Some general physical constants

Planck's constant	$h = 6.626\ 18(4) \times 10^{-34}$ joule second $\hbar = \dfrac{h}{2\pi} = 1.054\ 589(6) \times 10^{-34}$ joule second
Speed of light (in vacuum)	$c = 2.997\ 924\ 58(1) \times 10^{8}$ m/s
Electron charge	$q = -\ 1.602\ 189(5) \times 10^{-19}$ coulomb
Electron mass	$m_e = 9.109\ 53(5) \times 10^{-31}$ kg
Proton mass	$M_p = 1.672\ 65(1) \times 10^{-27}$ kg
Neutron mass	$M_n = 1.674\ 95(1) \times 10^{-27}$ kg
	$\dfrac{M_p}{m_e} = 1\ 836.1515(7)$
Electron Compton wavelength	$\lambda_c = h/m_e c = 2.426\ 309(4) \times 10^{-2}$ Å $\bar{\lambda}_c = \hbar/m_e c = 3.861\ 591(7) \times 10^{-3}$ Å
Fine structure constant (dimensionless)	$\alpha = \dfrac{q^2}{4\pi\varepsilon_0 \hbar c} = \dfrac{e^2}{\hbar c} = \dfrac{1}{137.036\ 0(1)}$
Bohr radius	$a_0 = \dfrac{\bar{\lambda}_c}{\alpha} = 0.529\ 177\ 1(5)$ Å
Hydrogen atom ionization energy (without proton recoil effect)	$-\ E_{I\infty} = \alpha^2 m_e c^2/2 = 13.605\ 80(5)$ eV
Rydberg's constant	$R_\infty = -\ E_{I\infty}/hc = 1.097\ 373\ 18(8) \times 10^{5}\ \text{cm}^{-1}$
"Classical" electron radius	$r_e = \dfrac{q^2}{4\pi\varepsilon_0 m_e c^2} = 2.817\ 938(7)$ fermi
Bohr magneton	$\mu_B = q\hbar/2m_e = -\ 9.274\ 08(4) \times 10^{-24}$ joule/tesla
Electron spin g factor	$g_e = 2 \times 1.001\ 159\ 657(4)$
Nuclear magneton	$\mu_n = -\ q\hbar/2M_p = 5.050\ 82(2) \times 10^{-27}$ joule/tesla
Boltzmann's constant	$k_B = 1.380\ 66(4) \times 10^{-23}$ joule/K
Avogadro's number	$N_A = 6.022\ 05(3) \times 10^{23}$

QUANTUM MECHANICS

Claude Cohen-Tannoudji
Bernard Diu
Franck Laloë

QUANTUM MECHANICS

Volume II

Translated from the French by **Susan Reid Hemley, Nicole Ostrowsky, Dan Ostrowsky**

A Wiley-Interscience Publication
JOHN WILEY & SONS
New York · London · Sydney · Toronto

HERMANN
Publishers in arts and science Paris

CLAUDE COHEN-TANNOUDJI, Professor at the Collège de France, was born in 1933. Since 1960 he has been involved in research at the École Normale in Paris with Professors Alfred Kastler and Jean Brossel; his principal research is in optical pumping and the interaction of radiation and matter.

BERNARD DIU, Professor at the University of Paris VII, was born in 1935. He is currently engaged in research at the Laboratory of Theoretical Physics and High Energy in Paris, in the field of strong-interaction particle physics.

FRANCK LALOË, born in 1940, was Lecturer at the University of Paris VI, then appointed to the C.N.R.S. Since 1964 he has been with Professors Kastler and Brossel at the École Normale; his research bears principally on optical pumping of rare gas ions and atoms.

Published by Hermann and by John Wiley & Sons, Inc. Printed in France.

Mécanique quantique was originally published in French by Hermann in 1973; second revised and enlarged edition 1977.

Library of Congress Cataloging in Publication Data:

Cohen-Tannoudji, C.
Quantum mechanics.

Translation of Mecanique quantique.
"A Wiley-Interscience publication."
Includes index.

1. Quantum theory. I. Diu, Bernard, joint author. II. Laloë, Franck, joint author. III. Title.

QC174.12.C6313 530.1'2 76-5874
ISBN 2-7056-5834-3 (Hermann)

ISBN 0-471-16434-8 (v. II) ISBN 0-471-16435-X (v. II) pbk.

10 9 8 7 6 5 4 3 2 1

Directions for Use

This book is made up of chapters and their complements :

– *The chapters* contain the fundamental concepts. Except for a few additions and variations, they correspond to a course given in the last year of a typical undergraduate physics program.

These fourteen chapters are *complete in themselves* and can be studied independently of the complements.

– *The complements* follow the appropriate chapter. They are listed at the end of each chapter in a "*reader's guide*" which discusses the difficulty and importance of every one of them. Each is labelled by a letter followed by a subscript which gives the number of the corresponding chapter (for example, the complements of chapter V are, in order, A_V, B_V, C_V...). They can be recognized immediately by the symbol ● which appears at the top of each of their pages.

The complements vary : some are intended to expand the treatment of the corresponding chapter or to provide more detailed discussion of certain points; others describe concrete examples or introduce various physical concepts. One of the complements (usually the last one) is a collection of exercises.

The *difficulty* of the complements varies. Some are very simple examples or extensions of the chapter, while others are more difficult (some are at graduate level); in any case, the reader should have studied the material in the chapter before using the complements.

The student should not try to study all the complements of a chapter at once. In accordance with his aims and interests, he should choose a small number of them (two or three, for example), plus a few exercises. The other complements can be left for later study.

Some passages within the book have been set in small type and these can be omitted on a first reading.

Table

VOLUME I

Chapter II The mathematical tools of quantum mechanics . . . 91

Complements of chapter II

Chapter III The postulates of quantum mechanics . . . 211

Complements of chapter III

Complements of chapter IV

Complements of chapter V

Complements of chapter VI

Complements of chapter VII

VOLUME II

Chapter VIII An elementary approach to the quantum theory of scattering by a potential 901

Complements of chapter VIII

Complements of chapter XI

Chapter XII An application of perturbation theory: the fine and hyperfine structure of the hydrogen atom . . . 1209

Complements of chapter XII

Complements of chapter XIII

Complements of chapter XIV

CHAPTER VIII

An elementary approach to the quantum theory of scattering by a potential

OUTLINE OF CHAPTER VIII

A. INTRODUCTION

1. Importancc of collision phenomena
2. Scattering by a potential
3. Definition of the scattering cross section
4. Organization of this chapter

B. STATIONARY SCATTERING STATES. CALCULATION OF THE CROSS SECTION

1. Definition of stationary scattering states
   a. Eigenvalue equation of the Hamiltonian
   b. Asymptotic form of the stationary scattering states. Scattering amplitude
2. Calculation of the cross section using probability currents
   a. "Probability fluid" associated with a stationary scattering state
   b. Incident current and scattered current
   c. Expression for the cross section
   d. Interference between the incident and the scattered waves
3. Integral scattering equation
4. The Born approximation
   a. Approximate solution of the integral scattering equation
   b. Interpretation of the formulas

B. SCATTERING BY A CENTRAL POTENTIAL. METHOD OF PARTIAL WAVES

1. Principle of the method of partial waves
2. Stationary states of a free particle
   a. Stationary states with well-defined momentum. Plane waves
   b. Stationary states with well-defined angular momentum. Free spherical waves
   c. Physical properties of free spherical waves
   d. Interference between the incident and the scattered waves
3. Partial waves in the potential $V(r)$
   a. Radial equation. Phase shifts
   b. Physical meaning of phase shifts
4. Expression of the cross section in terms of phase shifts
   a. Construction of the stationary scattering state from partial waves
   b. Calculation of the cross section

A. INTRODUCTION

1. Importance of collision phenomena

Many experiments in physics, especially in high energy physics, consist of directing a beam of particles (1) (produced for example, by an accelerator) onto a target composed of particles (2), and studying the resulting collisions : the various particles★ constituting the final state of the system – that is, the state after the collision (*cf.* fig. 1) – are detected and their characteristics (direction of emission, energy, etc.) are measured. Obviously, the aim of such a study is to determine the interactions that occur between the various particles entering into the collision.

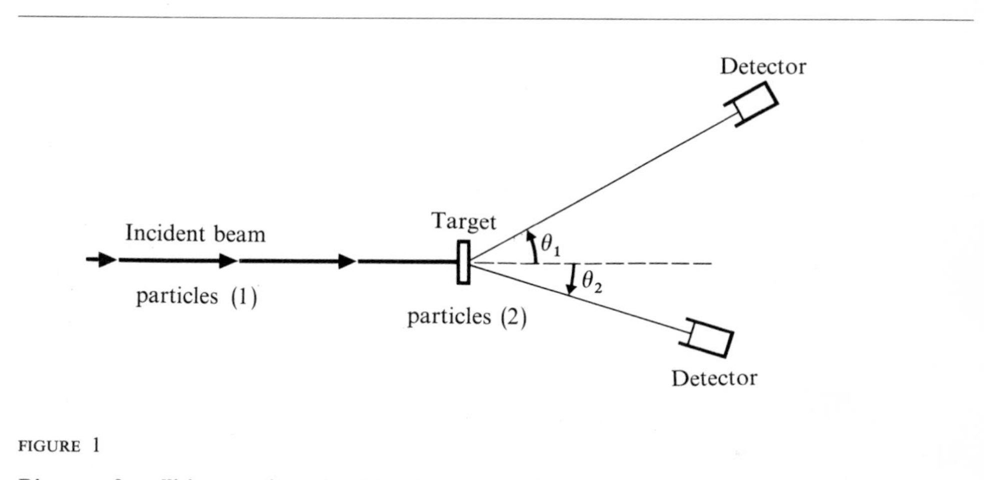

FIGURE 1

Diagram of a collision experiment involving the particles (1) of an incident beam and the particles (2) of a target. The two detectors represented in the figure measure the number of particles scattered through angles θ_1 and θ_2 with respect to the incident beam.

The phenomena observed are sometimes very complex. For example, if particles (1) and (2) are in fact composed of more elementary components (protons and neutrons in the case of nuclei), the latter can, during the collision, redistribute themselves amongst two or several final composite particles which are different from the initial particles; in this case, one speaks of "rearrangement collisions".

★ In practice, it is not always possible to detect all the particles emitted, and one must often be satisfied with partial information about the final system.

Moreover, at high energies, the relativistic possibility of the "materialization" of part of the energy appears: new particles are then created and the final state can include a great number of them (the higher the energy of the incident beam, the greater the number). Broadly speaking, one says that collisions give rise to *reactions*, which are described most often as in chemistry:

$$(1) + (2) \longrightarrow (3) + (4) + (5) + \dots \qquad \text{(A-1)}$$

Amongst all the reactions possible★ under given conditions, *scattering* reactions are defined as those in which the final state and the initial state are composed of the same particles (1) and (2). In addition, a scattering reaction is said to be elastic when none of the particles' internal states change during the collision.

2. Scattering by a potential

We shall confine ourselves in this chapter to the study of the elastic scattering of the incident particles (1) by the target particles (2). If the laws of classical mechanics were applicable, solving this problem would involve determining the deviations in the incident particles' trajectories due to the forces exerted by particles (2). For processes occurring on an atomic or nuclear scale, it is clearly out of the question to use classical mechanics to resolve the problem; we must study the evolution of the wave function associated with the incident particles under the influence of their interactions with the target particles [which is why we speak of the "scattering" of particles (1) by particles (2)]. Rather than attack this question in its most general form, we shall introduce the following simplifying hypotheses:

(*i*) We shall suppose that particles (1) and (2) have no spin. This simplifies the theory considerably but should not be taken to imply that the spin of particles is unimportant in scattering phenomena.

(*ii*) We shall not take into account the possible internal structure of particles (1) and (2). The following arguments are therefore not applicable to "inelastic" scattering phenomena, where part of the kinetic energy of (1) is absorbed in the final state by the internal degrees of freedom of (1) and (2) (*cf.* for example, the experiment of Franck and Hertz). We shall confine ourselves to the case of *elastic scattering*, which does not affect the internal structure of the particles.

(*iii*) We shall assume that the target is thin enough to enable us to neglect multiple scattering processes; that is, processes during which a particular incident particle is scattered several times before leaving the target.

(*iv*) We shall neglect any possibility of coherence between the waves scattered by the different particles which make up the target. This simplification is justified when the spread of the wave packets associated with particles (1) is small compared to the average distance between particles (2). Therefore we shall concern ourselves only with the elementary process of the scattering of a particle (1) of the beam by a particle (2) of the target. This excludes a certain number of phenomena which

★ Since the processes studied occur on a quantum level, it is not generally possible to predict with certainty what final state will result from a given collision; one merely attempts to predict the probabilities of the various possible states.

are nevertheless very interesting, such as coherent scattering by a crystal (Bragg diffraction) or scattering of slow neutrons by the phonons of a solid, which provide valuable information about the structure and dynamics of crystal lattices. When these coherence effects can be neglected, the flux of particles detected is simply the sum of the fluxes scattered by each of the $\mathcal{N}$ target particles, that is, $\mathcal{N}$ times the flux scattered by any one of them (the exact position of the scattering particle inside the target is unimportant since the target dimensions are much smaller than the distance between the target and the detector).

(v) We shall assume that the interactions between particles (1) and (2) can be described by a potential energy $V(\mathbf{r}_1 - \mathbf{r}_2)$, which depends only on the relative position $\mathbf{r} = \mathbf{r}_1 - \mathbf{r}_2$ of the particles. If we follow the reasoning of § B, chapter VII, then, in the center-of-mass reference frame★ of the two particles (1) and (2), the problem reduces to the study of *the scattering of a single particle by the potential* $V(\mathbf{r})$. The mass μ of this "relative particle" is related to the masses m_1 and m_2 of (1) and (2) by the formula:

$$\frac{1}{\mu} = \frac{1}{m_1} + \frac{1}{m_2} \tag{A-2}$$

3. Definition of the scattering cross section

Let Oz be the direction of the incident particles of mass μ (fig. 2). The potential $V(\mathbf{r})$ is localized around the origin O of the coordinate system [which is in fact

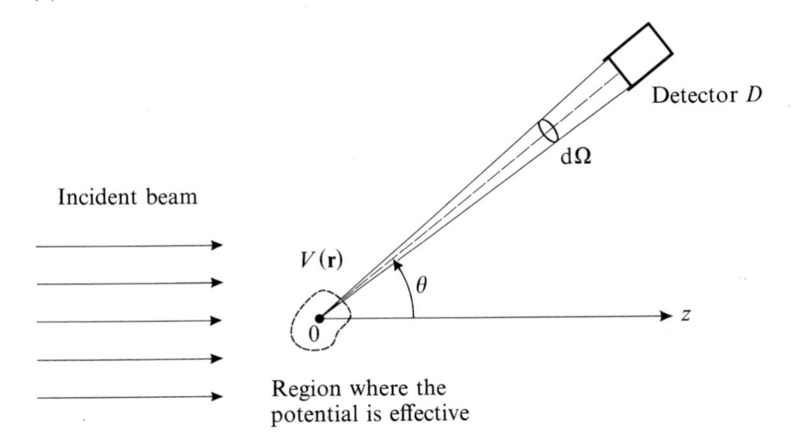

FIGURE 2

The incident beam, whose flux of particles is F_i, is parallel to the axis Oz; it is assumed to be much wider than the zone of influence of the potential $V(\mathbf{r})$, which is centered at O. Far from this zone of influence, a detector D measures the number $\mathrm{d}n$ of particles scattered per unit time into the solid angle $\mathrm{d}\Omega$, centered around the direction defined by the polar angles θ and φ. The number $\mathrm{d}n$ is proportional to F_i and to $\mathrm{d}\Omega$; the coefficient of proportionality $\sigma(\theta, \varphi)$ is, by definition, the scattering "cross section" in the direction (θ, φ).

★ In order to interpret the results obtained in scattering experiments, it is clearly necessary to return to the laboratory reference frame. Going from one frame of reference to another is a simple kinematic problem that we will not consider here. See for example Messiah (1.17), vol. I, chap. X, § 7.

the center of mass of the two real particles (1) and (2)]. We shall designate by F_i the flux of particles in the incident beam, that is, the number of particles per unit time which traverse a unit surface perpendicular to Oz in the region where z takes on very large negative values. (The flux F_i is assumed to be weak enough to allow us to neglect interactions between different particles of the incident beam.)

We place a detector far from the region under the influence of the potential and in the direction fixed by the polar angles θ and φ, with an opening facing O and subtending the solid angle $\mathrm{d}\Omega$ (the detector is situated at a distance from O which is large compared to the linear dimensions of the potential's zone of influence). We can thus count the number $\mathrm{d}n$ of particles scattered per unit time into the solid angle $\mathrm{d}\Omega$ about the direction (θ, φ).

$\mathrm{d}n$ is obviously proportional to $\mathrm{d}\Omega$ and to the incident flux F_i. We shall define $\sigma(\theta, \varphi)$ to be the coefficient of proportionality between $\mathrm{d}n$ and $F_i \, \mathrm{d}\Omega$:

$$\boxed{\mathrm{d}n = F_i \, \sigma(\theta, \varphi) \, \mathrm{d}\Omega} \tag{A-3}$$

The dimensions of $\mathrm{d}n$ and F_i are, respectively, T^{-1} and $(L^2T)^{-1}$. $\sigma(\theta, \varphi)$ therefore has the dimensions of a surface; it is called the *differential scattering cross section* in the direction (θ, φ). Cross sections are frequently measured in barns and submultiples of barns :

$$1 \text{ barn} = 10^{-24} \text{ cm}^2 \tag{A-4}$$

The definition (A-3) can be interpreted in the following way: the number of particles per unit time which reach the detector is equal to the number of particles which would cross a surface $\sigma(\theta, \varphi) \, \mathrm{d}\Omega$ placed perpendicular to Oz in the incident beam.

Similarly, the *total scattering cross section* σ is defined by the formula:

$$\sigma = \int \sigma(\theta, \varphi) \, \mathrm{d}\Omega \tag{A-5}$$

COMMENTS:

(*i*) Definition (A-3), in which $\mathrm{d}n$ is proportional to $\mathrm{d}\Omega$, implies that only the scattered particles are taken into consideration. The flux of these particles reaching a given detector D [of fixed surface and placed in the direction (θ, φ)] is inversely proportional to the square of the distance between D and O (this property is characteristic of a scattered flux). In practice, the incident beam is laterally bounded [although its width remains much larger than the extent of the zone of influence of $V(\mathbf{r})$], and the detector is placed outside its trajectory so that it receives only the scattered particles. Of course, such an arrangement does not permit the measurement of the cross section in the direction $\theta = 0$ (the forward direction), which can only be obtained by extrapolation from the values of $\sigma(\theta, \varphi)$ for small θ.

(*ii*) The concept of a cross section is not limited to the case of elastic scattering: reaction cross sections are defined in an analogous manner.

4. Organization of this chapter

§B is devoted to a brief study of scattering by an arbitrary potential $V(\mathbf{r})$ (decreasing however faster than $1/r$ as r tends toward infinity). First of all, in §B-1, we introduce the fundamental concepts of a stationary scattering state and a scattering amplitude. We then show, in §B-2, how knowledge of the asymptotic behavior of the wave functions associated with stationary scattering states enables us to obtain scattering cross sections. Afterwards, in §B-3, we discuss in a more precise way, using the integral scattering equation, the existence of these stationary scattering states. Finally (in §B-4), we derive an approximate solution of this equation, valid for weak potentials. This leads us to the Born approximation, in which the cross section is very simply related to the Fourier transform of the potential.

For a central potential $V(r)$, the general methods described in §B clearly remain applicable, but the method of partial waves, set forth in §C, is usually considered preferable. This method is based (§C-1) on the comparison of the stationary states with well-defined angular momentum in the presence of the potential $V(r)$ (which we shall call "partial waves") and their analogues in the absence of the potential ("free spherical waves"). Therefore, we begin by studying, in §C-2, the essential properties of the stationary states of a free particle, and more particularly those of free spherical waves. Afterwards (§C-3), we show that the difference between a partial wave in the potential $V(r)$ and a free spherical wave with the same angular momentum l is characterized by a "phase shift" δ_l. Thus, it is only necessary to know how stationary scattering states can be constructed from partial waves in order to obtain the expression of cross sections in terms of phase shifts (§C-4).

B. STATIONARY SCATTERING STATES. CALCULATION OF THE CROSS SECTION

In order to describe in quantum mechanical terms the scattering of a given incident particle by the potential $V(\mathbf{r})$, it is necessary to study the time evolution of the wave packet representing the state of the particle. The characteristics of this wave packet are assumed to be known for large negative values of the time t, when the particle is in the negative region of the Oz axis, far from and not yet affected by the potential $V(\mathbf{r})$. It is known that the subsequent evolution of the wave packet can be obtained immediately if it is expressed as a superposition of stationary states. This is why we are going to begin by studying the eigenvalue equation of the Hamiltonian:

$$H = H_0 + V(\mathbf{r}) \tag{B-1}$$

where:

$$H_0 = \frac{\mathbf{P}^2}{2\mu} \tag{B-2}$$

describes the particle's kinetic energy.

Actually, to simplify the calculations, we are going to base our reasoning directly on the stationary states and not on wave packets. We have already used this procedure in chapter I, in the study of "square" one-dimensional potentials (§ D-2 and complement H_I). It consists of considering a stationary state to represent a "probability fluid" in steady flow, and studying the structure of the corresponding probability currents. Naturally, this simplified reasoning is not rigorous: it remains to be shown that it leads to the same results as the correct treatment of the problem, which is based on wave packets. Assuming this will enable us to develop certain general ideas easily, without burying them in complicated calculations*.

1. Definition of stationary scattering states

a. EIGENVALUE EQUATION OF THE HAMILTONIAN

Schrödinger's equation describing the evolution of the particle in the potential $V(\mathbf{r})$ is satisfied by solutions associated with a well-defined energy E (stationary states):

$$\psi(\mathbf{r}, t) = \varphi(\mathbf{r})\, \mathrm{e}^{-iEt/\hbar} \tag{B-3}$$

where $\varphi(\mathbf{r})$ is a solution of the eigenvalue equation:

$$\left[-\frac{\hbar^2}{2\mu}\Delta + V(\mathbf{r}) \right] \varphi(\mathbf{r}) = E\, \varphi(\mathbf{r}) \tag{B-4}$$

We are going to assume that the potential $V(\mathbf{r})$ decreases faster than $1/r$ as r approaches infinity. Notice that this hypothesis excludes the Coulomb potential, which demands special treatment; we shall not consider it here.

We shall only be concerned with solutions of (B-4) associated with a positive energy E, equal to the kinetic energy of the incident particle before it reaches the zone of influence of the potential. Defining:

$$E = \frac{\hbar^2 k^2}{2\mu} \tag{B-5}$$

$$V(\mathbf{r}) = \frac{\hbar^2}{2\mu}\, U(\mathbf{r}) \tag{B-6}$$

enables us to write (B-4) in the form:

$$\left[\Delta + k^2 - U(\mathbf{r}) \right] \varphi(\mathbf{r}) = 0 \tag{B-7}$$

For each value of k (that is, of the energy E), equation (B-7) can be satisfied by an infinite number of solutions (the positive eigenvalues of the Hamiltonian H are

* The proof was given in complement J_I, for a particular one-dimensional problem; we verified that the same results are obtained by calculating the probability current associated with a stationary scattering state or by studying the evolution of a wave packet describing a particle which undergoes a collision.

infinitely degenerate). As in "square" one-dimensional potential problems (*cf.* chap. I, §D-2 and complement H_I), we must choose from amongst these solutions the one which corresponds to the physical problem being studied (for example, when we wanted to determine the probability that a particle with a given energy would cross a one-dimensional potential barrier, we chose the stationary state which, in the region on the other side of the barrier, was composed simply of a transmitted wave). Here, the choice proves to be more complicated, since the particle is moving in three-dimensional space and the potential $V(\mathbf{r})$ has, *a priori*, an arbitrary form. Therefore, we shall specify, using wave packet properties in an intuitive way, the conditions that must be imposed on the solutions of equation (B-7) if they are to be used in the description of a scattering process. We shall call the eigenstates of the Hamiltonian which satisfy these conditions *stationary scattering states*, and we shall designate by $v_k^{(\text{diff})}(\mathbf{r})$ the associated wave functions.

b. ASYMPTOTIC FORM OF STATIONARY SCATTERING STATES. SCATTERING AMPLITUDE

For large negative values of t, the incident particle is free [$V(\mathbf{r})$ is practically zero when one is sufficiently far from the point O], and its state is represented by a plane wave packet. Consequently, the stationary wave function that we are looking for must contain a term of the form e^{ikz}, where k is the constant which appears in equation (B-7). When the wave packet reaches the region which is under the influence of the potential $V(\mathbf{r})$, its structure is profoundly modified and its evolution complicated. Nevertheless, for large positive values of t, it has left this region and once more takes on a simple form: it is now split into a transmitted wave packet which continues to propagate along Oz in the positive direction (hence having the form e^{ikz}) and a scattered wave packet. Consequently, the wave function $v_k^{(\text{diff})}(\mathbf{r})$, representing the stationary scattering state associated with a given energy $E = \hbar^2k^2/2\mu$, will be obtained from the superposition of the plane wave e^{ikz} and a scattered wave (we are ignoring the problem of normalization).

The structure of the scattered wave obviously depends on the potential $V(\mathbf{r})$. Yet its asymptotic form (valid far from the zone of influence of the potential) is simple; reasoning by analogy with wave optics, we see that the scattered wave must present the following characteristics for large r:

(*i*) In a given direction (θ, φ), its radial dependence is of the form e^{ikr}/r. It is a divergent (or "outgoing") wave which has the same energy as the incident wave. The factor $1/r$ results from the fact that there are three spatial dimensions: $(\Delta + k^2)\,e^{ikr}$ is not zero, while:

$$(\Delta + k^2)\frac{e^{ikr}}{r} = 0 \quad \text{for } r \geqslant r_0 \text{ where } r_0 \text{ is any positive number} \tag{B-8}$$

(in optics, the factor $1/r$ insures that the total flux of energy passing through a sphere of radius r is independent of r for large r; in quantum mechanics, it is the probability flux passing through this sphere that does not depend on r).

(*ii*) Since scattering is not generally isotropic, the amplitude of the outgoing wave depends on the direction (θ, φ) being considered.

Finally, the wave function $v_k^{(\text{diff})}(\mathbf{r})$ associated with the stationary scattering state is, by definition, the solution of equation (B-7) whose asymptotic behavior is of the form :

$$v_k^{(\text{diff})}(\mathbf{r}) \underset{r \to \infty}{\sim} e^{ikz} + f_k(\theta, \varphi) \frac{e^{ikr}}{r} \tag{B-9}$$

In this expression, only the function $f_k(\theta, \varphi)$, which is called the *scattering amplitude*, depends on the potential $V(\mathbf{r})$. It can be shown (*cf.* §B-3) that equation (B-7) has indeed one and only one solution, for each value of k, that satisfies condition (B-9).

COMMENTS :

(*i*) We have already pointed out that in order to obtain simply the time evolution of the wave packet representing the state of the incident particle, it is necessary to expand it in terms of eigenstates of the total Hamiltonian H rather than in terms of plane waves. Therefore, let us consider a wave function of the form★:

$$\psi(\mathbf{r}, t) = \int_0^\infty dk\, g(k)\, v_k^{(\text{diff})}(\mathbf{r})\, e^{-iE_k t/\hbar} \tag{B-10}$$

where :

$$E_k = \frac{\hbar^2 k^2}{2\mu} \tag{B-11}$$

and where the function $g(k)$, taken to be real for the sake of simplicity, has a pronounced peak at $k = k_0$ and practically vanishes elsewhere. $\psi(\mathbf{r}, t)$ is a solution of Schrödinger's equation and therefore correctly describes the time evolution of the particle. It remains to be shown that this function indeed satisfies the boundary conditions imposed by the particular physical problem being considered. According to (B-9), it approaches asymptotically the sum of a plane wave packet and a scattered wave packet :

$$\psi(\mathbf{r}, t) \underset{r \to \infty}{\sim} \int_0^\infty dk\, g(k)\, e^{ikz}\, e^{-iE_k t/\hbar} + \int_0^\infty dk\, g(k)\, f_k(\theta, \varphi) \frac{e^{ikr}}{r}\, e^{-iE_k t/\hbar} \tag{B-12}$$

The position of the maximum of each of these packets can be obtained from the stationary phase condition (*cf.* chap. I, §C-2). A simple calculation then gives for the plane wave packet:

$$z_M(t) = v_G t \tag{B-13}$$

with:

$$v_G = \frac{\hbar k_0}{\mu} \tag{B-14}$$

★ Actually, it is also necessary to superpose the plane waves corresponding to wave vectors $\boldsymbol{k}$ having slightly different orientations, for the incident wave packet is limited in the direction perpendicular to Oz. For the sake of simplicity, we are concerning ourselves here only with the energy dispersion (which limits the spread of the wave packet along Oz).

As for the scattered wave packet, its maximum in the direction (θ, φ) is located at a distance from the point O given by:

$$r_M(\theta, \varphi\, ; t) = -\alpha'_{k_o}(\theta, \varphi) + v_G t \tag{B-15}$$

where $\alpha'_k(\theta, \varphi)$ is the derivative with respect to k of the argument of the scattering amplitude $f_k(\theta, \varphi)$. Note that formulas (B-13) and (B-15) are valid only in the asymptotic region (that is, for large $|t|$).

For large negative values of t, there is no scattered wave packet, as can be seen from (B-15). The waves of which it is composed interfere constructively only for negative values of r, and these values lie outside the domain permitted to r. Therefore, all that we find in this region is the plane wave packet, which, according to (B-13), is making its way towards the interaction region with a group velocity v_G. For large positive values of t, both packets are actually present; the first one moves off along the positive Oz axis, continuing along the path of the incident packet, and the second one diverges in all directions. The scattering process can thus be well described by the asymptotic condition (B-9).

(*ii*) The spatial extension Δz of the wave packet (B-10) is related to the momentum dispersion $\hbar \Delta k$ by the relation:

$$\Delta z \simeq \frac{1}{\Delta k} \tag{B-16}$$

We shall assume that Δk is small enough for Δz to be much larger than the linear dimensions of the potential's zone of influence. Under these conditions, the wave packet moving at a velocity v_G towards the point O (fig. 3) will take a time:

$$\Delta T \simeq \frac{\Delta z}{v_G} \simeq \frac{1}{v_G \Delta k} \tag{B-17}$$

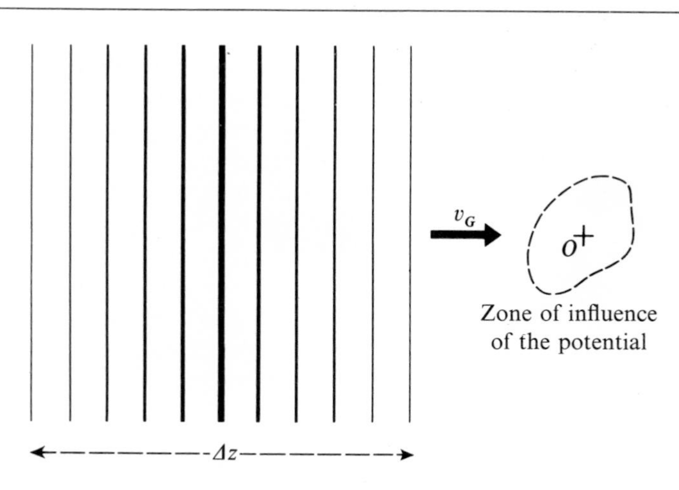

FIGURE 3

The incident wave packet of length Δz moves at a velocity v_G towards the potential $V(\mathbf{r})$; it interacts with the potential during a time of the order of $\Delta T = \Delta z / v_G$ (assuming the size of the potential's zone of influence to be negligible compared to Δz).

to cross this zone. Let us fix the time origin at the instant when the center of the incident wave packet reaches point O. Scattered waves exist only for $t \gtrsim -\Delta T/2$, i.e., after the forward edge of the incident wave packet has arrived at the potential's zone of influence. For $t = 0$, the most distant part of the scattered wave packet is at a distance of the order of $\Delta z/2$ from the point O.

Let us now consider an *a priori* different problem, where we have a time-dependent potential, obtained by multiplying $V(\mathbf{r})$ by a function $f(t)$ which increases slowly from 0 to 1 between $t = -\Delta T/2$ and $t = 0$. For t much less than $-\Delta T/2$, the potential is zero and we shall assume that the state of the particle is represented by a plane wave (extending throughout all space). This plane wave begins to be modified only at $t \simeq -\Delta T/2$, and at the instant $t = 0$ the scattered waves look like those in the preceding case.

Thus we see that there is a certain similarity between the two different problems that we have just described. On the one hand, we have scattering by a constant potential of an incident wave packet whose amplitude at the point O increases smoothly between the times $-\Delta T/2$ and zero; on the other hand, we have scattering of a plane wave of constant amplitude by a potential that is gradually "turned on" over the same time interval $[-\Delta T/2, 0]$.

If $\Delta k \longrightarrow 0$, the wave packet (B-10) tends toward a stationary scattering state [$g(k)$ tends toward $\delta(k - k_0)$]; in addition, according to (B-17), ΔT becomes infinite and the turning on of the potential associated with the function $f(t)$ becomes infinitely slow (for this reason, it is often said to be "adiabatic"). The preceding discussion, although very qualitative, thus makes it possible to describe a stationary scattering state as the result of adiabatically imposing a scattering potential on a free plane wave. We could make this interpretation more precise by studying in a more detailed way the evolution of the initial plane wave under the influence of the potential $f(t)V(\mathbf{r})$.

2. Calculation of the cross section using probability currents

a. PROBABILITY FLUID ASSOCIATED WITH A STATIONARY SCATTERING STATE

In order to determine the cross section, one should study the scattering of an incident wave packet by the potential $V(\mathbf{r})$. However, we can obtain the result much more simply by basing our reasoning on the stationary scattering states; we consider such a state to describe a *probability fluid in steady flow* and we calculate the cross section from the incident and scattered currents. As we have already pointed out, this method is analogous to the one we used in one-dimensional "square" barrier problems: in those problems, the ratio between the reflected (or transmitted) current and the incident current yielded the reflection (or transmission) coefficient directly.

Hence we shall calculate the contributions of the incident wave and the scattered wave to the probability current in a stationary scattering state. We recall that the expression for the current $\mathbf{J}(\mathbf{r})$ associated with a wave function $\varphi(\mathbf{r})$ is:

$$\mathbf{J}(\mathbf{r}) = \frac{1}{\mu} \mathrm{Re}\left[\varphi^*(\mathbf{r}) \frac{\hbar}{i} \nabla\varphi(\mathbf{r})\right] \qquad \text{(B-18)}$$

b. INCIDENT CURRENT AND SCATTERED CURRENT

The incident current $\mathbf{J}_i$ is obtained from (B-18) by replacing $\varphi(\mathbf{r})$ by the plane wave e^{ikz}; $\mathbf{J}_i$ is therefore directed along the Oz axis in the positive direction, and its modulus is:

$$|\mathbf{J}_i| = \frac{\hbar k}{\mu} \tag{B-19}$$

Since the scattered wave is expressed in spherical coordinates in formula (B-9), we shall calculate the components of the scattered current $\mathbf{J}_d$ along the local axes defined by this coordinate system. Recall that the corresponding components of the operator $\boldsymbol{\nabla}$ are:

$$\begin{aligned} (\boldsymbol{\nabla})_r &= \frac{\partial}{\partial r} \\ (\boldsymbol{\nabla})_\theta &= \frac{1}{r}\frac{\partial}{\partial \theta} \\ (\boldsymbol{\nabla})_\varphi &= \frac{1}{r \sin\theta}\frac{\partial}{\partial \varphi} \end{aligned} \tag{B-20}$$

If we replace $\varphi(\mathbf{r})$ in formula (B-18) by the function $f_k(\theta, \varphi) e^{ikr}/r$, we can easily obtain the scattered current in the asymptotic region:

$$\begin{aligned} (\mathbf{J}_d)_r &= \frac{\hbar k}{\mu}\frac{1}{r^2}|f_k(\theta, \varphi)|^2 \\ (\mathbf{J}_d)_\theta &= \frac{\hbar}{\mu}\frac{1}{r^3}\,\mathrm{Re}\left[\frac{1}{i} f_k^*(\theta, \varphi)\frac{\partial}{\partial \theta} f_k(\theta, \varphi)\right] \\ (\mathbf{J}_d)_\varphi &= \frac{\hbar}{\mu}\frac{1}{r^3 \sin\theta}\,\mathrm{Re}\left[\frac{1}{i} f_k^*(\theta, \varphi)\frac{\partial}{\partial \varphi} f_k(\theta, \varphi)\right] \end{aligned} \tag{B-21}$$

Since r is large, $(\mathbf{J}_d)_\theta$ and $(\mathbf{J}_d)_\varphi$ are negligible compared to $(\mathbf{J}_d)_r$, and the scattered current is practically radial.

c. EXPRESSION FOR THE CROSS SECTION

The incident beam is composed of independent particles, all of which are assumed to be prepared in the same way. Sending a great number of these particles amounts to repeating the same experiment a great number of times with one particle whose state is always the same. If this state is $v_k^{(\mathrm{diff})}(\mathbf{r})$, it is clear that the incident flux F_i (that is, the number of particles of the incident beam which cross a unit surface perpendicular to Oz per unit time) is proportional to the flux of the vector $\mathbf{J}_i$ across this surface; that is, according to (B-19):

$$F_i = C\,|\mathbf{J}_i| = C\frac{\hbar k}{\mu} \tag{B-22}$$

Similarly, the number dn of particles which strike the opening of the detector (fig. 2) per unit time is proportional to the flux of the vector $\mathbf{J}_d$ across the surface dS of this opening [the proportionality constant C is the same as in (B-22)]:

$$\begin{aligned} \mathrm{d}n &= C\,\mathbf{J}_d \,.\, \mathrm{d}\mathbf{S} = C\,(\mathbf{J}_d)_r\, r^2\, \mathrm{d}\Omega \\ &= C\,\frac{\hbar k}{\mu}\,|f_k(\theta, \varphi)|^2\, \mathrm{d}\Omega \end{aligned} \tag{B-23}$$

We see that dn is independent of r if r is sufficiently large.

If we substitute formulas (B-22) and (B-23) into the definition (A-3) of the differential cross section $\sigma(\theta, \varphi)$, we obtain:

$$\boxed{\sigma(\theta, \varphi) = |f_k(\theta, \varphi)|^2} \tag{B-24}$$

The differential cross section is thus simply the square of the modulus of the scattering amplitude.

d. INTERFERENCE BETWEEN THE INCIDENT AND THE SCATTERED WAVES

In the preceding sections, we have neglected a contribution to the current associated with $v_k^{(\text{diff})}(\mathbf{r})$ in the asymptotic region: the one which arises from interference between the plane

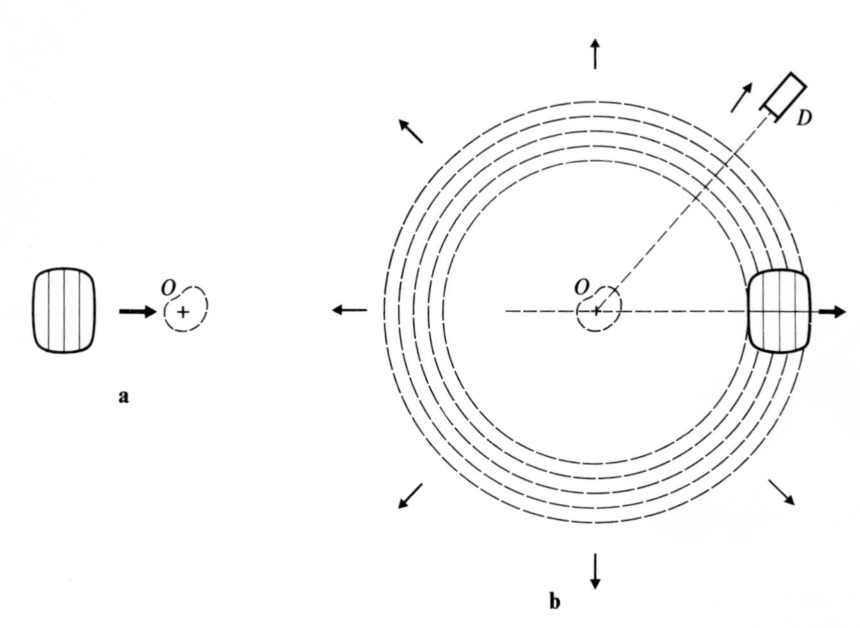

FIGURE 4

Before the collision (fig. a), the incident wave packet is moving towards the zone of influence of the potential. After the collision (fig. b), we observe a plane wave packet and a spherical wave packet scattered by the potential (dashed lines in the figure). The plane and scattered waves interfere in the forward direction in a destructive way (conservation of total probability); the detector *D* is placed in a lateral direction and can only see the scattered waves.

wave e^{ikz} and the scattered wave, and which is obtained by replacing $\varphi^*(\mathbf{r})$ in (B-18) by e^{-ikz} and $\varphi(\mathbf{r})$ by $f_k(\theta, \varphi)\, e^{ikr}/r$, and vice versa.

Nevertheless, we can convince ourselves that these interference terms do not appear when we are concerned with scattering in directions other than the forward direction ($\theta = 0$). In order to do so, let us go back to the description of the collision in terms of wave packets (fig. 4), and let us take into consideration the fact that in practice the wave packet always has a finite lateral spread. Initially, the incident wave packet is moving towards the zone of influence of $V(\mathbf{r})$ (fig. 4-a). After the collision (fig. 4-b), we find two wave packets : a plane one which results from the propagation of the incident wave packet (as if there were no scattering potential) and a scattered one moving away from the point O in all directions. The transmitted wave thus results from the interference between these two wave packets. In general, however, we place the detector D outside the beam, so that it is not struck by transmitted particles; thus we observe only the scattered wave packet and it is not necessary to take into consideration the interference terms that we have just mentioned.

Yet it follows from figure 4-b that interference between the plane and scattered wave packets cannot be neglected in the forward direction, where they occupy the same region of space. The transmitted wave packet results from this interference. It must have a smaller amplitude than the incident packet because of conservation of total probability (that is, conservation of the number of particles: particles scattered in all directions of space other than the forward direction leave the beam, whose intensity is thus attenuated after it has passed the target). It is thus the destructive interference between the plane and forward-scattered wave packets that insures the global conservation of the total number of particles.

3. Integral scattering equation

We propose to show now, in a more precise way than in § B-1-b, how one can demonstrate the existence of stationary wave functions whose asymptotic behavior is of the form (B-9). In order to do so, we shall introduce the integral scattering equation, whose solutions are precisely these stationary scattering state wave functions.

Let us go back to the eigenvalue equation of H [formula (B-7)] and put it in the form:

$$(\Delta + k^2)\,\varphi(\mathbf{r}) = U(\mathbf{r})\,\varphi(\mathbf{r}) \tag{B-25}$$

Suppose (we shall see later that this is in fact the case) that there exists a function $G(\mathbf{r})$ such that:

$$(\Delta + k^2)\, G(\mathbf{r}) = \delta(\mathbf{r}) \tag{B-26}$$

[$G(\mathbf{r})$ is called the "Green's function" of the operator $\Delta + k^2$]. Then any function $\varphi(\mathbf{r})$ which satisfies:

$$\varphi(\mathbf{r}) = \varphi_0(\mathbf{r}) + \int d^3r'\, G(\mathbf{r} - \mathbf{r}')\, U(\mathbf{r}')\, \varphi(\mathbf{r}') \tag{B-27}$$

where $\varphi_0(\mathbf{r})$ is a solution of the homogeneous equation:

$$(\Delta + k^2)\,\varphi_0(\mathbf{r}) = 0 \tag{B-28}$$

obeys the differential equation (B-25). To show this, we apply the operator $\Delta + k^2$ to both sides of equation (B-27); taking (B-28) into account, we obtain:

$$(\Delta + k^2)\,\varphi(\mathbf{r}) = (\Delta + k^2)\int d^3r'\, G(\mathbf{r} - \mathbf{r}')\, U(\mathbf{r}')\, \varphi(\mathbf{r}') \tag{B-29}$$

Assuming we can move the operator inside the integral, it will act only on the variable $\mathbf{r}$, and we shall have, according to (B-26):

$$\begin{aligned}(\Delta + k^2)\,\varphi(\mathbf{r}) &= \int d^3r'\, \delta(\mathbf{r} - \mathbf{r}')\, U(\mathbf{r}')\, \varphi(\mathbf{r}') \\ &= U(\mathbf{r})\, \varphi(\mathbf{r})\end{aligned} \tag{B-30}$$

Inversely, it can be shown that any solution of (B-25) satisfies (B-27)*. The differential equation (B-25) can thus be replaced by the integral equation (B-27).

We shall see that it is often easier to base our reasoning on the integral equation. Its principal advantage derives from the fact that by choosing $\varphi_0(\mathbf{r})$ and $G(\mathbf{r})$ correctly, one can incorporate into the equation the desired asymptotic behavior. Thus, one single integral equation, called the *integral scattering equation*, becomes the equivalent of the differential equation (B-25) and the asymptotic condition (B-9).

To begin with, let us consider (B-26). It implies that $(\Delta + k^2)G(\mathbf{r})$ must be identically equal to zero in any region which does not include the origin [which, according to (B-8), is the case when $G(\mathbf{r})$ is equal to e^{ikr}/r]. Moreover, according to formula (61) of appendix II, $G(\mathbf{r})$ must behave like $-1/4_\pi r$ when r approaches zero. In fact, it is easy to show that the functions:

$$G_\pm(\mathbf{r}) = -\frac{1}{4\pi}\frac{e^{\pm ikr}}{r} \tag{B-31}$$

are solutions of equation (B-26). We may write:

$$\begin{aligned}\Delta G_\pm(\mathbf{r}) = e^{\pm ikr}\,\Delta\left(-\frac{1}{4\pi r}\right) - \frac{1}{4\pi r}\,\Delta(e^{\pm ikr}) \\ + 2\left[\boldsymbol{\nabla}\left(-\frac{1}{4\pi r}\right)\right]\cdot\left[\boldsymbol{\nabla} e^{\pm ikr}\right]\end{aligned} \tag{B-32}$$

A simple calculation then gives (*cf.* appendix II):

$$\Delta G_\pm(\mathbf{r}) = -k^2 G_\pm(\mathbf{r}) + \delta(\mathbf{r}) \tag{B-33}$$

which is what we wished to prove. G_+ and G_- are called, respectively "outgoing and incoming Green's functions".

The actual form of the desired asymptotic behavior (B-9) suggests the choice of the incident plane wave e^{ikz} for $\varphi_0(\mathbf{r})$ and the choice of the outgoing Green's function

* This can be seen intuitively if one considers $U(\mathbf{r})\varphi(\mathbf{r})$ to be the right-hand side of a differential equation : the general solution of (B-25) is then obtained by adding to the general solution of the homogeneous equation a particular solution of the complete equation [second term of (B-27)].

$G_+(\mathbf{r})$ for $G(\mathbf{r})$. In fact, we are going to show that the integral scattering equation can be written :

$$v_k^{(\text{diff})}(\mathbf{r}) = e^{ikz} + \int d^3r' G_+(\mathbf{r} - \mathbf{r}')\, U(\mathbf{r}')\, v_k^{(\text{diff})}(\mathbf{r}') \tag{B-34}$$

whose solutions present the asymptotic behaviour given by (B-9).

To do this, let us place ourselves at a point M (position $\mathbf{r}$), very far from the various points P (position $\mathbf{r}'$) of the zone of influence of the potential, whose linear dimensions are of the order of L★ (fig. 5):

$$\begin{aligned} r &\gg L \\ r' &\lesssim L \end{aligned} \tag{B-35}$$

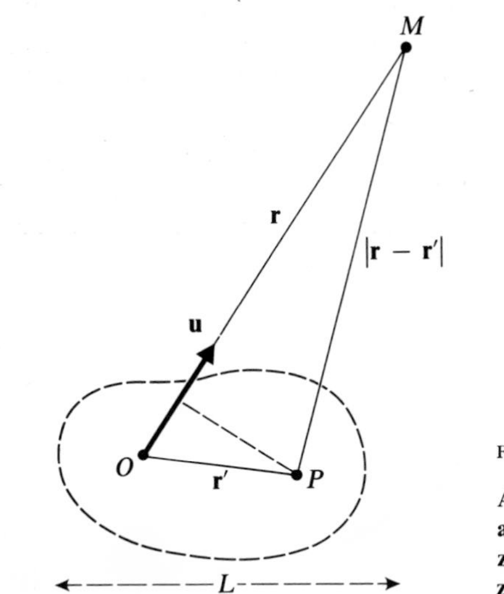

FIGURE 5

Approximate calculation of the distance $|\mathbf{r} - \mathbf{r}'|$ between a point M very far from O and a point P situated in the zone of influence of the potential (the dimensions of this zone of influence are of the order of L).

Since the angle between MO and MP is very small, the length MP (that is, $|\mathbf{r} - \mathbf{r}'|$) is equal, to within a good approximation, to the projection of MP on MO:

$$|\mathbf{r} - \mathbf{r}'| \simeq r - \mathbf{u} \cdot \mathbf{r}' \tag{B-36}$$

where $\mathbf{u}$ is the unit vector in the $\mathbf{r}$ direction. It follows that, for large r:

$$G_+(\mathbf{r} - \mathbf{r}') = -\frac{1}{4\pi}\frac{e^{ik|\mathbf{r}-\mathbf{r}'|}}{|\mathbf{r} - \mathbf{r}'|} \underset{r\to\infty}{\sim} -\frac{1}{4\pi}\frac{e^{ikr}}{r}\, e^{-ik\,\mathbf{u}.\mathbf{r}'} \tag{B-37}$$

★ Recall that we have explicitly assumed that $U(\mathbf{r})$ decreases at infinity faster than $1/r$.

Substituting this expression back into equation (B-34), we obtain the asymptotic behavior of $v_k^{(\text{diff})}(\mathbf{r})$:

$$v_k^{(\text{diff})}(\mathbf{r}) \underset{r \to \infty}{\sim} e^{ikz} - \frac{1}{4\pi} \frac{e^{ikr}}{r} \int d^3r' \, e^{-ik\,\mathbf{u}.\mathbf{r}'} U(\mathbf{r}') \, v_k^{(\text{diff})}(\mathbf{r}') \tag{B-38}$$

which is indeed of the form (B-9), since the integral is no longer a function of the distance $r = OM$ but only (through the unit vector $\mathbf{u}$) of the polar angles θ and φ which fix the direction of the vector $\mathbf{OM}$. Thus, by setting:

$$f_k(\theta, \varphi) = -\frac{1}{4\pi} \int d^3r' \, e^{-ik\,\mathbf{u}.\mathbf{r}'} U(\mathbf{r}') \, v_k^{(\text{diff})}(\mathbf{r}') \tag{B-39}$$

we are led to an expression which is identical to (B-9).

It is therefore clear that the solutions of the integral scattering equation (B-34) are indeed the stationary scattering states*.

COMMENT:

It is often convenient to define the *incident wave vector* $\mathbf{k}_i$ as a vector of modulus k directed along the Oz axis of the beam such that:

$$e^{ikz} = e^{i\mathbf{k}_i.\mathbf{r}} \tag{B-40}$$

In the same way, the vector $\mathbf{k}_d$ which has the same modulus k as the incident wave vector but whose direction is fixed by the angles θ and φ is called the *scattered wave vector* in the direction (θ, φ):

$$\mathbf{k}_d = k\,\mathbf{u} \tag{B-41}$$

Finally, the *scattering* (*or transferred*) *wave vector* in the direction (θ, φ) is the difference between $\mathbf{k}_d$ and $\mathbf{k}_i$ (fig. 6):

$$\mathbf{K} = \mathbf{k}_d - \mathbf{k}_i \tag{B-42}$$

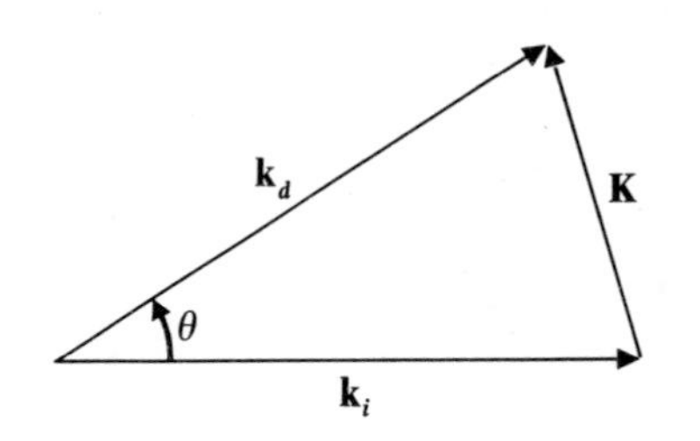

FIGURE 6

Incident wave vector $\mathbf{k}_i$, scattered wave vector $\mathbf{k}_d$ and transferred wave vector $\mathbf{K}$.

* In order to prove the existence of stationary scattering states rigorously, it would thus be sufficient to demonstrate that equation (B-34) admits a solution.

4. The Born approximation

a. APPROXIMATE SOLUTION OF THE INTEGRAL SCATTERING EQUATION

If we take (B-40) into account, we can write the integral scattering equation in the form:

$$v_k^{(\text{diff})}(\mathbf{r}) = e^{i\mathbf{k}_i\cdot\mathbf{r}} + \int d^3r' G_+(\mathbf{r} - \mathbf{r}')U(\mathbf{r}')\, v_k^{(\text{diff})}(\mathbf{r}') \tag{B-43}$$

We are going to try to solve this equation by iteration.

A simple change of notation ($\mathbf{r} \Longrightarrow \mathbf{r}'$; $\mathbf{r}' \Longrightarrow \mathbf{r}''$) permits us to write:

$$v_k^{(\text{diff})}(\mathbf{r}') = e^{i\mathbf{k}_i\cdot\mathbf{r}'} + \int d^3r'' G_+(\mathbf{r}' - \mathbf{r}'')\, U(\mathbf{r}'')\, v_k^{(\text{diff})}(\mathbf{r}'') \tag{B-44}$$

Inserting this expression in (B-43), we obtain:

$$\begin{aligned} v_k^{(\text{diff})}(\mathbf{r}) = e^{i\mathbf{k}_i\cdot\mathbf{r}} &+ \int d^3r' G_+(\mathbf{r} - \mathbf{r}')\, U(\mathbf{r}')\, e^{i\mathbf{k}_i\cdot\mathbf{r}'} \\ &+ \int d^3r' \int d^3r'' G_+(\mathbf{r} - \mathbf{r}')\, U(\mathbf{r}')\, G_+(\mathbf{r}' - \mathbf{r}'')\, U(\mathbf{r}'')\, v_k^{(\text{diff})}(\mathbf{r}'') \end{aligned} \tag{B-45}$$

The first two terms on the right-hand side of (B-45) are known; only the third one contains the unknown function $v_k^{(\text{diff})}(\mathbf{r})$. This procedure can be repeated : changing $\mathbf{r}$ to $\mathbf{r}''$ and $\mathbf{r}'$ to $\mathbf{r}'''$ in (B-43) gives $v_k^{(\text{diff})}(\mathbf{r}'')$, which can be reinserted in (B-45). We then have :

$$\begin{aligned} v_k^{(\text{diff})}(\mathbf{r}) = e^{i\mathbf{k}_i\cdot\mathbf{r}} &+ \int d^3r' G_+(\mathbf{r} - \mathbf{r}')\, U(\mathbf{r}')\, e^{i\mathbf{k}_i\cdot\mathbf{r}'} \\ &+ \int d^3r' \int d^3r'' G_+(\mathbf{r} - \mathbf{r}')\, U(\mathbf{r}')\, G_+(\mathbf{r}' - \mathbf{r}'')\, U(\mathbf{r}'')\, e^{i\mathbf{k}_i\cdot\mathbf{r}''} \\ &+ \int d^3r' \int d^3r'' \int d^3r''' G_+(\mathbf{r} - \mathbf{r}')\, U(\mathbf{r}')\, G_+(\mathbf{r}' - \mathbf{r}'')\, U(\mathbf{r}'') \\ &\qquad \times G_+(\mathbf{r}'' - \mathbf{r}''')\, U(\mathbf{r}''')\, v_k^{(\text{diff})}(\mathbf{r}''') \end{aligned} \tag{B-46}$$

where the first three terms are known; the unknown function $v_k^{(\text{diff})}(\mathbf{r})$ has been pushed back into the fourth term.

Thus we can construct, step by step, what is called the *Born expansion* of the stationary scattering wave function. Note that each term of this expansion brings in one higher power of the potential than the preceding one. Thus, if the potential is weak, each successive term is smaller than the preceding one. If we push the expansion far enough, we can neglect the last term on the right-hand side and thus obtain $v_k^{(\text{diff})}(\mathbf{r})$ entirely in terms of known quantities.

If we substitute this expansion of $v_k^{(\text{diff})}(\mathbf{r})$ into expression (B-39), we obtain the Born expansion of the scattering amplitude. In particular, if we limit ourselves to first order in U, all we need to do is replace $v_k^{(\text{diff})}(\mathbf{r}')$ by $e^{i\mathbf{k}_i \cdot \mathbf{r}'}$ on the right-hand side of (B-39). This is the *Born approximation*:

$$\begin{aligned} f_k^{(B)}(\theta, \varphi) &= -\frac{1}{4\pi} \int d^3r' \, e^{-ik \, \mathbf{u}.\mathbf{r}'} U(\mathbf{r}') \, e^{i\mathbf{k}_i.\mathbf{r}'} \\ &= -\frac{1}{4\pi} \int d^3r' \, e^{-i(\mathbf{k}_d - \mathbf{k}_i).\mathbf{r}'} U(\mathbf{r}') \\ &= -\frac{1}{4\pi} \int d^3r' \, e^{-i\mathbf{K}.\mathbf{r}'} U(\mathbf{r}') \end{aligned} \tag{B-47}$$

where $\mathbf{K}$ is the scattering wave vector defined in (B-42). The scattering cross section, in the Born approximation, is thus very simply related to the Fourier transform of the potential, since, using (B-24) and (B-6), (B-47) implies:

$$\sigma_k^{(B)}(\theta, \varphi) = \frac{\mu^2}{4\pi^2\hbar^4} \left| \int d^3r \, e^{-i\mathbf{K}.\mathbf{r}} V(\mathbf{r}) \right|^2 \tag{B-48}$$

According to figure 6, the direction and modulus of the scattering wave vector $\mathbf{K}$ depend both on the modulus k of $\mathbf{k}_i$ and $\mathbf{k}_d$ and on the scattering direction (θ, φ). Thus, for a given θ and φ, the Born cross section varies with k, that is, with the energy of the incident beam. Similarly, for a given energy, $\sigma^{(B)}$ varies with θ and φ. We thus see, within the simple framework of the Born approximation, how studying the variation of the differential cross section in terms of the scattering direction and the incident energy gives us information about the potential $V(\mathbf{r})$.

b. INTERPRETATION OF THE FORMULAS

We can give formula (B-45) a physical interpretation which brings out very clearly the formal analogy between quantum mechanics and wave optics.

Let us consider the zône of influence of the potential to be a scattering medium whose density is proportional to $U(\mathbf{r})$. The function $G_+(\mathbf{r} - \mathbf{r}')$ [formula (B-31)] represents the amplitude at the point $\mathbf{r}$ of a wave radiated by a point source situated at $\mathbf{r}'$. Consequently, the first two terms of formula (B-45) describe the total wave at the point $\mathbf{r}$ as the result of the superposition of the incident wave $e^{i\mathbf{k}_i.\mathbf{r}}$ and an infinite number of waves coming from *secondary sources* induced in the scattering medium by the incident wave. The amplitude of each of these sources is indeed proportional to the incident wave $(e^{i\mathbf{k}_i.\mathbf{r}'})$ and the density of the scattering material $[U(\mathbf{r}')]$, evaluated at the corresponding point $\mathbf{r}'$. This interpretation, symbolized by figure 7, recalls *Huygens' principle* in wave optics.

Actually, formula (B-45) includes a third term. However, we can interpret in an analogous fashion the successive terms of the Born expansion. Since the scattering medium extends over a certain area, a given secondary source is excited not only by the incident wave but also by scattered waves coming from other

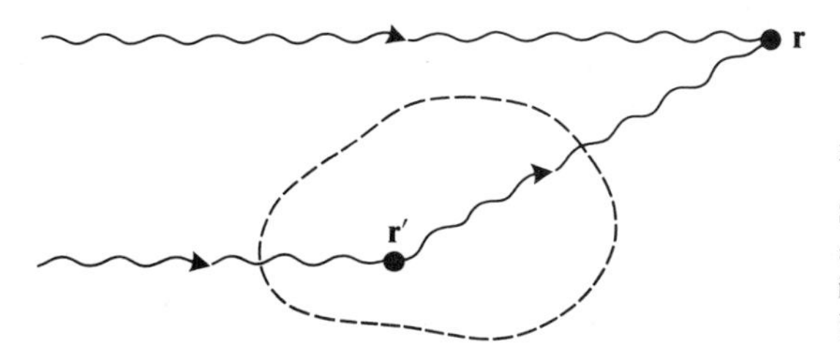

FIGURE 7

Schematic representation of the Born approximation: we only consider the incident wave and the waves scattered by one interaction with the potential.

secondary sources. Figure 8 represents symbolically the third term of the Born expansion [*cf.* formula (B-46)]. If the scattering medium has a very low density [$U(\mathbf{r})$ very small], we can neglect the influence of secondary sources on each other.

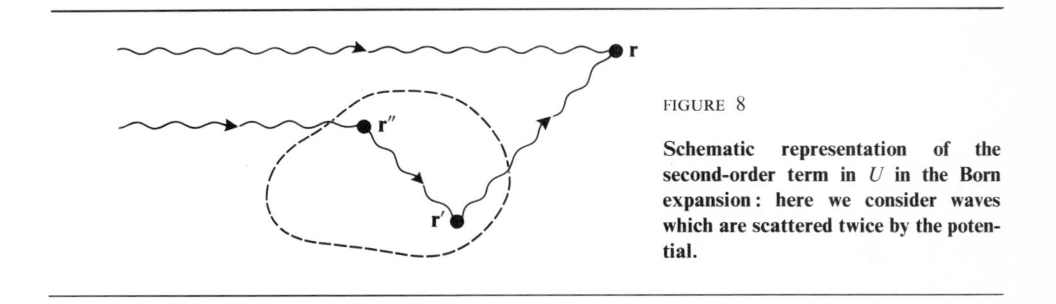

FIGURE 8

Schematic representation of the second-order term in U in the Born expansion: here we consider waves which are scattered twice by the potential.

COMMENT:

The interpretation that we have just given for higher-order terms in the Born expansion has nothing to do with the multiple scattering processes that can occur inside a thick target: we are only concerned, here, with the scattering of one particle of the beam by a single particle of the target, while multiple scattering brings in the successive interactions of the same incident particle with several different particles of the target.

C. SCATTERING BY A CENTRAL POTENTIAL. METHOD OF PARTIAL WAVES

1. Principle of the method of partial waves

In the special case of a central potential $V(r)$, the orbital angular momentum $\mathbf{L}$ of the particle is a constant of the motion. Therefore, there exist stationary states with well-defined angular momentum: that is, eigenstates common to H, $\mathbf{L}^2$ and L_z. We shall call the wave functions associated with these states *partial waves* and we shall write them $\varphi_{k,l,m}(\mathbf{r})$. The corresponding eigenvalues of H, $\mathbf{L}^2$ and L_z are, respectively, $\hbar^2k^2/2\mu$, $l(l+1)\hbar^2$ and $m\hbar$. Their angular dependence is always given by the spherical harmonics $Y_l^m(\theta, \varphi)$; the potential $V(r)$ influences only their radial dependence.

We expect that, for large r, the partial waves will be very close to the common eigenfunctions of H_0, $\mathbf{L}^2$ and L_z, where H_0 is the free Hamiltonian [formula (B-2)]. This is why we are first going to study, in §C-2, the stationary states of a free particle, and, in particular, those which have a well-defined angular momentum. The corresponding wave functions $\varphi_{k,l,m}^{(0)}(\mathbf{r})$ are *free spherical waves*: their angular dependence is, of course, that of a spherical harmonic and we shall see that the asymptotic expression for their radial function is the superposition of an incoming wave e^{-ikr}/r and an outgoing wave e^{ikr}/r with a well-determined phase difference.

The asymptotic expression for the partial wave $\varphi_{k,l,m}(\mathbf{r})$ in the potential $V(r)$ is also (§C-3) the superposition of an incoming wave and an outgoing wave. However, the phase difference between these two waves is different from the one which characterizes the corresponding free spherical wave: the potential $V(r)$ introduces a supplementary *phase shift* δ_l. This phase shift constitutes the only difference between the asymptotic behavior of $\varphi_{k,l,m}$ and that of $\varphi_{k,l,m}^{(0)}$. Consequently, for fixed k, the phase shifts δ_l for all values of l are all we need to know to be able to calculate the cross section.

In order to carry out this calculation, we shall express (§C-4) the stationary scattering state $v_k^{(\text{diff})}(\mathbf{r})$ as a linear combination of partial waves $\varphi_{k,l,m}(\mathbf{r})$ having the same energy but different angular momenta. Simple physical arguments suggest that the coefficients of this linear combination should be the same as those of the free spherical wave expansion of the plane wave e^{ikz}; this is in fact confirmed by an explicit calculation.

The use of partial waves thus permits us to express the scattering amplitude, and hence the cross section, in terms of the phase shifts δ_l. This method is particularly attractive when the range of the potential is not moch longer than the wavelength associated with the particle's motion, for, in this case, only a small number of phase shifts are involved (§C-3-b-β).

2. Stationary states of a free particle

In classical mechanics, a free particle of mass μ moves along a uniform linear trajectory. Its momentum $\mathbf{p}$, its energy $E = \mathbf{p}^2/2\mu$ and its angular momentum $\mathscr{L} = \mathbf{r} \times \mathbf{p}$ relative to the origin of the coordinate system are constants of the motion.

In quantum mechanics, the observables $\mathbf{P}$ and $\mathbf{L} = \mathbf{R} \times \mathbf{P}$ do not commute. Hence they represent incompatible quantities: it is impossible to measure the momentum and the angular momentum of a particle simultaneously.

The quantum mechanical Hamiltonian H_0 is written:

$$H_0 = \frac{1}{2\mu}\mathbf{P}^2 \tag{C-1}$$

H_0 does not constitute by itself a C.S.C.O. : its eigenvalues are infinitely degenerate (§2-a). On the other hand, the four observables :

$$H_0, P_x, P_y, P_z \tag{C-2}$$

form a C.S.C.O. Their common eigenstates are stationary states of well defined momentum. A free particle may also be considered as being placed in a zero central potential. The results of chap. VII then indicate that the three observables :

$$H_0, \mathbf{L}^2, L_z \tag{C-3}$$

form a C.S.C.O. The corresponding eigenstates are stationay states with well-defined angular momentum (more precisely, $\mathbf{L}^2$ and L_z have well-defined values, but L_x and L_y do not).

The bases of the state space defined by the C.S.C.O.'s (C-2) and (C-3) are distinct, since **P** and **L** are incompatible quantities. We are going to study these two bases and show how one can pass from one to the other.

a. STATIONARY STATES WITH WELL-DEFINED MOMENTUM. PLANE WAVES

We already know (*cf.* chap. II, E-2-d) that the three observables P_x, P_y and P_z form a C.S.C.O. (for a spinless particle). Their common eigenstates form a basis for the $\{ | \mathbf{p} \rangle \}$ representation:

$$\mathbf{P} \, | \, \mathbf{p} \rangle = \mathbf{p} \, | \, \mathbf{p} \rangle \tag{C-4}$$

Since H_0 commutes with these three observables, the states $| \, \mathbf{p} \rangle$ are necessarily eigenstates of H_0:

$$H_0 \, | \, \mathbf{p} \rangle = \frac{\mathbf{p}^2}{2\mu} \, | \, \mathbf{p} \rangle \tag{C-5}$$

The spectrum of H_0 is therefore continuous and includes all positive numbers and zero. Each of these eigenvalues is infinitely degenerate: to a fixed positive energy E there corresponds an infinite number of kets $| \, \mathbf{p} \rangle$ since there exists an infinite number of ordinary vectors **p** whose modulus satisfies:

$$|\mathbf{p}| = \sqrt{2\mu E} \tag{C-6}$$

The wave functions associated with the kets $| \, \mathbf{p} \rangle$ are the plane waves (*cf.* chap. II, §E-1-a):

$$\langle \, \mathbf{r} \, | \, \mathbf{p} \, \rangle = \left(\frac{1}{2\pi\hbar} \right)^{3/2} e^{i\mathbf{p}.\mathbf{r}/\hbar} \tag{C-7}$$

We shall introduce here the wave vector **k** to characterize a plane wave:

$$\mathbf{k} = \frac{\mathbf{p}}{\hbar} \tag{C-8}$$

and we shall define:

$$| \mathbf{k} \rangle = (\hbar)^{3/2} | \mathbf{p} \rangle \tag{C-9}$$

The kets $| \mathbf{k} \rangle$ are stationary states with well-defined momentum:

$$H_0 | \mathbf{k} \rangle = \frac{\hbar^2 \mathbf{k}^2}{2\mu} | \mathbf{k} \rangle \tag{C-10-a}$$

$$\mathbf{P} | \mathbf{k} \rangle = \hbar \mathbf{k} | \mathbf{k} \rangle \tag{C-10-b}$$

They are orthonormal in the extended sense:

$$\langle \mathbf{k} | \mathbf{k}' \rangle = \delta(\mathbf{k} - \mathbf{k}') \tag{C-11}$$

and form a basis in the state space:

$$\int d^3k \, | \mathbf{k} \rangle \langle \mathbf{k} | = 1 \tag{C-12}$$

The associated wave functions are the plane waves normalized, in a slightly different way:

$$\langle \mathbf{r} | \mathbf{k} \rangle = \left(\frac{1}{2\pi} \right)^{3/2} e^{i \mathbf{k}.\mathbf{r}} \tag{C-13}$$

b. STATIONARY STATES WITH WELL-DEFINED ANGULAR MOMENTUM. FREE SPHERICAL WAVES

In order to obtain the eigenfunctions common to H_0, $\mathbf{L}^2$ and L_z, all we have to do is solve the radial equation for an identically zero central potential. The detailed solution of this problem is given in complement $\mathrm{A}_{\mathrm{VIII}}$; we shall be satisfied here with giving the results.

Free spherical waves are the wave functions associated with the well-defined angular momentum stationary states $| \varphi_{k,l,m}^{(0)} \rangle$ of a free particle; they are written:

$$\varphi_{k,l,m}^{(0)}(\mathbf{r}) = \sqrt{\frac{2k^2}{\pi}} \, j_l(kr) \, Y_l^m(\theta, \varphi) \tag{C-14}$$

where j_l is a spherical Bessel function defined by:

$$j_l(\rho) = (-1)^l \rho^l \left(\frac{1}{\rho} \frac{\mathrm{d}}{\mathrm{d}\rho} \right)^l \frac{\sin \rho}{\rho} \tag{C-15}$$

The corresponding eigenvalues of H_0, $\mathbf{L}^2$ and L_z are, respectively, $\hbar^2 k^2/2\mu$, $l(l+1)\hbar^2$ and $m\hbar$.

The free spherical waves (C-14) are orthonormal in the extended sense:

$$\langle \varphi_{k,l,m}^{(0)} \mid \varphi_{k',l',m'}^{(0)} \rangle = \frac{2}{\pi} kk' \int_0^\infty j_l(kr)\, j_{l'}(k'r)\, r^2\, \mathrm{d}r \times \int \mathrm{d}\Omega\, Y_l^{m*}(\theta, \varphi)\, Y_{l'}^{m'}(\theta, \varphi)$$
$$= \delta(k - k')\, \delta_{ll'}\, \delta_{mm'} \tag{C-16}$$

and form a basis in the state space:

$$\int_0^\infty \mathrm{d}k \sum_{l=0}^{\infty} \sum_{m=-l}^{+l} \mid \varphi_{k,l,m}^{(0)} \rangle \langle \varphi_{k,l,m}^{(0)} \mid = 1 \tag{C-17}$$

c. PHYSICAL PROPERTIES OF FREE SPHERICAL WAVES

α. *Angular dependence*

The angular dependence of the free spherical wave $\varphi_{k,l,m}^{(0)}(\mathbf{r})$ is entirely given by the spherical harmonic $Y_l^m(\theta, \varphi)$. It is thus fixed by the eigenvalues of $\mathbf{L}^2$ and L_z (that is, by the indices l and m) and not by the energy. For example, a free $s(l = 0)$ spherical wave is always isotropic.

β. *Behavior in the neighborhood of the origin*

Let us consider an infinitesimal solid angle $\mathrm{d}\Omega_0$ about the direction (θ_0, φ_0); when the state of the particle is $\mid \varphi_{k,l,m}^{(0)} \rangle$, the probability of finding the particle in this solid angle between r and $r + \mathrm{d}r$ is proportional to:

$$r^2 j_l^2(kr) |Y_l^m(\theta_0, \varphi_0)|^2\, \mathrm{d}r\, \mathrm{d}\Omega_0 \tag{C-18}$$

It can be shown (complement A_{VIII}, § 2-c-α) that when ρ approaches zero:

$$j_l(\rho) \underset{\rho \to 0}{\sim} \frac{\rho^l}{(2l + 1)!!} \tag{C-19}$$

This result (which the discussion of chapter VII, § A-2-c would lead us to expect) implies that the probability (C-18) behaves like r^{2l+2} near the origin; hence, the larger l is, the more slowly it increases.

The shape of the function $\rho^2 j_l^2(\rho)$ is shown in figure 9. We see that this function remains small as long as:

$$\rho < \sqrt{l(l + 1)} \tag{C-20}$$

We may thus assume that the probability (C-18) is practically zero for:

$$r < \frac{1}{k}\sqrt{l(l + 1)} \tag{C-21}$$

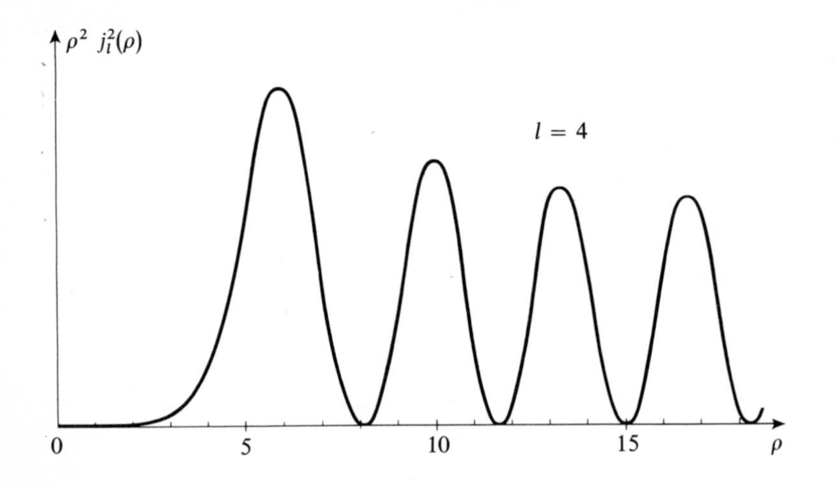

FIGURE 9

Graph of the function $\rho^2 j_l^2(\rho)$ giving the radial dependence of the probability of finding the particle in the state $|\varphi_{k,l,m}^{(0)}\rangle$. At the origin, this function behaves like ρ^{2l+2}; it remains practically zero as long as $\rho < \sqrt{l(l+1)}$.

This result is very important physically for it implies that a particle in the state $|\varphi_{k,l,m}^{(0)}\rangle$ is practically unaffected by what happens inside a sphere centered at O of radius:

$$b_l(k) = \frac{1}{k}\sqrt{l(l+1)} \tag{C-22}$$

We shall return to this point in §C-3-b-β.

COMMENT:

In classical mechanics, a free particle of momentum $\mathbf{p}$ and angular momentum $\mathscr{L}$ moves in a straight line whose distance b from the point O is given (fig. 10) by:

$$b = \frac{|\mathscr{L}|}{|\mathbf{p}|} \tag{C-23}$$

b is called the "collision" or "impact parameter" of the particle relative to O; the larger $|\mathscr{L}|$ is and the smaller the momentum (i.e. the energy), the larger b is. If $|\mathscr{L}|$ is replaced by $\hbar\sqrt{l(l+1)}$ and $|\mathbf{p}|$ by $\hbar k$ in (C-23), we again find expression (C-22) for $b_l(k)$, which can thus be interpreted semi-classically.

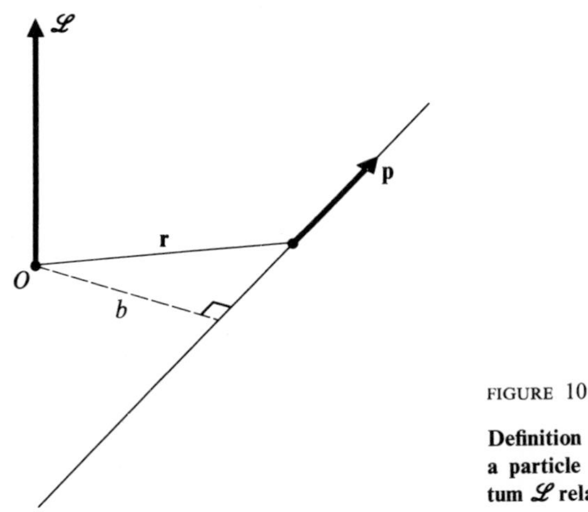

FIGURE 10

Definition of the classical impact parameter b of a particle of momentum p and angular momentum $\mathscr{L}$ relative to O.

γ. *Asymptotic behavior*

It can be shown (complement A_{VIII}, §2-c-β) that as ρ approaches infinity:

$$j_l(\rho) \underset{\rho \to \infty}{\simeq} \frac{1}{\rho} \sin\left(\rho - l\frac{\pi}{2}\right) \tag{C-24}$$

Consequently, the asymptotic behavior of the free spherical wave $\varphi^{(0)}_{k,l,m}(\mathbf{r})$ is:

$$\varphi^{(0)}_{k,l,m}(r, \theta, \varphi) \underset{r \to \infty}{\simeq} -\sqrt{\frac{2k^2}{\pi}}\, Y_l^m(\theta, \varphi)\, \frac{\mathrm{e}^{-ikr}\,\mathrm{e}^{il\frac{\pi}{2}} - \mathrm{e}^{ikr}\,\mathrm{e}^{-il\frac{\pi}{2}}}{2ikr} \tag{C-25}$$

At infinity, $\varphi^{(0)}_{k,l,m}$ therefore results from the *superposition of an incoming wave e^{-ikr}/r and an outgoing wave e^{ikr}/r*, whose amplitudes differ by a phase difference equal to $l\pi$.

COMMENT:

Suppose that we construct a packet of free spherical waves, all corresponding to the same values of l and m. A line of reasoning analogous to that of comment (i) of §B-1-b may be applied to it. The following conclusion results: for large negative values of t, only an incoming wave packet exists; while for large positive values of t, only an outgoing wave packet exists. Therefore, a free spherical wave may be thought of in the following manner: at first we have an incoming wave converging towards O; it becomes distorted as it approaches this point, retraces its steps when it is at a distance of the order of $b_l(k)$ [formula (C-22)], and gives rise to an outgoing wave with a phase shift of $l\pi$.

d. EXPANSION OF A PLANE WAVE IN TERMS OF FREE SPHERICAL WAVES

We thus have two distinct bases formed by eigenstates of H_0: the $\{ | \mathbf{k} \rangle \}$ basis associated with the plane waves and the $\{ | \varphi_{k,l,m}^{(0)} \rangle \}$ basis associated with the free spherical waves. It is possible to expand any ket of one basis in terms of vectors of the other one.

Let us consider in particular the ket $| 0, 0, k \rangle$, associated with a plane wave of wave vector $\mathbf{k}$ directed along Oz:

$$\langle \mathbf{r} | 0, 0, k \rangle = \left(\frac{1}{2\pi} \right)^{3/2} e^{ikz} \tag{C-26}$$

$| 0, 0, k \rangle$ represents a state of well-defined energy and momentum ($E = \hbar^2 k^2 / 2\mu$; $\mathbf{p}$ directed along Oz with modulus $\hbar k$). Now:

$$e^{ikz} = e^{ikr \cos \theta} \tag{C-27}$$

is independent of φ; since L_z corresponds to $\frac{\hbar}{i} \frac{\partial}{\partial \varphi}$ in the $\{ | \mathbf{r} \rangle \}$ representation, the ket $| 0, 0, k \rangle$ is also an eigenvector of L_z, with the eigenvalue zero:

$$L_z | 0, 0, k \rangle = 0 \tag{C-28}$$

Using the closure relation (C-17), we can write:

$$| 0, 0, k \rangle = \int_0^\infty dk' \sum_{l=0}^{\infty} \sum_{m=-l}^{+l} | \varphi_{k',l,m}^{(0)} \rangle \langle \varphi_{k',l,m}^{(0)} | 0, 0, k \rangle \tag{C-29}$$

Since $| 0, 0, k \rangle$ and $| \varphi_{k,l,m}^{(0)} \rangle$ are two eigenstates of H_0, they are orthogonal if the corresponding eigenvalues are different; their scalar product is therefore proportional to $\delta(k' - k)$. Similarly, they are both eigenstates of L_z and their scalar product is proportional to δ_{m0} [*cf.* relation (C-28)]. Formula (C-29) therefore takes on the form:

$$| 0, 0, k \rangle = \sum_{l=0}^{\infty} c_{k,l} | \varphi_{k,l,0}^{(0)} \rangle \tag{C-30}$$

The coefficients $c_{k,l}$ can be calculated explicitly (complement A_{VIII}, §3). Thus we obtain:

$$\boxed{e^{ikz} = \sum_{l=0}^{\infty} i^l \sqrt{4\pi(2l+1)}\, j_l(kr)\, Y_l^0(\theta)} \tag{C-31}$$

A state of well-defined linear momentum is therefore formed by the superposition of states corresponding to all possible angular momenta.

COMMENT:

The spherical harmonic $Y_l^0(\theta)$ is proportional to the Legendre polynomial $P_l(\cos\theta)$ (complement A_{VI}, § 2-e-α):

$$Y_l^0(\theta) = \sqrt{\frac{(2l+1)}{4\pi}}\, P_l(\cos\theta) \tag{C-32}$$

Hence the expansion (C-31) is often written in the form:

$$e^{ikz} = \sum_{l=0}^{\infty} i^l\,(2l+1)\, j_l(kr)\, P_l(\cos\theta) \tag{C-33}$$

3. Partial waves in the potential $V(r)$

We are now going to study the eigenfunctions common to H (the total Hamiltonian), $\mathbf{L}^2$ and L_z; that is, the partial waves $\varphi_{k,l,m}(\mathbf{r})$.

a. RADIAL EQUATION. PHASE SHIFTS

For any central potential $V(r)$, the partial waves $\varphi_{k,l,m}(\mathbf{r})$ are of the form:

$$\varphi_{k,l,m}(\mathbf{r}) = R_{k,l}(r)\, Y_l^m(\theta,\varphi) = \frac{1}{r}\, u_{k,l}(r)\, Y_l^m(\theta,\varphi) \tag{C-34}$$

where $u_{k,l}(r)$ is the solution of the radial equation:

$$\left[-\frac{\hbar^2}{2\mu}\frac{\mathrm{d}^2}{\mathrm{d}r^2} + \frac{l(l+1)\hbar^2}{2\mu r^2} + V(r)\right] u_{k,l}(r) = \frac{\hbar^2 k^2}{2\mu}\, u_{k,l}(r) \tag{C-35}$$

satisfying the condition at the origin:

$$u_{k,l}(0) = 0 \tag{C-36}$$

It is just as if we were dealing with a one-dimensional problem, where a particle of mass μ is under the influence of the potential (fig. 11):

$$\begin{aligned} V_{\text{eff}}(r) &= V(r) + \frac{l(l+1)\hbar^2}{2\mu r^2} && \text{for } r > 0 \\ V_{\text{eff}}(r) & \quad \text{infinite} && \text{for } r < 0 \end{aligned} \tag{C-37}$$

For large r, equation (C-35) reduces to:

$$\left[\frac{\mathrm{d}^2}{\mathrm{d}r^2} + k^2\right] u_{k,l}(r) \underset{r\to\infty}{\simeq} 0 \tag{C-38}$$

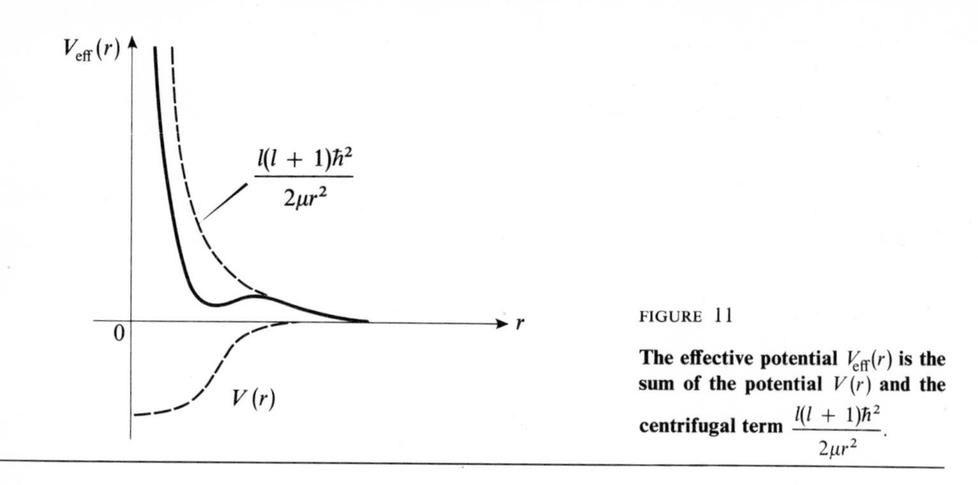

FIGURE 11

The effective potential $V_{\text{eff}}(r)$ is the sum of the potential $V(r)$ and the centrifugal term $\frac{l(l+1)\hbar^2}{2\mu r^2}$.

whose general solution is:

$$u_{k,l}(r) \underset{r \to \infty}{\sim} A\,\mathrm{e}^{ikr} + B\,\mathrm{e}^{-ikr} \tag{C-39}$$

Since $u_{k,l}(r)$ must satisfy condition (C-36), the constants A and B cannot be arbitrary. In the equivalent one-dimensional problem [formulas (C-37)], equation (C-36) is related to the fact that the potential is infinite for negative r, and expression (C-39) represents the superposition of an "incident" plane wave e^{-ikr} coming from the right (along the axis on which the fictitious particle being studied moves) and a "reflected" plane wave e^{ikr} propagating from left to right. Since there can be no "transmitted" wave [as $V(r)$ is infinite on the negative part of the axis], the "reflected" current must be equal to the "incident" current. Thus we see that condition (C-36) implies that, in the asymptotic expression (C-39):

$$|A| = |B| \tag{C-40}$$

Consequently:

$$u_{k,l}(r) \underset{r \to \infty}{\sim} |A| \left[\mathrm{e}^{ikr}\,\mathrm{e}^{i\varphi_A} + \mathrm{e}^{-ikr}\,\mathrm{e}^{i\varphi_B}\right] \tag{C-41}$$

which can be written in the form:

$$u_{k,l}(r) \underset{r \to \infty}{\sim} C \sin(kr - \beta_l) \tag{C-42}$$

The real phase β_l is completely determined by imposing continuity between (C-42) and the solution of (C-35) which goes to zero at the origin. In the case of an identically null potential $V(r)$, we saw in §C-2-c-γ that β_l is equal to $l\pi/2$. It is convenient to take this value as a point of reference, that is, to write:

$$u_{k,l}(r) \underset{r \to \infty}{\sim} C \sin\left(kr - l\frac{\pi}{2} + \delta_l\right) \tag{C-43}$$

The quantity δ_l defined in this way is called the *phase shift* of the partial wave $\varphi_{k,l,m}(\mathbf{r})$; it obviously depends on k, that is, on the energy.

b. PHYSICAL MEANING OF PHASE SHIFTS

α. *Comparison between partial waves and free spherical waves*

Taking (C-34) and (C-43) into account, we may write the expression for the asymptotic behavior of $\varphi_{k,l,m}(\mathbf{r})$ in the form:

$$\varphi_{k,l,m}(\mathbf{r}) \underset{r\to\infty}{\sim} C \frac{\sin(kr - l\pi/2 + \delta_l)}{r} Y_l^m(\theta, \varphi)$$

$$\underset{r\to\infty}{\sim} - C\, Y_l^m(\theta, \varphi) \frac{e^{-ikr} e^{i\left(l\frac{\pi}{2} - \delta_l\right)} - e^{ikr} e^{-i\left(l\frac{\pi}{2} - \delta_l\right)}}{2ir} \tag{C-44}$$

We see that the partial wave $\varphi_{k,l,m}(\mathbf{r})$, like a free spherical wave [formula (C-25)], results from the *superposition of an incoming wave and an outgoing wave.*

In order to develop the comparison between partial waves and free spherical waves in detail, we can modify the incoming wave of (C-44) so as to make it identical with the one in (C-25). To do this, we define a new partial wave $\tilde{\varphi}_{k,l,m}(\mathbf{r})$ by multiplying $\varphi_{k,l,m}(\mathbf{r})$ by $e^{i\delta_l}$ (this global phase factor has no physical importance) and by choosing the constant C in such a way that:

$$\tilde{\varphi}_{k,l,m}(\mathbf{r}) \underset{r\to\infty}{\sim} - Y_l^m(\theta, \varphi) \frac{e^{-ikr} e^{il\pi/2} - e^{ikr} e^{-il\pi/2} e^{2i\delta_l}}{2ikr} \tag{C-45}$$

This expression can then be interpreted in the following way (*cf.* the comment in §C-2-c-γ): at the outset, we have the same incoming wave as in the case of a free particle (aside from the normalization constant $\sqrt{2k^2/\pi}$). As this incoming wave approaches the zone of influence of the potential, it is more and more perturbed by this potential. When, after turning back, it is transformed into an outgoing wave, *it has accumulated a phase shift of* $2\delta_l$ relative to the free outgoing wave that would have resulted if the potential $V(r)$ had been identically zero. The factor $e^{2i\delta_l}$ (which varies with l and k) thus finally summarizes the total effect of the potential on a particle of angular momentum l.

COMMENT:

Actually, the preceding discussion is only valid if we base our reasoning on a wave packet formed by superposing partial waves $\varphi_{k,l,m}(\mathbf{r})$ with the same l and m, but slightly different k. For large negative values of t, we have only an incoming wave packet; it is the subsequent evolution of this wave packet directed towards the potential's zone of influence that we have analyzed above.

We could also adopt the point of view of comment (*ii*) of §B-1-b; that is, we could study the effect on a stationary free spherical wave of slowly "turning on" the potential $V(r)$. The same type of reasoning would then demonstrate that the partial wave $\varphi_{k,l,m}(\mathbf{r})$ can be obtained from a free spherical wave $\varphi_{k,l,m}^{(0)}(\mathbf{r})$ by adiabatically turning on the potential $V(r)$.

β. *Finite-range potentials*

Let us suppose that the potential $V(r)$ has a finite range r_0 ; that is, that :

$$V(r) = 0 \qquad \text{for} \quad r > r_0 \tag{C-46}$$

We pointed out earlier (§C-2-c-β) that the free spherical wave $\varphi_{k,l,m}^{(0)}$ scarcely penetrates a sphere centered at O of radius $b_l(k)$ [formula (C-22)]. Therefore, if we return to the interpretation of formula (C-45) that we have just given, we see that a potential satisfying (C-46) has virtually no effect on waves for which :

$$b_l(k) \gg r_0 \tag{C-47}$$

since the corresponding incoming wave turns back before reaching the zone of influence of $V(r)$. Thus, for each value of the energy, there exists a critical value l_M of the angular momentum, which, according to (C-22), is given approximately by:

$$\sqrt{l_M(l_M + 1)} \simeq kr_0 \tag{C-48}$$

The phase shifts δ_l are appreciable only for values of l less than or of the order of l_M.

The shorter the range of the potential and the lower the incident energy, the smaller l_M*. Therefore, it may happen that the only non-zero phase shifts are those corresponding to the first few partial waves : the $s(l = 0)$ wave at very low energy, followed by s and p waves for slightly greater energies, etc.

4. Expression of the cross section in terms of phase shifts

Phase shifts characterize the modifications, caused by the potential, of the asymptotic behavior of stationary states with well-defined angular momentum. Knowing them should therefore allow us to determine the cross section. In order to demonstrate this, all we must do is express the stationary scattering state $v_k^{(\text{diff})}(\mathbf{r})$ in terms of partial waves**, and calculate the scattering amplitude in this way.

a. CONSTRUCTION OF THE STATIONARY SCATTERING STATE FROM PARTIAL WAVES

We must find a linear superposition of partial waves whose asymptotic behavior is of the form (B-9). Since the stationary scattering state is an eigenstate of the Hamiltonian H, the expansion of $v_k^{(\text{diff})}(\mathbf{r})$ involves only partial waves having

* l_M is of the order of kr_0, which is the ratio between the range r_0 of the potential and the wavelength of the incident particle.

** If there exist bound states of the particle in the potential $V(r)$ (stationary states with negative energy), the system of partial waves does not constitute a basis of the state space; in order to form such a basis, it is necessary to add the wave functions of the bound states to the partial waves.

the same energy $\hbar^2k^2/2\mu$. Note also that, in the case of a central potential $V(r)$, the scattering problem we are studying is symmetrical with respect to rotation around the Oz axis defined by the incident beam. Consequently, the stationary scattering wave function $v_k^{(\text{diff})}(\mathbf{r})$ is independent of the azimuthal angle φ, so that its expansion includes only partial waves for which m is zero. Finally, we have an expression of the form :

$$v_k^{(\text{diff})}(\mathbf{r}) = \sum_{l=0}^{\infty} c_l \, \tilde{\varphi}_{k,l,0}(\mathbf{r}) \tag{C-49}$$

The problem thus consists of finding the coefficients c_l.

α. *Intuitive argument*

When $V(r)$ is identically zero, the function $v_k^{(\text{diff})}(\mathbf{r})$ reduces to the plane wave e^{ikz}, and the partial waves become free spherical waves $\varphi_{k,l,m}^{(0)}(\mathbf{r})$. In this case, we already know the expansion (C-49) : it is given by formula (C-31).

For non-zero $V(r)$, $v_k^{(\text{diff})}(\mathbf{r})$ includes a diverging scattered wave as well as a plane wave. Furthermore, we have seen that $\tilde{\varphi}_{k,l,0}(\mathbf{r})$ differs from $\varphi_{k,l,0}^{(0)}(\mathbf{r})$ in its asymptotic behavior only by the presence of the outgoing wave, which has the same radial dependence as the scattered wave. We should therefore expect that the coefficients c_l of the expansion (C-49) will be the same as those in formula (C-31)★, that is :

$$v_k^{(\text{diff})}(\mathbf{r}) = \sum_{l=0}^{\infty} i^l \sqrt{4\pi(2l+1)} \, \tilde{\varphi}_{k,l,0}(\mathbf{r}) \tag{C-50}$$

COMMENT :

We can also understand (C-50) in terms of the interpretation offered in comment (*ii*) of §B-1-b and the comment in §C-3-b-α. If we have a plane wave whose expansion is given by (C-31) and we turn on the potential $V(r)$ adiabatically, the wave is transformed into a stationary scattering state : the left-hand side of (C-31) must then be replaced by $v_k^{(\text{diff})}(\mathbf{r})$. In addition, each free spherical wave $j_l(kr)Y_l^0(\theta)$ appearing on the right-hand side of (C-31) is transformed into a partial wave $\tilde{\varphi}_{k,l,0}(\mathbf{r})$ when the potential is turned on. If we take into account the linearity of Schrödinger's equation, we finally obtain (C-50).

β. *Explicit derivation*

Let us now consider formula (C-50), which was suggested by a physical approach to the problem, and let us show that it does indeed supply the desired expansion.

First of all, the right-hand side of (C-50) is a superposition of eigenstates of H having the same energy $\hbar^2k^2/2\mu$; consequently, this superposition remains a stationary state.

★ Note that the expansion (C-31) brings in $j_l(kr)Y_l^0(\theta)$, that is, the free spherical wave $\varphi_{k,l,0}^{(0)}$ divided by the normalization factor $\sqrt{2k^2/\pi}$; this is why we defined $\tilde{\varphi}_{k,l,m}(\mathbf{r})$ [formula (C-45)] from expression (C-25) divided by this same factor.

Therefore, all we must do is make sure that the asymptotic behavior of the sum (C-50) is indeed of type (B-9). In order to do this, we use (C-45):

$$\sum_{l=0}^{\infty} i^l \sqrt{4\pi(2l+1)}\, \tilde{\varphi}_{k,l,0}(\mathbf{r}) \underset{r\to\infty}{\sim} - \sum_{l=0}^{\infty} i^l \sqrt{4\pi(2l+1)}\, Y_l^0(\theta) \times \frac{1}{2ikr}\left[e^{-ikr} e^{il\frac{\pi}{2}} - e^{ikr} e^{-il\frac{\pi}{2}} e^{2i\delta_l}\right] \tag{C-51}$$

In order to bring out the asymptotic behavior of expansion (C-31), we write:

$$e^{2i\delta_l} = 1 + 2i\, e^{i\delta_l} \sin \delta_l \tag{C-52}$$

and, regrouping the terms which are independent of δ_l, we have:

$$\sum_{l=0}^{\infty} i^l \sqrt{4\pi(2l+1)}\, \tilde{\varphi}_{k,l,0}(\mathbf{r}) \underset{r\to\infty}{\sim} - \sum_{l=0}^{\infty} i^l \sqrt{4\pi(2l+1)}\, Y_l^0(\theta) \times \left[\frac{e^{-ikr} e^{il\pi/2} - e^{ikr} e^{-il\pi/2}}{2ikr} - \frac{e^{ikr}}{r}\frac{1}{k} e^{-il\frac{\pi}{2}} e^{i\delta_l} \sin\delta_l\right] \tag{C-53}$$

Taking (C-25) and (C-31) into consideration, we recognize, in the first term of the right-hand side, the asymptotic expansion of the plane wave e^{ikz}, and we obtain finally:

$$\sum_{l=0}^{\infty} i^l \sqrt{4\pi(2l+1)}\, \tilde{\varphi}_{k,l,0}(\mathbf{r}) \underset{r\to\infty}{\sim} e^{ikz} + f_k(\theta)\frac{e^{ikr}}{r} \tag{C-54}$$

with*:

$$\boxed{f_k(\theta) = \frac{1}{k}\sum_{l=0}^{\infty} \sqrt{4\pi(2l+1)}\, e^{i\delta_l} \sin\delta_l\, Y_l^0(\theta)} \tag{C-55}$$

We have thus demonstrated that the expansion of (C-50) is correct and have found at the same time the expression for the scattering amplitude $f_k(\theta)$ in terms of the phase shifts δ_l.

b. CALCULATION OF THE CROSS SECTION

The differential scattering cross section is then given by formula (B-24):

$$\sigma(\theta) = |f_k(\theta)|^2 = \frac{1}{k^2}\left|\sum_{l=0}^{\infty} \sqrt{4\pi(2l+1)}\, e^{i\delta_l}\sin\delta_l\, Y_l^0(\theta)\right|^2 \tag{C-56}$$

* The factor i^l is compensated by $e^{-il\frac{\pi}{2}} = (-i)^l = \left(\frac{1}{i}\right)^l$.

from which we deduce the total scattering cross section by integrating over the angles:

$$\sigma = \int d\Omega \; \sigma(\theta) = \frac{1}{k^2} \sum_{l,l'} 4\pi \sqrt{(2l+1)(2l'+1)} \; e^{i(\delta_l - \delta_{l'})} \sin \delta_l \sin \delta_{l'} \times \int d\Omega \; Y_{l'}^{0*}(\theta) \, Y_l^0(\theta) \quad \text{(C-57)}$$

Since the spherical harmonics are orthonormal [formula (D-23) of chapter VI], we have finally:

$$\boxed{\sigma = \frac{4\pi}{k^2} \sum_{l=0}^{\infty} (2l+1) \sin^2 \delta_l} \quad \text{(C-58)}$$

Thus the terms resulting from interference between waves of different angular momenta disappear from the total cross section. For any potential $V(r)$, the contribution $\frac{4\pi}{k^2}(2l+1)\sin^2 \delta_l$ associated with a given value of l is positive and has an upper bound, for a given energy, of $\frac{4\pi}{k^2}(2l+1)$.

In theory, formulas (C-56) and (C-58) necessitate knowing all the phase shifts δ_l. Recall (*cf.* §C-3-a) that these phase shifts can be calculated from the radial equation if the potential $V(r)$ is known; this equation must be solved separately for each value of l (most of the time, moreover, this implies resorting to numerical techniques). In other words, the method of partial waves is attractive from a practical point of view only when there is a sufficiently small number of non-zero phase shifts. For a finite-range potential $V(r)$, we saw in §C-3-b-β that the phase shifts δ_l are negligible for $l > l_M$, the critical value l_M being defined by formula (C-48).

When the potential $V(r)$ is unknown at the outset, we attempt to reproduce the experimental curves which give the differential cross section at a fixed energy by introducing a small number of non-zero phase shifts. Furthermore, the very form of the θ-dependence of the cross section often suggests the minimum number of phase shifts needed. For example, if we limit ourselves to the *s*-wave, formula (C-56) gives an isotropic differential cross section (Y_0^0 is a constant). Thus if the experiments in fact imply a variation of $\sigma(\theta)$ with θ, it means that phase shifts other than that of the *s*-wave are not equal to zero. Once we have thereby determined, from experimental results corresponding to different energies, the phase shifts which do effectively contribute to the cross section, we can look for theoretical models of potentials which produce these phase shifts and their energy dependence.

COMMENT:

The dependence of cross sections on the energy $E = \hbar^2 k^2 / 2\mu$ of the incident particle is just as interesting as the θ-dependence of $\sigma(\theta)$. In particular, in certain cases, one observes rapid variations of the total cross section σ in the neighborhood of certain energy values. For example, if one of the phase shifts δ_l takes on the value $\pi/2$ for $E = E_0$, the corresponding contribution to σ reaches its upper limit and the cross section may show a sharp peak at $E = E_0$. This phenomenon is called "scattering resonance". We can compare it to the behavior described in chapter I (§ D-2-c-β) of the transmission coefficient of a "square" one-dimensional potential well.

References and suggestions for further reading:

Dicke and Wittke (1.14), chap. 16; Messiah (1.17), chap. X; Schiff (1.18), chaps. 5 and 9.

More advanced topics:

Coulomb scattering: Messiah (1.17), chap. XI; Schiff (1.18), § 21; Davydov (1.20), chap. XI, § 100.

Formal collision and S-matrix theory: Merzbacher (1.16), chap. 19; Roman (2.3), part II, chap. 4; Messiah (1.17), chap. XIX; Schweber (2.16), part 3, chap. 11.

Description of collisions in terms of wave packets: Messiah (1.17), chap. X, §§ 4, 5, 6; Goldberger and Watson (2.4), chaps. 3 and 4.

Determination of the potential from the phase shifts (the inverse problem): Wu and Ohmura (2.1), § G.

Applications: Davydov (1.20), chap. XI; Sobel'man (11.12), chap. 11; Mott and Massey (2.5); Martin and Spearman (16.18).

Scattering by multi-particle systems in the Born approximation and space-time correlation functions: Van Hove (2.39).

COMPLEMENTS OF CHAPTER VIII

Chapter VIII, at the upper limit of the program, is principally intended to serve as a reference for other courses, for example in nuclear physics. Physical applications of collision theory can be found in these courses.

A_{VIII} : THE FREE PARTICLE : STATIONARY STATES WITH WELL-DEFINED ANGULAR MOMENTUM

A_{VIII} : formal examination of stationary wave functions for a free particle with well-defined angular momentum. The use of the L_+ and L_- operators permits the introduction of spherical Bessel functions and the demonstration of a certain number of their properties which were used in §C of chapter VIII.

B_{VIII} : PHENOMENOLOGICAL DESCRIPTION OF COLLISIONS WITH ABSORPTION

B_{VIII} : permits the extension of the formalism of chapter VIII to collisions with absorption and establishes the "optical theorem". A phenomenological point of view is used whose principle is analogous to that of complement K_{III}. Not difficult if chapter VIII has been well assimilated.

C_{VIII} : SOME SIMPLE APPLICATIONS OF SCATTERING THEORY

C_{VIII} : illustration of the results of chapter VIII by several specific examples. Section 1 is recommended for a first reading, since it presents important physical results in a simple manner (Rutherford's formula). Section 2 can be considered as a worked example. Section 3 gives problems without their solutions.

Complement A_{VIII}

THE FREE PARTICLE: STATIONARY STATES WITH WELL-DEFINED ANGULAR MOMENTUM

We introduced, in §C-2 of chapter VIII, two distinct bases of stationary states of a free (spinless) particle whose Hamiltonian is written:

$$H_0 = \frac{\mathbf{P}^2}{2\mu} \tag{1}$$

The first of these bases is composed of the eigenstates common to H_0 and the three components of the momentum $\mathbf{P}$; the associated wave functions are the plane waves. The second consists of the stationary states with well-defined angular momentum, that is, the eigenstates common to H_0, $\mathbf{L}^2$ and L_z, whose principal properties we pointed out in §§ C-2-b, c and d of chapter VIII. We intend to study here this second basis in more detail. In particular, we wish to derive a certain number of results used in chapter VIII.

1. The radial equation

The Hamiltonian (1) commutes with the three components of the orbital angular momentum $\mathbf{L}$ of the particle:

$$[H_0, \mathbf{L}] = \mathbf{0} \tag{2}$$

Consequently, we can apply the general theory developped in §A of chapter VII to this particular problem. We therefore know that the free spherical waves (eigenfunctions common to H_0, $\mathbf{L}^2$ and L_z) are necessarily of the form:

$$\varphi^{(0)}_{\kappa,l,m}(\mathbf{r}) = R^{(0)}_{\kappa,l}(r)\, Y_l^m(\theta, \varphi) \tag{3}$$

The radial function $R^{(0)}_{\kappa,l}(r)$ is a solution of the equation:

$$\left[-\frac{\hbar^2}{2\mu}\frac{1}{r}\frac{\mathrm{d}^2}{\mathrm{d}r^2}r + \frac{l(l+1)\hbar^2}{2\mu r^2} \right] R^{(0)}_{\kappa,l}(r) = E_{\kappa,l}\, R^{(0)}_{\kappa,l}(r) \tag{4}$$

where $E_{\kappa,l}$ is the eigenvalue of H_0 corresponding to $\varphi^{(0)}_{\kappa,l,m}(\mathbf{r})$. If we set:

$$R^{(0)}_{\kappa,l}(r) = \frac{1}{r}\, u^{(0)}_{\kappa,l}(r) \tag{5}$$

the function $u^{(0)}_{\kappa,l}$ is given by the equation:

$$\left[\frac{\mathrm{d}^2}{\mathrm{d}r^2} - \frac{l(l+1)}{r^2} + \frac{2\mu E_{\kappa,l}}{\hbar^2}\right] u^{(0)}_{\kappa,l}(r) = 0 \tag{6}$$

to which we must add the condition :

$$u^{(0)}_{\kappa,l}(0) = 0 \tag{7}$$

It can be shown, first of all, that equations (6) and (7) enable us to find the spectrum of the Hamiltonian H_0, which we already know from the study of the plane waves [formula (C-5) of chapter VIII]. To do this, note that the minimum value of the potential (which is, in fact, identically zero) is zero and that consequently there cannot exist a stationary state with negative energy (*cf.* complement M_{III}). Consider, therefore, any positive value of the constant $E_{\kappa,l}$ appearing in equation (6), and set:

$$k = \frac{1}{\hbar}\sqrt{2\mu E_{\kappa,l}} \tag{8}$$

As r approaches infinity, the centrifugal term $l(l+1)/r^2$ becomes negligible compared to the constant term of equation (6), which can thus be approximated by:

$$\left[\frac{\mathrm{d}^2}{\mathrm{d}r^2} + k^2\right] u^{(0)}_{\kappa,l}(r) \underset{r\to\infty}{\simeq} 0 \tag{9}$$

Consequently, all solutions of equation (6) have an asymptotic behavior (linear combination of e^{ikr} and e^{-ikr}) which is physically acceptable. Therefore, the only restriction comes from condition (7): we know (*cf.* chap. VII, §A-3-b) that there exists, for a given value of $E_{\kappa,l}$, one and only one function (to within a constant factor) which satisfies (6) and (7). For any positive $E_{\kappa,l}$, the radial equation (6) thus has one and only one acceptable solution.

Thus, the spectrum of H_0 indeed includes all positive energies. Moreover, we see that the set of possible values of $E_{\kappa,l}$ does not depend on l; we shall therefore omit the index l for the energies. As for the index κ, we shall identify it with the constant defined in (8); this allows us to write :

$$E_k = \frac{\hbar^2 k^2}{2\mu} \quad ; \quad k \geqslant 0 \tag{10}$$

Each of these energies is infinitely degenerate. Indeed, for fixed k, there exists an acceptable solution $u^{(0)}_{k,l}(r)$ of the radial equation corresponding to the energy E_k for every value (positive integral or zero) of l. Moreover, formula (3) associates $(2l + 1)$ independent wave functions $\varphi_{k,l,m}^{(0)}(\mathbf{r})$ with a given radial function $u_{k,l}^{(0)}(r)$. Thus, we again find in this particular case the general result demonstrated in §A-3-b of chapter VII : H_0, $\mathbf{L}^2$ and L_z form a C.S.C.O. in $\mathscr{E}_\mathbf{r}$, and the specification of the three indices k, l and m gives sufficient information for the determination of a unique function in the corresponding basis.

2. Free spherical waves

The radial functions $R_{k,l}^{(0)}(r) = \frac{1}{r} u_{k,l}^{(0)}(r)$ can be found by solving equation (6) or equation (4) directly. The latter is easily reduced (comment of §2-c-β below) to a differential equation known as the "spherical Bessel equation" whose solutions are well-known. Instead of using these results directly, we are going to see how the various eigenfunctions common to H_0, $\mathbf{L}^2$ and L_2 can be simply deduced from those which correspond to the eigenvalue 0 of $\mathbf{L}^2$.

a. RECURRENCE RELATIONS

Let us define the operator:

$$P_+ = P_x + iP_y \tag{11}$$

in terms of the components P_x and P_y of the momentum $\mathbf{P}$. We know that $\mathbf{P}$ is a vectorial observable (*cf.* complement B_{VI}, §5-c), which implies the following commutation relations★ between its components and those of the angular momentum $\mathbf{L}$:

$$\begin{aligned} [L_x, P_x] &= 0 \\ [L_x, P_y] &= i\hbar P_z \\ [L_x, P_z] &= - i\hbar P_y \end{aligned} \tag{12}$$

and the equations which are deduced from these by cyclic permutation of the indices x, y, z. Using these relations, a simple algebraic calculation gives the commutators of L_z and $\mathbf{L}^2$ with the operator P_+ ; we find:

$$[L_z, P_+] = \hbar P_+ \tag{13-a}$$

$$[\mathbf{L}^2, P_+] = 2\hbar(P_+L_z - P_zL_+) + 2\hbar^2 P_+ \tag{13-b}$$

Consider therefore any eigenfunction $\varphi_{k,l,m}^{(0)}(\mathbf{r})$ common to H_0, $\mathbf{L}^2$ and L_z, the corresponding eigenvalues being E_k, $l(l + 1)\hbar^2$ and $m\hbar$. By applying the operators L_+ and L_-, we can obtain the $2l$ other eigenfunctions associated with the same energy E_k and the same value of l. Since H_0 commutes with $\mathbf{L}$, we have, for example:

$$H_0L_+ \varphi_{k,l,m}^{(0)}(\mathbf{r}) = L_+H_0 \varphi_{k,l,m}^{(0)}(\mathbf{r}) = E_kL_+ \varphi_{k,l,m}^{(0)}(\mathbf{r}) \tag{14}$$

★ These relations can be obtained directly from the definition $\mathbf{L} = \mathbf{R} \times \mathbf{P}$ and the canonical commutation rules.

and $L_+ \varphi^{(0)}_{k,l,m}(\mathbf{r})$ (which is not zero if m is different from l) is an eigenfunction of H_0 with the same eigenvalue as $\varphi^{(0)}_{k,l,m}(\mathbf{r})$. Therefore:

$$L_\pm \varphi^{(0)}_{k,l,m}(\mathbf{r}) \propto \varphi^{(0)}_{k,l,m\pm 1}(\mathbf{r}) \tag{15}$$

Let us now allow P_+ to act on $\varphi^{(0)}_{k,l,m}(\mathbf{r})$. First of all, since H_0 commutes with $\mathbf{P}$, we can repeat the preceding argument for $P_+ \varphi^{(0)}_{k,l,m}$. Moreover, from relation (13-a):

$$\begin{aligned} L_z P_+ \varphi^{(0)}_{k,l,m}(\mathbf{r}) &= P_+ L_z \varphi^{(0)}_{k,l,m} + \hbar P_+ \varphi^{(0)}_{k,l,m} \\ &= (m+1)\hbar\, P_+ \varphi^{(0)}_{k,l,m}(\mathbf{r}) \end{aligned} \tag{16}$$

$P_+ \varphi^{(0)}_{k,l,m}$ is therefore an eigenfunction of L_z with the eigenvalue $(m+1)\hbar$. If we use equation (13-b) in the same way, we see that the presence of the term $P_z L_+$ implies that $P_+ \varphi^{(0)}_{k,l,m}$ is not, in general, an eigenfunction of $\mathbf{L}^2$; nevertheless, if $m = l$, the contribution of this term is zero:

$$\begin{aligned} \mathbf{L}^2 P_+ \varphi^{(0)}_{k,l,l} &= P_+ \mathbf{L}^2 \varphi^{(0)}_{k,l,l} + 2\hbar P_+ L_z \varphi^{(0)}_{k,l,l} + 2\hbar^2 P_+ \varphi^{(0)}_{k,l,l} \\ &= [l(l+1) + 2l + 2]\hbar^2 P_+ \varphi^{(0)}_{k,l,l} \\ &= (l+1)(l+2)\hbar^2 P_+ \varphi^{(0)}_{k,l,l} \end{aligned} \tag{17}$$

Consequently, $P_+ \varphi^{(0)}_{k,l,l}$ is a common eigenfunction of H_0, $\mathbf{L}^2$ and L_z with the eigenvalues E_k, $(l+1)(l+2)\hbar^2$ and $(l+1)\hbar$ respectively. Since these three observables form a C.S.C.O. (§ 1), there exists only one eigenfunction (to within a constant factor*) associated with this set of eigenvalues:

$$P_+ \varphi^{(0)}_{k,l,l}(\mathbf{r}) \propto \varphi^{(0)}_{k,l+1,l+1}(\mathbf{r}) \tag{18}$$

We are going to use the recurrence relations (15) and (18) to construct the $\{ \varphi^{(0)}_{k,l,m}(\mathbf{r}) \}$ basis from the functions $\varphi^{(0)}_{k,0,0}(\mathbf{r})$ corresponding to zero eigenvalues for $\mathbf{L}^2$ and L_z**.

b. CALCULATION OF FREE SPHERICAL WAVES

α. *Solution of the radial equation for $l = 0$*

In order to determine the functions $\varphi^{(0)}_{k,0,0}(\mathbf{r})$, we return to the radial equation (6), in which we set $l = 0$; taking definition (10) into account, this equation can therefore be written:

$$\left[\frac{d^2}{dr^2} + k^2\right] u^{(0)}_{k,0}(r) = 0 \tag{19}$$

* Later (§ 2-b), we shall specify the coefficients which insure the orthonormalization of the $\{ \varphi^{(0)}_{k,l,m}(\mathbf{r}) \}$ basis (in the extended sense, since k is a continuous index).

** It must not be thought that the operator $P_- = P_x - iP_y$ allows one to "step down" from an arbitrary value of l to zero. It can easily be shown, by an argument analogous to the preceding one, that:

$$P_- \varphi^{(0)}_{k,l,-l}(\mathbf{r}) \propto \varphi^{(0)}_{k,l+1,-(l+1)}(\mathbf{r})$$

The solution which goes to zero at the origin [condition (7)] is of the form:

$$u_{k,0}^{(0)}(r) = a_k \sin kr \tag{20}$$

We choose the constant a_k such that the functions $\varphi_{k,0,0}^{(0)}(\mathbf{r})$ are orthonormal in the extended sense; that is:

$$\int d^3r \; \varphi_{k,0,0}^{(0)*}(\mathbf{r}) \, \varphi_{k',0,0}^{(0)}(\mathbf{r}) = \delta(k - k') \tag{21}$$

It is easy to show (see below) that condition (21) is satisfied if:

$$a_k = \sqrt{\frac{2}{\pi}} \tag{22}$$

which yields (Y_0^0 being equal to $1/\sqrt{4\pi}$):

$$\varphi_{k,0,0}^{(0)}(\mathbf{r}) = \sqrt{\frac{2k^2}{\pi}} \frac{1}{\sqrt{4\pi}} \frac{\sin kr}{kr} \tag{23}$$

Let us verify that the functions (23) satisfy the orthonormalization relation (21). To do this, it is sufficient to calculate:

$$\begin{aligned}\int d^3r \; \varphi_{k,0,0}^{(0)*}(\mathbf{r}) \, \varphi_{k',0,0}^{(0)}(\mathbf{r}) &= \frac{2}{\pi} kk' \frac{1}{4\pi} \int_0^\infty r^2 \, \mathrm{d}r \frac{\sin kr}{kr} \frac{\sin k'r}{k'r} \int \mathrm{d}\Omega \\ &= \frac{2}{\pi} \int_0^\infty \mathrm{d}r \sin kr \sin k'r \end{aligned} \tag{24}$$

Replacing the sines by complex exponentials and extending the interval of integration over the range $-\infty$ to $+\infty$, we obtain:

$$\frac{2}{\pi} \int_0^\infty \mathrm{d}r \sin kr \sin k'r = \frac{2}{\pi} \left(-\frac{1}{4} \right) \int_{-\infty}^{+\infty} \mathrm{d}r \left[\mathrm{e}^{i(k+k')r} - \mathrm{e}^{i(k-k')r} \right] \tag{25}$$

Since k and k' are both positive, $k + k'$ is always different from zero and the contribution of the first term within the brackets is always zero. According to formula (34) of appendix II, the second term yields finally:

$$\begin{aligned}\int d^3r \; \varphi_{k,0,0}^{(0)*}(\mathbf{r}) \, \varphi_{k',0,0}^{(0)}(\mathbf{r}) &= \frac{2}{\pi} \left(-\frac{1}{4} \right) (-2\pi) \, \delta(k - k') \\ &= \delta(k - k') \end{aligned} \tag{26}$$

β. *Construction of the other waves by recurrence*

Let us now apply the operator P_+ defined in (11) to the function $\varphi_{k,0,0}^{(0)}(\mathbf{r})$ that we have just found. According to relation (18):

$$\begin{aligned}\varphi_{k,1,1}^{(0)}(\mathbf{r}) &\propto P_+ \, \varphi_{k,0,0}^{(0)}(\mathbf{r}) \\ &\propto P_+ \frac{\sin kr}{kr} \end{aligned} \tag{27}$$

In the $\{ | \mathbf{r} \rangle \}$ representation, which we have been using throughout, P_+ is the differential operator:

$$P_+ = \frac{\hbar}{i} \left(\frac{\partial}{\partial x} + i \frac{\partial}{\partial y} \right) \tag{28}$$

In formula (27), it acts on a function of r alone. Now:

$$\begin{aligned} P_+ f(r) &= \frac{\hbar}{i} \left(\frac{x}{r} + i \frac{y}{r} \right) \frac{\mathrm{d}}{\mathrm{d}r} f(r) \\ &= \frac{\hbar}{i} \sin \theta \, \mathrm{e}^{i\varphi} \frac{\mathrm{d}}{\mathrm{d}r} f(r) \end{aligned} \tag{29}$$

Thus we obtain:

$$\varphi_{k,1,1}^{(0)}(\mathbf{r}) \propto \sin \theta \, \mathrm{e}^{i\varphi} \left[\frac{\cos kr}{kr} - \frac{\sin kr}{(kr)^2} \right] \tag{30}$$

We recognize the angular dependence of $Y_1^1(\theta, \varphi)$ [complement A_{VI}, formula (32)]; by applying L_-, $\varphi_{k,1,0}^{(0)}(\mathbf{r})$ and $\varphi_{k,1,-1}^{(0)}(\mathbf{r})$ can be calculated.

Although $\varphi_{k,1,1}^{(0)}(\mathbf{r})$ depends on θ and φ, the application of P_+ to this function remains very simple. The canonical commutation relations indicate immediately that:

$$[P_+, X + iY] = 0 \tag{31}$$

Consequently, $\varphi_{k,2,2}^{(0)}(\mathbf{r})$ is given by:

$$\begin{aligned} \varphi_{k,2,2}^{(0)}(\mathbf{r}) &\propto P_+^2 \frac{\sin kr}{kr} \\ &\propto P_+ \frac{x + iy}{r} \frac{\mathrm{d}}{\mathrm{d}r} \frac{\sin kr}{kr} \\ &\propto (x + iy) P_+ \frac{1}{r} \frac{\mathrm{d}}{\mathrm{d}r} \frac{\sin kr}{kr} \\ &\propto (x + iy)^2 \frac{1}{r} \frac{\mathrm{d}}{\mathrm{d}r} \left[\frac{1}{r} \frac{\mathrm{d}}{\mathrm{d}r} \frac{\sin kr}{kr} \right] \end{aligned} \tag{32}$$

In general:

$$\varphi_{k,l,l}^{(0)}(\mathbf{r}) \propto (x + iy)^l \left(\frac{1}{r} \frac{\mathrm{d}}{\mathrm{d}r} \right)^l \frac{\sin kr}{kr} \tag{33}$$

The angular dependence of $\varphi_{k,l,l}^{(0)}$ is contained in the factor:

$$(x + iy)^l = r^l (\sin \theta)^l \, \mathrm{e}^{il\varphi} \tag{34}$$

which is indeed proportional to $Y_l^l(\theta, \varphi)$.

Let us define:

$$j_l(\rho) = (-1)^l \rho^l \left(\frac{1}{\rho} \frac{\mathrm{d}}{\mathrm{d}\rho} \right)^l \frac{\sin \rho}{\rho} \tag{35}$$

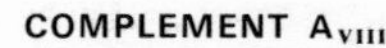

j_l, thus defined, is the *spherical Bessel function of order l*. The preceding calculation shows that $\varphi_{k,l,l}^{(0)}(\mathbf{r})$ is proportional to the product of $Y_l^l(\theta, \varphi)$ and $j_l(kr)$. We shall write (see below the problem of normalization):

$$R_{k,l}^{(0)}(r) = \sqrt{\frac{2k^2}{\pi}}\, j_l(kr) \tag{36}$$

The free spherical waves are then written:

$$\varphi_{k,l,m}^{(0)}(\mathbf{r}) = \sqrt{\frac{2k^2}{\pi}}\, j_l(kr)\, Y_l^m(\theta, \varphi) \tag{37}$$

They satisfy the orthonormalization relation:

$$\int d^3r\ \varphi_{k,l,m}^{(0)*}(\mathbf{r})\ \varphi_{k',l',m'}^{(0)}(\mathbf{r}) = \delta(k - k')\ \delta_{ll'}\ \delta_{mm'} \tag{38}$$

and the closure relation:

$$\int_0^\infty dk \sum_{l=0}^{\infty} \sum_{m=-l}^{+l} \varphi_{k,l,m}^{(0)}(\mathbf{r})\ \varphi_{k,l,m}^{(0)*}(\mathbf{r}') = \delta(\mathbf{r} - \mathbf{r}') \tag{39}$$

Let us now examine the normalization of the functions (37). To do so, let us begin by specifying the proportionality factors of the recurrence relations (15) and (18). For the first relation, we already know this factor from the properties of spherical harmonics (*cf.* complement A_{VI}):

$$L_\pm \varphi_{k,l,m}^{(0)}(\mathbf{r}) = \hbar\sqrt{l(l + 1) - m(m \pm 1)}\ \varphi_{k,l,m\pm 1}^{(0)}(\mathbf{r}) \tag{40}$$

As for relation (18), it can easily be shown, using the explicit expression for $Y_l^l(\theta, \varphi)$ [formulas (4) and (14) of complement A_{VI}], equations (31) and (29) and definition (35), that, if we take (37) into account, this relation can be written:

$$P_+ \varphi_{k,l,l}^{(0)}(\mathbf{r}) = \frac{\hbar k}{i}\sqrt{\frac{2l + 2}{2l + 3}}\ \varphi_{k,l+1,l+1}^{(0)}(\mathbf{r}) \tag{41}$$

In the orthonormalization relation (38), the factors $\delta_{ll'}\delta_{mm'}$ on the right-hand side arise from the angular integration and the orthonormality of the spherical harmonics. To establish relation (38), it is thus sufficient to show that the integral:

$$I_l(k, k') = \int d^3r\ \varphi_{k,l,l}^{(0)*}(\mathbf{r})\ \varphi_{k',l,l}^{(0)}(\mathbf{r}) \tag{42}$$

is equal to $\delta(k - k')$. We already know from (26) that $I_0(k, k')$ has this value. Consequently, we shall demonstrate that, if:

$$I_l(k, k') = \delta(k - k') \tag{43}$$

then the same is true for $I_{l+1}(k, k')$. Relation (41) permits us to write $I_{l+1}(k, k')$ in the form :

$$I_{l+1}(k, k') = \frac{1}{\hbar^2 kk'} \frac{2l+3}{2l+2} \int d^3r \, [P_+ \varphi_{k,l,l}^{(0)}(\mathbf{r})]^* \, [P_+ \varphi_{k',l,l}^{(0)}(\mathbf{r})]$$

$$= \frac{1}{\hbar^2 kk'} \frac{2l+3}{2l+2} \int d^3r \, \varphi_{k,l,l}^{(0)*}(\mathbf{r}) \, P_- P_+ \, \varphi_{k',l,l}^{(0)}(\mathbf{r}) \tag{44}$$

where $P_- = P_x - iP_y$ is the adjoint of P_+. Now:

$$P_- P_+ = P_x^2 + P_y^2 = \mathbf{P}^2 - P_z^2 \tag{45}$$

$\varphi_{k',l,l}^{(0)}$ is an eigenfunction of $\mathbf{P}^2$. Since, in addition, P_z is Hermitian, it results that :

$$I_{l+1}(k, k') = \frac{1}{\hbar^2 kk'} \frac{2l+3}{2l+2} \left\{ \hbar^2 k'^2 I_l(k, k') - \int d^3r \, [P_z \varphi_{k,l,l}^{(0)}(\mathbf{r})]^* \, [P_z \varphi_{k',l,l}^{(0)}(\mathbf{r})] \right\} \tag{46}$$

We must now calculate $P_z \varphi_{k,l,l}^{(0)}(\mathbf{r})$. Using the fact that $Y_l^l(\theta, \varphi)$ is proportional to $(x + iy)^l / r^l$, we easily find:

$$P_z \, \varphi_{k,l,l}^{(0)}(\mathbf{r}) = -\frac{\hbar k}{i} \sqrt{\frac{2k^2}{\pi}} \cos\theta \; Y_l^l(\theta, \varphi) \; j_{l+1}(kr)$$

$$= -\frac{\hbar k}{i} \frac{1}{\sqrt{2l+3}} \, \varphi_{k,l+1,l}^{(0)}(\mathbf{r}) \tag{47}$$

according to formula (35) of complement A_{VI}. Putting this result into (46), we finally obtain:

$$I_{l+1}(k, k') = \frac{2l+3}{2l+2} \frac{k'}{k} I_l(k, k') - \frac{1}{2l+2} I_{l+1}(k, k') \tag{48}$$

Hypothesis (43) thus implies:

$$I_{l+1}(k, k') = \delta(k - k') \tag{49}$$

which concludes the argument by recurrence.

c. PROPERTIES

α. *Behavior at the origin*

When ρ approaches zero, the function $j_l(\rho)$ behaves (see below) like:

$$j_l(\rho) \underset{\rho \to 0}{\simeq} \frac{\rho^l}{(2l+1)!!} \tag{50}$$

Consequently, $\varphi_{k,l,m}^{(0)}(\mathbf{r})$ is proportional to r^l in the neighborhood of the origin:

$$\varphi_{k,l,m}^{(0)}(\mathbf{r}) \underset{r \to 0}{\simeq} \sqrt{\frac{2k^2}{\pi}} \; Y_l^m(\theta, \varphi) \frac{(kr)^l}{(2l+1)!!} \tag{51}$$

To demonstrate formula (50), starting from definition (35), it is sufficient to expand $\sin\rho/\rho$ in a power series in ρ:

$$\frac{\sin\rho}{\rho} = \sum_{p=0}^{\infty} (-1)^p \frac{\rho^{2p}}{(2p+1)!} \tag{52}$$

We then apply the operator $\left(\frac{1}{\rho}\frac{\mathrm{d}}{\mathrm{d}\rho}\right)^l$, which yields:

$$\begin{aligned} j_l(\rho) &= (-1)^l \rho^l \left(\frac{1}{\rho}\frac{\mathrm{d}}{\mathrm{d}\rho}\right)^{l-1} \sum_{p=0}^{\infty} (-1)^p \frac{2p}{(2p+1)!} \rho^{2p-1-1} \\ &= (-1)^l \rho^l \sum_{p=0}^{\infty} (-1)^p \frac{2p(2p-2)(2p-4)\ldots[2p-2(l-1)]}{(2p+1)!} \rho^{2p-2l} \end{aligned} \tag{53}$$

The first l terms of the sum ($p = 0$ to $l - 1$) are zero, and the $(l + 1)$th is written:

$$j_l(\rho) \underset{\rho\to 0}{\sim} (-1)^l \rho^l (-1)^l \frac{2l(2l-2)(2l-4)\ldots 2}{(2l+1)!} \tag{54}$$

which proves (50).

β. *Asymptotic behavior*

When their argument approaches infinity, the spherical Bessel functions are related to the trigonometric functions in the following way:

$$j_l(\rho) \underset{\rho\to\infty}{\sim} \frac{1}{\rho} \sin\left(\rho - l\frac{\pi}{2}\right) \tag{55}$$

The asymptotic behavior of the free spherical waves is therefore:

$$\varphi_{k,l,m}^{(0)}(\mathbf{r}) \underset{r\to\infty}{\sim} \sqrt{\frac{2k^2}{\pi}}\, Y_l^m(\theta, \varphi) \frac{\sin(kr - l\pi/2)}{kr} \tag{56}$$

If we apply the operator $\frac{1}{\rho}\frac{\mathrm{d}}{\mathrm{d}\rho}$ once to $\frac{\sin\rho}{\rho}$, we can write $j_l(\rho)$ in the form:

$$j_l(\rho) = (-1)^l \rho^l \left(\frac{1}{\rho}\frac{\mathrm{d}}{\mathrm{d}\rho}\right)^{l-1} \left[\frac{\cos\rho}{\rho^2} - \frac{\sin\rho}{\rho^3}\right] \tag{57}$$

The second term inside the brackets is negligible compared to the first term when ρ approaches infinity. Moreover, when we apply $\frac{1}{\rho}\frac{\mathrm{d}}{\mathrm{d}\rho}$ a second time, the dominant term still comes from the derivative of the cosine. Thus we see that:

$$j_l(\rho) \underset{\rho\to\infty}{\sim} (-1)^l \rho^l \frac{1}{\rho^l}\frac{1}{\rho} \left(\frac{\mathrm{d}}{\mathrm{d}\rho}\right)^l \sin\rho \tag{58}$$

Since:

$$\left(\frac{\mathrm{d}}{\mathrm{d}\rho}\right)^l \sin \rho = (-1)^l \sin\left(\rho - l\frac{\pi}{2}\right) \tag{59}$$

the result is indeed (55).

COMMENT:

If we set:

$$kr = \rho \tag{60}$$

[k being defined by formula (10)], the radial equation (4) becomes:

$$\left[\frac{\mathrm{d}^2}{\mathrm{d}\rho^2} + \frac{2}{\rho}\frac{\mathrm{d}}{\mathrm{d}\rho} + \left(1 - \frac{l(l+1)}{\rho^2}\right)\right] R_l(\rho) = 0 \tag{61}$$

This is the spherical Bessel equation of order l. It has two linearly independent solutions, which can be distinguished, for example, by their behavior at the origin. One of them is then the spherical Bessel function $j_l(\rho)$, which satisfies (50) and (55). For the other, we can choose the "spherical Neumann function of order l", designated as $n_l(\rho)$, with the properties:

$$n_l(\rho) \underset{\rho \to 0}{\sim} \frac{(2l-1)!!}{\rho^{l+1}} \tag{62-a}$$

$$n_l(\rho) \underset{\rho \to \infty}{\sim} \frac{1}{\rho} \cos\left(\rho - l\frac{\pi}{2}\right) \tag{62-b}$$

3. Relation between free spherical waves and plane waves

We already know two distinct bases of eigenstates of H_0: the plane waves $v_{\mathbf{k}}^{(0)}(\mathbf{r})$ are eigenfunctions of the three components of the momentum $\mathbf{P}$; the free spherical waves $\varphi_{k,l,m}^{(0)}(\mathbf{r})$ are eigenfunctions of $\mathbf{L}^2$ and L_z. These two bases are different because $\mathbf{P}$ does not commute with $\mathbf{L}^2$ and L_z.

A given function of one of these bases can obviously be expanded in terms of the other basis. For example, we shall express a plane wave $v_{\mathbf{k}}^{(0)}(\mathbf{r})$ as a linear superposition of free spherical waves. Consider, therefore, a vector $\mathbf{k}$ in ordinary space. The plane wave $v_{\mathbf{k}}^{(0)}(\mathbf{r})$ that it characterizes is an eigenfunction of H_0 with the eigenvalue $\hbar^2\mathbf{k}^2/2\mu$. Therefore, its expansion will include only the $\varphi_{k,l,m}^{(0)}$ which correspond to this energy, that is those for which:

$$k = |\mathbf{k}| \tag{63}$$

This expansion will therefore be of the form:

$$v_{\mathbf{k}}^{(0)}(\mathbf{r}) = \sum_{l=0}^{\infty} \sum_{m=-l}^{+l} c_{l,m}(\mathbf{k})\, \varphi_{k,l,m}^{(0)}(\mathbf{r}) \tag{64}$$

the free indices **k** and k being related by equation (63). It is easy to show, using the properties of the spherical harmonics (*cf.* complement A_{VI}) and the spherical Bessel functions, that:

$$e^{i\mathbf{k}.\mathbf{r}} = 4\pi \sum_{l=0}^{\infty} \sum_{m=-l}^{+l} i^l \, Y_l^{m*}(\theta_k, \varphi_k) \, j_l(kr) \, Y_l^m(\theta, \varphi) \tag{65}$$

where θ_k and φ_k are the polar angles that fix the direction of the vector **k**. If **k** is directed along Oz, expansion (65) reduces to:

$$\begin{aligned} e^{ikz} &= \sum_{l=0}^{\infty} i^l \sqrt{4\pi(2l+1)} \, j_l(kr) \, Y_l^0(\theta) \\ &= \sum_{l=0}^{\infty} i^l \, (2l+1) \, j_l(kr) \, P_l(\cos\theta) \end{aligned} \tag{66}$$

where P_l is the Legendre polynomial of degree l [*cf.* equation (57) of complement A_{VI}].

Let us first demonstrate relation (66). To do this, let us assume that the vector **k** chosen is collinear with Oz :

$$k_x = k_y = 0 \tag{67}$$

and points in the same direction. In this case, equation (63) becomes:

$$k_z = k \tag{68}$$

and we want to expand the function:

$$e^{ikz} = e^{ikr\cos\theta} \tag{69}$$

in the $\{ \varphi_{k,l,m}^{(0)}(\mathbf{r}) \}$ basis. Since this function is independent of the angle φ, it is a linear combination of only those basis functions for which $m = 0$:

$$\begin{aligned} e^{ikr\cos\theta} &= \sum_{l=0}^{\infty} a_l \, \varphi_{k,l,0}^{(0)}(\mathbf{r}) \\ &= \sum_{l=0}^{\infty} c_l \, j_l(kr) \, Y_l^0(\theta) \end{aligned} \tag{70}$$

To calculate the numbers c_l, we can consider $e^{ikr\cos\theta}$ to be a function of the variable θ, with r playing the role of a parameter. Since the spherical harmonics form an orthonormal basis for functions of θ and φ, the "coefficient" $c_l j_l(kr)$ can be expressed as:

$$c_l \, j_l(kr) = \int d\Omega \; Y_l^{0*}(\theta) \, e^{ikr\cos\theta} \tag{71}$$

Replacing Y_l^0 by its expression in terms of $Y_l^l(\theta, \varphi)$ [formula (25) of complement A_{VI}], we obtain:

$$\begin{aligned} c_l \, j_l(kr) &= \frac{1}{\sqrt{(2l)!}} \int d\Omega \left[\left(\frac{L_-}{\hbar} \right)^l Y_l^l(\theta, \varphi) \right]^* e^{ikr\cos\theta} \\ &= \frac{1}{\sqrt{(2l)!}} \int d\Omega \, Y_l^{l*}(\theta, \varphi) \left[\left(\frac{L_+}{\hbar} \right)^l e^{ikr\cos\theta} \right] \end{aligned} \tag{72}$$

since L_+ is the adjoint operator of L_-. Formula (16) of complement A_{VI} then yields:

$$\left(\frac{L_+}{\hbar}\right)^l e^{ikr\cos\theta} = (-1)^l \, e^{il\varphi} \, (\sin\theta)^l \frac{d^l}{d(\cos\theta)^l} \, e^{ikr\cos\theta}$$
$$= (-1)^l \, e^{il\varphi} \, (\sin\theta)^l \, (ikr)^l \, e^{ikr\cos\theta} \tag{73}$$

Now $(\sin\theta)^l \, e^{il\varphi}$ is just, to within a constant factor, $Y_l^l(\theta, \varphi)$ [*cf.* formulas (4) and (14) of complement A_{VI}]. Consequently:

$$c_l \, j_l(kr) = (ikr)^l \frac{2^l \, l!}{\sqrt{(2l)!}} \sqrt{\frac{4\pi}{(2l+1)!}} \int d\Omega \, |Y_l^l(\theta, \varphi)|^2 \, e^{ikr\cos\theta} \tag{74}$$

It is therefore sufficient to choose a particular value of kr, for which we know the value of $j_l(kr)$, in order to calculate c_l. Allow, for example, kr to approach zero: we know that $j_l(kr)$ behaves like $(kr)^l$, and so, in fact, does the right-hand side of equation (74). More precisely, using (50), we find:

$$c_l \frac{1}{(2l+1)!!} = i^l \frac{2^l \, l!}{\sqrt{(2l)!}} \sqrt{\frac{4\pi}{(2l+1)!}} \int d\Omega \, |Y_l^l(\theta, \varphi)|^2 \tag{75}$$

that is, since Y_l^l is normalized to 1:

$$c_l = i^l \sqrt{4\pi(2l+1)} \tag{76}$$

This proves formula (66).

The general relation (65) can therefore be obtained as a consequence of the addition theorem for spherical harmonics [formula (70) of complement A_{VI}]. Whatever the direction of $\mathbf{k}$ (defined by the polar angles θ_k and φ_k), it is always possible, through a rotation of the system of axes, to return to the case we have just considered. Consequently, expansion (66) remains valid, providing that kz is replaced by $\mathbf{k} \cdot \mathbf{r}$ and $\cos\theta$ by $\cos\alpha$, where α is the angle between $\mathbf{k}$ and $\mathbf{r}$:

$$e^{i\mathbf{k}\cdot\mathbf{r}} = \sum_{l=0}^{\infty} i^l \, (2l+1) \, j_l(kr) \, P_l(\cos\alpha) \tag{77}$$

But the addition theorem for spherical harmonics permits the expression of $P_l(\cos\alpha)$ in terms of the angles (θ, φ) and (θ_k, φ_k), which yields finally formula (65).

The expansions (65) and (66) show that *a state of well-defined linear momentum involves all possible orbital angular momenta.*

To obtain the expansion of a given function $\varphi_{k,l,m}^{(0)}(\mathbf{r})$ in terms of plane waves, it is sufficient to invert formula (65), using the orthonormalization relation of spherical harmonics which are functions of θ_k and φ_k. This yields:

$$\int d\Omega_k \, Y_l^{m*}(\theta_k, \varphi_k) \, e^{i\mathbf{k}\cdot\mathbf{r}} = 4\pi i^l \, j_l(kr) \, Y_l^m(\theta, \varphi) \tag{78}$$

Thus we find:

$$\varphi_{k,l,m}^{(0)}(\mathbf{r}) = \frac{(-1)^l}{4\pi} i^l \sqrt{\frac{2k^2}{\pi}} \int \mathrm{d}\Omega_k \; Y_l^m(\theta_k, \varphi_k) \; \mathrm{e}^{i\mathbf{k}.\mathbf{r}} \tag{79}$$

An eigenfunction of $\mathbf{L}^2$ and L_z is therefore a linear superposition of all plane waves with the same energy : *a state of well-defined angular momentum involves all possible directions of the linear momentum.*

References:

Messiah (1.17), App. B, §6; Arfken (10.4), §11.6; Butkov (10.8), chap. 9, §9; see the subsection "Special functions and tables" of section 10 of the bibliography.

Complement B_{VIII}

PHENOMENOLOGICAL DESCRIPTION OF COLLISIONS WITH ABSORPTION

In chapter VIII, we confined ourselves to the study of the elastic scattering of particles by a potential. But we pointed out in the introduction of chap. VIII that collisions between particles can be inelastic and lead, under certain conditions, to numerous other reactions (creation or destruction of particles, etc...), particularly if the energy of the incident particles is high. When such reactions are possible, and one detects only elastically scattered particles, one observes that certain particles of the incident beam "disappear"; that is, they are not to be found either in the transmitted beam or amongst the elastically scattered particles. These particles are said to be "absorbed" during the interaction; in reality, they have taken part in reactions other than that of simple elastic scattering. If one is interested only in the elastic scattering, one seeks to describe the "absorption" globally, without going into detail about the other possible reactions. We are going to show here that the method of partial waves provides a convenient framework for such a phenomenological description.

1. Principle involved

We shall assume that the interactions responsible for the disappearance of the incident particles are invariant with respect to rotation about O. The scattering amplitude can therefore always be decomposed into partial waves, each of which corresponds to a fixed value of the angular momentum.

In this section, we shall see how the method of partial waves can be modified to take a possible absorption into consideration. To do this, let us return to the interpretation of partial waves that we gave in § C-3-b-α of chapter VIII. A free incoming wave penetrates the zone of influence of the potential and gives rise to an outgoing wave. The effect of the potential is to multiply this outgoing wave by $e^{2i\delta_l}$. Since the modulus of this factor is 1 (the phase shift δ_l is real), the amplitude of the outgoing wave is equal to that of the incoming wave. Consequently (see the calculation of §2-b below), the total flux of the incoming wave is equal to that of the outgoing wave: during the scattering, probability is conserved, that is, the total number of particles is constant. These considerations suggest that, in the cases where absorption phenomena occur, one can take them into account simply by giving the phase shift an imaginary part such that:

$$|e^{2i\delta_l}| < 1 \tag{1}$$

The amplitude of the outgoing wave with angular momentum l is thus smaller than that of the incoming wave from which it arises. The fact that the outgoing probability flux is smaller than the incoming flux expresses the "disappearance" of a certain number of particles.

We are going to make this idea more explicit and deduce from it the expressions for the scattering and absorption cross sections. However, we stress the fact that this is a purely phenomenological method: the parameters with which we shall characterize the absorption (modulus of $e^{2i\delta_l}$ for each partial wave) mask an often very complicated reality. Note also that if the total probability is no longer conserved it is impossible to describe the interaction by a simple potential. A correct treatment of the set of phenomena which can then arise during the collision would demand a more elaborate formalism than the one developed in chapter VIII.

2. Calculation of the cross sections

We return to the calculations of §C-4 of chapter VIII, setting:

$$\eta_l = e^{2i\delta_l} \tag{2}$$

Since the possibility of producing reactions other than that of elastic scattering is always expressed by a decrease in the number of elastically scattered particles, we must have:

$$|\eta_l| \leqslant 1 \tag{3}$$

(equality corresponding to cases where only elastic scattering is possible). The asymptotic form of the wave function which describes the elastic scattering is therefore [*cf.* formula (C-51) of chapter VIII]:

$$v_k^{(\text{diff})}(\mathbf{r}) \underset{r\to\infty}{\sim} -\sum_{l=0}^{\infty} i^l \sqrt{4\pi(2l+1)}\, Y_l^0(\theta)\, \frac{e^{-ikr}\, e^{il\frac{\pi}{2}} - \eta_l\, e^{ikr}\, e^{-il\frac{\pi}{2}}}{2ikr} \tag{4}$$

a. ELASTIC SCATTERING CROSS SECTION

The argument of §C-4-a of chapter VIII remains valid and gives the scattering amplitude $f_k(\theta)$ in the form:

$$f_k(\theta) = \frac{1}{k}\sum_{l=0}^{\infty} \sqrt{4\pi(2l+1)}\, Y_l^0(\theta)\, \frac{\eta_l - 1}{2i} \tag{5}$$

From this we deduce the differential elastic scattering cross section:

$$\sigma_{\text{el}}(\theta) = \frac{1}{k^2}\left|\sum_{l=0}^{\infty} \sqrt{4\pi(2l+1)}\, Y_l^0(\theta)\, \frac{\eta_l - 1}{2i}\right|^2 \tag{6}$$

and the total elastic scattering cross section:

$$\sigma_{\text{el}} = \frac{\pi}{k^2}\sum_{l=0}^{\infty} (2l+1)\, |1 - \eta_l|^2 \tag{7}$$

COMMENT:

According to the argument developed in § 1, the absorption of the wave (l) reaches a maximum when $|\eta_l|$ is zero, that is, when:

$$\eta_l = 0 \tag{8}$$

Formula (7) indicates however that, even in this limiting case, the contribution of the wave (l) to the elastic scattering cross section is not zero★. In other words, even if the interaction region is perfectly absorbing, it produces elastic scattering. This important phenomenon is a purely quantum effect. It can be compared to the behavior of a light wave which strikes an absorbing medium. Even if the absorption is total (perfectly black sphere or disc), a diffracted wave is observed (concentrated into a solid angle which becomes smaller as the surface of the disc becomes larger). Elastic scattering produced by a totally absorbing interaction is called, for this reason, *shadow scattering*.

b. ABSORPTION CROSS SECTION

Following the same principle as in §A-3 of chapter VIII, we define the absorption cross section σ_{abs}: it is the ratio between the number of particles absorbed per unit time and the incident flux.

To calculate this cross section, it is sufficient, as in §B-2 of chapter VIII, to evaluate the total amount of probability $\Delta\mathscr{P}$ which "disappears" per unit time. This probability can be obtained from the current $\mathbf{J}$ associated with the wave function (4). $\Delta\mathscr{P}$ is equal to the difference between the flux of the incoming waves across a sphere (S) of very large radius R_0 and that of the outgoing waves; it is therefore equal to minus the net flux of the vector $\mathbf{J}$ leaving this sphere. Thus:

$$\Delta\mathscr{P} = - \int_{(S)} \mathbf{J} \, . \, \mathrm{d}\mathbf{S} \tag{9}$$

with:

$$\mathbf{J} = \mathrm{Re}\left[v_k^{(diff)*}(\mathbf{r}) \, \frac{\hbar}{i\mu} \, \nabla v_k^{(diff)}(\mathbf{r}) \right] \tag{10}$$

Only the radial component J_r of the current contributes to the integral (9):

$$\Delta\mathscr{P} = - \int_{r = R_0} J_r \, r^2 \, \mathrm{d}\Omega \tag{11}$$

with:

$$J_r = \mathrm{Re}\left[v_k^{(diff)*}(\mathbf{r}) \, \frac{\hbar}{i\mu} \, \frac{\partial}{\partial r} \, v_k^{(diff)}(\mathbf{r}) \right] \tag{12}$$

★ This contribution is zero only if $\eta_l = 1$, that is, if the phase shift is real and equal to an integral multiple of π [this was already contained in formula (C-58) of chapter VIII].

In formula (12), the differentiation does not modify the angular dependence of the various terms which compose $v_k^{(\text{diff})}(\mathbf{r})$ [formula (4)]. Consequently, because of the orthogonality of the spherical harmonics, the cross terms between a partial wave (l) in $v_k^{(\text{diff})}(\mathbf{r})$ and a different wave (l') in $v_k^{(\text{diff})*}(\mathbf{r})$ make a zero contribution to integral (11). We have therefore:

$$\Delta\mathscr{P} = - \sum_{l=0}^{\infty} \int_{r=R_0} J_r^{(l)} \, r^2 \, d\Omega \tag{13}$$

where $J_r^{(l)}$ is the radial component of the current associated with the partial wave (l). A simple calculation gives:

$$J_r^{(l)} \underset{r\to\infty}{\sim} - \frac{\hbar k}{\mu} \frac{\pi(2l+1)}{k^2 r^2} [1 - |\eta_l|^2] \, |Y_l^0(\theta)|^2 \tag{14}$$

that is, finally, since $Y_l^0(\theta)$ is normalized:

$$\Delta\mathscr{P} = \frac{\hbar k}{\mu} \frac{\pi}{k^2} \sum_{l=0}^{\infty} (2l+1) [1 - |\eta_l|^2] \tag{15}$$

The absorption cross section σ_{abs} is therefore equal to the probability $\Delta\mathscr{P}$ divided by the incident current $\hbar k/\mu$:

$$\sigma_{\text{abs}} = \frac{\pi}{k^2} \sum_{l=0}^{\infty} (2l+1) [1 - |\eta_l|^2] \tag{16}$$

It is obvious that σ_{abs} is zero if all the η_l have a modulus of 1; that is, according to (2), if all the phase shifts are real. In this case, there is only elastic scattering, and the net flux of probability leaving a sphere of large radius R_0 is always zero. The total probability carried by the incoming waves is entirely transferred to the outgoing waves. On the other hand, when η_l is zero, the contribution of the wave (l) to the absorption cross section is maximum.

COMMENT:

The calculation of expression (15) shows that $\frac{\hbar k}{\mu} \frac{\pi}{k^2} (2l+1)$ is the amount of probability entering per unit time, and arising from the partial wave (l). Il we divide this quantity by the incident current $\hbar k/\mu$, we obtain a surface that can be called the "incoming cross section into the partial wave (l)":

$$\sigma_l = \frac{\pi}{k^2} (2l+1) \tag{17}$$

This formula can be interpreted classically. We can consider the incident plane wave as describing a beam of particles of uniform density, having a momentum $\hbar k$ parallel to Oz. What proportion of these particles reach the scattering potential, with an angular momentum $\hbar\sqrt{l(l+1)}$? We have already mentioned the link between angular momentum and the impact parameter in classical mechanics [*cf.* formula (C-23) of chapter VIII]:

$$|\mathscr{L}| = b \, |\mathbf{p}| = \hbar k \, b \tag{18}$$

All we must do, therefore, is to draw, in the plane passing through O and perpendicular to Oz, a circular ring centered at O, of average radius b_l such that:

$$\hbar\sqrt{l(l+1)} = \hbar k\, b_l \tag{19}$$

and of width Δb_l corresponding to $\Delta l = 1$ in formula (19) (fig. 1). All the particles crossing this surface reach the scattering potential with an angular momentum equal to $\hbar\sqrt{l(l+1)}$, to within $\hbar$. From (19) we derive:

$$b_l = \frac{1}{k}\sqrt{l(l+1)} \simeq \frac{1}{k}\left(l + \frac{1}{2}\right) \tag{20}$$

if $l \gg 1$, and consequently:

$$\Delta b_l = \frac{1}{k} \tag{21}$$

The area of the circular ring of figure 1 is therefore:

$$2\pi\, b_l\, \Delta b_l \simeq \frac{\pi}{k^2}(2l+1) \tag{22}$$

Thus we find again, very simply, σ_l.

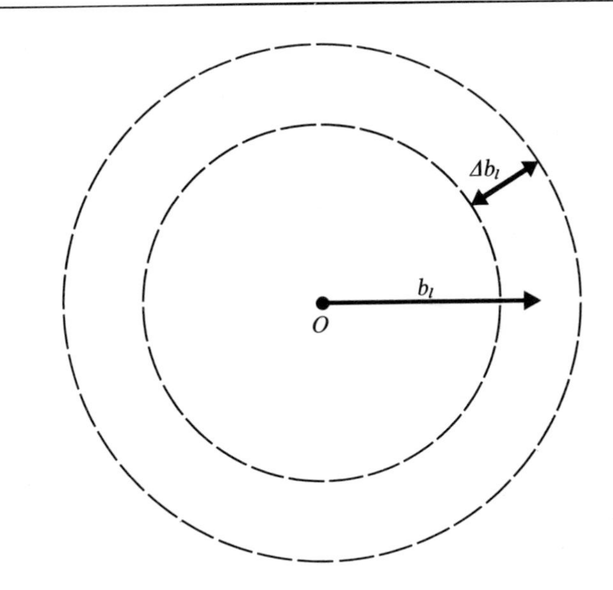

FIGURE 1

The incident particles must reach the potential with the impact parameter b_l to within Δb_l for their classical angular momentum to be $\hbar\sqrt{l(l+1)}$ to within $\hbar$.

c. TOTAL CROSS SECTION. OPTICAL THEOREM

When a collision can give rise to several different reactions or scattering phenomena, the total cross section σ_{tot} is defined as the sum of the cross sections (integrated over all the directions of space) corresponding to all these processes. The total cross section is thus the number of particles which, per unit time, participate in one or another of the possible reactions, divided by the incident flux.

If, as above, we treat globally all reactions other than elastic scattering, we have simply:

$$\sigma_{tot} = \sigma_{el} + \sigma_{abs} \tag{23}$$

Formulas (7) and (16) then give:

$$\sigma_{tot} = \frac{2\pi}{k^2} \sum_{l=0}^{\infty} (2l + 1)(1 - \text{Re}\,\eta_l) \tag{24}$$

Now $(1 - Re\,\eta_l)$ is the real part of $(1 - \eta_l)$, which appears in the elastic scattering amplitude [formula (5)]. Moreover, we know the value of $Y_l^0(\theta)$ for $\theta = 0$:

$$Y_l^0(0) = \sqrt{\frac{2l + 1}{4\pi}} \tag{25}$$

[*cf.* complement A_{VI}, formulas (57) and (60)]. Consequently, if we calculate from (5) the imaginary part of the elastic scattering amplitude in the forward direction, we find:

$$\text{Im}\, f_k(0) = \frac{1}{k} \sum_{l=0}^{\infty} (2l + 1) \frac{1 - \text{Re}\,\eta_l}{2} \tag{26}$$

Comparing this expression to formula (24), we see that:

$$\sigma_{tot} = \frac{4\pi}{k} \text{Im}\, f_k(0) \tag{27}$$

This relation between the total cross section and the imaginary part of the elastic scattering amplitude in the forward direction is valid in a very general sense; it constitutes what is called the *optical theorem*.

COMMENT:

The optical theorem is obviously valid in the case of purely elastic scattering ($\sigma_{abs} = 0$; $\sigma_{tot} = \sigma_{el}$). The fact that $f_k(0)$ – that is, the wave scattered in the forward direction – is related to the total cross section could have been predicted from the discussion in §B-2-d of chapter VIII. It is the interference in the forward direction between the incident plane wave and the scattered wave which accounts for the attenuation of the transmitted beam, due to the scattering of particles in all directions of space.

References and suggestions for further reading:

Optical model: Valentin (16.1), §X-3. High energy proton-proton collisions: Amaldi (16.31).

Complement C_{VIII}

SOME SIMPLE APPLICATIONS OF SCATTERING THEORY

There is no potential for which the scattering problem can be solved exactly★ by a simple analytical calculation. Therefore, in the examples that we are going to discuss here, we shall content ourselves with using the approximations that we introduced in chapter VIII.

1. The Born approximation for a Yukawa potential

Let us consider a potential of the form:

$$V(\mathbf{r}) = V_0 \frac{e^{-\alpha r}}{r} \tag{1}$$

where V_0 and α are real constants, with α positive. This potential is attractive or repulsive depending on whether V_0 is negative or positive. The larger $|V_0|$, the more intense the potential. Its range is characterized by the distance:

$$r_0 = \frac{1}{\alpha} \tag{2}$$

since, as figure 1 shows, $V(r)$ is practically zero when r exceeds $2r_0$ or $3r_0$.

The potential (1) bears the name of Yukawa, who had the idea of associating it with nuclear forces, whose range is of the order of a fermi. To explain the origin of this potential, Yukawa was led to predict the existence of the π-meson, which was indeed later discovered. Notice that for $\alpha = 0$ this potential becomes the Coulomb potential, which can thus be considered to be a Yukawa potential of infinite range.

★ Actually, we can rigorously treat the case of the Coulomb potential; however, as we pointed out in chapter VIII (§B-1), this necessitates a special method.

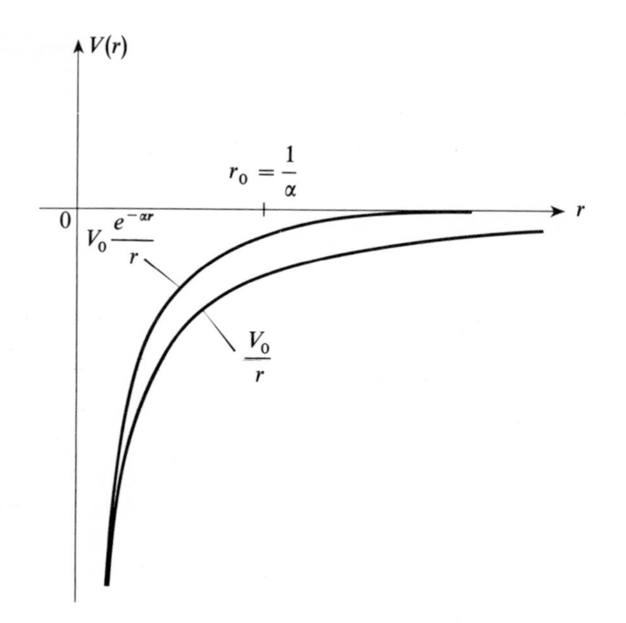

FIGURE 1

Yukawa potential and Coulomb potential. The presence of the term $e^{-\alpha r}$ causes the Yukawa potential to approach zero much more rapidly when $r \gg r_0 = 1/\alpha$ (range of the potential).

a. CALCULATION OF THE SCATTERING AND AMPLITUDE CROSS SECTION

We assume that $|V_0|$ is sufficiently small for the Born approximation (§ B-4 of chapter VIII) to be valid. According to formula (B-47) of chapter VIII, the scattering amplitude $f_k^{(B)}(\theta, \varphi)$ is then given by:

$$f_k^{(B)}(\theta, \varphi) = -\frac{1}{4\pi}\frac{2\mu V_0}{\hbar^2}\int d^3r \; e^{-i\mathbf{K}.\mathbf{r}} \frac{e^{-\alpha r}}{r} \tag{3}$$

where $\mathbf{K}$ is the momentum transferred in the direction (θ, φ) defined by relation (B-42) of chapter VIII.

Expression (3) involves the Fourier transform of the Yukawa potential. Since this potential depends only on the variable r, the angular integrations can easily be carried out (§2-e of appendix I), putting the scattering amplitude into the form:

$$f_k^{(B)}(\theta, \varphi) = -\frac{1}{4\pi}\frac{2\mu V_0}{\hbar^2}\frac{4\pi}{|\mathbf{K}|}\int_0^{\infty} r\, dr \sin|\mathbf{K}|r \; \frac{e^{-\alpha r}}{r} \tag{4}$$

After a simple calculation, we then find:

$$f_k^{(B)}(\theta, \varphi) = -\frac{2\mu V_0}{\hbar^2}\frac{1}{\alpha^2 + |\mathbf{K}|^2} \tag{5}$$

Figure 6 of chapter VIII shows that:

$$|\mathbf{K}| = 2k \sin\frac{\theta}{2} \tag{6}$$

where k is the modulus of the incident wave vector and θ is the scattering angle.

The differential scattering cross section is therefore written, in the Born approximation :

$$\sigma^{(B)}(\theta) = \frac{4\mu^2 V_0^2}{\hbar^4} \frac{1}{[\alpha^2 + 4k^2 \sin^2 \theta/2]^2} \tag{7}$$

It is independent of the azimuthal angle φ, as could have been foreseen from the fact that the problem of scattering by a central potential is symmetrical with respect to rotation about the direction of the incident beam. On the other hand, it depends, for a given energy (that is, for fixed k), on the scattering angle: in particular, the cross section in the forward direction ($\theta = 0$) is larger than the cross section in the backward direction ($\theta = \pi$). Finally, $\sigma^{(B)}(\theta)$, for fixed θ, is a decreasing function of the energy. Notice, moreover, that the sign of V_0 is of no importance in the scattering problem, at least in the Born approximation.

The total scattering cross section is easily obtained by integration:

$$\sigma^{(B)} = \int \mathrm{d}\Omega \; \sigma^{(B)}(\theta) = \frac{4\mu^2 V_0^2}{\hbar^4} \frac{4\pi}{\alpha^2(\alpha^2 + 4k^2)} \tag{8}$$

b. THE INFINITE-RANGE LIMIT

We noted above that the Yukawa potential approaches a Coulomb potential when α tends towards zero. What happens, in this limiting case, to the formulas that we have just established?

To obtain the Coulomb interaction potential between two particles having charges of $Z_1 q$ and $Z_2 q$ (q being the charge of the electron), we write:

$$\begin{aligned} \alpha &= 0 \\ V_0 &= Z_1 Z_2 e^2 \end{aligned} \tag{9}$$

with:

$$e^2 = \frac{q^2}{4\pi\varepsilon_0} \tag{10}$$

Formula (7) then gives:

$$\begin{aligned} \sigma^{(C)}(\theta) &= \frac{4\mu^2}{\hbar^4} \frac{Z_1^2 Z_2^2 e^4}{16\, k^4 \sin^4 \dfrac{\theta}{2}} \\ &= \frac{Z_1^2 Z_2^2 e^4}{16 E^2 \sin^4 \dfrac{\theta}{2}} \end{aligned} \tag{11}$$

(k has been replaced by its value in terms of the energy).

Expression (11) is indeed that of the Coulomb scattering cross section (*Rutherford's formula*). Of course, the way in which we have obtained it does not constitute a proof: the theory we have used is not applicable to the Coulomb potential. However, it is interesting to observe that the Born approximation for the Yukawa potential gives precisely Rutherford's formula for the limiting situation where the range of the potential approaches infinity.

COMMENT:

The total scattering cross section for a Coulomb potential is infinite since the corresponding integral diverges for small values of θ [expression (8) becomes infinite when α approaches zero]. This results from the infinite range of the Coulomb potential: even if the particle passes very far from the point O, it is affected by the potential. This suggests why the scattering cross section should be infinite. However, in reality, one never observes a rigorously pure Coulomb interaction over an infinite range. The potential created by a charged particle is always modified by the presence, in its more or less immediate neighborhood, of other particles of opposite charge (screening effect).

2. Low energy scattering by a hard sphere

Let us consider a central potential such that:

$$\begin{aligned} V(r) &= 0 \qquad \text{for} \quad r > r_0 \\ &= \infty \qquad \text{for} \quad r < r_0 \end{aligned} \tag{12}$$

In this case, we say that we are considering a "hard sphere" of radius r_0. We assume that the energy of the incident particle is sufficiently small for kr_0 to be much smaller than 1. We can then (§ C-3-b-β of chapter VIII and exercise 3-a below) neglect all the phase shifts except that of the s wave ($l = 0$). The scattering amplitude $f_k(\theta)$ is written, under these conditions:

$$f_k(\theta) = \frac{1}{k} e^{i\delta_0(k)} \sin \delta_0(k) \tag{13}$$

(since $Y_0^0 = 1/\sqrt{4\pi}$); the differential cross section is isotropic:

$$\sigma(\theta) = |f_k(\theta)|^2 = \frac{1}{k^2} \sin^2 \delta_0(k) \tag{14}$$

so that the total cross section is simply equal to:

$$\sigma = \frac{4\pi}{k^2} \sin^2 \delta_0(k) \tag{15}$$

To calculate the phase shift $\delta_0(k)$, it is necessary to solve the radial equation corresponding to $l = 0$. This equation is written [*cf.* formula (C-35) of chapter VIII]:

$$\left[\frac{d^2}{dr^2} + k^2\right] u_{k,0}(r) = 0 \qquad \text{for} \qquad r > r_0 \tag{16}$$

which must be completed by the condition:

$$u_{k,0}(r_0) = 0 \tag{17}$$

since the potential becomes infinite for $r = r_0$. The solution $u_{k,0}(r)$ of equations (16) and (17) is unique to within a constant factor:

$$\begin{aligned} u_{k,0}(r) &= C \sin k(r - r_0) \qquad \text{for} \qquad r > r_0 \\ &= 0 \qquad\qquad\qquad\quad \text{for} \qquad r < r_0 \end{aligned} \tag{18}$$

The phase shift δ_0 is, by definition, given by the asymptotic form of $u_{k,0}(r)$:

$$u_{k,0}(r) \underset{r \to \infty}{\sim} \sin(kr + \delta_0) \tag{19}$$

Thus, using solution (18), we find:

$$\delta_0(k) = -kr_0 \tag{20}$$

If we insert this value into expression (15) for the total cross section, we obtain:

$$\sigma = \frac{4\pi}{k^2}\sin^2 kr_0 \simeq 4\pi r_0^2 \tag{21}$$

since by hypothesis kr_0 is much smaller than 1. Therefore, σ is independent of the energy and equal to four times the apparent surface of the hard sphere seen by the particles of the incident beam. A calculation based on classical mechanics would give for the cross section the apparent surface πr_0^2 : only the particles which bounce elastically off the hard sphere would be deflected. In quantum mechanics, however, one studies the evolution of the wave associated with the incident particles, and the abrupt variation of $V(r)$ at $r = r_0$ produces a phenomenon analogous to the diffraction of a light wave.

COMMENT:

Even when the wavelength of the incident particles becomes negligible compared to r_0 ($kr_0 \gg 1$), the quantum cross section does not approach πr_0^2. It is possible, for very large k, to sum the series which gives the total cross section in terms of phase shifts [formula (C-58) of chapter VIII]; we then find:

$$\sigma \underset{k \to \infty}{\sim} 2\pi r_0^2 \tag{22}$$

Wave effects thus persist in the limiting case of very small wavelengths. This is due to the fact that the potential under study is discontinuous at $r = r_0$: it always varies appreciably within an interval which is smaller than the wavelength of the particles (*cf.* chap. I, §D-2-a).

3. Exercises

a. SCATTERING OF THE *p* WAVE BY A HARD SPHERE

We wish to study the phase shift $\delta_1(k)$ produced by a hard sphere on the p wave ($l = 1$). In particular, we want to verify that it becomes negligible compared to $\delta_0(k)$ at low energy.

α. Write the radial equation for the function $u_{k,1}(r)$ for $r > r_0$. Show that its general solution is of the form:

$$u_{k,1}(r) = C\left[\frac{\sin kr}{kr} - \cos kr + a\left(\frac{\cos kr}{kr} + \sin kr\right)\right]$$

where C and a are constants.

β. Show that the definition of $\delta_1(k)$ implies that:

$$a = \tan \delta_1(k)$$

γ. Determine the constant a from the condition imposed on $u_{k,1}(r)$ at $r = r_0$.

δ. Show that, as k approaches zero, $\delta_1(k)$ behaves like★ $(kr_0)^3$, which makes it negligible compared to $\delta_0(k)$.

b. "SQUARE SPHERICAL WELL": BOUND STATES AND SCATTERING RESONANCES

Consider a central potential $V(r)$ such that :

$$V(r) = -V_0 \quad \text{for} \quad r < r_0$$
$$= 0 \quad \text{for} \quad r > r_0$$

where V_0 is a positive constant. Set:

$$k_0 = \sqrt{\frac{2\mu V_0}{\hbar^2}}$$

We shall confine ourselves to the study of the s wave ($l = 0$).

α. *Bound states* ($E < 0$)

(i) Write the radial equation in the two regions $r > r_0$ and $r < r_0$, as well as the condition at the origin. Show that, if one sets:

$$\rho = \sqrt{\frac{-2\mu E}{\hbar^2}}$$

$$K = \sqrt{k_0^2 - \rho^2}$$

★ This result is true in general: for any potential of finite range r_0, the phase shift $\delta_l(k)$ behaves like $(kr_0)^{2l+1}$ at low energies.

the function $u_0(r)$ is necessarily of the form:

$$\begin{aligned} u_0(r) &= A\,e^{-\rho r} \qquad \text{for} \qquad r > r_0 \\ &= B \sin Kr \qquad \text{for} \qquad r < r_0 \end{aligned}$$

(ii) Write the matching conditions at $r = r_0$. Deduce from them that the only possible values for ρ are those which satisfy the equation :

$$\tan Kr_0 = -\frac{K}{\rho}$$

(iii) Discuss this equation: indicate the number of s bound states as a function of the depth of the well (for fixed r_0) and show, in particular, that there are no bound states if this depth is too small.

β. *Scattering resonances* ($E > 0$)

(i) Again write the radial equation, this time setting:

$$k = \sqrt{\frac{2\mu E}{\hbar^2}}$$

$$K' = \sqrt{k_0^2 + k^2}$$

Show that $u_{k,0}(r)$ is of the form:

$$\begin{aligned} u_{k,0}(r) &= A \sin (kr + \delta_0) \qquad \text{for} \qquad r > r_0 \\ &= B \sin K'r \qquad \text{for} \qquad r < r_0 \end{aligned}$$

(ii) Choose $A = 1$. Show, using the continuity conditions at $r = r_0$, that the constant B and the phase shift δ_0 are given by:

$$B^2 = \frac{k^2}{k^2 + k_0^2 \cos^2 K'r_0}$$

$$\delta_0 = -kr_0 + \alpha(k)$$

with:

$$\tan \alpha(k) = \frac{k}{K'} \tan K'r_0$$

(iii) Trace the curve representing B^2 as a function of k. This curve clearly shows resonances, for which B^2 is maximum. What are the values of k associated with these resonances? What is then the value of $\alpha(k)$? Show that, if there exists such a resonance for a small energy ($kr_0 \ll 1$), the corresponding contribution of the s wave to the total cross section is practically maximal.

γ. *Relation between bound states and scattering resonances*

Assume that $k_0 r_0$ is very close to $(2n + 1)\frac{\pi}{2}$, where n is an integer, and set:

$$k_0 r_0 = (2n + 1)\frac{\pi}{2} + \varepsilon \qquad \text{with} \qquad \varepsilon \ll 1$$

(*i*) Show that, if ε is positive, there exists a bound state whose binding energy $E = -\hbar^2\rho^2/2\mu$ is given by:

$$\rho \simeq \varepsilon k_0$$

(*ii*) Show that if, on the other hand, ε is negative, there exists a scattering resonance at energy $E = \hbar^2 k^2/2\mu$ such that:

$$k^2 \simeq -\frac{2k_0\varepsilon}{r_0}$$

(*iii*) Deduce from this that if the depth of the well is gradually decreased (for fixed r_0), the bound state which disappears when $k_0 r_0$ passes through an odd multiple of $\pi/2$ gives rise to a low energy scattering resonance.

References and suggestions for further reading :

Messiah (1.17), chap. IX, § 10 and chap. X, §§ III and IV; Valentin (16.1), Annexe II.

CHAPTER IX

Electron spin

OUTLINE OF CHAPTER IX

Until now, we have considered the electron to be a point particle possessing three degrees of freedom associated with its three coordinates x, y and z. Consequently, the quantum theory that we have developed is based on the hypothesis that an electron state, at a given time, is characterized by a wave function $\psi(x, y, z)$ which depends only on x, y and z. Within this framework, we have studied a certain number of physical systems: amongst others, the hydrogen atom (in chapter VII), which is particularly interesting because of the very precise experiments that can be performed on it. The results obtained in chapter VII actually describe the emission and absorption spectra of hydrogen very accurately. They give the energy levels correctly and make it possible to explain, using the corresponding wave functions, the selection rules (which indicate which frequencies, out of all the Bohr frequencies which are *a priori* possible, appear in the spectrum). Atoms with many electrons can be treated in an analogous fashion (by using approximations, however, since the complexity of the Schrödinger equation, even for the helium atom with two electrons, makes an exact analytic solution of the problem impossible). In this case as well, agreement between theory and experiment is satisfying.

However, when atomic spectra are studied in detail, certain phenomena appear, as we shall see, which cannot be interpreted within the framework of the theory that we have developed. This result is not surprising. It is clear that it is necessary to complete the preceding theory by a certain number of *relativistic corrections*: one must take into account the modifications brought in by *relativistic kinematics* (variation of mass with velocity, etc.) and *magnetic effects* which we have neglected. We know that these corrections are small (§ C-4-a of chapter VII): nevertheless, they do exist, and can be measured.

The *Dirac equation* gives a relativistic quantum mechanical description of the electron. Compared to the Schrödinger equation, it implies a profound modification in the quantum description of the properties of the electron; in addition to the corrections already pointed out concerning its position variables, a new characteristic of the electron appears: its *spin*. In a more general context, the structure of the Lorentz group (group of relativistic space-time transformations) reveals spin to be an intrinsic property of various particles, on the same footing, for example, as their rest mass★.

Historically, electron spin was discovered experimentally before the introduction of the Dirac equation. Furthermore, Pauli developed a theory which allowed

★ This does not mean that spin has a purely relativistic origin: it can be deduced from the structure of the non-relativistic transformation group (the Galilean group).

spin to be incorporated simply into non-relativistic quantum mechanics* through the addition of several supplementary postulates. Theoretical predictions for the atomic spectra are then obtained which are in excellent agreement with experimental results**.

It is Pauli's theory, which is much simpler than Dirac's, that we are going to develop in this chapter. We shall begin, in §A, by describing a certain number of experimental results, which revealed the existence of electron spin. Then we shall specify the postulates on which Pauli's theory is based. Afterwards, we shall examine, in §B, the special properties of an angular momentum 1/2. Finally, we shall show, in §C, how one can take into account simultaneously the position variables and the spin of a particle such as the electron.

A. INTRODUCTION OF ELECTRON SPIN

1. Experimental evidence

Experimental demonstrations of the existence of electron spin are numerous and appear in various important physical phenomena. For example, the magnetic properties of numerous substances, particularly of ferromagnetic metals, can only be explained if spin is taken into account. Here, however, we are going to confine ourselves to a certain number of simple phenomena observed experimentally in atomic physics: the fine structure of spectral lines, the Zeeman effect and, finally, the behavior of silver atoms in the Stern-Gerlach experiment.

a. FINE STRUCTURE OF SPECTRAL LINES

The precise experimental study of atomic spectral lines (for the hydrogen atom, for example) reveals a *fine structure*: each line is in fact made up of several components having nearly identical frequencies*** but which can be clearly distinguished by a device with good resolution. This means that there exist groups of atomic levels which are very closely spaced but distinct. In particular, the calculations of §C of chapter VII give the average energies of different groups of levels for the hydrogen atom but do not explain the splittings within each group.

* Pauli's theory can be obtained as a limiting case of Dirac's theory when the electron's speed is small compared to that of light.

** We shall see, for example in chapter XII where the general perturbation theory treated in chapter XI is used, how relativistic corrections and the existence of spin enable us to account quantitatively for the details of the hydrogen atomic spectrum (which would be inexplicable if we limited ourselves to the theory of chapter VII).

*** For example, the resonance line of the hydrogen atom ($2p \longleftrightarrow 1s$ transition) is actually double: the two components are separated by an interval of the order of 10^{-4} eV (that is, about 10^5 times smaller than the average $2p \longleftrightarrow 1s$ transition energy, which is equal to 10.2 eV).

b. « ANOMALOUS » ZEEMAN EFFECT

When an atom is placed in a uniform magnetic field, each of its lines (that is, each component of the fine structure) splits into a certain number of equidistant lines, the interval being proportional to the magnetic field : this is the *Zeeman effect*. The origin of the Zeeman effect can be easily understood by using the results of chapters VI and VII (complement D_{VII}). The theoretical explanation is based on the fact that a magnetic moment **M** is associated with the orbital angular momentum **L** of an electron:

$$\mathbf{M} = \frac{\mu_B}{\hbar} \mathbf{L} \tag{A-1}$$

where μ_B is the "Bohr magneton":

$$\mu_B = \frac{q\hbar}{2m_e} \tag{A-2}$$

However, while this theory is confirmed by experiment in certain cases (the so-called "normal" Zeeman effect), it is, in other cases, incapable of accounting quantitatively for the observed phenomena (the so-called "anomalous" Zeeman effect). The most striking "anomaly" appears for atoms with odd atomic number Z (in particular, for the hydrogen atom): their levels are divided into an *even number of Zeeman sub-levels*, while, according to the theory, this number should always be odd, being equal to $(2l + 1)$ with l an integer.

c. EXISTENCE OF HALF-INTEGRAL ANGULAR MOMENTA

We are confronted with the same difficulty in connection with the Stern-Gerlach experiment, which we described in chapter IV (§A-1); the beam of silver atoms is split symmetrically in two. These results suggest that *half-integral values of j* (which we saw in §C-2 of chapter VI to be *a priori* possible) do indeed exist. But this poses a serious problem, since we showed in §D-1-b of chapter VI that the orbital angular momentum of a particle such as an electron could only be integral (more precisely, it is the quantum number l which is integral). Even in atoms with several electrons, each of these has an integral orbital angular momentum, and we shall show in chapter X that, under these conditions, the total orbital angular momentum of the atom is necessarily integral. The existence of half-integral angular momenta thus cannot be explained without supplementary hypotheses.

COMMENT:

It is not possible to measure directly the angular momentum of the electron using the Stern-Gerlach apparatus. Unlike silver atoms, electrons possess an electric charge q, and the force due to the interaction between their magnetic moment and the inhomogeneous magnetic field would be completely masked by the Lorentz force $q\mathbf{v} \times \mathbf{B}$.

2. Quantum description: postulates of the Pauli theory

In order to resolve the preceding difficulties, Uehlenbeck and Goudsmit (1925) proposed the following hypothesis: the electron "spins" and this gives it an intrinsic angular momentum which is called the spin. To interpret the experimental results described above, one must also assume that a magnetic moment $\mathbf{M}_S$ is associated with this angular momentum $\mathbf{S}$★ :

$$\mathbf{M}_S = 2\frac{\mu_B}{\hbar}\mathbf{S} \tag{A-3}$$

Note that the coefficient of proportionality between the angular momentum and the magnetic moment is twice as large in (A-3) as in (A-1): one says that *the spin gyromagnetic ratio is twice the orbital gyromagnetic ratio.*

Pauli later stated this hypothesis more precisely and gave a quantum description of spin which is valid in the non-relativistic limit. To the general postulates of quantum mechanics that we set forth in chapter III must be added a certain number of postulates relating to spin.

Until now, we have studied the quantization of *orbital variables*. With the position $\mathbf{r}$ and the momentum $\mathbf{p}$ of a particle such as the electron, we associated the observables $\mathbf{R}$ and $\mathbf{P}$ acting in the state space $\mathscr{E}_{\mathbf{r}}$, which is isomorphic to the space $\mathscr{F}$ of wave functions. All physical quantities are functions of the fundamental variables $\mathbf{r}$ and $\mathbf{p}$, and the quantization rules enable us to associate with them observables acting in $\mathscr{E}_{\mathbf{r}}$. We shall call $\mathscr{E}_{\mathbf{r}}$ the *orbital state space.*

To these orbital variables we shall add *spin variables* which satisfy the following postulates :

(*i*) *The spin operator* $\mathbf{S}$ *is an angular momentum*. This means (§ B-2 of chapter VI) that its three components are observables which satisfy the commutation relations:

$$[S_x, S_y] = i\hbar S_z \tag{A-4}$$

and the two formulas which are deduced by cyclic permutation of the indices x, y, z.

(*ii*) The spin operators act in a new space, the "*spin state space*" $\mathscr{E}_s$, where $\mathbf{S}^2$ *and* S_z *constitute a C.S.C.O.* The space $\mathscr{E}_s$ is thus spanned by the set of eigenstates $| s, m \rangle$ common to $\mathbf{S}^2$ and S_z:

$$\mathbf{S}^2 \, | s, m \rangle = s(s+1)\hbar^2 \, | s, m \rangle \tag{A-5-a}$$

$$S_z \, | s, m \rangle = m\hbar \, | s, m \rangle \tag{A-5-b}$$

According to the general theory of angular momentum (§C of chapter VI), we know that s must be integral or half-integral and that m takes on all values included between $-s$ and $+s$ which differ from these two numbers by an integer (which may be zero). A given particle is characterized by *a unique value of* s: this particle is

★ Actually, when one takes into account the coupling of the electron with the quantized electromagnetic field (quantum electrodynamics), one finds that the coefficient of proportionality between $\mathbf{M}_S$ and $\mathbf{S}$ is not eactly $2\mu_B/\hbar$. The difference, which is of the order of 10^{-3} in relative value, is easily observable experimentally; it is often called the "anomalous magnetic moment" of the electron.

said to have a spin s. The spin state space $\mathscr{E}_s$ is therefore always of finite dimension $(2s + 1)$, and all spin states are eigenvectors of $\mathbf{S}^2$ with the same eigenvalue $s(s + 1)\hbar^2$.

(*iii*) *The state space* $\mathscr{E}$ of the particle being considered *is the tensor product of* $\mathscr{E}_\mathbf{r}$ *and* $\mathscr{E}_s$:

$$\mathscr{E} = \mathscr{E}_\mathbf{r} \otimes \mathscr{E}_s \tag{A-6}$$

Consequently (§F of chapter II), *all spin observables commute with all orbital observables.*

Except for the particular case where $s = 0$, it is therefore not sufficient to specify a ket of $\mathscr{E}_\mathbf{r}$ (that is, a square-integrable wave function) to characterize a state of the particle. In other words, the observables X, Y and Z do not constitute a C.S.C.O. in the space state $\mathscr{E}$ of the particle (no more than do P_x, P_y, P_z or any other C.S.C.O. of $\mathscr{E}_\mathbf{r}$). It is also necessary to know the spin state of the particle, that is, to add to the C.S.C.O. of $\mathscr{E}_\mathbf{r}$ a C.S.C.O. of $\mathscr{E}_s$ composed of spin observables, for example, $\mathbf{S}^2$ and S_z (or $\mathbf{S}^2$ and S_x). Every particle state is a linear combination of vectors which are tensor products of a ket of $\mathscr{E}_\mathbf{r}$ and a ket of $\mathscr{E}_s$ (see §C below).

(*iv*) *The electron is a spin 1/2 particle* ($s = 1/2$) and its *intrinsic magnetic moment* is given by formula (A-3). For the electron, the space $\mathscr{E}_s$ is therefore two-dimensional.

COMMENTS:

(*i*) The proton and the neutron, which are nuclear constituents, are also spin 1/2 particles, but their gyromagnetic ratios are different from that of the electron. At the present time we know of the existence of particles of spin 0, 1/2, 1, 3/2, 2, ... up to higher values such as 11/2.

(*ii*) In order to explain the existence of spin, we could imagine that a particle like the electron, instead of being a point, has a certain spatial extension. It would then be the rotation of the electron about its axis that would give rise to an intrinsic angular momentum. However, it is important to note that, in order to describe a structure that is more complex than a material point, it would be necessary to introduce more than three position variables. If, for example, the electron behaved like a solid body, six variables would be required : three coordinates to locate one of its points chosen once and for all, such as its center of gravity, and three angles to specify its orientation in space. The theory that we are considering here is radically different. It continues to treat the electron like a point (its position is fixed by three coordinates). The spin angular momentum is not derived from any position or momentum variable★. *Spin thus has no classical analogue.*

★ If it were, moreover, it would necessarily be integral.

B. SPECIAL PROPERTIES OF AN ANGULAR MOMENTUM 1/2

We shall restrict ourselves from now on to the case of the electron, which is a spin 1/2 particle. From the preceding chapters, we know how to handle its orbital variables. We are now going to study in more detail its spin degrees of freedom.

The spin state space $\mathscr{E}_s$ is two-dimensional. We shall take as a basis the orthonormal system $\{ | + \rangle, | - \rangle \}$ of eigenkets common to $\mathbf{S}^2$ and S_z which satisfy the equations:

$$\begin{cases} \mathbf{S}^2 \, | \pm \rangle = \dfrac{3}{4} \hbar^2 \, | \pm \rangle & \text{(B-1-a)} \\ S_z \, | \pm \rangle = \pm \dfrac{1}{2} \hbar \, | \pm \rangle & \text{(B-1-b)} \end{cases}$$

$$\begin{cases} \langle + | - \rangle = 0 & \text{(B-2-a)} \\ \langle + | + \rangle = \langle - | - \rangle = 1 & \text{(B-2-b)} \end{cases}$$

$$| + \rangle \langle + | + | - \rangle \langle - | = 1 \tag{B-3}$$

The most general spin state is described by an arbitrary vector of $\mathscr{E}_s$:

$$| \chi \rangle = c_+ \, | + \rangle + c_- \, | - \rangle \tag{B-4}$$

where c_+ and c_- are complex numbers. According to (B-1-a), all the kets of $\mathscr{E}_s$ are eigenvectors of $\mathbf{S}^2$ with the same eigenvalue $3\hbar^2/4$, which causes $\mathbf{S}^2$ to be proportional to the identity operator of $\mathscr{E}_s$:

$$\mathbf{S}^2 = \frac{3}{4} \hbar^2 \tag{B-5}$$

Since $\mathbf{S}$ is, by definition, an angular momentum, it possesses all the general properties derived in §C of chapter VI. The action of the operators:

$$S_\pm = S_x \pm i S_y \tag{B-6}$$

on the basis vectors $| + \rangle$ and $| - \rangle$ is given by the general formulas (C-50) of chapter VI when one sets $j = s = 1/2$:

$$\begin{array}{ll} S_+ \, | + \rangle = 0 & S_+ \, | - \rangle = \hbar \, | + \rangle \qquad \text{(B-7-a)} \\ S_- \, | + \rangle = \hbar \, | - \rangle & S_- \, | - \rangle = 0 \qquad \text{(B-7-b)} \end{array}$$

Any operator acting in $\mathscr{E}_s$ can be represented, in the $\{ | + \rangle, | - \rangle \}$ basis, by a 2×2 matrix. In particular, using (B-1-b) and (B-7), we find the matrices corresponding to S_x, S_y and S_z in the form:

$$(\mathbf{S}) = \frac{\hbar}{2} \boldsymbol{\sigma} \tag{B-8}$$

where $\boldsymbol{\sigma}$ designates the set of the three *Pauli matrices* :

$$\sigma_x = \begin{pmatrix} 0 & 1 \\ 1 & 0 \end{pmatrix} \qquad \sigma_y = \begin{pmatrix} 0 & -i \\ i & 0 \end{pmatrix} \qquad \sigma_z = \begin{pmatrix} 1 & 0 \\ 0 & -1 \end{pmatrix} \tag{B-9}$$

The Pauli matrices possess the following properties, which can easily be verified from their explicit form (B-9) (see also complement $\mathrm{A_{IV}}$):

$$\sigma_x^2 = \sigma_y^2 = \sigma_z^2 = 1 \tag{B-10-a}$$

$$\sigma_x\sigma_y + \sigma_y\sigma_x = 0 \tag{B-10-b}$$

$$[\sigma_x, \sigma_y] = 2i\,\sigma_z \tag{B-10-c}$$

$$\sigma_x\sigma_y = i\sigma_z \tag{B-10-d}$$

(to the last three formulas must be added those obtained through cyclic permutation of the x, y, z indices). It also follows from (B-9) that:

$$\mathrm{Tr}\,\sigma_x = \mathrm{Tr}\,\sigma_y = \mathrm{Tr}\,\sigma_z = 0 \tag{B-11-a}$$

$$\mathrm{Det}\,\sigma_x = \mathrm{Det}\,\sigma_y = \mathrm{Det}\,\sigma_z = -1 \tag{B-11-b}$$

Furthermore, any 2×2 matrix can be written as a linear combination, with complex coefficients, of the three Pauli matrices and the unit matrix. This is simply due to the fact that a 2×2 matrix has only four elements. Finally, it is easy to derive (see complement $\mathrm{A_{IV}}$) the following identity:

$$(\boldsymbol{\sigma}\,.\,\mathbf{A})(\boldsymbol{\sigma}\,.\,\mathbf{B}) = \mathbf{A}\,.\,\mathbf{B} + i\,\boldsymbol{\sigma}\,.\,(\mathbf{A} \times \mathbf{B}) \tag{B-12}$$

where **A** and **B** are two arbitrary vectors, or two vector operators whose three components commute with those of the spin **S**. If **A** and **B** do not commute with each other, the identity remains valid if **A** and **B** appear in the same order on the right-hand side as on the left-hand side.

The operators associated with electron spin have all the properties which follow directly from the general theory of angular momentum. They have, in addition, some specific properties related to their particular value of s (that is, of j), which is the smallest one possible (aside from zero). These specific properties can be deduced directly from (B-8) and formulas (B-10):

$$S_x^2 = S_y^2 = S_z^2 = \frac{\hbar^2}{4} \tag{B-13-a}$$

$$S_xS_y + S_yS_x = 0 \tag{B-13-b}$$

$$S_xS_y = \frac{i}{2}\hbar S_z \tag{B-13-c}$$

$$S_+^2 = S_-^2 = 0 \tag{B-13-d}$$

C. NON-RELATIVISTIC DESCRIPTION OF A SPIN 1/2 PARTICLE

We now know how to describe separately the external (orbital) and the internal (spin) degrees of freedom of the electron. In this section, we are going to assemble these different concepts into one formalism.

1. Observables and state vectors

a. STATE SPACE

When all its degrees of freedom are taken into account, the quantum state of an electron is characterized by a ket belonging to the space $\mathscr{E}$ which is the tensor product of $\mathscr{E}_{\mathbf{r}}$ and $\mathscr{E}_s$ (§ A-2).

We extend into $\mathscr{E}$, following the method described in § F-2-b of chapter II, both the operators originally defined in $\mathscr{E}_{\mathbf{r}}$ and those which initially acted in $\mathscr{E}_s$ (we shall continue to use the same notation for these extended operators as for the operators from which they are derived). We thus obtain a C.S.C.O. in $\mathscr{E}$ through the juxtaposition of a C.S.C.O. of $\mathscr{E}_{\mathbf{r}}$ and one of $\mathscr{E}_s$. For example, in $\mathscr{E}_s$, we can take $\mathbf{S}^2$ and S_z (or $\mathbf{S}^2$ and any component of $\mathbf{S}$). In $\mathscr{E}_{\mathbf{r}}$, we can choose $\{ X, Y, Z \}$, or $\{ P_x, P_y, P_z \}$, or, if H designates the Hamiltonian associated with a central potential, $\{ H, \mathbf{L}^2, L_z \}$ etc. From this we deduce various C.S.C.O. in $\mathscr{E}$:

$$\{ X, Y, Z, \mathbf{S}^2, S_z \} \tag{C-1-a}$$

$$\{ P_x, P_y, P_z, \mathbf{S}^2, S_z \} \tag{C-1-b}$$

$$\{ H, \mathbf{L}^2, L_z, \mathbf{S}^2, S_z \} \tag{C-1-c}$$

etc. Since all kets of $\mathscr{E}$ are eigenvectors of $\mathbf{S}^2$ with the same eigenvalue [formula (B-5)], we can omit $\mathbf{S}^2$ from the sets of observables.

We are going to use here the first of these C.S.C.O., (C-1-a). We shall take as a basis of $\mathscr{E}$ the set of vectors obtained from the tensor product of the kets $| \mathbf{r} \rangle \equiv | x, y, z \rangle$ of $\mathscr{E}_{\mathbf{r}}$ and the kets $| \varepsilon \rangle$ of $\mathscr{E}_s$:

$$| \mathbf{r}, \varepsilon \rangle \equiv | x, y, z, \varepsilon \rangle = | \mathbf{r} \rangle \otimes | \varepsilon \rangle \tag{C-2}$$

where the x, y, z, components of the vector $\mathbf{r}$, can vary from $-\infty$ to $+\infty$ (continuous indices), and ε is equal to $+$ or $-$ (discrete index). By definition, $| \mathbf{r}, \varepsilon \rangle$ is an eigenvector common to X, Y, Z, $\mathbf{S}^2$ and S_z:

$$\begin{aligned} X | \mathbf{r}, \varepsilon \rangle &= x | \mathbf{r}, \varepsilon \rangle \\ Y | \mathbf{r}, \varepsilon \rangle &= y | \mathbf{r}, \varepsilon \rangle \\ Z | \mathbf{r}, \varepsilon \rangle &= z | \mathbf{r}, \varepsilon \rangle \\ \mathbf{S}^2 | \mathbf{r}, \varepsilon \rangle &= \frac{3}{4} \hbar^2 | \mathbf{r}, \varepsilon \rangle \\ S_z | \mathbf{r}, \varepsilon \rangle &= \varepsilon \frac{\hbar}{2} | \mathbf{r}, \varepsilon \rangle \end{aligned} \tag{C-3}$$

Each ket $|\,\mathbf{r}, \varepsilon\,\rangle$ is unique to within a constant factor, since X, Y, Z, $\mathbf{S}^2$ and S_z constitute a C.S.C.O. The $\{\,|\,\mathbf{r}, \varepsilon\,\rangle\,\}$ system is orthonormal (in the extended sense), since the sets $\{\,|\,\mathbf{r}\,\rangle\,\}$ and $\{\,|\,+\,\rangle, |\,-\,\rangle\,\}$ are each orthonormal in $\mathscr{E}_{\mathbf{r}}$ and $\mathscr{E}_s$ respectively:

$$\langle\,\mathbf{r}', \varepsilon'\,|\,\mathbf{r}, \varepsilon\,\rangle = \delta_{\varepsilon'\varepsilon}\,\delta(\mathbf{r}' - \mathbf{r}) \tag{C-4}$$

($\delta_{\varepsilon'\varepsilon}$ is equal to 1 or 0 depending on whether ε' and ε are the same or different). Finally, it satisfies a closure relation in $\mathscr{E}$:

$$\sum_{\varepsilon}\int d^3r\,|\,\mathbf{r}, \varepsilon\,\rangle\langle\,\mathbf{r}, \varepsilon\,| = \int d^3r\,|\,\mathbf{r}, +\,\rangle\langle\,\mathbf{r}, +\,| + \int d^3r\,|\,\mathbf{r}, -\,\rangle\langle\,\mathbf{r}, -\,| = 1 \tag{C-5}$$

b. $\{\,|\,\mathbf{r}, \varepsilon\,\rangle\,\}$ REPRESENTATION

α. *State vectors*

Any state $|\,\psi\,\rangle$ of the space $\mathscr{E}$ can be expanded in the $\{\,|\,\mathbf{r}, \varepsilon\,\rangle\,\}$ basis. To do this, it suffices to use the closure relation (C-5):

$$|\,\psi\,\rangle = \sum_{\varepsilon}\int d^3r\,|\,\mathbf{r}, \varepsilon\,\rangle\langle\,\mathbf{r}, \varepsilon\,|\,\psi\,\rangle \tag{C-6}$$

The *vector* $|\,\psi\,\rangle$ can therefore be represented by the set of its coordinates in the $\{\,|\,\mathbf{r}, \varepsilon\,\rangle\,\}$ basis, that is, by the *numbers*:

$$\langle\,\mathbf{r}, \varepsilon\,|\,\psi\,\rangle = \psi_{\varepsilon}(\mathbf{r}) \tag{C-7}$$

which depend on the three continuous indices x, y, z (or, more succinctly, $\mathbf{r}$) and on the discrete index ε ($+$ or $-$). *In order to characterize completely the state of an electron, it is therefore necessary to specify two functions of the space variables x, y and z*:

$$\begin{aligned}\psi_+(\mathbf{r}) &= \langle\,\mathbf{r}, +\,|\,\psi\,\rangle\\ \psi_-(\mathbf{r}) &= \langle\,\mathbf{r}, -\,|\,\psi\,\rangle\end{aligned} \tag{C-8}$$

These two functions are often written in the form of a *two-component spinor*, which we shall write $[\psi](\mathbf{r})$:

$$[\psi](\mathbf{r}) = \begin{pmatrix}\psi_+(\mathbf{r})\\ \psi_-(\mathbf{r})\end{pmatrix} \tag{C-9}$$

The bra $\langle\,\psi\,|$ associated with the ket $|\,\psi\,\rangle$ is given by the adjoint of (C-6):

$$\langle\,\psi\,| = \sum_{\varepsilon}\int d^3r\,\langle\,\psi\,|\,\mathbf{r}, \varepsilon\,\rangle\langle\,\mathbf{r}, \varepsilon\,| \tag{C-10}$$

that is, taking (C-7) into account:

$$\langle \psi | = \sum_{\varepsilon} \int d^3r \, \psi_{\varepsilon}^*(\mathbf{r}) \langle \mathbf{r}, \varepsilon | \tag{C-11}$$

The bra $\langle \psi |$ is thus represented by the two functions $\psi_+^*(\mathbf{r})$ and $\psi_-^*(\mathbf{r})$, which can be written in the form of a spinor which is the adjoint of (C-9):

$$[\psi]^\dagger(\mathbf{r}) = \begin{pmatrix} \psi_+^*(\mathbf{r}) & \psi_-^*(\mathbf{r}) \end{pmatrix} \tag{C-12}$$

With this notation, the scalar product of two state vectors $| \psi \rangle$ and $| \varphi \rangle$, which, according to (C-5), is equal to:

$$\begin{aligned} \langle \psi | \varphi \rangle &= \sum_{\varepsilon} \int d^3r \, \langle \psi | \mathbf{r}, \varepsilon \rangle \langle \mathbf{r}, \varepsilon | \varphi \rangle \\ &= \int d^3r \, [\, \psi_+^*(\mathbf{r}) \, \varphi_+(\mathbf{r}) + \psi_-^*(\mathbf{r}) \, \varphi_-(\mathbf{r}) \,] \end{aligned} \tag{C-13}$$

can be written in the form:

$$\langle \psi | \varphi \rangle = \int d^3r \, [\psi]^\dagger(\mathbf{r}) \, [\varphi](\mathbf{r}) \tag{C-14}$$

This formula is very similar to the one which permitted the calculation of the scalar product of two kets of $\mathscr{E}_\mathbf{r}$ from the corresponding wave functions. However, it is important to note that here the matrix multiplication of the spinors $[\psi]^\dagger(\mathbf{r})$ and $[\varphi](\mathbf{r})$ must precede the spatial integration. In particular, the normalization of the vector $| \psi \rangle$ is expressed by:

$$\langle \psi | \psi \rangle = \int d^3r \, [\psi]^\dagger(\mathbf{r}) \, [\psi](\mathbf{r}) = \int d^3r \, [\, |\psi_+(\mathbf{r})|^2 + |\psi_-(\mathbf{r})|^2 \,] = 1 \tag{C-15}$$

Amongst the vectors of $\mathscr{E}$, some are the tensor products of a ket of $\mathscr{E}_\mathbf{r}$ and a ket of $\mathscr{E}_s$ (this is the case, for example, for the basis vectors). If the state vector under consideration is of this type:

$$| \psi \rangle = | \varphi \rangle \otimes | \chi \rangle \tag{C-16}$$

with:

$$\begin{aligned} | \varphi \rangle &= \int d^3r \, \varphi(\mathbf{r}) \, | \mathbf{r} \rangle \in \mathscr{E}_\mathbf{r} \\ | \chi \rangle &= c_+ \, | + \rangle + c_- \, | - \rangle \in \mathscr{E}_s \end{aligned} \tag{C-17}$$

the spinor associated with it takes on the simple form:

$$[\psi](\mathbf{r}) = \begin{pmatrix} \varphi(\mathbf{r})\, c_+ \\ \varphi(\mathbf{r})\, c_- \end{pmatrix} = \varphi(\mathbf{r}) \begin{pmatrix} c_+ \\ c_- \end{pmatrix} \tag{C-18}$$

This results from the definition of the scalar product in $\mathscr{E}$, and we have in this case:

$$\psi_+(\mathbf{r}) = \langle \mathbf{r}, + \mid \psi \rangle = \langle \mathbf{r} \mid \varphi \rangle \langle + \mid \chi \rangle = \varphi(\mathbf{r})\, c_+ \tag{C-19-a}$$

$$\psi_-(\mathbf{r}) = \langle \mathbf{r}, - \mid \psi \rangle = \langle \mathbf{r} \mid \varphi \rangle \langle - \mid \chi \rangle = \varphi(\mathbf{r})\, c_- \tag{C-19-b}$$

The square of the norm of $| \psi \rangle$ is then given by:

$$\langle \psi \mid \psi \rangle = \langle \varphi \mid \varphi \rangle \langle \chi \mid \chi \rangle = (|c_+|^2 + |c_-|^2) \int d^3r\, |\varphi(\mathbf{r})|^2 \tag{C-20}$$

β. *Operators*

Let $| \psi' \rangle$ be the ket obtained from the action of the linear operator A on the ket $| \psi \rangle$ of $\mathscr{E}$. According to the results of the preceding section, $| \psi' \rangle$ and $| \psi \rangle$ can be represented by the two-component spinors $[\psi'](\mathbf{r})$ and $[\psi](\mathbf{r})$. We are now going to show that one can associate with A a 2×2 matrix $[\![A]\!]$ such that:

$$[\psi'](\mathbf{r}) = [\![A]\!]\, [\psi](\mathbf{r}) \tag{C-21}$$

where the matrix elements remain in general differential operators with respect to the variable $\mathbf{r}$.

(*i*) *Spin operators.* These were initially defined in $\mathscr{E}_s$. Consequently, they act only on the ε index of the basis vectors $| \mathbf{r}, \varepsilon \rangle$, and their matrix form is the one stated in § B. We shall limit ourselves to one example, say that of the operator S_+. Its action on a vector $| \psi \rangle$ expanded as in (C-6) gives a vector $| \psi' \rangle$:

$$| \psi' \rangle = \hbar \int d^3r\, \psi_-(\mathbf{r})\, | \mathbf{r}, + \rangle \tag{C-22}$$

since S_+ annihilates all the $| \mathbf{r}, + \rangle$ kets and transforms $| \mathbf{r}, - \rangle$ into $\hbar\, | \mathbf{r}, + \rangle$. The components of $| \psi' \rangle$ in the $\{ | \mathbf{r}, \varepsilon \rangle \}$ basis are, according to (C-22):

$$\begin{aligned} \langle \mathbf{r}, + \mid \psi' \rangle &= \psi'_+(\mathbf{r}) = \hbar \psi_-(\mathbf{r}) \\ \langle \mathbf{r}, - \mid \psi' \rangle &= \psi'_-(\mathbf{r}) = 0 \end{aligned} \tag{C-23}$$

The spinor representing $| \psi' \rangle$ is therefore:

$$[\psi'](\mathbf{r}) = \hbar \begin{pmatrix} \psi_-(\mathbf{r}) \\ 0 \end{pmatrix} \tag{C-24}$$

This is indeed what is obtained if one performs the matrix multiplication of the spinor $[\psi](\mathbf{r})$ by:

$$[\![S_+]\!] = \frac{\hbar}{2}(\sigma_x + i\sigma_y) = \hbar \begin{pmatrix} 0 & 1 \\ 0 & 0 \end{pmatrix} \tag{C-25}$$

(*ii*) *Orbital operators.* Unlike the preceding operators, they always leave unchanged the ε index of the basis vector $| \mathbf{r}, \varepsilon \rangle$: their associated 2×2 matrices are always proportional to the unit matrix. On the other hand, they act on the $\mathbf{r}$-dependence of the spinors just as they act on ordinary wave functions. Consider, for example, the kets $| \psi' \rangle = X | \psi \rangle$ and $| \psi'' \rangle = P_x | \psi \rangle$. Their components in the $\{ | \mathbf{r}, \varepsilon \rangle \}$ basis are, respectively:

$$\psi'_\varepsilon(\mathbf{r}) = \langle \mathbf{r}, \varepsilon | X | \psi \rangle = x \, \psi_\varepsilon(\mathbf{r}) \tag{C-26-a}$$

$$\psi''_\varepsilon(\mathbf{r}) = \langle \mathbf{r}, \varepsilon | P_x | \psi \rangle = \frac{\hbar}{i} \frac{\partial}{\partial x} \psi_\varepsilon(\mathbf{r}) \tag{C-26-b}$$

The spinors $[\psi'](\mathbf{r})$ and $[\psi''](\mathbf{r})$ are thus obtained from $[\psi](\mathbf{r})$ by means of the 2×2 matrices:

$$[\![X]\!] = \begin{pmatrix} x & 0 \\ 0 & x \end{pmatrix} \tag{C-27-a}$$

$$[\![P_x]\!] = \frac{\hbar}{i} \begin{pmatrix} \dfrac{\partial}{\partial x} & 0 \\ 0 & \dfrac{\partial}{\partial x} \end{pmatrix} \tag{C-27-b}$$

(*iii*) *Mixed operators.* The most general operator acting in $\mathscr{E}$ is represented, in matrix notation, by a 2×2 matrix whose elements are differential operators with respect to the $\mathbf{r}$ variables. For example:

$$[\![L_z S_z]\!] = \frac{\hbar}{2} \begin{pmatrix} \dfrac{\hbar}{i} \dfrac{\partial}{\partial \varphi} & 0 \\ 0 & -\dfrac{\hbar}{i} \dfrac{\partial}{\partial \varphi} \end{pmatrix} \tag{C-28}$$

or:

$$[\![\mathbf{S} \cdot \mathbf{P}]\!] = \frac{\hbar}{2} (\sigma_x P_x + \sigma_y P_y + \sigma_z P_z) = \frac{\hbar^2}{2i} \begin{pmatrix} \dfrac{\partial}{\partial z} & \dfrac{\partial}{\partial x} - i \dfrac{\partial}{\partial y} \\ \dfrac{\partial}{\partial x} + i \dfrac{\partial}{\partial y} & -\dfrac{\partial}{\partial z} \end{pmatrix} \tag{C-29}$$

COMMENTS:

(*i*) The spinor representation $\{ | \mathbf{r}, \varepsilon \rangle \}$ is analogous to the $\{ | \mathbf{r} \rangle \}$ representation of $\mathscr{E}_\mathbf{r}$. The matrix element $\langle \psi | A | \varphi \rangle$ of any operator A of $\mathscr{E}$ is given by the formula:

$$\langle \psi | A | \varphi \rangle = \int d^3r \, [\psi]^\dagger(\mathbf{r}) \, [\![A]\!] \, [\varphi](\mathbf{r}) \tag{C-30}$$

where $[\![A]\!]$ designates the 2×2 matrix which represents the operator A (one first carries out the matrix multiplications and then integrates over all space).

This representation will only be used when it simplifies the reasoning and the calculations: as in $\mathscr{E}_{\mathbf{r}}$, the vectors and operators themselves will be used as much as possible.

(*ii*) Obviously, there also exists a $\{ | \mathbf{p}, \varepsilon \rangle \}$ representation, whose basis vectors are the eigenvectors common to the C.S.C.O. $\{ P_x, P_y, P_z, \mathbf{S}^2, S_z \}$. The definition of the scalar product in $\mathscr{E}$ yields:

$$\langle \mathbf{r}, \varepsilon | \mathbf{p}, \varepsilon' \rangle = \langle \mathbf{r} | \mathbf{p} \rangle \langle \varepsilon | \varepsilon' \rangle = \frac{1}{(2\pi\hbar)^{3/2}} \, e^{i\mathbf{p}.\mathbf{r}/\hbar} \delta_{\varepsilon\varepsilon'} \tag{C-31}$$

In the $\{ | \mathbf{p}, \varepsilon \rangle \}$ representation, one associates with each vector $| \psi \rangle$ of $\mathscr{E}$ a two-component spinor:

$$[\bar{\psi}](\mathbf{p}) = \begin{pmatrix} \bar{\psi}_+(\mathbf{p}) \\ \bar{\psi}_-(\mathbf{p}) \end{pmatrix} \tag{C-32}$$

with:

$$\begin{aligned} \bar{\psi}_+(\mathbf{p}) &= \langle \mathbf{p}, + | \psi \rangle \\ \bar{\psi}_-(\mathbf{p}) &= \langle \mathbf{p}, - | \psi \rangle \end{aligned} \tag{C-33}$$

According to (C-31), $\bar{\psi}_+(\mathbf{p})$ and $\bar{\psi}_-(\mathbf{p})$ are the Fourier transforms of $\psi_+(\mathbf{r})$ and $\psi_-(\mathbf{r})$:

$$\begin{aligned} \bar{\psi}_\varepsilon(\mathbf{p}) = \langle \mathbf{p}, \varepsilon | \psi \rangle &= \sum_{\varepsilon'} \int d^3r \, \langle \mathbf{p}, \varepsilon | \mathbf{r}, \varepsilon' \rangle \langle \mathbf{r}, \varepsilon' | \psi \rangle \\ &= \frac{1}{(2\pi\hbar)^{3/2}} \int d^3r \, e^{-i\mathbf{p}.\mathbf{r}/\hbar} \, \psi_\varepsilon(\mathbf{r}) \end{aligned} \tag{C-34}$$

The operators are still represented by 2×2 matrices, and those corresponding to the spin operators remain the same as in the $\{ | \mathbf{r}, \varepsilon \rangle \}$ representation.

2. Probability calculations for a physical measurement

Using the formalism we have just described, we can apply the postulates of chapter III to obtain predictions concerning the various measurements that one can imagine carrying out on an electron. We are going to give several examples.

First of all, consider the probabilistic interpretation of the components $\psi_+(\mathbf{r})$ and $\psi_-(\mathbf{r})$ of the state vector $| \psi \rangle$, which we assume to be normalized [formula (C-15)]. Imagine that we are simultaneously measuring the position of the electron and the component of its spin along Oz. Since X, Y, Z and S_z constitute a C.S.C.O., there exists only one state vector which corresponds to a given result : x, y, z and $\pm \hbar/2$. The probability $d^3\mathscr{P}(\mathbf{r}, +)$ of the electron being found in the infinitesimal volume d^3r around the point $\mathbf{r}(x, y, z)$ with its spin "up" (the component along Oz equal to $+ \hbar/2$) is equal to:

$$d^3\mathscr{P}(\mathbf{r}, +) = | \langle \mathbf{r}, + | \psi \rangle |^2 \, d^3r = |\psi_+(\mathbf{r})|^2 \, d^3r \tag{C-35}$$

In the same way:

$$d^3\mathscr{P}(\mathbf{r}, -) = |\langle \mathbf{r}, - | \psi \rangle|^2 \, d^3r = |\psi_-(\mathbf{r})|^2 \, d^3r \tag{C-36}$$

is the probability of the electron being found in the same volume as before but with its spin "down" (component along Oz equal to $-\hbar/2$).

If it is the component of the spin along Ox that is being measured at the same time as the position, all we need to do is use formulas (A-20) of chapter IV. The X, Y, Z and S_x operators also form a C.S.C.O.: to the measurement result $\{x, y, z, \pm\hbar/2\}$ corresponds a single state vector:

$$|\mathbf{r}\rangle|\pm\rangle_x = \frac{1}{\sqrt{2}}[|\mathbf{r}, +\rangle \pm |\mathbf{r}, -\rangle] \tag{C-37}$$

The probability of the electron being found in the volume d^3r around the point $\mathbf{r}$ with its spin in the positive direction of the Ox axis is then:

$$d^3r \times \left|\frac{1}{\sqrt{2}}[\langle \mathbf{r}, + | \psi \rangle + \langle \mathbf{r}, - | \psi \rangle]\right|^2 = \frac{1}{2}|\psi_+(\mathbf{r}) + \psi_-(\mathbf{r})|^2 d^3r \tag{C-38}$$

Obviously, one can measure the momentum of the electron instead of its position. One then uses the components of $|\psi\rangle$ relative to the vectors $|\mathbf{p}, \varepsilon\rangle$ [*cf.* comment (*ii*) of §1], that is, the Fourier transforms $\bar{\psi}_\pm(\mathbf{p})$ of $\psi_\pm(\mathbf{r})$. The probability $d^3\mathscr{P}(\mathbf{p}, \pm)$ of the momentum being $\mathbf{p}$ to within d^3p and of the spin component along Oz being $\pm\hbar/2$ is given by:

$$d^3\mathscr{P}(\mathbf{p}, \pm) = |\langle \mathbf{p}, \pm | \psi \rangle|^2 d^3p = |\bar{\psi}_\pm(\mathbf{p})|^2 d^3p \tag{C-39}$$

The various measurements that we have envisaged until now are all "complete" in the sense that they each relate to a C.S.C.O. For "incomplete measurements", several orthogonal states correspond to the same result, and it is necessary to sum the squares of the moduli of the corresponding probability amplitudes.

For example, if one does not seek to measure its spin, the probability $d^3\mathscr{P}(\mathbf{r})$ of finding the electron in the volume d^3r in the neighborhood of the point $\mathbf{r}$ is equal to:

$$d^3\mathscr{P}(\mathbf{r}) = [\,|\psi_+(\mathbf{r})|^2 + |\psi_-(\mathbf{r})|^2\,]\, d^3r \tag{C-40}$$

This is because two orthogonal state vectors, $|\mathbf{r}, +\rangle$ and $|\mathbf{r}, -\rangle$, are associated with the result $\{x, y, z\}$, their corresponding probability amplitudes being $\psi_+(\mathbf{r})$ and $\psi_-(\mathbf{r})$.

Finally, let us calculate the probability $\mathscr{P}_+$ that the spin component along Oz is $+\hbar/2$ (one is not seeking to measure the orbital variables). There exist an infinite number of orthogonal states, for example all the $|\mathbf{r}, +\rangle$ with arbitrary $\mathbf{r}$, which correspond to the result of the measurement. One must therefore sum over all possible values of $\mathbf{r}$ the squares of the moduli of the amplitudes $\langle \mathbf{r}, + | \psi \rangle = \psi_+(\mathbf{r})$, which gives:

$$\mathscr{P}_+ = \int d^3r \, |\psi_+(\mathbf{r})|^2 \tag{C-41}$$

Of course, if we are considering the component of the spin along Ox instead of along Oz, we integrate the result (C-38) over all space. These ideas generalize those of §B-2 of chapter IV, where we considered only the spin observables since the orbital variables could be treated classically.

References and suggestions for further reading:

History of the discovery of spin and references to original articles: Jammer (4.8), §3-4.

Evidence of spin in atomic physics: Eisberg and Resnick (1.3), chap. 8; Born (11.4), chap. VI; Kuhn (11.1), chap. III, §§A.5, A.6 and F; see references of chapter IV relating to the Stern-Gerlach experiment.

The spin magnetic moment of the electron: Cagnac and Pebay-Peyroula (11.2), chap. XII; Crane (11.16).

The Dirac equation: Schiff (1.18), chap. 13; Messiah (1.17), chap. XX; Bjorken and Drell (2.6), chaps. 1 to 4.

The Lorentz group: Omnes (16.13), chap. 4; Bacry (10.31), chaps. 7 and 8.

Spin 1 particles: Messiah (1.17), § XIII.21.

COMPLEMENTS OF CHAPTER IX

Several complements concerning spin $\frac{1}{2}$ properties can be found at the end of chapter IV. This is why chapter IX has only two complements.

A_{IX}: ROTATION OPERATORS FOR A SPIN 1/2 PARTICLE

A_{IX} is a continuation of complement B_{VI}. It studies in detail the relationship between the spin $\frac{1}{2}$ angular momentum and the geometric rotations of this spin. Moderately difficult. Can be omitted upon a first reading.

B_{IX}: EXERCISES

B_{IX}: exercise 4 is worked out in detail. It studies the polarization of a beam of spin $\frac{1}{2}$ particles caused by their reflection from a magnetized ferromagnetic material. This method is actually used in certain experiments.

Complement A_{IX}

ROTATION OPERATORS FOR A SPIN 1/2 PARTICLE

We are going to apply the ideas about rotation introduced in complement B_{VI} to the case of a spin 1/2 particle. First, we shall study the form that rotation operators take on in this case. We shall then examine the behavior, under rotation, of the ket representing the particle's state and of the two-component spinor associated with it.

1. Rotation operators in state space

a. TOTAL ANGULAR MOMENTUM

A spin 1/2 particle possesses an orbital angular momentum **L** and a spin angular momentum **S**. It is natural to define its total angular momentum as the sum of these two angular momenta:

$$\mathbf{J} = \mathbf{L} + \mathbf{S} \tag{1}$$

This definition is clearly consistent with the general considerations discussed in complement B_{VI}. It insures that not only **R** and **P**, but also **S**, be vectorial observables. (To test this, it is sufficient to calculate the commutators between the components of these observables and those of **J**; *cf.* § 5-c of complement B_{VI}).

b. DECOMPOSITION OF ROTATION OPERATORS INTO TENSOR PRODUCTS

In the state space of the particle under study, the rotation operator $R_{\mathbf{u}}(\alpha)$ is associated with the geometrical rotation $\mathscr{R}_{\mathbf{u}}(\alpha)$ through an angle α about the unit vector **u** (*cf.* complement B_{VI}, §4):

$$R_{\mathbf{u}}(\alpha) = e^{-\frac{i}{\hbar}\alpha\,\mathbf{J}.\mathbf{u}} \tag{2}$$

where **J** is the total angular momentum (1).

Since $\mathbf{L}$ acts only in $\mathscr{E}_{\mathbf{r}}$ and $\mathbf{S}$ only in $\mathscr{E}_s$ (which implies, in particular, that all components of $\mathbf{L}$ commute with all components of $\mathbf{S}$), we can write $R_{\mathbf{u}}(\alpha)$ in the form of a tensor product:

$$R_{\mathbf{u}}(\alpha) = {}^{(\mathbf{r})}R_{\mathbf{u}}(\alpha) \otimes {}^{(s)}R_{\mathbf{u}}(\alpha) \tag{3}$$

where:

$${}^{(\mathbf{r})}R_{\mathbf{u}}(\alpha) = \mathrm{e}^{-\frac{i}{\hbar}\alpha \mathbf{L}.\mathbf{u}} \tag{4}$$

and:

$${}^{(s)}R_{\mathbf{u}}(\alpha) = \mathrm{e}^{-\frac{i}{\hbar}\alpha \mathbf{S}.\mathbf{u}} \tag{5}$$

are the rotation operators associated with $\mathscr{R}_{\mathbf{u}}(\alpha)$ in $\mathscr{E}_{\mathbf{r}}$ and $\mathscr{E}_s$ respectively.

Consequently, if one performs the rotation $\mathscr{R}_{\mathbf{u}}(\alpha)$ on a spin 1/2 particle whose state is represented by a ket which is a tensor product:

$$|\psi\rangle = |\varphi\rangle \otimes |\chi\rangle \tag{6}$$

with:

$$\begin{aligned} &|\varphi\rangle \in \mathscr{E}_{\mathbf{r}} \\ &|\chi\rangle \in \mathscr{E}_s \end{aligned} \tag{7}$$

its state after rotation will be:

$$|\psi'\rangle = R_{\mathbf{u}}(\alpha)|\psi\rangle = [{}^{(\mathbf{r})}R_{\mathbf{u}}(\alpha)|\varphi\rangle] \otimes [{}^{(s)}R_{\mathbf{u}}(\alpha)|\chi\rangle] \tag{8}$$

The spin state of the particle is therefore also affected by the rotation. This is what we are going to study in more detail in § 2.

2. Rotation of spin states

We have already studied (§3 of complement B_{VI}) the rotation operators ${}^{(\mathbf{r})}R$ in the space $\mathscr{E}_{\mathbf{r}}$. Here we are interested in the operators ${}^{(S)}R$ which act in the spin state space $\mathscr{E}_s$.

a. EXPLICIT CALCULATION OF THE ROTATION OPERATORS IN $\mathscr{E}_s$

As in chapter IX, set:

$$\mathbf{S} = \frac{\hbar}{2}\boldsymbol{\sigma} \tag{9}$$

We want to calculate the operator:

$${}^{(s)}R_{\mathbf{u}}(\alpha) = \mathrm{e}^{-\frac{i}{\hbar}\alpha \mathbf{S}.\mathbf{u}} = \mathrm{e}^{-i\frac{\alpha}{2}\boldsymbol{\sigma}.\mathbf{u}} \tag{10}$$

To do this, let us use the definition of the exponential of an operator:

$${}^{(s)}R_{\mathbf{u}}(\alpha) = 1 - \frac{i\alpha}{2}\boldsymbol{\sigma}\,.\,\mathbf{u} + \frac{1}{2!}\left(-i\frac{\alpha}{2}\right)^2(\boldsymbol{\sigma}\,.\,\mathbf{u})^2 + \ldots + \frac{1}{n!}\left(-i\frac{\alpha}{2}\right)^n(\boldsymbol{\sigma}\,.\,\mathbf{u})^n + \ldots \tag{11}$$

Now, applying identity (B-12) of chapter IX, we immediately see that:

$$(\boldsymbol{\sigma} \cdot \mathbf{u})^2 = \mathbf{u}^2 = 1 \tag{12}$$

which leads to:

$$(\boldsymbol{\sigma} \cdot \mathbf{u})^n = \begin{cases} 1 & \text{if } n \text{ is even} \\ \boldsymbol{\sigma} \cdot \mathbf{u} & \text{if } n \text{ is odd} \end{cases} \tag{13}$$

Consequently, if we group together the even and odd terms respectively, expansion (11) can be written:

$$\begin{aligned} {}^{(s)}R_{\mathbf{u}}(\alpha) = & \left[1 - \frac{1}{2!}\left(\frac{\alpha}{2}\right)^2 + \ldots + \frac{(-1)^p}{(2p)!}\left(\frac{\alpha}{2}\right)^{2p} + \ldots\right] \\ & - i\boldsymbol{\sigma} \cdot \mathbf{u}\left[\frac{\alpha}{2} - \frac{1}{3!}\left(\frac{\alpha}{2}\right)^3 + \ldots + \frac{(-1)^p}{(2p+1)!}\left(\frac{\alpha}{2}\right)^{2p+1} + \ldots\right] \end{aligned} \tag{14}$$

that is, finally:

$$\boxed{{}^{(s)}R_{\mathbf{u}}(\alpha) = \cos\frac{\alpha}{2} - i\boldsymbol{\sigma} \cdot \mathbf{u} \sin\frac{\alpha}{2}} \tag{15}$$

It will be very easy to calculate the action of the operator ${}^{(S)}R$, in this form, on any spin state.

Using this formula, we can write the rotation matrix $R_{\mathbf{u}}^{(1/2)}(\alpha)$ explicitly in the $\{ |+\rangle, |-\rangle \}$ basis, since we already know [formulas (B-9) of chapter IX] the matrices which represent the σ_x, σ_y and σ_z operators. We find:

$$R_{\mathbf{u}}^{(1/2)}(\alpha) = \begin{pmatrix} \cos\frac{\alpha}{2} - iu_z \sin\frac{\alpha}{2} & (-iu_x - u_y)\sin\frac{\alpha}{2} \\ (-iu_x + u_y)\sin\frac{\alpha}{2} & \cos\frac{\alpha}{2} + iu_z \sin\frac{\alpha}{2} \end{pmatrix} \tag{16}$$

where u_x, u_y and u_z are the cartesian components of the vector $\mathbf{u}$.

b. OPERATOR ASSOCIATED WITH A ROTATION THROUGH AN ANGLE OF 2π

If we take 2π for the angle of rotation α, the geometrical rotation $\mathscr{R}_{\mathbf{u}}(2\pi)$ coincides, whatever the vector $\mathbf{u}$ may be, with the identity rotation. However, if we set $\alpha = 2\pi$ in formula (15), we see that:

$${}^{(s)}R_{\mathbf{u}}(2\pi) = -1 \tag{17}$$

whereas:

$${}^{(s)}R_{\mathbf{u}}(0) = 1 \tag{18}$$

The operator associated with a rotation through an angle of 2π is not the identity operator, but minus this operator. The group law is therefore conserved only locally in the correspondence between geometrical rotations and rotation operators in $\mathscr{E}_s$ [see discussion in complement B_{VI}, comment (*iii*) of §3-c-γ]. This is due to the half-integral value of the spin angular momentum of the particle which we are considering.

The fact that the spin state changes sign during a rotation through an angle of 2π is not disturbing, since two state vectors differing only by a global phase factor have the same physical properties. It is more important to study the way in which an observable A transforms during such a rotation. It is easy to show that:

$$A' = {}^{(S)}R_{\mathbf{u}}(2\pi)\, A \; {}^{(S)}R_{\mathbf{u}}^{\dagger}(2\pi) = A \tag{19}$$

This result is quite satisfying since a rotation through 2π cannot modify the measuring device associated with A. Consequently, the spectrum of A' must remain the same as that of A.

COMMENT:

We showed in complement B_{VI} [comment (*iii*) of §3-c-γ] that:

$${}^{(\mathbf{r})}R_{\mathbf{u}}(2\pi) = 1 \tag{20}$$

Consequently, in the global state space $\mathscr{E} = \mathscr{E}_{\mathbf{r}} \otimes \mathscr{E}_s$, as in $\mathscr{E}_s$, we have:

$$R_{\mathbf{u}}(2\pi) = {}^{(\mathbf{r})}R_{\mathbf{u}}(2\pi) \otimes {}^{(S)}R_{\mathbf{u}}(2\pi) = -1 \tag{21}$$

c. RELATIONSHIP BETWEEN THE VECTORIAL NATURE OF S AND THE BEHAVIOR OF A SPIN STATE UPON ROTATION

Consider an arbitrary spin state $|\chi\rangle$. We showed in chapter IV (§B-1-c) that there must exist angles θ and φ such that $|\chi\rangle$ can be written (except for a global phase factor which has no physical meaning):

$$|\chi\rangle = e^{-i\varphi/2}\cos\frac{\theta}{2}|+\rangle + e^{i\varphi/2}\sin\frac{\theta}{2}|-\rangle \tag{22}$$

$|\chi\rangle$ then appears as the eigenvector associated with the eigenvalue $+\hbar/2$ of the component $\mathbf{S}\cdot\mathbf{v}$ of the spin $\mathbf{S}$ along the unit vector $\mathbf{v}$ defined by the polar angles θ and φ. Now let us perform an arbitrary rotation on the state $|\chi\rangle$. Let us call $\mathbf{v}'$ the result of the transformation of $\mathbf{v}$ by the rotation being considered. Since $\mathbf{S}$ is a vectorial observable, the state $|\chi'\rangle$ after the rotation must be an eigenvector, with the eigenvalue $+\hbar/2$, of the component $\mathbf{S}\cdot\mathbf{v}'$ of $\mathbf{S}$ along the unit vector $\mathbf{v}'$ (*cf.* complement B_{VI}, §5):

$$|\chi\rangle = |+\rangle_v \Longrightarrow |\chi'\rangle = R|\chi\rangle \propto |+\rangle_{v'} \tag{23}$$

with:

$$\mathbf{v}' = \mathscr{R}\,\mathbf{v} \tag{24}$$

We shall be satisfied with verifying this for a specific case (*cf.* fig. 1). Choose for $\mathbf{v}$ the unit vector $\mathbf{e}_z$ of the Oz axis, and for $\mathbf{v}'$ an arbitrary unit vector, with polar angles θ and φ. $\mathbf{v}'$ is obtained from $\mathbf{v} = \mathbf{e}_z$ by a rotation through an angle θ about the unit vector $\mathbf{u}$, which is fixed by the polar angles:

$$\theta_u = \frac{\pi}{2}$$
$$\varphi_u = \varphi + \frac{\pi}{2} \tag{25}$$

Thus we must show that:

$$^{(s)}R_{\mathbf{u}}(\theta) \mid + \rangle \propto \mid + \rangle_{v'} \tag{26}$$

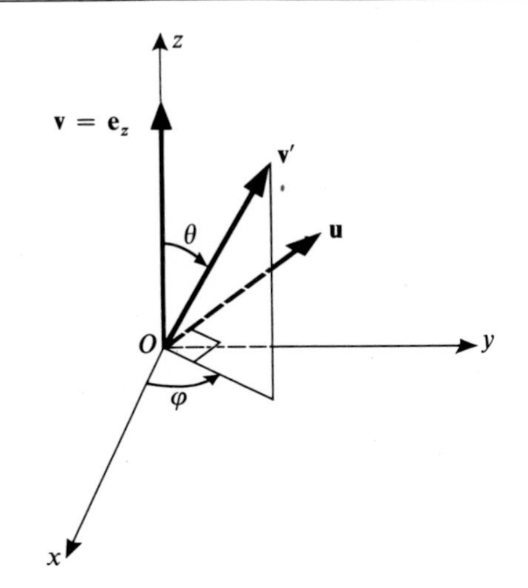

FIGURE 1

A rotation through an angle θ about u brings the vector $\mathbf{v} = \mathbf{e}_z$ onto the unit vector $\mathbf{v}'$, with polar angles θ and φ.

The cartesian components of the vector **u** are:

$$\begin{aligned} u_x &= - \sin \varphi \\ u_y &= \cos \varphi \\ u_z &= 0 \end{aligned} \tag{27}$$

so the operator $^{(s)}R_{\mathbf{u}}(\theta)$ can be written, using formula (15):

$$\begin{aligned} ^{(s)}R_{\mathbf{u}}(\theta) &= \cos\frac{\theta}{2} - i\boldsymbol{\sigma}\,.\,\mathbf{u}\sin\frac{\theta}{2} \\ &= \cos\frac{\theta}{2} - i\,(-\,\sigma_x \sin\varphi + \sigma_y \cos\varphi)\sin\frac{\theta}{2} \\ &= \cos\frac{\theta}{2} - \frac{1}{2}(\sigma_+\,\mathrm{e}^{-i\varphi} - \sigma_-\,\mathrm{e}^{i\varphi})\sin\frac{\theta}{2} \end{aligned} \tag{28}$$

with:

$$\sigma_{\pm} = \sigma_x \pm i\sigma_y \tag{29}$$

Now we know [*cf.* formulas (B-7) of chapter IX] that:

$$\begin{aligned} \sigma_+ \mid + \rangle &= 0 \\ \sigma_- \mid + \rangle &= 2 \mid - \rangle \end{aligned} \tag{30}$$

The result of the transformation of the ket $\mid + \rangle$ by the operator $^{(s)}R_{\mathbf{u}}(\theta)$ is therefore:

$$^{(s)}R_{\mathbf{u}}(\theta) \mid + \rangle = \cos\frac{\theta}{2} \mid + \rangle + \mathrm{e}^{i\varphi}\sin\frac{\theta}{2} \mid - \rangle \tag{31}$$

We recognize, to within a phase factor, the ket $| + \rangle_{v'}$ [*cf.* formula (22)]:

$$^{(s)}R_{\mathbf{u}}(\theta) \,|\, + \rangle = e^{i\varphi/2} \,|\, + \rangle_{v'} \tag{32}$$

3. Rotation of two-component spinors

We are now prepared to study the global behavior of a spin 1/2 particle under rotation. That is, we shall now take into account both its external and internal degrees of freedom.

Consider a spin 1/2 particle whose state is represented by the ket $| \psi \rangle$ of the state space $\mathscr{E} = \mathscr{E}_{\mathbf{r}} \otimes \mathscr{E}_s$. The ket $| \psi \rangle$ can be represented by the spinor $[\psi](\mathbf{r})$, having the components:

$$\psi_\varepsilon(\mathbf{r}) = \langle \mathbf{r}, \varepsilon \,|\, \psi \rangle \tag{33}$$

If we perform an arbitrary geometrical rotation $\mathscr{R}$ on this particle, its state then becomes:

$$| \psi' \rangle = R \,|\, \psi \rangle \tag{34}$$

where:

$$R = {}^{(\mathbf{r})}R \otimes {}^{(s)}R \tag{35}$$

is the operator associated, in $\mathscr{E}$, with the geometrical rotation $\mathscr{R}$. How is the spinor, $[\psi'](\mathbf{r})$, which corresponds to the state $| \psi' \rangle$, obtained from $[\psi](\mathbf{r})$?

In order to answer this question, let us write the components $\psi'_\varepsilon(\mathbf{r})$ of $[\psi']$:

$$\psi'_\varepsilon(\mathbf{r}) = \langle \mathbf{r}, \varepsilon \,|\, \psi' \rangle = \langle \mathbf{r}, \varepsilon \,|\, R \,|\, \psi \rangle \tag{36}$$

We can find the components of $[\psi](\mathbf{r})$ by inserting the closure relation relative to the $\{ | \mathbf{r}', \varepsilon' \rangle \}$ basis between R and $| \psi \rangle$:

$$\psi'_\varepsilon(\mathbf{r}) = \sum_{\varepsilon'} \int d^3r' \langle \mathbf{r}, \varepsilon \,|\, R \,|\, \mathbf{r}', \varepsilon' \rangle \langle \mathbf{r}', \varepsilon' \,|\, \psi \rangle \tag{37}$$

Now, since the vectors of the $\{ | \mathbf{r}, \varepsilon \rangle \}$ basis are tensor products, the matrix elements of the operator R in this basis can be decomposed in the following manner:

$$\langle \mathbf{r}, \varepsilon \,|\, R \,|\, \mathbf{r}', \varepsilon' \rangle = \langle \mathbf{r} \,|\, {}^{(\mathbf{r})}R \,|\, \mathbf{r}' \rangle \langle \varepsilon \,|\, {}^{(s)}R \,|\, \varepsilon' \rangle \tag{38}$$

We already know [*cf.* complement B_{VI}, formula (26)] that:

$$\langle \mathbf{r} \,|\, {}^{(\mathbf{r})}R \,|\, \mathbf{r}' \rangle = \langle \mathscr{R}^{-1}\mathbf{r} \,|\, \mathbf{r}' \rangle = \delta[\mathbf{r}' - (\mathscr{R}^{-1}\mathbf{r})] \tag{39}$$

Consequently, if we set:

$$\langle \varepsilon \,|\, {}^{(s)}R \,|\, \varepsilon' \rangle = R^{(1/2)}_{\varepsilon\varepsilon'} \tag{40}$$

formula (37) can finally be written:

$$\boxed{\psi'_\varepsilon(\mathbf{r}) = \sum_{\varepsilon'} R^{(1/2)}_{\varepsilon\varepsilon'} \, \psi_{\varepsilon'}(\mathscr{R}^{-1}\mathbf{r})} \tag{41}$$

that is, explicitly:

$$\begin{pmatrix} \psi'_+(\mathbf{r}) \\ \psi'_-(\mathbf{r}) \end{pmatrix} = \begin{pmatrix} R^{(1/2)}_{++} & R^{(1/2)}_{+-} \\ R^{(1/2)}_{-+} & R^{(1/2)}_{--} \end{pmatrix} \begin{pmatrix} \psi_+(\mathscr{R}^{-1}\mathbf{r}) \\ \psi_-(\mathscr{R}^{-1}\mathbf{r}) \end{pmatrix} \tag{42}$$

Thus we obtain the following result : each component of the new spinor $[\psi']$ at the point $\mathbf{r}$ is a linear combination of the two components of the original spinor $[\psi]$ evaluated at the point $\mathscr{R}^{-1}\mathbf{r}$ (that is, at the point that the rotation maps into $\mathbf{r}$)*. The coefficients of these linear combinations are the elements of the 2×2 matrix which represents $^{(s)}R$ in the $\{ | + \rangle, | - \rangle \}$ basis of $\mathscr{E}_s$ [*cf.* formula (16)].

References and suggestions for further reading:

Feynman III (1.2), chap. 6; chap. 18, §18-4 and added note 1; Messiah (1.17), App. C; Edmonds (2.21), chap. 4.

Rotation groups and SU(2): Bacry (10.31), chap. 6; Wigner (2.23), chap. 15; Meijer and Bauer (2.18), chap. 5.

Experiments dealing with rotations of a spin 1/2: article by Werner et al. (11.18).

* Note the close analogy between this behavior and that of a vector field under rotation.

Complement B_{IX}

EXERCISES

1. Consider a spin 1/2 particle. Call its spin **S**, its orbital angular momentum **L** and its state vector $| \psi \rangle$. The two functions $\psi_+(\mathbf{r})$ and $\psi_-(\mathbf{r})$ are defined by:

$$\psi_{\pm}(\mathbf{r}) = \langle \mathbf{r}, \pm | \psi \rangle$$

Assume that:

$$\psi_+(\mathbf{r}) = R(r)\left[Y_0^0(\theta, \varphi) + \frac{1}{\sqrt{3}} Y_1^0(\theta, \varphi) \right]$$

$$\psi_-(\mathbf{r}) = \frac{R(r)}{\sqrt{3}} [Y_1^1(\theta, \varphi) - Y_1^0(\theta, \varphi)]$$

where r, θ, φ, are the coordinates of the particle and $R(r)$ is a given function of r.

a. What condition must $R(r)$ satisfy for $| \psi \rangle$ to be normalized? √

b. S_z is measured with the particle in the state $| \psi \rangle$. What results can be found, and with what probabilities? Same question for L_z, then for S_x.

c. A measurement of $\mathbf{L}^2$, with the particle in the state $| \psi \rangle$, yielded zero. What state describes the particle just after this measurement? Same question if the measurement of $\mathbf{L}^2$ had given $2\hbar^2$.

2. Consider a spin 1/2 particle. **P** and **S** designate the observables associated with its momentum and its spin. We choose as the basis of the state space the orthonormal basis $| p_x, p_y, p_z, \pm \rangle$ of eigenvectors common to P_x, P_y, P_z and S_z (whose eigenvalues are, respectively, p_x, p_y, p_z and $\pm \hbar/2$).

We intend to solve the eigenvalue equation of the operator A which is defined by :

$$A = \mathbf{S} \cdot \mathbf{P}$$

a. Is A Hermitian?

b. Show that there exists a basis of eigenvectors of A which are also eigenvectors of P_x, P_y, P_z. In the subspace spanned by the kets $| p_x, p_y, p_z, \pm \rangle$, where p_x, p_y, p_z are fixed, what is the matrix representing A?

c. What are the eigenvalues of A, and what is their degree of degeneracy? Find a system of eigenvectors common to A and P_x, P_y, P_z.

3. The Pauli Hamiltonian

The Hamiltonian of an electron of mass m, charge q, spin $\frac{\hbar}{2}\boldsymbol{\sigma}$ (σ_x, σ_y, σ_z: Pauli matrices), placed in an electromagnetic field described by the vector potential $\mathbf{A}(\mathbf{r}, t)$ and the scalar potential $U(\mathbf{r}, t)$, is written:

$$H = \frac{1}{2m}[\mathbf{P} - q\mathbf{A}(\mathbf{R}, t)]^2 + qU(\mathbf{R}, t) - \frac{q\hbar}{2m}\boldsymbol{\sigma} \cdot \mathbf{B}(\mathbf{R}, t)$$

The last term represents the interaction between the spin magnetic moment $\frac{q\hbar}{2m}\boldsymbol{\sigma}$ and the magnetic field $\mathbf{B}(\mathbf{R}, t) = \boldsymbol{\nabla} \times \mathbf{A}(\mathbf{R}, t)$.

Show, using the properties of the Pauli matrices, that this Hamiltonian can also be written in the following form ("the Pauli Hamiltonian"):

$$H = \frac{1}{2m}\{\boldsymbol{\sigma} \cdot [\mathbf{P} - q\mathbf{A}(\mathbf{R}, t)]\}^2 + qU(\mathbf{R}, t)$$

4. We intend to study the reflection of a monoenergetic neutron beam which is perpendicularly incident on a block of a ferromagnetic material. We call Ox the direction of propagation of the incident beam and yOz the surface of the ferromagnetic material, which fills the entire $x > 0$ region (see figure). Let each incident neutron have an energy E and a mass m. The spin of the neutrons is $s = 1/2$ and their magnetic moment is written $\mathbf{M} = \gamma\mathbf{S}$ (γ is the gyromagnetic ratio and $\mathbf{S}$ is the spin operator).

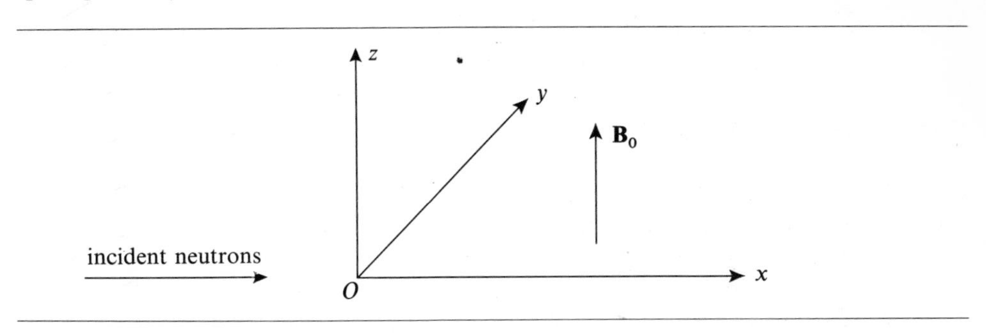

The potential energy of the neutrons is the sum of two terms:

– the first one corresponds to the interaction with the nucleons of the substance. Phenomenologically, it is represented by a potential $V(x)$, defined by $V(x) = 0$ for $x \leqslant 0$, $V(x) = V_0 > 0$ for $x > 0$.

– the second term corresponds to the interaction of the magnetic moment of each neutron with the internal magnetic field $\mathbf{B}_0$ of the material ($\mathbf{B}_0$ is assumed to be uniform and parallel to Oz). Thus we have $W = 0$ for $x \leqslant 0$, $W = \omega_0 S_z$ for $x > 0$ (with $\omega_0 = -\gamma B_0$). Throughout this exercise we shall confine ourselves to the case:

$$0 < \frac{\hbar\omega_0}{2} < V_0$$

a. Determine the stationary states of the particle which correspond to a positive incident momentum and a spin which is either parallel or antiparallel to Oz.

b. We assume in this question that $V_0 - \hbar\omega_0/2 < E < V_0 + \hbar\omega_0/2$. The incident neutron beam is unpolarized. Calculate the degree of polarization of the reflected beam. Can you imagine an application of this effect?

c. Now consider the general case where E has an arbitrary positive value. The spin of the incident neutrons points in the Ox direction. What is the direction of the spin of the reflected particles (there are three cases, depending on the relative values of E and $V_0 \pm \hbar\omega_0/2$)?

Solution of exercise 4

a. The Hamiltonian H of the particle is:

$$H = \frac{\mathbf{P}^2}{2m} + V(X) + W \tag{1}$$

$V(X)$, which acts only on the orbital variables, commutes with S_z. Since W is proportional to S_z, it also commutes with this operator. Furthermore, $V(X)$ commutes with P_y and P_z, as well as with W (obviously, since W acts only on the spin variables). We can therefore find a basis of eigenvectors common to H, S_z, P_y, P_z, which can be written:

$$| \varphi^{\pm}_{E,p_y,p_z} \rangle = | \varphi^{\pm}_E \rangle \otimes | p_y \rangle \otimes | p_z \rangle \otimes | \pm \rangle \tag{2}$$

with:

$$\begin{aligned} &| \varphi^{\pm}_E \rangle \in \mathscr{E}_x \\ &| p_y \rangle \in \mathscr{E}_y \,;\, P_y | p_y \rangle = p_y | p_y \rangle \\ &| p_z \rangle \in \mathscr{E}_z \,;\, P_z | p_z \rangle = p_z | p_z \rangle \\ &| \pm \rangle \in \mathscr{E}_s \,;\, S_z | \pm \rangle = \pm \frac{\hbar}{2} | \pm \rangle \end{aligned} \tag{3}$$

where the ket $| \varphi^{\pm}_E \rangle$ is a solution of the eigenvalue equation:

$$\left[\frac{P_x^2}{2m} + V(X) + \frac{1}{2m}(p_y^2 + p_z^2) \pm \frac{\hbar\omega_0}{2} \right] | \varphi^{\pm}_E \rangle = E \, | \varphi^{\pm}_E \rangle \tag{4}$$

We assume in the statement of the problem that the neutron beam is normally incident, so we can set $p_y = p_z = 0$. Let $\varphi^{\pm}_E(x) = \langle x | \varphi^{\pm}_E \rangle$ be the wave function associated with $| \varphi^{\pm}_E \rangle$; it satisfies the equation:

$$\left[-\frac{\hbar^2}{2m}\frac{\mathrm{d}^2}{\mathrm{d}x^2} + V(x) \pm \frac{\hbar\omega_0}{2} \right] \varphi^{\pm}_E(x) = E \, \varphi^{\pm}_E(x) \tag{5}$$

Thus the problem is reduced to that of a classical one-dimensional "square well": reflection from a "potential step" (*cf.* complement H_I).

In the $x < 0$ region, $V(x)$ is zero and the total energy E (which is positive) is greater than the potential energy. We know in this case that the wave function is a superposition of imaginary oscillatory exponentials:

$$\varphi_E^{\pm}(x) = A_{\pm}\, e^{ikx} + B_{\pm}\, e^{-ikx} \qquad \text{if} \quad x < 0 \tag{6}$$

with:

$$k = \sqrt{\frac{2m}{\hbar^2} E} \tag{7}$$

$A_{\pm}$ gives the amplitude of the wave associated with an incident particle having a spin either parallel or antiparallel to Oz. $B_{\pm}$ gives the amplitude of the wave associated with a reflected particle for the same two spin directions.

In the $x > 0$ region, $V(x)$ is equal to V_0 and, depending on the relative values of E and $V_0 \pm \hbar\omega_0/2$, the wave functions can behave like oscillatory or damped exponentials. We shall consider three cases:

(*i*) If $E > V_0 + \dfrac{\hbar\omega_0}{2}$, we set:

$$k'_{\pm} = \sqrt{\frac{2m}{\hbar^2}\left(E - V_0 \mp \frac{\hbar\omega_0}{2}\right)} \tag{8}$$

and the transmitted wave behaves like an oscillatory exponential:

$$\varphi_E^{\pm}(x) = C_{\pm}\, e^{ik'_{\pm}x} \qquad \text{if} \quad x > 0 \tag{9}$$

Moreover, the continuity conditions for the wave function and its derivative imply [*cf.* complement H_I, relations (13) and (14)]:

$$\frac{B_{\pm}}{A_{\pm}} = \frac{k - k'_{\pm}}{k + k'_{\pm}} \qquad\qquad \frac{C_{\pm}}{A_{\pm}} = \frac{2k}{k + k'_{\pm}} \tag{10}$$

(*ii*) If, on the other hand, $E < V_0 - \dfrac{\hbar\omega_0}{2}$, we must introduce the quantities $\rho_{\pm}$:

$$\rho_{\pm} = \sqrt{\frac{2m}{\hbar^2}\left(V_0 \pm \frac{\hbar\omega_0}{2} - E\right)} \tag{11}$$

and the wave in the $x > 0$ region is a real, damped exponential (evanescent wave):

$$\varphi_E^{\pm}(x) = D_{\pm}\, e^{-\rho_{\pm}x} \qquad \text{if} \quad x > 0 \tag{12}$$

with, in this case [*cf.* complement H_I, equations (22) and (23)]:

$$\frac{B_{\pm}}{A_{\pm}} = \frac{k - i\rho_{\pm}}{k + i\rho_{\pm}}\,; \qquad \frac{D_{\pm}}{A_{\pm}} = \frac{2k}{k + i\rho_{\pm}} \tag{13}$$

(*iii*) Finally, in the intermediate case $V_0 - \dfrac{\hbar\omega_0}{2} < E < V_0 + \dfrac{\hbar\omega_0}{2}$, we have:

$$\varphi_E^{+}(x) = D_{+}\, e^{-\rho_{+}x} \qquad \text{if} \quad x > 0 \tag{14-a}$$

$$\varphi_E^{-}(x) = C_{-}\, e^{ik'_{-}x} \qquad \text{if} \quad x > 0 \tag{14-b}$$

[definitions (8) and (11) of k'_- and ρ_+ are still valid]. Depending on the spin orientation, the wave is either a damped or an oscillatory exponential. We then have:

$$\frac{B_+}{A_+} = \frac{k - i\rho_+}{k + i\rho_+}\,; \qquad \frac{D_+}{A_+} = \frac{2k}{k + i\rho_+} \tag{15-a}$$

$$\frac{B_-}{A_-} = \frac{k - k'_-}{k + k'_-}\,; \qquad \frac{C_-}{A_-} = \frac{2k}{k + k'_-} \tag{15-b}$$

b. When $V_0 - \frac{\hbar\omega_0}{2} < E < V_0 + \frac{\hbar\omega_0}{2}$, we are in the situation of case (*iii*) above. If the projection onto Oz of the incident neutron spin is equal to $\hbar/2$, the corresponding reflection coefficient is:

$$R_+ = \left|\frac{B_+}{A_+}\right|^2 = \left|\frac{k - i\rho_+}{k + i\rho_+}\right|^2 = 1 \tag{16}$$

On the other hand, if the projection of the spin onto Oz is equal to $-\hbar/2$, the reflection coefficient is no longer 1, since it is given by:

$$R_- = \left|\frac{B_-}{A_-}\right|^2 = \left(\frac{k - k'_-}{k + k'_-}\right)^2 < 1 \tag{17}$$

Thus we see how the reflected beam can be polarized since, depending on the direction of its spin, the neutron has a different probability of being reflected. An unpolarized incident beam can be considered to be formed of neutrons whose spins have a probability 1/2 of being in the state $|+\rangle$ and a probability 1/2 of being in the state $|-\rangle$. Taking (16) and (17) into account, we see that the probability that a particle of the reflected beam will have its spin in the state $|+\rangle$ is $\frac{1}{1 + R_-}$, while for the state $|-\rangle$ it is $\frac{R_-}{1 + R_-}$. Therefore, the degree of polarization of the reflected beam is:

$$T = \frac{1 - R_-}{1 + R_-} = \frac{2kk'_-}{k^2 + k'^2_-} \tag{18}$$

In practice, reflection from a saturated ferromagnetic substance is actually used in the laboratory to obtain beams of polarized neutrons. To increase the degree of polarization obtained, the beam is made to fall obliquely on the surface of the ferromagnetic mirror; thus, the theoretical results obtained here are not directly applicable. However, the principle of the experiment is the same. The ferromagnetic substance chosen is often cobalt. When cobalt is magnetized to saturation, one can obtain high degrees of polarization T ($T \gtrsim 80\,\%$). Note, furthermore, that the same neutron beam reflection device can serve as an "analyzer" as well as a "polarizer" for spin directions. This possibility has been exploited in precision measurements of the magnetic moment of the neutron.

c. Consider a neutron whose momentum, of magnitude $p = \hbar k$, is parallel to Ox. Assume that the projection $\langle S_x \rangle$ of its spin is equal to $\hbar/2$. Its state is [*cf*. chap. IV, relation (A-20)]:

$$| \psi \rangle = | p \rangle \otimes \frac{1}{\sqrt{2}} [| + \rangle + | - \rangle] \tag{19}$$

with:

$$\langle \mathbf{r} | p \rangle = \frac{1}{(2\pi\hbar)^{3/2}} e^{ipx/\hbar} \tag{20}$$

How can we construct a stationary state of the particle in which the incident wave has the form (19)? We simply have to consider the state:

$$| \psi_S \rangle = \frac{1}{\sqrt{2}} [| \varphi^+_{E,0,0} \rangle + | \varphi^-_{E,0,0} \rangle] \tag{21}$$

which is a linear combination of two eigenkets of H defined in (2), associated with the same eigenvalue $E = p^2/2m$. The part of the ket $| \psi_S \rangle$ which describes the reflected wave is then:

$$| - p \rangle \otimes \frac{1}{\sqrt{2}} [B_+ | + \rangle + B_- | - \rangle] \tag{22}$$

where B_+ and B_- are given, depending on the case, by (10), (13) or (15) (A_+ and A_- being replaced by 1). Let us calculate, for a state such as (22), the mean value $\langle \mathbf{S} \rangle$. Since this state is a tensor product, the spin variables and the orbital variables are not correlated. Therefore, $\langle \mathbf{S} \rangle$ can easily be obtained from the spin state vector $B_+ | + \rangle + B_- | - \rangle$, which gives:

$$\langle S_x \rangle = \frac{\hbar}{2} \frac{B_+^* B_- + B_-^* B_+}{|B_+|^2 + |B_-|^2} \tag{23-a}$$

$$\langle S_y \rangle = \frac{\hbar}{2} \frac{i(B_-^* B_+ - B_+^* B_-)}{|B_+|^2 + |B_-|^2} \tag{23-b}$$

$$\langle S_z \rangle = \frac{\hbar}{2} \frac{|B_+|^2 - |B_-|^2}{|B_+|^2 + |B_-|^2} \tag{23-c}$$

Three cases can then be distinguished:

(*i*) If $E > V_0 + \hbar\omega_0/2$, we see from (10) that B_+ and B_- are real. Formulas (23) then show that $\langle S_x \rangle$ and $\langle S_z \rangle$ are not zero but that $\langle S_y \rangle = 0$. Upon reflection of the neutron, the spin has thus undergone a rotation about Oy. Physically, it is the difference between the degrees of reflection of neutrons whose spin is parallel to Oz and those whose spin is antiparallel to Oz which explains why the $\langle S_z \rangle$ component becomes positive.

(*ii*) If $E < V_0 - \hbar\omega_0/2$, equations (13) show that B_+ and B_- are not real: they are two complex numbers having different phases but the same modulus. According to (23), we have, in this case, $\langle S_z \rangle = 0$ but $\langle S_x \rangle \neq 0$ and $\langle S_y \rangle \neq 0$. Upon reflection of the neutron, the spin thus undergoes a rotation about Oz. The physical origin of this rotation is the following : because of the existence of the evanescent wave, the neutron spends a certain time in the $x > 0$ region; the Larmor precession about $\mathbf{B}_0$ that it undergoes during this time accounts for the rotation of its spin.

(*iii*) If $V_0 - \hbar\omega_0/2 < E < V_0 + \hbar\omega_0/2$, B_+ is a complex number while B_- is a real number, and their moduli are different. None of the spin components, $\langle S_x \rangle$, $\langle S_y \rangle$ or $\langle S_z \rangle$, is then zero. This rotation of the spin upon reflection of the neutron is explained by a combination of the effects pointed out in (*i*) and (*ii*).

CHAPTER X

Addition of angular momenta

OUTLINE OF CHAPTER X

A. INTRODUCTION	1. Total angular momentum in classical mechanics 2. The importance of total angular momentum in quantum mechanics
B. ADDITION OF TWO SPIN 1/2'S. ELEMENTARY METHOD	1. Statement of the problem *a.* State space *b.* Total spin **S**. Commutation relations *c.* The basis change to be performed 2. The eigenvalues of S_z and their degrees of degeneracy 3. Diagonalization of $\mathbf{S}^2$ *a.* Calculation of the matrix representing $\mathbf{S}^2$ *b.* Eigenvalues and eigenvectors of $\mathbf{S}^2$ 4. Results : triplet and singlet
C. ADDITION OF TWO ARBITRARY ANGULAR MOMENTA. GENERAL METHOD	1. Review of the general theory of angular momentum 2. Statement of the problem *a.* State space *b.* Total angular momentum. Commutation relations *c.* The basis change to be performed 3. Eigenvalues of $\mathbf{J}^2$ and J_z *a.* Special case of two spin 1/2's *b.* The eigenvalues of J_z and their degrees of degeneracy *c.* The eigenvalues of $\mathbf{J}^2$ 4. Common eigenvectors of $\mathbf{J}^2$ and J_z *a.* Special case of two spin 1/2's *b.* General case (arbitrary j_1 and j_2) *c.* Clebsch-Gordan coefficients

A. INTRODUCTION

1. Total angular momentum in classical mechanics

Consider a system of N classical particles. The total angular momentum $\mathcal{L}$ of this system with respect to a fixed point O is the vector sum of the individual angular momenta of the N particles with respect to this point O:

$$\mathcal{L} = \sum_{i=1}^{N} \mathcal{L}_i \tag{A-1}$$

with:

$$\mathcal{L}_i = \mathbf{r}_i \times \mathbf{p}_i \tag{A-2}$$

The time derivative of $\mathcal{L}$ is equal to the moment with respect to O of the external forces. Consequently, when the external forces are zero (an isolated system) or all directed towards the same center, the total angular momentum of the system (with respect to any point in the first case and with respect to the center of force in the second one) is a constant of the motion. This is not the case for each of the individual angular momenta $\mathcal{L}_i$ if there are internal forces, that is, if the various particles of the system interact.

We shall illustrate this point with an example. Consider a system composed of two particles, (1) and (2), subject to the same central force field (which can be created by a third particle assumed to be heavy enough to remain motionless at the origin). If these two particles exert no force on each other, their angular momenta $\mathcal{L}_1$ and $\mathcal{L}_2$ with respect to the center of force O are both constants of the motion. The only force then acting on particle (1), for example, is directed towards O; its moment with respect to this point is therefore zero, as is $\frac{\mathrm{d}}{\mathrm{d}t}\mathcal{L}_1$. On the other hand, if particle (1) is also subject to a force due to the presence of particle (2), the moment with respect to O of this force is not generally zero, and, consequently, $\mathcal{L}_1$ is no longer a constant of the motion. However, if the interaction between the two particles obeys the principle of action and reaction, the moment of the force exerted by (1) on (2) with respect to O exactly compensates that of the force exerted by (2) on (1): the total angular momentum $\mathcal{L}$ is conserved over time.

Therefore, in a system of interacting particles, *only the total angular momentum is a constant of the motion*: forces inside the system induce a transfer of angular momentum from one particle to the other. Thus we see why it is useful to study the properties of the total angular momentum.

2. The importance of total angular momentum in quantum mechanics

Let us treat the preceding example quantum mechanically. In the case of two non-interacting particles, the Hamiltonian of the system is given simply, in the $\{ | \mathbf{r}_1, \mathbf{r}_2 \rangle \}$ representation:

$$H_0 = H_1 + H_2 \tag{A-3}$$

with:

$$\begin{aligned} H_1 &= -\frac{\hbar^2}{2\mu_1} \Delta_1 + V(r_1) \\ H_2 &= -\frac{\hbar^2}{2\mu_2} \Delta_2 + V(r_2) \end{aligned} \tag{A-4}$$

[μ_1 and μ_2 are the masses of the two particles, $V(r)$ is the central potential to which they are subject, and Δ_1 and Δ_2 denote the Laplacian operators relative to the coordinates of particles (1) and (2) respectively]. We know from chapter VII (§ A-2-a) that the three components of the operator $\mathbf{L}_1$ associated with the angular momentum $\mathscr{L}_1$ of particle (1) commute with H_1:

$$[\mathbf{L}_1, H_1] = \mathbf{0} \tag{A-5}$$

Also, all observables relating to one of the particles commute with all those corresponding to the other one; in particular:

$$[\mathbf{L}_1, H_2] = \mathbf{0} \tag{A-6}$$

From (A-5) and (A-6), we see that the three components of $\mathbf{L}_1$ are constants of the motion. An analogous argument is obviously valid for $\mathbf{L}_2$.

Now assume that the two particles interact, and that the corresponding potential energy $v(|\mathbf{r}_1 - \mathbf{r}_2|)$ depends only on the distance between them $|\mathbf{r}_1 - \mathbf{r}_2|$*:

$$|\mathbf{r}_1 - \mathbf{r}_2| = \sqrt{(x_1 - x_2)^2 + (y_1 - y_2)^2 + (z_1 - z_2)^2} \tag{A-7}$$

In this case, the Hamiltonian of the system is:

$$H = H_1 + H_2 + v(|\mathbf{r}_1 - \mathbf{r}_2|) \tag{A-8}$$

where H_1 and H_2 are given by (A-4). According to (A-5) and (A-6), the commutator of $\mathbf{L}_1$ with H reduces to:

$$[\mathbf{L}_1, H] = [\mathbf{L}_1, v(|\mathbf{r}_1 - \mathbf{r}_2|)] \tag{A-9}$$

that is, for the component L_{1z}, for example:

$$[L_{1z}, H] = [L_{1z}, v(|\mathbf{r}_1 - \mathbf{r}_2|)] = \frac{\hbar}{i}\left(x_1 \frac{\partial v}{\partial y_1} - y_1 \frac{\partial v}{\partial x_1} \right) \tag{A-10}$$

* The corresponding classical forces then necessarily obey the principle of action and reaction.

Expression (A-10) is generally not zero: $\mathbf{L}_1$ is no longer a constant of the motion. On the other hand, if we define the *total angular momentum operator* $\mathbf{L}$ by an expression similar to (A-1):

$$\mathbf{L} = \mathbf{L}_1 + \mathbf{L}_2 \tag{A-11}$$

we obtain an operator whose three components are constants of the motion. For example, we find:

$$[L_z, H] = [L_{1z} + L_{2z}, H] \tag{A-12}$$

According to (A-10), this commutator is equal to:

$$\begin{aligned}[L_z, H] &= [L_{1z} + L_{2z}, H] \\ &= \frac{\hbar}{i}\left(x_1 \frac{\partial v}{\partial y_1} - y_1 \frac{\partial v}{\partial x_1} + x_2 \frac{\partial v}{\partial y_2} - y_2 \frac{\partial v}{\partial x_2}\right)\end{aligned} \tag{A-13}$$

But, since v depends only on $|\mathbf{r}_1 - \mathbf{r}_2|$, given by (A-7), we have:

$$\frac{\partial v}{\partial x_1} = v' \frac{\partial |\mathbf{r}_1 - \mathbf{r}_2|}{\partial x_1} = v' \frac{x_1 - x_2}{|\mathbf{r}_1 - \mathbf{r}_2|} \tag{A-14-a}$$

$$\frac{\partial v}{\partial x_2} = v' \frac{\partial |\mathbf{r}_1 - \mathbf{r}_2|}{\partial x_2} = v' \frac{x_2 - x_1}{|\mathbf{r}_1 - \mathbf{r}_2|} \tag{A-14-b}$$

and analogous expressions for $\frac{\partial v}{\partial y_1}, \frac{\partial v}{\partial y_2}, \frac{\partial v}{\partial z_1}$ and $\frac{\partial v}{\partial z_2}$ (v' is the derivative of v, considered as a function of a single variable). Substituting these values into (A-13):

$$\begin{aligned}[L_z, H] &= \frac{\hbar}{i} \frac{v'}{|\mathbf{r}_1 - \mathbf{r}_2|} \Big\{ x_1(y_1 - y_2) - y_1(x_1 - x_2) \\ &\qquad + x_2(y_2 - y_1) - y_2(x_2 - x_1) \Big\} \\ &= 0\end{aligned} \tag{A-15}$$

We therefore arrive at the same conclusion as in classical mechanics.

Until now we have implicitly assumed that the particles being studied had no spin. Now let us examine another important example: that of a single particle with spin. First, we assume that this particle is subject only to a central potential $V(r)$. Its Hamiltonian is then the one studied in § A of chapter VII. We know that the three components of the orbital angular momentum $\mathbf{L}$ commute with this Hamiltonian. In addition, since the spin operators commute with the orbital observables, the three components of the spin $\mathbf{S}$ are also constants of the motion. But we shall see in chapter XII that relativistic corrections introduce into the Hamiltonian a *spin-orbit coupling* term of the form:

$$H_{SO} = \xi(r)\mathbf{L} \cdot \mathbf{S} \tag{A-16}$$

where $\xi(r)$ is a known function of the single variable r (the physical meaning of this coupling will be explained in chapter XII). When this term is taken into account,

L and **S** no longer commute with the total Hamiltonian. For example★:

$$\begin{aligned}[L_z, H_{SO}] &= \xi(r)\,[L_z, L_xS_x + L_yS_y + L_zS_z] \\ &= \xi(r)\,(i\hbar L_yS_x - i\hbar L_xS_y)\end{aligned} \qquad \text{(A-17)}$$

and, similarly:

$$\begin{aligned}[S_z, H_{SO}] &= \xi(r)\,[S_z, L_xS_x + L_yS_y + L_zS_z] \\ &= \xi(r)\,(i\hbar L_xS_y - i\hbar L_yS_x)\end{aligned} \qquad \text{(A-18)}$$

However, if we set:

$$\mathbf{J} = \mathbf{L} + \mathbf{S} \qquad \text{(A-19)}$$

the three components of **J** *are constants of the motion.* To see this, we can simply add equations (A-17) and (A-18):

$$[J_z, H_{SO}] = [L_z + S_z, H_{SO}] = 0 \qquad \text{(A-20)}$$

(an analogous proof could be given for the other components of **J**). The operator **J** defined by (A-19) is said to be the total angular momentum of a particle with spin.

In the two cases just described, we have two partial angular momenta $\mathbf{J}_1$ and $\mathbf{J}_2$, which commute. We know a basis of the state space composed of eigenvectors common to $\mathbf{J}_1^2$, J_{1z}, $\mathbf{J}_2^2$, J_{2z}. However, $\mathbf{J}_1$ and $\mathbf{J}_2$ are not constants of the motion, while the components of the total angular momentum:

$$\mathbf{J} = \mathbf{J}_1 + \mathbf{J}_2 \qquad \text{(A-21)}$$

commute with the Hamiltonian of the system. We shall therefore try to construct, using the preceding basis, *a new basis formed by eigenvectors of* $\mathbf{J}^2$ *and* J_z. The problem thus posed in general terms is that of the *addition (or composition) of two angular momenta* $\mathbf{J}_1$ *and* $\mathbf{J}_2$.

The importance of this new basis, formed of eigenvectors of $\mathbf{J}^2$ and J_z, is easy to understand. To determine the stationary states of the system, that is, the eigenstates of H, it is simpler to diagonalize the matrix which represents H in this new basis. Since H commutes with $\mathbf{J}^2$ and J_z, this matrix can be broken down into as many blocks as there are eigensubspaces associated with the various sets of eigenvalues of $\mathbf{J}^2$ and J_z (*cf.* chap. II, § D-3-a). Its structure is much simpler than that of the matrix which represents H in the basis of eigenvectors common to $\mathbf{J}_1^2$, J_{1z}, $\mathbf{J}_2^2$, J_{2z}, since neither J_{1z} nor J_{2z} generally commutes with H.

We shall leave aside for now the problem of the diagonalization of H (whether exact or approximate) in the basis of eigenstates of $\mathbf{J}^2$ and J_z. Rather, we shall concentrate on the construction of this new basis from the one formed by the eigenstates of $\mathbf{J}_1^2$, J_{1z}, $\mathbf{J}_2^2$, J_{2z}. A certain number of physical applications (many-electron atoms, fine and hyperfine line structure, etc.) will be considered after we have studied perturbation theory (complements of chapter XI and chapter XII).

★ To establish (A-17) and (A-18), one uses the fact that **L**, which acts only on the angular variables θ and φ, commutes with $\xi(r)$, which depends only on r.

We shall begin (§B) with an elementary treatment of a simple case, in which the two partial angular momenta we wish to add are spin 1/2's. This will allow us to familiarize ourselves with various aspects of the problem, before we treat, in §C, the addition of two arbitrary angular momenta.

B. ADDITION OF TWO SPIN 1/2'S. ELEMENTARY METHOD

1. Statement of the problem

We shall consider a system of two spin 1/2 particles (electrons or silver atoms in the ground state, for example), and we shall be concerned only with their spin degrees of freedom. Let $\mathbf{S}_1$ and $\mathbf{S}_2$ be the spin operators of the two particles.

a. STATE SPACE

We have already defined the state space of such a system. Recall that it is a four-dimensional space, obtained by taking the tensor product of the individual spin spaces of the two particles. We know an orthonormal basis of this space, which we shall denote by $\{ | \varepsilon_1, \varepsilon_2 \rangle \}$, that is, explicitly:

$$\{ | \varepsilon_1, \varepsilon_2 \rangle \} = \{ | +, + \rangle , | +, - \rangle , | -, + \rangle , | -, - \rangle \} \tag{B-1}$$

These vectors are eigenstates of the four observables $\mathbf{S}_1^2$, S_{1z}, $\mathbf{S}_2^2$, S_{2z} (which are actually the extensions, into the tensor product space, of operators, defined in each of the spin spaces):

$$\mathbf{S}_1^2 | \varepsilon_1, \varepsilon_2 \rangle = \mathbf{S}_2^2 | \varepsilon_1, \varepsilon_2 \rangle = \frac{3}{4}\hbar^2 | \varepsilon_1, \varepsilon_2 \rangle \tag{B-2-a}$$

$$S_{1z} | \varepsilon_1, \varepsilon_2 \rangle = \varepsilon_1 \frac{\hbar}{2} | \varepsilon_1, \varepsilon_2 \rangle \tag{B-2-b}$$

$$S_{2z} | \varepsilon_1, \varepsilon_2 \rangle = \varepsilon_2 \frac{\hbar}{2} | \varepsilon_1, \varepsilon_2 \rangle \tag{B-2-c}$$

$\mathbf{S}_1^2$, $\mathbf{S}_2^2$, S_{1z} and S_{2z} constitute a C.S.C.O. (the first two observables are actually multiples of the identity operator, and the set of operators remains complete even if they are omitted).

b. TOTAL SPIN S. COMMUTATION RELATIONS

We define the total spin $\mathbf{S}$ of the system by:

$$\mathbf{S} = \mathbf{S}_1 + \mathbf{S}_2 \tag{B-3}$$

It is simple, knowing that $\mathbf{S}_1$ and $\mathbf{S}_2$ are angular momenta, to show that $\mathbf{S}$ is as well. We can calculate, for example, the commutator of S_x and S_y:

$$\begin{aligned} [S_x, S_y] &= [S_{1x} + S_{2x}, S_{1y} + S_{2y}] \\ &= [S_{1x}, S_{1y}] + [S_{2x}, S_{2y}] \\ &= i\hbar S_{1z} + i\hbar S_{2z} \\ &= i\hbar S_z \end{aligned} \tag{B-4}$$

The operator $\mathbf{S}^2$ can be obtained by taking the (scalar) square of equation (B-3):

$$\mathbf{S}^2 = (\mathbf{S}_1 + \mathbf{S}_2)^2 = \mathbf{S}_1^2 + \mathbf{S}_2^2 + 2\mathbf{S}_1 \cdot \mathbf{S}_2 \tag{B-5}$$

since $\mathbf{S}_1$ and $\mathbf{S}_2$ commute. The scalar product $\mathbf{S}_1 \cdot \mathbf{S}_2$ can be expressed in terms of the operators $S_{1\pm}$, S_{1z} and $S_{2\pm}$, S_{2z}; it is easy to show that:

$$\begin{aligned} \mathbf{S}_1 \cdot \mathbf{S}_2 &= S_{1x}S_{2x} + S_{1y}S_{2y} + S_{1z}S_{2z} \\ &= \frac{1}{2}(S_{1+}S_{2-} + S_{1-}S_{2+}) + S_{1z}S_{2z} \end{aligned} \tag{B-6}$$

Note that, since $\mathbf{S}_1$ and $\mathbf{S}_2$ each commute with $\mathbf{S}_1^2$ and $\mathbf{S}_2^2$, so do the three components of $\mathbf{S}$. In particular, $\mathbf{S}^2$ and S_z commute with $\mathbf{S}_1^2$ and $\mathbf{S}_2^2$:

$$[S_z, \mathbf{S}_1^2] = [S_z, \mathbf{S}_2^2] = 0 \tag{B-7-a}$$

$$[\mathbf{S}^2, \mathbf{S}_1^2] = [\mathbf{S}^2, \mathbf{S}_2^2] = 0 \tag{B-7-b}$$

In addition, S_z obviously commutes with S_{1z} and S_{2z}:

$$[S_z, S_{1z}] = [S_z, S_{2z}] = 0 \tag{B-8}$$

However, $\mathbf{S}^2$ *commutes with neither* S_{1z} *nor* S_{2z} since, according to (B-5):

$$\begin{aligned} [\mathbf{S}^2, S_{1z}] &= [\mathbf{S}_1^2 + \mathbf{S}_2^2 + 2\mathbf{S}_1 \cdot \mathbf{S}_2, S_{1z}] \\ &= 2[\mathbf{S}_1 \cdot \mathbf{S}_2, S_{1z}] \\ &= 2[S_{1x}S_{2x} + S_{1y}S_{2y}, S_{1z}] \\ &= 2i\hbar(- S_{1y}S_{2x} + S_{1x}S_{2y}) \end{aligned} \tag{B-9}$$

[this calculation is analogous to the one performed in (A-17) and (A-18)]. The commutator of $\mathbf{S}^2$ with S_{2z} is, of course, equal and opposite to the preceding one, so that $S_z = S_{1z} + S_{2z}$ commutes with $\mathbf{S}^2$.

c. THE BASIS CHANGE TO BE PERFORMED

The basis (B-1), as we have seen, is composed of eigenvectors common to the C.S.C.O.:

$$\{ \mathbf{S}_1^2, \mathbf{S}_2^2, S_{1z}, S_{2z} \} \tag{B-10}$$

Also, we have just shown that the four observables:

$$\mathbf{S}_1^2, \mathbf{S}_2^2, \mathbf{S}^2, S_z \tag{B-11}$$

commute. We shall see in what follows that they also form a C.S.C.O.

Adding the two spins $\mathbf{S}_1$ and $\mathbf{S}_2$ amounts to constructing the orthonormal system of eigenvectors common to the set (B-11). This system will be different from (B-1), since $\mathbf{S}^2$ does not commute with S_{1z} and S_{2z}. We shall write the vectors of this new basis $| S, M \rangle$, with the eigenvalues of $\mathbf{S}_1^2$ and $\mathbf{S}_2^2$ (which remain the same) implicit. The vectors $| S, M \rangle$ therefore satisfy the equations:

$$\mathbf{S}_1^2 \, | S, M \rangle = \mathbf{S}_2^2 \, | S, M \rangle = \frac{3}{4}\hbar^2 \, | S, M \rangle \tag{B-12-a}$$

$$\mathbf{S}^2 \, | S, M \rangle = S(S+1)\hbar^2 | S, M \rangle \tag{B-12-b}$$

$$S_z \, | S, M \rangle = M\hbar \, | S, M \rangle \tag{B-12-c}$$

We know that $\mathbf{S}$ is an angular momentum. Consequently, S must be a positive integer or half-integer, and M varies by one-unit jumps between $-S$ and $+S$. The problem is therefore to find what values S and M can actually have, and to express the basis vectors $| S, M \rangle$ in terms of those of the known basis.

In this section, we shall confine ourselves to solving this problem by the elementary method involving the calculation and diagonalization of the 4×4 matrices representing $\mathbf{S}^2$ and S_z in the $\{ | \varepsilon_1, \varepsilon_2 \rangle \}$ basis. In §C, we shall use another, more elegant, method, and generalize it to the case of two arbitrary angular momenta.

2. The eigenvalues of S_z and their degrees of degeneracy

The observables $\mathbf{S}_1^2$ and $\mathbf{S}_2^2$ are easy to deal with: all vectors of the state space are their eigenvectors, with the same eigenvalue $3\hbar^2/4$. Consequently, equations (B-12-a) are automatically satisfied for all kets $| S, M \rangle$.

We have already noted [formulas (B-7) and (B-8)] that S_z commutes with the four observables of the C.S.C.O. (B-10). We should therefore expect *the basis vectors* $\{ | \varepsilon_1, \varepsilon_2 \rangle \}$ *to be automatically eigenvectors of* S_z. We can indeed show, using (B-2-b) and (B-2-c), that:

$$S_z \, | \varepsilon_1, \varepsilon_2 \rangle = (S_{1z} + S_{2z}) | \varepsilon_1, \varepsilon_2 \rangle = \frac{1}{2}(\varepsilon_1 + \varepsilon_2)\,\hbar \, | \varepsilon_1, \varepsilon_2 \rangle \tag{B-13}$$

$| \varepsilon_1, \varepsilon_2 \rangle$ is therefore an eigenstate of S_z with the eigenvalue:

$$M = \frac{1}{2}(\varepsilon_1 + \varepsilon_2) \tag{B-14}$$

Since ε_1 and ε_2 can each be equal to ± 1, we see that *M can take on the values $+1$, 0 and -1.*

The values $M = 1$ and $M = -1$ are not degenerate. Only one eigenvector corresponds to each of them : $| +, + \rangle$ for the first one and $| -, - \rangle$ for the second one. *On the other hand, $M = 0$ is two-fold degenerate:* two orthogonal eigenvectors are associated with it, $| +, - \rangle$ and $| -, + \rangle$. Any linear combination of these two vectors is an eigenstate of S_z, with the eigenvalue 0.

These results appear clearly in the matrix which represents S_z in the $\{ | \varepsilon_1, \varepsilon_2 \rangle \}$ basis. Choosing the basis vectors in the order indicated in (B-1), that matrix can be written :

$$(S_z) = \hbar \begin{pmatrix} 1 & 0 & 0 & 0 \\ 0 & 0 & 0 & 0 \\ 0 & 0 & 0 & 0 \\ 0 & 0 & 0 & -1 \end{pmatrix} \tag{B-15}$$

3. Diagonalization of $\mathbf{S}^2$

All that remains to be done is to find and then diagonalize the matrix which represents $\mathbf{S}^2$ in the $\{ | \varepsilon_1, \varepsilon_2 \rangle \}$ basis. We know in advance that it is not diagonal, since $\mathbf{S}^2$ does not commute with S_{1z} and S_{2z}.

a. CALCULATION OF THE MATRIX REPRESENTING $\mathbf{S}^2$

We are going to apply $\mathbf{S}^2$ to each of the basis vectors. To do this, we shall use formulas (B-5) and (B-6):

$$\mathbf{S}^2 = \mathbf{S}_1^2 + \mathbf{S}_2^2 + 2S_{1z}S_{2z} + S_{1+}S_{2-} + S_{1-}S_{2+} \tag{B-16}$$

The four vectors $| \varepsilon_1, \varepsilon_2 \rangle$ are eigenvectors of $\mathbf{S}_1^2$, $\mathbf{S}_2^2$, S_{1z} and S_{2z} [formulas (B-2)], and the action of the operators $S_{1\pm}$ and $S_{2\pm}$ can be derived from formulas (B-7) of chapter IX. We therefore find:

$$\begin{aligned} \mathbf{S}^2 | +, + \rangle &= \left(\frac{3}{4}\hbar^2 + \frac{3}{4}\hbar^2\right) | +, + \rangle + \frac{1}{2}\hbar^2 | +, + \rangle \\ &= 2\hbar^2 | +, + \rangle \end{aligned} \tag{B-17-a}$$

$$\begin{aligned} \mathbf{S}^2 | +, - \rangle &= \left(\frac{3}{4}\hbar^2 + \frac{3}{4}\hbar^2\right) | +, - \rangle - \frac{1}{2}\hbar^2 | +, - \rangle + \hbar^2 | -, + \rangle \\ &= \hbar^2 [| +, - \rangle + | -, + \rangle] \end{aligned} \tag{B-17-b}$$

$$\begin{aligned} \mathbf{S}^2 | -, + \rangle &= \left(\frac{3}{4}\hbar^2 + \frac{3}{4}\hbar^2\right) | -, + \rangle - \frac{1}{2}\hbar^2 | -, + \rangle + \hbar^2 | +, - \rangle \\ &= \hbar^2 [| -, + \rangle + | +, - \rangle] \end{aligned} \tag{B-17-c}$$

$$\begin{aligned} \mathbf{S}^2 | -, - \rangle &= \left(\frac{3}{4}\hbar^2 + \frac{3}{4}\hbar^2\right) | -, - \rangle + \frac{1}{2}\hbar^2 | -, - \rangle \\ &= 2\hbar^2 | -, - \rangle \end{aligned} \tag{B-17-d}$$

The matrix representing $\mathbf{S}^2$ in the basis of the four vectors $| \varepsilon_1, \varepsilon_2 \rangle$, arranged in the order given in (B-1), is therefore:

$$(\mathbf{S}^2) = \hbar^2 \begin{pmatrix} 2 & 0 & 0 & 0 \\ 0 & 1 & 1 & 0 \\ 0 & 1 & 1 & 0 \\ 0 & 0 & 0 & 2 \end{pmatrix} \tag{B-18}$$

COMMENT:

The zeros appearing in this matrix were to be expected. $\mathbf{S}^2$ commutes with S_z, and therefore has non-zero matrix elements only between eigenvectors of S_z associated with the same eigenvalue. According to the results of § 2, the only non-diagonal elements of $\mathbf{S}^2$ which could be different from zero are those which relate $| +, - \rangle$ to $| -, + \rangle$.

b. EIGENVALUES AND EIGENVECTORS OF $\mathbf{S}^2$

Matrix (B-18) can be broken down into three submatrices (as shown by the dotted lines). Two of them are one-dimensional: *the vectors* $| +, + \rangle$ and $| -, - \rangle$ *are eigenvectors of* $\mathbf{S}^2$, as is also shown by relations (B-17-a) and (B-17-d). The associated eigenvalues are both equal to $2\hbar^2$.

We must now diagonalize the 2×2 submatrix:

$$(\mathbf{S}^2)_0 = \hbar^2 \begin{pmatrix} 1 & 1 \\ 1 & 1 \end{pmatrix} \tag{B-19}$$

which represents $\mathbf{S}^2$ inside the two-dimensional subspace spanned by $| +, - \rangle$ and $| -, + \rangle$, that is, the eigensubspace of S_z corresponding to $M = 0$. The eigenvalues $\lambda\hbar^2$ of matrix (B-19) can be obtained by solving the characteristic equation:

$$(1 - \lambda)^2 - 1 = 0 \tag{B-20}$$

The roots of this equation are $\lambda = 0$ and $\lambda = 2$. This yields the last two eigenvalues of $\mathbf{S}^2$: 0 and $2\hbar^2$. An elementary calculation yields the corresponding eigenvectors:

$$\frac{1}{\sqrt{2}}[| +, - \rangle + | -, + \rangle] \qquad \text{for the eigenvalue } 2\hbar^2 \tag{B-21-a}$$

$$\frac{1}{\sqrt{2}}[| +, - \rangle - | -, + \rangle] \qquad \text{for the eigenvalue } 0 \tag{B-21-b}$$

(of course, they are defined only to within a global phase factor; the coefficients $1/\sqrt{2}$ insure their normalization).

The operator $\mathbf{S}^2$ therefore possesses two distinct eigenvalues : 0 and $2\hbar^2$. The first one is non-degenerate and corresponds to vector (B-21-b). The second one is three-fold degenerate, and the vectors $| +, + \rangle$, $| -, - \rangle$ and (B-21-a) form an orthonormal basis in the associated eigensubspace.

4. Results: triplet and singlet

Thus we have obtained the eigenvalues of $\mathbf{S}^2$ and S_z, as well as a system of eigenvectors common to these two observables. We shall summarize these results by expressing them in the notation of equations (B-12).

The quantum number S of (B-12-b) can take on two values: 0 and 1. The first one is associated with a single vector, (B-21-b), which is also an eigenvector of S_z with the eigenvalue 0, since it is a linear combination of $| +, - \rangle$ and $| -, + \rangle$; we shall therefore denote this vector by $| 0, 0 \rangle$:

$$| 0, 0 \rangle = \frac{1}{\sqrt{2}}[| +, - \rangle - | -, + \rangle] \tag{B-22}$$

Three vectors which differ by their values of M are associated with the value $S = 1$:

$$\begin{cases} | 1, 1 \rangle = | +, + \rangle \\ | 1, 0 \rangle = \dfrac{1}{\sqrt{2}}[| +, - \rangle + | -, + \rangle] \\ | 1, -1 \rangle = | -, - \rangle \end{cases} \tag{B-23}$$

It can easily be shown that the four vectors $| S, M \rangle$ given in (B-22) and (B-23) form an orthonormal basis. Specification of S and M suffices to define uniquely a vector of this basis. From this, it can be shown that $\mathbf{S}^2$ and S_z constitute a C.S.C.O. (which could include $\mathbf{S}_1^2$ and $\mathbf{S}_2^2$, although it is not necessary here).

Therefore, *when two spin 1/2's* ($s_1 = s_2 = 1/2$) *are added, the number* S which characterizes the eigenvalues $S(S + 1)\hbar^2$ of the observable $\mathbf{S}^2$ *can be equal either to* 1 *or to* 0. *With each of these two values of S is associated a family of* $(2S + 1)$ *orthogonal vectors* (three for $S = 1$, one for $S = 0$) *corresponding to the* $(2S + 1)$ *values of M which are compatible with S.*

COMMENTS:

(*i*) The family (B-23) of the three vectors $| 1, M \rangle$ ($M = 1, 0, -1$) constitutes what is called a *triplet*; the vector $| 0, 0 \rangle$ is called a *singlet* state.

(*ii*) The triplet states are *symmetric* with respect to an exchange of two spins, whereas the singlet state is *antisymmetric*. This means that if each vector $| \varepsilon_1, \varepsilon_2 \rangle$ is replaced by the vector $| \varepsilon_2, \varepsilon_1 \rangle$, expressions (B-23) remain invariant, while (B-22) changes sign. We shall see in chapter XIV the importance of this property when the two particles whose spins are added are identical. Furthermore, it enables us to find the right linear combination of $| +, - \rangle$ and $| -, + \rangle$ which must be associated with $| +, + \rangle$ and $| -, - \rangle$ (clearly symmetric) in order to complete the triplet. The singlet state, on the other hand, is the antysymmetric linear combination of $| +, - \rangle$ and $| -, + \rangle$, which is orthogonal to the preceding one.

C. ADDITION OF TWO ARBITRARY ANGULAR MOMENTA. GENERAL METHOD

1. Review of the general theory of angular momentum

Consider an arbitrary system, whose state space is $\mathscr{E}$, and an angular momentum **J** relative to this system (**J** can be either a partial angular momentum or the total angular momentum of the system). We showed in chapter VI (§C-3) that it is always possible to construct a standard basis $\{ | k, j, m \rangle \}$ composed of eigenvectors common to $\mathbf{J}^2$ and J_z:

$$\mathbf{J}^2 \, | k, j, m \rangle = j(j + 1)\hbar^2 \, | k, j, m \rangle \tag{C-1-a}$$

$$J_z \, | k, j, m \rangle = m\hbar \, | k, j, m \rangle \tag{C-1-b}$$

such that the action of the operators J_+ and J_- obeys the relations:

$$J_{\pm} \, | k, j, m \rangle = \hbar\sqrt{j(j + 1) - m(m \pm 1)} \, | k, j, m \pm 1 \rangle \tag{C-2}$$

We denote by $\mathscr{E}(k, j)$ the vector space spanned by the set of vectors of the standard basis which correspond to fixed values of k and j. There are $(2j + 1)$ of these vectors, and, according to (C-1) and (C-2), they can be transformed into each other by $\mathbf{J}^2$, J_z, J_+ and J_-. The state space can be considered to be a direct sum of orthogonal subspaces $\mathscr{E}(k, j)$ which possess the following properties:

(*i*) $\mathscr{E}(k, j)$ is $(2j + 1)$-dimensional.

(*ii*) $\mathscr{E}(k, j)$ is globally invariant under the action of $\mathbf{J}^2$, J_z, $J_{\pm}$, and, more generally, of any function $F(\mathbf{J})$. In other words, these operators have non-zero matrix elements only inside each of the subspaces $\mathscr{E}(k, j)$.

(*iii*) Inside a subspace $\mathscr{E}(k, j)$, the matrix elements of any function $F(\mathbf{J})$ of the angular momentum **J** are independent of k.

COMMENT:

As we pointed out in §C-3-a of chapter VI, we can give the index k a concrete physical meaning by choosing for the standard basis the system of eigenvectors common to $\mathbf{J}^2$, J_z and one or several observables which commute with the three components of **J** and form a C.S.C.O. with $\mathbf{J}^2$ and J_z. If, for example:

$$[A, \mathbf{J}] = \mathbf{0} \tag{C-3}$$

and if the set $\{ A, \mathbf{J}^2, J_z \}$ is a C.S.C.O., we require the vectors $| k, j, m \rangle$ to be eigenvectors of A:

$$A \, | k, j, m \rangle = a_{k,j} \, | k, j, m \rangle \tag{C-4}$$

Relations (C-1), (C-2) and (C-4) determine the standard basis $\{ | k, j, m \rangle \}$ in this case. Each of the $\mathscr{E}(k, j)$ is an eigensubspace of A, and the index k distinguishes between the various eigenvalues $a_{k,j}$ associated with each value of j.

2. Statement of the problem

a. STATE SPACE

Consider a physical system formed by the union of two subsystems (for example, a two-particle system). We shall use indices 1 and 2 to label quantities relating to the two subsystems.

We shall assume that we know, in the state space $\mathscr{E}_1$ of subsystem (1), a standard basis $\{ | k_1, j_1, m_1 \rangle \}$ composed of common eigenvectors of $\mathbf{J}_1^2$ and J_{1z}, where $\mathbf{J}_1$ is the angular momentum operator of subsystem (1):

$$\mathbf{J}_1^2 | k_1, j_1, m_1 \rangle = j_1(j_1 + 1)\hbar^2 | k_1, j_1, m_1 \rangle \tag{C-5-a}$$

$$J_{1z} | k_1, j_1, m_1 \rangle = m_1\hbar | k_1, j_1, m_1 \rangle \tag{C-5-b}$$

$$J_{1\pm} | k_1, j_1, m_1 \rangle = \hbar\sqrt{j_1(j_1 + 1) - m_1(m_1 \pm 1)} \, | k_1, j_1, m_1 \pm 1 \rangle \tag{C-5-c}$$

Similarly, the state space $\mathscr{E}_2$ of subsystem (2) is spanned by a standard basis $\{ | k_2, j_2, m_2 \rangle \}$:

$$\mathbf{J}_2^2 | k_2, j_2, m_2 \rangle = j_2(j_2 + 1)\hbar^2 | k_2, j_2, m_2 \rangle \tag{C-6-a}$$

$$J_{2z} | k_2, j_2, m_2 \rangle = m_2\hbar | k_2, j_2, m_2 \rangle \tag{C-6-b}$$

$$J_{2\pm} | k_2, j_2, m_2 \rangle = \hbar\sqrt{j_2(j_2 + 1) - m_2(m_2 \pm 1)} \, | k_2, j_2, m_2 \pm 1 \rangle \tag{C-6-c}$$

The state space of the global system is the tensor product of $\mathscr{E}_1$ and $\mathscr{E}_2$:

$$\mathscr{E} = \mathscr{E}_1 \otimes \mathscr{E}_2 \tag{C-7}$$

We know a basis of the global system, formed by taking the tensor product of the bases chosen in $\mathscr{E}_1$ and $\mathscr{E}_2$. We shall denote by $| k_1, k_2; j_1, j_2; m_1, m_2 \rangle$ the vectors of this basis:

$$| k_1, k_2; j_1, j_2; m_1, m_2 \rangle = | k_1, j_1, m_1 \rangle \otimes | k_2, j_2, m_2 \rangle \tag{C-8}$$

The spaces $\mathscr{E}_1$ and $\mathscr{E}_2$ can be considered to be the direct sums of the subspaces $\mathscr{E}_1(k_1, j_1)$ and $\mathscr{E}_2(k_2, j_2)$, which possess the properties recalled in §C-1:

$$\mathscr{E}_1 = \sum_{\oplus} \mathscr{E}_1(k_1, j_1) \tag{C-9-a}$$

$$\mathscr{E}_2 = \sum_{\oplus} \mathscr{E}_2(k_2, j_2) \tag{C-9-b}$$

Consequently, $\mathscr{E}$ is the direct sum of the subspaces $\mathscr{E}(k_1, k_2; j_1, j_2)$ obtained by taking the tensor product of a space $\mathscr{E}_1(k_1, j_1)$ and a space $\mathscr{E}_2(k_2, j_2)$:

$$\mathscr{E} = \sum_{\oplus} \mathscr{E}(k_1, k_2; j_1, j_2) \tag{C-10}$$

with:

$$\mathscr{E}(k_1, k_2; j_1, j_2) = \mathscr{E}_1(k_1, j_1) \otimes \mathscr{E}_2(k_2, j_2) \tag{C-11}$$

The dimension of the subspace $\mathscr{E}(k_1, k_2; j_1, j_2)$ is $(2j_1 + 1)(2j_2 + 1)$. This subspace is globally invariant under the action of any function of $\mathbf{J}_1$ and $\mathbf{J}_2$ ($\mathbf{J}_1$ and $\mathbf{J}_2$ here denote the extensions into $\mathscr{E}$ of the angular momentum operators originally defined in $\mathscr{E}_1$ and $\mathscr{E}_2$ respectively).

b. TOTAL ANGULAR MOMENTUM. COMMUTATION RELATIONS

The total angular momentum of the system under consideration is defined by:

$$\mathbf{J} = \mathbf{J}_1 + \mathbf{J}_2 \tag{C-12}$$

where $\mathbf{J}_1$ and $\mathbf{J}_2$, extensions of operators acting in the different spaces $\mathscr{E}_1$ and $\mathscr{E}_2$, commute. Of course, the components of $\mathbf{J}_1$, on the one hand, and of $\mathbf{J}_2$, on the other, satisfy the commutation relations which characterize angular momenta. It is easy to verify that the components of $\mathbf{J}$ also satisfy such relations [the calculation is the same as in (B-4)].

Since $\mathbf{J}_1$ and $\mathbf{J}_2$ each commute with $\mathbf{J}_1^2$ and $\mathbf{J}_2^2$, so does $\mathbf{J}$. In particular, $\mathbf{J}^2$ and J_z commute with $\mathbf{J}_1^2$ and $\mathbf{J}_2^2$:

$$[J_z, \mathbf{J}_1^2] = [J_z, \mathbf{J}_2^2] = 0 \tag{C-13-a}$$

$$[\mathbf{J}^2, \mathbf{J}_1^2] = [\mathbf{J}^2, \mathbf{J}_2^2] = 0 \tag{C-13-b}$$

Furthermore, J_{1z} and J_{2z} obviously commute with J_z:

$$[J_{1z}, J_z] = [J_{2z}, J_z] = 0 \tag{C-14}$$

but not with $\mathbf{J}^2$ since this last operator can be written in terms of $\mathbf{J}_1$ and $\mathbf{J}_2$ in the form:

$$\mathbf{J}^2 = \mathbf{J}_1^2 + \mathbf{J}_2^2 + 2\mathbf{J}_1 \cdot \mathbf{J}_2 \tag{C-15}$$

and, as in (B-9), J_{1z} and J_{2z} do not commute with $\mathbf{J}_1 \cdot \mathbf{J}_2$. We can also transform the expression for $\mathbf{J}^2$ into:

$$\mathbf{J}^2 = \mathbf{J}_1^2 + \mathbf{J}_2^2 + 2J_{1z}J_{2z} + J_{1+}J_{2-} + J_{1-}J_{2+} \tag{C-16}$$

c. THE BASIS CHANGE TO BE PERFORMED

A vector $| k_1, k_2; j_1, j_2; m_1, m_2 \rangle$ of basis (C-8) is a simultaneous eigenstate of the observables:

$$\mathbf{J}_1^2, \mathbf{J}_2^2, J_{1z}, J_{2z} \tag{C-17}$$

with the respective eigenvalues $j_1(j_1 + 1)\hbar^2$, $j_2(j_2 + 1)\hbar^2$, $m_1\hbar$, $m_2\hbar$. *Basis* (C-8) *is well adapted to the study of the individual angular momenta* $\mathbf{J}_1$ *and* $\mathbf{J}_2$ of the two subsystems.

According to (C-13), the observables:

$$\mathbf{J}_1^2, \mathbf{J}_2^2, \mathbf{J}^2, J_z \tag{C-18}$$

also commute. We are going to construct an orthonormal system of common eigenvectors of these observables : *this new basis will be well adapted to the study of the total angular momentum* of the system. Note that this basis will be different from the preceding one, since $\mathbf{J}^2$ does not commute with J_{1z} and J_{2z} (§b above).

COMMENT:

To give a physical meaning to the indices k_1 and k_2, let us assume (comment of § C-1) that we know, in $\mathscr{E}_1$, a C.S.C.O., $\{ A_1, \mathbf{J}_1^2, J_{1z} \}$, where A_1 commutes with the three components of $\mathbf{J}_1$, and, in $\mathscr{E}_2$, a C.S.C O., $\{ A_2, \mathbf{J}_2^2, J_{2z} \}$, where A_2 commutes with the three components of $\mathbf{J}_2$. We can choose for a standard basis $\{ | k_1, j_1, m_1 \rangle \}$ the orthonormal system of eigenvectors common to A_1, $\mathbf{J}_1^2$, and J_{1z}, and for $\{ | k_2, j_2, m_2 \rangle \}$ the orthonormal system of eigenvectors common to A_2, $\mathbf{J}_2^2$ and J_{2z}. The set:

$$\{ A_1, A_2 ; \mathbf{J}_1^2, \mathbf{J}_2^2 ; J_{1z}, J_{2z} \} \tag{C-19}$$

then constitutes a C.S.C.O. in $\mathscr{E}$, whose eigenvectors are the kets (C-8). Since the observable A_1 commutes separately with the components of $\mathbf{J}_1$ and with those of $\mathbf{J}_2$, it also commutes with $\mathbf{J}$ and, in particular, with $\mathbf{J}^2$ and J_z. The same is, of course, true of A_2. Consequently, the observables:

$$A_1, A_2, \mathbf{J}_1^2, \mathbf{J}_2^2, \mathbf{J}^2, J_z \tag{C-20}$$

commute. We shall see that they in fact form a C.S.C.O.; the new basis we are trying to find is the orthonormal system of eigenvectors of this C.S.C.O.

The subspace $\mathscr{E}(k_1, k_2; j_1, j_2)$ of $\mathscr{E}$ defined in (C-11) is globally invariant under the action of any operator which is a function of $\mathbf{J}_1$ and $\mathbf{J}_2$, and, therefore, under the action of any function of the total angular momentum $\mathbf{J}$. It follows that the observables $\mathbf{J}^2$ and J_z, which we want to diagonalize, have non-zero matrix elements only between vectors belonging to the same subspace $\mathscr{E}(k_1, k_2; j_1, j_2)$. The matrices (which are, in general, infinite) representing $\mathbf{J}^2$ and J_z in the basis (C-8) are "block diagonal", that is, they can be broken down into a series of submatrices, each of which corresponds to a particular subspace $\mathscr{E}(k_1, k_2; j_1, j_2)$. *The problem therefore reduces to a change of basis inside each of the subspaces* $\mathscr{E}(k_1, k_2, j_1, j_2)$, which are of finite dimension $(2j_1 + 1)(2j_2 + 1)$.

Moreover, the matrix elements in the basis (C-8) of any function of $\mathbf{J}_1$ and $\mathbf{J}_2$ are independent of k_1 and k_2. This is therefore true of those of $\mathbf{J}^2$ and J_z. Consequently, *the problem of the diagonalization of* $\mathbf{J}^2$ *and* J_z *is the same inside all the subspaces* $\mathscr{E}(k_1, k_2; j_1, j_2)$ *which correspond to the same values of* j_1 *and* j_2. It is for this reason that one usually speaks of *adding angular momenta* j_1 *and* j_2 without specifying the other quantum numbers. To simplify the notation, we shall henceforth omit the indices k_1 and k_2. We shall denote by $\mathscr{E}(j_1, j_2)$ the subspace $\mathscr{E}(k_1, k_2; j_1, j_2)$, and by $| j_1, j_2; m_1, m_2 \rangle$, the vectors of basis (C-8) belonging to this subspace:

$$\mathscr{E}(j_1, j_2) \equiv \mathscr{E}(k_1, k_2; j_1, j_2) \tag{C-21-a}$$

$$| j_1, j_2; m_1, m_2 \rangle \equiv | k_1, k_2; j_1, j_2; m_1, m_2 \rangle \tag{C-21-b}$$

Since $\mathbf{J}$ is an angular momentum and $\mathscr{E}(j_1, j_2)$ is globally invariant under the action of any function of $\mathbf{J}$, the results of chapter VI recalled above (§C-1) are applicable. Consequently, $\mathscr{E}(j_1, j_2)$ is a direct sum of orthogonal subspaces $\mathscr{E}(k, J)$, each of which is globally invariant under the action of $\mathbf{J}^2$, J_z, J_+ and J_-:

$$\mathscr{E}(j_1, j_2) = \sum_{\oplus} \mathscr{E}(k, J) \tag{C-22}$$

Thus, finally, we are left with the following double problem:

(i) Given j_1 and j_2, *what are the values of* J which appear in (C-22), and *how many distinct subspaces* $\mathscr{E}(k, J)$ are associated with each of them?

(ii) *How can the eigenvectors of* $\mathbf{J}^2$ *and* J_z belonging to $\mathscr{E}(j_1, j_2)$ *be expanded on the* $\{ | j_1, j_2; m_1, m_2 \rangle \}$ *basis*?

§C-3 supplies the answer to the first question, and §C-4 to the second.

COMMENTS:

(i) We have introduced $\mathbf{J}_1$ and $\mathbf{J}_2$ as the angular momenta of two distinct subspaces. In fact, we know (§A-2) that we may want to add the orbital and spin angular momenta of the same particle. All the discussions and results of this section are applicable to this case, with $\mathscr{E}_1$ and $\mathscr{E}_2$ simply being replaced by $\mathscr{E}_\mathbf{r}$ and $\mathscr{E}_s$.

(ii) In order to add several angular momenta, one first adds the first two, then the angular momentum so obtained to the third one, and so on until the last one has been added.

3. Eigenvalues of $\mathbf{J}^2$ and J_z

a. SPECIAL CASE OF TWO SPIN 1/2'S

First of all, let us again take up the simple problem treated in §B. The spaces $\mathscr{E}_1$ and $\mathscr{E}_2$ each contain, in this case, a single invariant subspace, and the tensor product space $\mathscr{E}$, a single subspace $\mathscr{E}(j_1, j_2)$, for which $j_1 = j_2 = 1/2$.

The results recalled in §C-1 make it very simple to find the values of the quantum number S associated with the total spin. The space $\mathscr{E} = \mathscr{E}(1/2, 1/2)$ must be a direct sum of $(2S + 1)$-dimensional subspaces $\mathscr{E}(k, S)$. Each of these subspaces contains one and only one eigenvector of S_z corresponding to each of the values of M such that $|M| \leqslant S$. Now, we know (*cf.* §B-2) that the only values taken on by M are 1, -1 and 0, the first two being non-degenerate and the third, two-fold degenerate. From this, the following conclusions can be deduced directly:

(i) Values of S greater than 1 are excluded. For example, for $S = 2$ to be possible there would have to exist at least one eigenvector of S_z of eigenvalue $2\hbar$.

(ii) $S = 1$ occurs (since $M = 1$ does) only once : $M = 1$ is not degenerate.

(iii) This is also true for $S = 0$. The subspace characterized by $S = 1$ includes only one vector for which $M = 0$, and this value of M is doubly degenerate in the space $\mathscr{E}(1/2, 1/2)$.

The four-dimensional space $\mathscr{E}(1/2, 1/2)$ can therefore be broken down into a subspace associated with $S = 1$ (which is three-dimensional) and a subspace associated with $S = 0$ (which is one-dimensional).

Using a completely analogous argument, we shall determine the possible values of J in the general case in which j_1 and j_2 are arbitrary.

b. THE EIGENVALUES OF J_z AND THEIR DEGREES OF DEGENERACY

In accordance with the conclusions of §2-c, we shall consider a well-defined subspace $\mathscr{E}(j_1, j_2)$, of dimension $(2j_1 + 1)(2j_2 + 1)$. We shall assume that j_1 and j_2 are labeled such that:

$$j_1 \geqslant j_2 \tag{C-23}$$

The vectors $| j_1, j_2 ; m_1, m_2 \rangle$ are already eigenstates of J_z:

$$\begin{aligned} J_z \, | j_1, j_2; m_1, m_2 \rangle &= (J_{1z} + J_{2z}) \, | j_1, j_2; m_1, m_2 \rangle \\ &= (m_1 + m_2)\hbar \, | j_1, j_2; m_1, m_2 \rangle \end{aligned} \tag{C-24}$$

and the corresponding eigenvalues $M\hbar$ are such that:

$$M = m_1 + m_2 \tag{C-25}$$

Consequently, M takes on the following values:

$$j_1 + j_2 \quad , j_1 + j_2 - 1 \quad , j_1 + j_2 - 2 \quad , \ldots, \; - (j_1 + j_2) \tag{C-26}$$

To find the degree of degeneracy $g_{j_1, j_2}(M)$ of these values, we can use the following geometrical procedure. In a two-dimensional diagram, we associate with each vector $| j_1, j_2 ; m_1, m_2 \rangle$ the point whose abscissa is m_1 and whose ordinate is m_2. All these points are situated inside, or on the sides of, the rectangle whose corners are at (j_1, j_2), $(j_1, -j_2)$, $(-j_1, -j_2)$ and $(-j_1, j_2)$. Figure 1 represents the 15 points associated with the basis vectors in the case in which $j_1 = 2$ and $j_2 = 1$ (the values of m_1 and m_2 are shown beside each point). All points situated on the same dashed line (of slope -1) correspond to the same value of $M = m_1 + m_2$. The number of such points is therefore equal to the degeneracy $g_{j_1, j_2}(M)$ of this value of M.

Now consider the various values of M, in decreasing order, tracing the line defined by each of them (fig. 1). $M = j_1 + j_2$ is not degenerate, since the line it characterizes passes only through the upper right-hand corner, whose coordinates are (j_1, j_2):

$$g_{j_1, j_2}(j_1 + j_2) = 1 \tag{C-27}$$

$M = j_1 + j_2 - 1$ is doubly degenerate, since the corresponding line contains the points $(j_1, j_2 - 1)$ and $(j_1 - 1, j_2)$:

$$g_{j_1, j_2}(j_1 + j_2 - 1) = 2 \tag{C-28}$$

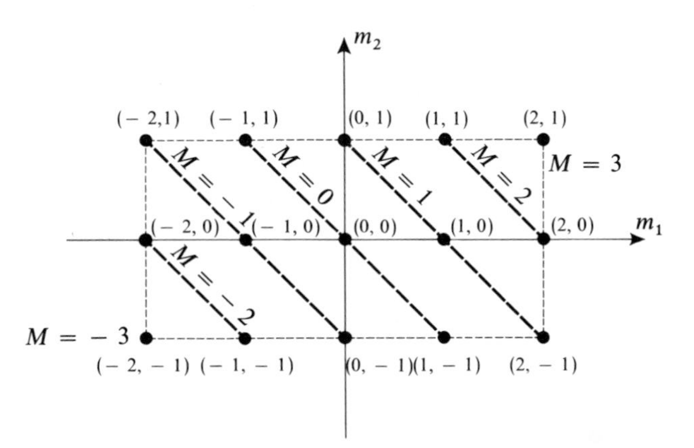

FIGURE 1

Pairs of possible values (m_1, m_2) for the kets $| j_1, j_2; m_1, m_2 \rangle$. We have chosen the case in which $j_1 = 2$ and $j_2 = 1$. The points associated with a given value of $M = m_1 + m_2$ are situated on a straight line of slope -1 (dashed lines).

The degree of degeneracy thus increases by one when M decreases by one, until we reach the lower right-hand corner of the rectangle ($m_1 = j_1, m_2 = -j_2$), that is, the value $M = j_1 - j_2$. The number of points on the line is then at a maximum and is equal to:

$$g_{j_1,j_2}(j_1 - j_2) = 2j_2 + 1 \tag{C-29}$$

When M falls below $j_1 - j_2$, $g_{j_1,j_2}(M)$ first remains constant and equal to its maximum value as long as the line associated with M cuts across the entire width

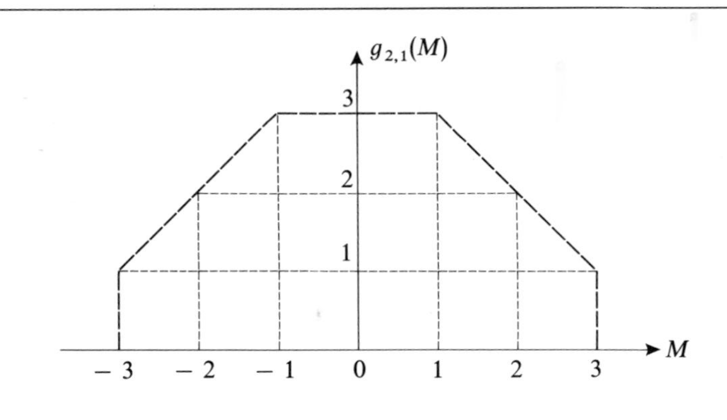

FIGURE 2

Value of the degree of degeneracy $g_{j_1,j_2}(M)$ as a function of M. As in figure 1, we have shown the case in which $j_1 = 2$ and $j_2 = 1$. The degree of degeneracy $g_{j_1\,j_2}(M)$ is simply obtained by counting the number of points on the corresponding dashed line of figure 1.

of the rectangle, that is, until it passes through the upper left-hand corner of the rectangle ($m_1 = -j_1, m_2 = j_2$):

$$g_{j_1,j_2}(M) = 2j_2 + 1 \quad \text{for} \quad -(j_1 - j_2) \leqslant M \leqslant j_1 - j_2 \tag{C-30}$$

Finally, for M less than $-(j_1 - j_2)$, the corresponding line no longer intersects with the upper horizontal side of the rectangle, and $g_{j_1,j_2}(M)$ steadily decreases by one each time M decreases by one, again reaching 1 when $M = -(j_1 + j_2)$ (lower left-hand corner of the rectangle). Consequently:

$$g_{j_1,j_2}(-M) = g_{j_1,j_2}(M) \tag{C-31}$$

These results are summarized, for $j_1 = 2$ and $j_2 = 1$, in figure 2, which gives $g_{2,1}(M)$ as a function of M.

c. THE EIGENVALUES OF $\mathbf{J}^2$

Note, first of all, that the values (C-26) of M are all integral if j_1 and j_2 are both integral or both half-integral, and all half-integral if one of them is integral and the other half-integral. Consequently, the corresponding values of J will also be all integral in the first case and all half-integral in the second.

Since the maximum value attained by M is $j_1 + j_2$, none of the values of J greater than $j_1 + j_2$ is found in $\mathscr{E}(j_1, j_2)$ and therefore none appears in the direct sum (C-22). With $J = j_1 + j_2$ is associated one invariant subspace (since $M = j_1 + j_2$ exists) and only one (since $M = j_1 + j_2$ is not degenerate). In this subspace $\mathscr{E}(J = j_1 + j_2)$, there is one and only one vector which corresponds to $M = j_1 + j_2 - 1$; now this value of M is two-fold degenerate in $\mathscr{E}(j_1, j_2)$; therefore, $J = j_1 + j_2 - 1$ also occurs, and to it corresponds a single invariant subspace $\mathscr{E}(J = j_1 + j_2 - 1)$.

More generally, we shall denote by $p_{j_1,j_2}(J)$ the number of subspaces $\mathscr{E}(k, J)$ of $\mathscr{E}(j_1, j_2)$ associated with a given value of J, that is, the number of different values of k for this value of J (j_1 and j_2 having been fixed at the beginning). $p_{j_1,j_2}(J)$ and $g_{j_1,j_2}(M)$ are very simply related. Consider a particular value of M. To it corresponds one and only one vector in each subspace $\mathscr{E}(k, J)$ such that $J \geqslant |M|$. Its degree of degeneracy $g_{j_1,j_2}(M)$ in $\mathscr{E}(j_1, j_2)$ can therefore be written:

$$\begin{aligned} g_{j_1,j_2}(M) = p_{j_1,j_2}(J = |M|) &+ p_{j_1,j_2}(J = |M| + 1) \\ &+ p_{j_1,j_2}(J = |M| + 2) + \ldots \end{aligned} \tag{C-32}$$

Inverting, we obtain $p_{j_1,j_2}(J)$ in terms of $g_{j_1,j_2}(M)$:

$$\begin{aligned} p_{j_1,j_2}(J) &= g_{j_1,j_2}(M = J) - g_{j_1,j_2}(M = J + 1) \\ &= g_{j_1,j_2}(M = -J) - g_{j_1,j_2}(M = -J - 1) \end{aligned} \tag{C-33}$$

The results of §C-3-b then enable us to determine simply the values of the quantum number J which actually occur in $\mathscr{E}(j_1, j_2)$ and the number of invariant subspaces $\mathscr{E}(k, J)$ which are associated with them. First of all, we have, obviously:

$$p_{j_1,j_2}(J) = 0 \quad \text{for} \quad J > j_1 + j_2 \tag{C-34}$$

since $g_{j_1,j_2}(M)$ is zero for $|M| > j_1 + j_2$. Furthermore, according to (C-27) and (C-28):

$$p_{j_1,j_2}(J = j_1 + j_2) = g_{j_1,j_2}(M = j_1 + j_2) = 1 \tag{C-35-a}$$

$$p_{j_1,j_2}(J = j_1 + j_2 - 1) = g_{j_1,j_2}(M = j_1 + j_2 - 1) - g_{j_1,j_2}(M = j_1 + j_2) = 1 \tag{C-35-b}$$

Thus, by iteration, we find all the values of $p_{j_1,j_2}(J)$:

$$p_{j_1,j_2}(J = j_1 + j_2 - 2) = 1, \ldots, \tag{C-36-a}$$

$$\ldots, p_{j_1,j_2}(J = j_1 - j_2) = 1 \tag{C-36-b}$$

and, finally, according to (C-30):

$$p_{j_1,j_2}(J) = 0 \qquad \text{for} \quad J < j_1 - j_2 \tag{C-37}$$

Therefore, for fixed j_1 and j_2, that is, inside a given space $\mathscr{E}(j_1, j_2)$, the eigenvalues of $\mathbf{J}^2$ are such that★:

$$J = j_1 + j_2 \quad , j_1 + j_2 - 1 \quad , j_1 + j_2 - 2 \quad , \ldots, |j_1 - j_2| \tag{C-38}$$

With each of these values is associated *a single* invariant subspace $\mathscr{E}(J)$, so that the index k which appears in (C-22) is actually unnecessary. This means, in particular, that if we fix a value of J belonging to the set (C-38) and a value of M which is compatible with it, there corresponds to them one and only one vector in $\mathscr{E}(j_1, j_2)$: the specification of J suffices for the determination of the subspace $\mathscr{E}(J)$, in which the specification of M then defines one and only one vector. In other words, $\mathbf{J}^2$ and J_z form a C.S.C.O. in $\mathscr{E}(j_1, j_2)$.

COMMENT:

It can be shown that the number of pairs (J, M) found in $\mathscr{E}(j_1, j_2)$ is indeed equal to the dimension $(2j_1 + 1)(2j_2 + 1)$ of this space. This number (if, for example, $j_1 \geqslant j_2$) is equal to:

$$\sum_{J=j_1-j_2}^{j_1+j_2} (2J + 1) \tag{C-39}$$

If we set:

$$J = j_1 - j_2 + i \tag{C-40}$$

it is easy to calculate the sum (C-39):

$$\begin{aligned}\sum_{J=j_1-j_2}^{j_1+j_2} (2J + 1) &= \sum_{i=0}^{2j_2} [2(j_1 - j_2 + i) + 1] \\ &= [2(j_1 - j_2) + 1](2j_2 + 1) + 2\,\frac{2j_2(2j_2 + 1)}{2} \\ &= (2j_2 + 1)(2j_1 + 1)\end{aligned} \tag{C-41}$$

★ Thus far, we have assumed $j_1 \geqslant j_2$, but it is simple to extend the discussion to the opposite case $j_1 < j_2$: all we need to do is invert indices 1 and 2.

4. Common eigenvectors of $\mathbf{J}^2$ and J_z

We shall denote by $| J, M \rangle$ the common eigenvectors of $\mathbf{J}^2$ and J_z belonging to the space $\mathscr{E}(j_1, j_2)$. To be completely rigorous, we should have to recall the values of j_1 and j_2 in this notation, but we shall not write them explicitly, since they are the same as in the vectors (C-21-b) of which the $| J, M \rangle$ are linear combinations. Of course, the indices J and M refer to the eigenvalues of $\mathbf{J}^2$ and J_z:

$$\mathbf{J}^2 \; | J, M \rangle = J(J + 1)\hbar^2 \; | J, M \rangle \tag{C-42-a}$$

$$J_z \; | J, M \rangle = M\hbar \; | J, M \rangle \tag{C-42-b}$$

and the vectors $| J, M \rangle$, like all those of the space $\mathscr{E}(j_1, j_2)$, are eigenvectors of $\mathbf{J}_1^2$ and $\mathbf{J}_2^2$ with eigenvalues $j_1(j_1 + 1)\hbar^2$ and $j_2(j_2 + 1)\hbar^2$ respectively.

a. SPECIAL CASE OF TWO SPIN 1/2'S

First of all, we shall show how use of the general results concerning angular momenta leads us to the expression for the vectors $| S, M \rangle$ established in § B-3. It will not be necessary to diagonalize the matrix which represents $\mathbf{S}^2$. By generalizing this method, we shall then construct (§ 4-b) the vectors $| J, M \rangle$ for the case of arbitrary j_1 and j_2.

α. *The subspace $\mathscr{E}(S = 1)$*

The ket $| +, + \rangle$ is, in the state space $\mathscr{E} = \mathscr{E}(1/2, 1/2)$, the only eigenvector of S_z associated with $M = 1$. Since $\mathbf{S}^2$ and S_z commute, and the value $M = 1$ is not degenerate, $| +, + \rangle$ must also be an eigenvector of $\mathbf{S}^2$ (§ D-3-a of chapter II). According to the reasoning of § C-3-a, the corresponding value of S must be 1. Therefore, we can choose the phase of the vector $| S = 1, M = 1 \rangle$ such that:

$$| 1, 1 \rangle = | +, + \rangle \tag{C-43}$$

It is then easy to find the other states of the triplet, since we know from the general theory of angular momentum that:

$$\begin{aligned} S_- \, | 1, 1 \rangle &= \hbar\sqrt{1(1 + 1) - 1(1 - 1)} \, | 1, 0 \rangle \\ &= \hbar\sqrt{2} \, | 1, 0 \rangle \end{aligned} \tag{C-44}$$

Consequently:

$$| 1, 0 \rangle = \frac{1}{\hbar\sqrt{2}} S_- \, | +, + \rangle \tag{C-45}$$

To calculate $| 1, 0 \rangle$ explicitly in the $\{ | \varepsilon_1, \varepsilon_2 \rangle \}$ basis, it suffices to recall that definition (B-3) of the total spin $\mathbf{S}$ implies:

$$S_- = S_{1-} + S_{2-} \tag{C-46}$$

We then obtain:

$$
\begin{aligned}
| 1, 0 \rangle &= \frac{1}{\hbar\sqrt{2}}(S_{1-} + S_{2-})| +, + \rangle \\
&= \frac{1}{\hbar\sqrt{2}}[\hbar | -, + \rangle + \hbar | +, - \rangle] \\
&= \frac{1}{\sqrt{2}}[| -, + \rangle + | +, - \rangle]
\end{aligned}
\tag{C-47}
$$

Finally, we can again apply S_- to $| 1, 0 \rangle$, that is, $(S_{1-} + S_{2-})$ to expression (C-47). This yields:

$$
\begin{aligned}
| 1, -1 \rangle &= \frac{1}{\hbar\sqrt{2}} S_- | 1, 0 \rangle \\
&= \frac{1}{\hbar\sqrt{2}}(S_{1-} + S_{2-})\frac{1}{\sqrt{2}}[| -, + \rangle + | +, - \rangle] \\
&= \frac{1}{2\hbar}[\hbar | -, - \rangle + \hbar | -, - \rangle] \\
&= | -, - \rangle
\end{aligned}
\tag{C-48}
$$

Of course, this last result could have been obtained directly, using an argument analogous to the one applied above to $| +, + \rangle$. However, the preceding calculation has a slight advantage : it enables us, in accordance with the general conventions set forth in §C-3-a of chapter VI, to fix the phase factors which could appear in $| 1, 0 \rangle$ and $| 1, -1 \rangle$ with respect to the one chosen for $| 1, 1 \rangle$ in (C-43).

β. *The state* $| S = 0, M = 0 \rangle$

The only vector $| S = 0, M = 0 \rangle$ of the subspace $\mathscr{E}(S = 0)$ is determined, to within a constant factor, by the condition that it must be orthogonal to the three vectors $| 1, M \rangle$ which we have just constructed.

Since it is orthogonal to $| 1, 1 \rangle = | +, + \rangle$ and $| 1, -1 \rangle = | -, - \rangle$, $| 0, 0 \rangle$ must be a linear combination of $| +, - \rangle$ and $| -, + \rangle$:

$$
| 0, 0 \rangle = \alpha | +, - \rangle + \beta | -, + \rangle \tag{C-49}
$$

which will be normalized if:

$$
\langle 0, 0 | 0, 0 \rangle = |\alpha|^2 + |\beta|^2 = 1 \tag{C-50}
$$

We now insist that its scalar product with $| 1, 0 \rangle$ [*cf.* (C-47)] be zero :

$$
\frac{1}{\sqrt{2}}(\alpha + \beta) = 0 \tag{C-51}
$$

The coefficients α and β are therefore equal in absolute value and of opposite sign. With (C-50) taken into account, this fixes them to within a phase factor:

$$\alpha = -\beta = \frac{1}{\sqrt{2}} e^{i\chi} \tag{C-52}$$

where χ is any real number. We shall choose $\chi = 0$, which yields:

$$| 0, 0 \rangle = \frac{1}{\sqrt{2}} [| +, - \rangle - | -, + \rangle] \tag{C-53}$$

Thus we have calculated the four vectors $| S, M \rangle$ without explicitly having had to write the matrix which represents $\mathbf{S}^2$ in the $\{ | \varepsilon_1, \varepsilon_2 \rangle \}$ basis.

b. GENERAL CASE (ARBITRARY j_1 AND j_2)

We showed in §C-3-c that the decomposition of $\mathscr{E}(j_1, j_2)$ into a direct sum of invariant subspaces $\mathscr{E}(J)$ is:

$$\mathscr{E}(j_1, j_2) = \mathscr{E}(j_1 + j_2) \oplus \mathscr{E}(j_1 + j_2 - 1) \oplus ... \oplus \mathscr{E}(|j_1 - j_2|) \tag{C-54}$$

We shall now see how to determine the vectors $| J, M \rangle$ which span these subspaces.

α. *The subspace* $\mathscr{E}(J = j_1 + j_2)$

The ket $| j_1, j_2; m_1 = j_1, m_2 = j_2 \rangle$ is, in $\mathscr{E}(j_1, j_2)$, the only eigenvector of J_z associated with $M = j_1 + j_2$. Since $\mathbf{J}^2$ and J_z commute, and the value $M = j_1 + j_2$ is not degenerate, $| j_1, j_2; m_1 = j_1, m_2 = j_2 \rangle$ must also be an eigenvector of $\mathbf{J}^2$. According to (C-54), the corresponding value of J can only be $j_1 + j_2$. We can choose the phase of the vector:

$$| J = j_1 + j_2, M = j_1 + j_2 \rangle$$

such that:

$$| j_1 + j_2, j_1 + j_2 \rangle = | j_1, j_2; j_1, j_2 \rangle \tag{C-55}$$

Repeated application of the operator J_- on this expression enables us to complete the family of vectors $| J, M \rangle$ for which $J = j_1 + j_2$. Thus, according to the general formulas (C-50) of chapter VI:

$$J_- | j_1 + j_2, j_1 + j_2 \rangle = \hbar \sqrt{2(j_1 + j_2)} \, | j_1 + j_2, j_1 + j_2 - 1 \rangle \tag{C-56}$$

We can therefore calculate the vector corresponding to $J = j_1 + j_2$ and $M = j_1 + j_2 - 1$ by applying $J_- = J_{1-} + J_{2-}$ to the vector $| j_1, j_2; j_1, j_2 \rangle$:

$$
\begin{aligned}
| j_1 + j_2, j_1 + j_2 - 1 \rangle &= \frac{1}{\hbar\sqrt{2(j_1 + j_2)}} J_- \, | j_1 + j_2, j_1 + j_2 \rangle \\
&= \frac{1}{\hbar\sqrt{2(j_1 + j_2)}} (J_{1-} + J_{2-}) \, | j_1, j_2 ; j_1, j_2 \rangle \\
&= \frac{1}{\hbar\sqrt{2(j_1 + j_2)}} [\hbar\sqrt{2j_1} \, | j_1, j_2 ; j_1 - 1, j_2 \rangle \\
&\qquad + \hbar\sqrt{2j_2} \, | j_1, j_2 ; j_1, j_2 - 1 \rangle]
\end{aligned}
\tag{C-57}
$$

that is:

$$
\begin{aligned}
| j_1 + j_2, j_1 + j_2 - 1 \rangle &= \sqrt{\frac{j_1}{j_1 + j_2}} \, | j_1, j_2 ; j_1 - 1, j_2 \rangle \\
&\quad + \sqrt{\frac{j_2}{j_1 + j_2}} \, | j_1, j_2 ; j_1, j_2 - 1 \rangle
\end{aligned}
\tag{C-58}
$$

Indeed, note that we obtain in this way a linear combination of the two basis vectors which correspond to $M = j_1 + j_2 - 1$, and that this combination is directly normalized.

We then repeat the procedure: we construct $| j_1 + j_2, j_1 + j_2 - 2 \rangle$ by letting J_- act on both sides of (C-58) (for the right-hand side, we take this operator in the form $J_{1-} + J_{2-}$), and so on, through $| j_1 + j_2, -(j_1 + j_2) \rangle$, which is found to be equal to $| j_1, j_2 ; -j_1, -j_2 \rangle$.

We therefore know how to calculate the first $[2(j_1 + j_2) + 1]$ vectors of the $\{ | J, M \rangle \}$ basis, which correspond to $J = j_1 + j_2$ and $M = j_1 + j_2$, $j_1 + j_2 - 1, ..., -(j_1 + j_2)$ and span the subspace $\mathscr{E}(J = j_1 + j_2)$ of $\mathscr{E}(j_1, j_2)$.

β. *The other subspaces* $\mathscr{E}(J)$

Now consider the space $\mathscr{S}(j_1 + j_2)$, the supplement of $\mathscr{E}(j_1 + j_2)$ in $\mathscr{E}(j_1, j_2)$. According to (C-54), $\mathscr{S}(j_1 + j_2)$ can be broken down into:

$$
\mathscr{S}(j_1 + j_2) = \mathscr{E}(j_1 + j_2 - 1) \oplus \mathscr{E}(j_1 + j_2 - 2) \oplus ... \oplus \mathscr{E}(|j_1 - j_2|) \tag{C-59}
$$

We can therefore apply to it the same reasoning as was used in §α.

In $\mathscr{S}(j_1 + j_2)$, the degree of degeneracy $g'_{j_1,j_2}(M)$ of a given value of M is smaller by one than $g_{j_1,j_2}(M)$, since $\mathscr{E}(j_1 + j_2)$ possesses one and only one vector associated with this value of M:

$$
g'_{j_1,j_2}(M) = g_{j_1,j_2}(M) - 1 \tag{C-60}
$$

This means, in particular, that $M = j_1 + j_2$ no longer exists in $\mathscr{S}(j_1 + j_2)$, and that the new maximum value $M = j_1 + j_2 - 1$ is not degenerate. From this we see, as in §α, that the corresponding vector must be proportional to $| J = j_1 + j_2 - 1, M = j_1 + j_2 - 1 \rangle$. It is easy to find its expansion on the $\{ | j_1, j_2 ; m_1, m_2 \rangle \}$ basis, since, because of the value of M, it is surely of the form:

$$
\begin{aligned}
| j_1 + j_2 - 1, j_1 + j_2 - 1 \rangle &= \alpha \, | j_1, j_2 ; j_1, j_2 - 1 \rangle \\
&\quad + \beta \, | j_1, j_2 ; j_1 - 1, j_2 \rangle
\end{aligned}
\tag{C-61}
$$

with:

$$|\alpha|^2 + |\beta|^2 = 1 \tag{C-62}$$

to insure its normalization. It must also be orthogonal to $| j_1 + j_2, j_1 + j_2 - 1 \rangle$, which belongs to $\mathscr{E}(j_1 + j_2)$ and for which the expression is given by (C-58). The coefficients α and β must therefore satisfy:

$$\alpha \sqrt{\frac{j_2}{j_1 + j_2}} + \beta \sqrt{\frac{j_1}{j_1 + j_2}} = 0 \tag{C-63}$$

Relations (C-62) and (C-63) determine α and β to within a phase factor. We shall choose α and β to be real and, for example, α positive. With these conventions:

$$| j_1 + j_2 - 1, j_1 + j_2 - 1 \rangle = \sqrt{\frac{j_1}{j_1 + j_2}} | j_1, j_2; j_1, j_2 - 1 \rangle - \sqrt{\frac{j_2}{j_1 + j_2}} | j_1, j_2; j_1 - 1, j_2 \rangle \tag{C-64}$$

This vector is the first of a new family, characterized by $J = j_1 + j_2 - 1$. As in § α, we can derive the others by applying J_- as many times as necessary. Thus we obtain $[2(j_1 + j_2 - 1) + 1]$ vectors $| J, M \rangle$ corresponding to

$$J = j_1 + j_2 - 1 \text{ and } M = j_1 + j_2 - 1, j_1 + j_2 - 2, \ldots, -(j_1 + j_2 - 1),$$

and spanning the subspace $\mathscr{E}(J = j_1 + j_2 - 1)$.

Now consider the space $\mathscr{S}(j_1 + j_2, j_1 + j_2 - 1)$, the supplement of the direct sum $\mathscr{E}(j_1 + j_2) \oplus \mathscr{E}(j_1 + j_2 - 1)$ in $\mathscr{E}(j_1, j_2)$★:

$$\mathscr{S}(j_1 + j_2, j_1 + j_2 - 1) = \mathscr{E}(j_1 + j_2 - 2) \oplus \ldots \oplus \mathscr{E}(|j_1 - j_2|) \tag{C-65}$$

In $\mathscr{S}(j_1 + j_2, j_1 + j_2 - 1)$, the degeneracy of each value of M is again decreased by one with respect to what it was in $\mathscr{S}(j_1 + j_2)$. In particular, the maximum value is now $M = j_1 + j_2 - 2$, and it is not degenerate. The corresponding vector of $\mathscr{S}(j_1 + j_2, j_1 + j_2 - 1)$ must therefore be $| J = j_1 + j_2 - 2, M = j_1 + j_2 - 2 \rangle$. To calculate it in the $\{ | j_1, j_2; m_1, m_2 \rangle \}$ basis, it is sufficient to note that it is a linear combination of the three vectors $| j_1, j_2; j_1, j_2 - 2 \rangle$, $| j_1, j_2; j_1 - 1, j_2 - 1 \rangle$, $| j_1, j_2; j_1 - 2, j_2 \rangle$. The coefficients of this combination are fixed to within a phase factor by the triple condition that it be normalized and orthogonal to $| j_1 + j_2, j_1 + j_2 - 2 \rangle$ and $| j_1 + j_2 - 1, j_1 + j_2 - 2 \rangle$ (which are already known). Finally, the use of J_- enables us to find the other vectors of this third family, thus defining $\mathscr{E}(j_1 + j_2 - 2)$.

The procedure can be repeated without difficulty until we have exhausted all values of M greater than or equal to $|j_1 - j_2|$ [and, consequently, according to (C-31), also all those less than or equal to $-|j_1 - j_2|$]. We then know all the desired $| J, M \rangle$ vectors. This method will be illustrated by two examples in complement A_X.

★ Of course, $\mathscr{S}(j_1 + j_2, j_1 + j_2 - 1)$ exists only if $j_1 + j_2 - 2$ is not less than $|j_1 - j_2|$.

c. CLEBSCH-GORDAN COEFFICIENTS

In each space $\mathscr{E}(j_1, j_2)$, the eigenvectors of $\mathbf{J}^2$ and J_z are linear combinations of vectors of the initial $\{ | j_1, j_2 ; m_1, m_2 \rangle \}$ basis:

$$| J, M \rangle = \sum_{m_1 = -j_1}^{j_1} \sum_{m_2 = -j_2}^{j_2} | j_1, j_2 ; m_1, m_2 \rangle \langle j_1, j_2 ; m_1, m_2 | J, M \rangle \qquad \text{(C-66)}$$

The coefficients $\langle j_1, j_2 ; m_1, m_2 | J, M \rangle$ of these expansions are called *Clebsch-Gordan coefficients*.

COMMENT:

To be completely rigorous, we should write the vectors $| j_1, j_2 ; m_1, m_2 \rangle$ and $| J, M \rangle$ as $| k_1, k_2 ; j_1, j_2 ; m_1, m_2 \rangle$ and $| k_1, k_2 ; j_1, j_2 ; J, M \rangle$ respectively [the values of k_1 and k_2, like those of j_1 and j_2, would then be the same on both sides of relations (C-66)]. However, we shall not write k_1 and k_2 in the symbols which represent the Clebsch-Gordan coefficients, since we know that these coefficients are independent of k_1 and k_2 (§C-2-c).

It is not possible to give a general expression for the Clebsch-Gordan coefficients, but the method presented in §C-4-b enables us to calculate them by iteration for any values of j_1 and j_2. For practical applications, there are *numerical tables* of Clebsch-Gordan coefficients.

Actually, to determine the Clebsch-Gordan coefficients uniquely, a certain number of *phase conventions* must be chosen. [We mentioned this fact when we wrote expressions (C-55) and (C-64)]. Clebsch-Gordan coefficients are always chosen to be real. The choice then bears on the signs of some of them (obviously, the relative signs of the coefficients appearing in the expansion of the same vector $| J, M \rangle$ are fixed; only the global sign of the expansion can be chosen arbitrarily).

The results of § C-4-b imply that $\langle j_1, j_2 ; m_1, m_2 | J, M \rangle$ is different from zero only if :

$$M = m_1 + m_2 \qquad \text{(C-67-a)}$$

$$|j_1 - j_2| \leqslant J \leqslant j_1 + j_2 \qquad \text{(C-67-b)}$$

where J is of the same type (integral or half-integral) as $j_1 + j_2$ and $|j_1 - j_2|$. Condition (C-67-b) is often called the "triangle rule": one must be able to form a triangle with three line segments of lengths j_1, j_2 and J.

Since the vectors $| J, M \rangle$ also form an orthonormal basis of the space $\mathscr{E}(j_1, j_2)$, the expressions which are the inverse of (C-66) can be written:

$$| j_1, j_2 ; m_1, m_2 \rangle = \sum_{J = j_1 - j_2}^{j_1 + j_2} \sum_{M = -J}^{J} | J, M \rangle \langle J, M | j_1, j_2 ; m_1, m_2 \rangle \qquad \text{(C-68)}$$

Since the Clebsch-Gordan coefficients have all been chosen to be real, the scalar products appearing in (C-68) are such that:

$$\langle J, M | j_1, j_2 ; m_1, m_2 \rangle = \langle j_1, j_2 ; m_1, m_2 | J, M \rangle \qquad \text{(C-69)}$$

The Clebsch-Gordan coefficients therefore enable us to express the vectors of the old basis $\{ \mid j_1, j_2 ; m_1, m_2 \rangle \}$, in terms of those of the new basis $\{ \mid J, M \rangle \}$.

The Clebsch-Gordan coefficients possess interesting properties, some of which will be studied in complement B_X.

References and suggestions for further reading:

Messiah (1.17), chap. XIII, §V; Rose (2.19), chap. III. Edmonds (2.21), chaps. 3 and 6.

Relation with group theory : Meijer and Bauer (2.18), chap. 5, §5 and App. III of that chapter; Bacry (10.31), chap. 6; Wigner (2.23), chaps. 14 and 15.

Vectorial spherical harmonics: Edmonds (2.21), §5-10; Jackson (7.5), chap. 16; Berestetskii et al. (2.8), §§6 and 7; Akhiezer and Berestetskii (2.14), §4.

COMPLEMENTS OF CHAPTER X

A_X : EXAMPLES OF ADDITION OF ANGULAR MOMENTA

A_X : illustrates the results of chapter X by the simplest cases not treated in detail in this chapter : two angular momenta equal to 1, and an integral angular momentum l with a spin $\frac{1}{2}$. Easy, recommended as an exercise illustrating methods of addition of angular momenta.

B_X : CLEBSCH-GORDAN COEFFICIENTS

C_X : ADDITION OF SPHERICAL HARMONICS

B_X, C_X : technical complements intended to demonstrate certain useful mathematical results; can be used as references.

B_X : study of Clebsch-Gordan coefficients, which frequently appear in physical problems involving angular momentum and rotational invariance.

C_X : proof of an expression concerning the product of spherical harmonics; useful for certain subsequent complements and exercises.

D_X : VECTOR OPERATORS : THE WIGNER-ECKART THEOREM

E_X : ELECTRIC MULTIPOLE MOMENTS

D_X, E_X : introduction of physical concepts (vector observables, multipole moments) which play important roles in numerous fields.

D_X : study of vector operators; proof of the Wigner-Eckart theorem, which establishes proportionality rules between the matrix elements of these operators. Rather theoretical, but recommended for its numerous applications. Can be helpful in an atomic physics course (the vector model, calculation of Landé factors, etc.).

E_X : definition and properties of electric multipole moments of a classical or quantum mechanical system; study of their selection rules (these multipole moments are frequently used in atomic and nuclear physics). Moderately difficult.

F_X : EVOLUTION OF TWO ANGULAR MOMENTA $\mathbf{J}_1$ AND $\mathbf{J}_2$ COUPLED BY AN INTERACTION $a\mathbf{J}_1 . \mathbf{J}_2$

F_X : can be considered to be a worked exercise, treating a problem fundamental to the vector model of the atom : the time evolution of two angular momenta $\mathbf{J}_1$ and $\mathbf{J}_2$ coupled by an interaction $W = a\mathbf{J}_1 . \mathbf{J}_2$. This dynamical point of view completes, as it were, the results of chapter X concerning the eigenstates of W. Fairly simple.

(continued on the next page)

G_X : EXERCISES

G_X : exercises 7 to 10 are more difficult than the others. Exercices 7, 8, 9 are extensions of complements D_X and F_X (concept of a standard component and that of an irreductible tensor operator, the Wigner-Eckart theorem). Exercise 10 takes up the problem of the various ways of coupling three angular momenta.

Complement A_X

EXAMPLES OF ADDITION OF ANGULAR MOMENTA

To illustrate the general method of addition of angular momenta described in chapter X, we shall apply it here to two examples.

1. Addition of $j_1 = 1$ and $j_2 = 1$

First consider the case in which $j_1 = j_2 = 1$. This is the case, for example, for a two-particle system in which both orbital angular momenta are equal to 1. Since each of the two particles is then in a p state, this is said to be a "p^2 configuration".

The space $\mathscr{E}(1, 1)$ with which we are concerned has $3 \times 3 = 9$ dimensions. We assume the basis composed of common eigenstates of $\mathbf{J}_1^2$, $\mathbf{J}_2^2$, J_{1z} and J_{2z} to be known:

$$\{ | 1, 1; m_1, m_2 \rangle \}, \qquad \text{with} \quad m_1, m_2 = 1, 0, -1 \tag{1}$$

and we want to determine the $\{ | J, M \rangle \}$ basis of common eigenvectors of $\mathbf{J}_1^2$, $\mathbf{J}_2^2$, $\mathbf{J}^2$ and J_z, where $\mathbf{J}$ is the total angular momentum.

According to §C-3 of chapter X, the possible values of the quantum number J are:

$$J = 2, 1, 0. \tag{2}$$

We must therefore construct three families of vectors $| J, M \rangle$, containing, respectively, five, three and one vectors of the new basis.

a. THE SUBSPACE $\mathscr{E}(J = 2)$

The ket $| J = 2, M = 2 \rangle$ can be written simply:

$$| 2, 2 \rangle = | 1, 1; 1, 1 \rangle \tag{3}$$

Applying J_- to it, we find the vector $|\,J = 2, M = 1\,\rangle$:

$$\begin{aligned} |\,2, 1\,\rangle &= \frac{1}{2\hbar} J_- \,|\,2, 2\,\rangle \\ &= \frac{1}{2\hbar}(J_{1-} + J_{2-})\,|\,1, 1; 1, 1\,\rangle \\ &= \frac{1}{2\hbar}[\hbar\sqrt{2}\,|\,1, 1; 0, 1\,\rangle + \hbar\sqrt{2}\,|\,1, 1; 1, 0\,\rangle] \\ &= \frac{1}{\sqrt{2}}[|\,1, 1; 1, 0\,\rangle + |\,1, 1; 0, 1\,\rangle] \end{aligned} \tag{4}$$

We use J_- again to calculate $|\,J = 2, M = 0\,\rangle$. After a simple calculation, we find:

$$|\,2, 0\,\rangle = \frac{1}{\sqrt{6}}[|\,1, 1; 1, -1\,\rangle + 2\,|\,1, 1; 0, 0\,\rangle + |\,1, 1; -1, 1\,\rangle] \tag{5}$$

then:

$$|\,2, -1\,\rangle = \frac{1}{\sqrt{2}}[|\,1, 1; 0, -1\,\rangle + |\,1, 1; -1, 0\,\rangle] \tag{6}$$

and, finally:

$$|\,2, -2\rangle = |\,1, 1; -1, -1\rangle \tag{7}$$

b. THE SUBSPACE $\mathscr{E}(J = 1)$

We shall now proceed to the subspace $\mathscr{E}(J = 1)$. The vector $|\,J = 1, M = 1\,\rangle$ must be a linear combination of the two basis kets $|\,1, 1; 1, 0\,\rangle$ and $|\,1, 1; 0, 1\,\rangle$ (the only ones for which $M = 1$):

$$|\,1, 1\rangle = \alpha\,|\,1, 1; 1, 0\rangle + \beta\,|\,1, 1; 0, 1\,\rangle \tag{8}$$

with:

$$|\alpha|^2 + |\beta|^2 = 1 \tag{9}$$

For it to be orthogonal to the vector $|\,2, 1\,\rangle$, it is necessary [*cf.* (4)] that:

$$\alpha + \beta = 0 \tag{10}$$

We choose α and β to be real, and choose, by convention, α positive*. Under these conditions:

$$|\,1, 1\,\rangle = \frac{1}{\sqrt{2}}[|\,1, 1; 1, 0\,\rangle - |\,1, 1; 0, 1\,\rangle] \tag{11}$$

* In general, the component of the ket $|\,J, J\,\rangle$ on the ket $|\,j_1, j_2; m_1 = j_1, m_2 = J - j_1\,\rangle$ is always chosen to be real and positive (*cf.* complement B_X, § 2).

Application of J_- here again enables us to deduce $| 1, 0 \rangle$ and $| 1, -1 \rangle$. We easily find, using the same technique as above:

$$| 1, 0 \rangle = \frac{1}{\sqrt{2}} [| 1, 1; 1, -1 \rangle - | 1, 1; -1, 1 \rangle] \tag{12}$$

$$| 1, -1 \rangle = \frac{1}{\sqrt{2}} [| 1, 1; 0, -1 \rangle - | 1, 1; -1, 0 \rangle] \tag{13}$$

It is interesting to note that expansion (12) does not contain the vector $| 1, 1; 0, 0 \rangle$, although it also corresponds to $M = 0$. It so happens that the corresponding Clebsch-Gordan coefficient is zero:

$$\langle 1, 1; 0, 0 | 1, 0 \rangle = 0 \tag{14}$$

c. THE VECTOR $| J = 0, M = 0 \rangle$

We are left with the calculation of the last vector of the $\{ | J, M \rangle \}$ basis, associated with $J = M = 0$. This vector is a linear combination of the three basis kets for which $M = 0$:

$$| 0, 0 \rangle = a | 1, 1; 1, -1 \rangle + b | 1, 1; 0, 0 \rangle + c | 1, 1; -1, 1 \rangle \tag{15}$$

with :

$$|a|^2 + |b|^2 + |c|^2 = 1 \tag{16}$$

It must also be orthogonal to $| 2, 0 \rangle$ [formula (5)] and $| 1, 0 \rangle$ [formula (12)]. This gives the two conditions:

$$a + 2b + c = 0 \tag{17-a}$$

$$a - c = 0 \tag{17-b}$$

These relations imply:

$$a = -b = c \tag{18}$$

We again choose a, b and c real, and agree to choose a positive (see note, p. 1028). We then obtain, using (16) and (18):

$$| 0, 0 \rangle = \frac{1}{\sqrt{3}} [| 1, 1; 1, -1 \rangle - | 1, 1; 0, 0 \rangle + | 1, 1; -1, 1 \rangle] \tag{19}$$

This completes the construction of the $\{ | J, M \rangle \}$ basis for the case $j_1 = j_2 = 1$.

COMMENT:

If the physical problem under study is that of a p^2 configuration of a two-particle system, the wave functions which represent the states of the initial basis are of the form:

$$\langle \mathbf{r}_1, \mathbf{r}_2 | 1, 1; m_1, m_2 \rangle = R_{k_1,1}(r_1) R_{k_2,1}(r_2) Y_1^{m_1}(\theta_1, \varphi_1) Y_1^{m_2}(\theta_2, \varphi_2) \tag{20}$$

where $\mathbf{r}_1(r_1, \theta_1, \varphi_1)$ and $\mathbf{r}_2(r_2, \theta_2, \varphi_2)$ give the positions of the two particles. Since the radial functions are independent of the quantum numbers m_1 and m_2, the linear combinations which give the wave functions associated with the kets $| J, M \rangle$ are functions only of the angular dependence. For example, in the $\{ | \mathbf{r}_1, \mathbf{r}_2 \rangle \}$ representation, equation (19) can be written:

$$\langle \mathbf{r}_1, \mathbf{r}_2 | 0, 0 \rangle = R_{k_1,1}(r_1) R_{k_2,1}(r_2) \frac{1}{\sqrt{3}} [Y_1^1(\theta_1, \varphi_1) Y_1^{-1}(\theta_2, \varphi_2) - Y_1^0(\theta_1, \varphi_1) Y_1^0(\theta_2, \varphi_2) + Y_1^{-1}(\theta_1, \varphi_1) Y_1^1(\theta_2, \varphi_2)] \tag{21}$$

2. Addition of an integral orbital angular momentum l and a spin 1/2

Now consider the addition of an orbital angular momentum ($j_1 = l$, an integer) and a spin 1/2 ($j_2 = 1/2$). This problem is encountered, for example, whenever one wants to study the total angular momentum of a spin 1/2 particle such as the electron.

The space $\mathscr{E}(l, 1/2)$ which we are considering here is $2(2l + 1)$-dimensional. We already know a basis of this space*:

$$\{ | l, 1/2; m, \varepsilon \rangle \} \quad \text{with} \quad m = l, l - 1, ..., - l \text{ and } \varepsilon = \pm \tag{22}$$

formed of eigenstates of the observables $\mathbf{L}^2$, $\mathbf{S}^2$, L_z and S_z, where $\mathbf{L}$ and $\mathbf{S}$ are the orbital angular momentum and spin under consideration. We want to construct the eigenvectors $| J, M \rangle$ of $\mathbf{J}^2$ and J_z, where $\mathbf{J}$ is the total angular momentum of the system:

$$\mathbf{J} = \mathbf{L} + \mathbf{S} \tag{23}$$

First of all, note that if l is zero, the solution to the problem is obvious. It is easy to show in this case that the vectors $| 0, 1/2; 0, \varepsilon \rangle$ are also eigenvectors of $\mathbf{J}^2$ and J_z with eigenvalues such as $J = 1/2$ and $M = \varepsilon/2$. On the other hand, if l is not zero, there are two possible values of J:

$$J = l + \frac{1}{2} \quad , \quad l - \frac{1}{2} \tag{24}$$

a. THE SUBSPACE $\mathscr{E}(J = l + 1/2)$

The $(2l + 2)$ vectors $| J, M \rangle$ spanning the subspace $\mathscr{E}(J = l + 1/2)$ can be obtained by using the general method of chapter X. We have, first of all:

$$| l + \frac{1}{2}, l + \frac{1}{2} \rangle = | l, \frac{1}{2}; l, + \rangle \tag{25}$$

* If we wanted to conform strictly to the notation of chapter X, we should have to write $\pm 1/2$, and not ε, in the basis kets. But we agreed in chapters IV and IX to denote the eigenvectors of S_z in the spin state space by $| + \rangle$ and $| - \rangle$.

Through the action of J_-(★), we obtain $| l + \frac{1}{2}, l - \frac{1}{2} \rangle$:

$$
\begin{aligned}
| l + \frac{1}{2}, l - \frac{1}{2} \rangle &= \frac{1}{\hbar\sqrt{2l+1}} J_- | l + \frac{1}{2}, l + \frac{1}{2} \rangle \\
&= \frac{1}{\hbar\sqrt{2l+1}} (L_- + S_-) | l, \frac{1}{2}; l, + \rangle \\
&= \frac{1}{\hbar\sqrt{2l+1}} \left[\hbar\sqrt{2l} \, | l, \frac{1}{2}; l-1, + \rangle + \hbar \, | l, \frac{1}{2}; l, - \rangle \right] \\
&= \sqrt{\frac{2l}{2l+1}} \, | l, \frac{1}{2}; l-1, + \rangle + \frac{1}{\sqrt{2l+1}} \, | l, \frac{1}{2}; l, - \rangle
\end{aligned}
\tag{26}
$$

We apply J_- again. An analogous calculation yields:

$$
\begin{aligned}
| l + \frac{1}{2}, l - \frac{3}{2} \rangle = \frac{1}{\sqrt{2l+1}} \Big[& \sqrt{2l-1} \, | l, \frac{1}{2}; l-2, + \rangle \\
& + \sqrt{2} \, | l, \frac{1}{2}; l-1, - \rangle \Big]
\end{aligned}
\tag{27}
$$

More generally, the vector $| l + 1/2, M \rangle$ will be a linear combination of the only two basis vectors associated with M: $| l, 1/2; M - 1/2, + \rangle$ and $| l, 1/2; M + 1/2, - \rangle$ (M is, of course, half-integral). Comparing (25), (26) and (27), we can guess that this linear combination should be the following one:

$$
\begin{aligned}
| l + \frac{1}{2}, M \rangle = \frac{1}{\sqrt{2l+1}} \Big[& \sqrt{l + M + \frac{1}{2}} \, | l, \frac{1}{2}; M - \frac{1}{2}, + \rangle \\
& + \sqrt{l - M + \frac{1}{2}} \, | l, \frac{1}{2}; M + \frac{1}{2}, - \rangle \Big]
\end{aligned}
\tag{28}
$$

with:

$$
M = l + \frac{1}{2} \;, l - \frac{1}{2} \;, l - \frac{3}{2} \;, \ldots, -l + \frac{1}{2} \;, -\left(l + \frac{1}{2}\right)
\tag{29}
$$

★ To find the numerical coefficients appearing in the following equations, we can simply use the relation: $j(j+1) - m(m-1) = (j+m)(j-m+1)$.

Reasoning by recurrence, we can show this to be true, since application of J_- to both sides of (28) yields:

$$|\, l + \frac{1}{2}, M - 1 \rangle = \frac{1}{\hbar\sqrt{\left(l + M + \frac{1}{2}\right)\left(l - M + \frac{3}{2}\right)}} J_- \,|\, l + \frac{1}{2}, M \rangle$$

$$= \frac{1}{\hbar\sqrt{\left(l + M + \frac{1}{2}\right)\left(l - M + \frac{3}{2}\right)}} \frac{1}{\sqrt{2l + 1}} \times$$

$$\times \left[\sqrt{l + M + \frac{1}{2}}\,\hbar\sqrt{\left(l + M - \frac{1}{2}\right)\left(l - M + \frac{3}{2}\right)}\;|\, l, \frac{1}{2}; M - \frac{3}{2}, + \rangle\right.$$

$$+ \sqrt{l + M + \frac{1}{2}}\,\hbar\;|\, l, \frac{1}{2}; M - \frac{1}{2}, - \rangle$$

$$\left. + \sqrt{l - M + \frac{1}{2}}\,\hbar\sqrt{\left(l + M + \frac{1}{2}\right)\left(l - M + \frac{1}{2}\right)}\;|\, l, \frac{1}{2}; M - \frac{1}{2}, - \rangle\right]$$

$$= \frac{1}{\sqrt{2l + 1}}\left[\sqrt{l + M - \frac{1}{2}}\;|\, l, \frac{1}{2}; M - \frac{3}{2}, + \rangle\right.$$

$$\left. + \sqrt{l - M + \frac{3}{2}}\;|\, l, \frac{1}{2}; M - \frac{1}{2}, - \rangle\right] \quad (30)$$

We indeed obtain the same expression as in (28), with M changed to $M - 1$.

b. THE SUBSPACE $\mathscr{E}(J = l - 1/2)$

We shall now try to determine the expression for the $2l$ vectors $|\, J, M \rangle$ associated with $J = l - 1/2$. The one which corresponds to the maximum value $l - 1/2$ of M is a normalized linear combination of $|\, l, 1/2; l - 1, + \rangle$ and $|\, l, 1/2; l, - \rangle$, and it must be orthogonal to $|\, l + 1/2, l - 1/2 \rangle$ [formula (26)]. Choosing the coefficient of $|\, l, 1/2; l, - \rangle$ real and positive (*cf.* note p. 1028), we easily find:

$$|\, l - \frac{1}{2}, l - \frac{1}{2} \rangle = \frac{1}{\sqrt{2l + 1}}\left[\sqrt{2l}\,|\, l, \frac{1}{2}; l, - \rangle - |\, l, \frac{1}{2}; l - 1, + \rangle\right] \quad (31)$$

The operator J_- enables us to deduce successively all the other vectors of the family characterized by $J = l - 1/2$. Since there are only two basis vectors with a given value of M, and since $|\, l - 1/2, M \rangle$ is orthogonal to $|\, l + 1/2, M \rangle$, (28) leads us to expect that:

$$|\, l - \frac{1}{2}, M \rangle = \frac{1}{\sqrt{2l + 1}}\left[\sqrt{l + M + \frac{1}{2}}\;|\, l, \frac{1}{2}; M + \frac{1}{2}, - \rangle\right.$$

$$\left. - \sqrt{l - M + \frac{1}{2}}\;|\, l, \frac{1}{2}; M - \frac{1}{2}, + \rangle\right] \quad (32)$$

for:

$$M = l - \frac{1}{2} \ , l - \frac{3}{2} \ , ..., - l + \frac{3}{2} \ , - \left(l - \frac{1}{2}\right) \tag{33}$$

By an argument analogous to the one in §2-a, this formula can also be proved by recurrence.

COMMENTS:

(*i*) The states $| \, l, 1/2; m, \varepsilon \rangle$ of a spin 1/2 particle can be represented by two-component spinors of the form:

$$[\psi_{l,\frac{1}{2};m,+}](\mathbf{r}) = R_{k,l}(r) \, Y_l^m(\theta, \varphi) \begin{pmatrix} 1 \\ 0 \end{pmatrix} \tag{34-a}$$

$$[\psi_{l,\frac{1}{2};m,-}](\mathbf{r}) = R_{k,l}(r) \, Y_l^m(\theta, \varphi) \begin{pmatrix} 0 \\ 1 \end{pmatrix} \tag{34-b}$$

The preceding calculations then show that the spinors associated with the states $| \, J, M \rangle$ can be written:

$$[\psi_{l+\frac{1}{2},M}](\mathbf{r}) = \frac{1}{\sqrt{2l+1}} R_{k,l}(r) \begin{pmatrix} \sqrt{l + M + \frac{1}{2}} \; Y_l^{M-\frac{1}{2}}(\theta, \varphi) \\ \sqrt{l - M + \frac{1}{2}} \; Y_l^{M+\frac{1}{2}}(\theta, \varphi) \end{pmatrix} \tag{35-a}$$

$$[\psi_{l-\frac{1}{2},M}](\mathbf{r}) = \frac{1}{\sqrt{2l+1}} R_{k,l}(r) \begin{pmatrix} -\sqrt{l - M + \frac{1}{2}} \; Y_l^{M-\frac{1}{2}}(\theta, \varphi) \\ \sqrt{l + M + \frac{1}{2}} \; Y_l^{M+\frac{1}{2}}(\theta, \varphi) \end{pmatrix} \tag{35-b}$$

(*ii*) In the particular case $l = 1$, formulas (25), (28), (31) and (32) yield:

$$\begin{aligned}
|\tfrac{3}{2}, \tfrac{3}{2}\rangle &= |\, 1, \tfrac{1}{2}; 1, + \rangle \\
|\tfrac{3}{2}, \tfrac{1}{2}\rangle &= \sqrt{\tfrac{2}{3}}\,|\, 1, \tfrac{1}{2}; 0, + \rangle + \tfrac{1}{\sqrt{3}}\,|\, 1, \tfrac{1}{2}; 1, - \rangle \\
|\tfrac{3}{2}, -\tfrac{1}{2}\rangle &= \tfrac{1}{\sqrt{3}}\,|\, 1, \tfrac{1}{2}; -1, + \rangle + \sqrt{\tfrac{2}{3}}\,|\, 1, \tfrac{1}{2}; 0, - \rangle \\
|\tfrac{3}{2}, -\tfrac{3}{2}\rangle &= |\, 1, \tfrac{1}{2}; -1, - \rangle
\end{aligned} \tag{36-a}$$

and:

$$| \frac{1}{2}, \frac{1}{2} \rangle = \sqrt{\frac{2}{3}} | 1, \frac{1}{2}; 1, - \rangle - \frac{1}{\sqrt{3}} | 1, \frac{1}{2}; 0, + \rangle$$

$$| \frac{1}{2}, - \frac{1}{2} \rangle = \frac{1}{\sqrt{3}} | 1, \frac{1}{2}; 0, - \rangle - \sqrt{\frac{2}{3}} | 1, \frac{1}{2}; - 1, + \rangle \qquad (36\text{-b})$$

References and suggestions for further reading:

Addition of an angular momentum l and an angular momentum $S = 1$: see "vectorial spherical harmonics" in the references of chapter X.

Complement B_X

CLEBSCH-GORDAN COEFFICIENTS

Clebsch-Gordan coefficients were introduced in chapter X [*cf.* relation (C-66)]: they are the coefficients $\langle j_1, j_2; m_1, m_2 \mid J, M \rangle$ involved in the expansion of the ket $\mid J, M \rangle$ on the $\{ \mid j_1, j_2; m_1, m_2 \rangle \}$ basis:

$$\mid J, M \rangle = \sum_{m_1=-j_1}^{j_1} \sum_{m_2=-j_2}^{j_2} \langle j_1, j_2; m_1, m_2 \mid J, M \rangle \mid j_1, j_2; m_1, m_2 \rangle \tag{1}$$

In this complement, we shall derive some interesting properties of Clebsch-Gordan coefficients, some of which were simply stated in chapter X.

Note that, to define the $\langle j_1, j_2; m_1, m_2 \mid J, M \rangle$ completely, equation (1) is not sufficient. The normalized vector $\mid J, M \rangle$ is fixed only to within a phase factor by the corresponding eigenvalues $J(J+1)\hbar^2$ and $M\hbar$, and a phase convention must be chosen in order to complete the definition. In chapter X, we used the action of the J_- and J_+ operators to fix the relative phase of the $(2J+1)$ kets $\mid J, M \rangle$ associated with the same value of J. In this complement, we shall complete this choice of phase by adopting a convention for the phase of the kets $\mid J, J \rangle$. This will enable us to show that all the Clebsch-Gordan coefficients are then real.

However, before approaching, in § 2, the problem of the choice of the phase of the $\langle j_1, j_2; m_1, m_2 \mid J, M \rangle$, we shall, in §1, study some of their most useful properties which do not depend on this phase convention. Finally, §3 presents various relations which will be of use in other complements.

1. General properties of Clebsch-Gordan coefficients

a. SELECTION RULES

Two important selection rules, which follow directly from the results of chapter X concerning the addition of angular momenta, have already been given

in this chapter [*cf.* relations (C-67-a) and (C-67-b)]. We shall simply restate them here : the Clebsch-Gordan coefficient $\langle j_1, j_2 ; m_1, m_2 \mid J, M \rangle$ is necessarily zero if the following two conditions are not simultaneously satisfied:

$$M = m_1 + m_2 \tag{2}$$

$$|j_1 - j_2| \leqslant J \leqslant j_1 + j_2 \tag{3-a}$$

Inequality (3-a) is often called the "triangle selection rule", since it means that a triangle can be formed with three line segments of lengths j_1, j_2 and J (*cf.* fig. 1). These three numbers therefore play symmetrical roles here, and (3-a) can also be written in the form:

$$|J - j_1| \leqslant j_2 \leqslant J + j_1 \tag{3-b}$$

or:

$$|J - j_2| \leqslant j_1 \leqslant J + j_2 \tag{3-c}$$

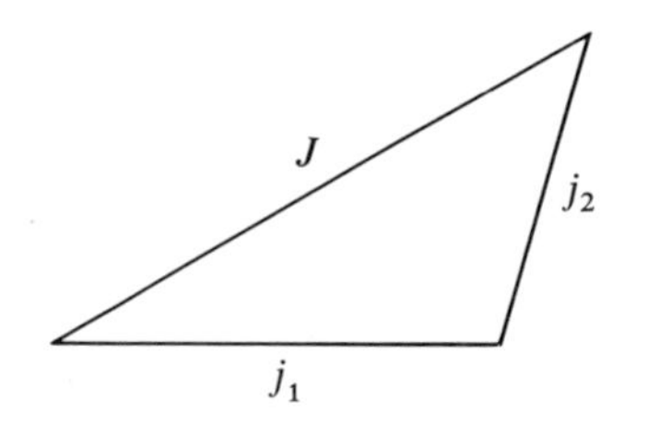

FIGURE 1

Triangle selection rule : the coefficient $\langle j_1, j_2 ; m_1, m_2 | J, M \rangle$ can be different from zero only if it is possible to form a triangle with three line segments of lengths j_1, j_2, J.

Moreover, the general properties of angular momentum require that the ket $| J, M \rangle$ and, therefore, the coefficient $\langle j_1, j_2 ; m_1, m_2 \mid J, M \rangle$, exist only if M takes on one of the values:

$$M = J \;, J - 1 \;, J - 2 \;, ..., - J \tag{4-a}$$

Similarly, it is necessary that:

$$m_1 = j_1, j_1 - 1, ..., - j_1 \tag{4-b}$$

$$m_2 = j_2, j_2 - 1, ..., - j_2 \tag{4-c}$$

If this is not the case, the Clebsch-Gordan coefficients are not defined. However, in what follows, it will be convenient to assume that they exist for all m_1, m_2 and M, but that they are zero if at least one of conditions (4) is not satisfied. These relations thus play the role of new selection rules for the Clebsch-Gordan coefficients.

b. ORTHOGONALITY-RELATIONS

Inserting the closure relation★:

$$\sum_{m_1=-j_1}^{j_1} \sum_{m_2=-j_2}^{j_2} | j_1, j_2; m_1, m_2 \rangle \langle j_1, j_2; m_1, m_2 | = 1 \tag{5}$$

in the orthogonality relation of the kets $| J, M \rangle$:

$$\langle J, M | J', M' \rangle = \delta_{JJ'} \delta_{MM'} \tag{6}$$

we obtain:

$$\sum_{m_1=-j_1}^{j_1} \sum_{m_2=-j_2}^{j_2} \langle J, M | j_1, j_2; m_1, m_2 \rangle \langle j_1, j_2; m_1, m_2 | J', M' \rangle = \delta_{JJ'} \delta_{MM'} \tag{7-a}$$

We shall see later [*cf.* relation (18-b)] that the Clebsch-Gordan coefficients are real, which enables us to write this relation in the form:

$$\sum_{m_1=-j_1}^{j_1} \sum_{m_2=-j_2}^{j_2} \langle j_1, j_2; m_1, m_2 | J, M \rangle \langle j_1, j_2; m_1, m_2 | J', M' \rangle = \delta_{JJ'} \delta_{MM'} \tag{7-b}$$

Thus we obtain a first "orthogonality relation" for the Clebsch-Gordan coefficients. We note, moreover, that the summation which appears in it is performed, in fact, over only one index : for the coefficients of the left-hand side to be different from zero, m_1 and m_2 must be related by (2).

Similarly, we insert the closure relation:

$$\sum_{J=|j_1-j_2|}^{j_1+j_2} \sum_{M=-J}^{J} | J, M \rangle \langle J, M | = 1 \tag{8}$$

in the orthogonality relation of the kets $| j_1, j_2; m_1, m_2 \rangle$; we obtain:

$$\sum_{J=|j_1-j_2|}^{j_1+j_2} \sum_{M=-J}^{J} \langle j_1, j_2; m_1, m_2 | J, M \rangle \langle J, M | j_1, j_2; m'_1, m'_2 \rangle = \delta_{m_1m'_1} \delta_{m_2m'_2} \tag{9-a}$$

that is, with (18-b) taken into account:

$$\sum_{J=|j_1-j_2|}^{j_1+j_2} \sum_{M=-J}^{J} \langle j_1, j_2; m_1, m_2 | J, M \rangle \langle j_1, j_2; m'_1, m'_2 | J, M \rangle = \delta_{m_1m'_1} \delta_{m_2m'_2} \tag{9-b}$$

Again, the summation is performed over only one index: since we must have $M = m_1 + m_2$, the summation over M reduces to a single term.

★ This closure relation is valid for a given subspace $\mathscr{E}(k_1, k_2; j_1, j_2)$ (*cf.* chap. X, §C-2).

c. RECURRENCE RELATIONS

In this section, we shall use the fact that the kets $|j_1, j_2; m_1, m_2\rangle$ form a standard basis. Thus:

$$J_{1\pm}\,|j_1, j_2; m_1, m_2\rangle = \hbar\sqrt{j_1(j_1+1) - m_1(m_1 \pm 1)}\,|j_1, j_2; m_1 \pm 1, m_2\rangle$$
$$J_{2\pm}\,|j_1, j_2; m_1, m_2\rangle = \hbar\sqrt{j_2(j_2+1) - m_2(m_2 \pm 1)}\,|j_1, j_2; m_1, m_2 \pm 1\rangle \tag{10}$$

Similarly, by construction, the kets $|J, M\rangle$ satisfy:

$$J_\pm\,|J, M\rangle = \hbar\sqrt{J(J+1) - M(M \pm 1)}\,|J, M \pm 1\rangle \tag{11}$$

We shall therefore apply the J_- operator to relation (1). Since $J_- = J_{1-} + J_{2-}$, we obtain (if $M > -J$):

$$\sqrt{J(J+1) - M(M-1)}\,|J, M-1\rangle = \sum_{m_1'=-j_1}^{j_1}\sum_{m_2'=-j_2}^{j_2}\langle j_1, j_2; m_1', m_2'\,|\,J, M\rangle$$
$$\times\Big[\sqrt{j_1(j_1+1) - m_1'(m_1'-1)}\,|j_1, j_2; m_1'-1, m_2'\rangle + \sqrt{j_2(j_2+1) - m_2'(m_2'-1)}\,|j_1, j_2; m_1', m_2'-1\rangle\Big] \tag{12}$$

Multiplying this relation by the bra $\langle j_1, j_2; m_1, m_2|$, we find:

$$\sqrt{J(J+1) - M(M-1)}\,\langle j_1, j_2; m_1, m_2\,|\,J, M-1\rangle$$
$$= \sqrt{j_1(j_1+1) - m_1(m_1+1)}\,\langle j_1, j_2; m_1+1, m_2\,|\,J, M\rangle$$
$$+ \sqrt{j_2(j_2+1) - m_2(m_2+1)}\,\langle j_1, j_2; m_1, m_2+1\,|\,J, M\rangle \tag{13}$$

If the value of M is equal to $-J$, we have $J_-\,|J, -J\rangle = 0$, and relation (13) remains valid if we use the convention, given above in § 1-b, according to which $\langle j_1, j_2; m_1, m_2\,|\,J, M\rangle$ is zero if $|M| > J$.

Analogously, application of the operator $J_+ = J_{1+} + J_{2+}$ to relation (1) leads to:

$$\sqrt{J(J+1) - M(M+1)}\,\langle j_1, j_2; m_1, m_2\,|\,J, M+1\rangle$$
$$= \sqrt{j_1(j_1+1) - m_1(m_1-1)}\,\langle j_1, j_2; m_1-1, m_2\,|\,J, M\rangle$$
$$+ \sqrt{j_2(j_2+1) - m_2(m_2-1)}\,\langle j_1, j_2; m_1, m_2-1\,|\,J, M\rangle \tag{14}$$

(the left-hand side of this relation is zero if $M = J$); (13) and (14) are recurrence relations for the Clebsch-Gordan coefficients.

2. Phase conventions. Reality of Clebsch-Gordan coefficients

As we have seen, expressions (12) fix the relative phases of the kets $| J, M \rangle$ associated with the same value of J. To complete the definition of the Clebsch-Gordan coefficients involved in (1), we must choose the phase of the various kets $| J, J \rangle$. To this end, we shall begin by studying some properties of the coefficients $\langle j_1, j_2 ; m_1, m_2 | J, J \rangle$.

a. THE COEFFICIENTS $\langle j_1, j_2 ; m_1, m_2 | J, J \rangle$; PHASE OF THE KET $| J, J \rangle$

In the coefficient $\langle j_1, j_2 ; m_1, m_2 | J, J \rangle$, the maximum value of m_1 is $m_1 = j_1$. According to selection rule (2), m_2 is then equal to $J - j_1$ [whose modulus is well below j_2, according to (3-b)]. As m_1 decreases from this maximum value j_1, one unit at a time, m_2 increases until it reaches its maximum value $m_2 = j_2$ [m_1 is then equal to $J - j_2$, whose modulus is well below j_1, according to (3-c)]. In theory, therefore, $(j_1 + j_2 - J + 1)$ non-zero Clebsch-Gordan coefficients $\langle j_1, j_2 ; m_1, m_2 | J, J \rangle$ can exist. We are going to show that, in fact, none of them is ever zero.

If we set $M = J$ in (14), we obtain:

$$\langle j_1, j_2 ; m_1 - 1, m_2 | J, J \rangle = - \sqrt{\frac{j_2(j_2 + 1) - m_2(m_2 - 1)}{j_1(j_1 + 1) - m_1(m_1 - 1)}} \langle j_1, j_2 ; m_1, m_2 - 1 | J, J \rangle \tag{15}$$

The radical on the right-hand side of this relation is never zero, nor is it infinite, so long as the Clebsch-Gordan coefficients appearing there satisfy rules (4-b) and (4-c). Relation (15) therefore shows that if $\langle j_1, j_2 ; j_1, J - j_1 | J, J \rangle$ were equal to zero, $\langle j_1, j_2 ; j_1 - 1, J - j_1 + 1 | J, J \rangle$ would be zero as well, as would be all the succeeding coefficients $\langle j_1, j_2 ; m_1, J - m_1 | J, J \rangle$. Now, this is impossible, since the ket $| J, J \rangle$, which is normalized, cannot be zero. Therefore, all the coefficients $\langle j_1, j_2 ; m_1, J - m_1 | J, J \rangle$ (with $j_1 \geqslant m_1 \geqslant J - j_2$) are different from zero.

In particular, the coefficient $\langle j_1, j_2 ; j_1, J - j_1 | J, J \rangle$, in which m_1 takes on its maximum value, is not zero. To fix the phase of the ket $| J, J \rangle$, we shall require this coefficient to satisfy the condition:

$$\langle j_1, j_2 ; j_1, J - j_1 | J, J \rangle \quad \text{real and positive} \tag{16}$$

Relation (15) then implies by recurrence that all the coefficients

$$\langle j_1, j_2 ; m_1, J - m_1 | J, J \rangle$$

are real [their sign being $(-1)^{j_1 - m_1}$].

COMMENT:

The phase convention we have chosen for the ket $| J, J \rangle$ gives the two angular momenta $\mathbf{J}_1$ and $\mathbf{J}_2$ asymmetrical roles. It actually depends on the order in which the quantum numbers j_1 and j_2 are arranged in the Clebsch-Gordan coefficients: if j_1 and j_2 are permuted, the phase of the ket $| J, J \rangle$ is fixed by the condition:

$$\langle j_2, j_1 ; j_2, J - j_2 | J, J \rangle \quad \text{real and positive} \tag{17}$$

which is not necessarily equivalent, *a priori*, to (16) [(16) and (17) may define different phases for the ket $| J, J \rangle$]. We shall return to this point in §3-b.

b. OTHER CLEBSCH-GORDAN COEFFICIENTS

Relation (13) enables us to express, in terms of the $\langle j_1, j_2; m_1, m_2 \mid J, J \rangle$, all the coefficients $\langle j_1, j_2; m_1, m_2 \mid J, J-1 \rangle$, then all the coefficients $\langle j_1, j_2; m_1, m_2 \mid J, J-2 \rangle$, etc. This relation, in which no imaginary numbers are involved, requires that all Clebsch-Gordan coefficients be real:

$$\langle j_1, j_2; m_1, m_2 \mid J, M \rangle^* = \langle j_1, j_2; m_1, m_2 \mid J, M \rangle \tag{18-a}$$

which can also be written:

$$\langle j_1, j_2; m_1, m_2 \mid J, M \rangle = \langle J, M \mid j_1, j_2; m_1, m_2 \rangle \tag{18-b}$$

However, the signs of the $\langle j_1, j_2; m_1, m_2 \mid J, M \rangle$ do not obey any simple rule for $M \neq J$.

3. Some useful relations

In this section, we give some useful relations, which complement those given in §1. To prove them, we shall begin by studying the signs of a certain number of Clebsch-Gordan coefficients.

a. THE SIGNS OF SOME COEFFICIENTS

α. *The coefficients* $\langle j_1, j_2; m_1, m_2 \mid j_1 + j_2, M \rangle$

Convention (16) requires the coefficient $\langle j_1, j_2; j_1, j_2 \mid j_1 + j_2, j_1 + j_2 \rangle$ to be real and positive; it is, moreover, equal to 1 (*cf.* chap. X, § C-4-b-α). Setting $M = J = j_1 + j_2$ in (13), we then see that the coefficients $\langle j_1, j_2; m_1, m_2 \mid j_1 + j_2, j_1 + j_2 - 1 \rangle$ are positive. By recurrence, it is then easy to prove that:

$$\langle j_1, j_2; m_1, m_2 \mid j_1 + j_2, M \rangle \geqslant 0 \tag{19}$$

β. *Coefficients in which m_1 has its maximum value*

Consider the coefficient $\langle j_1, j_2; m_1, m_2 \mid J, M \rangle$. In theory, the maximum value of m_1 is $m_1 = j_1$. However, we then have $m_2 = M - j_1$, which, according to (4-c), is possible only if $M - j_1 \geqslant -j_2$, that is:

$$M \geqslant j_1 - j_2 \tag{20}$$

If, on the other hand:

$$M \leqslant j_1 - j_2 \tag{21}$$

the maximum value of m_1 corresponds to the minimum value of m_2 ($m_2 = -j_2$), and is therefore equal to $m_1 = M + j_2$.

Let us show that all Clebsch-Gordan coefficients for which m_1 has its maximum value are non-zero and positive. To do so, we set $m_1 = j_1$ in (13); we find:

$$\sqrt{J(J+1) - M(M-1)}\,\langle j_1, j_2; j_1, m_2 \mid J, M-1 \rangle$$
$$= \sqrt{j_2(j_2+1) - m_2(m_2+1)}\,\langle j_1, j_2; j_1, m_2 + 1 \mid J, M \rangle \tag{22}$$

Using this relation, an argument by recurrence starting with (16) shows that all the coefficients $\langle j_1, j_2; j_1, M - j_1 \mid J, M \rangle$ are positive [and non-zero if M satisfies (20)]. Analogously, setting $m_2 = -j_2$ in (14), we could prove that all the coefficients $\langle j_1, j_2; M + j_2, -j_2 \mid J, M \rangle$ are positive [if M satisfies (21)].

γ. *The coefficients* $\langle j_1, j_2; m_1, m_2 \mid J, J \rangle$ *and* $\langle j_1, j_2; m_1, m_2 \mid J, -J \rangle$

We saw in § 2-a that the sign of $\langle j_1, j_2; m_1, m_2 \mid J, J \rangle$ is $(-1)^{j_1 - m_1}$. In particular:

$$\text{the sign of } \langle j_1, j_2; J - j_2, j_2 \mid J, J \rangle = (-1)^{j_1 + j_2 - J} \tag{23}$$

To determine the sign of $\langle j_1, j_2; m_1, m_2 \mid J, -J \rangle$, we can set $M = -J$ in (13), whose left-hand side then goes to zero. We therefore see that the sign of $\langle j_1, j_2; m_1, m_2 \mid J, -J \rangle$ changes whenever m_1 (or m_2) varies by ± 1. Since, according to § β, $\langle j_1, j_2; j_2 - J, -j_2 \mid J, -J \rangle$ is positive, it follows that the sign of $\langle j_1, j_2; m_1, m_2 \mid J, -J \rangle$ is $(-1)^{m_2 + j_2}$, and, in particular:

$$\text{the sign of } \langle j_1, j_2; -j_1, -J + j_1 \mid J, -J \rangle = (-1)^{j_1 + j_2 - J} \tag{24}$$

b. CHANGING THE ORDER OF j_1 AND j_2

With the conventions we have chosen, the phase of the ket $\mid J, J \rangle$ depends on the order in which the two angular momenta j_1 and j_2 are arranged in the Clebsch-Gordan coefficients (*cf.* comment of § 2-a). If they are taken in the order j_1, j_2, the component of $\mid J, J \rangle$ along $\mid j_1, j_2; j_1, J - j_1 \rangle$ is positive, which means that the sign of the component along $\mid j_1, j_2; J - j_2, j_2 \rangle$ is $(-1)^{j_1 + j_2 - J}$, as is indicated by (23). On the other hand, if we pick the order j_2, j_1, relation (17) shows that the latter component is positive. Therefore, if we invert j_1 and j_2, the ket $\mid J, J \rangle$ is multiplied by $(-1)^{j_1 + j_2 - J}$. The same is true for the kets $\mid J, M \rangle$, which are constructed from $\mid J, J \rangle$ by the action of J_- in such a way that the order of j_1 and j_2 plays no role. Finally, the exchange of j_1 and j_2 leads to the relation :

$$\langle j_2, j_1; m_2, m_1 \mid J, M \rangle = (-1)^{j_1 + j_2 - J} \langle j_1, j_2; m_1, m_2 \mid J, M \rangle \tag{25}$$

c. CHANGING THE SIGN OF M, m_1 AND m_2

In chapter X and in this complement, we have constructed all the kets $\mid J, M \rangle$ (and, therefore, the Clebsch-Gordan coefficients) from the kets $\mid J, J \rangle$, by applying the operator J_-. We can take the opposite point of view, and start with the kets $\mid J, -J \rangle$, using the operator J_+. The reasoning which follows is exactly the same, and we find for the kets $\mid J, -M \rangle$ the same expansion coefficients on the kets $\mid j_1, j_2; -m_1, -m_2 \rangle$ as for the $\mid J, M \rangle$ on the $\mid j_1, j_2; m_1, m_2 \rangle$. The only differences that can appear are related to the phase conventions for the kets $\mid J, M \rangle$, since the analogue of (16) then requires $\langle j_1, j_2; -j_1, -J + j_1 \mid J, -J \rangle$ to be real and positive. Now, according to (24), the sign of this coefficient is, in reality, $(-1)^{j_1 + j_2 - J}$. Consequently:

$$\langle j_1, j_2; -m_1, -m_2 \mid J, -M \rangle = (-1)^{j_1 + j_2 - J} \langle j_1, j_2; m_1, m_2 \mid J, M \rangle \tag{26}$$

In particular, if we set $m_1 = m_2 = 0$, we see that the coefficient $\langle j_1, j_2; 0, 0 \mid J, 0 \rangle$ is zero when $j_1 + j_2 - J$ is an odd number.

d. THE COEFFICIENTS $\langle j, j; m, -m \mid 0, 0 \rangle$

According to (3-a), J can be zero only if j_1 and j_2 are equal. We therefore substitute the values $j_1 = j_2 = j$, $m_1 = m$, $m_2 = -m - 1$ and $J = M = 0$ into (13); we obtain:

$$\langle j, j; m + 1, -(m + 1) \mid 0, 0 \rangle = -\langle j, j; m, -m \mid 0, 0 \rangle \tag{27}$$

All the coefficients $\langle j, j; m, -m \mid 0, 0 \rangle$ are therefore equal in modulus. Their signs change whenever m varies by one, and, since $\langle j, j; j, -j \mid 0, 0 \rangle$ is positive, it is given by $(-1)^{j-m}$. Taking into account orthogonality relation (7-b), which indicates that:

$$\sum_{m=-j}^{j} \langle j, j; m, -m \mid 0, 0 \rangle^2 = 1 \tag{28}$$

we find:

$$\langle j, j; m, -m \mid 0, 0 \rangle = \frac{(-1)^{j-m}}{\sqrt{2j+1}} \tag{29}$$

References:

Messiah (1.17), app. C; Rose (2.19), chap. III and app. I; Edmonds (2.21), chap. 3; Sobel'man (11.12), chap. 4, § 13.

Tables of Clebsch-Gordan coefficients: Condon and Shortley (11.13), chap. III, §14; Bacry (10.34), app. C.

Tables of $3j$ and $6j$ coefficients : Edmonds (2.21), Table 2; Rotenberg et al. (10.48).

Complement C_X

ADDITION OF SPHERICAL HARMONICS

In this complement, we use the properties of Clebsch-Gordan coefficients to prove relations that will be of use to us later, especially in complements E_X and A_{XIII}: the spherical harmonic addition relations. With this aim in mind, we shall begin by introducing and studying the functions of two sets of polar angles Ω_1 and Ω_2, the $\Phi_J^M(\Omega_1\,;\,\Omega_2)$.

1. The functions $\Phi_J^M(\Omega_1\,;\,\Omega_2)$

Consider two particles (1) and (2), of state spaces $\mathscr{E}_{\mathbf{r}}^1$ and $\mathscr{E}_{\mathbf{r}}^2$ and orbital angular momenta $\mathbf{L}_1$ and $\mathbf{L}_2$. We choose for the space $\mathscr{E}_{\mathbf{r}}^1$ a standard basis, formed by the kets $\{\,|\,\varphi_{k_1,l_1,m_1}\rangle\,\}$, whose wave functions are:

$$\varphi_{k_1,l_1,m_1}(\mathbf{r}_1) = R_{k_1,l_1}(r_1)\,Y_{l_1}^{m_1}(\Omega_1) \tag{1}$$

(Ω_1 denotes the set of polar angles $\{\,\theta_1,\varphi_1\,\}$ of the first particle). Similarly, we choose for $\mathscr{E}_{\mathbf{r}}^2$ a standard basis, $\{\,|\,\varphi_{k_2,l_2,m_2}\rangle\,\}$. In all that follows, we shall confine the states of the two particles to the subspaces $\mathscr{E}(k_1,l_1)$ and $\mathscr{E}(k_2,l_2)$, where k_1, l_1, k_2 and l_2 are fixed, and the radial functions $R_{k_1,l_1}(r_1)$ and $R_{k_2,l_2}(r_2)$ play no role.

The angular momentum of the total system (1) + (2) is:

$$\mathbf{J} = \mathbf{L}_1 + \mathbf{L}_2 \tag{2}$$

According to the results of chapter X, we can construct a basis of $\mathscr{E}(k_1,l_1)\otimes\mathscr{E}(k_2,l_2)$ of eigenvectors $|\,\Phi_J^M\rangle$ common to $\mathbf{J}^2$ [eigenvalue $J(J+1)\hbar^2$] and J_z (eigenvalue $M\hbar$). These vectors are of the form:

$$|\,\Phi_J^M\rangle = \sum_{m_1=-l_1}^{l_1}\sum_{m_2=-l_2}^{l_2}\langle\,l_1,l_2\,;m_1,m_2\,|\,J,M\,\rangle\,|\,\varphi_{k_1,l_1,m_1}(1)\rangle\otimes|\,\varphi_{k_2,l_2,m_2}(2)\rangle \tag{3-a}$$

the inverse change of basis being given by:

$$|\,\varphi_{k_1,l_1,m_1}(1)\rangle\otimes|\,\varphi_{k_2,l_2,m_2}(2)\rangle = \sum_{J=|l_1-l_2|}^{l_1+l_2}\sum_{M=-J}^{J}\langle\,l_1,l_2\,;m_1,m_2\,|\,J,M\,\rangle\,|\,\Phi_J^M\rangle \tag{3-b}$$

Relation (3-a) shows that the angular dependence of the states $| \Phi_J^M \rangle$ is described by the functions:

$$\Phi_J^M(\Omega_1 ; \Omega_2) = \sum_{m_1} \sum_{m_2} \langle l_1, l_2 ; m_1, m_2 | J, M \rangle \, Y_{l_1}^{m_1}(\Omega_1) \, Y_{l_2}^{m_2}(\Omega_2) \tag{4-a}$$

Similarly, relation (3-b) implies that:

$$Y_{l_1}^{m_1}(\Omega_1) \, Y_{l_2}^{m_2}(\Omega_2) = \sum_{J=|l_1-l_2|}^{l_1+l_2} \sum_{M=-J}^{J} \langle l_1, l_2 ; m_1, m_2 | J, M \rangle \, \Phi_J^M(\Omega_1 ; \Omega_2) \tag{4-b}$$

To the observables $\mathbf{L}_1$ and $\mathbf{L}_2$ correspond, for the wave functions, differential operators acting on the variables $\Omega_1 = \{ \theta_1, \varphi_1 \}$ and $\Omega_2 = \{ \theta_2, \varphi_2 \}$; in particular:

$$L_{1z} \Longrightarrow \frac{\hbar}{i} \frac{\partial}{\partial \varphi_1} \tag{5-a}$$

$$L_{2z} \Longrightarrow \frac{\hbar}{i} \frac{\partial}{\partial \varphi_2} \tag{5-b}$$

Since, by construction, the ket $| \Phi_J^M \rangle$ is an eigenvector of $J_z = L_{1z} + L_{2z}$, we can write:

$$\frac{\hbar}{i} \left(\frac{\partial}{\partial \varphi_1} + \frac{\partial}{\partial \varphi_2} \right) \Phi_J^M(\theta_1, \varphi_1 ; \theta_2, \varphi_2) = M\hbar \, \Phi_J^M(\theta_1, \varphi_1 ; \theta_2, \varphi_2) \tag{6}$$

Similarly, we have:

$$J_{\pm} | \Phi_J^M \rangle = \hbar \sqrt{J(J+1) - M(M \pm 1)} \, | \Phi_J^{M \pm 1} \rangle \tag{7}$$

which implies, with formulas (D-6) of chapter VI taken into account:

$$\left\{ e^{\pm i\varphi_1} \left[\pm \frac{\partial}{\partial \theta_1} + i \cot \theta_1 \frac{\partial}{\partial \varphi_1} \right] + e^{\pm i\varphi_2} \left[\pm \frac{\partial}{\partial \theta_2} + i \cot \theta_2 \frac{\partial}{\partial \varphi_2} \right] \right\} \Phi_J^M(\theta_1, \varphi_1 ; \theta_2, \varphi_2)$$
$$= \sqrt{J(J+1) - M(M \pm 1)} \, \Phi_J^{M \pm 1}(\theta_1, \varphi_1 ; \theta_2, \varphi_2) \tag{8}$$

2. The functions $F_l^m(\Omega)$

We now introduce the function F_l^m defined by:

$$F_l^m(\theta, \varphi) \equiv F_l^m(\Omega) = \Phi_{J=l}^{M=m}(\Omega_1 = \Omega ; \Omega_2 = \Omega) \tag{9}$$

F_l^m is a function of a single pair of polar angles $\Omega = \{ \theta, \varphi \}$, and can therefore characterize the angular dependence of a wave function associated with a single particle, of state space $\mathscr{E}_{\mathbf{r}}$ and angular momentum $\mathbf{L}$. In fact, we shall see that F_l^m is not a new function, but is simply proportional to the spherical harmonic Y_l^m.

To demonstrate this, we shall show that F_l^m is an eigenfunction of $\mathbf{L}^2$ and L_z with the eigenvalues $l(l+1)\hbar^2$ and $m\hbar$. We therefore begin by calculating the action of L_z on F_l^m. According to (9), F_l^m depends on θ and φ by way of $\Omega_1 = \{ \theta_1, \varphi_1 \}$ and $\Omega_2 = \{ \theta_2, \varphi_2 \}$, which are both taken equal to Ω. If we apply the differentiation theorem for functions of functions, we find:

$$
\begin{aligned}
L_z F_l^m(\theta, \varphi) &= \frac{\hbar}{i} \frac{\partial}{\partial \varphi} F_l^m(\theta, \varphi) \\
&= \frac{\hbar}{i} \left\{ \left[\frac{\partial}{\partial \varphi_1} + \frac{\partial}{\partial \varphi_2} \right] \Phi_{J=l}^{M=m}(\Omega_1 ; \Omega_2) \right\}_{\Omega_1 = \Omega_2 = \Omega}
\end{aligned} \tag{10}
$$

Relation (6) then yields:

$$
L_z F_l^m(\theta, \varphi) = m\hbar \, F_l^m(\theta, \varphi) \tag{11}
$$

which proves part of the result being sought. To calculate the action of $\mathbf{L}^2$ on F_l^m, we use the fact that:

$$
\mathbf{L}^2 = \frac{1}{2}(L_+ L_- + L_- L_+) + L_z^2 \tag{12}
$$

Now, by using an argument analogous to the one which enabled us to write (10) and (11), relation (8) leads to:

$$
L_\pm F_l^m(\theta, \varphi) = \hbar \sqrt{l(l+1) - m(m \pm 1)} \, F_l^{m \pm 1}(\theta, \varphi) \tag{13}
$$

With this, (12) then yields:

$$
\begin{aligned}
\mathbf{L}^2 F_l^m(\theta, \varphi) &= \frac{\hbar^2}{2} \Big\{ [l(l+1) - m(m-1)] \\
&\qquad + [l(l+1) - m(m+1)] + 2m^2 \Big\} F_l^m(\theta, \varphi) \\
&= l(l+1)\hbar^2 \, F_l^m(\theta, \varphi)
\end{aligned} \tag{14}
$$

F_l^m, which, according to (11), is an eigenfunction of L_z with the eigenvalue $m\hbar$, is therefore also an eigenfunction of $\mathbf{L}^2$ with the eigenvalue $l(l+1)\hbar^2$. Since $\mathbf{L}^2$ and L_z form a C.S.C.O. in the space of functions of θ and φ alone, F_l^m is necessarily proportional to the spherical harmonic Y_l^m. Relation (13) enables us to show easily that the proportionality coefficient does not depend on m, and we find:

$$
F_l^m(\theta, \varphi) = \lambda(l) \, Y_l^m(\theta, \varphi) \tag{15}
$$

We must now calculate this proportionality coefficient $\lambda(l)$. To do so, we shall choose a particular direction in space, the Oz direction ($\theta = 0$, φ indeterminate). In this direction, all the spherical harmonics Y_l^m are zero, except those corresponding to $m = 0$ [since Y_l^m is proportional to $e^{im\varphi}$, they must be zero for the value of Y_l^m in the Oz direction to be defined uniquely; to see this, set $\theta = 0$ in (66), (67)

and (69) of complement A_{VI}]. When $m = 0$, the spherical harmonic $Y_l^m(\theta = 0, \varphi)$ is given by [*cf.* complement A_{VI}, relations (57) and (60)]:

$$Y_l^0(\theta = 0, \varphi) = \sqrt{\frac{2l + 1}{4\pi}} \tag{16}$$

Substituting these results into (4-a) and (9), we find:

$$F_l^{m=0}(\theta = 0, \varphi) = \langle l_1, l_2 ; 0, 0 \mid l, 0 \rangle \frac{\sqrt{(2l_1 + 1)(2l_2 + 1)}}{4\pi} \tag{17}$$

Furthermore, according to (15) and (16):

$$F_l^{m=0}(\theta = 0, \varphi) = \lambda(l)\sqrt{\frac{2l + 1}{4\pi}} \tag{18}$$

We therefore have:

$$\lambda(l) = \sqrt{\frac{(2l_1 + 1)(2l_2 + 1)}{4\pi(2l + 1)}} \langle l_1, l_2 ; 0, 0 \mid l, 0 \rangle \tag{19}$$

3. Expansion of a product of spherical harmonics ; the integral of a product of three spherical harmonics

With (9), (15) and (19) taken into account, relations (4-a) and (4-b) imply that:

$$Y_l^m(\Omega) = \left[\sqrt{\frac{(2l_1 + 1)(2l_2 + 1)}{4\pi(2l + 1)}} \langle l_1, l_2 ; 0, 0 \mid l, 0 \rangle\right]^{-1} \times \sum_{m_1}\sum_{m_2} \langle l_1, l_2 ; m_1, m_2 \mid l, m \rangle \, Y_{l_1}^{m_1}(\Omega)\, Y_{l_2}^{m_2}(\Omega) \tag{20}$$

and:

$$Y_{l_1}^{m_1}(\Omega) Y_{l_2}^{m_2}(\Omega) = \sum_{l=|l_1-l_2|}^{l_1+l_2} \sum_{m=-l}^{l} \sqrt{\frac{(2l_1 + 1)(2l_2 + 1)}{4\pi(2l + 1)}} \langle l_1, l_2 ; 0, 0 \mid l, 0 \rangle \times \langle l_1, l_2 ; m_1, m_2 \mid l, m \rangle \, Y_l^m(\Omega) \tag{21}$$

This last relation (in which the summation over m is actually unnecessary, since the only non-zero terms necessarily satisfy $m = m_1 + m_2$) is called *the spherical harmonic addition relation*★. According to formula (26) of complement B_X, the

★ In the particular case in which $l_2 = 1$, $m_2 = 0$[$Y_1^0(\theta, \varphi) \propto \cos\theta$], it yields formula (35) of complement A_{VI}.

Clebsch-Gordan coefficient $\langle l_1, l_2; 0, 0 \mid l, 0 \rangle$ is different from zero only if $l_1 + l_2 - l$ is even. The product $Y_{l_1}^{m_1}(\Omega) Y_{l_2}^{m_2}(\Omega)$ can therefore be expanded only in terms of spherical harmonics of orders:

$$l = l_1 + l_2 \quad , l_1 + l_2 - 2 \quad , l_1 + l_2 - 4 \quad , \dots, |l_1 - l_2| \tag{22}$$

In (21), the parity $(-1)^l$ of all the terms of the expansion on the right-hand side is thus indeed equal to $(-1)^{l_1 + l_2}$, the parity of the product which constitutes the left-hand side.

We can use the spherical harmonic addition relation to calculate the integral:

$$I = \int Y_{l_1}^{m_1}(\Omega)\, Y_{l_2}^{m_2}(\Omega)\, Y_{l_3}^{m_3}(\Omega)\, \mathrm{d}\Omega \tag{23}$$

Substituting (21) into (23), we find expressions of the type:

$$K(l, m; l_3, m_3) = \int Y_l^m(\Omega)\, Y_{l_3}^{m_3}(\Omega)\, \mathrm{d}\Omega \tag{24}$$

which, with the spherical harmonic complex conjugation relations and orthogonality relations taken into account [*cf.* complement A_{VI}, relations (55) and (45)], are equal to:

$$K(l, m; l_3, m_3) = (-1)^m \delta_{l l_3} \delta_{m, -m_3} \tag{25}$$

The value of I is therefore:

$$\int Y_{l_1}^{m_1}(\Omega)\, Y_{l_2}^{m_2}(\Omega)\, Y_{l_3}^{m_3}(\Omega)\, \mathrm{d}\Omega = (-1)^{m_3} \sqrt{\frac{(2l_1 + 1)(2l_2 + 1)}{4\pi(2l_3 + 1)}}$$

$$\times \langle l_1, l_2; 0, 0 \mid l_3, 0 \rangle \langle l_1, l_2; m_1, m_2 \mid l_3, -m_3 \rangle \tag{26}$$

This integral is, consequently, different from zero only if:

(*i*) $m_1 + m_2 + m_3 = 0$, as could have been predicted directly, since the integral over φ in (23) is $\int_0^{2\pi} \mathrm{d}\varphi\, \mathrm{e}^{i(m_1 + m_2 + m_3)\varphi} = \delta_{0, m_1 + m_2 + m_3}$.

(*ii*) a triangle can be formed with three line segments of lengths l_1, l_2 and l_3.

(*iii*) $l_1 + l_2 - l_3$ is even (necessary for $\langle l_1, l_2; 0, 0 \mid l_3, 0 \rangle$ to be different from zero), that is, if the product of the three spherical harmonics $Y_{l_1}^{m_1}$, $Y_{l_2}^{m_2}$ and $Y_{l_3}^{m_3}$ is an even function (obviously a necessary condition for its integral over all directions of space to be different from zero).

Relation (26) expresses, for the particular case of the spherical harmonics, a more general theorem, called the Wigner-Eckart theorem.

Complement D_X

VECTOR OPERATORS : THE WIGNER-ECKART THEOREM

In complement B_{VI} (*cf.* §5-b), we defined the concept of a scalar operator: it is an operator A which commutes with the angular momentum **J** of the system under study. An important property of these operators was then given (*cf.* §6-c-β of that complement): in a standard basis, $\{ | k, j, m \rangle \}$, the non-zero matrix elements $\langle k, j, m | A | k', j', m' \rangle$ of a scalar operator must satisfy the conditions $j = j'$ and $m = m'$; in addition, these elements do not depend on m*, which allows us to write:

$$\langle k, j, m | A | k', j', m' \rangle = a_j(k, k')\delta_{jj'}\delta_{mm'} \tag{1}$$

In particular, if the values of k and j are fixed, which amounts to considering the "restriction" of A (*cf.* complement B_{II}, § 3) to the subspace $\mathscr{E}(k, j)$ spanned by the $(2j + 1)$ kets $| k, j, m \rangle$ $(m = -j, -j + 1, ..., +j)$, we obtain a very simple $(2j + 1) \times (2j + 1)$ matrix: it is diagonal and all its elements are equal.

Now consider another scalar operator B. The matrix corresponding to it in the subspace $\mathscr{E}(k, j)$ possesses the same property: it is proportional to the unit matrix. Therefore, the matrix corresponding to B can easily be obtained from the one associated with A, by multiplying all the (diagonal) elements by the same constant. We therefore see that the restrictions of two scalar operators A and B to a subspace $\mathscr{E}(k, j)$ are always proportional. Denoting by $P(k, j)$ the projector onto the subspace $\mathscr{E}(k, j)$, we can write this result in the form**:

$$P(k, j) B P(k, j) = \lambda(k, j) P(k, j) A P(k, j) \tag{2}$$

The aim of this complement is to study another type of operator which possesses properties analogous to the ones just recalled : the vector operator. We shall see that if **V** and **V**′ are vectorial, their matrix elements also obey selection rules, which we shall establish. Moreover, we shall show that the restrictions of **V** and **V**′ to $\mathscr{E}(k, j)$ are always proportional:

$$P(k, j) \mathbf{V}' P(k, j) = \mu(k, j) P(k, j) \mathbf{V} P(k, j) \tag{3}$$

These results constitute the Wigner-Eckart theorem for vector operators.

* The proof of these properties was outlined in complement B_{VI}. We shall return to this point in this complement (§3-a) when we study the matrix elements of a scalar Hamiltonian.

** For two given operators A and B, the proportionality coefficient generally depends on the subspace $\mathscr{E}(k, j)$ chosen; this is why we write $\lambda(k, j)$.

COMMENT:

Actually, the Wigner-Eckart theorem is much more general. For example, it enables us to obtain selection rules for the matrix elements of **V** between two kets belonging to two different subspaces $\mathscr{E}(k, j)$ and $\mathscr{E}(k', j')$, or to relate these elements to the corresponding elements of **V**′. The Wigner-Eckart theorem can also be applied to a whole class of operators, of which scalars and vectors merely represent special cases: the irreducible tensor operators (*cf.* exercise 8 of complement G_X), which we shall not treat here.

1. Definition of vector operators; examples

In §5-c of complement B_{VI}, we showed that an observable **V** is a vector if its three components V_x, V_y and V_z in an orthonormal frame $Oxyz$ satisfy the following commutation relations:

$$[J_x, V_x] = 0 \tag{4-a}$$

$$[J_x, V_y] = i\hbar V_z \tag{4-b}$$

$$[J_x, V_z] = -i\hbar V_y \tag{4-c}$$

as well as those obtained by cyclic permutation of the indices x, y and z.

To give an idea of what this means, we shall give some examples of vector operators.

(*i*) The angular momentum **J** is itself a vector; replacing **V** by **J** in formulas (4), we simply obtain the relations which define an angular momentum (*cf.* chap. VI).

(*ii*) For a spinless particle whose state space is $\mathscr{E}_{\mathbf{r}}$, we have $\mathbf{J} = \mathbf{L}$. It is then simple to show that **R** and **P** are vector operators. We have, for example:

$$\begin{aligned} [L_x, X] &= [YP_z - ZP_y, X] = 0 \\ [L_x, Y] &= [-ZP_y, Y] = i\hbar Z \\ [L_x, Z] &= [YP_z, Z] = -i\hbar Y \end{aligned} \tag{5}$$

(*iii*) For a particle of spin **S**, whose state space is $\mathscr{E}_{\mathbf{r}} \otimes \mathscr{E}_s$, **J** is given by $\mathbf{J} = \mathbf{L} + \mathbf{S}$. In this case, the operators **L**, **S**, **R**, **P** are vectors. If we take into account the fact that all the spin operators (which act only in $\mathscr{E}_s$) commute with the orbital operators (which act only in $\mathscr{E}_{\mathbf{r}}$), the proof of these properties follows immediately from (*i*) and (*ii*).

On the other hand, operators of the type $\mathbf{L}^2$, $\mathbf{L} \cdot \mathbf{S}$, etc., are not vectors, but scalars [*cf.* comment (*i*) of complement B_{VI}, §5-c]. Other vector operators could, however, be constructed from those we have mentioned: $\mathbf{R} \times \mathbf{S}$, $(\mathbf{L} \cdot \mathbf{S})\mathbf{P}$, etc.

(*iv*) Consider the system (1) + (2), formed by the union of two systems: (1), of state space $\mathscr{E}_1$, and (2), of state space $\mathscr{E}_2$. If $\mathbf{V}(1)$ is an operator which acts only in $\mathscr{E}_1$, and if this operator is a vector [that is, satisfies commutation relations (4) with the angular momentum $\mathbf{J}_1$ of the first system], then the extension of $\mathbf{V}(1)$ into $\mathscr{E}_1 \otimes \mathscr{E}_2$ is also a vector. For example, for a two-electron system, the operators $\mathbf{L}_1$, $\mathbf{R}_1$, $\mathbf{S}_2$, etc. are vectors.

2. The Wigner-Eckart theorem for vector operators

a. NON-ZERO MATRIX ELEMENTS OF V IN A STANDARD BASIS

We introduce the operators V_+, V_-, J_+ and J_- defined by:

$$V_\pm = V_x \pm iV_y$$
$$J_\pm = J_x \pm iJ_y \tag{6}$$

Using relations (4), we can easily show that:

$$[J_x, V_\pm] = \mp \hbar V_z \tag{7-a}$$
$$[J_y, V_\pm] = -i\hbar V_z \tag{7-b}$$
$$[J_z, V_\pm] = \pm \hbar V_\pm \tag{7-c}$$

from which we can deduce the commutation relations of $J_\pm$ and $V_\pm$:

$$[J_+, V_+] = 0 \tag{8-a}$$
$$[J_+, V_-] = 2\hbar V_z \tag{8-b}$$
$$[J_-, V_+] = -2\hbar V_z \tag{8-c}$$
$$[J_-, V_-] = 0 \tag{8-d}$$

Now consider the matrix elements of **V** in a standard basis. We shall see that the fact that **V** is a vector implies that a large number of them are zero. First of all, we shall show that the matrix elements $\langle k, j, m \mid V_z \mid k', j', m' \rangle$ are necessarily zero whenever m is different from m'. It suffices to note that V_z and J_z commute [which follows, after cyclic permutation of the indices x, y and z, from relation (4-a)]. Therefore, the matrix elements of V_z between two vectors $\mid k, j, m \rangle$ corresponding to different eigenvalues $m\hbar$ of J_z are zero (*cf.* chap. II, § D-3-a-β).

For the matrix elements $\langle k, j, m \mid V_\pm \mid k', j', m' \rangle$ of $V_\pm$, we shall show that they are different from zero only if $m - m' = \pm 1$. Equation (7-c) indicates that:

$$J_z V_\pm = V_\pm J_z \pm \hbar V_\pm \tag{9}$$

Applying both sides of this relation to the ket $\mid k', j', m' \rangle$, we obtain:

$$\begin{aligned} J_z(V_\pm \mid k', j', m' \rangle) &= V_\pm J_z \mid k', j', m' \rangle \pm \hbar V_\pm \mid k', j', m' \rangle \\ &= (m' \pm 1)\hbar \, V_\pm \mid k', j', m' \rangle \end{aligned} \tag{10}$$

This relation indicates that $V_\pm \mid k', j', m' \rangle$ is an eigenvector of J_z with the eigenvalue $(m' \pm 1)\hbar$★. Since two eigenvectors of the Hermitian operator J_z associated

★ It should not be concluded that $V_\pm \mid k, j, m \rangle$ is necessarily proportional to $\mid k, j, m \pm 1 \rangle$. In fact, the argument we have given shows only that:

$$V_\pm \mid k, j, m \rangle = \sum_{k'} \sum_{j'} c_{k',j'} \mid k', j', m \pm 1 \rangle.$$

For us to be able to omit, for example, the summation over j', it would be necessary for $V_\pm$ to commute with $\mathbf{J}^2$, which is not generally the case.

with different eigenvalues are orthogonal, it follows that the scalar product $\langle k, j, m \mid V_{\pm} \mid k', j', m' \rangle$ is zero if $m \neq m' \pm 1$.

Summing up, the selection rules obtained for the matrix elements of **V** are as follows:

$$V_z \Longrightarrow \Delta m = m - m' = 0 \tag{11-a}$$
$$V_+ \Longrightarrow \Delta m = m - m' = +1 \tag{11-b}$$
$$V_- \Longrightarrow \Delta m = m - m' = -1 \tag{11-c}$$

From these results, we can easily deduce the forms of the matrices which represent the restrictions of the components of **V** inside a subspace $\mathscr{E}(k, j)$. The one associated with V_z is diagonal, and those associated with $V_{\pm}$ have matrix elements only just above and just below the principal diagonal.

b. PROPORTIONALITY BETWEEN THE MATRIX ELEMENTS OF J AND V INSIDE A SUBSPACE $\mathscr{E}(k, j)$

α. *Matrix elements of V_+ and V_-*

Expressing the fact that the matrix element of the commutator (8-a) between the bra $\langle k, j, m + 2 \mid$ and the ket $\mid k, j, m \rangle$ is zero, we have:

$$\langle k, j, m + 2 \mid J_+ V_+ \mid k, j, m \rangle = \langle k, j, m + 2 \mid V_+ J_+ \mid k, j, m \rangle \tag{12}$$

On both sides of this relation and between the operators J_+ and V_+, we insert the closure relation:

$$\sum_{k', j', m'} \mid k', j', m' \rangle \langle k', j', m' \mid = 1 \tag{13}$$

We thus obtain the matrix elements $\langle k, j, m \mid J_+ \mid k', j', m' \rangle$ of J_+; by the very construction of the standard basis $\{ \mid k, j, m \rangle \}$, they are different from zero only if $k = k', j = j'$ and $m = m' + 1$. The summations over k', j' and m' are therefore unnecessary in this case, and (12) can be written:

$$\begin{aligned} &\langle k, j, m + 2 \mid J_+ \mid k, j, m + 1 \rangle \langle k, j, m + 1 \mid V_+ \mid k, j, m \rangle \\ &\quad = \langle k, j, m + 2 \mid V_+ \mid k, j, m + 1 \rangle \langle k, j, m + 1 \mid J_+ \mid k, j, m \rangle \end{aligned} \tag{14}$$

that is:

$$\frac{\langle k, j, m + 1 \mid V_+ \mid k, j, m \rangle}{\langle k, j, m + 1 \mid J_+ \mid k, j, m \rangle} = \frac{\langle k, j, m + 2 \mid V_+ \mid k, j, m + 1 \rangle}{\langle k, j, m + 2 \mid J_+ \mid k, j, m + 1 \rangle} \tag{15}$$

(as long as the bras and kets appearing in this relation exist, that is, as long as $j - 2 \geqslant m \geqslant -j$, we can show immediately that neither of the denominators can

go to zero). Writing the relation thus obtained for $m = -j, -j+1, ..., j-2$, we get:

$$\frac{\langle k, j, -j+1 \mid V_+ \mid k, j, -j \rangle}{\langle k, j, -j+1 \mid J_+ \mid k, j, -j \rangle} = \frac{\langle k, j, -j+2 \mid V_+ \mid k, j, -j+1 \rangle}{\langle k, j, -j+2 \mid J_+ \mid k, j, -j+1 \rangle} = \cdots$$
$$= \frac{\langle k, j, m+1 \mid V_+ \mid k, j, m \rangle}{\langle k, j, m+1 \mid J_+ \mid k, j, m \rangle} = \cdots$$
$$= \frac{\langle k, j, j \mid V_+ \mid k, j, j-1 \rangle}{\langle k, j, j \mid J_+ \mid k, j, j-1 \rangle} \tag{16}$$

that is, if we call $\alpha_+(k, j)$ the common value of these ratios:

$$\langle k, j, m+1 \mid V_+ \mid k, j, m \rangle = \alpha_+(k, j) \langle k, j, m+1 \mid J_+ \mid k, j, m \rangle \tag{17}$$

where $\alpha_+(k, j)$ depends on k and on j, but not on m.

In addition, selection rule (11-b) implies that all the matrix elements $\langle k, j, m \mid V_+ \mid k, j, m' \rangle$ and $\langle k, j, m \mid J_+ \mid k, j, m' \rangle$ are zero if $\Delta m = m - m' \neq +1$. Therefore, whatever m and m', we have:

$$\langle k, j, m \mid V_+ \mid k, j, m' \rangle = \alpha_+(k, j) \langle k, j, m \mid J_+ \mid k, j, m' \rangle \tag{18-a}$$

This result expresses the fact that all the matrix elements of V_+ inside $\mathscr{E}(k, j)$ are proportional to those of J_+.

An analogous argument can be made by taking the matrix element of the commutator (8-d) between the bra $\langle k, j, m-2 \mid$ and the ket $\mid k, j, m \rangle$ to be zero. We are thus led to:

$$\langle k, j, m \mid V_- \mid k, j, m' \rangle = \alpha_-(k, j) \langle k, j, m \mid J_- \mid k, j, m' \rangle \tag{18-b}$$

an equation which expresses the fact that the matrix elements of V_- and J_- inside $\mathscr{E}(k, j)$ are proportional.

β. *Matrix elements of V_z*

To relate the matrix elements of V_z to those of J_z, we now place relation (8-c) between the bra $\langle k, j, m \mid$ and the ket $\mid k, j, m \rangle$:

$$\begin{aligned} -2\hbar \langle k, j, m \mid V_z \mid k, j, m \rangle &= \langle k, j, m \mid (J_- V_+ - V_+ J_-) \mid k, j, m \rangle \\ &= \hbar \sqrt{j(j+1) - m(m+1)} \langle k, j, m+1 \mid V_+ \mid k, j, m \rangle \\ &\quad - \hbar \sqrt{j(j+1) - m(m-1)} \langle k, j, m \mid V_+ \mid k, j, m-1 \rangle \end{aligned} \tag{19}$$

Using (18-a), we get :

$$\begin{aligned}\langle k, j, m \mid V_z \mid k, j, m \rangle &= -\frac{1}{2}\alpha_+(k, j)\Big\{\sqrt{j(j+1) - m(m+1)}\,\langle k, j, m+1 \mid J_+ \mid k, j, m \rangle \\ &\qquad - \sqrt{j(j+1) - m(m-1)}\,\langle k, j, m \mid J_+ \mid k, j, m-1 \rangle\Big\} \\ &= -\frac{\hbar}{2}\alpha_+(k, j)\,\{ j(j+1) - m(m+1) - j(j+1) + m(m-1) \}\end{aligned} \tag{20}$$

that is:

$$\langle k, j, m \mid V_z \mid k, j, m \rangle = m\hbar\, \alpha_+(k, j) \tag{21}$$

Similarly, an analogous argument based on (8-b) and (18-b) leads to:

$$\langle k, j, m \mid V_z \mid k, j, m \rangle = m\hbar\, \alpha_-(k, j) \tag{22}$$

Relations (21) and (22) show that $\alpha_+(k, j)$ and $\alpha_-(k, j)$ are necessarily equal; from now on, we shall call their common value $\alpha(k, j)$:

$$\alpha(k, j) = \alpha_+(k, j) = \alpha_-(k, j) \tag{23}$$

In addition, these relations imply that:

$$\langle k, j, m \mid V_z \mid k, j, m' \rangle = \alpha(k, j)\,\langle k, j, m \mid J_z \mid k, j, m' \rangle \tag{24}$$

γ. *Generalization to an arbitrary component of* **V**

Any component of **V** is a linear combination of V_+, V_- and V_z. Consequently, using relation (23), we can summarize (18-a), (18-b) and (24) by writing :

$$\langle k, j, m \mid \mathbf{V} \mid k, j, m' \rangle = \alpha(k, j)\,\langle k, j, m \mid \mathbf{J} \mid k, j, m' \rangle \tag{25}$$

Therefore, *inside $\mathscr{E}(k, j)$, all the matrix elements of* **V** *are proportional to those of* **J**. This result expresses the Wigner-Eckart theorem, for a special case. Introducing the "restrictions" of **V** and **J** to $\mathscr{E}(k, j)$ (*cf.* complement B_{II}, §3), we can also write it :

$$P(k, j)\,\mathbf{V}\,P(k, j) = \alpha(k, j)\,P(k, j)\,\mathbf{J}\,P(k, j) \tag{26}$$

COMMENT:

J commutes with $P(k, j)$ [*cf.* (27)]; since, moreover

$$[P(k, j)]^2 = P(k, j)$$

we can omit either one of the two projectors $P(k, j)$ on the right-hand side of (26).

c. CALCULATION OF THE PROPORTIONALITY CONSTANT; THE PROJECTION THEOREM

Consider the operator $\mathbf{J} \cdot \mathbf{V}$; its restriction to $\mathscr{E}(k, j)$ is $P(k, j)\mathbf{J} \cdot \mathbf{V}P(k, j)$. To transform this expression, we can use the fact that:

$$[\mathbf{J}, P(k, j)] = \mathbf{0} \tag{27}$$

a relation that can easily be verified by showing that the action of the commutators $[J_z, P(k, j)]$ and $[J_\pm, P(k, j)]$ on any ket of the $\{ | k, j, m \rangle \}$ basis yields zero. Using (26), we then get:

$$\begin{aligned} P(k,j)\, \mathbf{J} \cdot \mathbf{V}\, P(k,j) &= \mathbf{J} \cdot [P(k,j)\, \mathbf{V}\, P(k,j)] \\ &= \alpha(k,j)\, \mathbf{J}^2 P(k,j) \\ &= \alpha(k,j)\, j(j+1)\hbar^2\, P(k,j) \end{aligned} \tag{28}$$

The restriction to the space $\mathscr{E}(k, j)$ of the operator $\mathbf{J} \cdot \mathbf{V}$ is therefore equal to the identity operator★ multiplied by $\alpha(k, j)j(j + 1)\hbar^2$. Therefore, if $| \psi_{k,j} \rangle$ denotes an arbitrary normalized state belonging to the subspace $\mathscr{E}(k, j)$, the mean value $\langle \mathbf{J} \cdot \mathbf{V} \rangle_{k,j}$ of $\mathbf{J} \cdot \mathbf{V}$ is independent of the ket $| \psi_{k,j} \rangle$ chosen, since:

$$\langle \mathbf{J} \cdot \mathbf{V} \rangle_{k,j} = \langle \psi_{k,j} | \mathbf{J} \cdot \mathbf{V} | \psi_{k,j} \rangle = \alpha(k,j)\, j(j+1)\hbar^2 \tag{29}$$

If we substitute this relation into (26), we see that, *inside the subspace* $\mathscr{E}(k, j)$★★:

$$\mathbf{V} = \frac{\langle \mathbf{J} \cdot \mathbf{V} \rangle_{k,j}}{\langle \mathbf{J}^2 \rangle_{k,j}} \mathbf{J} = \frac{\langle \mathbf{J} \cdot \mathbf{V} \rangle_{k,j}}{j(j+1)\hbar^2} \mathbf{J} \tag{30}$$

This result is often called the "projection theorem". Whatever the physical system being studied, as long as we are concerned only with states belonging to the same subspace $\mathscr{E}(k, j)$, we can assume that all vector operators are proportional to $\mathbf{J}$.

We can give the following classical physical interpretation of this property: if $\mathbf{j}$ denotes the total angular momentum of any isolated physical system, all the physical quantities attached to the system rotate about $\mathbf{j}$, which is a constant vector (*cf.* fig. 1). In particular, for a vector quantity $\mathbf{v}$, all that remains after averaging over time is its projection $\mathbf{v}_{//}$ onto $\mathbf{j}$, that is, a vector parallel to $\mathbf{j}$, given by:

$$\mathbf{v}_{//} = \frac{\mathbf{j} \cdot \mathbf{v}}{\mathbf{j}^2} \mathbf{j} \tag{31}$$

a formula which is indeed analogous to (30).

★ Since $\mathbf{J} \cdot \mathbf{V}$ is a scalar, the fact that its restriction is proportional to the identity operator was to be expected.

★★ We shall say that an operator relation is valid only inside a given subspace when it is actually valid only for the restrictions of the operators being considered to this subspace. To be completely rigorous, we should therefore have to place both sides of relation (30) between two projectors $P(k, j)$.

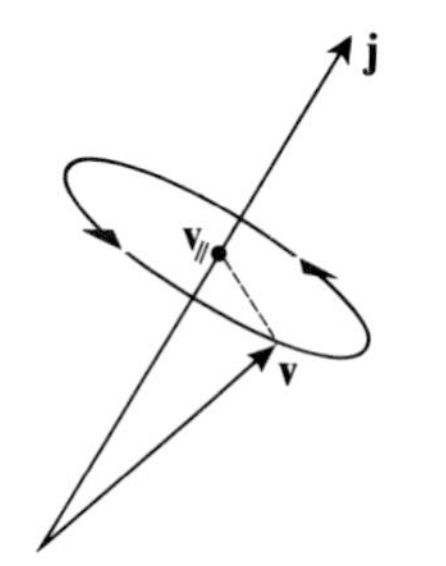

FIGURE 1

Classical interpretation of the projection theorem: since the vector v rotates very rapidly about the total angular momentum j, only its static component $\mathbf{v}_{//}$ should be taken into account.

COMMENTS:

(*i*) It cannot be deduced from (30) that, in the total state space [the direct sum of all the subspaces $\mathscr{E}(k, j)$], **V** and **J** are proportional. It must be noted that the proportionality constant $\alpha(k, j)$ (or $\langle \mathbf{J} \cdot \mathbf{V} \rangle_{k,j}$) depends on the subspace $\mathscr{E}(k, j)$ chosen. Moreover, any vector operator **V** may possess non-zero matrix elements between kets belonging to different subspaces $\mathscr{E}(k, j)$, while the corresponding elements of **J** are always zero.

(*ii*) Consider a second vector operator **W**. Its restriction inside $\mathscr{E}(k, j)$ is proportional to **J**, and therefore also to the restriction of **V**. Therefore, *inside a subspace $\mathscr{E}(k, j)$, all vector operators are proportional.*

However, to calculate the proportionality coefficient between **V** and **W**, we cannot simply replace **J** by **W** in (30) (which would give the value $\langle \mathbf{V} \cdot \mathbf{W} \rangle_{k,j} / \langle \mathbf{W}^2 \rangle_{k,j}$). In the proof leading to relation (30), we used the fact that **J** commutes with $P(k, j)$ in (28), which is not generally the case for **W**. To calculate this proportionality coefficient correctly, we note that, inside the subspace $\mathscr{E}(k, j)$:

$$\mathbf{W} = \frac{\langle \mathbf{J} \cdot \mathbf{W} \rangle_{k,j}}{\langle \mathbf{J}^2 \rangle_{k,j}} \mathbf{J} \tag{32}$$

This yields, with (30) taken into account:

$$\mathbf{V} = \frac{\langle \mathbf{J} \cdot \mathbf{V} \rangle_{k,j}}{\langle \mathbf{J} \cdot \mathbf{W} \rangle_{k,j}} \mathbf{W} \tag{33}$$

3. Application: calculation of the Landé g_J factor of an atomic level

In this section, we shall apply the Wigner-Eckart theorem to the calculation of the effect of a magnetic field **B** on the energy levels of an atom. We shall see that this theorem considerably simplifies the calculations and enables us to predict, in a very general way, that the magnetic field removes degeneracies, causing equidistant levels to appear (to first order in B). The energy difference of these states is proportional to B and to a constant g_J (the Landé factor) which we shall calculate.

Let **L** be the total orbital angular momentum of the electrons of an atom (the sum of their individual orbital angular momenta $\mathbf{L}_i$), and let **S** be their total spin angular momentum (the sum of their individual spins $\mathbf{S}_i$). The total internal angular momentum of the atom (assuming the spin of the nucleus to be zero) is:

$$\mathbf{J} = \mathbf{L} + \mathbf{S} \tag{34}$$

In the absence of a magnetic field, we call H_0 the Hamiltonian of the atom; H_0 commutes with $\mathbf{J}$★. We shall assume that H_0, $\mathbf{L}^2$, $\mathbf{S}^2$, $\mathbf{J}^2$ and J_z form a C.S.C.O., and we shall call $| E_0, L, S, J, M \rangle$ their common eigenvectors, of eigenvalues E_0, $L(L+1)\hbar^2$, $S(S+1)\hbar^2$, $J(J+1)\hbar^2$ and $M\hbar$, respectively.

This hypothesis is valid for a certain number of light atoms for which the angular momentum coupling is of the $\mathbf{L}\,.\,\mathbf{S}$ type (*cf.* complement B_{XIV}). However, for other atoms, which have a different type of coupling (for example, the rare gases other than helium), this is not the case. Calculations based on the Wigner-Eckart theorem, similar to those presented here, can then be performed, and the central physical ideas remain the same. For the sake of simplicity, we shall confine ourselves here to the case in which L and S are actually good quantum numbers for the atomic state under study.

a. ROTATIONAL DEGENERACY; MULTIPLETS

Consider the ket $J_\pm | E_0, L, S, J, M \rangle$. According to the hypotheses set forth above, $J_\pm$ commutes with H_0; therefore, $J_\pm | E_0, L, S, J, M \rangle$ is an eigenvector of H_0 with the eigenvalue E_0. Furthermore, in accordance with the general properties of angular momenta and their addition, we have:

$$J_\pm | E_0, L, S, J, M \rangle = \hbar\sqrt{J(J+1) - M(M \pm 1)}\, | E_0, L, S, J, M \pm 1 \rangle \tag{35}$$

This relation shows that, starting with a state $| E_0, L, S, J, M \rangle$, we can construct others with the same energy: those for which $-J \leqslant M \leqslant J$. It follows that the eigenvalue E_0 is necessarily at least $(2J+1)$-fold degenerate. This is an essential degeneracy, since it is related to the rotational invariance of H_0 (an accidental degeneracy may also be present). In atomic physics, the corresponding $(2J+1)$-fold degenerate energy level is called a multiplet. The eigensubspace associated with it, spanned by the kets $| E_0, L, S, J, M \rangle$ with $M = J, J-1, ..., -J$, will be written $\mathscr{E}(E_0, L, S, J)$.

b. REMOVAL OF THE DEGENERACY BY A MAGNETIC FIELD; ENERGY DIAGRAM

In the presence of a magnetic field $\mathbf{B}$ parallel to Oz, the Hamiltonian becomes (*cf.* complement D_{VII}):

$$H = H_0 + H_1 \tag{36}$$

★ This general property follows from the invariance of the energy of the atom under a rotation of all the electrons, performed about an axis passing through the origin (which is the position of the nucleus, assumed to be motionless). H_0, which is invariant under rotation, therefore commutes with $\mathbf{J}$ (H_0 is a scalar operator; *cf.* complement B_{VI}, §5-b).

with:

$$H_1 = \omega_L(L_z + 2S_z) \tag{37}$$

(the factor 2 before S_z arises from the electron spin gyromagnetic ratio). The "Larmor angular frequency" ω_L of the electron is defined in terms of its mass m and its charge q by:

$$\omega_L = -\frac{qB}{2m} = -\frac{\mu_B}{\hbar}B \tag{38}$$

(where $\mu_B = q\hbar/2m$ is the Bohr magneton).

To calculate the effect of the magnetic field on the energy levels of the atom, we shall consider only the matrix elements of H_1 inside the subspace $\mathscr{E}(E_0, L, S, J)$ associated with the multiplet under study. Perturbation theory, which will be explained in chapter XI, justifies this procedure when B is not too large.

Inside the subspace $\mathscr{E}(E_0, L, S, J)$, we have, according to the projection theorem (§ 2-c):

$$\mathbf{L} = \frac{\langle \mathbf{L}\,.\,\mathbf{J} \rangle_{E_0,L,S,J}}{J(J+1)\hbar^2}\,\mathbf{J} \tag{39-a}$$

$$\mathbf{S} = \frac{\langle \mathbf{S}\,.\,\mathbf{J} \rangle_{E_0,L,S,J}}{J(J+1)\hbar^2}\,\mathbf{J} \tag{39-b}$$

where $\langle \mathbf{L}\,.\,\mathbf{J} \rangle_{E_0,L,S,J}$ and $\langle \mathbf{S}\,.\,\mathbf{J} \rangle_{E_0,L,S,J}$ denote respectively the mean values of the operators $\mathbf{L}\,.\,\mathbf{J}$ and $\mathbf{S}\,.\,\mathbf{J}$ for the states of the system belonging to $\mathscr{E}(E_0, L, S, J)$. Now, we can write:

$$\mathbf{L}\,.\,\mathbf{J} = \mathbf{L}\,.\,(\mathbf{L} + \mathbf{S}) = \mathbf{L}^2 + \frac{1}{2}(\mathbf{J}^2 - \mathbf{L}^2 - \mathbf{S}^2) \tag{40-a}$$

as well as:

$$\mathbf{S}\,.\,\mathbf{J} = \mathbf{S}\,.\,(\mathbf{L} + \mathbf{S}) = \mathbf{S}^2 + \frac{1}{2}(\mathbf{J}^2 - \mathbf{L}^2 - \mathbf{S}^2) \tag{40-b}$$

It follows that:

$$\langle \mathbf{L}\,.\,\mathbf{J} \rangle_{E_0,L,S,J} = L(L+1)\hbar^2 + \frac{\hbar^2}{2}[J(J+1) - L(L+1) - S(S+1)] \tag{41-a}$$

and:

$$\langle \mathbf{S}\,.\,\mathbf{J} \rangle_{E_0,L,S,J} = S(S+1)\hbar^2 + \frac{\hbar^2}{2}[J(J+1) - L(L+1) - S(S+1)] \tag{41-b}$$

Relations (41), substituted into (39) and then into (37), show that, inside the subspace $\mathscr{E}(E_0, L, S, J)$, the operator H_1 is given by:

$$H_1 = g_J \omega_L J_z \tag{42}$$

where the Landé g_J factor of the multiplet under consideration is equal to:

$$g_J = \frac{3}{2} + \frac{S(S+1) - L(L+1)}{2J(J+1)} \tag{43}$$

Relation (42) implies that the eigenstates of H_1 inside the eigensubspace $\mathscr{E}(E_0, L, S, J)$ are simply the basis vectors $| E_0, L, S, J, M \rangle$, with the eigenvalues:

$$E_1(M) = g_J M \hbar \omega_L \tag{44}$$

We see that the magnetic field completely removes the degeneracy of the multiplet. As is shown by the diagram in figure 2, a set of $(2J + 1)$ equidistant levels appears, each one corresponding to one of the possible values of M. Such a diagram permits generalization of our earlier study of the polarization and frequency of optical lines emitted by a fictitious atom with a single spinless electron (the "normal" Zeeman effect; *cf.* complement D_{VII}), to the case of atoms with several electrons whose spins must be taken into account.

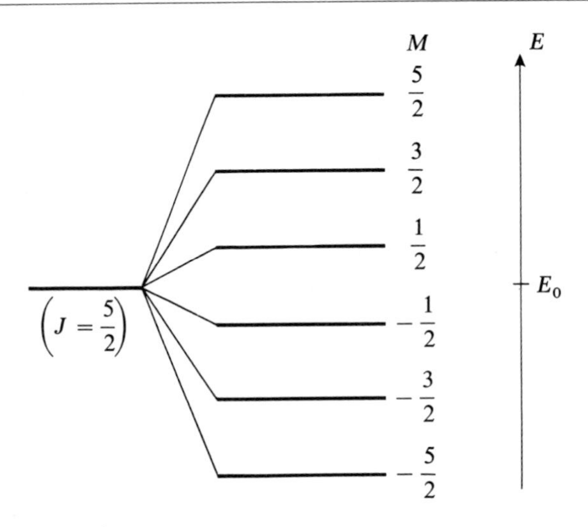

FIGURE 2

Energy diagram showing the removal of the $(2J + 1)$-fold degeneracy of a multiplet (here $J = 5/2$) by a static magnetic field B. The distance between two adjacent levels is proportional to $|\mathbf{B}|$ and to the Landé g_J factor.

References and suggestions for further reading :

Tensor operators: Schiff (1.18), §28; Messiah (1.17), chap. XIII, §VI; Edmonds (2.21), chap. 5; Rose (2.19), chap. 5; Meijer and Bauer (2.18), chap. 6.

Complement E_X

ELECTRIC MULTIPOLE MOMENTS

Consider a system $\mathscr{S}$ composed of N charged particles placed in a given electrostatic potential $U(\mathbf{r})$. We shall show in this complement how to calculate the interaction energy of the system $\mathscr{S}$ with the potential $U(\mathbf{r})$ by introducing the electric multipole moments of $\mathscr{S}$. First of all, we shall begin by recalling how these moments are introduced in classical physics. Then we shall construct the corresponding quantum mechanical operators, and we shall see how, in a large number of cases, their use considerably simplifies the study of the electrical properties of a quantum mechanical system. This is because these operators possess general properties which are independent of the system being studied, satisfying in particular certain selection rules. For example, if the state of the system $\mathscr{S}$ being studied has an angular momentum j [i.e. is an eigenvector of $\mathbf{J}^2$ with the eigenvalue $j(j + 1)\hbar^2$], we shall see that the mean values of all multipole operators of order higher than $2j$ are necessarily zero.

1. Definition of multipole moments

a. EXPANSION OF THE POTENTIAL ON THE SPHERICAL HARMONICS

For the sake of simplicity, we begin by studying a system $\mathscr{S}$ composed of a single particle, of charge q and position $\mathbf{r}$, placed in the potential $U(\mathbf{r})$. We shall then generalize the results obtained to N-particle systems.

α. *Case of a single particle*

In classical physics, the potential energy of the particle is:

$$V(\mathbf{r}) = qU(\mathbf{r}) \tag{1}$$

Since the spherical harmonics form a basis for functions of θ and φ, we can expand $U(\mathbf{r})$ in the form:

$$U(\mathbf{r}) = \sum_{l=0}^{\infty} \sum_{m=-l}^{l} f_{l,m}(r)\, Y_l^m(\theta, \varphi) \tag{2}$$

We shall assume the charges creating the electrostatic potential to be placed outside the region of space in which the particle being studied can be found. In this whole region, we then have:

$$\Delta U(\mathbf{r}) = 0 \tag{3}$$

Now, we know [*cf.* relation (A-15) of chapter VII] that the Laplacian Δ is related to the differential operator $\mathbf{L}^2$ acting on the angular variables θ and φ by:

$$\Delta = \frac{1}{r}\frac{\partial^2}{\partial r^2} r - \frac{\mathbf{L}^2}{\hbar^2 r^2} \tag{4}$$

Also, the very definition of the spherical harmonics implies that:

$$\mathbf{L}^2 Y_l^m(\theta, \varphi) = l(l+1)\hbar^2 Y_l^m(\theta, \varphi) \tag{5}$$

It is therefore easy to calculate the Laplacian of expansion (2). If we write, using (3), that each of the terms thus obtained is zero, we get:

$$\left[\frac{1}{r}\frac{\partial^2}{\partial r^2} r - \frac{l(l+1)}{r^2}\right] f_{l,m}(r) = 0 \tag{6}$$

This equation has two linearly independent solutions, r^l and $r^{-(l+1)}$. Since $U(\mathbf{r})$ is not infinite for $r = 0$, we must choose:

$$f_{l,m}(r) = \sqrt{\frac{4\pi}{2l+1}}\, c_{l,m}\, r^l \tag{7}$$

where the $c_{l,m}$ are coefficients which depend on the potential under consideration (the factor $\sqrt{4\pi/(2l+1)}$ is introduced for convenience, as will be seen later).

We can therefore write (2) in the form:

$$V(\mathbf{r}) = qU(\mathbf{r}) = \sum_{l=0}^{\infty} \sum_{m=-l}^{l} c_{l,m}\, \mathcal{Q}_l^m(\mathbf{r}) \tag{8}$$

where the functions $\mathcal{Q}_l^m(\mathbf{r})$ are defined by their expressions in spherical coordinates:

$$\mathcal{Q}_l^m(\mathbf{r}) = q\sqrt{\frac{4\pi}{2l+1}}\, r^l\, Y_l^m(\theta, \varphi) \tag{9}$$

In quantum mechanics, the same type of expansion is possible; the potential energy operator of the particle is $V(\mathbf{R}) = qU(\mathbf{R})$, whose matrix elements in the $\{ | \mathbf{r} \rangle \}$ representation are (*cf.* complement B_{II}, §4-b):

$$\langle \mathbf{r} | qU(\mathbf{R}) | \mathbf{r}' \rangle = qU(\mathbf{r})\, \delta(\mathbf{r} - \mathbf{r}') \tag{10}$$

Expansion (8) then yields:

$$V(\mathbf{R}) = qU(\mathbf{R}) = \sum_{l=0}^{\infty} \sum_{m=-l}^{l} c_{l,m}\, Q_l^m \tag{11}$$

where the operators Q_l^m are defined by:

$$\langle \mathbf{r} \mid Q_l^m \mid \mathbf{r}' \rangle = \mathcal{Q}_l^m(\mathbf{r})\, \delta(\mathbf{r} - \mathbf{r}') = q \sqrt{\frac{4\pi}{2l+1}}\, r^l\, Y_l^m(\theta, \varphi)\, \delta(\mathbf{r} - \mathbf{r}') \tag{12}$$

The Q_l^m are called "electric multipole operators".

β. *Generalization to N particles*

Now consider N particles, with positions $\mathbf{r}_1, \mathbf{r}_2, ..., \mathbf{r}_N$ and charges $q_1, q_2, ..., q_N$. Their coupling energy with the external potential $U(\mathbf{r})$ is:

$$V(\mathbf{r}_1, \mathbf{r}_2, ..., \mathbf{r}_N) = \sum_{n=1}^{N} q_n U(\mathbf{r}_n) \tag{13}$$

The argument of the preceding section can immediately be generalized to show that:

$$V(\mathbf{r}_1, \mathbf{r}_2, ..., \mathbf{r}_N) = \sum_{l=0}^{\infty} \sum_{m=-l}^{l} c_{l,m}\, \mathcal{Q}_l^m(\mathbf{r}_1, \mathbf{r}_2, ..., \mathbf{r}_N) \tag{14}$$

where the coefficients $c_{l,m}$ [which depend on the potential $U(\mathbf{r})$] have the same values as in the preceding section, and the functions $\mathcal{Q}_l^m$ are defined by their values in polar coordinates:

$$\mathcal{Q}_l^m(\mathbf{r}_1, \mathbf{r}_2, ..., \mathbf{r}_N) = \sqrt{\frac{4\pi}{2l+1}} \sum_{n=1}^{N} q_n\, (r_n)^l\, Y_l^m(\theta_n, \varphi_n) \tag{15}$$

(θ_n and φ_n are the polar angles of $\mathbf{r}_n$). The multipole moments of the total system are therefore simply the sums of the moments associated with each of the particles.

Similarly, in quantum mechanics, the coupling energy of the N particles with the external potential is described by the operator:

$$V(\mathbf{R}_1, \mathbf{R}_2, ..., \mathbf{R}_N) = \sum_{l=0}^{\infty} \sum_{m=-l}^{l} c_{l,m}\, Q_l^m \tag{16}$$

with:

$$\langle \mathbf{r}_1, \mathbf{r}_2, ..., \mathbf{r}_N \mid Q_l^m \mid \mathbf{r}_1', \mathbf{r}_2', ..., \mathbf{r}_N' \rangle = \mathcal{Q}_l^m(\mathbf{r}_1, \mathbf{r}_2, ..., \mathbf{r}_N)\, \delta(\mathbf{r}_1 - \mathbf{r}_1')\, \delta(\mathbf{r}_2 - \mathbf{r}_2') \ldots \delta(\mathbf{r}_N - \mathbf{r}_N') \tag{17}$$

b. PHYSICAL INTERPRETATION OF MULTIPOLE OPERATORS

α. *The operator Q_0^0; the total charge of the system*

Since Y_0^0 is a constant ($Y_0^0 = 1/\sqrt{4\pi}$), definition (15) implies that:

$$\mathcal{Q}_0^0 = \sum_{n=1}^{N} q_n \tag{18}$$

The operator Q_0^0 is therefore a constant which is equal to the total charge of the system.

The first term of expansion (14) therefore gives the coupling energy of the system with the potential $U(\mathbf{r})$, assuming all the particles to be situated at the origin O. This is obviously a good approximation if $U(\mathbf{r})$ does not vary very much in relative value over distances comparable to those separating the various particles from O (if the system $\mathscr{S}$ is centered at O, this distance is of the order of the dimensions of $\mathscr{S}$). Furthermore, there exists a special case in which expansion (14) is rigorously given by its first term : the case where the potential $U(\mathbf{r})$ is uniform, and therefore proportional to the spherical harmonic $l = 0$.

β. *The operators Q_1^m; the electric dipole moment*

According to (15) and the expression for the spherical harmonics Y_1^m [*cf.* complement A_{VI}, equations (32)], we have:

$$\begin{cases} \mathscr{Q}_1^1 = -\dfrac{1}{\sqrt{2}} \sum_n q_n\,(x_n + iy_n) \\ \mathscr{Q}_1^0 = \sum_n q_n\, z_n \\ \mathscr{Q}_1^{-1} = \dfrac{1}{\sqrt{2}} \sum_n q_n\,(x_n - iy_n) \end{cases} \tag{19}$$

These three quantities can be considered to be the components of a vector on the complex basis of three vectors $\mathbf{e}_1$, $\mathbf{e}_0$ and $\mathbf{e}_{-1}$:

$$\boldsymbol{\mathscr{D}} = -\,\mathscr{Q}_1^{-1}\mathbf{e}_1 + \mathscr{Q}_1^0\,\mathbf{e}_0 - \mathscr{Q}_1^1\,\mathbf{e}_{-1} \tag{20}$$

with:

$$\mathbf{e}_1 = -\frac{1}{\sqrt{2}}(\mathbf{e}_x + i\mathbf{e}_y)\,; \quad \mathbf{e}_0 = \mathbf{e}_z\,; \quad \mathbf{e}_{-1} = \frac{1}{\sqrt{2}}(\mathbf{e}_x - i\mathbf{e}_y) \tag{21}$$

(where $\mathbf{e}_x$, $\mathbf{e}_y$, $\mathbf{e}_z$ are the unit vectors of the Ox, Oy and Oz axes). The components of this vector $\boldsymbol{\mathscr{D}}$ on the $Oxyz$ axes are then:

$$\begin{aligned} \mathscr{Q}_1^x &= \frac{1}{\sqrt{2}}[\mathscr{Q}_1^{-1} - \mathscr{Q}_1^1] = \sum_n q_n\, x_n \\ \mathscr{Q}_1^y &= \frac{i}{\sqrt{2}}[\mathscr{Q}_1^{-1} + \mathscr{Q}_1^1] = \sum_n q_n\, y_n \\ \mathscr{Q}_1^z &= \mathscr{Q}_1^0 \qquad\qquad\qquad = \sum_n q_n\, z_n \end{aligned} \tag{22}$$

We recognize the three components of the total electric dipole moment of the system $\mathscr{S}$ with respect to the origin O:

$$\boldsymbol{\mathscr{D}} = \sum_{n=1}^{N} q_n\,\mathbf{r}_n \tag{23}$$

The operators Q_1^m are therefore actually the components of the electric dipole $\mathbf{D} = \sum_n q_n \mathbf{R}_n$.

Relations (19) enable us, moreover, to write the $l = 1$ terms of expansion (14) in the form:

$$\sum_{m=-1}^{+1} c_{1,m}\, \mathcal{Q}_1^m = -\frac{1}{\sqrt{2}}(c_{1,1} - c_{1,-1}) \sum_n q_n\, x_n - \frac{i}{\sqrt{2}}(c_{1,1} + c_{1,-1}) \sum_n q_n\, y_n + c_{1,0} \sum_n q_n\, z_n \quad (24)$$

We shall now show that the combinations of the coefficients $c_{1,m}$ which appear in this expression are none other than the components of the gradient of the potential $U(\mathbf{r})$ at $\mathbf{r} = \mathbf{0}$. If we take the gradient of expansion (8) of $U(\mathbf{r})$, the $l = 0$ term (which is constant) disappears; the $l = 1$ term can be put into a form analogous to (24) and gives:

$$[\boldsymbol{\nabla} U(\mathbf{r})]_{\mathbf{r}=0} = -\frac{1}{\sqrt{2}}(c_{1,1} - c_{1,-1})\, \mathbf{e}_x - \frac{i}{\sqrt{2}}(c_{1,1} + c_{1,-1})\, \mathbf{e}_y + c_{1,0}\, \mathbf{e}_z \quad (25)$$

As for the $l > 1$ terms of (8), they are polynomials in x, y, z of degree higher than 1 (*cf.* §§ γ and δ below) which make no contribution to the gradient at $\mathbf{r} = \mathbf{0}$. The $l = 1$ term of expansion (14) can therefore be written, using (23) and (25):

$$\left(\sum_{n=1}^{N} q_n \mathbf{r}_n \right) \cdot (\boldsymbol{\nabla} U)_{\mathbf{r}=0} = -\, \mathcal{D} \,.\, \mathcal{E}(\mathbf{r} = \mathbf{0}) \quad (26)$$

where:

$$\mathcal{E}(\mathbf{r}) = -\, \boldsymbol{\nabla}\, U(\mathbf{r}) \quad (27)$$

is the electric field at the point $\mathbf{r}$. Thus we recognize (26) as the well-known expression for the coupling energy between an electric dipole and the field $\mathcal{E}$.

COMMENTS:

(*i*) In physics, we often deal with systems whose total charge is zero (atoms, for example). $\mathcal{Q}_0^0$ is then equal to zero, and the first multipole operator entering into expansion (14) is the electric dipole moment. This expansion can often be limited to the $l = 1$ terms [hence expression (26)], since the terms for which $l \geqslant 2$ are generally much smaller (this is the case, for example, if the electric field varies little over distances comparable to the distances of the particles from the origin; the $l \geqslant 2$ terms are, furthermore, rigorously zero in a special case: the case in which the electric field is uniform [*cf.* §§γ and δ below)].

(*ii*) For a system $\mathscr{S}$ composed of two particles of opposite charge $+q$ and $-q$ (an electric dipole), the dipole moment $\mathscr{D}$ is:

$$\mathscr{D} = q(\mathbf{r}_1 - \mathbf{r}_2) \tag{28}$$

Its value, which is related to the position of the "relative particle" (*cf.* chap. VII, § B) associated with the system $\mathscr{S}$, therefore does not depend on the choice of the origin O. Actually this is a more general property: it is simple to show that the electric dipole moment of any electrically neutral system $\mathscr{S}$ is independent of the origin O chosen.

γ. *The operators Q_2^m; the electric quadrupole moment*

Using the explicit expression for the Y_2^m [*cf.* complement A_{VI}, relations (33)], we could show without difficulty that:

$$\begin{cases} \mathscr{Q}_2^{\pm 2} = \dfrac{\sqrt{6}}{4} \displaystyle\sum_n q_n \, (x_n \pm i y_n)^2 \\ \mathscr{Q}_2^{\pm 1} = \mp \dfrac{\sqrt{6}}{2} \displaystyle\sum_n q_n \, z_n (x_n \pm i y_n) \\ \mathscr{Q}_2^{0} = \dfrac{1}{2} \displaystyle\sum_n q_n \, (3 z_n^2 - r_n^2) \end{cases} \tag{29}$$

In this way, we obtain the five components of the electric quadrupole moment of the system $\mathscr{S}$. While the total charge of $\mathscr{S}$ is a scalar, and its dipole moment is a vector $\mathscr{D}$, it can be shown that the quadrupole moment is a second-rank tensor. In addition, an argument similar to the one in § β would enable us to write the $l = 2$ terms of expansion (14) in the form:

$$\sum_{m=-2}^{+2} c_{2,m} \, \mathscr{Q}_2^m = \sum_{i,j} \left[\frac{\partial^2 U}{\partial x^i \, \partial x^j} \right]_{\mathbf{r}=\mathbf{0}} \sum_{n=1}^{N} q_n \, x_n^i x_n^j \tag{30}$$

(with $x^i, x^j = x, y$ or z). These terms describe the coupling between the electric quadrupole moment of the system $\mathscr{S}$ and the gradient of the field $\mathscr{E}(\mathbf{r})$ at the point $\mathbf{r} = \mathbf{0}$.

δ. *Generalization: the electric l-pole moment*

We could generalize the preceding arguments and show from the general expression for the spherical harmonics [*cf.* complement A_{VI}, relations (26) or (30)] that:

– the quantities $\mathscr{Q}_l^m$ are polynomials (which are homogeneous in x, y and z) of degree l.

– the contribution to expansion (14) of the l terms involves lth order derivatives of the potential $U(\mathbf{r})$, evaluated at $\mathbf{r} = \mathbf{0}$.

Expression (14) for the potential can thus be seen to be a Taylor series expansion in the neighborhood of the origin. As the variation of the potential $U(\mathbf{r})$

in the region about $\mathscr{S}$ becomes more complicated, higher order terms must be retained in the expansion. For example, if $U(\mathbf{r})$ is constant, we have seen that the $l = 0$ term is the only one involved. If the field $\mathscr{E}(\mathbf{r})$ is uniform, the $l = 1$ terms must be added to the expansion. If it is the gradient of the field $\mathscr{E}$ which is uniform, we must have $l \leqslant 2$, and so forth.

c. PARITY OF MULTIPOLE OPERATORS

Finally, we shall consider the parity of the Q_l^m. We know that the parity of Y_l^m is $(-1)^l$ [*cf.* chap. VI, relation (D-28)]. Therefore (*cf.* complement F_{II}, §2-a), the electric multipole operator Q_l^m has a definite parity, equal to $(-1)^l$, independent of m. This property will prove useful in what follows.

d. ANOTHER WAY TO INTRODUCE MULTIPOLE MOMENTS

We shall consider the same system of N charged particles as in §1-a. However, instead of considering the interaction energy of this system with a given external potential $U(\mathbf{r})$, we shall try to calculate the potential $W(\boldsymbol{\rho})$ created by these charges at a distant point $\boldsymbol{\rho}$ (*cf.* fig. 1). For the sake of simplicity, we shall use classical mechanics to treat this problem. The potential $W(\boldsymbol{\rho})$ is then:

$$W(\boldsymbol{\rho}) = \frac{1}{4\pi\varepsilon_0} \sum_{n=1}^{N} \frac{q_n}{|\boldsymbol{\rho} - \mathbf{r}_n|} \tag{31}$$

Now, when $|\boldsymbol{\rho}| \gg |\mathbf{r}_n|$, it can be shown that:

$$\frac{1}{|\boldsymbol{\rho} - \mathbf{r}_n|} = \frac{1}{\rho} \sum_{l=0}^{\infty} \left(\frac{r_n}{\rho}\right)^l P_l(\cos \alpha_n) \tag{32}$$

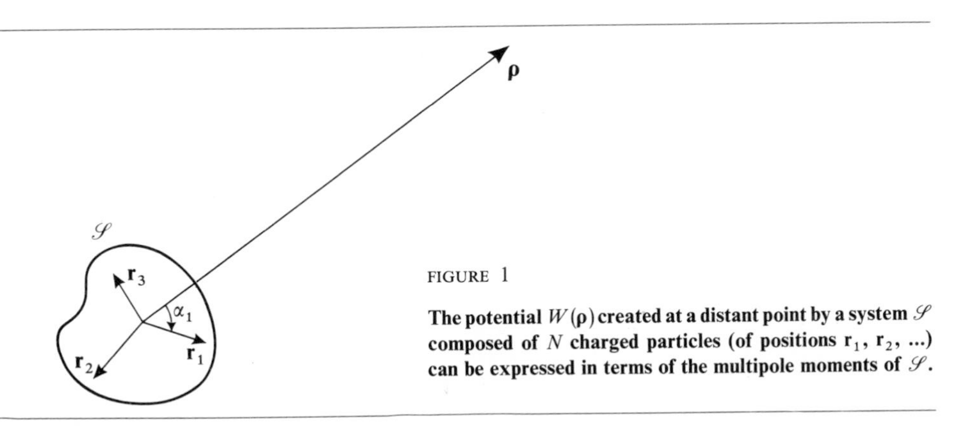

FIGURE 1

The potential $W(\boldsymbol{\rho})$ created at a distant point by a system $\mathscr{S}$ composed of N charged particles (of positions $\mathbf{r}_1$, $\mathbf{r}_2$, ...) can be expressed in terms of the multipole moments of $\mathscr{S}$.

where α_n denotes the angle $(\boldsymbol{\rho}, \mathbf{r}_n)$, and P_l is the lth-order Legendre polynomial. Using the spherical harmonic addition theorem (*cf.* complement A_{VI}, §2-e-γ), we can write :

$$P_l(\cos \alpha_n) = \frac{4\pi}{2l+1} \sum_{m=-l}^{+l} (-1)^m \, Y_l^{-m}(\theta_n, \varphi_n) Y_l^m(\Theta, \Phi) \tag{33}$$

(where Θ and Φ denote the polar angles of $\boldsymbol{\rho}$). Substituting (32) and (33) into (31), we finally obtain:

$$W(\boldsymbol{\rho}) = \frac{1}{4\pi\varepsilon_0} \sum_{l=0}^{\infty} \sum_{m=-l}^{l} \sqrt{\frac{4\pi}{2l+1}} (-1)^m \, \mathcal{Q}_l^{-m} \frac{1}{\rho^{l+1}} Y_l^m(\Theta, \Phi) \tag{34}$$

where $\mathcal{Q}_l^m(\mathbf{r}_1, \mathbf{r}_2, ..., \mathbf{r}_N)$ is defined by relation (15).

Relation (34) shows that the specification of the $\mathcal{Q}_l^m$ perfectly defines the potential created by the particle system in regions of space outside the system $\mathscr{S}$. This potential $W(\boldsymbol{\rho})$ can thus be seen to be the sum of an infinite number of terms:

(*i*) The $l = 0$ term gives the contribution of the total charge of the system. This term is isotropic (it does not depend on Θ and Φ) and can be written:

$$W_0(\boldsymbol{\rho}) = \frac{1}{4\pi\varepsilon_0} \frac{1}{\rho} \sum_n q_n \tag{35}$$

This is the $1/\rho$ potential which would be created by the charges if they were all situated at O. It is zero if the system is globally neutral.

(*ii*) The $l = 1$ term gives the contribution of the electric dipole moment $\mathcal{D}$ of the system. By performing transformations analogous to those in §b-β, it can be shown that this contribution can be written:

$$W_1(\boldsymbol{\rho}) = \frac{1}{4\pi\varepsilon_0} \frac{\mathcal{D} \cdot \boldsymbol{\rho}}{\rho^3} \tag{36}$$

This potential decreases like $1/\rho^2$ when ρ increases.

(*iii*) The $l = 2, 3, ...$ terms give, in the same way, the contributions to the potential $W(\boldsymbol{\rho})$ of the successive multipole moments of the system under study. When ρ increases, each of these contributions decreases like $1/\rho^{l+1}$, and its angular dependence is described by an lth-order spherical harmonic. Moreover, we see from (34) and definition (15) that the potential due to the multipole moment $\mathcal{Q}_l$ is at most of the order of magnitude of $W_0(\rho) \times (d/\rho)^l$, where d is the maximum distance of the various particles of the system $\mathscr{S}$ from the origin. Therefore, if we are concerned with the potential at a point $\boldsymbol{\rho}$ such that $\rho \gg d$ (the potential at a distant point), the $W_l(\boldsymbol{\rho})$ terms decrease very rapidly when l increases, and we do not make a large error by retaining only the lowest values of l in (34).

COMMENT:

If we wanted to calculate the magnetic field created by a system of moving charges, we could introduce the magnetic multipole moments of the system in an analogous way : the magnetic dipole moment*, the magnetic quadrupole moment, etc... The parities of the magnetic moments are the opposite of those of the corresponding electric moments: the magnetic dipole moment is even, the magnetic quadrupole moment is odd, and so on. This property arises from the fact that the electric field is a polar vector while the magnetic field is an axial vector.

* There is no magnetic multipole moment of order $l = 0$ (magnetic monopole). This result is related to the fact that the magnetic field, whose divergence is zero according to Maxwell's equations, has a conservative flux.

2. Matrix elements of electric multipole moments

We shall again consider, for the sake of simplicity, a system composed of a single spinless particle. However, generalization to N-particle systems presents no theoretical difficulty.

The state space $\mathscr{E}_{\mathbf{r}}$ of the particle is spanned by an orthonormal basis, $\{ | \chi_{n,l,m} \rangle \}$, of common eigenvectors of $\mathbf{L}^2$ [eigenvalue $l(l+1)\hbar^2$] and L_z (eigenvalue $m\hbar$). We shall evaluate the matrix elements of a multipole operator Q_l^m in such a basis.

a. GENERAL EXPRESSION FOR THE MATRIX ELEMENTS

α. *Expansion of the matrix elements*

From the general results of chapter VII, we know that the wave functions associated with the states $| \chi_{n,l,m} \rangle$ are necessarily of the form:

$$\chi_{n,l,m}(\mathbf{r}) = R_{n,l}(r)\, Y_l^m(\theta, \varphi) \tag{37}$$

The matrix element of the operator Q_l^m can therefore be written, using (12):

$$\begin{aligned}
\langle \chi_{n_1,l_1,m_1} | Q_l^m | \chi_{n_2,l_2,m_2} \rangle &= \\
&= \int_0^\infty r^2\, \mathrm{d}r \int_0^\pi \sin\theta\, \mathrm{d}\theta \int_0^{2\pi} \mathrm{d}\varphi\, \chi^*_{n_1,l_1,m_1}(r,\theta,\varphi)\, \mathscr{Q}_l^m(r,\theta,\varphi)\, \chi_{n_2,l_2,m_2}(r,\theta,\varphi) \\
&= q\sqrt{\frac{4\pi}{2l+1}} \int_0^\infty r^2\, \mathrm{d}r\, R^*_{n_1,l_1}(r)\, R_{n_2,l_2}(r)\, r^l \int_0^\pi \sin\theta\, \mathrm{d}\theta \\
&\qquad \times \int_0^{2\pi} \mathrm{d}\varphi\, Y_{l_1}^{m_1 *}(\theta,\varphi)\, Y_l^m(\theta,\varphi)\, Y_{l_2}^{m_2}(\theta,\varphi)
\end{aligned} \tag{38}$$

Thus, in the matrix element under consideration, we have a radial integral and an angular integral. The latter, furthermore, can be simplified; using the complex conjugation relation for the spherical harmonics [*cf.* chap. VI, relation (D-29)] and relation (26) of complement C_X (the Wigner-Eckart theorem for spherical harmonics), we can show that it can be written:

$$\begin{aligned}
(-1)^{m_1} \int_0^\pi \sin\theta\, \mathrm{d}\theta \int_0^{2\pi} \mathrm{d}\varphi\, Y_{l_1}^{-m_1}(\theta,\varphi)\, Y_l^m(\theta,\varphi)\, Y_{l_2}^{m_2}(\theta,\varphi) &= \\
= \sqrt{\frac{(2l+1)(2l_2+1)}{4\pi(2l_1+1)}}\, \langle l_2, l; 0, 0 | l_1, 0 \rangle \langle l_2, l; m_2, m | l_1, m_1 \rangle
\end{aligned} \tag{39}$$

Finally, we obtain:

$$\begin{aligned}
\langle \chi_{n_1,l_1,m_1} | Q_l^m | \chi_{n_2,l_2,m_2} \rangle &= \\
&= \frac{1}{\sqrt{2l_1+1}} \langle \chi_{n_1,l_1} \| Q_l \| \chi_{n_2,l_2} \rangle \langle l_2, l; m_2, m | l_1, m_1 \rangle
\end{aligned} \tag{40}$$

Where the "reduced matrix element" $\langle \chi_{n_1,l_1} \| Q_l \| \chi_{n_2,l_2} \rangle$ of the lth-order electric multipole operator is defined by:

$$\langle \chi_{n_1,l_1} \|Q_l\| \chi_{n_2,l_2} \rangle = q\sqrt{2l_2 + 1} \langle l_2, l; 0, 0 | l_1, 0 \rangle \times \int_0^\infty \mathrm{d}r\, r^{l+2} R^*_{n_1,l1}(r) R_{n_2,l_2}(r) \quad (41)$$

Relation (40) expresses, in the particular case of electric multipole operators, a general theorem whose application in the case of vector operators has already been illustrated (*cf.* complement D_X): the Wigner-Eckart theorem.

COMMENT:

We have confined ourselves here to a system $\mathscr{S}$ composed of a single spinless particle. Nevertheless, if we consider a system of N particles which may have spins, we can generalize the results we have obtained. To do so, we must introduce the total angular momentum $\mathbf{J}$ of the system (the sum of the orbital and spin angular momenta of the N particles), and denote by $| \chi_{n,j,m} \rangle$ the eigenvectors common to $\mathbf{J}^2$ and J_z. We can then derive a relation similar to (40), in which l_1 and l_2 are replaced by j_1 and j_2 (*cf.* complement G_X, exercise 8). However, the quantum numbers j_1, j_2, m_1 and m_2 can then be either integral or half-integral, depending on the physical system being considered.

β. *The reduced matrix element*

The reduced matrix element $\langle \chi_{n_1,l_1} \| Q_l \| \chi_{n_2,l_2} \rangle$ is independent of m, m_1 and m_2. It involves the radial part $R_{n,l}(r)$ of the wave functions $\chi_{n,l,m}(r, \theta, \varphi)$. Its value therefore depends on the $\{ | \chi_{n,l,m} \rangle \}$ basis chosen, and general properties can hardly be attributed to it. However, it can be noted that the Clebsch-Gordan coefficient $\langle l_2, l; 0, 0 | l_1, 0 \rangle$ involved in (41) is zero if $l_1 + l_2 + l$ is odd (*cf.* complement B_X, §3-c); this implies that the reduced matrix element has the same property.

COMMENT:

This property is related to the $(-1)^l$ parity of the electric multipole operators Q_l^m. For the magnetic multipole operators, we have already pointed out that their parity is $(-1)^{l+1}$; therefore it is when $l_1 + l_2 + l$ is even that their matrix elements are zero.

γ. *The angular part of the matrix element*

In (40), the Clebsch-Gordan coefficient $\langle l_2, l; m_2, m | l_1, m_1 \rangle$ arises solely from the angular integral appearing in the matrix element of Q_l^m [*cf.* (38)]. This coefficient depends only on the quantum numbers associated with the angular momenta of the states being considered and does not involve the radial dependence $R_{n,l}(r)$ of the wave functions. This is why it appears in the matrix elements of multipole operators whenever one chooses a basis of eigenvectors common to $\mathbf{L}^2$ and L_z (or $\mathbf{J}^2$ and J_z for a system of N particles which may have spins;

cf. comment of §α above). Now, we know that such bases are frequently used in quantum mechanics, and, in particular, that the stationary states of a particle in a central potential $W(r)$ can be chosen in this form. The radial functions $R_{n,l}(r)$ associated with the stationary states thus depend on the potential $W(r)$ chosen; this is therefore also true for the reduced matrix element $\langle \chi_{n_1,l_1} \| Q_l \| \chi_{n_2,l_2} \rangle$. On the other hand, this is not the case for the angular dependence of the wave functions, and the same Clebsch-Gordan coefficient appears for all $W(r)$; this is why it plays a universal role.

b. SELECTION RULES

According to the properties of Clebsch-Gordan coefficients (*cf.* complement B_X, §1), $\langle l_2, l; m_2, m \mid l_1, m_1 \rangle$ can be different from zero only if we have both:

$$m_1 = m_2 + m \tag{42}$$

$$|l_1 - l_2| \leqslant l \leqslant l_1 + l_2 \tag{43}$$

Therefore, relation (40) implies that if at least one of these conditions is not met, the matrix element $\langle \chi_{n_1,l_1,m_1} \mid Q_l^m \mid \chi_{n_2,l_2,m_2} \rangle$ is necessarily zero. We thus obtain selection rules which enable us, without calculations, to simplify considerably our search for the matrix which represents any multipole operator Q_l^m.

Furthermore, we saw in § 2-a-β that the reduced matrix element of a multipole operator obeys another selection rule:

– for an electric multipole operator:

$$l_1 + l_2 + l = \text{an even number} \tag{44-a}$$

– for a magnetic multipole operator:

$$l_1 + l_2 + l = \text{an odd number} \tag{44-b}$$

c. PHYSICAL CONSEQUENCES

α. *The mean value of a multipole operator in a state of well-defined angular momentum*

Assume that the state $| \psi \rangle$ of the particle is one of the basis states $| \chi_{n_1,l_1,m_1} \rangle$. The mean value $\langle Q_l^m \rangle$ of the operator Q_l^m is then:

$$\langle Q_l^m \rangle = \langle \chi_{n_1,l_1,m_1} \mid Q_l^m \mid \chi_{n_1,l_1,m_1} \rangle \tag{45}$$

Conditions (42) and (43) are written here:

$$m = 0 \tag{46}$$

$$0 \leqslant l \leqslant 2l_1 \tag{47}$$

Thus we obtain the following important rules:

– the mean values, in a state $| \chi_{n_1,l_1,m_1} \rangle$, of all the operators Q_l^m are zero if $m \neq 0$:

$$\langle Q_l^m \rangle = 0 \qquad \text{if} \quad m \neq 0 \tag{48}$$

— *the mean values, in a state* $| \chi_{n_1, l_1, m_1} \rangle$, *of all operators of order* l *higher than* $2l_1$ *are zero:*

$$\langle Q_l^m \rangle = 0 \qquad \text{if} \quad l > 2l_1 \tag{49}$$

If we now assume that the state $| \psi \rangle$, instead of being a state $| \chi_{n_1, l_1, m_1} \rangle$, is any superposition of such states, all corresponding to the same value of l_1, it is not difficult to show that rule (49) remains valid [but not rule (48), since, in general, matrix elements for which $m_1 \neq m_2$ then contribute to the mean value $\langle Q_l^m \rangle$]. Relation (49) is therefore a very general one and can be applied whenever the system is in an eigenstate of $\mathbf{L}^2$.

Furthermore, relations (44) imply that the mean value of an lth-order multipole operator can be different from zero only if:

— for an electric multipole operator:

$$l = \text{an even number} \tag{50-a}$$

— for a magnetic multipole operator:

$$l = \text{an odd number} \tag{50-b}$$

The preceding rules enable us to obtain, conveniently and without calculations, some simple physical results. For example, in an $l = 0$ state (like the ground state of the hydrogen atom), the dipole moments (electric or magnetic), quadrupole moments (electric or magnetic), etc. are always zero. For an $l = 1$ state, only the 0th-, 1st- and 2nd-order multipole operators can be non-zero; parity rules (50) indicate that they are the total charge and electric quadrupole of the system, as well as its magnetic dipole.

COMMENT:

The predictions obtained can be generalized to more complex systems (many-electron atoms for example). If the angular momentum of such a system is j (integral or half-integral) one can show that it suffices to replace, in (49), l_1 by j.

We shall apply, for example, rules (49) and (50) to the study of the electromagnetic properties of an atomic nucleus. We know that such a nucleus is a bound system composed of protons and neutrons, interacting through nuclear forces. If, in the ground state*, the eigenvalue of the square of the angular momentum is $I(I + 1)\hbar^2$, the quantum number I is called the nuclear spin. The rules we have stated indicate that:

— if $I = 0$, the electromagnetic interactions of the nucleus are characterized by its total charge, all the other multipole moments being zero. This is the case, for example, for ^4He nuclei ("α-particles"), ^{20}Ne nuclei, etc.

* In atomic physics, one generally consider the nucleus to be in its ground state: the energies involved, although high enough to excite the electronic cloud of the atom, are much too small to excite the nucleus.

– if $I = 1/2$, the nucleus has an electric charge and a magnetic dipole moment [parity rule (50-a) excludes an electric dipole moment]. This is the case for the ^{3}He nucleus and the ^{1}H nucleus (i.e., the proton), as well as for all spin 1/2 particles (electrons, muons, neutrons, etc.).

– if $I = 1$, we must add the electric quadrupole moment to the charge and the magnetic dipole moment. This is the case for ^{2}H (deuterium), ^{6}Li, etc.

This argument can be generalized to any value of I. Actually, very few nuclei have spins greater than 3 or 4.

β. *Matrix elements between states of different quantum numbers*

For arbitrary l_1, l_2, m_1 and m_2, the selection rules must be applied in their general forms, (42), (43) and (44). Consider, for example, a particle of charge q subjected to a central potential $V_0(r)$, whose stationary states are the states $| \chi_{n,l,m} \rangle$. Assume that we then add an additional electric field $\mathscr{E}$, uniform and parallel to Oz. In the corresponding coupling Hamiltonian, the only non-zero term is the electric dipole term (*cf.* § 1-b-β):

$$\begin{aligned} V(\mathbf{R}) &= - \mathbf{D} \cdot \mathscr{E} \\ &= - D_z \mathscr{E} \end{aligned} \tag{51}$$

As we saw in (22), the operator D_z is equal to the operator Q_1^0. Selection rules (42) and (43) then indicate that :

– the states $| \chi_{n,l,m} \rangle$ coupled by the additional Hamiltonian $V(\mathbf{R})$ necessarily correspond to the same value of m.

– the l-values of two states necessarily differ by ± 1 [they cannot be equal, according to (44-a)]. We can predict without calculation that a large number of matrix elements of $V(\mathbf{R})$ are zero. This considerably simplifies, for example, the study of the Stark effect (*cf.* complement E_{XII}), and that of the selection rules governing the emission spectrum of atoms (*cf.* complement A_{XIII}).

References and suggestions for further reading:

Cagnac and Pebay-Peyroula (11.2), annexe IV; Valentin (16.1), chap. VIII; Jackson (7.5), chaps. 4 and 16.

Complement F_X

EVOLUTION OF TWO ANGULAR MOMENTA J_1 AND J_2 COUPLED BY AN INTERACTION $aJ_1 . J_2$

In a physical system, we must often consider the effect of a coupling between two partial angular momenta $\mathbf{J}_1$ and $\mathbf{J}_2$. These can, for example, be the angular momenta of two electrons of an atom, or the orbital and spin angular momenta of an electron. In the presence of such a coupling, $\mathbf{J}_1$ and $\mathbf{J}_2$ are no longer constants of the motion; only:

$$\mathbf{J} = \mathbf{J}_1 + \mathbf{J}_2 \tag{1}$$

commutes with the total Hamiltonian of the system.

We shall assume that the term of the Hamiltonian which introduces a coupling between $\mathbf{J}_1$ and $\mathbf{J}_2$ has the simple form:

$$W = a\,\mathbf{J}_1 . \mathbf{J}_2 \tag{2}$$

where a is a real constant. Such a situation is frequently encountered in atomic physics. We shall see numerous examples in chapter XII, when we use perturbation theory to study the effect on the hydrogen atom spectrum of interactions involving electron or proton spins. When the coupling has the form (2), classical theory predicts that the classical angular momenta $\mathcal{J}_1$ and $\mathcal{J}_2$ will precess about their resultant $\mathcal{J}$ with an angular velocity proportional to the constant a (*cf.* §1 below). The "vector model" of the atom, which played a very important role in the history of the development of atomic physics, is founded on this result. In this complement, we shall show how, with the knowledge of the common eigenstates of $\mathbf{J}^2$ and J_z, one can study the motion of the mean values $\langle \mathbf{J}_1 \rangle$ and $\langle \mathbf{J}_2 \rangle$, and again derive, at least partially, the results of the vector model of the atom (§§ 2 and 3). In addition, this study will enable us to specify in simple cases the polarization of the electromagnetic waves emitted or absorbed in magnetic dipole transitions. Finally, (§ 4), we shall take up the case in which the two angular momenta $\mathbf{J}_1$ and $\mathbf{J}_2$ are coupled only during a collision but not permanently. This case will serve as a simple illustration of the important concept of correlation between two systems.

1. Classical review

a. EQUATIONS OF MOTION

If θ is the angle between the classical angular momenta $\mathcal{J}_1$ and $\mathcal{J}_2$ (fig. 1), the coupling energy can be written:

$$\mathcal{W} = a\,\mathcal{J}_1 \cdot \mathcal{J}_2 = a\,\mathcal{J}_1 \mathcal{J}_2 \cos\theta \tag{3}$$

Let $\mathcal{H}_0$ be the energy of the total system in the absence of coupling [$\mathcal{H}_0$ can represent, for example, the sum of the rotational kinetic energies of systems (1) and (2)]. We shall assume:

$$\mathcal{W} \ll \mathcal{H}_0 \tag{4}$$

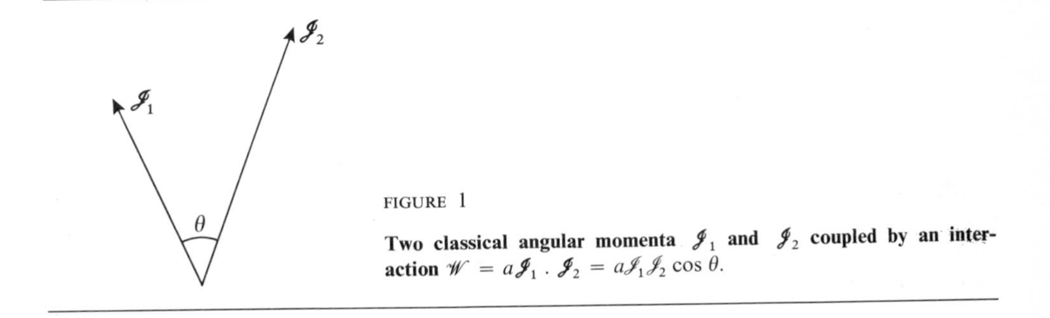

FIGURE 1

Two classical angular momenta $\mathcal{J}_1$ and $\mathcal{J}_2$ coupled by an interaction $\mathcal{W} = a\mathcal{J}_1 \cdot \mathcal{J}_2 = a\mathcal{J}_1\mathcal{J}_2 \cos\theta$.

Let us calculate the moment $\mathcal{M}_1$ of the forces acting on system (1). Let $\mathbf{u}$ be a unit vector and $\mathrm{d}\mathcal{W}$, the variation of the coupling energy when the system (1) is rotated through an angle $\mathrm{d}\alpha$ about $\mathbf{u}$. We know (the theorem of virtual work) that:

$$\mathcal{M}_1 \cdot \mathbf{u} = -\frac{\mathrm{d}\mathcal{W}}{\mathrm{d}\alpha} \tag{5}$$

Starting with (3) and (5), we then obtain, via a simple calculation:

$$\mathcal{M}_1 = -\,a\mathcal{J}_1 \times \mathcal{J}_2 \tag{6-a}$$
$$\mathcal{M}_2 = -\,a\mathcal{J}_2 \times \mathcal{J}_1 \tag{6-b}$$

and, consequently:

$$\frac{\mathrm{d}\mathcal{J}_1}{\mathrm{d}t} = -\,a\mathcal{J}_1 \times \mathcal{J}_2 \tag{7-a}$$

$$\frac{\mathrm{d}\mathcal{J}_2}{\mathrm{d}t} = -\,a\mathcal{J}_2 \times \mathcal{J}_1 \tag{7-b}$$

b. **MOTION OF $\mathcal{J}_1$ AND $\mathcal{J}_2$**

Adding (7-a) and (7-b), we obtain:

$$\frac{\mathrm{d}}{\mathrm{d}t}(\mathcal{J}_1 + \mathcal{J}_2) = \mathbf{0} \tag{8}$$

which shows that the total angular momentum $\mathcal{J}_1 + \mathcal{J}_2$ is indeed a constant of the motion. Furthermore, it can easily be deduced from (7-a) and (7-b) that:

$$\mathcal{J}_1 \cdot \left(\frac{\mathrm{d}\mathcal{J}_1}{\mathrm{d}t}\right) = \mathcal{J}_2 \cdot \left(\frac{\mathrm{d}\mathcal{J}_2}{\mathrm{d}t}\right) = 0 \tag{9}$$

and:

$$\mathcal{J}_1 \cdot \left(\frac{\mathrm{d}}{\mathrm{d}t}\mathcal{J}_2\right) + \left(\frac{\mathrm{d}}{\mathrm{d}t}\mathcal{J}_1\right) \cdot \mathcal{J}_2 = \frac{\mathrm{d}}{\mathrm{d}t}(\mathcal{J}_1 \cdot \mathcal{J}_2) = 0 \tag{10}$$

The angle θ between $\mathcal{J}_1$ and $\mathcal{J}_2$, as well as the moduli of $\mathcal{J}_1$ and $\mathcal{J}_2$, therefore remain constant over time. Finally :

$$\frac{\mathrm{d}}{\mathrm{d}t}\mathcal{J}_1 = a\mathcal{J}_2 \times \mathcal{J}_1 = a(\mathcal{J} - \mathcal{J}_1) \times \mathcal{J}_1 = a\mathcal{J} \times \mathcal{J}_1 \tag{11}$$

Since $\mathcal{J} = \mathcal{J}_1 + \mathcal{J}_2$ is constant, the preceding equation shows that $\mathcal{J}_1$ precesses about $\mathcal{J}$ with an angular velocity equal to $a|\mathcal{J}|$ (fig. 2).

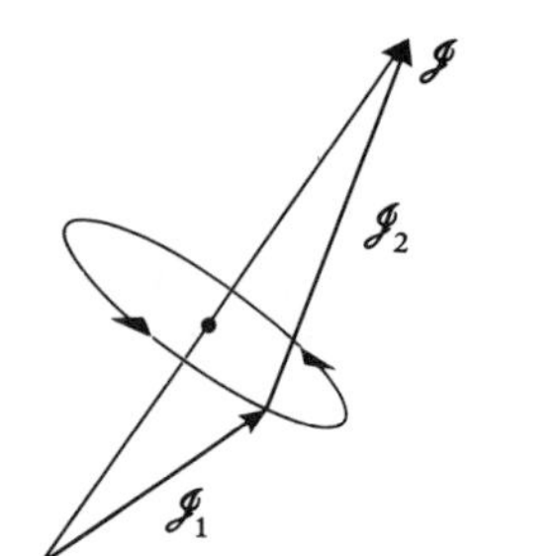

FIGURE 2

Under the effect of the coupling $\mathcal{W} = a\,\mathcal{J}_1 \cdot \mathcal{J}_2$, the angular momenta $\mathcal{J}_1$ and $\mathcal{J}_2$ precess about their resultant $\mathcal{J}$, which is a constant of the motion.

Under the effect of the coupling, $\mathcal{J}_1$ and $\mathcal{J}_2$ therefore precess about their resultant $\mathcal{J}$ with an angular velocity proportional to $|\mathcal{J}|$ and to the coupling constant a.

2. Quantum mechanical evolution of the mean values $\langle \mathbf{J}_1 \rangle$ and $\langle \mathbf{J}_2 \rangle$

a. **CALCULATION OF $\frac{\mathrm{d}}{\mathrm{d}t}\langle \mathbf{J}_1 \rangle$ AND $\frac{\mathrm{d}}{\mathrm{d}t}\langle \mathbf{J}_2 \rangle$**

Recall, first of all, that if A is an observable of a quantum mechanical system of Hamiltonian H, we have (*cf.* chap. III, §D-1-d):

$$\frac{\mathrm{d}}{\mathrm{d}t}\langle A \rangle(t) = \frac{1}{i\hbar}\langle [A, H] \rangle(t) \tag{12}$$

In the present case, the Hamiltonian H is equal to:

$$H = H_0 + W \tag{13}$$

where H_0 is the sum of the energies of systems (1) and (2), and W is the coupling between $\mathbf{J}_1$ and $\mathbf{J}_2$ given in (2). In the absence of such a coupling, $\mathbf{J}_1$ and $\mathbf{J}_2$ are constants of the motion (they commute with H_0). Therefore, in the presence of the coupling, we have simply:

$$\frac{\mathrm{d}}{\mathrm{d}t}\langle \mathbf{J}_1 \rangle = \frac{1}{i\hbar}\langle [\mathbf{J}_1, W] \rangle = \frac{a}{i\hbar}\langle [\mathbf{J}_1, \mathbf{J}_1 \cdot \mathbf{J}_2] \rangle \tag{14}$$

and an analogous expression for $\frac{\mathrm{d}}{\mathrm{d}t}\langle \mathbf{J}_2 \rangle$. The calculation of the commutator appearing in formula (14) does not present any difficulty. We have, for example:

$$\begin{aligned} [J_{1x}, \mathbf{J}_1 \cdot \mathbf{J}_2] &= [J_{1x}, J_{1y}J_{2y}] + [J_{1x}, J_{1z}J_{2z}] \\ &= i\hbar J_{1z}J_{2y} - i\hbar J_{1y}J_{2z} \\ &= -\, i\hbar(\mathbf{J}_1 \times \mathbf{J}_2)_x \end{aligned} \tag{15}$$

From this, we see finally that:

$$\frac{\mathrm{d}}{\mathrm{d}t}\langle \mathbf{J}_1 \rangle = -\, a \langle \mathbf{J}_1 \times \mathbf{J}_2 \rangle \tag{16-a}$$

$$\frac{\mathrm{d}}{\mathrm{d}t}\langle \mathbf{J}_2 \rangle = -\, a \langle \mathbf{J}_2 \times \mathbf{J}_1 \rangle \tag{16-b}$$

b. DISCUSSION

Note the close analogy between formulas (7-a) and (7-b) on the one hand and formulas (16-a) and (16-b) on the other. Adding (16-a) and (16-b), we again find that **J** is a constant of the motion, since:

$$\frac{\mathrm{d}}{\mathrm{d}t}\langle \mathbf{J}_1 \rangle + \frac{\mathrm{d}}{\mathrm{d}t}\langle \mathbf{J}_2 \rangle = \frac{\mathrm{d}}{\mathrm{d}t}\langle \mathbf{J} \rangle = \mathbf{0} \tag{17}$$

However, we must recall that, in general:

$$\langle \mathbf{J}_1 \times \mathbf{J}_2 \rangle \neq \langle \mathbf{J}_1 \rangle \times \langle \mathbf{J}_2 \rangle \tag{18}$$

The motion of the mean values is therefore not necessarily identical to the classical motion. To examine this point in greater detail, we shall now study a special case: that in which $\mathbf{J}_1$ and $\mathbf{J}_2$ are two spin 1/2's, which we shall denote by $\mathbf{S}_1$ and $\mathbf{S}_2$.

3. The special case of two spin 1/2's

The evolution of a quantum mechanical system can easily be calculated in the basis of eigenstates of the Hamiltonian of this system. Therefore, we shall begin by determining the stationary states of the two-spin system.

a. STATIONARY STATES OF THE TWO-SPIN SYSTEM

Let

$$\mathbf{S} = \mathbf{S}_1 + \mathbf{S}_2 \tag{19}$$

be the total spin. Squaring both sides of (19), we obtain:

$$\mathbf{S}^2 = \mathbf{S}_1^2 + \mathbf{S}_2^2 + 2\mathbf{S}_1 \cdot \mathbf{S}_2 \tag{20}$$

which enables us to write W in the form:

$$W = a\ \mathbf{S}_1 \cdot \mathbf{S}_2 = \frac{a}{2}[\mathbf{S}^2 - \mathbf{S}_1^2 - \mathbf{S}_2^2] = \frac{a}{2}\left[\mathbf{S}^2 - \frac{3}{2}\hbar^2\right] \tag{21}$$

(all vectors of the state space are eigenvectors of $\mathbf{S}_1^2$ and $\mathbf{S}_2^2$ with the eigenvalue $3\hbar^2/4$).

In the absence of coupling, the Hamiltonian H_0 of the system is diagonal in the $\{ \mid \varepsilon_1, \varepsilon_2 \rangle \}$ basis (with $\varepsilon_1 = \pm, \varepsilon_2 = \pm$) of eigenstates of S_{1z} and S_{2z}, as well as in the $\{ \mid S, M \rangle \}$ basis (with $S = 0$ or 1, $-S \leqslant M \leqslant +S$) of eigenstates of $\mathbf{S}^2$ and S_z. The various vectors $\mid \varepsilon_1, \varepsilon_2 \rangle$ or $\mid S, M \rangle$ are eigenvectors of H_0 with the same eigenvalue, which we shall take to be the energy origin.

When we take the coupling W into account, we see from formula (21) that the total Hamiltonian $H = H_0 + W$ is no longer diagonal in the $\{ \mid \varepsilon_1, \varepsilon_2 \rangle \}$ basis. However, we may write:

$$(H_0 + W) \mid S, M \rangle = \frac{a\hbar^2}{2}\left[S(S+1) - \frac{3}{2}\right] \mid S, M \rangle \tag{22}$$

The stationary states of the two-spin system therefore separate into two levels (fig. 3) : the $S = 1$ level, three-fold degenerate with energy $E_1 = a\hbar^2/4$, and the

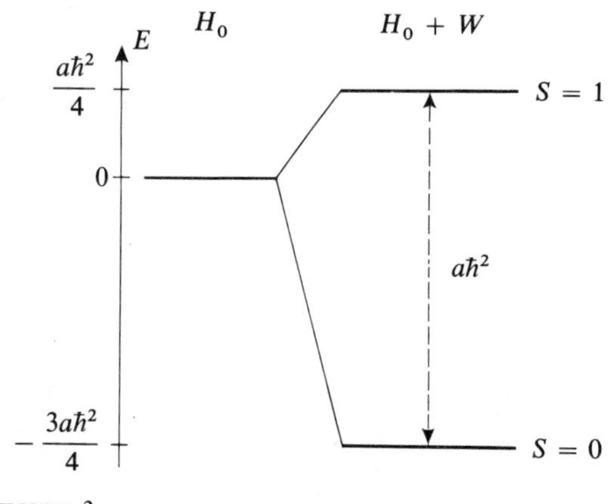

FIGURE 3

Energy levels of a system of two spin 1/2's. On the left-hand side of the figure, the coupling is assumed to be zero, and we obtain a single level which is four-fold degenerate. The coupling $W = a\mathbf{S}_1 \cdot \mathbf{S}_2$ splits it into two distinct levels, separated by an energy of $a\hbar^2$: the triplet level ($S = 1$, three-fold degenerate) and the singlet level ($S = 0$, non-degenerate).

non-degenerate $S = 0$ level, with energy $E_0 = -3a\hbar^2/4$. The splitting between the two levels is equal to $a\hbar^2$. If we set:

$$a\hbar^2 = \hbar\Omega \tag{23}$$

$\Omega/2\pi$ is the only non-zero Bohr frequency of the two-spin system.

b. CALCULATION OF $\langle \mathbf{S}_1 \rangle(t)$

In order to find the evolution of $\langle \mathbf{S}_1 \rangle(t)$, we must first calculate the matrices representing S_{1x}, S_{1y} and S_{1z} (or, more simply, S_{1z} and $S_{1+} = S_{1x} + iS_{1y}$) in the $\{ | S, M \rangle \}$ basis of stationary states. If we use expressions (B-22) and (B-23) of chapter X, which give the expansion of the states $| S, M \rangle$ on the $\{ | \varepsilon_1, \varepsilon_2 \rangle \}$ basis, it is possible to easily calculate the action of S_{1z} or S_{1+} on the kets $| S, M \rangle$. Thus we find:

$$\left\{ \begin{aligned} S_{1z} | 1, 1 \rangle &= \frac{\hbar}{2} | 1, 1 \rangle \\ S_{1z} | 1, 0 \rangle &= \frac{\hbar}{2} | 0, 0 \rangle \\ S_{1z} | 1, -1 \rangle &= -\frac{\hbar}{2} | 1, -1 \rangle \\ S_{1z} | 0, 0 \rangle &= \frac{\hbar}{2} | 1, 0 \rangle \end{aligned} \right. \tag{24}$$

and:

$$\left\{ \begin{aligned} S_{1+} | 1, 1 \rangle &= 0 \\ S_{1+} | 1, 0 \rangle &= \frac{\hbar}{\sqrt{2}} | 1, 1 \rangle \\ S_{1+} | 1, -1 \rangle &= \frac{\hbar}{\sqrt{2}} (| 1, 0 \rangle + | 0, 0 \rangle) \\ S_{1+} | 0, 0 \rangle &= -\frac{\hbar}{\sqrt{2}} | 1, 1 \rangle \end{aligned} \right. \tag{25}$$

From this, we can immediately derive the matrices representing S_{1z} and S_{1+} in the basis of the four states $| S, M \rangle$ arranged in the order $| 1, 1 \rangle$, $| 1, 0 \rangle$, $| 1, -1 \rangle$ and $| 0, 0 \rangle$:

$$(S_{1z}) = \frac{\hbar}{2} \begin{pmatrix} 1 & 0 & 0 & 0 \\ 0 & 0 & 0 & 1 \\ 0 & 0 & -1 & 0 \\ 0 & 1 & 0 & 0 \end{pmatrix} \tag{26}$$

$$(S_{1+}) = \frac{\hbar}{\sqrt{2}} \begin{pmatrix} 0 & 1 & 0 & -1 \\ 0 & 0 & 1 & 0 \\ 0 & 0 & 0 & 0 \\ 0 & 0 & 1 & 0 \end{pmatrix} \tag{27}$$

COMMENT:

It can easily be shown that the restrictions of the S_{1z} and S_{1+} matrices to the $S = 1$ subspace are proportional respectively (with the same proportionality coefficient) to the matrices representing S_z and S_+ in the same subspace. This result could have been expected, in view of the Wigner-Eckart theorem relative to vector operators (*cf.* complement D_X).

Let

$$| \psi(0) \rangle = \alpha \, | 0, 0 \rangle + \beta_{-1} \, | 1, -1 \rangle + \beta_0 \, | 1, 0 \rangle + \beta_1 \, | 1, 1 \rangle \tag{28}$$

be the state of the system at the instant $t = 0$. From this we deduce the expression for $| \psi(t) \rangle$ (to within the factor $e^{3ia\hbar t/4}$):

$$| \psi(t) \rangle = \alpha \, | 0, 0 \rangle + [\beta_{-1} \, | 1, -1 \rangle + \beta_0 \, | 1, 0 \rangle + \beta_1 \, | 1, 1 \rangle] \, e^{-i\Omega t} \tag{29}$$

It is then easy to obtain, using (26) and (27):

$$\begin{aligned} \langle S_{1z} \rangle(t) &= \langle \psi(t) \, | S_{1z} \, | \psi(t) \rangle \\ &= \frac{\hbar}{2} [\, |\beta_1|^2 - |\beta_{-1}|^2 + e^{i\Omega t} \, \alpha\beta_0^* + e^{-i\Omega t} \, \alpha^*\beta_0] \end{aligned} \tag{30}$$

$$\begin{aligned} \langle S_{1+} \rangle(t) &= \langle \psi(t) \, | S_{1+} \, | \psi(t) \rangle \\ &= \frac{\hbar}{\sqrt{2}} [\, \beta_1^*\beta_0 + \beta_0^*\beta_{-1} - e^{i\Omega t} \, \beta_1^*\alpha + e^{-i\Omega t} \, \alpha^*\beta_{-1}] \end{aligned} \tag{31}$$

$\langle S_{1x} \rangle(t)$ and $\langle S_{1y} \rangle(t)$ can be expressed in terms of $\langle S_{1+} \rangle(t)$:

$$\langle S_{1x} \rangle(t) = \text{Re} \, \langle S_{1+} \rangle(t) \tag{32}$$

$$\langle S_{1y} \rangle(t) = \text{Im} \, \langle S_{1+} \rangle(t) \tag{33}$$

Analogous calculations enable us to obtain the three components of $\langle \mathbf{S}_2 \rangle(t)$.

c. DISCUSSION. POLARIZATION OF MAGNETIC DIPOLE TRANSITIONS

Studying the motion of $\langle \mathbf{S}_1 \rangle(t)$ does more than compare the vector model of the atom with the predictions of quantum mechanics. It also enables us to specify the polarization of the electromagnetic waves emitted due to the motion of $\langle \mathbf{S}_1 \rangle(t)$.

The Bohr frequency $\Omega/2\pi$ appears in the evolution of $\langle \mathbf{S}_1 \rangle(t)$ because of the existence of non-zero matrix elements of S_{1x}, S_{1y}, or S_{1z} between the state $| 0, 0 \rangle$ and one of the states $| 1, M \rangle$ (with $M = -1, 0, +1$). In (28) or (29), we shall begin by assuming that, with α non-zero, only one of the three coefficients β_{-1}, β_0 or β_1 is different from zero. The examination of the motion of $\langle \mathbf{S}_1 \rangle(t)$ in the three corresponding cases thus will enable us to specify the polarization of the radiation associated with the three magnetic dipole transitions :

$$| 0, 0 \rangle \longleftrightarrow | 1, 0 \rangle, \quad | 0, 0 \rangle \longleftrightarrow | 1, 1 \rangle \quad \text{and} \quad | 0, 0 \rangle \longleftrightarrow | 1, -1 \rangle$$

We can always choose α to be real; we shall set:

$$\beta_M = |\beta_M| \, e^{i\varphi_M} \qquad (M = -1, 0, 1) \tag{34}$$

COMMENT:

Actually, the electromagnetic waves are radiated by the magnetic moments $\mathbf{M}_1$ and $\mathbf{M}_2$ associated with $\mathbf{S}_1$ and $\mathbf{S}_2$ (hence the name, magnetic dipole transitions). $\mathbf{M}_1$ and $\mathbf{M}_2$ are proportional respectively to $\mathbf{S}_1$ and $\mathbf{S}_2$. To be completely rigorous, we should then study the evolution of $\langle \mathbf{M}_1 + \mathbf{M}_2 \rangle(t)$. Here we shall assume $\langle \mathbf{M}_1 \rangle \gg \langle \mathbf{M}_2 \rangle$. Such a situation is found, for example, in the ground state of the hydrogen atom : the hyperfine structure of this state is due to the coupling between the spin of the electron and that of the proton (*cf.* chap. XII, § D). But the magnetic moment of the electron spin is much larger than that of the proton, so that the emission and absorption of electromagnetic waves at the hyperfine transition frequency are essentially governed by the motion of the electron spin. Taking $\langle \mathbf{M}_2 \rangle$ into account as well would complicate the calculations without modifying the conclusions.

α. *The* $|0, 0\rangle \longleftrightarrow |1, 0\rangle$ *transition* $(\beta_1 = \beta_{-1} = 0)$

If we take $\beta_1 = \beta_{-1} = 0$ in (30), (31), (32) and (33), we get:

$$\begin{aligned} \langle S_{1x} \rangle(t) &= \langle S_{1y} \rangle(t) = 0 \\ \langle S_{1z} \rangle(t) &= \hbar\alpha \, |\beta_0| \cos(\Omega t - \varphi_0) \end{aligned} \tag{35}$$

Furthermore, it can easily be seen that:

$$\langle S_x \rangle(t) = \langle S_y \rangle(t) = \langle S_z \rangle(t) = 0 \tag{36}$$

$\langle \mathbf{S}_1 \rangle(t)$ and $\langle \mathbf{S}_2 \rangle(t)$ are then permanently of opposite direction and vibrate along Oz at the frequency $\Omega/2\pi$ (fig. 4).

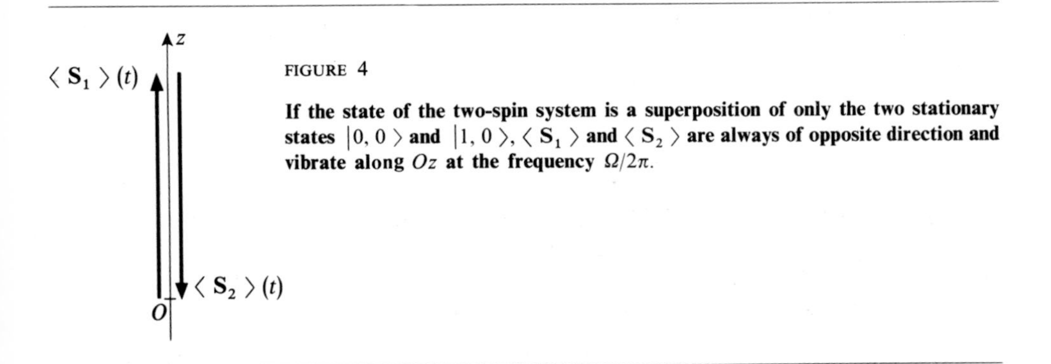

FIGURE 4

If the state of the two-spin system is a superposition of only the two stationary states $|0, 0\rangle$ and $|1, 0\rangle$, $\langle \mathbf{S}_1 \rangle$ and $\langle \mathbf{S}_2 \rangle$ are always of opposite direction and vibrate along Oz at the frequency $\Omega/2\pi$.

The electromagnetic waves emitted by $\langle \mathbf{S}_1 \rangle$ therefore have a magnetic field★ linearly polarized along Oz ("π polarization").

★ Since these are magnetic dipole transitions, we are concerned with the magnetic field vector of the radiated wave. In the case of an electric dipole transition (*cf.* complement D_{VII}, § 2-c), on the other hand, we would be concerned with the radiated electric field.

We see in this example that $(\langle \mathbf{S}_1 \rangle)^2$ varies over time and is therefore not equal to $\langle \mathbf{S}_1^2 \rangle$ (which is constant and equal to $3\hbar^2/4$). This represents an important difference with the classical situation studied in § 1, in which $\mathcal{J}_1$ maintains a constant length over time.

β. *The* $|\, 0, 0 \rangle \longleftrightarrow |\, 1, 1 \rangle$ *transition* $\quad (\beta_0 = \beta_{-1} = 0)$

We find in this case:

$$\begin{cases} \langle S_{1z} \rangle(t) = \dfrac{\hbar}{2} |\beta_1|^2 \\ \langle S_{1x} \rangle(t) = -\dfrac{\hbar}{\sqrt{2}} \alpha \, |\beta_1| \cos(\Omega t - \varphi_1) \\ \langle S_{1y} \rangle(t) = -\dfrac{\hbar}{\sqrt{2}} \alpha \, |\beta_1| \sin(\Omega t - \varphi_1) \end{cases} \tag{37}$$

Furthermore, it can easily be verified that:

$$\begin{cases} \langle S_z \rangle(t) = \hbar \, |\beta_1|^2 \\ \langle S_x \rangle(t) = \langle S_y \rangle(t) = 0 \end{cases} \tag{38}$$

From this, it can be seen (fig. 5) that $\langle \mathbf{S}_1 \rangle(t)$ and $\langle \mathbf{S}_2 \rangle(t)$ precess counterclockwise at an angular velocity Ω about their resultant $\langle \mathbf{S} \rangle$, which is parallel to Oz. The electromagnetic waves emitted by $\langle \mathbf{S}_1 \rangle(t)$ in this case therefore have a right-hand circular polarization ("σ_+ polarization").

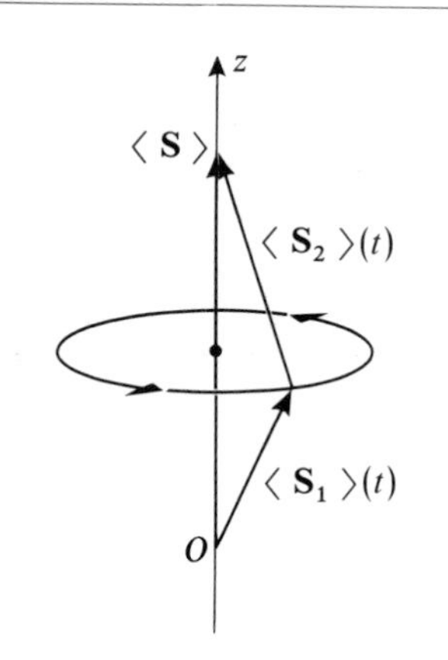

FIGURE 5

If the state of the two-spin system is a superposition of only the stationary states $|\, 0, 0 \rangle$ and $|\, 1, 1 \rangle$, $\langle \mathbf{S}_1 \rangle$ and $\langle \mathbf{S}_2 \rangle$ precess counterclockwise about their resultant $\langle \mathbf{S} \rangle$, with the angular velocity Ω.

Note that here the motion obtained for the mean values $\langle \mathbf{S}_1 \rangle$ and $\langle \mathbf{S}_2 \rangle$ is the classical motion.

γ. *The* $|\, 0, 0 \rangle \longleftrightarrow |\, 1, -1 \rangle$ *transition* $\quad (\beta_0 = \beta_1 = 0)$

The calculations are closely analogous to those of the preceding section and lead to the following result (fig. 6): $\langle \mathbf{S}_1 \rangle(t)$ and $\langle \mathbf{S}_2 \rangle(t)$ precess about Oz, again at the angular velocity Ω, but in the clockwise direction. It must be noted that $\langle S_z \rangle = -\hbar|\beta_{-1}|^2$ is now negative, so that while the direction of the precession of $\langle \mathbf{S}_1 \rangle$ and $\langle \mathbf{S}_2 \rangle$ about Oz is different from what it was in the preceding case, it remains the same relative to $\langle \mathbf{S} \rangle$. The electromagnetic waves emitted by $\langle \mathbf{S}_1 \rangle$ are now left-hand circularly polarized ("σ_- polarization").

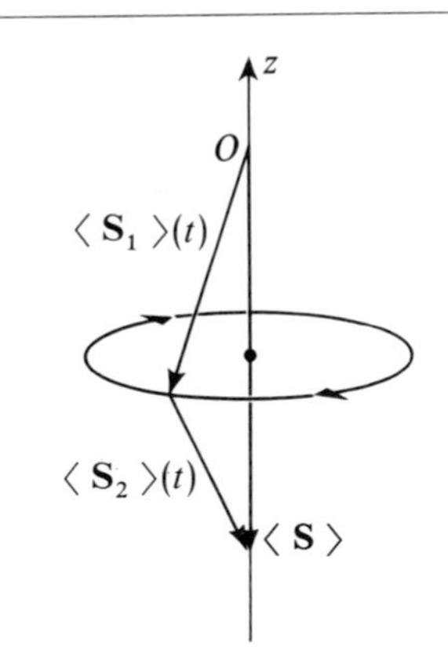

FIGURE 6

If the state of the two-spin system is a superposition of only the stationary states $|0, 0\rangle$ and $|1, -1\rangle$, $\langle \mathbf{S}_1 \rangle$ and $\langle \mathbf{S}_2 \rangle$ still precess in the counterclockwise direction with the angular velocity Ω about their resultant $\langle \mathbf{S} \rangle$; however, the latter is now directed opposite to Oz.

δ. *General case*

In the general case (any α, β_{-1}, β_0 and β_1), we see from (30), (31), (32) and (33) that the components of $\langle \mathbf{S}_1 \rangle(t)$ on the three axes contain a static part and a part modulated at the frequency $\Omega/2\pi$. Since these three projected motions are sinusoidal motions of the same frequency, the tip of $\langle \mathbf{S}_1 \rangle(t)$ describes an ellipse in space. Since the sum

$$\langle \mathbf{S}_1 \rangle(t) + \langle \mathbf{S}_2 \rangle(t) = \langle \mathbf{S} \rangle$$

remains constant, the tip of $\langle \mathbf{S}_2 \rangle(t)$ also describes an ellipse (fig. 7).

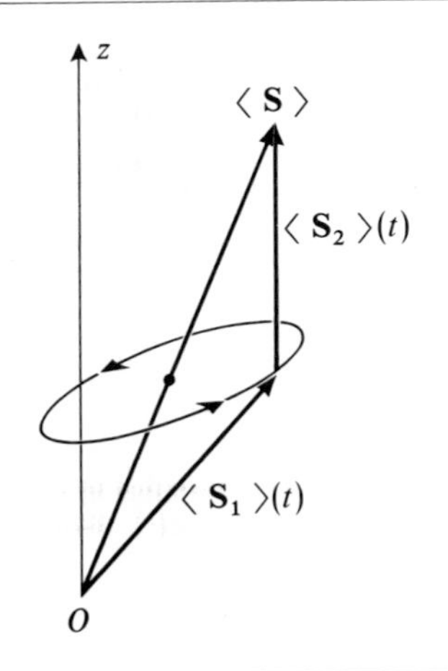

FIGURE 7

Motion of $\langle \mathbf{S}_1 \rangle(t)$ and $\langle \mathbf{S}_2 \rangle(t)$ in the general case, in which the state of the two-spin system is a superposition of the four stationary states $|0, 0\rangle$, $|1, 1\rangle$, $|1, 0\rangle$ and $|1, -1\rangle$. The resultant $\langle \mathbf{S} \rangle$ is still constant but is not necessarily directed along Oz. $\langle \mathbf{S}_1 \rangle$ and $\langle \mathbf{S}_2 \rangle$ no longer have constant lengths, and their tips describe ellipses.

Thus we find for the general case only part of the results of the vector model of the atom. We do find that, the larger the coupling constant a, the more rapidly $\langle \mathbf{S}_1 \rangle(t)$ and $\langle \mathbf{S}_2 \rangle(t)$ precess about $\langle \mathbf{S} \rangle$. However, as we saw clearly in the special case α studied above, $|\langle \mathbf{S}_1 \rangle(t)|$ is not constant, and the tip of $\langle \mathbf{S}_1 \rangle(t)$ does not describe a circle in the general case.

4. Study of a simple model for the collision of two spin 1/2 particles

a. DESCRIPTION OF THE MODEL

Consider two spin 1/2 particles, whose external degrees of freedom we shall treat classically and whose spin degrees of freedom we shall treat quantum mechanically. We shall assume that their trajectories are rectilinear (fig. 8) and that the interaction between the two spins $\mathbf{S}_1$ and $\mathbf{S}_2$ is of the form $W = a\, \mathbf{S}_1 \cdot \mathbf{S}_2$, where the coupling constant a is a rapidly decreasing function of the distance r separating the two particles.

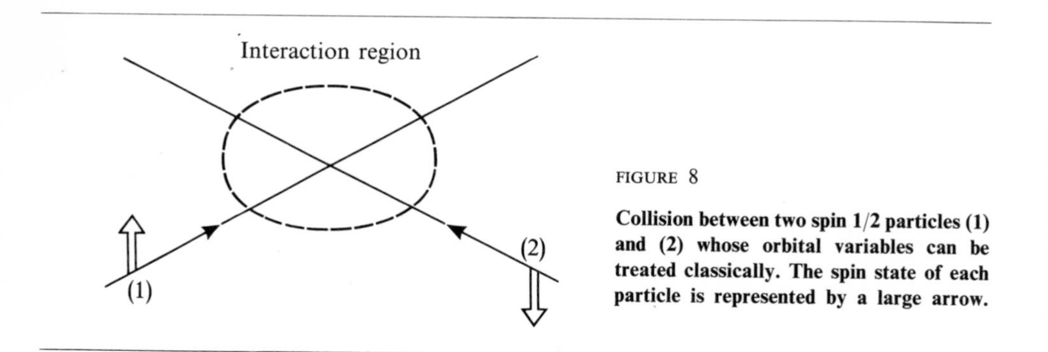

FIGURE 8

Collision between two spin 1/2 particles (1) and (2) whose orbital variables can be treated classically. The spin state of each particle is represented by a large arrow.

Since r varies over time, so does a. The shape of the variation of a with respect to t is shown in figure 9. The maximum corresponds to the time when the distance between the two particles is at a minimum. To simplify the reasoning, we shall replace the curve in figure 9 by the one in figure 10.

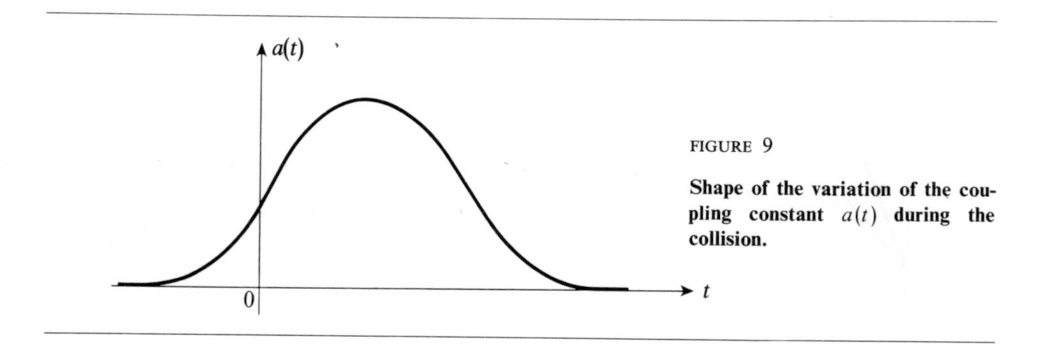

FIGURE 9

Shape of the variation of the coupling constant $a(t)$ during the collision.

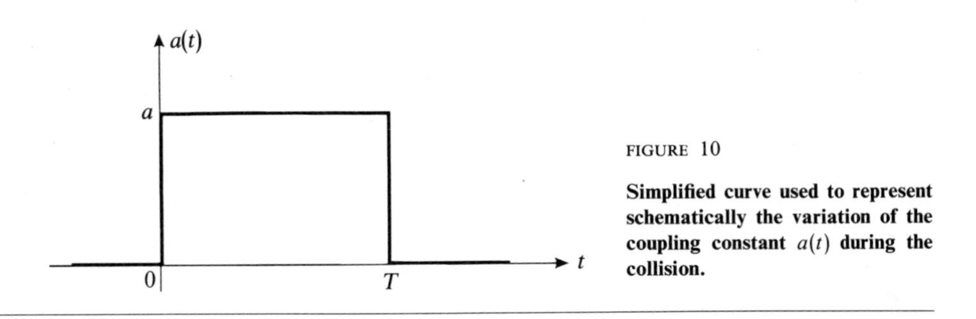

FIGURE 10

Simplified curve used to represent schematically the variation of the coupling constant $a(t)$ during the collision.

The problem we have here is the following: before the collision, that is, at $t = -\infty$, the spin state of the two-particle system is:

$$| \psi(-\infty) \rangle = | +, - \rangle \tag{39}$$

What is the state $| \psi(+\infty) \rangle$ of the system after the collision?

b. THE STATE OF THE SYSTEM AFTER COLLISION

Since the Hamiltonian is zero for $t < 0$, we have:

$$\begin{aligned} | \psi(0) \rangle &= | \psi(-\infty) \rangle = | +, - \rangle \\ &= \frac{1}{\sqrt{2}} [| 1, 0 \rangle + | 0, 0 \rangle] \end{aligned} \tag{40}$$

The results of the preceding section concerning the eigenstates and eigenvalues of $W = a \mathbf{S}_1 . \mathbf{S}_2$ are applicable between times 0 and T and enable us to calculate $| \psi(T) \rangle$:

$$| \psi(T) \rangle = \frac{1}{\sqrt{2}} [| 1, 0 \rangle e^{-iE_1 T/\hbar} + | 0, 0 \rangle e^{-iE_0 T/\hbar}] \tag{41}$$

Multiplying (41) by the global phase factor $e^{i(E_0 + E_1)T/2\hbar}$ (of no physical importance), setting $E_1 - E_0 = \hbar\Omega$ [*cf.* formula (23)], and returning to the $\{ | \varepsilon_1, \varepsilon_2 \rangle \}$ basis, we find:

$$| \psi(T) \rangle = \cos\frac{\Omega T}{2} | +, - \rangle - i \sin\frac{\Omega T}{2} | -, + \rangle \tag{42}$$

Finally, since the Hamiltonian is zero for $t > T$, we have:

$$| \psi(+\infty) \rangle = | \psi(T) \rangle \tag{43}$$

COMMENT:

The calculation could be performed for an arbitrary function $a(t)$ of the type shown in figure 9. It would then be found necessary to replace, in the preceding formula, $aT = \frac{\Omega T}{\hbar}$ by $\int_{-\infty}^{+\infty} a(t) \, dt$ (*cf.* exercise 2 of complement E_{XIII}).

c. DISCUSSION. CORRELATION INTRODUCED BY THE COLLISION

If the condition:

$$\frac{\Omega T}{2} = \frac{\pi}{2} + k\pi \quad , \quad k \text{ an integer} \gtrless 0 \tag{44}$$

is satisfied, we see from (42) that:

$$| \psi(+\infty) \rangle = | -, + \rangle \tag{45}$$

The orientation of the two spins, in this case, is exchanged during the collision.

On the other hand, if:

$$\frac{\Omega T}{2} = k\pi \quad , \quad k \text{ an integer} \gtrless 0 \tag{46}$$

we find that:

$$| \psi(+\infty) \rangle = | +, - \rangle = | \psi(-\infty) \rangle \tag{47}$$

In this case, the collision has no effect on the orientation of the spins.

For other values of T, we have:

$$| \psi(+\infty) \rangle = \alpha | +, - \rangle + \beta | -, + \rangle \tag{48}$$

with α and β simultaneously non-zero. The state of the two-spin system has been transformed by the collision into a linear superposition of the two states $| +, - \rangle$ and $| -, + \rangle$. $| \psi(+\infty) \rangle$ *is therefore no longer a tensor product*, although $| \psi(-\infty) \rangle$ was one: the interaction of the two spins has introduced *correlations* between them.

To see this, we shall analyze an experiment in which, after the collision, an observer [observer (1)] measures S_{1z}. According to formula (48) for $| \psi(+\infty) \rangle$, he has the probability $|\alpha|^2$ of finding $+\hbar/2$ and $|\beta|^2$ of finding $-\hbar/2$ [according to (42), $|\alpha|^2 + |\beta|^2 = 1$]. Assume that he finds $-\hbar/2$. Immediately after this measurement, the state of the total system is, according to the wave packet reduction postulate, $| -, + \rangle$. If, at this moment, a second observer [observer (2)] measures S_{2z}, he will always find $+\hbar/2$. Similarly, it can easily be shown that if observer (1) finds the result $+\hbar/2$, observer (2) will then always find $-\hbar/2$. Thus, the result obtained by observer (1) critically influences the result that observer (2) will obtain later, even if at the time of these two measurements, the particles are extremely far apart. This apparently paradoxical result (the Einstein-Podolsky-Rosen paradox) reflects the existence of a strong correlation between the two spins, which has appeared because of their interaction during the collision.

Note, finally, that if we are concerned with only one of the two spins, it is impossible to describe its state after the collision by a state vector, since, according to formula (48), $| \psi(+\infty) \rangle$ is not a tensor product. Spin (1), for example, can be described in this case only by a density operator (*cf.* complement E_{III}). Let

$$\rho = | \psi(+\infty) \rangle \langle \psi(+\infty) | \tag{49}$$

be the density operator of the total two-spin system. According to the results of complement E_{III} (§ 5-b), the density operator of spin (1) can be obtained by taking the partial trace of ρ with respect to the spin variables of particle (2):

$$\rho(1) = \mathrm{Tr}_2\, \rho \tag{50}$$

Similarly:

$$\rho(2) = \mathrm{Tr}_1\, \rho \tag{51}$$

It is easy to calculate, from expression (48) for $|\,\psi(+\infty)\,\rangle$, the matrix representing ρ in the four-state basis, $\{\,|\,+,+\,\rangle, |\,+,-\,\rangle, |\,-,+\,\rangle, |\,-,-\,\rangle\,\}$, arranged in this order. We find:

$$\rho = \begin{pmatrix} 0 & 0 & 0 & 0 \\ 0 & |\alpha|^2 & \alpha\beta^* & 0 \\ 0 & \beta\alpha^* & |\beta|^2 & 0 \\ 0 & 0 & 0 & 0 \end{pmatrix} \tag{52}$$

Applying (50) and (51), we then find:

$$\rho(1) = \begin{pmatrix} |\alpha|^2 & 0 \\ 0 & |\beta|^2 \end{pmatrix} \tag{53}$$

$$\rho(2) = \begin{pmatrix} |\beta|^2 & 0 \\ 0 & |\alpha|^2 \end{pmatrix} \tag{54}$$

Starting with expressions (53) and (54), we can form:

$$\rho' = \rho(1) \otimes \rho(2) \tag{55}$$

whose matrix representation can be written:

$$\rho' = \begin{pmatrix} |\alpha|^2\,|\beta|^2 & 0 & 0 & 0 \\ 0 & |\alpha|^4 & 0 & 0 \\ 0 & 0 & |\beta|^4 & 0 \\ 0 & 0 & 0 & |\alpha|^2\,|\beta|^2 \end{pmatrix} \tag{56}$$

We see that ρ' is different from ρ, reflecting the existence of correlations between the two spins.

References and suggestions for further readings

The vector model of the atom: Eisberg and Resnick (1.3), chap. 8, §5; Cagnac and Pebay-Peyroula (11.2), chaps. XVI, §3B and XVII, §§3E and 4C.

The Einstein-Podolsky-Rosen paradox: see references of complement D_{III}.

Complement G_X

EXERCISES

1. Consider a deuterium atom (composed of a nucleus of spin $I = 1$ and an electron). The electronic angular momentum is $\mathbf{J} = \mathbf{L} + \mathbf{S}$, where $\mathbf{L}$ is the orbital angular momentum of the electron and $\mathbf{S}$ is its spin. The total angular momentum of the atom is $\mathbf{F} = \mathbf{J} + \mathbf{I}$, where $\mathbf{I}$ is the nuclear spin. The eigenvalues of $\mathbf{J}^2$ and $\mathbf{F}^2$ are $J(J+1)\hbar^2$ and $F(F+1)\hbar^2$ respectively.

a. What are the possible values of the quantum numbers J and F for a deuterium atom in the $1s$ ground state?

b. Same question for deuterium in the $2p$ excited state.

2. The hydrogen atom nucleus is a proton of spin $I = 1/2$.

a. In the notation of the preceding exercise, what are the possible values of the quantum numbers J and F for a hydrogen atom in the $2p$ level?

b. Let $\{ | n, l, m \rangle \}$ be the stationary states of the Hamiltonian H_0 of the hydrogen atom studied in §C of chapter VII.

Let $\{ | n, l, s, J, M_J \rangle \}$ be the basis obtained by adding $\mathbf{L}$ and $\mathbf{S}$ to form $\mathbf{J}$ ($M_J\hbar$ is the eigenvalue of J_z); and let $\{ | n, l, s, J, I, F, M_F \rangle \}$ be the basis obtained by adding $\mathbf{J}$ and $\mathbf{I}$ to form $\mathbf{F}$ ($M_F\hbar$ is the eigenvalue of F_z).

The magnetic moment operator of the electron is :

$$\mathbf{M} = \mu_B(\mathbf{L} + 2\mathbf{S})/\hbar$$

In each of the subspaces $\mathscr{E}(n = 2, l = 1, s = 1/2, J, I = 1/2, F)$ arising from the $2p$ level and subtended by the $2F + 1$ vectors

$$| n = 2, l = 1, s = \frac{1}{2}, J, I = \frac{1}{2}, F, M_F \rangle$$

corresponding to fixed values of J and F, the projection theorem (*cf.* complement D_X, §§ 2-c and 3) enables us to write:

$$\mathbf{M} = g_{JF}\,\mu_B\mathbf{F}/\hbar$$

Calculate the various possible values of the Landé factors g_{JF} corresponding to the $2p$ level.

3. Consider a system composed of two spin 1/2 particles whose orbital variables are ignored. The Hamiltonian of the system is:

$$H = \omega_1 S_{1z} + \omega_2 S_{2z}$$

where S_{1z} and S_{2z} are the projections of the spins $\mathbf{S}_1$ and $\mathbf{S}_2$ of the two particles onto Oz, and ω_1 and ω_2 are real constants.

a. The initial state of the system, at time $t = 0$, is:

$$| \psi(0) \rangle = \frac{1}{\sqrt{2}} [| + - \rangle + | - + \rangle]$$

(with the notation of § B of chapter X). At time t, $\mathbf{S}^2 = (\mathbf{S}_1 + \mathbf{S}_2)^2$ is measured. What results can be found, and with what probabilities?

b. If the initial state of the system is arbitrary, what Bohr frequencies can appear in the evolution of $\langle \mathbf{S}^2 \rangle$? Same question for $S_x = S_{1x} + S_{2x}$.

4. Consider a particle (a) of spin 3/2 which can disintegrate into two particles, (b) of spin 1/2 and (c) of spin 0. We place ourselves in the rest frame of (a). Total angular momentum is conserved during the disintegration.

a. What values can be taken on by the relative orbital angular momentum of the two final particles? Show that there is only one possible value if the parity of the relative orbital state is fixed. Would this result remain valid if the spin of particle (a) were greater than 3/2?

b. Assume that particle (a) is initially in the spin state characterized by the eigenvalue $m_a \hbar$ of its spin component along Oz. We know that the final orbital state has a definite parity. Is it possible to determine this parity by measuring the probabilities of finding particle (b) either in the state $| + \rangle$ or in the state $| - \rangle$ (you may use the general formulas of complement A_X, §2)?

5. Let $\mathbf{S} = \mathbf{S}_1 + \mathbf{S}_2 + \mathbf{S}_3$ be the total angular momentum of three spin 1/2 particles (whose orbital variables will be ignored). Let $| \varepsilon_1, \varepsilon_2, \varepsilon_3 \rangle$ be the eigenstates common to S_{1z}, S_{2z}, S_{3z}, of respective eigenvalues $\varepsilon_1 \hbar/2$, $\varepsilon_2 \hbar/2$, $\varepsilon_3 \hbar/2$. Give a basis of eigenvectors common to $\mathbf{S}^2$ and S_z, in terms of the kets $| \varepsilon_1, \varepsilon_2, \varepsilon_3 \rangle$. Do these two operators form a C.S.C.O.? (Begin by adding two of the spins, then add the partial angular momentum so obtained to the third one.)

6. Let $\mathbf{S}_1$ and $\mathbf{S}_2$ be the intrinsic angular momenta of two spin 1/2 particles, $\mathbf{R}_1$ and $\mathbf{R}_2$, their position observables, and m_1 and m_2, their masses $\left(\text{with } \mu = \dfrac{m_1 m_2}{m_1 + m_2}, \text{ the reduced mass}\right)$. Assume that the interaction W between the two particles is of the form:

$$W = U(R) + V(R) \frac{\mathbf{S}_1 \cdot \mathbf{S}_2}{\hbar^2}$$

where $U(R)$ and $V(R)$ depend only on the distance $R = |\mathbf{R}_1 - \mathbf{R}_2|$ between the particles.

a. Let $\mathbf{S} = \mathbf{S}_1 + \mathbf{S}_2$ be the total spin of the two particles.

α. Show that:

$$P_1 = \frac{3}{4} + \frac{\mathbf{S}_1 \cdot \mathbf{S}_2}{\hbar^2}$$

$$P_0 = \frac{1}{4} - \frac{\mathbf{S}_1 \cdot \mathbf{S}_2}{\hbar^2}$$

are the projectors onto the total spin states $S = 1$ and $S = 0$ respectively.

β. Show from this that $W = W_1(R)P_1 + W_0(R)P_0$, where $W_1(R)$ and $W_0(R)$ are two functions of R, to be expressed in terms of $U(R)$ and $V(R)$.

b. Write the Hamiltonian H of the "relative particle" in the center of mass frame; $\mathbf{P}$ denotes the momentum of this relative particle. Show that H commutes with $\mathbf{S}^2$ and does not depend on S_z. Show from this that it is possible to study separately the eigenstates of H corresponding to $S = 1$ and $S = 0$.

Show that one can find eigenstates of H, with eigenvalue E, of the form:

$$|\psi_E\rangle = \lambda_{00}\,|\varphi_E^0\rangle\,|S = 0, M = 0\rangle + \sum_{M=-1}^{+1} \lambda_{1M}\,|\varphi_E^1\rangle\,|S = 1, M\rangle$$

where λ_{00} and λ_{1M} are constants, and $|\varphi_E^0\rangle$ and $|\varphi_E^1\rangle$ are kets of the state space $\mathscr{E}_\mathbf{r}$ of the relative particle ($M\hbar$ is the eigenvalue of S_z). Write the eigenvalue equations satisfied by $|\varphi_E^0\rangle$ and $|\varphi_E^1\rangle$.

c. We want to study collisions between the two particles under consideration. Let $E = \hbar^2k^2/2\mu$ be the energy of the system in the center of mass frame. We assume in all that follows that, before the collision, one of the particles is in the $|+\rangle$ spin state, and the other one, in the $|-\rangle$ spin state. Let $|\psi_k^{\downarrow\uparrow}\rangle$ be the corresponding stationary scattering state (*cf*. chap. VIII, § B). Show that:

$$|\psi_k^{\uparrow\downarrow}\rangle = \frac{1}{\sqrt{2}}|\varphi_k^0\rangle\,|S = 0, M = 0\rangle + \frac{1}{\sqrt{2}}|\varphi_k^1\rangle\,|S = 1, M = 0\rangle$$

where $|\varphi_k^0\rangle$ and $|\varphi_k^1\rangle$ are the stationary scattering states for a spinless particle of mass μ, scattered respectively by the potentials $W_0(R)$ and $W_1(R)$.

d. Let $f_0(\theta)$ and $f_1(\theta)$ be the scattering amplitudes which correspond to $|\varphi_k^0\rangle$ and $|\varphi_k^1\rangle$. Calculate, in terms of $f_0(\theta)$ and $f_1(\theta)$, the scattering cross section $\sigma_b(\theta)$ of the two particles in the θ direction, with simultaneous flip of the two spins (the spin which was in the $|+\rangle$ state goes into the $|-\rangle$ state, and vice versa).

e. Let δ_l^0 and δ_l^1 be the phase shifts of the l partial waves associated respectively with $W_0(R)$ and $W_1(R)$ (*cf*. chap. VIII, §C-3). Show that the total scattering cross section σ_b, with simultaneous flip of the two spins, is equal to:

$$\sigma_b = \frac{\pi}{k^2}\sum_{l=0}^{\infty}(2l + 1)\sin^2(\delta_l^1 - \delta_l^0)$$

7. We define the standard components of a vector operator **V** as the three operators:

$$\begin{cases} V_1^{(1)} = -\dfrac{1}{\sqrt{2}}(V_x + iV_y) \\ V_0^{(1)} = V_z \\ V_{-1}^{(1)} = \dfrac{1}{\sqrt{2}}(V_x - iV_y) \end{cases}$$

Using the standard components $V_p^{(1)}$ and $W_q^{(1)}$ of the two vector operators **V** and **W**, we construct the operators:

$$[V^{(1)} \otimes W^{(1)}]_M^{(K)} = \sum_p \sum_q \langle 1, 1; p, q \mid K, M \rangle \, V_p^{(1)} W_q^{(1)}$$

where the $\langle 1, 1; p, q \mid K, M \rangle$ are the Clebsch-Gordan coefficients entering into the addition of two angular momenta 1 (these coefficients can be obtained from the results of §1 of complement A_X).

a. Show that $[V^{(1)} \otimes W^{(1)}]_0^{(0)}$ is proportional to the scalar product $\mathbf{V} \cdot \mathbf{W}$ of the two vector operators.

b. Show that the three operators $[V^{(1)} \otimes W^{(1)}]_M^{(1)}$ are proportional to the three standard components of the vector operator $\mathbf{V} \times \mathbf{W}$.

c. Express the five components $[V^{(1)} \otimes W^{(1)}]_M^{(2)}$ in terms of V_z, $V_\pm = V_x \pm iV_y$, W_z, $W_\pm = W_x \pm iW_y$.

d. We choose $\mathbf{V} = \mathbf{W} = \mathbf{R}$, where **R** is the position observable of a particle. Show that the five operators $[R^{(1)} \otimes R^{(1)}]_M^{(2)}$ are proportional to the five components Q_2^M of the electric quadrupole moment operator of this particle [*cf.* formula (29) of complement E_X].

e. We choose $\mathbf{V} = \mathbf{W} = \mathbf{L}$, where **L** is the orbital angular momentum of the particle. Express the five operators $[L^{(1)} \otimes L^{(1)}]_M^{(2)}$ in terms of L_z, L_+, L_-. What are the selection rules satisfied by these five operators in a standard basis $\{ \mid k, l, m \rangle \}$ of eigenstates common to $\mathbf{L}^2$ and L_z (in other words, on what conditions is the matrix element

$$\langle k, l, m \mid [L^{(1)} \otimes L^{(1)}]_M^{(2)} \mid k', l', m' \rangle$$

non-zero)?

8. Irreducible tensor operators; Wigner-Eckart theorem

The $2K + 1$ operators $T_Q^{(K)}$, with K an integer $\geqslant 0$ and

$Q = -K, -K + 1, ..., +K,$

are, by definition, the $2K + 1$ components of an irreducible tensor operator of rank K if they satisfy the following commutation relations with the total angular momentum **J** of the physical system:

$$[J_z, T_Q^{(K)}] = \hbar Q T_Q^{(K)} \tag{1}$$

$$[J_+, T_Q^{(K)}] = \hbar\sqrt{K(K+1) - Q(Q+1)}\, T_{Q+1}^{(K)} \tag{2}$$

$$[J_-, T_Q^{(K)}] = \hbar\sqrt{K(K+1) - Q(Q-1)}\, T_{Q-1}^{(K)} \tag{3}$$

a. Show that a scalar operator is an irreducible tensor operator of rank $K = 0$, and that the three standard components of a vector operator (*cf.* exercise 7) are the components of an irreducible tensor operator of rank $K = 1$.

b. Let $\{ | k, J, M \rangle \}$ be a standard basis of common eigenstates of $\mathbf{J}^2$ and J_z. By taking both sides of (1) to have the same matrix elements between $| k, J, M \rangle$ and $| k', J', M' \rangle$, show that $\langle k, J, M | T_Q^{(K)} | k', J', M' \rangle$ is zero if M is not equal to $Q + M'$.

c. Proceeding in the same way with relations (2) and (3), show that the $(2J+1)(2K+1)(2J'+1)$ matrix elements $\langle k, J, M | T_Q^{(K)} | k', J', M' \rangle$ corresponding to fixed values of k, J, K, k', J' satisfy recurrence relations identical to those satisfied by the $(2J+1)(2K+1)(2J'+1)$ Clebsch-Gordan coefficients $\langle J', K; M', Q | J, M \rangle$ (*cf.* complement B_X, §§ 1-c and 2) corresponding to fixed values of J, K, J'.

d. Show that:

$$\langle k, J, M | T_Q^{(K)} | k', J', M' \rangle = \alpha \langle J', K; M', Q | J, M \rangle \tag{4}$$

where α is a constant depending only on k, J, K, k', J', which is usually written in the form:

$$\alpha = \frac{1}{\sqrt{2J+1}} \langle k, J \| T^{(K)} \| k', J' \rangle$$

e. Show that, conversely, if $(2K+1)$ operators $T_Q^{(K)}$ satisfy relation (4) for all $| k, J, M \rangle$ and $| k', J', M' \rangle$, they satisfy relations (1), (2) and (3), that is, they constitute the $(2K+1)$ components of an irreducible tensor operator of rank K.

f. Show that, for a spinless particle, the electric multipole moment operators Q_l^m introduced in complement E_X are irreducible tensor operators of rank l in the state space $\mathscr{E}_\mathbf{r}$ of this particle. Show that, in addition, when the spin degrees of freedom are taken into account, the operators Q_l^m remain irreducible tensor operators in the state space $\mathscr{E}_\mathbf{r} \otimes \mathscr{E}_s$ (where $\mathscr{E}_s$ is the spin state space).

g. Derive the selection rules satisfied by the Q_l^m in a standard basis $\{ | k, l, J, M_J \rangle \}$ obtained by adding the orbital angular momentum $\mathbf{L}$ and the spin $\mathbf{S}$ of the particle to form the total angular momentum $\mathbf{J} = \mathbf{L} + \mathbf{S}$ [$l(l+1)\hbar^2$, $J(J+1)\hbar^2$, $M_J\hbar$ are the eigenvalues of $\mathbf{L}^2$, $\mathbf{J}^2$, J_z respectively].

9. Let $A_{Q_1}^{(K_1)}$ be an irreducible tensor operator (exercise 8) of rank K_1 acting in a state space $\mathscr{E}_1$, and $B_{Q_2}^{(K_2)}$, an irreducible tensor operator of rank K_2 acting in a state space $\mathscr{E}_2$. With $A_{Q_1}^{(K_1)}$ and $B_{Q_2}^{(K_2)}$, we construct the operator :

$$C_Q^{(K)} = [A^{(K_1)} \otimes B^{(K_2)}]_Q^{(K)} = \sum_{Q_1 Q_2} \langle K_1, K_2; Q_1, Q_2 | K, Q \rangle A_{Q_1}^{(k_1)} B_{Q_1}^{(K_2)}$$

a. Using the recurrence relations for Clebsch-Gordan coefficients (*cf.* complement B_X), show that the $C_Q^{(K)}$ satisfy commutation relations (1), (2) and (3) of exercise 8 with the total angular momentum $\mathbf{J} = \mathbf{J}_1 + \mathbf{J}_2$ of the system. Show that the $C_Q^{(K)}$ are the components of an irreducible tensor operator of rank K.

b. Show that the operator $\sum_Q (-1)^Q A_Q^{(K)} B_{-Q}^{(K)}$ is a scalar operator (you may use the results of §3-d of complement B_X).

10. Addition of three angular momenta

Let $\mathscr{E}(1)$, $\mathscr{E}(2)$, $\mathscr{E}(3)$ be the state spaces of three systems, (1), (2) and (3), of angular momenta $\mathbf{J}_1$, $\mathbf{J}_2$, $\mathbf{J}_3$. We shall write $\mathbf{J} = \mathbf{J}_1 + \mathbf{J}_2 + \mathbf{J}_3$ for the total angular momentum. Let $\{ | k_a, j_a, m_a \rangle \}$, $\{ | k_b, j_b, m_b \rangle \}$, $\{ | k_c, j_c, m_c \rangle \}$ be the standard bases of $\mathscr{E}(1)$, $\mathscr{E}(2)$, $\mathscr{E}(3)$, respectively. To simplify the notation, we shall omit the indices k_a, k_b, k_c, as we did in chapter X.

We are interested in the eigenstates and eigenvalues of the total angular momentum in the subspace $\mathscr{E}(j_a, j_b, j_c)$ subtended by the kets:

$$\{ | j_a m_a \rangle | j_b m_b \rangle | j_c m_c \rangle \}$$
$$- j_a \leqslant m_a \leqslant j_a, \; - j_b \leqslant m_b \leqslant j_b, \; - j_c \leqslant m_c \leqslant j_c \tag{1}$$

We want to add j_a, j_b, j_c to form an eigenstate of $\mathbf{J}^2$ and J_z characterized by the quantum numbers j_f and m_f. We shall denote by:

$$| j_a, (j_b j_c) j_e ; j_f m_f \rangle \tag{2}$$

such a normalized eigenstate obtained by first adding j_b to j_c to form an angular momentum j_e, then adding j_a to j_e to form the state $| j_f m_f \rangle$. One could also add j_a and j_b to form j_g and then add j_c to j_g to form the normalized state $| j_f m_f \rangle$, written:

$$| (j_a j_b) j_g, j_c ; j_f m_f \rangle \tag{3}$$

a. Show that the system of kets (2), corresponding to the various possible values of j_e, j_f, m_f, forms an orthonormal basis in $\mathscr{E}(j_a, j_b, j_c)$. Same question for the system of kets (3), corresponding to the various values of j_g, j_f, m_f.

b. Show, by using the operators $J_\pm$, that the scalar product $\langle (j_a j_b) j_g, j_c ; j_f m_f | j_a, (j_b j_c) j_e ; j_f m_f \rangle$ does not depend on m_f, denoting such a scalar product by $\langle (j_a j_b) j_g, j_c ; j_f | j_a, (j_b j_c) j_e ; j_f \rangle$.

c. Show that:

$$| j_a, (j_b j_c) j_e ; j_f m_f \rangle = \sum_{j_g} \langle (j_a j_b) j_g, j_c ; j_f | j_a, (j_b j_c) j_e ; j_f \rangle \, | (j_a j_b) j_g, j_c ; j_f m_f \rangle \tag{4}$$

d. Using the Clebsch-Gordan coefficients, write the expansions for vectors (2) and (3) in the basis (1). Show that:

$$\sum_{m_e} \langle j_b, j_c ; m_b, m_c \mid j_e, m_e \rangle \langle j_a, j_e ; m_a, m_e \mid j_f, m_f \rangle =$$

$$\sum_{j_g m_g} \langle j_a, j_b ; m_a, m_b \mid j_g, m_g \rangle \langle j_g, j_c ; m_g, m_c \mid j_f, m_f \rangle$$

$$\times \langle (j_a j_b) j_g, j_c ; j_f \mid j_a, (j_b j_c) j_e ; j_f \rangle \quad (5)$$

e. Starting with relation (5), prove, using the Clebsch-Gordan coefficient orthogonality relations, the following relations:

$$\sum_{m_a m_b m_e} \langle j_b, j_c ; m_b, m_c \mid j_e, m_e \rangle \langle j_a, j_e ; m_a, m_e \mid j_f, m_f \rangle \langle j_d, m_d \mid j_a, j_b ; m_a, m_b \rangle$$

$$= \langle j_d, j_c ; m_d, m_c \mid j_f, m_f \rangle \langle (j_a j_b) j_d, j_c ; j_f \mid j_a, (j_b j_c) j_e ; j_f \rangle \quad (6)$$

as well as:

$$\langle (j_a j_b) j_d, j_c ; j_f \mid j_a, (j_b j_c) j_e ; j_f \rangle = \frac{1}{2j_f + 1} \sum_{m_a m_b m_c m_d m_e m_f} \langle j_b, j_c ; m_b, m_c \mid j_e, m_e \rangle$$

$$\times \langle j_a, j_e ; m_a, m_e \mid j_f, m_f \rangle \langle j_d, m_d \mid j_a, j_b ; m_a, m_b \rangle \langle j_f, m_f \mid j_d, j_c ; m_d, m_c \rangle \quad (7)$$

Exercises 8 and 9:

References : see references of complement D_X

Exercise 10:

References: Edmonds (2.21), chap. 6; Messiah (1.17), §XIII-29 and App. C; Rose (2.19), App. I.

CHAPTER XI

Stationary perturbation theory

OUTLINE OF CHAPTER XI

The quantum mechanical study of conservative physical systems (that is, systems whose Hamiltonians are not explicitly time-dependent) is based on the eigenvalue equation of the Hamiltonian operator. We have already encountered two important examples of physical systems (the harmonic oscillator and the hydrogen atom) whose Hamiltonians are simple enough for their eigenvalue equations to be solved exactly. However, this happens in only a very small number of problems. In general, the equation is too complicated for us to be able to find its solutions in an analytic form★. For example, we do not know how to treat many-electron atoms, even helium, exactly. Besides, the hydrogen atom theory explained in chapter VII (§C) takes into account only the electrostatic interaction between the proton and the electron; when relativistic corrections (such as magnetic forces) are added to this principal interaction, the equation obtained for the hydrogen atom can no longer be solved analytically. We must then resort to solving it numerically, frequently with a computer. There exist, however, *approximation methods* which enable us to obtain analytically approximate solutions of the basic eigenvalue equation in certain cases. In this chapter, we shall study one of these methods, known as "stationary perturbation theory"★★. (In chapter XIII, we shall describe "time-dependent perturbation theory", which is used to treat systems whose Hamiltonians contain explicitly time-dependent terms.)

Stationary perturbation theory is very widely used in quantum physics, since it corresponds very well to the physicist's usual approach to problems. In studying a phenomenon or a physical system, one begins by isolating the principal effects which are responsible for the main features of this phenomenon or this system. When they have been understood, one tries to explain the "finer" details by taking into account less important effects that were neglected in the first approximation. It is in treating these secondary effects that one commonly uses perturbation theory. In chapter XII, we shall see, for example, the importance of perturbation theory in atomic physics: it will enable us to calculate the relativistic corrections in the case of the hydrogen atom. Similarly, complement B_{XIV}, which is devoted to the helium atom, indicates how perturbation theory allows us to treat many-electron atoms. Numerous other applications of perturbation theory are given in the complements of this chapter and the following ones.

★ Of course, this phenomenon is not limited to the domain of quantum mechanics. In all fields of physics, there are very few problems which can be treated completely analytically.

★★ Perturbation theory also exists in classical mechanics, where it is, in principle, entirely analogous to the one we shall describe here.

Let us mention, finally, another often used approximation method, the variational method, which we shall present in complement E_{XI}. We shall briefly examine its applications in solid state physics (complement F_{XI}) and a molecular physics (complement G_{XI}).

A. DESCRIPTION OF THE METHOD

1. Statement of the problem

Perturbation theory is applicable when the Hamiltonian H of the system being studied can be put in the form:

$$H = H_0 + W \tag{A-1}$$

where the eigenstates and eigenvalues of H_0 are known, and where W is much smaller than H_0. The operator H_0, which is time-independent, is called the "unperturbed Hamiltonian" and W, the "perturbation". If W is not time-dependent, we say that we are dealing with a "stationary perturbation"; this is the case we are considering in this chapter (the case of time-dependent perturbations will be studied in chapter XIII). The problem is then to find the modifications produced in the energy levels of the system and in its stationary states by the addition of the perturbation W.

When we say that W is much smaller than H_0, this means that the matrix elements of W are much smaller than those of H_0★. To make this more explicit, we shall assume that W is proportional to a real parameter λ which is dimensionless and much smaller than 1:

$$\begin{aligned} W &= \lambda \hat{W} \\ \lambda &\ll 1 \end{aligned} \tag{A-2}$$

(where $\hat{W}$ is an operator whose matrix elements are comparable to those of H_0). Perturbation theory consist of expanding the eigenvalues and eigenstates of H in powers of λ, keeping only a finite number of terms (often only one or two) of these expansions.

We shall assume the eigenstates and eigenvalues of the unperturbed Hamiltonian H_0 to be known. In addition, we shall assume that *the unperturbed energies form a discrete spectrum*, and we shall label them by an integral index p: E_p^0. The corresponding eigenstates will be denoted by $| \varphi_p^i \rangle$, the additional index i permitting us, in the case of a degenerate eigenvalue E_p^0, to distinguish between the various vectors of an orthonormal basis of the associated eigensubspace. We therefore have:

$$H_0 | \varphi_p^i \rangle = E_p^0 | \varphi_p^i \rangle \tag{A-3}$$

★ More precisely, the important point is that the matrix elements of W are much smaller than the differences between eigenvalues of H_0 (*cf.* comment of § B-1-b).

where the set of vectors $| \varphi_p^i \rangle$ forms an orthonormal basis of the state space:

$$\langle \varphi_p^i | \varphi_{p'}^{i'} \rangle = \delta_{pp'} \delta_{ii'} \tag{A-4-a}$$

$$\sum_p \sum_i | \varphi_p^i \rangle \langle \varphi_p^i | = 1 \tag{A-4-b}$$

If we substitute (A-2) into (A-1), we can consider the Hamiltonian of the system to be continuously dependent on the parameter λ characterizing the intensity of the perturbation:

$$H(\lambda) = H_0 + \lambda \hat{W} \tag{A-5}$$

When λ is equal to zero, $H(\lambda)$ is equal to the unperturbed Hamiltonian H_0. The eigenvalues $E(\lambda)$ of $H(\lambda)$ generally depend on λ, and figure 1 represents possible forms of their variations with respect to λ.

An eigenvector of $H(\lambda)$ is associated with each curve of figure 1. For a given value of λ, these vectors form a basis of the state space [$H(\lambda)$ is an observable]. When λ is much smaller than 1, the eigenvalues $E(\lambda)$ and the eigenvectors $| \psi(\lambda) \rangle$ of $H(\lambda)$ remain very close to those of $H_0 = H(\lambda = 0)$, which they approach when $\lambda \longrightarrow 0$.

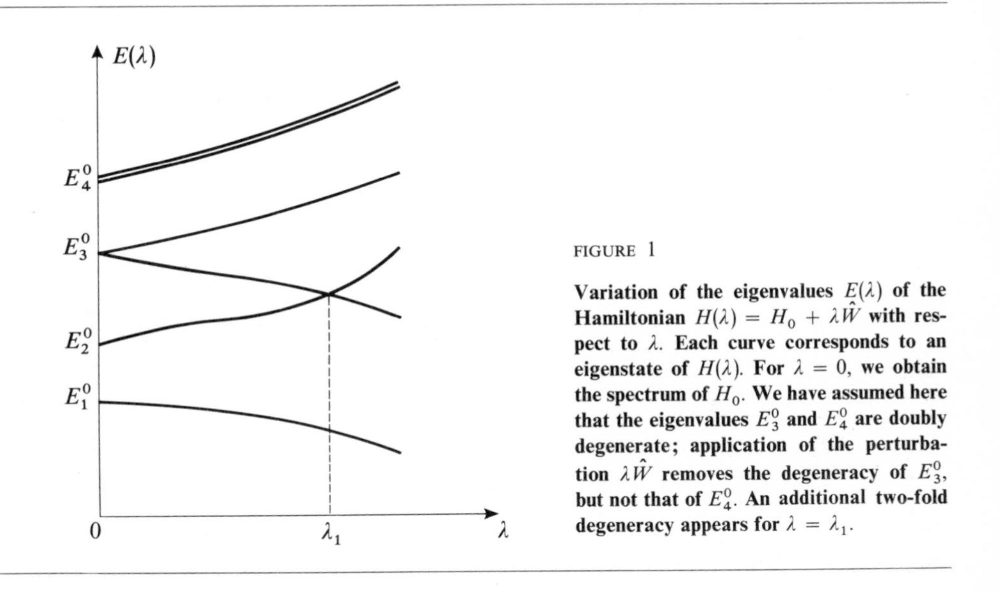

FIGURE 1

Variation of the eigenvalues $E(\lambda)$ of the Hamiltonian $H(\lambda) = H_0 + \lambda \hat{W}$ with respect to λ. Each curve corresponds to an eigenstate of $H(\lambda)$. For $\lambda = 0$, we obtain the spectrum of H_0. We have assumed here that the eigenvalues E_3^0 and E_4^0 are doubly degenerate; application of the perturbation $\lambda \hat{W}$ removes the degeneracy of E_3^0, but not that of E_4^0. An additional two-fold degeneracy appears for $\lambda = \lambda_1$.

$H(\lambda)$ may, of course, have one or several degenerate eigenvalues. For example, in figure 1, the double curve represents a doubly degenerate energy (the one which approaches E_4^0 when $\lambda \longrightarrow 0$), which corresponds, for all λ, to a two-dimensional eigensubspace. It is also possible for several distinct eigenvalues $E(\lambda)$ to approach the same unperturbed energy E_p^0 when $\lambda \longrightarrow 0$★ (this is the case for E_3^0 in figure 1).

★ It is not impossible that additional degeneracies may appear for particular non-zero values of λ (crossing at $\lambda = \lambda_1$ in figure 1). We shall assume here that λ is small enough to avoid such a situation.

We then say that the effect of the perturbation is to remove the degeneracy of the corresponding eigenvalue of H_0.

In the following section, we shall give an approximate solution of the eigenvalue equation of $H(\lambda)$ for $\lambda \ll 1$.

2. Approximate solution of the $H(\lambda)$ eigenvalue equation

We are looking for the eigenstates $| \psi(\lambda) \rangle$ and eigenvalues $E(\lambda)$ of the Hermitian operator $H(\lambda)$:

$$H(\lambda) | \psi(\lambda) \rangle = E(\lambda) | \psi(\lambda) \rangle \tag{A-6}$$

We shall assume★ that $E(\lambda)$ and $| \psi(\lambda) \rangle$ can be expanded in powers of λ in the form:

$$E(\lambda) = \varepsilon_0 + \lambda \varepsilon_1 + \ldots + \lambda^q \varepsilon_q + \ldots \tag{A-7-a}$$

$$| \psi(\lambda) \rangle = | 0 \rangle + \lambda | 1 \rangle + \ldots + \lambda^q | q \rangle + \ldots \tag{A-7-b}$$

We then substitute these expansions, as well as definition (A-5) of $H(\lambda)$, into equation (A-6):

$$(H_0 + \lambda \hat{W}) \left[\sum_{q=0}^{\infty} \lambda^q | q \rangle \right] = \left[\sum_{q'=0}^{\infty} \lambda^{q'} \varepsilon_{q'} \right] \left[\sum_{q=0}^{\infty} \lambda^q | q \rangle \right] \tag{A-8}$$

We require this equation to be satisfied for λ small but arbitrary. We must therefore equate the coefficients of successive powers of λ on both sides. Thus, we obtain:

– for 0th-order terms in λ:

$$H_0 | 0 \rangle = \varepsilon_0 | 0 \rangle \tag{A-9}$$

– for 1st-order terms:

$$(H_0 - \varepsilon_0) | 1 \rangle + (\hat{W} - \varepsilon_1) | 0 \rangle = 0 \tag{A-10}$$

– for 2nd-order terms:

$$(H_0 - \varepsilon_0) | 2 \rangle + (\hat{W} - \varepsilon_1) | 1 \rangle - \varepsilon_2 | 0 \rangle = 0 \tag{A-11}$$

– for qth-order terms:

$$(H_0 - \varepsilon_0) | q \rangle + (\hat{W} - \varepsilon_1) | q - 1 \rangle - \varepsilon_2 | q - 2 \rangle \ldots - \varepsilon_q | 0 \rangle = 0 \tag{A-12}$$

★ This is not obvious from a mathematical point of view, the basic problem being the convergence of the series (A-7).

We shall confine ourselves here to the study of the first three equations, that is, we shall neglect, in expansions (A-7), terms of orders higher than 2 in λ.

We know that the eigenvalue equation (A-6) defines $|\psi(\lambda)\rangle$ only to within a constant factor. We can therefore choose the norm of $|\psi(\lambda)\rangle$ and its phase: *we shall require $|\psi(\lambda)\rangle$ to be normalized, and we shall choose its phase such that the scalar product $\langle 0 | \psi(\lambda)\rangle$ is real.* To 0th order, this implies that the vector denoted by $|0\rangle$ must be normalized:

$$\langle 0 | 0 \rangle = 1 \tag{A-13}$$

Its phase, however, remains arbitrary ; we shall see in §§B and C how it can be chosen in each particular case. To 1st order, the square of the norm of $|\psi(\lambda)\rangle$ can be written:

$$\begin{aligned} \langle \psi(\lambda) | \psi(\lambda)\rangle &= [\langle 0 | + \lambda \langle 1 |][| 0 \rangle + \lambda | 1 \rangle] + O(\lambda^2) \\ &= \langle 0 | 0 \rangle + \lambda[\langle 1 | 0 \rangle + \langle 0 | 1 \rangle] + O(\lambda^2) \end{aligned} \tag{A-14}$$

(where the symbol $O(\lambda^p)$ signifies all the terms of order higher than or equal to p). Using (A-13), we see that this expression is equal to 1 to first order if the λ term is zero. But the choice of phase indicates that the scalar product $\langle 0 | 1 \rangle$ is real (since λ is real). We therefore obtain:

$$\langle 0 | 1 \rangle = \langle 1 | 0 \rangle = 0 \tag{A-15}$$

An analogous argument, for 2nd order in λ, yields:

$$\langle 0 | 2 \rangle = \langle 2 | 0 \rangle = -\frac{1}{2}\langle 1 | 1 \rangle \tag{A-16}$$

and, for qth order:

$$\begin{aligned} \langle 0 | q \rangle &= \langle q | 0 \rangle \\ &= -\frac{1}{2}[\langle q-1 | 1 \rangle + \langle q-2 | 2 \rangle + \ldots \\ &\qquad + \langle 2 | q-2 \rangle + \langle 1 | q-1 \rangle] \end{aligned} \tag{A-17}$$

When we confine ourselves to second order in λ, the perturbation equations are therefore (A-9), (A-10) and (A-11). With the conventions we have set, we must add conditions (A-13), (A-15) and (A-16).

Equation (A-9) expresses the fact that $|0\rangle$ is an eigenvector of H_0 with the eigenvalue ε_0. ε_0 therefore belongs to the spectrum of H_0. This was to be expected, since each eigenvalue of $H(\lambda)$, when $\lambda \longrightarrow 0$, approaches one of the unperturbed energies. We then choose a particular value of ε_0, that is, an eigenvalue E_n^0 of H_0. As figure 1 shows, there can exist one or several different energies $E(\lambda)$ of $H(\lambda)$ which approach E_n^0 when $\lambda \longrightarrow 0$.

Consider the set of eigenstates of $H(\lambda)$ corresponding to the various eigenvalues $E(\lambda)$ which approach E_n^0 when $\lambda \longrightarrow 0$. They span a vector subspace whose dimension clearly cannot vary discontinuously when λ varies in the neighborhood of zero. This dimension is consequently equal to the degeneracy g_n

of E_n^0. In particular, if E_n^0 is non-degenerate, it can give rise only to a single energy $E(\lambda)$, and this energy is non-degenerate.

To study the influence of the perturbation W, we shall consider separately the case of non-degenerate, and degenerate levels of H_0.

B. PERTURBATION OF A NON-DEGENERATE LEVEL

Consider a particular non-degenerate eigenvalue E_n^0 of the unperturbed Hamiltonian H_0. Associated with it is an eigenvector $| \varphi_n \rangle$ which is unique to within a constant factor. We want to determine the modifications in this unperturbed energy and in the corresponding stationary state produced by the addition of the perturbation W to the Hamiltonian.

To do so, we shall use perturbation equations (A-9) through (A-12), as well as conditions (A-13) and (A-15) through (A-17). For the eigenvalue of $H(\lambda)$ which approaches E_n^0 when $\lambda \longrightarrow 0$, we have:

$$\varepsilon_0 = E_n^0 \tag{B-1}$$

which, according to (A-9), implies that $| 0 \rangle$ must be proportional to $| \varphi_n \rangle$. The vectors $| 0 \rangle$ and $| \varphi_n \rangle$ are both normalized [*cf.* (A-13)], and we shall choose:

$$| 0 \rangle = | \varphi_n \rangle \tag{B-2}$$

Thus, when $\lambda \longrightarrow 0$, we again find the unperturbed state $| \varphi_n \rangle$ with the same phase.

We call $E_n(\lambda)$ the eigenvalue of $H(\lambda)$ which, when $\lambda \longrightarrow 0$, approaches the eigenvalue E_n^0 of H_0. We shall assume λ small enough for this eigenvalue to remain non-degenerate, that is, for a unique eigenvector $| \psi_n(\lambda) \rangle$ to correspond to it (in the case of the $n = 2$ level of figure 1, this is satisfied if $\lambda < \lambda_1$). We shall now calculate the first terms of the expansion of $E_n(\lambda)$ and $| \psi_n(\lambda) \rangle$ in powers of λ.

1. First-order corrections

We shall begin by determining ε_1 and the vector $| 1 \rangle$ from equation (A-10) and condition (A-15).

a. ENERGY CORRECTION

Projecting equation (A-10) onto the vector $| \varphi_n \rangle$, we obtain:

$$\langle \varphi_n | (H_0 - \varepsilon_0) | 1 \rangle + \langle \varphi_n | (\hat{W} - \varepsilon_1) | 0 \rangle = 0 \tag{B-3}$$

The first term is zero, since $| \varphi_n \rangle = | 0 \rangle$ is an eigenvector of the Hermitian operator H_0 with the eigenvalue $E_n^0 = \varepsilon_0$. With (B-2) taken into account, equation (B-3) then yields:

$$\varepsilon_1 = \langle \varphi_n | \hat{W} | 0 \rangle = \langle \varphi_n | \hat{W} | \varphi_n \rangle \tag{B-4}$$

In the case of a non-degenerate state E_n^0, the eigenvalue $E_n(\lambda)$ of H which corresponds to E_n^0 can be written, to first order in the perturbation $W = \lambda \hat{W}$:

$$\boxed{E_n(\lambda) = E_n^0 + \langle \varphi_n | W | \varphi_n \rangle + O(\lambda^2)} \tag{B-5}$$

The first-order correction to a non-degenerate energy E_n^0 is simply equal to the mean value of the perturbation term W in the unperturbed state $| \varphi_n \rangle$.

b. EIGENVECTOR CORRECTION

The projection (B-3) obviously does not exhaust all the information contained in perturbation equation (A-10). We must now project this equation onto all the vectors of the $\{ | \varphi_p^i \rangle \}$ basis other than $| \varphi_n \rangle$. We obtain, using (B-1) and (B-2):

$$\langle \varphi_p^i | (H_0 - E_n^0) | 1 \rangle + \langle \varphi_p^i | (\hat{W} - \varepsilon_1) | \varphi_n \rangle = 0 \qquad (p \neq n) \tag{B-6}$$

(since the eigenvalues E_p^0 other than E_n^0 can be degenerate, we must retain the degeneracy index i here). Since the eigenvectors of H_0 associated with different eigenvalues are orthogonal, the last term, $\varepsilon_1 \langle \varphi_p^i | \varphi_n \rangle$, is zero. Furthermore, in the first term, we can let H_0 act on the left on $\langle \varphi_p^i |$. (B-6) then becomes:

$$(E_p^0 - E_n^0) \langle \varphi_p^i | 1 \rangle + \langle \varphi_p^i | \hat{W} | \varphi_n \rangle = 0 \tag{B-7}$$

which gives the coefficients of the desired expansion of the vector $| 1 \rangle$ on all the unperturbed basis states, except $| \varphi_n \rangle$:

$$\langle \varphi_p^i | 1 \rangle = \frac{1}{E_n^0 - E_p^0} \langle \varphi_p^i | \hat{W} | \varphi_n \rangle \qquad (p \neq n) \tag{B-8}$$

The last coefficient which we lack, $\langle \varphi_n | 1 \rangle$, is actually zero, according to condition (A-15), which we have not yet used [$| \varphi_n \rangle$, according to (B-2), coincides with $| 0 \rangle$]:

$$\langle \varphi_n | 1 \rangle = 0 \tag{B-9}$$

We therefore know the vector $| 1 \rangle$ since we know its expansion on the $\{ | \varphi_p^i \rangle \}$ basis:

$$| 1 \rangle = \sum_{p \neq n} \sum_i \frac{\langle \varphi_p^i | \hat{W} | \varphi_n \rangle}{E_n^0 - E_p^0} | \varphi_p^i \rangle \tag{B-10}$$

Consequently, to first order in the perturbation $W = \lambda \hat{W}$, the eigenvector $| \psi_n(\lambda) \rangle$ of H corresponding to the unperturbed state $| \varphi_n \rangle$ can be written:

$$| \psi_n(\lambda) \rangle = | \varphi_n \rangle + \sum_{p \neq n} \sum_i \frac{\langle \varphi_p^i | W | \varphi_n \rangle}{E_n^0 - E_p^0} | \varphi_p^i \rangle + O(\lambda^2) \tag{B-11}$$

The first-order correction of the state vector is a linear superposition of all the unperturbed states other than $| \varphi_n \rangle$: the perturbation W is said to produce a "mixing" of the state $| \varphi_n \rangle$ with the other eigenstates of H_0. The contribution of a given state $| \varphi_p^i \rangle$ is zero if the perturbation W has no matrix element between $| \varphi_n \rangle$ and $| \varphi_p^i \rangle$. In general, the stronger the coupling induced by W between $| \varphi_n \rangle$ and $| \varphi_p^i \rangle$ (characterized by the matrix element $\langle \varphi_p^i | W | \varphi_n \rangle$), and the closer the level E_p^0 to the level E_n^0 under study, the greater the mixing with $| \varphi_p^i \rangle$.

COMMENT:

We have assumed that the perturbation W is much smaller than the unperturbed Hamiltonian H_0, that is, that the matrix elements of W are much smaller than those of H_0. It appears here that this hypothesis is not sufficient: the first order correction of the state vector is small only if *the non-diagonal matrix elements of W are much smaller than the corresponding unperturbed energy differences.*

2. Second-order corrections

The second-order corrections can be extracted from perturbation equation (A-11) by the same method as above, with the addition of condition (A-16).

a. ENERGY CORRECTION

To calculate ε_2, we project equation (A-11) onto the vector $| \varphi_n \rangle$, using (B-1) and (B-2):

$$\langle \varphi_n | (H_0 - E_n^0) | 2 \rangle + \langle \varphi_n | (\hat{W} - \varepsilon_1) | 1 \rangle - \varepsilon_2 \langle \varphi_n | \varphi_n \rangle = 0 \qquad \text{(B-12)}$$

For the same reason as in § B-1-a, the first term is zero. This is also the case for $\varepsilon_1 \langle \varphi_n | 1 \rangle$, since, according to (B-9), $| 1 \rangle$ is orthogonal to $| \varphi_n \rangle$. We then get:

$$\varepsilon_2 = \langle \varphi_n | \hat{W} | 1 \rangle \qquad \text{(B-13)}$$

that is, substituting expression (B-10) for the vector $| 1 \rangle$:

$$\varepsilon_2 = \sum_{p \neq n} \sum_i \frac{|\langle \varphi_p^i | \hat{W} | \varphi_n \rangle|^2}{E_n^0 - E_p^0} \qquad \text{(B-14)}$$

This result enables us to write the energy $E_n(\lambda)$, to second order in the perturbation $W = \lambda \hat{W}$, in the form:

$$E_n(\lambda) = E_n^0 + \langle \varphi_n | W | \varphi_n \rangle + \sum_{p \neq n} \sum_i \frac{|\langle \varphi_p^i | W | \varphi_n \rangle|^2}{E_n^0 - E_p^0} + O(\lambda^3) \qquad \text{(B-15)}$$

COMMENT:

The second-order energy correction for the state $| \varphi_n \rangle$ due to the presence of the state $| \varphi_p^i \rangle$ has the sign of $E_n^0 - E_p^0$. We can therefore say that, to second order, the closer the state $| \varphi_p^i \rangle$ to the state $| \varphi_n \rangle$, and the stronger the "coupling" $| \langle \varphi_p^i | W | \varphi_n \rangle |$, the more these two levels "repel" each other.

b. EIGENVECTOR CORRECTION

By projecting equation (A-11) onto the set of basis vectors $| \varphi_p^i \rangle$ different from $| \varphi_n \rangle$, and by using conditions (A-16), we could obtain the expression for the ket $| 2 \rangle$, and therefore the eigenvector to second order. Such a calculation presents no theoretical difficulties, and we shall not give it here.

COMMENT:

In (B-4), the first-order energy correction is expressed in terms of the zeroeth-order eigenvector. Similarly, in (B-13), the second-order energy correction involves the first-order eigenvector [which explains a certain similarity of formulas (B-10) and (B-14)]. This is a general result : by projecting (A-12) onto $| \varphi_n \rangle$, one makes the first term go to zero, which gives ε_q in terms of the corrections of order $q - 1$, $q - 2$, ... of the eigenvector. This is why we generally retain one more term in the energy expansion than in that of the eigenvector : for example, the energy is given to second order and the eigenvector to first order.

c. UPPER LIMIT OF ε_2

If we limit the energy expansion to first order in λ, we can obtain an approximate idea of the error involved by evaluating the second-order term which is simple to obtain.

Consider expression (B-14) for ε_2. It contains a sum (which is generally infinite) of terms whose numerators are positive or zero. We denote by ΔE the absolute value of the difference between the energy E_n^0 of the level being studied and that of the closest level. For all n, we obviously have:

$$|E_n^0 - E_p^0| \geqslant \Delta E \tag{B-16}$$

This gives us an upper limit for the absolute value of ε_2:

$$|\varepsilon_2| \leqslant \frac{1}{\Delta E} \sum_{p \neq n} \sum_i | \langle \varphi_p^i | \hat{W} | \varphi_n \rangle |^2 \tag{B-17}$$

which can be written:

$$\begin{aligned} |\varepsilon_2| &\leqslant \frac{1}{\Delta E} \sum_{p \neq n} \sum_i \langle \varphi_n | \hat{W} | \varphi_p^i \rangle \langle \varphi_p^i | \hat{W} | \varphi_n \rangle \\ &\leqslant \frac{1}{\Delta E} \langle \varphi_n | \hat{W} \left[\sum_{p \neq n} \sum_i | \varphi_p^i \rangle \langle \varphi_p^i | \right] \hat{W} | \varphi_n \rangle \end{aligned} \tag{B-18}$$

The operator which appears inside the brackets differs from the identity operator only by the projector onto the state $| \varphi_n \rangle$, since the basis of unperturbed states satisfies the closure relation:

$$| \varphi_n \rangle \langle \varphi_n | + \sum_{p \neq n} \sum_i | \varphi_p^i \rangle \langle \varphi_p^i | = 1 \tag{B-19}$$

Inequality (B-18) therefore becomes simply:

$$\begin{aligned} |\varepsilon_2| &\leqslant \frac{1}{\Delta E} \langle \varphi_n | \hat{W} [1 - | \varphi_n \rangle \langle \varphi_n |] \hat{W} | \varphi_n \rangle \\ &\leqslant \frac{1}{\Delta E} [\langle \varphi_n | \hat{W}^2 | \varphi_n \rangle - (\langle \varphi_n | \hat{W} | \varphi_n \rangle)^2] \end{aligned} \tag{B-20}$$

Multiplying both sides of (B-20) by λ^2, we obtain an upper limit for the second-order term in the expansion of $E_n(\lambda)$, in the form:

$$|\lambda^2 \varepsilon_2| \leqslant \frac{1}{\Delta E} (\Delta W)^2 \tag{B-21}$$

where ΔW is the root-mean-square deviation of the perturbation W in the unperturbed state $| \varphi_n \rangle$. This indicates the order of magnitude of the error committed by taking only the first-order correction into account.

C. PERTURBATION OF A DEGENERATE STATE

Now assume that the level E_n^0 whose perturbation we want to study is g_n-fold degenerate (where g_n is greater than 1, but finite). We denote by $\mathscr{E}_n^0$ the corresponding eigensubspace of H_0. In this case, the choice:

$$\varepsilon_0 = E_n^0 \tag{C-1}$$

does not suffice to determine the vector $| 0 \rangle$, since equation (A-9) can theoretically be satisfied by any linear combination of the g_n vectors $| \varphi_n^i \rangle$ ($i = 1, 2, ..., g_n$). We know only that $| 0 \rangle$ belongs to the eigensubspace $\mathscr{E}_n^0$ spanned by them.

We shall see that, this time, under the action of the perturbation W, the level E_n^0 generally gives rise to several distinct "sublevels". Their number, f_n, is between 1 and g_n. If f_n is less than g_n, some of these sublevels are degenerate, since the total number of orthogonal eigenvectors of H associated with the f_n sublevels is always equal to g_n. To calculate the eigenvalues and eigenstates of the total Hamiltonian H, we shall limit ourselves, as usual, to first order in λ for the energies and to zeroeth order for the eigenvectors.

To determine ε_1 and $| 0 \rangle$, we can project equation (A-10) onto the g_n basis vectors $| \varphi_n^i \rangle$. Since the $| \varphi_n^i \rangle$ are eigenvectors of H_0 with the eigenvalue $E_n^0 = \varepsilon_0$, we obtain the g_n relations:

$$\langle \varphi_n^i | \hat{W} | 0 \rangle = \varepsilon_1 \langle \varphi_n^i | 0 \rangle \tag{C-2}$$

We now insert, between the operator $\hat{W}$ and the vector $| 0 \rangle$, the closure relation for the $\{ | \varphi_p^i \rangle \}$ basis :

$$\sum_p \sum_{i'} \langle \varphi_n^i | \hat{W} | \varphi_p^{i'} \rangle \langle \varphi_p^{i'} | 0 \rangle = \varepsilon_1 \langle \varphi_n^i | 0 \rangle \tag{C-3}$$

The vector $| 0 \rangle$, which belongs to the eigensubspace associated with E_n^0, is orthogonal to all the basis vectors $| \varphi_p^{i'} \rangle$ for which p is different from n. Consequently, on the left-hand side of (C-3), the sum over the index p reduces to a single term ($p = n$), which gives :

$$\sum_{i'=1}^{g_n} \langle \varphi_n^i | \hat{W} | \varphi_n^{i'} \rangle \langle \varphi_n^{i'} | 0 \rangle = \varepsilon_1 \langle \varphi_n^i | 0 \rangle \tag{C-4}$$

We arrange the g_n^2 numbers $\langle \varphi_n^i | \hat{W} | \varphi_n^{i'} \rangle$ (where n is fixed and $i, i' = 1, 2, ..., g_n$) in a $g_n \times g_n$ matrix of row index i and column index i'. This square matrix, which we shall denote by $(\hat{W}^{(n)})$ is, so to speak, cut out of the matrix which represents $\hat{W}$ in the $\{ | \varphi_p^i \rangle \}$ basis : $(\hat{W}^{(n)})$ is the part which corresponds to $\mathscr{E}_n^0$. Equations (C-4) then show that the column vector of elements $\langle \varphi_n^i | 0 \rangle$ ($i = 1, 2, ..., g_n$) is an eigenvector of $(\hat{W}^{(n)})$ with the eigenvalue ε_1.

System (C-4) can, moreover, be transformed into a *vector equation inside* $\mathscr{E}_n^0$. All we need to do is define the operator $\hat{W}^{(n)}$, the *restriction of* $\hat{W}$ *to the subspace* $\mathscr{E}_n^0$. $\hat{W}^{(n)}$ acts only in $\mathscr{E}_n^0$, and it is represented in this subspace by the matrix of elements $\langle \varphi_n^i | \hat{W} | \varphi_n^{i'} \rangle$, that is, by $(\hat{W}^{(n)})$*. System (C-4) is thus equivalent to the vector equation:

$$\hat{W}^{(n)} | 0 \rangle = \varepsilon_1 | 0 \rangle \tag{C-5}$$

[We stress the fact that the operator $\hat{W}^{(n)}$ is different from the operator $\hat{W}$ of which it is the restriction : equation (C-5) is an eigenvalue equation inside $\mathscr{E}_n^0$, and not in all space].

Therefore, *to calculate the eigenvalues (to first order) and the eigenstates (to zeroeth order) of the Hamiltonian corresponding to a degenerate unperturbed state* E_n^0, *diagonalize the matrix* $(W^{(n)})$, *which represents the perturbation*** W, *inside the eigensubspace* $\mathscr{E}_n^0$ *associated with* E_n^0.

Let us examine more closely the first-order effect of the perturbation W on the degenerate state E_n^0. Let ε_1^j ($j = 1, 2, ..., f_n^{(1)}$) be the various distinct roots of the characteristic equation of $(\hat{W}^{(n)})$. Since $(\hat{W}^{(n)})$ is Hermitian, its eigenvalues are all real, and the sum of their degrees of degeneracy is equal to g_n ($g_n \geqslant f_n^{(1)}$). Each eigenvalue introduces a different energy correction. Therefore, under the influence of the perturbation $W = \lambda \hat{W}$, the degenerate level splits, to first order, into $f_n^{(1)}$ distinct sublevels, whose energies can be written :

$$E_{n,j}(\lambda) = E_n^0 + \lambda \varepsilon_1^j \qquad j = 1, 2, ... f_n^{(1)} \leqslant g_n \tag{C-6}$$

* If P_n is the projector onto the subspace $\mathscr{E}_n^0$, $\hat{W}^{(n)}$ can be written (complement B_{II}, § 3):

$\hat{W}^{(n)} = P_n \hat{W} P_n$

** $(W^{(n)})$ is simply equal to $\lambda(\hat{W}^{(n)})$; this is why its eigenvalues yield directly the corrections $\lambda\varepsilon_1$.

If $f_n^{(1)} = g_n$, we say that, to first order, the perturbation W completely removes the degeneracy of the level E_n^0. If $f_n^{(1)} < g_n$, the degeneracy, to first order, is only partially removed (or not at all if $f_n^{(1)} = 1$).

We shall now choose an eigenvalue ε_1^j of $\hat{W}^{(n)}$. If this eigenvalue is non-degenerate, the corresponding eigenvector $| 0 \rangle$ is uniquely determined (to within a phase factor) by (C-5) [or by the equivalent system (C-4)]. There then exists a single eigenvalue $E(\lambda)$ of $H(\lambda)$ which is equal to $E_n^0 + \lambda \varepsilon_1^j$, to first order, and this eigenvalue is non-degenerate★. On the other hand, if the eigenvalue ε_1^j of $\hat{W}^{(n)}$ being considered presents a q-fold degeneracy, (C-5) indicates only that $| 0 \rangle$ belongs to the corresponding q-dimensional subspace $\mathscr{F}_j^{(1)}$.

This property of ε_1^j can, actually, reflect two very different situations. One could distinguish between them by pursuing the perturbation calculation to higher orders of λ, and seeing whether the remaining degeneracy is removed. These two situations are the following :

(i) Suppose that there is only one exact energy $E(\lambda)$ which is equal, to first order, to $E_n^0 + \lambda\varepsilon_1^j$, and that this energy is q-fold degenerate [in figure 1, for example, the energy $E(\lambda)$ which approaches E_4^0 when $\lambda \longrightarrow 0$ is two-fold degenerate, for any value of λ]. A q-dimensional eigensubspace then corresponds to the eigenvalue $E(\lambda)$, whatever λ, so that the degeneracy of the approximate eigenvalues will never be removed, to any order of λ.

In this case, the zeroeth-order eigenvector $| 0 \rangle$ of $H(\lambda)$ cannot be completely specified, since the only condition imposed on $| 0 \rangle$ is that of belonging to a subspace which is the limit, when $\lambda \longrightarrow 0$, of the q-dimensional eigensubspace of $H(\lambda)$ corresponding to $E(\lambda)$. This limit is none other than the eigensubspace $\mathscr{F}_j^{(1)}$ of $(\hat{W}^{(n)})$ associated with the eigenvalue ε_1^j chosen.

This first case often arises when H_0 and W possess common symmetry properties, implying an essential degeneracy for $H(\lambda)$. Such a degeneracy then remains to all orders in perturbation theory.

(ii) It may also happen that several different energies $E(\lambda)$ are equal, to first order, to $E_n^0 + \lambda\varepsilon_1^j$ (the difference between these energies then appears in a calculation at second or higher orders). In this case, the subspace $\mathscr{F}_j^{(1)}$ obtained to first order is only the direct sum of the limits, for $\lambda \longrightarrow 0$, of several eigensubspaces associated with these various energies $E(\lambda)$. In other words, all the eigenvectors of $H(\lambda)$ corresponding to these energies certainly approach kets of $\mathscr{F}_j^{(1)}$, but, inversely, a particular ket of $\mathscr{F}_j^{(1)}$ is not necessarily the limit $| 0 \rangle$ of an eigenket of $H(\lambda)$.

In this situation, going to higher order terms allows one, not only to improve the accuracy of the energies, but also to determine the zeroeth-order kets $| 0 \rangle$. However, in practice, the partial information contained in equation (C-5) is often considered sufficient.

COMMENTS:

(i) When we use the perturbation method to treat all the energies★★ of the spectrum of H_0, we must diagonalize the perturbation W inside each of the eigensubspaces $\mathscr{E}_n^0$ corresponding to these energies. It must be understood that this problem is much simpler than the initial problem, which is the complete diagonalization of the Hamiltonian in the entire state space. Perturbation theory enables us to ignore completely the matrix elements of W between vectors belonging to different subspaces $\mathscr{E}_n^0$. Therefore, instead of having to

★ The proof of this point is analogous to the one which shows that a non-degenerate level of H_0 gives rise to a non-degenerate level of $H(\lambda)$ (*cf.* end of § A-2).

★★ The perturbation of a non-degenerate state, studied in § B, can be seen as a special case of that of a degenerate state.

diagonalize a generally infinite matrix, we need only diagonalize, for each of the energies E_n^0 in which we are interested, a matrix of smaller dimensions, generally finite.

(*ii*) The matrix $(\hat{W}^{(n)})$ clearly depends on the $\{ | \varphi_n^i \rangle \}$ basis initially chosen in this subspace $\mathscr{E}_n^0$ (although the eigenvalues and eigenkets of $\hat{W}^{(n)}$ obviously do not depend on it). Therefore, before we begin the perturbation calculation, it is advantageous to find a basis which simplifies as much as possible the form of $(W^{(n)})$ for this subspace, and, consequently, the search for its eigenvalues and eigenvectors (the simplest situation is obviously the one in which this matrix is obtained directly in a diagonal form). To find such a basis, we often use observables which commute both with H_0 and W★. Assume that we have an observable A which commutes with H_0 and W. Since H_0 and A commute, we can choose for the basis vectors $| \varphi_n^i \rangle$ eigenstates common to H_0 and A. Furthermore, since W commutes with A, its matrix elements are zero between eigenvectors of A associated with different eigenvalues. The matrix $(W^{(n)})$ then contains numerous zeros, which facilitates its diagonalization.

(*iii*) Just as for non-degenerate levels (*cf.* comment of §B-1-b), the method described in this section is valid only if the matrix elements of the perturbation W are much smaller than the differences between the energy of the level under study and those of the other levels (this conclusion would have been evident if we had calculated higher-order corrections). However, it is possible to extend this method to the case of a group of unperturbed levels which are very close to each other (but distinct) and very far from all the other levels of the system being considered. This means, of course, that the matrix elements of the perturbation W are of the same order of magnitude as the energy differences inside the group, but are negligible compared to the separation between a level in the group and one outside. We can then approximately determine the influence of the perturbation W by diagonalizing the matrix which represents $H = H_0 + W$ inside this group of levels. It is by relying on an approximation of this type that we can, in certain cases, reduce the study of a physical problem to that of a two-level system, such as those described in chapter IV (§C).

References and suggestions for further reading:

For other perturbation methods, see, for example:
Brillouin-Wigner series (an expansion which is simple for all orders but which involves the perturbed energies in the energy denominators): Ziman (2.26), §3.1.
The resolvent method (an operator method which is well suited to the calculation of higher-order corrections): Messiah (1.17), chap. XVI, §III; Roman (2.3), §4-5-d.
Method of Dalgarno and Lewis (which replaces the summations over the intermediate states by differential equations): Borowitz (1.7), §14-5; Schiff (1.18), chap. 8, §33. Original references: (2.34), (2.35), (2.36).

★ Recall that this does not imply that H_0 and W commute.

The W.K.B. method, applicable to quasi-classical situations: Landau and Lifshitz (1.19), chap. 7; Messiah (1.17), chap. VI, §II; Merzbacher (1.16), chap. VII; Schiff (1.18), §34; Borowitz (1.7), chaps. 8 and 9.

The Hartree and Hartree-Fock methods: Messiah (1.17), chap. XVIII, §II; Slater (11.8), chap. 8 and 9 (Hartree) and 17 (Hartree-Fock); Bethe and Jackiw (1.21), chap. 4. See also references of complement A_{XIV}.

COMPLEMENTS OF CHAPTER XI

A_{XI} : A ONE-DIMENSIONAL HARMONIC OSCILLATOR SUBJECTED TO A PERTURBING POTENTIAL IN x, x^2, x^3

A_{XI}, B_{XI}, C_{XI}, D_{XI}: illustrations of stationary perturbation theory using simple and important examples.

A_{XI}: study of a one-dimensional harmonic oscillator perturbed by a potential in x, x^2, x^3. Simple, advised for a first reading. The last example (perturbing potential in x^3) permits the study of the anharmonicity in the vibration of a diatomic molecule(a refinement of the model presented in complement A_V).

B_{XI} : INTERACTION BETWEEN THE MAGNETIC DIPOLES OF TWO SPIN 1/2 PARTICLES

B_{XI}: can be considered as a worked example, illustrating perturbation theory for non-degenerate as well as degenerate states. Familiarizes the reader with the dipole-dipole interaction between magnetic moments of two spin 1/2 particles. Simple.

C_{XI} : VAN DER WAALS FORCES

C_{XI}: study of the long-distance forces between two neutral atoms using perturbation theory (Van der Waals forces). The accent is placed on the physical interpretation of the results. A little less simple than the two preceding complements; can be reserved for later study.

D_{XI} : THE VOLUME EFFECT : THE INFLUENCE OF THE SPATIAL EXTENSION OF THE NUCLEUS ON THE ATOMIC LEVELS

D_{XI}: study of the influence of the nuclear volume on the energy levels of hydrogen-like atoms. Simple. Can be considered as a sequel of complement A_{VII}.

E_{XI} : THE VARIATIONAL METHOD

E_{XI}: presentation of another approximation method, the variational method. Important, since the applications of the variational method are very numerous.

F_{XI} : ENERGY BANDS OF ELECTRONS IN SOLIDS : A SIMPLE MODEL

F_{XI}, G_{XI}: two important applications of the variational method.

F_{XI}: introduction, using the strong-bonding approximation, of the concept of an allowed energy band for the electrons of a solid. Essential, because of its numerous applications. Moderately difficult. The accent is placed on the interpretation of the results. The view point adopted is different from that of complement O_{III} and somewhat simpler.

G_{XI} : A SIMPLE EXAMPLE OF THE CHEMICAL BOND : THE H_2^+ ION

G_{XI}: studies the phenomenon of the chemical bond for the simplest possible case, that of the (ionized) H_2^+ molecule. Shows how quantum mechanics explains the attractive forces between two atoms whose electronic wave functions overlap. Includes a proof of the virial theorem. Essential from the point of view of chemical physics. Moderately difficult.

H_{XI} : EXERCISES

Complement A_{XI}

A ONE-DIMENSIONAL HARMONIC OSCILLATOR SUBJECTED TO A PERTURBING POTENTIAL IN x, x^2, x^3

In order to illustrate the general considerations of chapter XI by a simple example, we shall use stationary perturbation theory to study the effect of a perturbing potential in x, x^2 or x^3 on the energy levels of a one-dimensional harmonic oscillator (none of these levels is degenerate, *cf.* chap. V).

The first two cases (a perturbing potential in x and in x^2) are exactly soluble. Consequently, we shall be able to verify in these two examples that the perturbation expansion coincides with the limited expansion of the exact solution with respect to the parameter which characterizes the strength of the perturbation. The last case (a perturbing potential in x^3) is very important in practice for the following reason. Consider a potential $V(x)$ which has a minimum at $x = 0$. To a first approximation, $V(x)$ can be replaced by the first term (in x^2) of its Taylor series expansion, in which case we are considering a harmonic oscillator and, therefore, an exactly soluble problem. The next term of the expansion of $V(x)$, which is proportional to x^3, then constitutes the first correction to this approximation. Calculation of the effect of the term in x^3, consequently, is necessary whenever we want to study the anharmonicity of the vibrations of a physical system. It permits us, for example, to evaluate the deviations of the vibrational spectrum of diatomic molecules from the predictions of the (purely harmonic) model of complement A_V.

1. Perturbation by a linear potential

We shall use the notation of chapter V. Let:

$$H_0 = \frac{P^2}{2m} + \frac{1}{2} m\omega^2 X^2 \tag{1}$$

be the Hamiltonian of a one-dimensional harmonic oscillator of eigenvectors $| \varphi_n \rangle$ and eigenvalues★:

★ To specify that it is the unperturbed Hamiltonian that we are considering, we add, as in chapter XI, the index 0 to the eigenvalue of H_0.

$$E_n^0 = \left(n + \frac{1}{2}\right)\hbar\omega \tag{2}$$

with $n = 0, 1, 2, ...$

We add to this Hamiltonian the perturbation:

$$W = \lambda\hbar\omega\hat{X} \tag{3}$$

where λ is a real dimensionless constant much smaller than 1, and $\hat{X}$ is given by formula (B-1) of chapter V (since $\hat{X}$ is of the order of 1, $\hbar\omega\hat{X}$ is of the order of H_0 and plays the role of the operator $\hat{W}$ of chapter XI). The problem consists of finding the eigenstates $| \psi_n \rangle$ and eigenvalues E_n of the Hamiltonian:

$$H = H_0 + W \tag{4}$$

a. THE EXACT SOLUTION

We have already studied an example of a linear perturbation in X : when the oscillator, assumed to be charged, is placed in a uniform electric field $\mathscr{E}$, we must add to H_0 the electrostatic Hamiltonian:

$$W = - q\mathscr{E}X = - q\mathscr{E}\sqrt{\frac{\hbar}{m\omega}}\,\hat{X} \tag{5}$$

where q is the charge of the oscillator. The effect of such a term on the stationary states of the harmonic oscillator was studied in detail in complement F_V. It is therefore possible to use the results of this complement to determine the eigenstates and eigenvalues of the Hamiltonian H given by (4) if we perform the substitution:

$$\lambda\hbar\omega \longleftrightarrow - q\mathscr{E}\sqrt{\frac{\hbar}{m\omega}} \tag{6}$$

Expression (39) of F_V thus yields immediately:

$$E_n = \left(n + \frac{1}{2}\right)\hbar\omega - \frac{\lambda^2}{2}\hbar\omega \tag{7}$$

Similarly, we see from (40) of F_V (after having replaced P by its expression in terms of the creation and annihilation operators $a^\dagger$ and a):

$$| \psi_n \rangle = \mathrm{e}^{-\frac{\lambda}{\sqrt{2}}(a^\dagger - a)}\,| \varphi_n \rangle \tag{8}$$

The limited expansion of the exponential then yields:

$$\begin{aligned} | \psi_n \rangle &= \left[1 - \frac{\lambda}{\sqrt{2}}(a^\dagger - a) + ...\right]| \varphi_n \rangle \\ &= | \varphi_n \rangle - \lambda\sqrt{\frac{n+1}{2}}\,| \varphi_{n+1} \rangle + \lambda\sqrt{\frac{n}{2}}\,| \varphi_{n-1} \rangle + ... \end{aligned} \tag{9}$$

b. THE PERTURBATION EXPANSION

We replace $\hat{X}$ by $\frac{1}{\sqrt{2}}(a^\dagger + a)$ in (3) [*cf.* formula (B-7-a) of chapter V]. We obtain:

$$W = \lambda \frac{\hbar\omega}{\sqrt{2}}(a^\dagger + a) \tag{10}$$

W then mixes the state $| \varphi_n \rangle$ only with the two states $| \varphi_{n+1} \rangle$ and $| \varphi_{n-1} \rangle$. The only non-zero matrix elements of W are, consequently:

$$\begin{aligned} \langle \varphi_{n+1} | W | \varphi_n \rangle &= \lambda \sqrt{\frac{n+1}{2}}\, \hbar\omega \\ \langle \varphi_{n-1} | W | \varphi_n \rangle &= \lambda \sqrt{\frac{n}{2}}\, \hbar\omega \end{aligned} \tag{11}$$

According to general expression (B-15) of chapter XI, we have:

$$E_n = E_n^0 + \langle \varphi_n | W | \varphi_n \rangle + \sum_{n' \neq n} \frac{|\langle \varphi_{n'} | W | \varphi_n \rangle|^2}{E_n^0 - E_{n'}^0} + \ldots \tag{12}$$

Substituting (11) into (12) and replacing $E_n^0 - E_{n'}^0$ by $(n - n')\hbar\omega$, we immediately obtain:

$$\begin{aligned} E_n &= E_n^0 + 0 - \frac{\lambda^2(n+1)}{2}\hbar\omega + \frac{\lambda^2 n}{2}\hbar\omega + \ldots \\ &= \left(n + \frac{1}{2}\right)\hbar\omega - \frac{\lambda^2}{2}\hbar\omega + \ldots \end{aligned} \tag{13}$$

This shows that the perturbation expansion of the eigenvalue to second order in λ coincides* with the exact solution (7).

Similarly, general formula (B-11) of chapter XI:

$$| \psi_n \rangle = | \varphi_n \rangle + \sum_{n' \neq n} \frac{\langle \varphi_{n'} | W | \varphi_n \rangle}{E_n^0 - E_{n'}^0} | \varphi_{n'} \rangle + \ldots \tag{14}$$

yields here:

$$| \psi_n \rangle = | \varphi_n \rangle - \lambda \sqrt{\frac{n+1}{2}} | \varphi_{n+1} \rangle + \lambda \sqrt{\frac{n}{2}} | \varphi_{n-1} \rangle + \ldots \tag{15}$$

an expression which is identical to expansion (9) of the exact solution.

* It can be shown that all terms of order higher than 2 in the perturbation expansion are zero.

2. Perturbation by a quadratic potential

We now assume W to have the following form:

$$W = \frac{1}{2}\rho\,\hbar\omega\hat{X}^2 = \frac{1}{2}\rho\,m\omega^2X^2 \tag{16}$$

where ρ is a real dimensionless parameter much smaller than 1. H can then be written:

$$H = H_0 + W = \frac{P^2}{2m} + \frac{1}{2}m\omega^2(1+\rho)X^2 \tag{17}$$

In this case, the effect of the perturbation is simply to change the spring constant of the harmonic oscillator. If we set:

$$\omega'^2 = \omega^2(1+\rho) \tag{18}$$

we see that H is still a harmonic oscillator Hamiltonian, whose angular frequency has become ω'.

In this section, we shall confine ourselves to the study of the eigenvalues of H. According to (17) and (18), they can be written simply:

$$E_n = \left(n+\frac{1}{2}\right)\hbar\omega' = \left(n+\frac{1}{2}\right)\hbar\omega\sqrt{1+\rho} \tag{19}$$

that is, expanding the radical:

$$E_n = \left(n+\frac{1}{2}\right)\hbar\omega\left[1+\frac{\rho}{2}-\frac{\rho^2}{8}+\ldots\right] \tag{20}$$

Let us now find result (20) by using stationary perturbation theory. Expression (16) can also be written:

$$W = \frac{1}{4}\rho\,\hbar\omega(a^\dagger+a)^2 = \frac{1}{4}\rho\,\hbar\omega(a^{\dagger 2}+a^2+aa^\dagger+a^\dagger a)$$

$$= \frac{1}{4}\rho\,\hbar\omega[a^{\dagger 2}+a^2+2a^\dagger a+1] \tag{21}$$

From this, it can be seen that the only non-zero matrix elements of W associated with $|\varphi_n\rangle$ are:

$$\langle\varphi_n|W|\varphi_n\rangle = \frac{1}{2}\rho\left(n+\frac{1}{2}\right)\hbar\omega$$

$$\langle\varphi_{n+2}|W|\varphi_n\rangle = \frac{1}{4}\rho[(n+1)(n+2)]^{1/2}\,\hbar\omega \tag{22}$$

$$\langle\varphi_{n-2}|W|\varphi_n\rangle = \frac{1}{4}\rho[n(n-1)]^{1/2}\,\hbar\omega$$

When we use this result to evaluate the various terms of (12), we find:

$$E_n = E_n^0 + \frac{\rho}{2}\left(n + \frac{1}{2}\right)\hbar\omega - \frac{\rho^2}{16}(n+1)(n+2)\frac{\hbar\omega}{2} + \frac{\rho^2}{16}n(n-1)\frac{\hbar\omega}{2} + \ldots$$

$$= E_n^0 + \left(n + \frac{1}{2}\right)\hbar\omega\frac{\rho}{2} - \left(n + \frac{1}{2}\right)\hbar\omega\frac{\rho^2}{8} + \ldots$$

$$= \left(n + \frac{1}{2}\right)\hbar\omega\left[1 + \frac{\rho}{2} - \frac{\rho^2}{8} + \ldots\right] \tag{23}$$

which indeed coincides with expansion (20).

3. Perturbation by a potential in x^3

We now add to H_0 the perturbation:

$$W = \sigma\hbar\omega\hat{X}^3 \tag{24}$$

where σ is a real dimensionless number much smaller than 1.

a. THE ANHARMONIC OSCILLATOR

Figure 1 represents the variation with respect to x of the total potential $\frac{1}{2}m\omega^2x^2 + W(x)$ in which the particle is moving. The dashed line gives the parabolic potential $\frac{1}{2}m\omega^2x^2$ of the "unperturbed" harmonic oscillator. We have chosen $\sigma < 0$, so that the total potential (the solid curve in the figure) increases less rapidly for $x > 0$ than for $x < 0$.

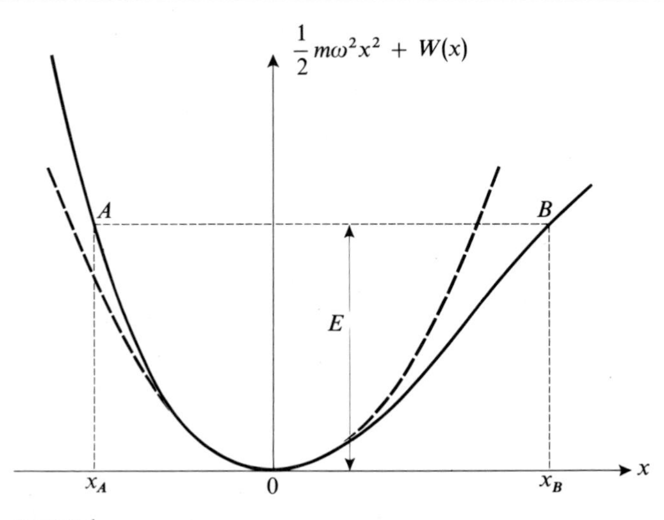

FIGURE 1

Variation of the potential associated with an anharmonic oscillator with respect to x. We treat the difference between the real potential (solid line) and the harmonic potential (dashed line) of the unperturbed Hamiltonian as a perturbation (x_A and x_B are the limits of the classical motion of energy E).

When the problem is treated in classical mechanics, the particle with total energy E is found to oscillate between two points, x_A and x_B (fig. 1), which are no longer symmetrical with respect to O. This motion, while it remains periodic, is no longer sinusoidal: there appears, in the Fourier series expansion of $x(t)$, a whole series of harmonics of the fundamental frequency. This is why such a system is called an "anharmonic oscillator" (its motion is no longer harmonic). Finally, let us point out that the period of the motion is no longer independent of the energy E, as was the case for the harmonic oscillator.

b. THE PERTURBATION EXPANSION

α. *Matrix elements of the perturbation W*

We replace $\hat{X}$ by $\frac{1}{\sqrt{2}}(a^\dagger + a)$ in (24). Using relations (B-9) and (B-17) of chapter V, we obtain, after a simple calculation:

$$W = \frac{\sigma\hbar\omega}{2^{3/2}}\left[a^{\dagger 3} + a^3 + 3Na^\dagger + 3(N+1)a\right] \tag{25}$$

where $N = a^\dagger a$ was defined in chapter V [formula (B-13)].

From this can immediately be deduced the only non-zero matrix elements of W associated with $|\varphi_n\rangle$:

$$\langle \varphi_{n+3} | W | \varphi_n \rangle = \sigma\left[\frac{(n+3)(n+2)(n+1)}{8}\right]^{\frac{1}{2}} \hbar\omega$$

$$\langle \varphi_{n-3} | W | \varphi_n \rangle = \sigma\left[\frac{n(n-1)(n-2)}{8}\right]^{\frac{1}{2}} \hbar\omega$$

$$\langle \varphi_{n+1} | W | \varphi_n \rangle = 3\sigma\left(\frac{n+1}{2}\right)^{\frac{3}{2}} \hbar\omega$$

$$\langle \varphi_{n-1} | W | \varphi_n \rangle = 3\sigma\left(\frac{n}{2}\right)^{\frac{3}{2}} \hbar\omega \tag{26}$$

β. *Calculation of the energies*

We substitute results (26) into the perturbation expansion of E_n (formula 12). Since the diagonal element of W is zero, there is no first-order correction. The four matrix elements (26) enter, however, into the second-order correction. A simple calculation thus yields:

$$E_n = \left(n + \frac{1}{2}\right)\hbar\omega - \frac{15}{4}\sigma^2\left(n + \frac{1}{2}\right)^2 \hbar\omega - \frac{7}{16}\sigma^2\hbar\omega + ... \tag{27}$$

The effect of W is therefore to lower the levels (whatever the sign of σ). The larger n, the greater the shift (fig. 2). The difference between two adjacent levels is equal to:

$$E_n - E_{n-1} = \hbar\omega\left[1 - \frac{15}{2}\sigma^2 n\right] \tag{28}$$

It is no longer independent of n, as it was for the harmonic oscillator. The energy states are no longer equidistant and move closer together as n increases.

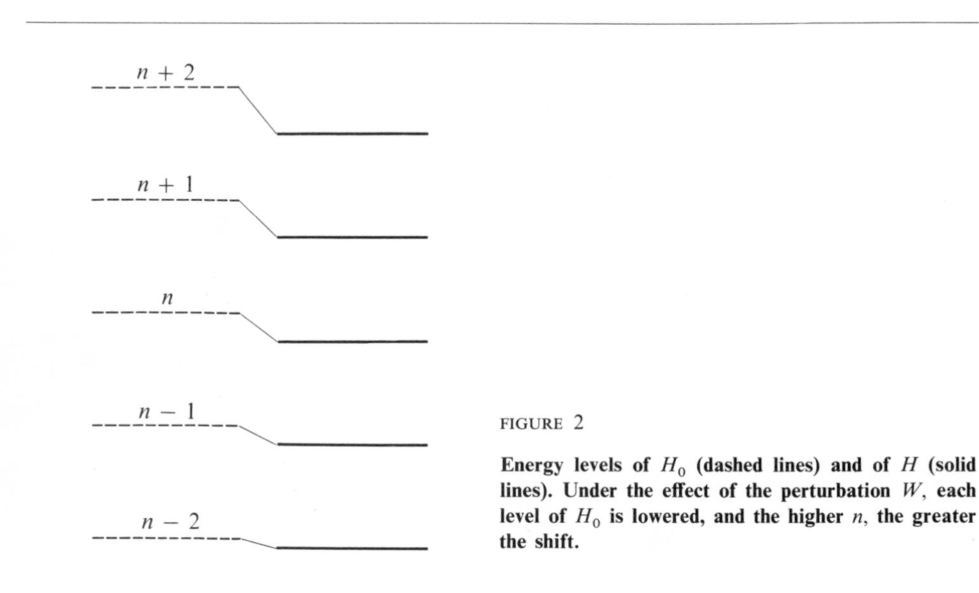

FIGURE 2

Energy levels of H_0 (dashed lines) and of H (solid lines). Under the effect of the perturbation W, each level of H_0 is lowered, and the higher n, the greater the shift.

γ. *Calculation of the eigenstates*

Substituting relations (26) into expansion (14), we easily obtain:

$$\begin{aligned} |\psi_n\rangle = |\varphi_n\rangle &- 3\sigma\left(\frac{n+1}{2}\right)^{\frac{3}{2}}|\varphi_{n+1}\rangle + 3\sigma\left(\frac{n}{2}\right)^{\frac{3}{2}}|\varphi_{n-1}\rangle \\ &- \frac{\sigma}{3}\left[\frac{(n+3)(n+2)(n+1)}{8}\right]^{\frac{1}{2}}|\varphi_{n+3}\rangle \\ &+ \frac{\sigma}{3}\left[\frac{n(n-1)(n-2)}{8}\right]^{\frac{1}{2}}|\varphi_{n-3}\rangle + \ldots \end{aligned} \tag{29}$$

Under the effect of the perturbation W, the state $|\varphi_n\rangle$ is therefore mixed with the states $|\varphi_{n+1}\rangle$, $|\varphi_{n-1}\rangle$, $|\varphi_{n+3}\rangle$ and $|\varphi_{n-3}\rangle$.

c. APPLICATION: THE ANHARMONICITY OF THE VIBRATIONS OF A DIATOMIC MOLECULE

In complement A_V, we showed that a heteropolar diatomic molecule could absorb or emit electromagnetic waves whose frequency coincides with the vibrational frequency of the two nuclei of the molecule about their equilibrium position. If we denote by x the displacement $r - r_e$ of the two nuclei from their equilibrium position r_e, the electric dipole moment of the molecule can be written:

$$D(x) = d_0 + d_1 x + \ldots \tag{30}$$

The vibrational frequencies of this dipole are therefore the Bohr frequencies which can appear in the expression for $\langle X \rangle(t)$. For a harmonic oscillator, the selection rules satisfied by X are such that only one Bohr frequency can be involved, the frequency $\omega/2\pi$ (*cf.* complement A_V).

When we take the perturbation W into account, the states $|\varphi_n\rangle$ of the oscillator are "mixed" [*cf.* expression (29)], and X can connect states $|\psi_n\rangle$ and $|\psi_{n'}\rangle$ for which $n' - n \neq \pm 1$: new frequencies can thus be absorbed or emitted by the molecule.

To analyze this phenomenon more closely, we shall assume that the molecule is initially in its vibrational ground state $|\psi_0\rangle$ (this is practically always the case at ordinary temperatures T since, in general, $\hbar\omega \gg kT$). By using expression (29), we can calculate, to first order★ in σ, the matrix elements of $\hat{X}$ between the state $|\psi_0\rangle$ and an arbitrary state $|\psi_n\rangle$. A simple calculation thus yields the following matrix elements (all the others are zero to first order in σ):

$$\langle \psi_1 | \hat{X} | \psi_0 \rangle = \frac{1}{\sqrt{2}} \tag{31-a}$$

$$\langle \psi_2 | \hat{X} | \psi_0 \rangle = \frac{1}{\sqrt{2}}\sigma \tag{31-b}$$

$$\langle \psi_0 | \hat{X} | \psi_0 \rangle = -\frac{3}{2}\sigma \tag{31-c}$$

From this, we can find the transition frequencies observable in the absorption spectrum of the ground state. We naturally find the frequency:

$$\nu_1 = \frac{E_1 - E_0}{h} \tag{32-a}$$

which appears with the greatest intensity since, according to (31-a), $\langle \psi_1 | \hat{X} | \psi_0 \rangle$ is of zeroeth-order in σ. Then, with a much smaller intensity [*cf.* formula (31-b)], we find the frequency:

$$\nu_2 = \frac{E_2 - E_0}{h} \tag{32-b}$$

★ It would not be correct to keep terms of order higher than 1 in the calculation, since expansion (29) is valid only to first order in σ.

which is often called the second harmonic (although it is not rigorously equal to twice ν_1).

COMMENT:

Result (31-c) means that the mean value of $\hat{X}$ is not zero in the ground state. This can easily be understood from figure 1, since the oscillatory motion is no longer symmetrical about O. If σ is negative (the case in figure 1), the oscillator spends more time in the $x > 0$ region than in the $x < 0$ region, and the mean value of X must be positive. We thus understand the sign appearing in (31-c).

The preceding calculation thus reveals only one new line in the absorption spectrum. Actually, the perturbation calculation could be pursued to higher orders in σ, taking into account higher order terms in expansion (30) of the dipole moment $D(x)$, as well as terms in x^4, x^5, ... in the expansion of the potential in the neighborhood of $x = 0$. All the frequencies:

$$\nu_n = \frac{E_n - E_0}{h} \tag{33}$$

with $n = 3, 4, 5, \ldots$, would then be found to be present in the absorption spectrum of the molecule (with intensities decreasing very rapidly with n). This would finally give, for this spectrum, the form shown in figure 3. This is what is actually

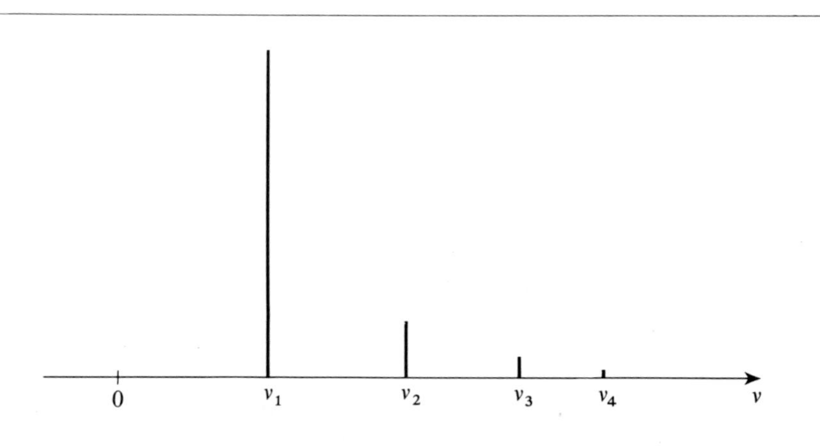

FIGURE 3

Form of the vibrational spectrum of a heteropolar diatomic molecule. Because of the anharmonicity of the potential and higher order terms in the power series expansion in x (the distance between the two atoms) of the molecular dipole moment $D(x)$, a series of "harmonic" frequencies ν_2, ν_3, ..., ν_n, ... appear in addition to the fundamental frequency ν_1. Note that the corresponding lines are not quite equidistant and that their intensity decreases rapidly with n.

observed. Note that the various spectral lines of figure 3 are not equidistant since, according to formula (28):

$$\nu_1 - 0 = \frac{E_1 - E_0}{h} = \frac{\omega}{2\pi}\left(1 - \frac{15}{2}\sigma^2\right) \quad (34)$$

$$\nu_2 - \nu_1 = \frac{E_2 - E_1}{h} = \frac{\omega}{2\pi}(1 - 15\,\sigma^2) \quad (35)$$

$$\nu_3 - \nu_2 = \frac{E_3 - E_2}{h} = \frac{\omega}{2\pi}\left(1 - \frac{45}{2}\sigma^2\right) \quad (36)$$

which gives the relation:

$$(\nu_2 - \nu_1) - \nu_1 = (\nu_3 - \nu_2) - (\nu_2 - \nu_1) = -\frac{15\omega}{4\pi}\sigma^2 \quad (37)$$

Thus we see that the study of the precise positions of the lines of the absorption spectrum makes it possible to find the parameter σ.

COMMENTS:

(*i*) The constant ξ appearing in (52) of complement F_{VII} can be evaluated by using formula (27) of this complement. Comparing these two expressions and replacing n by v in (27), we obtain:

$$\xi = -\frac{15}{4}\sigma^2 \quad (38)$$

Now, the perturbing potential in F_{VII} is equal to $-gx^3$, while here we have chosen it equal to $\sigma\hbar\omega\hat{x}^3$, that is, equal to:

$$\sigma\left(\frac{m^3\omega^5}{\hbar}\right)^{\frac{1}{2}} x^3 \quad (39)$$

We therefore have:

$$\sigma = -g\left(\frac{\hbar}{m^3\omega^5}\right)^{\frac{1}{2}} \quad (40)$$

which, substituted into (38), finally yields:

$$\xi = -\frac{15}{4}\frac{g^2\hbar}{m^3\omega^5} \quad (41)$$

(*ii*) In the expansion of the potential in the neighborhood of $x = 0$, the term in x^4 is much smaller than the term in x^3 but it corrects the energies to first order, while the term in x^3 enters only in second order (*cf.* §3-b-β above). It is therefore necessary to evaluate these two corrections simultaneously (they may be comparable) when the spectrum of figure 3 is studied more precisely.

References and suggestions for further reading:

Anharmonicity of the vibrations of a diatomic molecule: Herzberg (12.4), vol. I, chap. III, §2.

Complement B_{XI}

INTERACTION BETWEEN THE MAGNETIC DIPOLES OF TWO SPIN 1/2 PARTICLES

In this complement, we intend to use stationary perturbation theory to study the energy levels of a system of two spin 1/2 particles placed in a static field $\mathbf{B}_0$ and coupled by a magnetic dipole-dipole interaction.

Such systems do exist. For example, in a gypsum monocrystal ($CaSO_4$, $2H_2O$), the two protons of each crystallization water molecule occupy fixed positions, and the dipole-dipole interaction between them leads to a fine structure in the nuclear magnetic resonance spectrum.

In the hydrogen atom, there also exists a dipole-dipole interaction between the electron spin and the proton spin. In this case, however, the two particles are moving relative to each other, and we shall see that the effect of the dipole-dipole interaction vanishes due to the symmetry of the $1s$ ground state. The hyperfine structure observed in this state is thus due to other interactions (contact interaction; *cf.* chap. XII, §§ B-2 and D-2 and complement A_{XII}).

1. The interaction Hamiltonian W

a. THE FORM OF THE HAMILTONIAN W. PHYSICAL INTERPRETATION

Let $\mathbf{S}_1$ and $\mathbf{S}_2$ be the spins of particles (1) and (2), and $\mathbf{M}_1$ and $\mathbf{M}_2$, their corresponding magnetic moments:

$$\begin{aligned} \mathbf{M}_1 &= \gamma_1 \mathbf{S}_1 \\ \mathbf{M}_2 &= \gamma_2 \mathbf{S}_2 \end{aligned} \tag{1}$$

[where γ_1 and γ_2 are the gyromagnetic ratios of (1) and (2)].

We call W the interaction of the magnetic moment $\mathbf{M}_2$ with the field created by $\mathbf{M}_1$ at (2). If $\mathbf{n}$ denotes the unit vector of the line joining the two particles and r, the distance between them (fig. 1), W can be written:

$$W = \frac{\mu_0}{4\pi} \gamma_1 \gamma_2 \frac{1}{r^3} [\mathbf{S}_1 \cdot \mathbf{S}_2 - 3(\mathbf{S}_1 \cdot \mathbf{n})(\mathbf{S}_2 \cdot \mathbf{n})] \tag{2}$$

The calculation which enables us to obtain expression (2) is in every way analogous to the one which will be presented in complement C_{XI} and which leads to the expression for the interaction between two electric dipoles (*cf.* p. 1132).

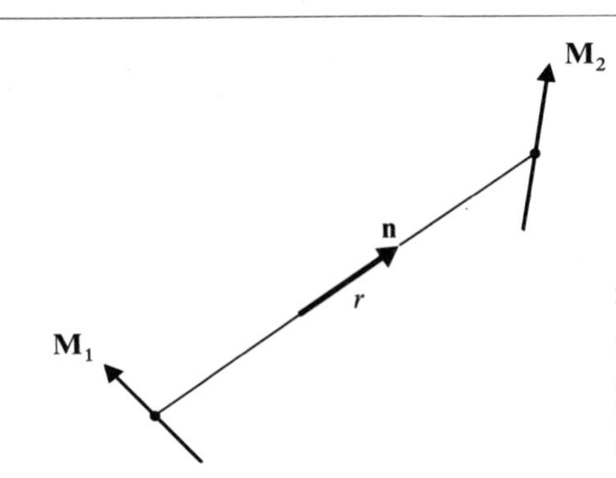

FIGURE 1

Relative disposition of the magnetic moments $\mathbf{M}_1$ and $\mathbf{M}_2$ of particles (1) and (2) (r is the distance between the two particles, and $\mathbf{n}$ is the unit vector of the straight line between them).

b. AN EQUIVALENT EXPRESSION FOR W

Let θ and φ be the polar angles of $\mathbf{n}$. If we set:

$$\xi(r) = -\frac{\mu_0}{4\pi}\frac{\gamma_1\gamma_2}{r^3} \tag{3}$$

we get:

$$\begin{aligned} W &= \xi(r)\{\,3[S_{1z}\cos\theta + \sin\theta\,(S_{1x}\cos\varphi + S_{1y}\sin\varphi)] \\ &\qquad\qquad \times [S_{2z}\cos\theta + \sin\theta\,(S_{2x}\cos\varphi + S_{2y}\sin\varphi)] - \mathbf{S}_1\cdot\mathbf{S}_2\,\} \\ &= \xi(r)\left\{3\left[S_{1z}\cos\theta + \frac{1}{2}\sin\theta\,(S_{1+}\,e^{-i\varphi} + S_{1-}\,e^{i\varphi})\right]\right. \\ &\qquad\qquad \left.\times\left[S_{2z}\cos\theta + \frac{1}{2}\sin\theta\,(S_{2+}\,e^{-i\varphi} + S_{2-}\,e^{i\varphi})\right] - \mathbf{S}_1\cdot\mathbf{S}_2\right\} \end{aligned} \tag{4}$$

that is:

$$W = \xi(r)\,[T_0 + T'_0 + T_1 + T_{-1} + T_2 + T_{-2}] \tag{5}$$

where:

$$\left\{\begin{aligned} T_0 &= (3\cos^2\theta - 1)\,S_{1z}S_{2z} \\ T'_0 &= -\frac{1}{4}(3\cos^2\theta - 1)(S_{1+}S_{2-} + S_{1-}S_{2+}) \\ T_1 &= \frac{3}{2}\sin\theta\cos\theta\,e^{-i\varphi}\,(S_{1z}S_{2+} + S_{1+}S_{2z}) \\ T_{-1} &= \frac{3}{2}\sin\theta\cos\theta\,e^{i\varphi}\,(S_{1z}S_{2-} + S_{1-}S_{2z}) \\ T_2 &= \frac{3}{4}\sin^2\theta\,e^{-2i\varphi}\,S_{1+}S_{2+} \\ T_{-2} &= \frac{3}{4}\sin^2\theta\,e^{2i\varphi}\,S_{1-}S_{2-} \end{aligned}\right. \tag{6}$$

Each of the terms T_q (or T'_q) appearing in (5) is, according to (6), the product of a function of θ and φ proportional to the second-order spherical harmonic Y_2^q and an operator acting only on the spin degrees of freedom [the space and spin operators appearing in (6) are second-rank tensors; W, for this reason, is often called the "tensor interaction"].

c. SELECTION RULES

r, θ and φ are the spherical coordinates of the relative particle associated with the system of two particles (1) and (2). The operator W acts only on these variables and on the spin degrees of freedom of the two particles. Let $\{ | \varphi_{n,l,m} \rangle \}$ be a standard basis in the state space $\mathscr{E}_r$ of the relative particle, and $\{ | \varepsilon_1, \varepsilon_2 \rangle \}$, the basis of eigenvectors common to S_{1z} and S_{2z} in the spin state space ($\varepsilon_1 = \pm, \varepsilon_2 = \pm$). The state space in which W acts is spanned by the $\{ | \varphi_{n,l,m} \rangle \otimes | \varepsilon_1, \varepsilon_2 \rangle \}$ basis, in which it is very easy, using expressions (5) and (6), to find the selection rules satisfied by the matrix elements of W.

α. *Spin degrees of freedom*

– T_0 changes neither ε_1 nor ε_2.

– T'_0 "flips" both spins:

$$| +, - \rangle \longrightarrow | -, + \rangle \quad \text{and} \quad | -, + \rangle \longrightarrow | +, - \rangle$$

– T_1 flips one of the two spins up:

$$| -, \varepsilon_2 \rangle \longrightarrow | +, \varepsilon_2 \rangle \quad \text{or} \quad | \varepsilon_1, - \rangle \longrightarrow | \varepsilon_1, + \rangle$$

– Similarly, T_{-1} flips one of the two spins down:

$$| +, \varepsilon_2 \rangle \longrightarrow | -, \varepsilon_2 \rangle \quad \text{or} \quad | \varepsilon_1, + \rangle \longrightarrow | \varepsilon_1, - \rangle$$

– Finally, T_2 and T_{-2} flip both spins up and down, respectively:

$$| -, - \rangle \longrightarrow | +, + \rangle \quad \text{and} \quad | +, + \rangle \longrightarrow | -, - \rangle$$

β. *Orbital degrees of freedom*

When we calculate the matrix element of $\xi(r) T_q$ between the state $| \varphi_{n,l,m} \rangle$ and the state $| \varphi_{n',l',m'} \rangle$, the following angular integral appears:

$$\int Y_{l'}^{m'*}(\theta, \varphi) \, Y_2^q(\theta, \varphi) \, Y_l^m(\theta, \varphi) \, d\Omega \tag{7}$$

which, according to the results of complement C_X, is different from zero only for:

$$l' = l, l - 2, l + 2 \tag{8-a}$$

$$m' = m + q \tag{8-b}$$

Note that the case $l = l' = 0$, although not in contradiction with (8), is excluded because we must always be able to form a triangle with l, l' and 2, which is impossible when $l = l' = 0$. We must have then:

$$l, l' \geqslant 1 \tag{8-c}$$

2. Effects of the dipole-dipole interaction on the Zeeman sublevels of two fixed particles

In this section, we shall assume the two particles to be fixed in space. We shall therefore quantize only the spin degrees of freedom, considering the quantities r, θ and φ as given parameters.

The two particles are placed in a static field $\mathbf{B}_0$ parallel to Oz. The Zeeman Hamiltonian H_0, describing the interaction of the two spin magnetic moments with $\mathbf{B}_0$, can then be written:

$$H_0 = \omega_1 S_{1z} + \omega_2 S_{2z} \tag{9}$$

with:

$$\begin{aligned} \omega_1 &= -\gamma_1 B_0 \\ \omega_2 &= -\gamma_2 B_0 \end{aligned} \tag{10}$$

In the presence of the dipole-dipole interaction W, the total Hamiltonian H of the system becomes:

$$H = H_0 + W \tag{11}$$

We shall assume the field B_0 to be large enough for us to be able to treat W like a perturbation.

a. CASE WHERE THE TWO PARTICLES HAVE DIFFERENT MAGNETIC MOMENTS

α. *Zeeman levels and the magnetic resonance spectrum in the absence of interaction*

According to (9), we have:

$$H_0 \,|\, \varepsilon_1, \varepsilon_2 \rangle = \frac{\hbar}{2}(\varepsilon_1\omega_1 + \varepsilon_2\omega_2) \,|\, \varepsilon_1, \varepsilon_2 \rangle \tag{12}$$

Figure 2-a represents the energy levels of the two-spin system in the absence of the dipole-dipole interaction (we have assumed $\omega_1 > \omega_2 > 0$). Since $\omega_1 \neq \omega_2$, these levels are all non-degenerate.

If we apply a radio-frequency field $\mathbf{B}_1 \cos \omega t$ parallel to Ox, we obtain a series of magnetic resonance lines. The frequencies of these resonances correspond to the various Bohr frequencies which can appear in the evolution of $\langle \gamma_1 S_{1x} + \gamma_2 S_{2x} \rangle$ (the radio-frequency field interacts with the component along Ox of the total magnetic moment). The solid-line (dashed-line) arrows of figure 2-a join levels between which $S_{1x}(S_{2x})$ has a non-zero matrix element. Thus we see that

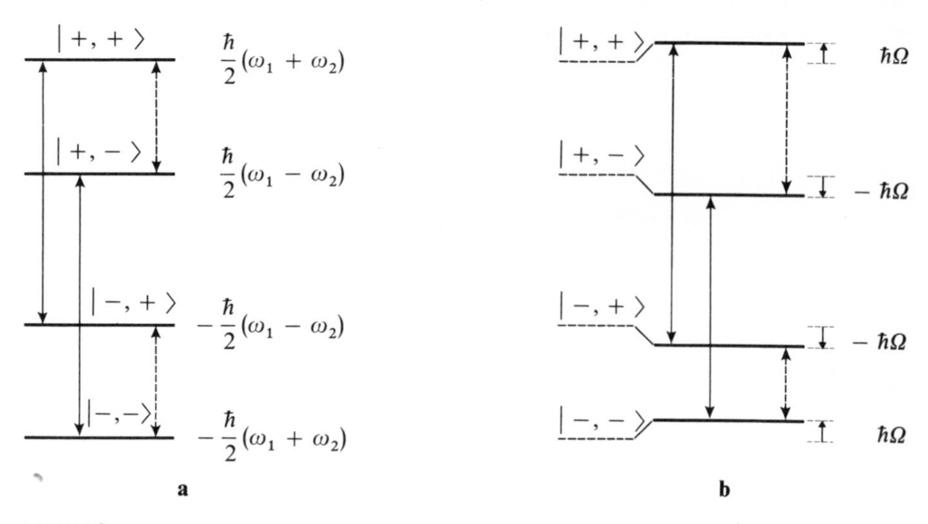

FIGURE 2

Energy levels of two spin 1/2 particles, placed in a static field $\mathbf{B}_0$ parallel to Oz. The two Larmor angular frequencies, $\omega_1 = -\gamma_1 B_0$ and $\omega_2 = -\gamma_2 B_0$, are assumed to be different.
For figure a, the energy levels are calculated without taking account of the dipole-dipole interaction W between the two spins.
For figure b, we take this interaction into account. The levels undergo a shift whose approximate value, to first order in W, is indicated on the right-hand side of the figure. The solid-line arrows join the levels between which S_{1x} has a non-zero matrix element, and the dashed-line arrows, those for which S_{2x} does.

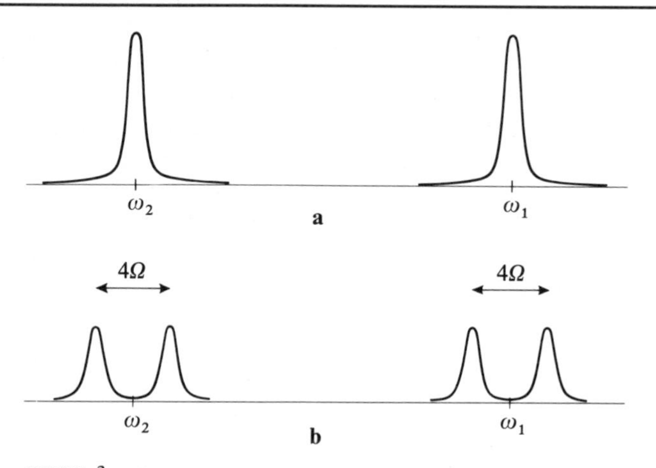

FIGURE 3

The Bohr frequencies which appear in the evolution of $\langle S_{1x} \rangle$ and $\langle S_{2x} \rangle$ give the positions of the magnetic resonance lines which can be observed for the two-spin system (the transitions corresponding to the arrows of figure 2).
In the absence of a dipole-dipole interaction, two resonances are obtained, each one corresponding to one of the two spins (fig. a). The dipole-dipole interaction is expressed by a splitting of each of the two preceding lines (fig. b).

there are two distinct Bohr angular frequencies, equal to ω_1 and ω_2 (fig. 3-a), which correspond simply to the resonances of the individual spins, (1) and (2).

β. *Modifications created by the interaction*

Since all the levels of figure 2-a are non-degenerate, the effect of W can be obtained to first order by calculating the diagonal elements of W, $\langle \varepsilon_1, \varepsilon_2 \mid W \mid \varepsilon_1, \varepsilon_2 \rangle$. It is clear from expressions (5) and (6) that only the term T_0 makes a non-zero contribution to this matrix element, which is then equal to:

$$\langle \varepsilon_1, \varepsilon_2 \mid W \mid \varepsilon_1, \varepsilon_2 \rangle = \xi(r)\,(3\cos^2\theta - 1)\frac{\varepsilon_1\varepsilon_2\hbar^2}{4} = \varepsilon_1\varepsilon_2\,\hbar\Omega \tag{13}$$

with:

$$\Omega = \frac{\hbar}{4}\,\xi(r)\,(3\cos^2\theta - 1) = \frac{-\hbar\mu_0}{16\pi}\frac{\gamma_1\gamma_2}{r^3}\,(3\cos^2\theta - 1) \tag{14}$$

Since W is much smaller than H_0, we have:

$$\Omega \ll \omega_1 - \omega_2 \tag{15}$$

From this we can immediately deduce the level shifts to first order in W: $\hbar\Omega$ for $\mid +, + \rangle$ and $\mid -, - \rangle$, and $-\hbar\Omega$ for $\mid +, - \rangle$ and for $\mid -, + \rangle$ (fig. 2-b).

What now happens to the magnetic resonance spectrum of figure 3-a? If we are concerned only with lines whose intensities are of zeroeth order in W (that is, those which approach the lines of figure 2-a when W approaches zero), then to calculate the Bohr frequencies which appear in $\langle S_{1x} \rangle$ and $\langle S_{2x} \rangle$ we simply use the zeroeth-order expressions for the eigenvectors*. It is then the same transitions which are involved (compare the arrows of figures 2-a and 2-b). We see, however, that the two lines which correspond to the frequency ω_1 in the absence of coupling (solid-line arrows) now have different frequencies: $\omega_1 + 2\Omega$ and $\omega_1 - 2\Omega$. Similarly, the two lines corresponding to ω_2 (dashed-line arrows) now have frequencies of $\omega_2 + 2\Omega$ and $\omega_2 - 2\Omega$. The magnetic resonance spectrum is therefore now composed of two "doublets" centered at ω_1 and ω_2, the interval between the two components of each doublet being equal to 4Ω (fig. 3-b).

Thus, the dipole-dipole interaction leads to a fine structure in the magnetic resonance spectrum, for which we can give a simple physical interpretation. The magnetic moment $\mathbf{M}_1$ associated with $\mathbf{S}_1$ creates a "local field" $\mathbf{b}$ at particle (2). Since we assume $\mathbf{B}_0$ to be very large, $\mathbf{S}_1$ precesses very rapidly about Oz, so we can consider only the S_{1z} component (the local field created by the other components oscillates too rapidly to have a significant effect). The local field $\mathbf{b}$ therefore has a different direction depending on whether the spin is in the state $\mid + \rangle$ or $\mid - \rangle$, that is, depending on whether it points up or down. It follows that the total field

* If we used higher-order expressions for the eigenvectors, we would see other lines of lower intensity appear (they disappear when $W \longrightarrow 0$).

"seen" by particle **(2)**, which is the sum of $\mathbf{B}_0$ and $\mathbf{b}$, can take on two possible values★. This explains the appearance of two resonance frequencies for the spin (2). The same argument would obviously enable us to understand the origin of the doublet centered at ω_1.

b. CASE WHERE THE TWO PARTICLES HAVE EQUAL MAGNETIC MOMENTS

α. *Zeeman levels and the magnetic resonance spectrum in the absence of the interaction*

Formula (12) remains valid if we choose ω_1 and ω_2 to be equal. We shall therefore set:

$$\omega_1 = \omega_2 = \omega = -\gamma B_0 \tag{16}$$

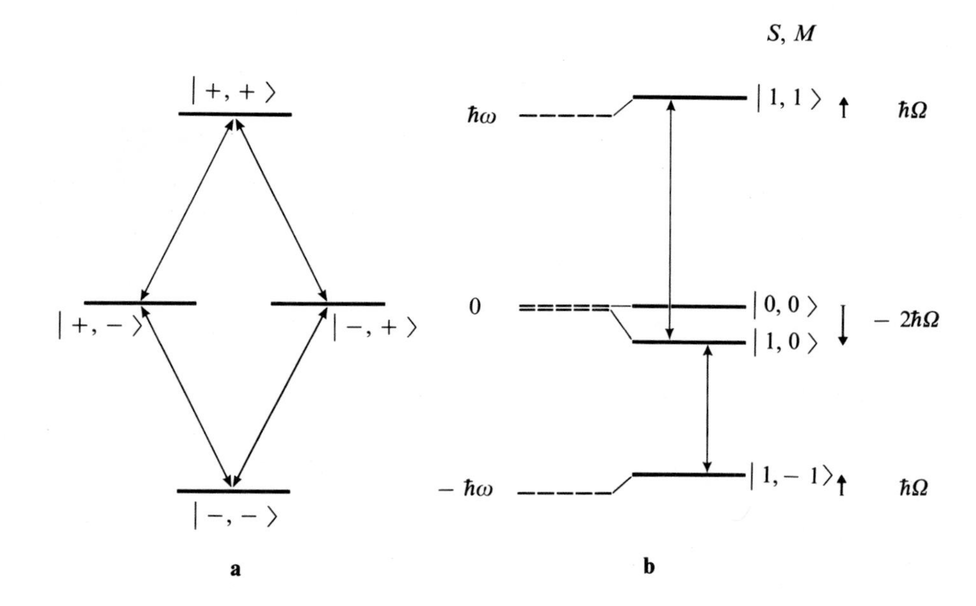

FIGURE 4

The two spin 1/2 particles are assumed to have the same magnetic moment and, consequently, the same Larmor angular frequency $\omega = -\gamma B_0$.

In the absence of a dipole-dipole interaction, we obtain three levels, one of which is two-fold degenerate (fig. a). Under the effect of the dipole-dipole interaction (fig. b), these levels undergo shifts whose approximate values (to first order in W) are indicated on the right-hand side of the figure. To zeroeth-order in W, the stationary states are the eigenstates $| S, M \rangle$ of the total spin. The arrows join the levels between which $S_{1x} + S_{2x}$ has a non-zero matrix element.

★ Actually, since $|\mathbf{B}_0| \gg |\mathbf{b}|$, it is only the component of $\mathbf{b}$ along $\mathbf{B}_0$ which is involved.

The energy levels are shown in figure (4-a). The upper level, $| +, + \rangle$, and the lower level, $| -, - \rangle$, of energies $\hbar\omega$ and $-\hbar\omega$, are non-degenerate. On the other hand, the intermediate level, of energy 0, is two-fold degenerate: to it correspond the two eigenstates $| +, - \rangle$ and $| -, + \rangle$.

The frequencies of the magnetic resonance lines can be obtained by finding the Bohr frequencies involved in the evolution of $\langle S_{1x} + S_{2x} \rangle$ (the total magnetic moment is now proportional to the total spin $\mathbf{S} = \mathbf{S}_1 + \mathbf{S}_2$). We easily obtain the four transitions represented by the arrows in figure 4-a, which correspond to a single angular frequency ω. This finally yields the spectrum of figure 5-a.

FIGURE 5

Shape of the magnetic resonance spectrum which can be observed for a system of two spin 1/2 particles, with the same gyromagnetic ratio, placed in a static field B_0.
In the absence of a dipole-dipole interaction, we observe a single resonance (fig. a). In the presence of a dipole-dipole interaction (fig. b), the preceding line splits. The separation 6Ω between the two components of the doublet is proportional to $3\cos^2\theta - 1$, where θ is the angle between the static-field B_0 and the straight line joining the two particles.

β. *Modifications created by the interaction*

The shifts of the non-degenerate levels $| +, + \rangle$ and $| -, - \rangle$ can be obtained as they were before, and are both equal to $\hbar\Omega$ [we must replace, however, γ_1 and γ_2 by γ in expression (14) for Ω].

Since the intermediate level is two-fold degenerate, the effect of W on this level can now be obtained by diagonalizing the matrix which represents the restriction of W to the subspace $\{ | +, - \rangle, | -, + \rangle \}$. The calculation of the diagonal elements is performed as above and yields:

$$\langle +, - | W | +, - \rangle = \langle -, + | W | -, + \rangle = -\hbar\,\Omega \tag{17}$$

As for the non-diagonal element $\langle +, - | W | -, + \rangle$, we easily see from expressions (5) and (6) that only the term T'_0 contributes to it:

$$\begin{aligned}\langle +, - | W | -, + \rangle &= -\frac{\xi(r)}{4}(3\cos^2\theta - 1) \\ &\qquad \times \langle +, - | (S_{1+}S_{2-} + S_{1-}S_{2+}) | -, + \rangle \\ &= -\xi(r)\frac{\hbar^2}{4}(3\cos^2\theta - 1) = -\hbar\Omega\end{aligned} \tag{18}$$

We are then led to the diagonalization of the matrix:

$$- \hbar\Omega\begin{pmatrix} 1 & 1 \\ 1 & 1 \end{pmatrix} \tag{19}$$

whose eigenvalues are $-2\hbar\Omega$ and 0, associated respectively with the eigenvectors $|\psi_1\rangle = \frac{1}{\sqrt{2}}(|+,-\rangle + |-,+\rangle)$ and $|\psi_2\rangle = \frac{1}{\sqrt{2}}(|+,-\rangle - |-,+\rangle)$.

Figure 4-b represents the energy levels of the system of two coupled spins. The energies, to first order in W, are given by the eigenstates to zeroeth order.

Note that these eigenstates are none other than the eigenstates $|S, M\rangle$ common to $\mathbf{S}^2$ and S_z, where $\mathbf{S} = \mathbf{S}_1 + \mathbf{S}_2$ is the total spin. Since the operator S_x commutes with $\mathbf{S}^2$, it can couple only the triplet states, that is, $|1, 0\rangle$ to $|1, 1\rangle$ and $|1, 0\rangle$ to $|1, -1\rangle$. This gives the two transitions represented by the arrows in figure 4-b, and to which correspond the Bohr frequencies $\omega + 3\Omega$ and $\omega - 3\Omega$. The magnetic resonance spectrum is therefore composed of a doublet centered at ω, the separation between the two components of the doublet being equal to 6Ω (fig. 5-b).

c. EXAMPLE : THE MAGNETIC RESONANCE SPECTRUM OF GYPSUM

The case studied in § b above corresponds to that of two protons of a crystallization water molecule in a gypsum monocrystal ($CaSO_4$, $2H_2O$). These two protons have identical magnetic moments and can be considered to occupy fixed positions in the crystal. Moreover, they are much closer to each other than to other protons (belonging to other water molecules). Since the dipole-dipole interaction decreases very quickly with distance ($1/r^3$ law), we can neglect interactions between protons belonging to other water molecules.

The magnetic resonance spectrum is indeed observed to contain a doublet* whose separation depends on the angle θ between the field $\mathbf{B}_0$ and the straight line joining the two protons. If we rotate the crystal with respect to the field $\mathbf{B}_0$, this angle θ varies, and the separation between the two components of the doublet changes. Thus, by studying the variations of this separation, we can determine the positions of the water molecules relative to the crystal axes.

When the sample under study is not a monocrystal, but rather a powder composed of small, randomly oriented monocrystals, θ takes on all possible values. We then observe a wide band, due to the superposition of doublets having different separations.

3. Effects of the interaction in a bound state

We shall now assume that the two particles, (1) and (2), are not fixed, but can move with respect to each other.

Consider, for example, the case of the hydrogen atom (a proton and an electron). When we take only the electrostatic forces into account, the ground state

* Actually, in a Gypsum monocrystal, there are two different orientations for the water molecules, and, consequently, two doublets corresponding to the two possible values of θ.

of this atom (in the center of mass frame) is described by the ket $| \varphi_{1,0,0} \rangle$, labeled by the quantum numbers $n = 1$, $l = 0$, $m = 0$ (*cf.* chap. VII). The proton and the electron are spin 1/2 particles. The ground state is therefore four-fold degenerate, and a possible basis in the corresponding subspace is made up of the four vectors:

$$\{ | \varphi_{1,0,0} \rangle \otimes | \varepsilon_1, \varepsilon_2 \rangle \} \tag{20}$$

where ε_1 and ε_2, equal to + or −, represent respectively the eigenvalues of S_z and I_z (**S** and **I**: the electron and proton spins).

What is the effect on this ground state of the dipole-dipole interaction between **S** and **I**? The matrix elements of W are much smaller than the energy difference between the 1*s* level and the excited levels, so that it is possible to treat the effect of W by perturbation theory. To first order, it can be evaluated by diagonalizing the 4×4 matrix of elements $\langle \varphi_{1,0,0}\varepsilon_1'\varepsilon_2' | W | \varphi_{1,0,0}\varepsilon_1\varepsilon_2 \rangle$. The calculation of these matrix elements, according to (5) and (6), involves angular integrals of the form:

$$\int Y_0^{0*}(\theta, \varphi)\, Y_2^q(\theta, \varphi)\, Y_0^0(\theta, \varphi)\, \mathrm{d}\Omega \tag{21}$$

which are equal to zero, according to the selection rules established in §1-c above [in this particular case, it can be shown very simply that integral (21) is equal to zero: since Y_0^0 is a constant, expression (21) is proportional to the scalar product of Y_2^q and Y_0^0, which is equal to zero because of the spherical harmonic orthogonality relations].

The dipole-dipole interaction does not modify the energy of the ground state to first order. It enters, however, into the (hyperfine) structure of the excited levels with $l \geqslant 1$. We must then calculate the matrix elements $\langle \varphi_{n,l,m'}\varepsilon_1'\varepsilon_2' | W | \varphi_{n,l,m}\varepsilon_1\varepsilon_2 \rangle$, that is, the integrals:

$$\int Y_l^{m'*}(\theta, \varphi)\, Y_2^q(\theta, \varphi)\, Y_l^m(\theta, \varphi)\, \mathrm{d}\Omega$$

which, according to (8-c), become non-zero as soon as $l \geqslant 1$.

References and suggestions for further reading:

Evidence in nuclear magnetic resonance experiments of the magnetic dipole interactions between two spins in a rigid lattice: Abragam (14.1), chap. IV, §II and chap. VII, §IA; Slichter (14.2), chap. 3; Pake (14.6).

Complement C_{XI}

VAN DER WAALS FORCES

The character of the forces exerted between two neutral atoms changes with the order of magnitude of the distance R separating these two atoms.

Consider, for example, two hydrogen atoms. When R is of the order of atomic dimensions (that is, of the order of the Bohr radius a_0), the electronic wave functions overlap, and the two atoms attract each other, since they tend to form an H_2 molecule. The potential energy of the system has a minimum* for a certain value R_e of the distance R between the atoms. The physical origin of this attraction (and therefore of the chemical bond) lies in the fact that the electrons can oscillate between the two atoms (*cf.* §§C-2-c and C-3-d of chapter IV). The stationary wave functions of the two electrons are no longer localized about only one of the nuclei; this lowers the energy of the ground state (*cf.* complement G_{XI}).

At greater distances, the phenomena change completely. The electrons can no longer move from one atom to the other, since the probability amplitude of such a process decreases with the decreasing overlap of the wave functions, that is, exponentially with the distance. The preponderant effect is then the electrostatic interaction between the electric dipole moments of the two neutral atoms. This gives rise to a total energy which is attractive and which decreases, not exponentially, but with $1/R^6$. This is the origin of the *Van der Waals forces*, which we intend to study in this complement by using stationary perturbation theory (confining ourselves, for the sake of simplicity, to the case of two hydrogen atoms).

It should be clearly understood that the fundamental nature of Van der Waals forces is the same as that of the forces responsible for the chemical bond : the basic Hamiltonian is electrostatic in both cases. Only the variation of the energies of the quantum stationary states of the two-atom system with respect to R allows us to define and differentiate these two types of forces.

* At very short distances, the repulsive forces between the nuclei always dominate.

Van der Waals forces play an important role in physical chemistry, especially when the two atoms under consideration have no valence electrons (forces between rare gas atoms, stable molecules, etc.). They are partially responsible for the differences between the behavior of a real gas and that of an ideal gas. Finally, as we have already said, these are long-range forces, and are therefore involved in the stability of colloids.

We shall begin by determining the expression for the dipole-dipole interaction Hamiltonian between two neutral hydrogen atoms (§1). This will then enable us to study the Van der Waals forces between two atoms in the $1s$ state (§2), or between an atom in the $2p$ state and an atom in the $1s$ state (§3). Finally, we shall show (§4) that a hydrogen atom in the $1s$ state is attracted by its electrical mirror image in a perfectly conducting wall.

1. The electrostatic interaction Hamiltonian for two hydrogen atoms

a. NOTATION

The two protons of the two hydrogen atoms are assumed to remain motionless at points A and B (fig. 1). We shall set:

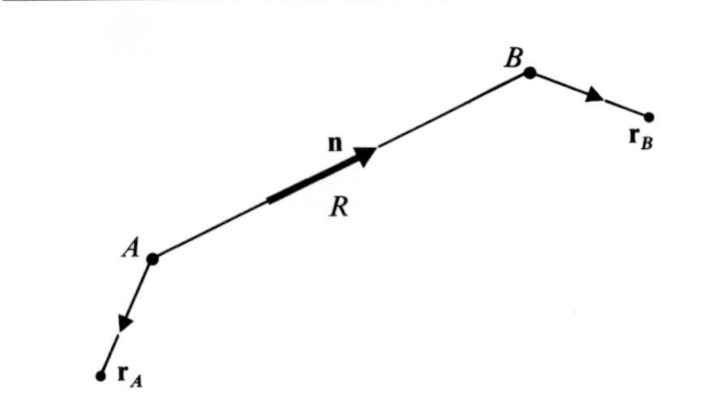

FIGURE 1

Relative position of the two hydrogen atoms. R is the distance between the two protons, which are situated at A and B, and n is the unit vector on the line joining them. $\mathbf{r}_A$ and $\mathbf{r}_B$ are the position vectors of the two electrons with respect to points A and B respectively.

$$\mathbf{R} = \mathbf{OB} - \mathbf{OA} \tag{1}$$

$$R = |\mathbf{R}| \tag{2}$$

$$\mathbf{n} = \frac{\mathbf{R}}{|\mathbf{R}|} \tag{3}$$

R is the distance between the two atoms, and $\mathbf{n}$ is the unit vector on the line which joins them. Let $\mathbf{r}_A$ be the position vector of the electron attached to atom (A) with respect to point A, and $\mathbf{r}_B$, the position vector of the electron attached to atom B

with respect to B. We call:

$$\mathscr{D}_A = q\mathbf{r}_A \tag{4}$$

$$\mathscr{D}_B = q\mathbf{r}_B \tag{5}$$

the electric dipole moments of the two atoms (q is the electron charge).

We shall assume throughout this complement that:

$$R \gg |\mathbf{r}_A|, |\mathbf{r}_B| \tag{6}$$

Although they are identical, the electrons of the two atoms are well separated, and their wave functions do not overlap. It is therefore not necessary to apply the symmetrization postulate (*cf.* chap. XIV, §D-2-b).

b. CALCULATION OF THE ELECTROSTATIC INTERACTION ENERGY

Atom (A) creates at (B) an electrostatic potential U with which the charges of (B) interact. This gives rise to an interaction energy $\mathscr{W}$.

We saw in complement E_X that U can be calculated in terms of R, $\mathbf{n}$ and the multipole moments of atom (A). Since (A) is neutral, the most important contribution to U is that of the electric dipole moment $\mathscr{D}_A$. Similarly, since (B) is neutral, the most important term in $\mathscr{W}$ comes from the interaction between the dipole moment $\mathscr{D}_B$ of (B) and the electric field $\mathbf{E} = -\nabla U$ which is essentially created by $\mathscr{D}_A$. This explains the name of "dipole-dipole interaction" given to the dominant term of $\mathscr{W}$. There exist, of course, smaller terms (dipole-quadrupole, quadrupole-quadrupole, etc.), and $\mathscr{W}$ is written:

$$\mathscr{W} = \mathscr{W}_{dd} + \mathscr{W}_{dq} + \mathscr{W}_{qd} + \mathscr{W}_{qq} + \dots \tag{7}$$

To calculate $\mathscr{W}_{dd}$, we shall start with the expression for the electrostatic potential created by $\mathscr{D}_A$ at (B):

$$U(\mathbf{R}) = \frac{1}{4\pi\varepsilon_0} \frac{\mathscr{D}_A \cdot \mathbf{R}}{R^3} \tag{8}$$

from which we see:

$$\mathbf{E} = -\nabla_{\mathbf{R}} U = -\frac{q}{4\pi\varepsilon_0} \frac{1}{R^3} [\mathbf{r}_A - 3(\mathbf{r}_A \cdot \mathbf{n})\mathbf{n}] \tag{9}$$

and, consequently:

$$\mathscr{W}_{dd} = -\mathbf{E} \cdot \mathscr{D}_B = \frac{e^2}{R^3} [\mathbf{r}_A \cdot \mathbf{r}_B - 3(\mathbf{r}_A \cdot \mathbf{n})(\mathbf{r}_B \cdot \mathbf{n})] \tag{10}$$

We have set $e^2 = q^2/4\pi\varepsilon_0$, and we have used expressions (4) and (5) for $\mathscr{D}_A$ and $\mathscr{D}_B$. In this complement, we shall choose the Oz axis parallel to $\mathbf{n}$, so that (10) can be written:

$$\mathscr{W}_{dd} = \frac{e^2}{R^3} (x_A x_B + y_A y_B - 2z_A z_B) \tag{11}$$

In quantum mechanics, $\mathscr{W}_{dd}$ becomes the operator W_{dd}, which can be obtained by replacing in (11) x_A, y_A, ..., z_B by the corresponding observables X_A, Y_A, ..., Z_B, which act in the state spaces $\mathscr{E}_A$ and $\mathscr{E}_B$ of the two hydrogen atoms*:

$$W_{dd} = \frac{e^2}{R^3}(X_A X_B + Y_A Y_B - 2Z_A Z_B) \tag{12}$$

2. Van der Waals forces between two hydrogen atoms in the 1*s* ground state

a. EXISTENCE OF A $-C/R^6$ ATTRACTIVE POTENTIAL

α. *Principle of the calculation*

The Hamiltonian of the system is:

$$H = H_{0A} + H_{0B} + W_{dd} \tag{13}$$

where H_{0A} and H_{0B} are the energies of atoms (A) and (B) when they are isolated.

In the absence of W_{dd}, the eigenstates of H are given by the equation:

$$(H_{0A} + H_{0B}) \mid \varphi^A_{n,l,m} ; \varphi^B_{n',l',m'} \rangle = (E_n + E_{n'}) \mid \varphi^A_{n,l,m} ; \varphi^B_{n',l',m'} \rangle \tag{14}$$

where the $\mid \varphi_{n,l,m} \rangle$ and the E_n were calculated in §C of chapter VII. In particular, the ground state of $H_{0A} + H_{0B}$ is $\mid \varphi^A_{1,0,0} ; \varphi^B_{1,0,0} \rangle$, of energy $-2E_I$. It is non-degenerate (we do not take spins into account).

The problem is to evaluate the shift in this ground state due to W_{dd} and, in particular, its R-dependence. This shift represents, so to speak, the interaction potential energy of the two atoms in the ground state.

Since W_{dd} is much smaller than H_{0A} and H_{0B}, we can calculate this effect by stationary perturbation theory.

β. *First-order effect of the dipole-dipole interaction*

Let us show that the first-order correction:

$$\varepsilon_1 = \langle \varphi^A_{1,0,0} ; \varphi^B_{1,0,0} \mid W_{dd} \mid \varphi^A_{1,0,0} ; \varphi^B_{1,0,0} \rangle \tag{15}$$

is zero. ε_1 involves, according to expression (12) for W_{dd}, products of the form $\langle \varphi^A_{1,0,0} \mid X_A \mid \varphi^A_{1,0,0} \rangle \langle \varphi^B_{1,0,0} \mid X_B \mid \varphi^B_{1,0,0} \rangle$ (and analogous quantities in which X_A is replaced by Y_A, Z_A and X_B by Y_B, Z_B), which are zero, since, in a stationary state of the atom, the mean values of the components of the position operator are zero.

* The translational external degrees of freedom of the two atoms are not quantized : for the sake of simplicity, we assume the two protons to be infinitely heavy and motionless. In (12), R is therefore a parameter and not an observable.

COMMENT:

The other terms, W_{dq}, W_{qd}, W_{qq}..., of expansion (7) involve products of two multipole moments, one relative to (A) and the other one to (B), at least one of which is of order higher than 1. Their contributions are also zero to first order: they are expressed in terms of mean values in the ground state of multipole operators of order greater than or equal to one, and we know (*cf.* complement E_X, §2-c) that such mean values are zero in an $l = 0$ state (triangle rule of Clebsch-Gordan-coefficients). Therefore we must find the second-order effect of W_{dd}, which then constitutes the most important energy correction.

γ. *Second-order effect of the dipole-dipole interaction*

According to the results of chapter XI, the second-order energy correction can be written:

$$\varepsilon_2 = \sum_{\substack{nlm \\ n'l'm'}}' \frac{|\langle \varphi^A_{n,l,m} ; \varphi^B_{n',l',m'} | W_{dd} | \varphi^A_{1,0,0} ; \varphi^B_{1,0,0} \rangle|^2}{-2E_I - E_n - E_{n'}} \tag{16}$$

where the notation Σ' means that the state $| \varphi^A_{1,0,0} ; \varphi^B_{1,0,0} \rangle$ is excluded from the summation★.

Since W_{dd} is proportional to $1/R^3$, ε_2 is proportional to $1/R^6$. Furthermore, all the energy denominators are negative, since we are starting from the ground state. Therefore, *the dipole-dipole interaction gives rise to a negative energy proportional to* $1/R^6$:

$$\varepsilon_2 = -\frac{C}{R^6} \tag{17}$$

Van der Waals forces are therefore attractive and vary with $1/R^7$.

Finally, let us calculate the expansion of the ground state to first order in W_{dd}. We find, according to formula (B-11) of chapter XI:

$$| \psi_0 \rangle = | \varphi^A_{1,0,0} ; \varphi^B_{1,0,0} \rangle + \sum_{\substack{nlm \\ n'l'm'}}' | \varphi^A_{n,l,m} ; \varphi^B_{n',l',m'} \rangle \frac{\langle \varphi^A_{n,l,m} ; \varphi^B_{n',l',m'} | W_{dd} | \varphi^A_{1,0,0} ; \varphi^B_{1,0,0} \rangle}{-2E_I - E_n - E_{n'}} + \ldots \tag{18}$$

COMMENT:

The matrix elements appearing in expressions (16) and (18) involve the quantities $\langle \varphi^A_{n,l,m} | X_A | \varphi^A_{1,0,0} \rangle \langle \varphi^B_{n',l',m'} | X_B | \varphi^B_{1,0,0} \rangle$ (and analogous quantities in which X_A and

★ This summation is performed not only over the bound states, but also over the continuous spectrum of $H_{0A} + H_{0B}$

X_B are replaced by Y_A and Y_B or Z_A and Z_B), which are different from zero only if $l = 1$ and $l' = 1$. These quantities are indeed proportional to products of angular integrals

$$\left[\int Y_l^{m*}(\Omega_A)\, Y_1^q(\Omega_A)\, Y_0^0(\Omega_A)\, \mathrm{d}\Omega_A\right] \times \left[\int Y_{l'}^{m'*}(\Omega_B)\, Y_1^{q'}(\Omega_B)\, Y_0^0(\Omega_B)\, \mathrm{d}\Omega_B\right]$$

which, according to the results of complement C_X, are zero if $l \neq 1$ or $l' \neq 1$. We can therefore, in (16) and (18), replace l and l' by 1.

b. APPROXIMATE CALCULATION OF THE CONSTANT C

According to (16) and (12), the constant C appearing in (17) is given by:

$$C = e^4 \sum_{\substack{nlm \\ n'l'm'}}' \frac{|\langle \varphi_{n,l,m}^A ; \varphi_{n',l',m'}^B | (X_A X_B + Y_A Y_B - 2Z_A Z_B) | \varphi_{1,0,0}^A ; \varphi_{1,0,0}^B \rangle|^2}{2E_I + E_n + E_{n'}} \tag{19}$$

We must have $n \geqslant 2$ and $n' \geqslant 2$. For bound states, $|E_n| = E_I/n^2$ is smaller than E_I, and the error is not significant if we replace in (19) E_n and $E_{n'}$ by 0. For states in the continuous spectrum, E_n varies between 0 and $+\infty$. The matrix elements of the numerator become small, however, as soon as the size of E_n becomes appreciable, since the spatial oscillations of the wave function are then numerous in the region in which $\varphi_{1,0,0}(\mathbf{r})$ is non-zero.

To have an idea of the order of magnitude of C, we can therefore replace all the energy denominators of (19) by $2E_I$. Using the closure relation and the fact that the diagonal element of W_{dd} is zero (§2-a-β), we then get:

$$C \simeq \frac{e^4}{2E_I} \langle \varphi_{1,0,0}^A ; \varphi_{1,0,0}^B | (X_A X_B + Y_A Y_B - 2Z_A Z_B)^2 | \varphi_{1,0,0}^A ; \varphi_{1,0,0}^B \rangle \tag{20}$$

This expression is simple to calculate: because of the spherical symmetry of the $1s$ state, the mean values of the cross terms of the type $X_A Y_A$, $X_B Y_B$, ..., are zero. Furthermore, and for the same reason, the various quantities:

$$\langle \varphi_{1,0,0}^A | X_A^2 | \varphi_{1,0,0}^A \rangle, \langle \varphi_{1,0,0}^A | Y_A^2 | \varphi_{1,0,0}^A \rangle \; ..., \langle \varphi_{1,0,0}^B | Z_B^2 | \varphi_{1,0,0}^B \rangle$$

are all equal to one third of the mean value of $\mathbf{R}_A^2 = X_A^2 + Y_A^2 + Z_A^2$. We finally obtain, therefore, using the expression for the wave function $\varphi_{1,0,0}(\mathbf{r})$:

$$C \simeq \frac{e^4}{2E_I} \times 6 \left| \langle \varphi_{1,0,0}^A | \frac{\mathbf{R}_A^2}{3} | \varphi_{1,0,0}^A \rangle \right|^2 = 6\, e^2 a_0^5 \tag{21}$$

(where a_0 is the Bohr radius) and, consequently:

$$\varepsilon_2 \simeq -6e^2 \frac{a_0^5}{R^6} = -6 \frac{e^2}{R} \left(\frac{a_0}{R}\right)^5 \tag{22}$$

The preceding calculation is valid only if $a_0 \ll R$ (no overlapping of the wave functions). Thus we see that ε_2 is of the order of the electrostatic interaction between two charges q and $-q$, multiplied by the reduction factor $(a_0/R)^5 \ll 1$.

c. DISCUSSION

α. *"Dynamical" interpretation of Van der Waals forces*

At any given instant, the electric dipole moment (we shall say, more simply, the dipole) of each atom has a mean value of zero in the ground state $| \varphi_{1,0,0}^A \rangle$ or $| \varphi_{1,0,0}^B \rangle$. This does not mean that any individual measurement of a component of this dipole will yield zero. If we make such a measurement, we generally find a non-zero value; however, we have the same probability of finding the opposite value. The dipole of a hydrogen atom in the ground state should therefore be thought of as constantly undergoing random fluctuations.

We shall begin by neglecting the influence of one dipole on the motion of the other one. Since the two dipoles are then fluctuating randomly and independently, their mean interaction is zero: this explains the fact that W_{dd} has no first-order effect.

However, the two dipoles are not really independent. Consider the electrostatic field created by dipole (A) at (B). This field follows the fluctuations of dipole (A). The dipole it induces at (B) is therefore correlated with dipole (A), so the electrostatic field which "returns" to (A) is no longer uncorrelated with the motion of dipole (A). Thus, although the motion of dipole (A) is random, its interaction with its own field, which is "reflected" to it by (B), does not have a mean value of zero. This is the physical interpretation of the second-order effect of W_{dd}.

The dynamical aspect is therefore useful for understanding the origin of Van der Waals forces. If we were to think of the two hydrogen atoms in the ground state as two spherical and "static" clouds of negative electricity (with a positive point charge at the center of each one), we would be led to a rigorously zero interaction energy.

β. *Correlations between the two dipole moments*

Let us show more precisely that there exists a correlation between the two dipoles.

When we take W_{dd} into account, the ground state of the system is no longer $| \varphi_{1,0,0}^A ; \varphi_{1,0,0}^B \rangle$, but $| \psi_0 \rangle$ [*cf.* expression (18)]. A simple calculation then yields:

$$\langle \psi_0 | X_A | \psi_0 \rangle = \dots = \langle \psi_0 | Z_B | \psi_0 \rangle = 0 \tag{23}$$

to first order in W_{dd}.

Consider, for example, $\langle \psi_0 | X_A | \psi_0 \rangle$. The zeroeth-order term,

$$\langle \varphi_{1,0,0}^A ; \varphi_{1,0,0}^B | X_A | \varphi_{1,0,0}^A ; \varphi_{1,0,0}^B \rangle$$

is zero, since it is equal to the mean value of X_A in the ground state $| \varphi_{1,0,0}^A \rangle$. To first order, the summation appearing in formula (18) must be included. Since W_{dd} contains only products of the form $X_A X_B$, the coefficients of the kets $| \varphi^A_{1,0,0} ; \varphi^B_{n',l',m'} \rangle$ and $| \varphi^A_{n,l,m} ; \varphi^B_{1,0,0} \rangle$ in this summation are zero. The first-order terms which could be different from zero are therefore proportional to

$$\langle \varphi^A_{n,l,m} ; \varphi^B_{n',l',m'} | X_A | \varphi^A_{1,0,0} ; \varphi^B_{1,0,0} \rangle, \qquad \text{with} \quad l \neq 0 \quad \text{and} \quad l' \neq 0 ;$$

These terms are all zero since X_A does not act on $| \varphi^B_{1,0,0} \rangle$ and $\langle \varphi^B_{n',l',m'} | \varphi^B_{1,0,0} \rangle = 0$ for $l' \neq 0$.

Thus, even in the presence of an interaction, the mean values of the components of each dipole are zero. This is not surprising : in the interpretation of § 2-c-α, the dipole induced in (*B*) by the field of dipole (*A*) fluctuates randomly, like this field, and has, consequently, a mean value of zero.

Let us show, on the other hand, that the two dipoles are correlated, by evaluating the mean value of a product of two components, one relative to dipole (*A*) and the other, to dipole (*B*). We shall calculate $\langle \psi_0 | (X_A X_B + Y_A Y_B - 2Z_A Z_B) | \psi_0 \rangle$, for example, which, according to (12), is nothing more than $\frac{R^3}{e^2} \langle \psi_0 | W_{dd} | \psi_0 \rangle$. Using (18), we immediately find, taking (15) and (16) into account, that:

$$\langle \psi_0 | (X_A X_B + Y_A Y_B - 2Z_A Z_B) | \psi_0 \rangle = 2\varepsilon_2 \frac{R^3}{e^2} \neq 0 \tag{24}$$

Thus, the mean values of the products $X_A X_B$, $Y_A Y_B$ and $Z_A Z_B$ are not zero, as would be the products of mean values $\langle X_A \rangle \langle X_B \rangle$, $\langle Y_A \rangle \langle Y_B \rangle$, $\langle Z_A \rangle \langle Z_B \rangle$ according to (23). This proves the existence of a correlation between the two dipoles.

γ. *Long-range modification of Van der Waals forces*

The description of §2-c-α above enables us to understand that the preceding calculations are no longer valid if the two atoms are too far apart. The field produced by (*A*) and "reflected" by (*B*) returns to (*A*) with a time lag due to the propagation (*A*) ⟶ (*B*) ⟶ (*A*), and we have argued as if the interactions were instantaneous.

We can see that this propagation time can no longer be neglected when it becomes of the order of the characteristic times of the atom's evolution, that is, of the order of $2\pi/\omega_{n1}$, where $\omega_{n1} = (E_n - E_1)/\hbar$ denotes a Bohr angular frequency. In other words, the calculations performed in this complement assume that the distance R between the two atoms is much smaller than the wavelengths $2\pi c/\omega_{n1}$ of the spectrum of these atoms (about 1 000 Å).

A calculation which takes propagation effects into account gives an interaction energy which, at large distances, decreases as $1/R^7$. The $1/R^6$ law which we have found therefore applies to an intermediate range of distances, neither too large (because of the time lag) nor too small (to avoid overlapping of the wave functions).

3. Van der Waals forces between a hydrogen atom in the 1s state and a hydrogen atom in the 2p state

a. ENERGIES OF THE STATIONARY STATES OF THE TWO-ATOM SYSTEM. RESONANCE EFFECT

The first excited level of the unperturbed Hamiltonian $H_{0A} + H_{0B}$ is eight-fold degenerate. The associated eigensubspace is spanned by the eight states $\{ | \varphi^A_{1,0,0}; \varphi^B_{2,0,0} \rangle; | \varphi^A_{2,0,0}; \varphi^B_{1,0,0} \rangle; | \varphi^A_{1,0,0}; \varphi^B_{2,1,m} \rangle$, with $m = -1, 0, +1$; $| \varphi^A_{2,1,m'}; \varphi^B_{1,0,0} \rangle$, with $m' = -1, 0, +1 \}$, which correspond to a situation in which one of the two atoms is in the ground state, while the other one is in a state of the $n = 2$ level.

According to perturbation theory for a degenerate state, we must diagonalize the 8×8 matrix representing the restriction of W_{dd} to the eigensubspace to obtain the first-order effect of W_{dd}. We shall show that the only non-zero matrix elements of W_{dd} are those which connect a state $| \varphi^A_{1,0,0}; \varphi^B_{2,1,m} \rangle$ to the state $| \varphi^A_{2,1,m}; \varphi^B_{1,0,0} \rangle$. The operators X_A, Y_A, Z_A appearing in the expression for W_{dd} are odd and can therefore couple $| \varphi^A_{1,0,0} \rangle$ only to one of the $| \varphi^A_{2,1,m} \rangle$; an analogous argument is valid for X_B, Y_B, Z_B. Finally, the dipole-dipole interaction is invariant under a rotation of the two atoms about the Oz axis which joins them; therefore W_{dd} commutes with $L_{Az} + L_{Bz}$ and then can only join two states for which the sum of the eigenvalues of L_{Az} and L_{Bz} is the same.

Therefore, the preceding 8×8 matrix can be broken down into four 2×2 matrices. One of them is entirely zero (the one which concerns the $2s$ states), and the other three are of the form:

$$\begin{pmatrix} 0 & k_m/R^3 \\ k_m/R^3 & 0 \end{pmatrix} \tag{25}$$

where we have set:

$$\langle \varphi^A_{1,0,0}; \varphi^B_{2,1,m} | W_{dd} | \varphi^A_{2,1,m}; \varphi^B_{1,0,0} \rangle = \frac{k_m}{R^3} \tag{26}$$

k_m is a calculable constant, of the order of $e^2 a_0^2$, which we shall not treat here.

We can immediately diagonalize matrix (25), obtaining the eigenvalues $+ k_m/R^3$ and $- k_m/R^3$, associated respectively with the eigenstates:

$$\frac{1}{\sqrt{2}}(| \varphi^A_{1,0,0}; \varphi^B_{2,1,m} \rangle + | \varphi^A_{2,1,m}; \varphi^B_{1,0,0} \rangle)$$

and

$$\frac{1}{\sqrt{2}}(| \varphi^A_{1,0,0}; \varphi^B_{2,1,m} \rangle - | \varphi^A_{2,1,m}; \varphi^B_{1,0,0} \rangle)$$

This reveals the following important results:

– The interaction energy varies with $1/R^3$ and not with $1/R^6$, since W_{dd} now modifies the energies to first order. The Van der Waals forces are therefore more important than they are between two hydrogen atoms in the $1s$ state (resonance effect between two different states of the total system with the same unperturbed energy).

– The sign of the interaction can be positive or negative (eigenvalues $+ k_{m}/R^3$ and $- k_{m}/R^3$). There therefore exist states of the two-atom system for which there is attraction, and others for which there is repulsion.

b. TRANSFER OF THE EXCITATION FROM ONE ATOM TO THE OTHER

The two states $| \varphi^A_{1,0,0} ; \varphi^B_{2,1,m} \rangle$ and $| \varphi^A_{2,1,m} ; \varphi^B_{1,0,0} \rangle$ have the same unperturbed energy and are coupled by a non-diagonal perturbation. According to the general results of §C of chapter IV (two-level system), we know that there is oscillation of the system from one level to the other with a frequency proportional to the coupling.

Therefore, if the system starts in the state $| \varphi^A_{1,0,0} ; \varphi^B_{2,1,m} \rangle$ at $t = 0$, it arrives, after a certain time (the larger R, the longer the time), in the state $| \varphi^A_{2,1,m} ; \varphi^B_{1,0,0} \rangle$. The excitation thus passes from (B) to (A), then returns to (B), and so on.

COMMENT:

If the two atoms are not fixed but, for example, undergo collision, R varies over time and the passage of the excitation from one atom to the other is no longer periodic. The corresponding collisions, called resonant collisions, play an important role in the broadening of spectral lines

4. Interaction of a hydrogen atom in the ground state with a conducting wall

We shall now consider a single hydrogen atom (A) situated at a distance d from a wall which is assumed to be perfectly conducting. The Oz axis is taken along the perpendicular to the wall passing through A (fig. 2). The distance d is assumed to be much larger than the atomic dimensions, so that the atomic structure of the wall can be ignored, and we can assume that the atom interacts with its electrical image on the other side of this wall (that is, with a symmetrical atom with opposite charges). The dipole interaction energy between the atom and the wall can easily be obtained from expression (12) for W_{dd} by making the following substitutions:

$$\begin{cases} e^2 \longrightarrow -e^2 \\ R \longrightarrow 2d \\ X_B \longrightarrow X'_A = X_A \\ Y_B \longrightarrow Y'_A = Y_A \\ Z_B \longrightarrow Z'_A = -Z_A \end{cases} \tag{27}$$

(the change of e^2 to $-e^2$ is due to the sign difference of the image charges).

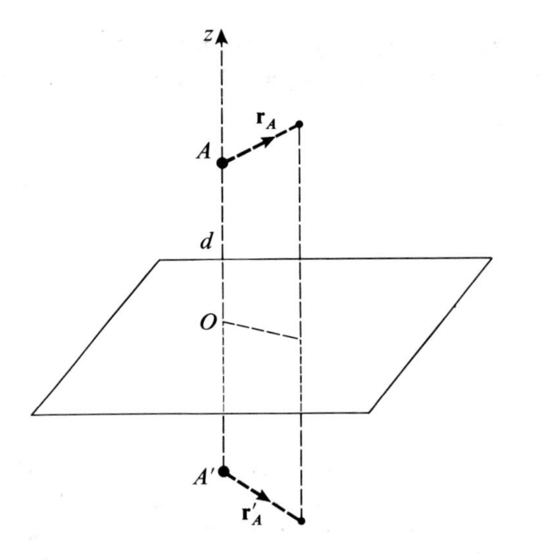

FIGURE 2

To calculate the interaction energy of a hydrogen atom with a perfectly conducting wall, we can assume that the electric dipole moment $q\mathbf{r}_A$ of the atom interacts with its electrical image $-q\mathbf{r}'_A$ (d is the distance between the proton A and the wall).

We then get:

$$W = -\frac{e^2}{8d^3}(X_A^2 + Y_A^2 + 2Z_A^2) \tag{28}$$

which represents the interaction energy of the atom with the wall [W acts only on the degrees of freedom of (A)].

If the atom is in its ground state, the energy correction to first order in W is then:

$$\varepsilon'_1 = \langle \varphi_{1,0,0} | W | \varphi_{1,0,0} \rangle \tag{29}$$

Using the spherical symmetry of the $1s$ state, we obtain:

$$\varepsilon'_1 = -\frac{e^2}{8d^3} 4 \langle \varphi_{1,0,0} | \frac{\mathbf{R}_A^2}{3} | \varphi_{1,0,0} \rangle = -\frac{e^2 a_0^2}{2d^3} \tag{30}$$

We see that the atom is attracted by the wall; the attraction energy varies with $1/d^3$, and, therefore, the force of attraction varies with $1/d^4$.

The fact that W has an effect even to first order can easily be understood in terms of the discussion of § 2-c above. In this case, there is a perfect correlation between the two dipoles, since they are images of each other.

References and suggestions for further reading:

Kittel (13.2), chap. 3, p. 82; Davydov (1.20), chap. XII, §§124 and 125; Langbein (12.9).

For a discussion of retardation effects, see: Power (2.11), §§7.5 and 8.4 (quantum electrodynamic approach); Landau and Lifshitz (7.12), chap. XIII, §90 (electromagnetic fluctuation approach).

See also Derjaguin's article (12.12).

Complement D_{XI}

THE VOLUME EFFECT: THE INFLUENCE OF THE SPATIAL EXTENSION OF THE NUCLEUS ON THE ATOMIC LEVELS

The energy levels and the stationary states of the hydrogen atom were studied in chapter VII by assuming the proton to be a charged point particle, which creates an electrostatic $1/r$ Coulomb potential. Actually, this is not quite true. The proton is not strictly a point charge; its charge fills a volume which has a certain size (of the order of 1 fermi $= 10^{-13}$ cm). When an electron is extremely close to the center of the proton, it "sees" a potential which no longer varies with $1/r$ and which depends on the spatial charge distribution associated with the proton. This is true, furthermore, for all atoms: inside the volume of the nucleus, the electrostatic potential depends on how the charges are distributed. We thus expect the atomic energy levels, which are determined by the potential to which the electrons are subject at all points of space, to be affected by this distribution: this is what is called the "volume effect". The experimental and theoretical study of such an effect is therefore important, since it can supply information about the internal structure of nuclei.

In this complement, we shall give a simplified treatment of the volume effect of hydrogen-like atoms. To have an idea of the order of magnitude of the energy shifts it causes, we shall confine ourselves to a model in which the nucleus is represented by a sphere of radius ρ_0, in which the charge $-Zq$ is uniformly distributed. In this model, the potential created by the nucleus is (*cf.* complement A_V, § 4-b):

$$V(r) = \begin{cases} -\dfrac{Ze^2}{r} & \text{for } r \geqslant \rho_0 \\[2ex] \dfrac{Ze^2}{2\rho_0}\left[\left(\dfrac{r}{\rho_0}\right)^2 - 3\right] & \text{for } r \leqslant \rho_0 \end{cases} \tag{1}$$

(we have set $e^2 = q^2/4\pi\varepsilon_0$). The shape of the variation of $V(r)$ with respect to r is shown in figure 1.

The exact solution of the Schrödinger equation for an electron subject to such a potential poses a complicated problem. Therefore, we shall content ourselves with an approximate solution, based on perturbation theory. In a first approximation, we shall consider the potential to be a Coulomb potential [which amounts to

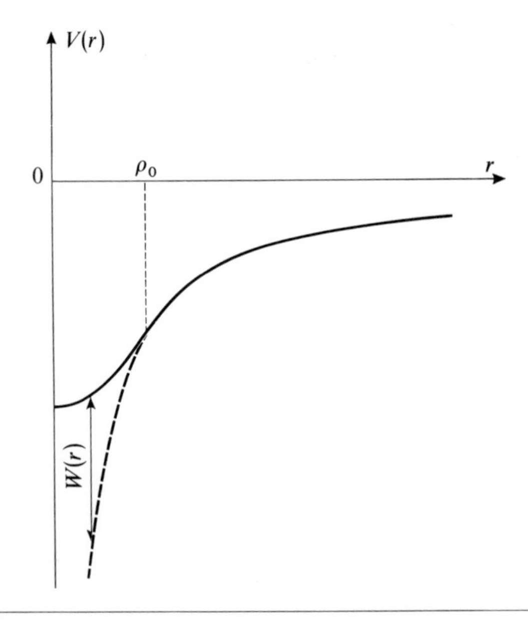

FIGURE 1

Variation with respect to r of the electrostatic potential $V(r)$ created by the charge distribution $-Zq$ of the nucleus, assumed to be uniformly distributed inside a sphere of radius ρ_0. For $r \leqslant \rho_0$, the potential is parabolic. For $r \geqslant \rho_0$, it is a Coulomb potential [the extension of this Coulomb potential into the $r \leqslant \rho_0$ zone is represented by the dashed line; $W(r)$ is the difference between $V(r)$ and the Coulomb potential].

setting $\rho_0 = 0$ in (1)]. The energy levels of the hydrogen atom are then the ones found in § C of chapter VII. We shall treat the difference $W(r)$ between the potential $V(r)$ written in (1) and the Coulomb potential as a perturbation. This difference is zero when r is greater than the radius ρ_0 of the nucleus. It is therefore reasonable that it should cause a small shift in the atomic levels (the corresponding wave functions extend over dimensions of the order of $a_0 \gg \rho_0$), which justifies a treatment by first-order perturbation theory.

1. First-order energy correction

a. CALCULATION OF THE CORRECTION

By definition, $W(r)$ is equal to:

$$W(r) = \begin{cases} \dfrac{Ze^2}{2\rho_0}\left[\left(\dfrac{r}{\rho_0}\right)^2 + \dfrac{2\rho_0}{r} - 3\right] & \text{if } 0 \leqslant r \leqslant \rho_0 \\ 0 & \text{if } r \geqslant \rho_0 \end{cases} \tag{2}$$

Let $| \varphi_{n,l,m} \rangle$ be the stationary states of the hydrogen-like atom in the absence of the perturbation W. To evaluate the effect of W to first order, we must calculate the matrix elements:

$$\langle \varphi_{n,l,m} | W | \varphi_{n,l',m'} \rangle = \int d\Omega\, Y_l^{m*}(\Omega)\, Y_{l'}^{m'}(\Omega) \times \int_0^\infty r^2\, dr\, R_{n,l}^*(r)\, R_{n,l'}(r)\, W(r) \tag{3}$$

In this expression, the angular integral simply gives $\delta_{ll'}\delta_{mm'}$. To simplify the radial integral, we shall make an approximation and assume* that:

$$\rho_0 \ll a_0 \tag{4}$$

that is, that the $r \leqslant \rho_0$ region, in which $W(r)$ is not zero, is much smaller than the spatial extent of the functions $R_{n,l}(r)$. When $r \leqslant \rho_0$, we then have:

$$R_{n,l}(r) \simeq R_{n,l}(0) \tag{5}$$

The radial integral can therefore be written:

$$I = \frac{Ze^2}{2\rho_0} |R_{n,l}(0)|^2 \int_0^{\rho_0} r^2 \, dr \left[\left(\frac{r}{\rho_0} \right)^2 + \frac{2\rho_0}{r} - 3 \right] \tag{6}$$

which gives:

$$I = \frac{Ze^2}{10} \rho_0^2 \, |R_{n,l}(0)|^2 \tag{7}$$

and:

$$\langle \varphi_{n,l,m} \mid W \mid \varphi_{n,l',m'} \rangle = \frac{Ze^2}{10} \rho_0^2 \, |R_{n,l}(0)|^2 \, \delta_{ll'} \, \delta_{mm'} \tag{8}$$

We see that the matrix representing W in the subspace $\mathscr{E}_n$ corresponding to the nth level of the unperturbed Hamiltonian is diagonal. Therefore, the first-order energy correction associated with each state $| \varphi_{n,l,m} \rangle$ can be written simply:

$$\Delta E_{n,l} = \frac{Ze^2}{10} \rho_0^2 \, |R_{n,l}(0)|^2 \tag{9}$$

This correction does not depend on m**. Furthermore, since $R_{n,l}(0)$ is zero unless $l = 0$ (*cf.* chap. VII, § C-4-c), only the s states ($l = 0$ states) are shifted, by a quantity which is equal to:

$$\begin{aligned} \Delta E_{n,0} &= \frac{Ze^2}{10} \rho_0^2 \, |R_{n,0}(0)|^2 \\ &= \frac{2\pi Ze^2}{5} \rho_0^2 \, |\varphi_{n,0,0}(0)|^2 \end{aligned} \tag{10}$$

(we have used the fact that $Y_0^0 = 1/\sqrt{4\pi}$).

* This is certainly the case for the hydrogen atom. In §2, we shall examine condition (4) in greater detail.

** This result could have been expected, since the perturbation W, which is invariant under rotation, is a scalar (*cf.* complement B_{VI}, §5-b).

b. DISCUSSION

$\Delta E_{n,0}$ can be written:

$$\Delta E_{n,0} = \frac{3}{10} w \mathscr{P} \tag{11}$$

where:

$$w = \frac{Ze^2}{\rho_0} \tag{12}$$

is the potential energy of the electron at a distance ρ_0 from the center of the nucleus, and:

$$\mathscr{P} = \frac{4}{3} \pi \rho_0^3 \, |\varphi_{n,0,0}(0)|^2 \tag{13}$$

is the probability of finding the electron inside the nucleus. $\mathscr{P}$ and w enter into (11) because the effect of the perturbation $W(r)$ is felt only inside the nucleus.

For the method which led us to (10) and (11) to be consistent, the correction $\Delta E_{n,0}$ must be much smaller than the energy differences between unperturbed levels. Since w is very large (an electron and a proton attract each other very strongly when they are very close), $\mathscr{P}$ must therefore be extremely small. Before taking up the more precise calculation in §2, we shall evaluate the order of magnitude of these quantities. Let:

$$a_0(Z) = \frac{\hbar^2}{Zme^2} \tag{14}$$

be the Bohr radius when the total charge of the nucleus is $-Zq$. If n is not too high, the wave functions $\varphi_{n,0,0}(\mathbf{r})$ are practically localized inside a region of space whose volume is approximately $[a_0(Z)]^3$. As for the nucleus, its volume is of the order of ρ_0^3, so:

$$\mathscr{P} \simeq \left[\frac{\rho_0}{a_0(Z)}\right]^3 \tag{15}$$

Relation (11) then yields:

$$\begin{aligned} \Delta E_{n,0} &\simeq \frac{Ze^2}{\rho_0}\left[\frac{\rho_0}{a_0(Z)}\right]^3 \\ &= \frac{Ze^2}{a_0(Z)}\left[\frac{\rho_0}{a_0(Z)}\right]^2 \end{aligned} \tag{16}$$

Now, $Ze^2/a_0(Z)$ is of the order of magnitude of the binding energy $E_I(Z)$ of the unperturbed atom. The relative value of the correction is therefore equal to:

$$\frac{\Delta E_{n,0}}{E_I(Z)} \simeq \left[\frac{\rho_0}{a_0(Z)}\right]^2 \tag{17}$$

If condition (4) is met, this correction will indeed be very small. We shall now calculate it more precisely in some special cases.

2. Application to some hydrogen-like systems

a. THE HYDROGEN ATOM AND HYDROGEN-LIKE IONS

For the ground state of the hydrogen atom, we have [*cf.* chap. VII, relation (C-39-a)]:

$$R_{1,0}(r) = 2(a_0)^{-3/2}\, e^{-r/a_0} \tag{18}$$

[where a_0 is obtained by setting $Z = 1$ in (14)]. Formula (10) then gives:

$$\Delta E_{1,0} = \frac{2}{5}\frac{e^2}{a_0}\left(\frac{\rho_0}{a_0}\right)^2 = \frac{4}{5}E_I\left(\frac{\rho_0}{a_0}\right)^2 \tag{19}$$

Now, we know that, for hydrogen:

$$a_0 \simeq 0.53\ \text{Å} = 5.3 \times 10^{-11}\ \text{m} \tag{20}$$

Furthermore, the radius ρ_0 of the proton is of the order of:

$$\rho_0(\text{proton}) \simeq 1\ \text{F} = 10^{-15}\ \text{m} \tag{21}$$

If we substitute these numerical values into (19), we obtain:

$$\Delta E_{1,0} \simeq 4.5 \times 10^{-10}\ E_I \simeq 6 \times 10^{-9}\ \text{eV} \tag{22}$$

The result is therefore very small.

For a hydrogen-like ion, the nucleus has a charge of $-Zq$. We can then apply (10), which amounts to replacing e^2 in (19) by Ze^2, and a_0 by $a_0(Z) = a_0/Z$. We obtain:

$$\Delta E_{1,0}(Z) = \frac{2}{5}\frac{Z^2e^2}{a_0}\left[\frac{\rho_0(A, Z)}{a_0} \times Z\right]^2 \tag{23}$$

where $\rho_0(A, Z)$ is the radius of the nucleus, composed of A nucleons (protons or neutrons), Z of which are protons. In practice, the number of nucleons of a nucleus is not very different from $2Z$; in addition, the "nuclear density saturation" property is expressed by the approximate relation:

$$\rho_0(A, Z) \propto A^{1/3} \propto Z^{1/3} \tag{24}$$

The variation of the energy correction with respect to Z is then given by:

$$\Delta E_{1,0}(Z) \propto Z^{14/3} \tag{25}$$

or:

$$\frac{\Delta E_{1,0}(Z)}{E_I(Z)} \propto Z^{8/3} \tag{26}$$

$\Delta E_{1,0}(Z)$ therefore varies very rapidly with Z, under the effect of several concordant factors: when Z increases, a_0 decreases and ρ_0 increases. The volume effect is therefore significantly larger for heavy hydrogen-like ions than for hydrogen.

COMMENT:

The volume effect also exists for all the other atoms. It is responsible for an isotopic shift of the lines of the emission spectrum. For two distinct isotopes of the same chemical element, the number Z of protons of the nucleus is the same, but the number $A - Z$ of neutrons is different; the spatial distributions of the nuclear charges are therefore not identical for the two nuclei.

Actually, for light atoms, the isotopic shift is caused principally by the nuclear finite mass effect (*cf.* complement A_{VII}, §1-a-α). On the other hand, for heavy atoms (for which the reduced mass varies very little from one isotope to another), the finite mass effect is small; however, the volume effect increases with Z and becomes preponderant.

b. MUONIC ATOMS

We have already discussed some simple properties of muonic atoms (*cf.* complement A_V, §4 and A_{VII}, §2-a). In particular, we have pointed out that the Bohr radius associated with them is distinctly smaller than for ordinary atoms (this is caused by the fact that the mass of the μ^- muon is approximately equal to 207 times that of the electron). From the qualitative discussion of §1-b, we may therefore expect an important volume effect for muonic atoms. We shall evaluate it by choosing two limiting cases: a light muonic atom (hydrogen) and a heavy one (lead).

α. *The muonic hydrogen atom*

The Bohr radius is then:

$$a_0(\mu^-, p^+) \simeq \frac{a_0}{207} \tag{27}$$

that is, of the order of 250 fermi. It therefore remains, in this case, distinctly greater than ρ_0. If we replace a_0 by $a_0/207$ in (19), we find:

$$\Delta E_{1,0}(\mu^-, p^+) \simeq 1.9 \times 10^{-5} \times E_I(\mu^-, p^+) \simeq 5 \times 10^{-2} \text{ eV} \tag{28}$$

Although the volume effect is much larger than for the ordinary hydrogen atom, it still yields only a small correction to the energy levels.

β. *The muonic lead atom*

The Bohr radius of the muonic lead atom is [*cf.* complement A_V, relation (25)]:

$$a_0(\mu^-, \text{Pb}) \simeq 3 \text{ F} = 3 \times 10^{-15} \text{ m} \tag{29}$$

The μ^- muon is now very close to the lead nucleus; it is therefore practically unaffected by the repulsion of the atomic electrons which are located at distinctly greater distances. This could lead us to believe that (10), which was proven for hydrogen-like atoms and ions, is directly applicable to this case. Actually, this is not true, since the radius of the lead nucleus is equal to:

$$\rho_0(\text{Pb}) \simeq 8.5 \text{ F} = 8.5 \times 10^{-15} \text{ m} \tag{30}$$

which is not small compared to $a_0(\mu^-, \text{Pb})$. Equation (10) would therefore lead to large corrections (several MeV), of the same order of magnitude as the energy $E_I(\mu^-, \text{Pb})$. We therefore see that, in this case, the volume effect can no longer be treated as a perturbation (see discussion of §4 of complement A_V). To calculate the energy levels, it is necessary to know the potential $V(r)$ exactly and to solve the corresponding Schrödinger equation.

The muon is therefore more inside the nucleus than outside, that is, according to (1), in a region in which the potential is parabolic. In a first approximation, we could consider the potential to be parabolic everywhere (as is done in complement A_V) and then treat as a perturbation the difference which exists for $r \geqslant \rho_0$ between the real potential and the parabolic potential. However, the extension of the wave function corresponding to such a potential is not sufficiently smaller than ρ_0 for such an approximation to lead to precise results, and the only valid method consists of solving the Schrödinger equation correspon-ding to the real potential.

References and suggestions for further reading:

The isotopic volume effect: Kuhn (11.1), chap. VI, §C-3; Sobel'man (11.12), chap. 6, §24.

Muonic atoms (sometimes called mesic atoms): Cagnac and Pebay-Peyroula (11.2), chap. XIX, §7-C; De Benedetti (11.21); Wiegand (11.22); Weissenberg (16.19), §4-2.

Complement E_{XI}

THE VARIATIONAL METHOD

The perturbation theory studied in chapter XI is not the only general approximation method applicable to conservative systems. We shall give a concise description here of another of these methods, which also has numerous applications, especially in atomic and molecular physics, nuclear physics, and solid state physics. First of all, we shall indicate, in §1, the principle of the variational method. Then we shall use the simple example of the one-dimensional harmonic oscillator to bring out its principal features (§2), which we shall briefly discuss in §3. Complements F_{XI} and G_{XI} apply the variational method to simple models which enable us to understand the behavior of electrons in a solid and the chemical bond.

1. Principle of the method

Consider an arbitrary physical system whose Hamiltonian H is time-independent. To simplify the notation, we shall assume that the entire spectrum of H is discrete and non-degenerate:

$$H \mid \varphi_n \rangle = E_n \mid \varphi_n \rangle ; \; n = 0, 1, 2, \ldots \tag{1}$$

Although the Hamiltonian H is known, this is not necessarily the case for its eigenvalues E_n and the corresponding eigenstates $\mid \varphi_n \rangle$. The variational method is, of course, most useful in the cases in which we do not know how to diagonalize H exactly.

a. A PROPERTY OF THE GROUND STATE OF A SYSTEM

Choose an arbitrary ket $\mid \psi \rangle$ of the state space of the system. The mean value of the Hamiltonian H in the state $\mid \psi \rangle$ is such that:

$$\langle H \rangle = \frac{\langle \psi \mid H \mid \psi \rangle}{\langle \psi \mid \psi \rangle} \geqslant E_0 \tag{2}$$

(where E_0 is the smallest eigenvalue of H), equality occuring if and only if $\mid \psi \rangle$ is an eigenvector of H with the eigenvalue E_0.

To prove inequality (2), we expand the ket $| \psi \rangle$ on the basis of eigenstates of H:

$$| \psi \rangle = \sum_n c_n | \varphi_n \rangle \tag{3}$$

We then have:

$$\langle \psi | H | \psi \rangle = \sum_n |c_n|^2 E_n \geqslant E_0 \sum_n |c_n|^2 \tag{4}$$

with, of course:

$$\langle \psi | \psi \rangle = \sum_n |c_n|^2 \tag{5}$$

which proves (2). For inequality (4) to become an equality, it is necessary and sufficient that all the coefficients c_n be zero, with the exception of c_0; $| \psi \rangle$ is then an eigenvector of H with the eigenvalue E_0.

This property is the basis for a method of approximate determination of E_0. We choose (in theory, arbitrarily, but in fact, by using physical criteria) a family of kets $| \psi(\alpha) \rangle$ which depend on a certain number of parameters which we symbolize by α. We calculate the mean value $\langle H \rangle(\alpha)$ of the Hamiltonian H in these states, and we minimize $\langle H \rangle(\alpha)$ with respect to the parameters α. The minimal value so obtained constitutes an approximation of the ground state E_0 of the system. The kets $| \psi(\alpha) \rangle$ are called *trial kets*, and the method itself, the *variational method*.

COMMENT:

The preceding proof can easily be generalized to cases in which the spectrum of H is degenerate or includes a continuous part.

b. GENERALIZATION: THE RITZ THEOREM

We shall show that, more generally, *the mean value of the Hamiltonian H is stationary in the neighborhood of its discrete eigenvalues.*

Consider the mean value of H in the state $| \psi \rangle$:

$$\langle H \rangle = \frac{\langle \psi | H | \psi \rangle}{\langle \psi | \psi \rangle} \tag{6}$$

as a functional of the state vector $| \psi \rangle$, and calculate its increment $\delta \langle H \rangle$ when $| \psi \rangle$ becomes $| \psi \rangle + | \delta\psi \rangle$, where $| \delta\psi \rangle$ is assumed to be infinitely small. To do so, it is useful to write (6) in the form:

$$\langle H \rangle \langle \psi | \psi \rangle = \langle \psi | H | \psi \rangle \tag{7}$$

and to differentiate both sides of this relation:

$$\begin{aligned} \langle \psi | \psi \rangle \delta \langle H \rangle + \langle H \rangle [\langle \psi | \delta\psi \rangle + \langle \delta\psi | \psi \rangle] \\ = \langle \psi | H | \delta\psi \rangle + \langle \delta\psi | H | \psi \rangle \end{aligned} \tag{8}$$

that is, since $\langle H \rangle$ is a number:

$$\langle \psi | \psi \rangle \delta \langle H \rangle = \langle \psi | [H - \langle H \rangle] | \delta\psi \rangle + \langle \delta\psi | [H - \langle H \rangle] | \psi \rangle \quad (9)$$

The mean value $\langle H \rangle$ will be stationary if:

$$\delta \langle H \rangle = 0 \quad (10)$$

which, according to (9), means that:

$$\langle \psi | [H - \langle H \rangle] | \delta\psi \rangle + \langle \delta\psi | [H - \langle H \rangle] | \psi \rangle = 0 \quad (11)$$

We set:

$$| \varphi \rangle = [H - \langle H \rangle] | \psi \rangle \quad (12)$$

Relation (11) can then be written simply:

$$\langle \varphi | \delta\psi \rangle + \langle \delta\psi | \varphi \rangle = 0 \quad (13)$$

This last relation must be satisfied for any infinitesimal ket $| \delta\psi \rangle$. In particular, if we choose:

$$| \delta\psi \rangle = \delta\lambda | \varphi \rangle \quad (14)$$

(where $\delta\lambda$ is an infinitely small real number), (13) becomes:

$$2 \langle \varphi | \varphi \rangle \delta\lambda = 0 \quad (15)$$

The norm of the ket $| \varphi \rangle$ is therefore zero, and $| \varphi \rangle$ must consequently be zero. With definition (12) taken into account, this means that:

$$H | \psi \rangle = \langle H \rangle | \psi \rangle \quad (16)$$

Consequently, the mean value $\langle H \rangle$ is stationary if and only if the state vector $| \psi \rangle$ to which it corresponds is an eigenvector of H, and the stationary values of $\langle H \rangle$ are the eigenvalues of the Hamiltonian.

The variational method can therefore be generalized and applied to the approximate determination of the eigenvalues of the Hamiltonian H. If the function $\langle H \rangle(\alpha)$ obtained from the trial kets $| \psi(\alpha) \rangle$ has several extrema, they give the approximate values of some of its energies E_n (*cf.* exercise 10 of complement H_{XI}).

c. A SPECIAL CASE WHERE THE TRIAL FUNCTIONS FORM A SUBSPACE

Assume that we choose for the trial kets the set of kets belonging to a vector subspace $\mathscr{F}$ of $\mathscr{E}$. In this case, the variational method reduces to the *resolution of the eigenvalue equation of the Hamiltonian H inside* $\mathscr{F}$, and no longer in all of $\mathscr{E}$.

To see this, we simply apply the argument of § 1-b, limiting it to the kets $| \psi \rangle$ of the subspace $\mathscr{F}$. The maxima and minima of $\langle H \rangle$, characterized by $\delta \langle H \rangle = 0$, are obtained when $| \psi \rangle$ is an eigenvector of H in $\mathscr{F}$. The corresponding eigenvalues constitute the variational method approximation for the true eigenvalues of H in $\mathscr{E}$.

We stress the fact that the restriction of the eigenvalue equation of H to a subspace $\mathscr{F}$ of the state space $\mathscr{E}$ can considerably simplify its solution. However, if $\mathscr{F}$ is badly chosen, it can also yield results which are rather far from the true eigenvalues and eigenvectors of H in $\mathscr{E}$ (*cf.* § 3). The subspace $\mathscr{F}$ must therefore be chosen so as to simplify the problem enough to make it soluble, without too greatly altering the physical reality. In certain cases, it is possible to reduce the study of a complex system to that of a two-level system (*cf.* chap. IV), or at least, to that of a system of a limited number of levels. Another important example of this procedure is the method of the *linear combination of atomic orbitals*, widely used in molecular physics. This method consists essentially (*cf.* complement G_{XI}) of the determination of the wave functions of electrons in a molecule in the form of linear combinations of eigenfunctions associated with the various atoms which constitute the molecule, treated as if they were isolated. It therefore limits the search for the molecular states to a subspace chosen using physical criteria. Similarly, in complement F_{XI}, we shall choose as a trial wave function for an electron in a solid a linear combination of atomic orbitals relative to the various ions which constitute this solid.

COMMENT:

Note that first-order perturbation theory fits into this special case of the variational method: $\mathscr{F}$ is then an eigensubspace of the unperturbed Hamiltonian H_0.

2. Application to a simple example

To illustrate the discussion of §1 and to give an idea of the validity of the approximations obtained with the help of the variational method, we shall apply this method to the one-dimensional harmonic oscillator, whose eigenvalues and eigenstates we know (*cf.* chap. V). We shall consider the Hamiltonian:

$$H = -\frac{\hbar^2}{2m}\frac{\mathrm{d}^2}{\mathrm{d}x^2} + \frac{1}{2}m\omega^2 x^2 \tag{17}$$

and we shall solve its eigenvalue equation approximately by variational calculations.

a. EXPONENTIAL TRIAL FUNCTIONS

Since the Hamiltonian (17) is even, it can easily be shown that its ground state is necessarily represented by an even wave function. To determine the characteristics of this ground state, we shall therefore choose even trial functions. We take, for example, the one-parameter family:

$$\psi_\alpha(x) = \mathrm{e}^{-\alpha x^2} \quad ; \quad \alpha > 0 \tag{18}$$

The square of the norm of the ket $| \psi_\alpha \rangle$ is equal to:

$$\langle \psi_\alpha | \psi_\alpha \rangle = \int_{-\infty}^{+\infty} \mathrm{d}x \, \mathrm{e}^{-2\alpha x^2} \tag{19}$$

and we find:

$$\begin{aligned} \langle \psi_\alpha | H | \psi_\alpha \rangle &= \int_{-\infty}^{+\infty} \mathrm{d}x \, \mathrm{e}^{-\alpha x^2} \left[-\frac{\hbar^2}{2m} \frac{\mathrm{d}^2}{\mathrm{d}x^2} + \frac{1}{2} m\omega^2 x^2 \right] \mathrm{e}^{-\alpha x^2} \\ &= \left[\frac{\hbar^2}{2m} \alpha + \frac{1}{8} m\omega^2 \frac{1}{\alpha} \right] \int_{-\infty}^{+\infty} \mathrm{d}x \, \mathrm{e}^{-2\alpha x^2} \end{aligned} \tag{20}$$

so that:

$$\langle H \rangle(\alpha) = \frac{\hbar^2}{2m} \alpha + \frac{1}{8} m\omega^2 \frac{1}{\alpha} \tag{21}$$

The derivative of the function $\langle H \rangle(\alpha)$ goes to zero for:

$$\alpha = \alpha_0 = \frac{1}{2} \frac{m\omega}{\hbar} \tag{22}$$

and we then have:

$$\langle H \rangle(\alpha_0) = \frac{1}{2} \hbar\omega \tag{23}$$

The minimum value of $\langle H \rangle(\alpha)$ is therefore exactly equal to the energy of the ground state of the harmonic oscillator. This result is due to the simplicity of the problem that we are studying: the wave function of the ground state happens to be precisely one of the functions of the trial family (18), the one which corresponds to value (22) of the parameter α. The variational method, in this case, gives the exact solution of the problem (this illustrates the theorem proven in §1-a).

If we want to calculate (approximately, in theory) the first excited state E_1 of the Hamiltonian (17), we should choose trial functions which are orthogonal to the wave function of the ground state. This follows from the discussion of §1-a, which shows that $\langle H \rangle$ has a lower bound of E_1, and no longer of E_0, if the coefficient c_0 is zero. We therefore choose the trial family of odd functions:

$$\psi_\alpha(x) = x \, \mathrm{e}^{-\alpha x^2} \tag{24}$$

In this case:

$$\langle \psi_\alpha | \psi_\alpha \rangle = \int_{-\infty}^{+\infty} \mathrm{d}x \, x^2 \, \mathrm{e}^{-2\alpha x^2} \tag{25}$$

and:

$$\langle \psi_\alpha | H | \psi_\alpha \rangle = \left[\frac{\hbar^2}{2m} \times 3\alpha + \frac{1}{2} m\omega^2 \times \frac{3}{4\alpha} \right] \int_{-\infty}^{+\infty} \mathrm{d}x \, x^2 \, \mathrm{e}^{-2\alpha x^2} \tag{26}$$

which yields:

$$\langle H \rangle(\alpha) = \frac{3\hbar^2}{2m}\alpha + \frac{3}{8} m\omega^2 \frac{1}{\alpha} \tag{27}$$

This function, for the same value α_0 as above [formula (22)], presents a minimum equal to:

$$\langle H \rangle(\alpha_0) = \frac{3}{2}\hbar\omega \tag{28}$$

Here again, we find exactly the energy E_1 and the associated eigenstate because the trial family includes the correct wave function.

b. RATIONAL WAVE FUNCTIONS

The calculations of §2-a enabled us to familiarize ourselves with the variational method, but they do not really allow us to judge its effectiveness as a method of approximation, since the families chosen always included the exact wave function. Therefore, we shall now choose trial functions of a totally different type, for example★:

$$\psi_a(x) = \frac{1}{x^2 + a} \quad ; \quad a > 0 \tag{29}$$

A simple calculation then yields:

$$\langle \psi_a | \psi_a \rangle = \int_{-\infty}^{+\infty} \frac{\mathrm{d}x}{(x^2 + a)^2} = \frac{\pi}{2a\sqrt{a}} \tag{30}$$

and, finally:

$$\langle H \rangle(a) = \frac{\hbar^2}{4m}\frac{1}{a} + \frac{1}{2} m\omega^2 a \tag{31}$$

The minimum value of this function is obtained for:

$$a = a_0 = \frac{1}{\sqrt{2}}\frac{\hbar}{m\omega} \tag{32}$$

and is equal to:

$$\langle H \rangle(a_0) = \frac{1}{\sqrt{2}}\hbar\omega \tag{33}$$

This minimum value is therefore equal to $\sqrt{2}$ times the exact ground state energy $\hbar\omega/2$. To measure the error committed, we can calculate the ratio of $\langle H \rangle(a_0) - \hbar\omega/2$ to the energy quantum $\hbar\omega$:

$$\frac{\langle H \rangle(a_0) - \frac{1}{2}\hbar\omega}{\hbar\omega} = \frac{\sqrt{2} - 1}{2} \simeq 20\,\% \tag{34}$$

★ Our choice here is dictated by the fact that we want the necessary integrals to be analytically calculable. Of course, in most real cases, one resorts to numerical integration.

3. Discussion

The example of §2-b shows that it is easy to obtain the ground state energy of a system, without significant error, starting with arbitrarily chosen trial kets. This is one of the principal advantages of the variational method. Since the exact eigenvalue is a minimum of the mean value $\langle H \rangle$, it is not surprising that $\langle H \rangle$ does not vary very much near this minimum.

On the other hand, as the same reasoning shows, the "approximate" state can be rather different from the true eigenstate. Thus, in the example of §2-b, the wave function $1/(x^2 + a_0)$ [where a_0 is given by formula (32)] decreases too rapidly for small values of x and much too slowly when x becomes large. Table I gives quantitative support for this qualitative assertion. It gives, for various values of x^2, the values of the exact normalized eigenfunction:

$$\varphi_0(x) = (2\alpha_0/\pi)^{1/4}\, e^{-\alpha_0 x^2}$$

[where α_0 was defined in (22)] and of the approximate normalized eigenfunction:

$$\sqrt{\frac{2}{\pi}}(a_0)^{3/4}\psi_{a_0}(x) = \sqrt{\frac{2}{\pi}}\frac{(a_0)^{3/4}}{x^2 + a_0} = \sqrt{\frac{2}{\pi}}(2\sqrt{2}\alpha_0)^{1/4}\frac{1}{1 + 2\sqrt{2}\alpha_0 x^2} \tag{35}$$

$x\sqrt{\alpha_0}$	$\left(\frac{2}{\pi}\right)^{1/4} e^{-\alpha_0 x^2}$	$\sqrt{\frac{2}{\pi}}\frac{(2\sqrt{2})^{1/4}}{1 + 2\sqrt{2}\alpha_0 x^2}$
0	0.893	1.034
1/2	0.696	0.605
1	0.329	0.270
3/2	0.094	0.140
2	0.016	0.083
5/2	0.002	0.055
3	0.000 1	0.039

TABLE I

It is therefore necessary to be very careful when physical properties other than the energy of the system are calculated using the approximate state obtained from the variational method. The validity of the result obtained varies enormously depending on the physical quantity under consideration. In the particular problem which we are studying here, we find, for example, that the approximate mean value of the operator X^2★ is not very different from the exact value:

$$\frac{\langle \psi_{a_0} | X^2 | \psi_{a_0} \rangle}{\langle \psi_{a_0} | \psi_{a_0} \rangle} = \frac{1}{\sqrt{2}}\frac{\hbar}{m\omega} \tag{36}$$

★ The mean value of X is automatically zero, as is correct since we have chosen even trial functions.

which is to be compared with $\hbar/2m\omega$. On the other hand, the mean value of X^4 is infinite for the wave function (35), while it is, of course, finite for the real wave function. More generally, table I shows that the approximation will be very poor for all properties which depend strongly on the behavior of the wave function for $x \gtrsim 2/\sqrt{\alpha_0}$.

The drawback we have just mentioned is all the more serious as it is very difficult, if not impossible, to evaluate the error in a variational calculation if we do not know the exact solution of the problem (and, of course, if we use the variational method, it is because we do not know this exact solution).

The variational method is therefore a very flexible approximation method, which can be adapted to very diverse situations and which gives great scope to physical intuition in the choice of trial kets. It gives good values for the energy rather easily, but the approximate state vectors may present certain completely unpredictable erroneous features, and we cannot check these errors. This method is particularly valuable when physical arguments give us an idea of the qualitative or semi-quantitative form of the solutions.

References and suggestions for further reading:

The Hartree-Fock method, often used in physics, is an application of the variational method. See references of chapter XI.

The variational method is of fundamental importance in molecular physics. See references of complement G_{XI}.

For a simple presentation of the use of variational principles in physics, see Feynman II (7.2), chap. 19.

Complement F_{XI}

ENERGY BANDS OF ELECTRONS IN SOLIDS : A SIMPLE MODEL

A crystal is composed of atoms evenly distributed in space so as to form a three-dimensional periodic lattice. The theoretical study of the properties of a crystal, which brings into play an extremely large number of particles (nuclei and electrons), poses a problem which is so complicated that it is out of the question to treat it rigorously. We must therefore resort to approximations.

The first of these is of the same type as the Born-Oppenheimer approximation (which we encountered in § 1 of complement A_V). It consists of considering, first of all, the positions of the nuclei as fixed, which enables us to study the stationary states of the electrons subjected to the potential created by the nuclei. The motion of the nuclei is not treated until later, using the knowledge of the electronic energies*. In this complement, we shall concern ourselves only with the first step of this calculation, and we shall assume the nuclei to be motionless at the nodes of the crystalline lattice.

This problem still remains extremely complicated. It is necessary to calculate the energies of a system of electrons subjected to a periodic potential and interacting with each other. We then make a second approximation : we assume that each electron, at a position $\mathbf{r}_i$, is subjected to the influence of a potential $V(\mathbf{r}_i)$ which takes into account the attraction exerted by the nuclei and the average effect of the repulsion of all the other electrons**. The problem is thus reduced to one involving independent particles, moving in a potential which has the periodicity of the crystalline lattice.

The physical characteristics of a crystal therefore depend, in a first approximation, on the behavior of independent electrons subjected to a periodic potential. We could be led to think that each electron remains bound to a given nucleus, as happens in isolated atoms. We shall see that, in reality, the situation is completely

* Recall that the study of the motion of the nuclei leads to the introduction of the normal vibrational modes of the crystal: phonons (*cf.* complement J_V).

** This approximation is of the same type as the "central field" approximation for isolated atoms (*cf.* complement A_{XIV}, § 1).

different. Even if an electron is initially in the neighborhood of a particular nucleus, it can move into the zone of attraction of an adjacent nucleus by the tunnel effect, then into another, and so on. Actually, the stationary states of the electrons are not localized in the neighborhood of any nucleus, but are completely delocalized: the probability density associated with them is uniformly distributed over all the nuclei*. Thus, the properties of an electron placed in a periodic potential resemble those of an electron free to move throughout the crystal more than they do those of an electron bound to a particular atom. Such a phenomenon could not exist in classical mechanics: the direction of a particle traveling through a crystal would change constantly under the influence of the potential variations (for example, upon skirting an ion). In quantum mechanics, the interference of the waves scattered by the different nuclei permit the propagation of an electron inside the crystal.

In §1, we shall study very qualitatively how the energy levels of isolated atoms are modified when they are brought gradually closer together to form a linear chain. Then, in §2, still confining ourselves, for simplicity, to the case of a linear chain, we shall calculate the energies and wave functions of stationary states a little more precisely. We shall perform the calculation in the "strong bonding approximation": when the electron is in one site, it can move to one of two neighboring sites via the tunnel effect. The strong bonding approximation is equivalent to assuming that the probability of its tunneling is small. We shall, in this way, establish a certain number of results (the delocalization of stationary states, the appearance of allowed and forbidden energy bands, the form of Bloch functions) which remain valid in more realistic models (three-dimensional crystals, bonds of arbitrary strength).

The "perturbation" approach which we shall adopt here constructs the stationary states of the electrons from atomic wave functions localized about the various ions. It has the advantage of showing how atomic levels change gradually to energy bands in a solid. Note, however, that the existence of energy bands can be directly established from the periodic nature of the structure in which the electron is placed (see, for example, complement O_{III}, in which we study quantization of the energy levels in a one-dimensional periodic potential).

Finally, we stress the fact that we are concerned here only with the properties of the individual stationary states of the electrons. To construct the stationary state of a system of N electrons from these individual states, it is necessary to apply the symmetrization postulate (*cf.* chap. XIV), since we are dealing with a system of identical particles. We shall treat this problem again in complement C_{XIV}, when we shall describe the spectacular consequences of Pauli's exclusion principle on the physical behavior of the electrons in a solid.

* This phenomenon is analogous to the one we encountered in the study of the ammonia molecule (*cf.* complement G_{IV}). There, since the nitrogen atom can move from one side of the plane of the hydrogen atoms to the other, by the tunnel effect, the stationary states give an equal probability of finding it in each of the two corresponding positions.

1. A first approach to the problem: qualitative discussion

We return to the example of the ionized H_2^+ molecule, studied in §§C-2-c and C-3-d of chapter IV. Consider, therefore, two protons P_1 and P_2, whose positions are fixed, and an electron which is subject to their electrostatic attraction. This electron sees a potential $V(\mathbf{r})$, which has the form indicated in figure 1. In terms of the distance R between P_1 and P_2 (considered as a parameter) what are the possible energies and the corresponding stationary states?

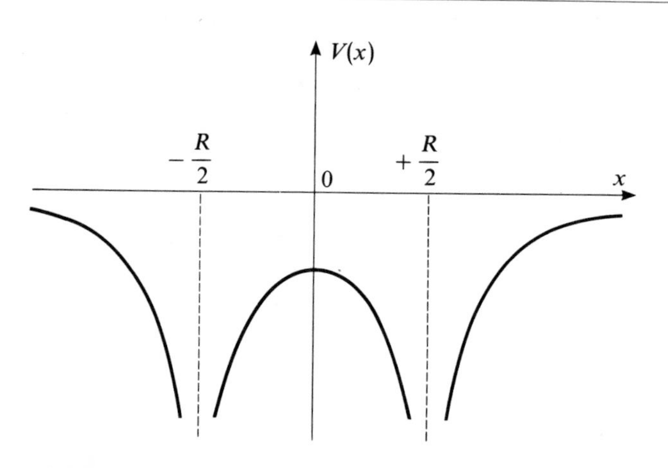

FIGURE 1

The potential seen by the electron in the ionized H_2^+ molecule as it moves along the Ox axis defined by the two protons. We obtain two wells separated by a barrier. If, at any instant, the electron is localized in one of the two wells, it can move into the other well via the tunnel effect.

We shall begin by considering the limiting case in which $R \gg a_0$ (where a_0 is the Bohr radius of the hydrogen atom). The ground state is then two-fold degenerate : the electron can form a hydrogen atom either with P_1 or with P_2; it is practically unaffected by the attraction of the other proton, which is very far away. In other words, the coupling between the states $| \varphi_1 \rangle$ and $| \varphi_2 \rangle$ considered in chapter IV (localized states in the neighborhood of P_1 or P_2; *cf.* fig. 13 of chapter IV) is then negligible, so that $| \varphi_1 \rangle$ and $| \varphi_2 \rangle$ are practically stationary states.

If we now choose a value of R comparable to a_0, it is no longer possible to neglect the attraction of one or the other of the protons. If, at $t = 0$, the electron is localized in the neighborhood of one of them, and even if its energy is lower than the height of the potential barrier situated between P_1 and P_2 (*cf.* fig. 1), it can move to the other proton by the tunnel effect. In chapter IV we studied the effect of coupling of the states $| \varphi_1 \rangle$ and $| \varphi_2 \rangle$, and we showed that it produces an oscillation of the system between these two states (the dynamical aspect). We have also seen (the static aspect) that this coupling removes the degeneracy of the ground state and that the corresponding stationary states are "delocalized" (for these states, the probability of finding the electron in the neighborhood of P_1 or P_2 is the same). Figure 2 shows the form of the variation with respect to R of the possible energies of the system★.

Two effects appear when we decrease the distance R between P_1 and P_2. On the one hand, an $R = \infty$ energy value gives rise to two distinct energies when R decreases (when the distance R is fixed at a given value R_0, the stronger the coupling between the states $| \varphi_1 \rangle$ and $| \varphi_2 \rangle$, the greater the difference between these two energies). On the other hand, the stationary states are delocalized.

It is easy to imagine what will happen if the electron is subject to the influence, not of two, but of three identical attractive particles (protons or positive ions), arranged, for example, in a straight line at intervals of R. When R is very large, the energy levels are triply degenerate, and the stationary states of the electron can be chosen to be localized in the neighborhood of any one of the fixed particles.

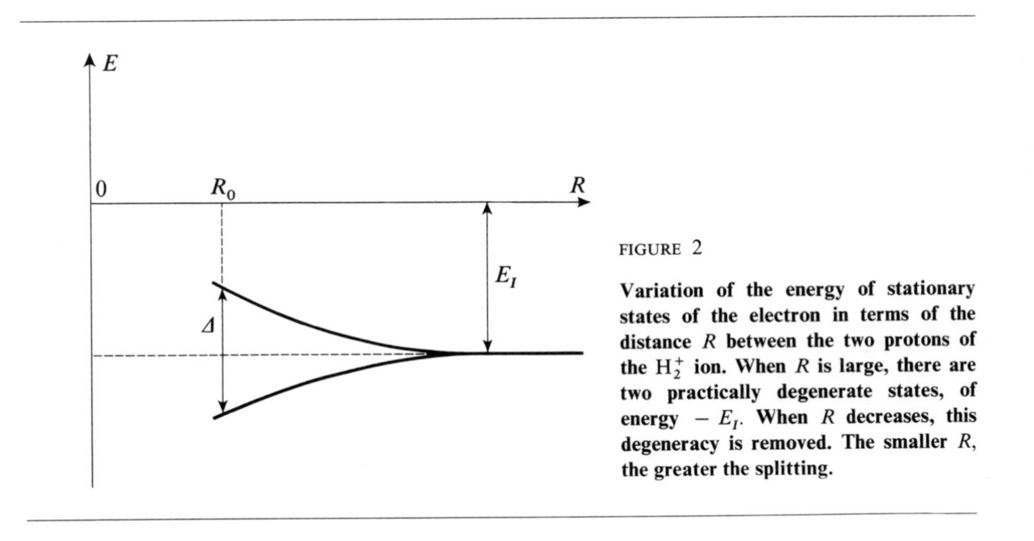

FIGURE 2

Variation of the energy of stationary states of the electron in terms of the distance R between the two protons of the H_2^+ ion. When R is large, there are two practically degenerate states, of energy $-E_I$. When R decreases, this degeneracy is removed. The smaller R, the greater the splitting.

If R is decreased, each energy gives rise to three generally distinct energies and, in a stationary state, the probabilities of finding the electron in the three wells are comparable. Moreover, if, at the initial instant, the electron is localized in the right-hand well, for example, it moves into the other wells during its subsequent evolution★★.

The same ideas remain valid for a chain composed of an arbitrary number $\mathcal{N}$ of ions which attract an electron. The potential seen by the electron is then composed of $\mathcal{N}$ regularly spaced identical wells (in the limit in which $\mathcal{N} \longrightarrow \infty$, it is a periodic potential). When the distance R between the ions is large, the energy levels are $\mathcal{N}$-fold degenerate. This degeneracy disappears if the ions are moved

★ A detailed study of the H_2^+ ion is presented in complement G_{XI}.

★★ See exercise 8 of complement J_{IV}.

closer together: each level gives rise to distinct levels, which are distributed, as shown in figure 3, in an energy interval of width Δ. What now happens if the value of $\mathcal{N}$ is very large? In each of the intervals Δ, the possible energies are so close that they practically form a continuum: "allowed energy bands" are thus obtained, separated by "forbidden bands". Each allowed band contains $\mathcal{N}$ levels (actually $2\mathcal{N}$ if the electron spin is taken into account). The stronger the coupling causing the electron to pass from one potential well to the next one, the greater the band width. (Consequently, we expect the lowest energy bands to be the narrowest since

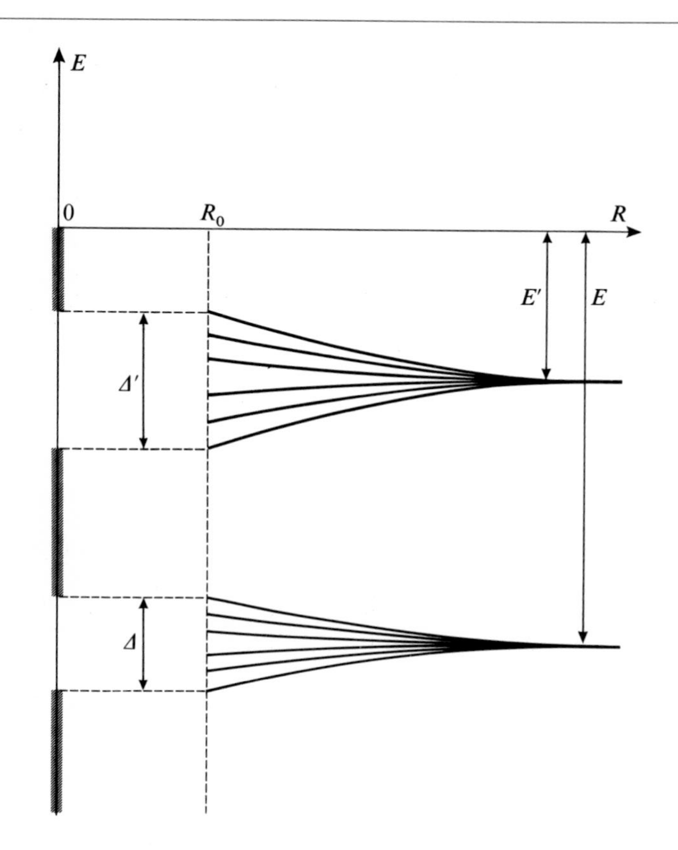

FIGURE 3

Energy levels of an electron subject to the action of $\mathcal{N}$ regularly spaced identical ions. When R is very large, the wave functions are localized about the various ions, and the energy levels are the atomic levels, $\mathcal{N}$-fold degenerate (the electron can form an atom with any one of the $\mathcal{N}$ ions). In the figure, two of these levels are shown, of energies $-E$ and $-E'$. When R decreases, the electron can pass from one ion to another by the tunnel effect, and the degeneracy of the levels is removed. The smaller R, the greater the splitting. For the value R_0 of R found in a crystal, each of the two original atomic levels is therefore broken down into $\mathcal{N}$ very close levels. If $\mathcal{N}$ is very large, these levels are so close that they yield energy bands, of widths Δ and Δ', separated by a forbidden band.

the tunnel effect which is responsible for this passage is less probable when the energy is smaller). The stationary states of the electron are all delocalized. The analogue here of figure 3 of complement M_{III} is figure 4, which represents the energy levels and gives an idea of the spatial extension of the associated wave functions.

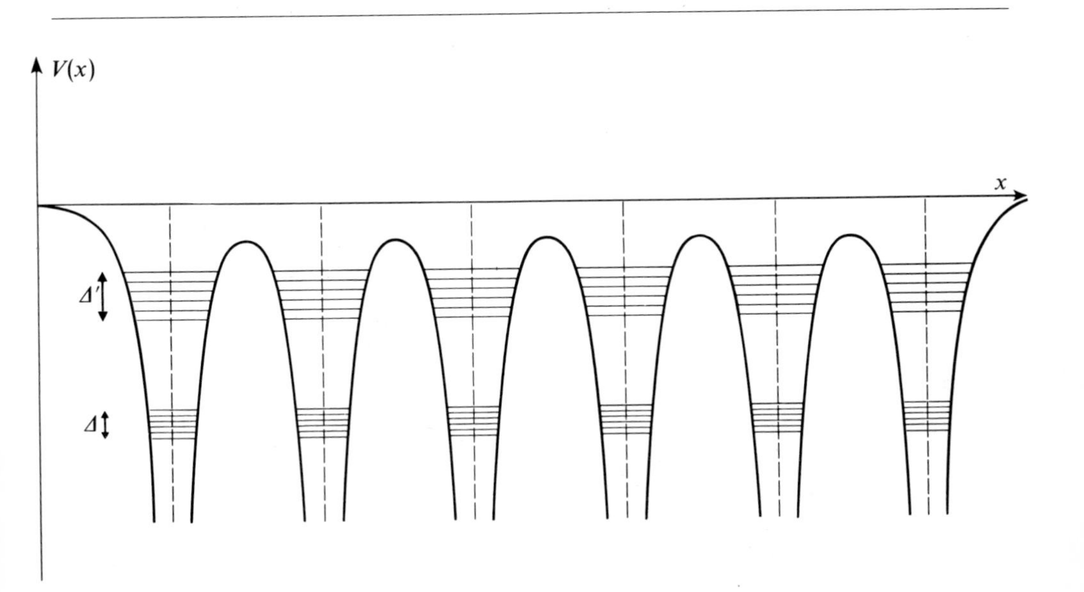

FIGURE 4

Energy levels for a potential composed of several regularly spaced wells. Two bands are shown in this figure, one of width Δ and the other of width Δ'. The deeper the band, the more narrow it is, since crossing the barrier by the tunnel effect is then more difficult.

Finally, note that if, at $t = 0$, the electron is localized at one end of the chain, it propagates along the chain during its subsequent evolution.

2. A more precise study using a simple model

a. CALCULATION OF THE ENERGIES AND STATIONARY STATES

To complete the qualitative considerations of the preceding section, we shall discuss the problem more precisely, using a simple model. We shall perform calculations analogous to those of §C of chapter IV, but adapted to the case in which the system under consideration contains an infinite number of ions (instead of two), regularly spaced in a linear chain.

α. *Description of the model; simplifying hypotheses*

Consider, therefore, an infinite linear chain of regularly spaced positive ions. As in chapter IV, we shall assume that the electron, when it is bound to a given ion, has only one possible state: we shall denote by $| v_n \rangle$ the state of the electron when it forms an atom with the *n*th ion of the chain. For the sake of simplicity, we shall neglect the mutual overlap of the wave functions $v_n(x)$ associated with neighboring atoms, and we shall assume the $\{ | v_n \rangle \}$ basis to be orthonormal:

$$\langle v_n | v_p \rangle = \delta_{np} \tag{1}$$

Moreover, we shall confine ourselves to the subspace of the state space spanned by the kets $| v_n \rangle$. It is obvious that by restricting the state space accessible to the electron in this way, we are making an approximation. This can be justified by using the variational method (*cf.* complement E_{XI}) : by diagonalizing the Hamiltonian H, not in the total space, but in the one spanned by the $| v_n \rangle$, it can be shown that we obtain a good approximation for the true energies of the electron.

We shall now write the matrix representing the Hamiltonian H in the $\{ | v_n \rangle \}$ basis. Since the ions all play equivalent roles, the matrix elements $\langle v_n | H | v_n \rangle$ are necessarily all equal to the same energy E_0. In addition to these diagonal elements, H also has non-diagonal elements $\langle v_n | H | v_p \rangle$ (coupling between the various states $| v_n \rangle$, which expresses the possibility for an electron to move from one ion to another). This coupling is obviously very weak for distant ions; this is why we shall take into account only the matrix elements $\langle v_n | H | v_{n \pm 1} \rangle$, which we shall choose equal to a real constant $-A$. Under these conditions, the (infinite) matrix which represents H can be written:

$$(H) = \begin{pmatrix} \ddots & & & & \\ & E_0 & -A & 0 & 0 \\ & -A & E_0 & -A & 0 \\ & 0 & -A & E_0 & -A \\ & 0 & 0 & -A & E_0 \\ & & & & & \ddots \end{pmatrix} \tag{2}$$

To find the possible energies and the corresponding stationary states, we must diagonalize this matrix.

β. *Possible energies; the concept of an energy band*

Let $| \varphi \rangle$ be an eigenvector of H; we shall write it in the form:

$$| \varphi \rangle = \sum_{q=-\infty}^{+\infty} c_q | v_q \rangle \tag{3}$$

Using (2), the eigenvalue equation:

$$H | \varphi \rangle = E | \varphi \rangle \tag{4}$$

projected onto $| v_q \rangle$, yields:

$$E_0 \, c_q - A \, c_{q+1} - A \, c_{q-1} = E \, c_q \tag{5}$$

When q takes on all positive or negative integral values, we thus obtain an infinite system of coupled linear equations which, in certain ways, recall the coupled equations (5) of complement J_V. As in that complement, we shall look for simple solutions of the form:

$$c_q = e^{ikql} \tag{6}$$

where l is the distance between two adjacent ions, and k is a constant whose dimensions are those of an inverse length. We require k to belong to the "first Brillouin zone", that is, to satisfy:

$$-\frac{\pi}{l} \leqslant k < +\frac{\pi}{l} \tag{7}$$

This is always possible, because two values of k differing by $2\pi/l$ give all the coefficients c_q the same value. Substituting (6) into (5), we obtain:

$$E_0 \, e^{ikql} - A[e^{ik(q+1)l} + e^{ik(q-1)l}] = E \, e^{ikql} \tag{8}$$

that is, dividing by e^{ikql}:

$$E = E(k) = E_0 - 2A \cos kl \tag{9}$$

If this condition is satisfied, the ket $| \varphi \rangle$ given by (3) and (6) is an eigenket of H; its energy depends on the parameter k, as is indicated by (9).

Figure 5 represents the variation of E with respect to k. It shows that the possible energies are situated in the interval $[E_0 - 2A, E_0 + 2A]$. We therefore obtain an allowed energy band, whose width $4A$ is proportional to the strength of the coupling.

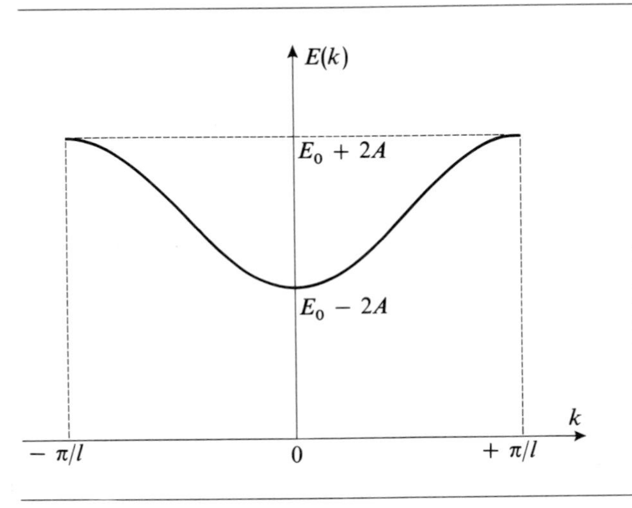

FIGURE 5

Possible energies of the electron in terms of the parameter k (k varies within the first Brillouin zone). An energy band therefore appears, with a width $4A$ which is proportional to the coupling between neighbouring atoms.

γ. *Stationary states; Bloch functions*

Let us calculate the wave function $\varphi_k(x) = \langle x \mid \varphi_k \rangle$ associated with the stationary state $\mid \varphi_k \rangle$ of energy $E(k)$. Relations (3) and (6) give:

$$\mid \varphi_k \rangle = \sum_{q=-\infty}^{+\infty} e^{ikql} \mid v_q \rangle \tag{10-a}$$

that is:

$$\varphi_k(x) = \sum_{q=-\infty}^{+\infty} e^{ikql} \, v_q(x) \tag{10-b}$$

where:

$$v_q(x) = \langle x \mid v_q \rangle \tag{11}$$

is the wave function associated with the state $\mid v_q \rangle$. Since the state $\mid v_q \rangle$ can be obtained from the state $\mid v_0 \rangle$ by a translation of ql, we have:

$$v_q(x) = v_0(x - ql) \tag{12}$$

so that (10-b) can be written:

$$\varphi_k(x) = \sum_{q=-\infty}^{+\infty} e^{ikql} \, v_0(x - ql) \tag{13}$$

We now calculate $\varphi_k(x + l)$:

$$\begin{aligned} \varphi_k(x + l) &= \sum_{q=-\infty}^{+\infty} e^{ikql} \, v_0[x - (q - 1)l] \\ &= e^{ikl} \sum_{q=-\infty}^{+\infty} e^{ik(q-1)l} \, v_0[x - (q - 1)l] \\ &= e^{ikl} \, \varphi_k(x) \end{aligned} \tag{14}$$

To express this remarkable property simply, we set:

$$\varphi_k(x) = e^{ikx} \, u_k(x) \tag{15}$$

The function $u_k(x)$ so defined then satisfies:

$$u_k(x + l) = u_k(x) \tag{16}$$

Therefore, the wave function $\varphi_k(x)$ is the product of e^{ikx} and a periodic function which has the period l of the lattice. A function of type (15) is called a *Bloch function*. Note that, if n is any integer:

$$|\varphi_k(x + nl)|^2 = |\varphi_k(x)|^2 \tag{17}$$

a result which demonstrates the delocalization of the electron : the probability density of finding the electron at any point on the x-axis is a periodic function of x.

COMMENT:

Expressions (15) and (16) have been proven here for a simple model. Actually, this result is more general and can be proven directly from the symmetries of the

Hamiltonian H (Bloch's theorem). To show this, let us call $S(a)$ the unitary operator associated with a translation of a along Ox (*cf.* complement E_{II}, §3). Since the system is invariant under any translation which leaves the ion chain unchanged, we must have:

$$[H, S(l)] = 0 \tag{18-a}$$

We can therefore construct a basis of eigenvectors common to the operator $S(l)$ and H. Now, equation (14) is simply the one which defines the eigenfunctions of $S(-l)$ [since this operator is unitary, its eigenvalues can always be written in the form e^{ikl}, where k satisfies condition (7); *cf.* complement C_{II}, §1-d]. It is then simple to get, as before, (15) and (16) from (14).

Note that, for any a, we have, in general:

$$[H, S(a)] \neq 0 \tag{18-b}$$

unlike the situation of a free particle (or one subject to the influence of a constant potential). For a free particle, since H commutes with all operators $S(a)$ (that is, with the momentum P_x; *cf.* complement E_{II}, § 3), the stationary wave functions are of the form:

$$w_k(x) \propto e^{ikx} \tag{19}$$

The fact that, in our case, (18-b) is satisfied only for certain values of a explains why form (15) is less restrictive than (19).

δ. *Periodic boundary conditions*

To each value of k in the interval $[-\pi/l, +\pi/l]$ therefore corresponds an eigenstate $|\varphi\rangle$ of H, with the coefficients c_q appearing in expansion (3) of $|\varphi\rangle$ given by equation (6). We thus obtain an infinite continuum of stationary states. This is due to the fact that we have considered a linear chain containing an infinite number of ions. What happens when we consider a finite linear chain, of length L, composed of a large number $\mathcal{N}$ of ions?

The qualitative considerations of § 1 show that there must then be $\mathcal{N}$ levels in the band ($2\mathcal{N}$ if spin is taken into account). The exact determination of the corresponding $\mathcal{N}$ stationary states is a difficult problem, since it is necessary to take into account the boundary conditions at the ends of the chain. It is clear, however, that the behavior of electrons sufficiently far from the ends are little affected by the "edge effects"*. This is why one generally prefers, in solid state physics, to substitute for the real boundary conditions, new boundary conditions, which, despite their artificial character, have the advantage of leading to much simpler calculations, while conserving the most important properties necessary for the comprehension of effects other than the edge effects.

These new boundary conditions, called periodic boundary conditions, or "Born-Von Karman conditions" (B.V.K. conditions), require the wave function to take on the same value at both ends of the chain. We can also imagine that we are placing an infinite number of identical chains, all of length L, end to end. We then

* For a three-dimensional crystal, this amounts to establishing a distinction between "bulk effects" and "surface effects".

require the wave function of the electron to be periodic, with a period L. Equations (5) remain valid, as does their solution (6), but the periodicity of the wave function now implies:

$$e^{ikL} = 1 \tag{20}$$

Consequently, the only possible values of k are of the form:

$$k_n = n\frac{2\pi}{L} \tag{21}$$

where n is a positive or negative integer or zero. Let us now verify that the B.V.K. conditions give the correct result for the number of stationary states contained in the band. To do so, we must calculate the number of allowed values k_n included in the first Brillouin zone. We obtain this number by dividing the width $2\pi/l$ of this zone by the interval $2\pi/L$ between two adjacent values of k, which indeed gives us:

$$\frac{2\pi}{l}\Big/\frac{2\pi}{L} = \frac{L}{l} = \mathcal{N} - 1 \simeq \mathcal{N} \tag{22}$$

We should also show that the $\mathcal{N}$ stationary states obtained with the B.V.K. conditions are distributed in the allowed band with the same density* $\rho(E)$ as the true stationary states (associated with the real boundary conditions). As the density of states $\rho(E)$ plays a very important role in the comprehension of the physical properties of a solid (we shall discuss this point in complement C_{XIV}), it is important for the new boundary conditions to leave it unchanged. That the B.V.K. conditions give the correct density of states will be proven in complement C_{XIV} (§1-c) for the simple example of a free electron gas enclosed in a "rigid box". In this case, the true stationary states can be calculated and compared with those obtained by using the periodic boundary conditions on the walls of the box (see also §3 of complement O_{III}).

b. DISCUSSION

Starting with a discrete non-degenerate level for an isolated atom (for example, the ground level) we have obtained a series of possible energies, grouped in an allowed band of width $4A$ for the chain of ions being considered. If we had started with another level of the atom (for example, the first excited level), we would have obtained another energy band, and so on. Each atomic level yields one energy band, as figure 6 shows, and there appears a series of allowed bands, separated by forbidden bands.

Relation (6) shows that, for a stationary state, the probability amplitude of finding the electron in the state $| v_q \rangle$ is an oscillating function of q, whose modulus does not depend on q. This recalls the properties of phonons, the normal vibrational modes of an infinite number of coupled oscillators for which all the oscillators participate in the collective vibration with the same amplitude, but with a certain phase shift (*cf.* complement J_V).

* $\rho(E)\,dE$ is the number of distinct stationary states with energies included between E and $E + dE$.

How can we obtain states in which the electron is not completely delocalized? For a free electron, we saw in chapter I that we must superpose plane waves so as to form a free "wave packet":

$$\hat{\psi}(x, t) = \frac{1}{\sqrt{2\pi}} \int \mathrm{d}k \; \hat{g}(k) \; \mathrm{e}^{i[kx - E(k)t/\hbar]} \tag{23}$$

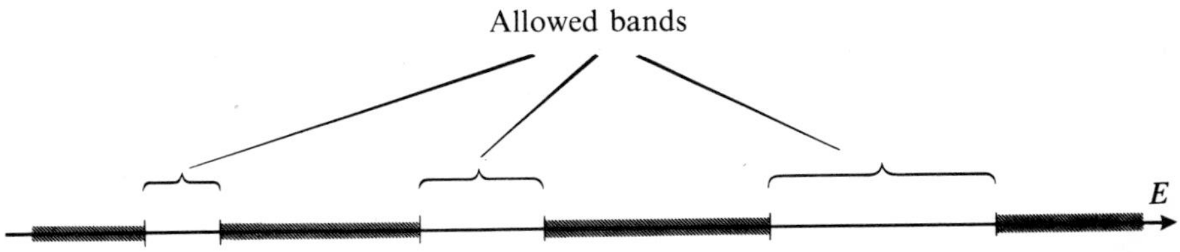

FIGURE 6

Allowed bands and forbidden bands on the energy axis.

The maximum of this wave packet propagates at the group velocity (*cf.* chap I, § C):

$$\hat{V}_G = \frac{1}{\hbar}\left[\frac{\mathrm{d}E}{\mathrm{d}k}\right]_{k = k_0} = \frac{\hbar k_0}{m} \tag{24}$$

[where k_0 is the value of k for which the function $\hat{g}(k)$ presents a peak]. Here, we must superpose wave functions of type (15), and the corresponding ket can be written:

$$| \psi(t) \rangle = \frac{1}{\sqrt{2\pi}} \int \mathrm{d}k \; g(k) \; \mathrm{e}^{-iE(k)t/\hbar} \, | \varphi_k \rangle \tag{25}$$

where $g(k)$ is a function of k which has the form of a peak about $k = k_0$. We shall calculate the probability amplitude of finding the electron in the state $| v_q \rangle$. Using (10-a) and (1), we can write:

$$\langle v_q | \psi(t) \rangle = \frac{1}{\sqrt{2\pi}} \int \mathrm{d}k \; g(k) \; \mathrm{e}^{i[kql - E(k)t/\hbar]} \tag{26}$$

Replacing ql by x in this relation, we obtain a function of x:

$$\chi(x, t) = \frac{1}{\sqrt{2\pi}} \int \mathrm{d}k \; g(k) \; \mathrm{e}^{i[kx - E(k)t/\hbar]} \tag{27}$$

Only the values at the points $x = 0, \pm ql, \pm 2ql$, etc... of this function are really significant and yield the desired probability amplitudes.

Relation (27) is entirely analogous to (23). By applying (24), it can be shown that $\chi(x, t)$ takes on significant values only in a limited domain of the x-axis whose center moves at the velocity:

$$V_G = \frac{1}{\hbar}\left[\frac{\mathrm{d}E(k)}{\mathrm{d}k}\right]_{k=k_0} \tag{28}$$

It follows that the probability amplitude $\langle v_q | \psi(t) \rangle$ is large only for certain values of q: therefore, the electron is no longer delocalized, but moves in the crystal at the velocity V_G given by (28).

Equation (9) enables us to calculate this velocity explicitly:

$$V_G = \frac{2Al}{\hbar} \sin k_0 l \tag{29}$$

This function is shown in figure (7). It is zero when $k_0 = 0$, that is, when the energy is minimal; this is also a property of the free electron. However, when k_0 takes on non-zero values, important departures from the behavior of a free electron occur. For example, as soon as $k_0 > \pi/2l$, the group velocity is no longer an increasing function of the energy. It even goes to zero when $k_0 = \pm \pi/l$ (at the borders of the first Brillouin zone). This indicates that an electron cannot move in the crystal if its energy is too close to the maximum value $E_0 + 2A$ appearing in figure 5. The optical analogy of this situation is Bragg reflection. X rays whose wavelength is equal to the unit edge of the crystalline lattice cannot propagate in it: interference of the waves scattered by each of the ions lead to total reflection.

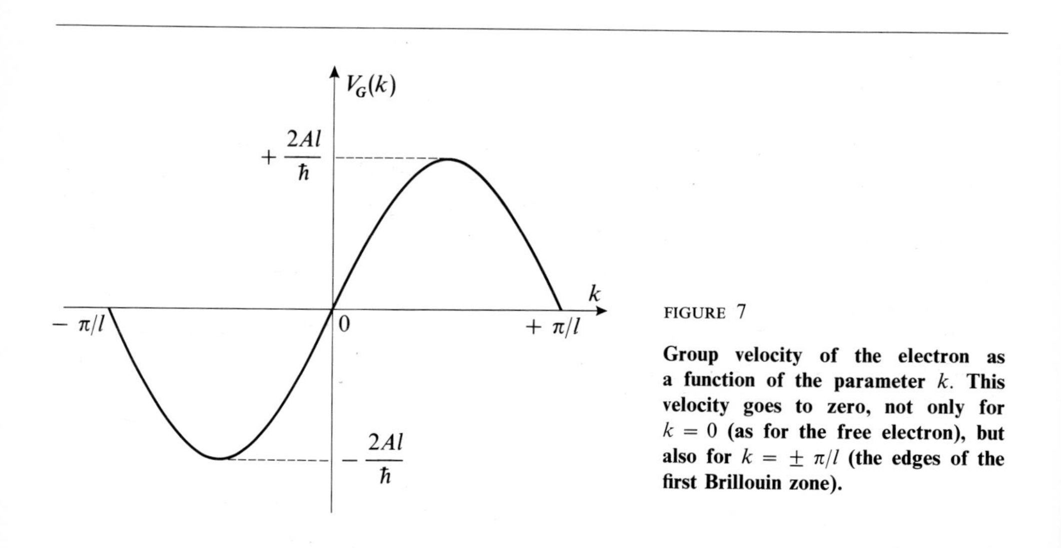

FIGURE 7

Group velocity of the electron as a function of the parameter k. This velocity goes to zero, not only for $k = 0$ (as for the free electron), but also for $k = \pm \pi/l$ (the edges of the first Brillouin zone).

References and suggestions for further reading:

Feynman III (1.2), chap. 13; Mott and Jones (13.7), chap. II, §4; references of section 13 of the bibliography.

Complement G_{XI}

A SIMPLE EXAMPLE OF THE CHEMICAL BOND: THE H_2^+ ION

1. Introduction

In this complement, we intend to show how quantum mechanics enables us to understand the existence and properties of the *chemical bond*, which is responsible for the formation of more or less complex molecules from isolated atoms. Our aim is to explain the basic nature of these phenomena and not, of course, to enter into details which could only be covered in a specialized book on molecular physics. This is why we shall study the simplest molecule possible, the H_2^+ ion, which is composed of two protons and a single electron. We have already discussed certain aspects of this problem, in chapter IV (§C-2-c) and in exercise 5 of complement K_I; we shall consider it here in a more realistic and systematic fashion.

a. GENERAL METHOD

When the two protons are very far from each other, the electron forms a hydrogen atom with one of them, and the other one remains isolated, in the form of an H^+ ion. If the two protons are brought closer together, the electron will be able to "jump" from one to the other. This radically modifies the situation (*cf.* chap. IV, §C-2). We shall therefore study the variation of the energies of the stationary states of the system with respect to the distance between the two protons. We shall see that the energy of the ground state reaches a minimum for a certain value of this distance, which explains the stability of the H_2^+ molecule.

In order to treat the problem exactly, it would be necessary to write the Hamiltonian of the three-particle system and solve its eigenvalue equation. However, it is possible to simplify this problem considerably by using the *Born-Oppenheimer approximation* (*cf.* complement A_V, §1-a). Since the motion of the electron in the molecule is considerably more rapid than that of the protons, the latter can be neglected in a first approximation. The problem is then reduced to the resolution of the eigenvalue equation of the Hamiltonian of the electron subject to the attraction of two protons which are assumed to be fixed. In other words, the distance R between the two protons is treated, not like a quantum mechanical variable, but like a *parameter*, on which the electronic Hamiltonian and total energy of the system depend.

In the case of the H_2^+ ion, it so happens that the equation simplified in this way is exactly soluble for all values of R. However, this is not true for other, more complex, molecules. The *variational method*, described in complement E_{XI}, must then be used. Although we are confining ourselves here to the study of the H_2^+ ion, we shall use the variational method, since it can be generalized to the case of other molecules.

b. NOTATION

We shall call R the distance between the two protons, situated at P_1 and P_2, and r_1 and r_2 the distances of the electron to each of the two protons (fig. 1). We shall relate these distances to a natural atomic unit, the Bohr radius a_0 (*cf.* chap. VII, § C-2), by setting:

$$\rho = R/a_0$$
$$\rho_1 = r_1/a_0 \qquad \rho_2 = r_2/a_0 \tag{1}$$

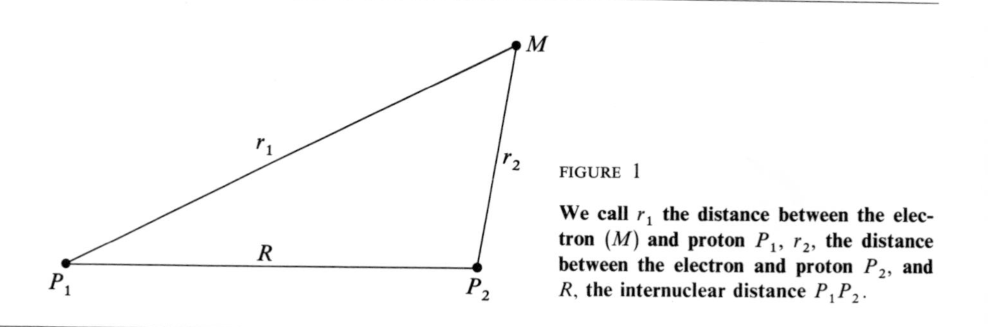

FIGURE 1

We call r_1 the distance between the electron (M) and proton P_1, r_2, the distance between the electron and proton P_2, and R, the internuclear distance P_1P_2.

The normalized wave function associated with the ground state $1s$ of the hydrogen atom formed around proton P_1 can then be written:

$$\varphi_1 = \frac{1}{\sqrt{\pi a_0^3}} e^{-\rho_1} \tag{2}$$

Similarly, we express the energies in terms of the natural unit $E_I = e^2/2a_0$; E_I is the ionization energy of the hydrogen atom.

It will sometimes be convenient in what follows to use a system of elliptic coordinates, in which a point M of space (here, the electron) is defined by:

$$\mu = \frac{r_1 + r_2}{R} = \frac{\rho_1 + \rho_2}{\rho}$$

$$\nu = \frac{r_1 - r_2}{R} = \frac{\rho_1 - \rho_2}{\rho} \qquad (3)$$

and the angle φ which fixes the orientation of the MP_1P_2 plane about the P_1P_2 axis (this angle also enters into the system of polar coordinates whose Oz axis coincides with P_1P_2). If we fix μ and ν, and if φ varies between 0 and 2π, the point M describes a circle about the P_1P_2 axis. If μ (or ν) and φ are fixed, M describes an ellipse (or a hyperbola) of foci P_1 and P_2 when ν (or μ) varies. It can easily be shown that the volume element in this system is:

$$d^3r = \frac{R^3}{8}(\mu^2 - \nu^2)\, d\mu\, d\nu\, d\varphi \qquad (4)$$

To do so, we simply calculate the Jacobian J of the transformation:

$$\{ x, y, z \} \Longrightarrow \{ \mu, \nu, \varphi \} \qquad (5)$$

We see immediately that, if P_1P_2 is chosen as the Oz axis, with the origin O in the middle of P_1P_2:

$$r_1^2 = x^2 + y^2 + \left(z - \frac{R}{2}\right)^2$$

$$r_2^2 = x^2 + y^2 + \left(z + \frac{R}{2}\right)^2$$

$$\tan \varphi = \frac{y}{x} \qquad (6)$$

We can then find:

$$\frac{\partial \mu}{\partial x} = \frac{1}{R}\left(\frac{\partial r_1}{\partial x} + \frac{\partial r_2}{\partial x}\right) = \frac{1}{R}\left(\frac{x}{r_1} + \frac{x}{r_2}\right) = \frac{\mu x}{r_1 r_2}$$

$$\frac{\partial \nu}{\partial x} = \frac{1}{R}\left(\frac{\partial r_1}{\partial x} - \frac{\partial r_2}{\partial x}\right) = -\frac{\nu x}{r_1 r_2}$$

$$\frac{\partial \mu}{\partial y} = \frac{\mu y}{r_1 r_2}$$

$$\frac{\partial \nu}{\partial y} = -\frac{\nu y}{r_1 r_2}$$

$$\frac{\partial \mu}{\partial z} = \frac{1}{R}\left[\frac{z - R/2}{r_1} + \frac{z + R/2}{r_2}\right] = \frac{\mu z + \nu R/2}{r_1 r_2}$$

$$\frac{\partial \nu}{\partial z} = \frac{1}{R}\left[\frac{z - R/2}{r_1} - \frac{z + R/2}{r_2}\right] = -\frac{\nu z + \mu R/2}{r_1 r_2}$$

$$\frac{\partial \varphi}{\partial x} = -\frac{y}{x^2 + y^2} \qquad \frac{\partial \varphi}{\partial y} = \frac{x}{x^2 + y^2} \qquad \frac{\partial \varphi}{\partial z} = 0 \qquad (7)$$

The Jacobian J can therefore be written:

$$J = \frac{1}{(r_1 r_2)^2} \begin{vmatrix} \mu x & \mu y & \mu z + \nu R/2 \\ -\nu x & -\nu y & -\nu z - \nu R/2 \\ -y/(x^2+y^2) & x/(x^2+y^2) & 0 \end{vmatrix} = \frac{1}{(r_1 r_2)^2} \frac{R}{2} (\mu^2 - \nu^2) \qquad (8)$$

Since:

$$\mu^2 - \nu^2 = \frac{4 r_1 r_2}{R^2} \qquad (9)$$

we get, finally:

$$J = \frac{8}{R^3(\mu^2 - \nu^2)} \qquad (10)$$

c. PRINCIPLE OF THE EXACT CALCULATION

In the Born-Oppenheimer approximation, the equation to be solved in order to find the energy levels of the electron in the Coulomb field of the two fixed protons can be written:

$$\left[-\frac{\hbar^2}{2m}\Delta - \frac{e^2}{r_1} - \frac{e^2}{r_2} + \frac{e^2}{R} \right] \varphi(\mathbf{r}) = E\, \varphi(\mathbf{r}) \qquad (11)$$

If we go into the elliptical coordinates defined in (3), we can separate the variables μ, ν and φ. Solving the equations so obtained, we find a discrete spectrum of possible energies for each value of R. We shall not perform this calculation here, but shall merely represent (the solid-line curve in figure 2) the variation of the ground state energy with respect to R. This will enable us to compare the results we shall obtain by the variational method with the values given by the exact solution of equation (11).

2. The variational calculation of the energies

a. CHOICE OF THE TRIAL KETS

Assume R to be much larger than a_0. If we are concerned with values of r_1 of the order of a_0, we have, practically:

$$\frac{e^2}{r_2} \simeq \frac{e^2}{R} \quad \text{for } R, r_2 \gg a_0 \qquad (12)$$

The Hamiltonian:

$$H = \frac{\mathbf{P}^2}{2m} - \frac{e^2}{r_1} - \frac{e^2}{r_2} + \frac{e^2}{R} \qquad (13)$$

is then very close to that of a hydrogen atom centered at proton P_1. Analogous conclusions are, of course, obtained for R much larger than a_0, and r_2 of the order of a_0. Therefore, when the two protons are very far apart, the eigenfunctions of the Hamiltonian (13) are practically the stationary wave functions of hydrogen atoms.

This is, of course, no longer true when a_0 is not negligible compared to R. We see, however, that it is convenient, for all R, to choose a family of trial kets constructed from atomic states centered at each of the two protons. This choice constitutes the application to the special case of the H_2^+ ion of a general method known as the *method of linear combination of atomic orbitals*. More precisely, we shall call $| \varphi_1 \rangle$ and $| \varphi_2 \rangle$ the kets which describe the $1s$ states of the two hydrogen atoms:

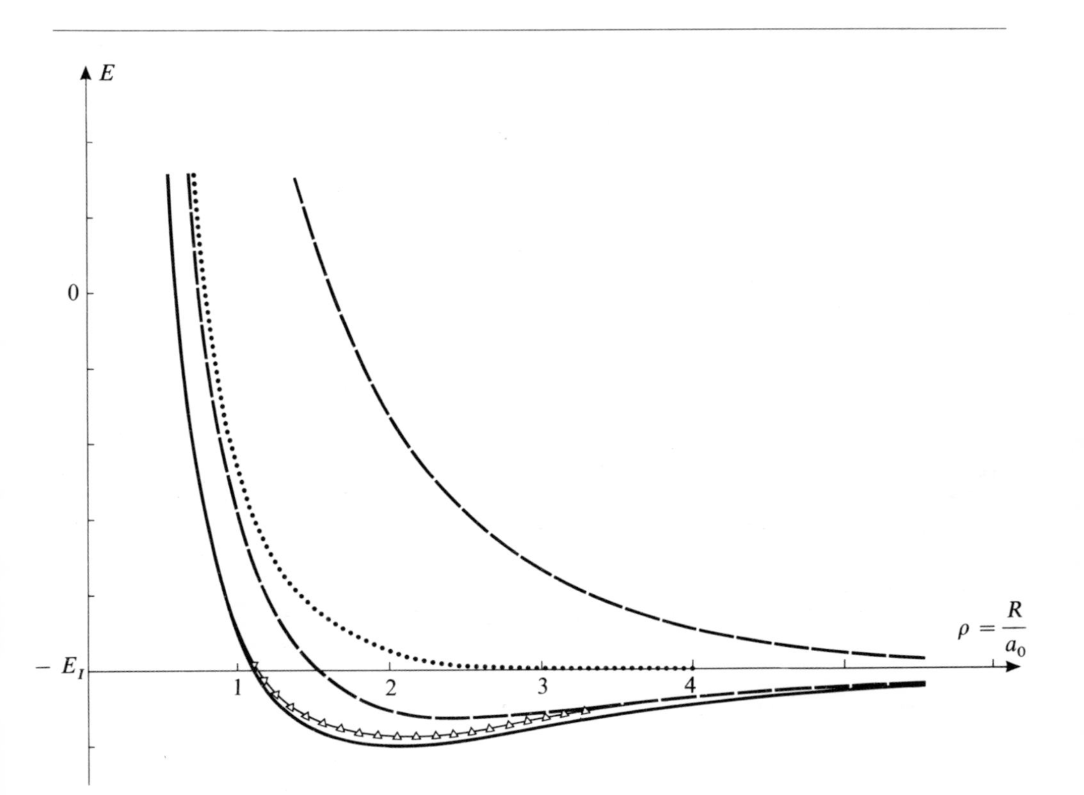

FIGURE 2

Variation of the energy E of the molecular ion H_2^+ with respect to the distance R between the two protons.

. solid line : the exact total energy of the ground state (the stability of the H_2^+ ion is due to the existence of a minimum in this curve).

. dotted line : the diagonal matrix element $H_{11} = H_{22}$ of the Hamiltonian H (the variation of this matrix element cannot explain the chemical bond).

. dashed line : the results of the simple variational calculation of §2 for the bonding and antibonding states (though approximate, this calculation explains the stability of the H_2^+ ion).

. triangles : the results of the more elaborate variational calculation of § 3-a (taking atomic orbitals of adjustable radius considerably improves the accuracy, especially at small distances).

$$\langle \mathbf{r} \mid \varphi_1 \rangle = \frac{1}{\sqrt{\pi a_0^3}} e^{-\rho_1}$$
$$\langle \mathbf{r} \mid \varphi_2 \rangle = \frac{1}{\sqrt{\pi a_0^3}} e^{-\rho_2} \tag{14}$$

We shall choose as trial kets all the kets belonging to the vector subspace $\mathscr{F}$ spanned by these two kets, that is, the set of kets $| \psi \rangle$ such that:

$$| \psi \rangle = c_1 | \varphi_1 \rangle + c_2 | \varphi_2 \rangle \tag{15}$$

The variational method (complement E_{XI}) then consists of finding the stationary values of:

$$\langle H \rangle = \frac{\langle \psi | H | \psi \rangle}{\langle \psi | \psi \rangle} \tag{16}$$

within this subspace. Since this is a vector subspace, the mean value $\langle H \rangle$ is minimal or maximal when $| \psi \rangle$ is an eigenvector of H inside this subspace $\mathscr{F}$, and the corresponding eigenvalue constitutes an approximation of a true eigenvalue of H in the total state space.

b. THE EIGENVALUE EQUATION OF THE HAMILTONIAN H IN THE TRIAL KET VECTOR SUBSPACE $\mathscr{F}$

The resolution of the eigenvalue equation of H within the subspace $\mathscr{F}$ is slightly complicated by the fact that $| \varphi_1 \rangle$ and $| \varphi_2 \rangle$ are not orthogonal.

Any vector $| \psi \rangle$ of $\mathscr{F}$ is of the form (15). For it to be an eigenvector of H in $\mathscr{F}$ with the eigenvalue E, it is necessary and sufficient that:

$$\langle \varphi_i | H | \psi \rangle = E \langle \varphi_i | \psi \rangle \qquad i = 1, 2 \tag{17}$$

that is:

$$\sum_{j=1}^{2} c_j \langle \varphi_i | H | \varphi_j \rangle = E \sum_{j=1}^{2} c_j \langle \varphi_i | \varphi_j \rangle \tag{18}$$

We set:

$$S_{ij} = \langle \varphi_i | \varphi_j \rangle$$
$$H_{ij} = \langle \varphi_i | H | \varphi_j \rangle \tag{19}$$

We must solve a system of two linear homogeneous equations:

$$(H_{11} - ES_{11}) c_1 + (H_{12} - ES_{12}) c_2 = 0$$
$$(H_{21} - ES_{21}) c_1 + (H_{22} - ES_{22}) c_2 = 0 \tag{20}$$

This system has a non-zero solution only if:

$$\begin{vmatrix} H_{11} - ES_{11} & H_{12} - ES_{12} \\ H_{21} - ES_{21} & H_{22} - ES_{22} \end{vmatrix} = 0 \tag{21}$$

The possible eigenvalues E are therefore the roots of a second-degree equation.

c. OVERLAP, COULOMB AND RESONANCE INTEGRALS

$| \varphi_1 \rangle$ and $| \varphi_2 \rangle$ are normalized; consequently:

$$S_{11} = S_{22} = 1 \tag{22}$$

On the other hand, $| \varphi_1 \rangle$ and $| \varphi_2 \rangle$ are not orthogonal. Since the wave-functions (14) associated with these two kets are real, we have:

$$S_{12} = S_{21} = S \tag{23}$$

with:

$$S = \langle \varphi_1 | \varphi_2 \rangle = \int d^3r \; \varphi_1(\mathbf{r}) \; \varphi_2(\mathbf{r}) \tag{24}$$

S is called an *overlap integral*, since it receives contributions only from points of space at which the atomic wave functions φ_1 and φ_2 are both different from zero (such points exist if the two atomic orbitals partially "overlap"). A simple calculation gives:

$$S = e^{-\rho}\left[1 + \rho + \frac{1}{3}\rho^2\right] \tag{25}$$

To find this result, we can use elliptic coordinates (3), since:

$$\rho_1 = \frac{\mu + \nu}{2}\rho$$

$$\rho_2 = \frac{\mu - \nu}{2}\rho \tag{26}$$

According to expression (14) for the wave functions and the one for the volume element, (4), we must calculate:

$$\begin{aligned} S &= \frac{1}{\pi a_0^3}\int_1^{+\infty} d\mu \int_{-1}^{+1} d\nu \int_0^{2\pi} d\varphi \, \frac{\rho^3 a_0^3}{8}(\mu^2 - \nu^2)\, e^{-\mu\rho} \\ &= \frac{\rho^3}{2}\int_1^{+\infty} d\mu \left(\mu^2 - \frac{1}{3}\right) e^{-\mu\rho} \end{aligned} \tag{27}$$

which easily yields (25).

By symmetry:

$$H_{11} = H_{22} \tag{28}$$

According to expression (13) for the Hamiltonian H, we obtain:

$$H_{11} = \langle \varphi_1 | \left[\frac{\mathbf{P}^2}{2m} - \frac{e^2}{r_1} \right] | \varphi_1 \rangle - \langle \varphi_1 | \frac{e^2}{r_2} | \varphi_1 \rangle + \frac{e^2}{R} \langle \varphi_1 | \varphi_1 \rangle \tag{29}$$

Now, $| \varphi_1 \rangle$ is a normalized eigenket of $\frac{\mathbf{P}^2}{2m} - \frac{e^2}{r_1}$. The first term of (29) is therefore equal to the energy $- E_I$ of the ground state of the hydrogen atom, and the third term is equal to e^2/R; we thus have:

$$H_{11} = - E_I + \frac{e^2}{R} - C \tag{30}$$

with:

$$C = \langle \varphi_1 | \frac{e^2}{r_2} | \varphi_1 \rangle = \int d^3r \; \frac{e^2}{r_2} \left[\varphi_1(\mathbf{r}) \right]^2 \tag{31}$$

C is called a *Coulomb integral*. It describes (to within a change of sign) the electrostatic interaction between the proton P_2 and the charge distribution associated with the electron when it is in the $1s$ atomic state around the proton P_1. We find:

$$C = E_I \times \frac{2}{\rho} \left[1 - \mathrm{e}^{-2\rho}(1 + \rho) \right] \tag{32}$$

To find this result, we use elliptic coordinates again:

$$\begin{aligned} C &= \frac{e^2}{a_0 \rho} \frac{1}{\pi a_0^3} \frac{\rho^3 a_0^3}{8} \int (\mu^2 - \nu^2) \, \mathrm{d}\mu \, \mathrm{d}\nu \, \mathrm{d}\varphi \frac{2}{\mu - \nu} \mathrm{e}^{-(\mu + \nu)\rho} \\ &= E_I \, \rho^2 \int_1^{+\infty} \mathrm{d}\mu \int_{-1}^{+1} \mathrm{d}\nu \, (\mu + \nu) \, \mathrm{e}^{-(\mu + \nu)\rho} \end{aligned} \tag{33}$$

Elementary integrations then lead to result (32).

In formula (30), C can be considered to be a modification of the repulsive energy e^2/R of the two protons: when the electron is in the state $| \varphi_1 \rangle$, the corresponding charge distribution "screens" the proton P_1. Since $|\varphi_1(\mathbf{r})|^2$ is spherically symmetric about P_1, if the proton P_2 was far enough from it this charge distribution would appear to P_2 like a negative point charge e situated at its center P_1, (so that the charge of the proton P_1 would be totally cancelled). This does not actually happen unless R is much larger than a_0:

$$\lim_{R \to \infty} \left[\frac{e^2}{R} - C \right] = 0 \tag{34}$$

For finite R, the screening effect can only be partial, and we must have:

$$\frac{e^2}{R} - C > 0 \tag{35}$$

The variation of the energy $\frac{e^2}{R} - C$ with respect to R is shown in figure 2 by the dotted line. It is clear that the variation of H_{11} (or H_{22}) with respect to R cannot explain the chemical bond, since this curve has no minimum.

Finally, let us calculate H_{12} and H_{21}. Since the wave functions $\varphi_1(\mathbf{r})$ and $\varphi_2(\mathbf{r})$ are real, we have:

$$H_{12} = H_{21} \tag{36}$$

Expression (13) for the Hamiltonian gives:

$$H_{12} = \langle \varphi_1 | \left[\frac{\mathbf{P}^2}{2m} - \frac{e^2}{r_2} \right] | \varphi_2 \rangle + \frac{e^2}{R} \langle \varphi_1 | \varphi_2 \rangle - \langle \varphi_1 | \frac{e^2}{r_1} | \varphi_2 \rangle \tag{37}$$

that is, according to definition (24) of S:

$$H_{12} = - E_I S + \frac{e^2}{R} S - A \tag{38}$$

with:

$$A = \langle \varphi_1 | \frac{e^2}{r_1} | \varphi_2 \rangle = \int d^3r \, \varphi_1(\mathbf{r}) \frac{e^2}{r_1} \varphi_2(\mathbf{r}) \tag{39}$$

We shall call A the *resonance integral*★. It is equal to:

$$A = E_I \times 2\,\mathrm{e}^{-\rho}(1 + \rho) \tag{40}$$

The use of elliptic coordinates enables us to write A in the form:

$$\begin{aligned} A &= \frac{e^2}{a_0} \frac{1}{\pi a_0^3} \frac{\rho^3 a_0^3}{8} \int (\mu^2 - \nu^2)\, \mathrm{d}\mu\, \mathrm{d}\nu\, \mathrm{d}\varphi \frac{2\,\mathrm{e}^{-\mu\rho}}{(\mu + \nu)\rho} \\ &= \rho^2 E_I \int_1^{+\infty} \mathrm{d}\mu\, 2\mu\, \mathrm{e}^{-\mu\rho} \end{aligned} \tag{41}$$

The fact that H_{12} is different from zero expresses the possibility of the electron "jumping" from the neighborhood of one of the protons to that of the other one. If, at some time, the electron is in the state $| \varphi_1 \rangle$ (or $| \varphi_2 \rangle$), it oscillates in time between the two sites, under the influence of the non-diagonal matrix element H_{12}. H_{12} is therefore responsible for the phenomenon of *quantum resonance*, which we described qualitatively in §C-2-c of chapter IV (hence the name of integral A).

To sum up, the parameters which are functions of R and which are involved in equation (21) for the approximate energies E are:

$$\begin{aligned} S_{11} &= S_{22} = 1 \\ S_{12} &= S_{21} = S \\ H_{11} &= H_{22} = - E_I + \frac{e^2}{R} - C \\ H_{12} &= H_{21} = \left(- E_I + \frac{e^2}{R} \right) S - A \end{aligned} \tag{42}$$

★ Certain authors call A an "exchange integral". We prefer to restrict the use of this term to another type of integral which is encountered in many-particle systems (complement B_{IV}, § 2-c-β).

where S, C and A are given by (25), (32) and (40), and are shown in figure 3. Note that the non-diagonal elements of determinant (21) take on significant values only if the orbitals $\varphi_1(\mathbf{r})$ and $\varphi_2(\mathbf{r})$ partially overlap, since the product $\varphi_1(\mathbf{r})\varphi_2(\mathbf{r})$ appears in definition (39) of A, as well as in that of S.

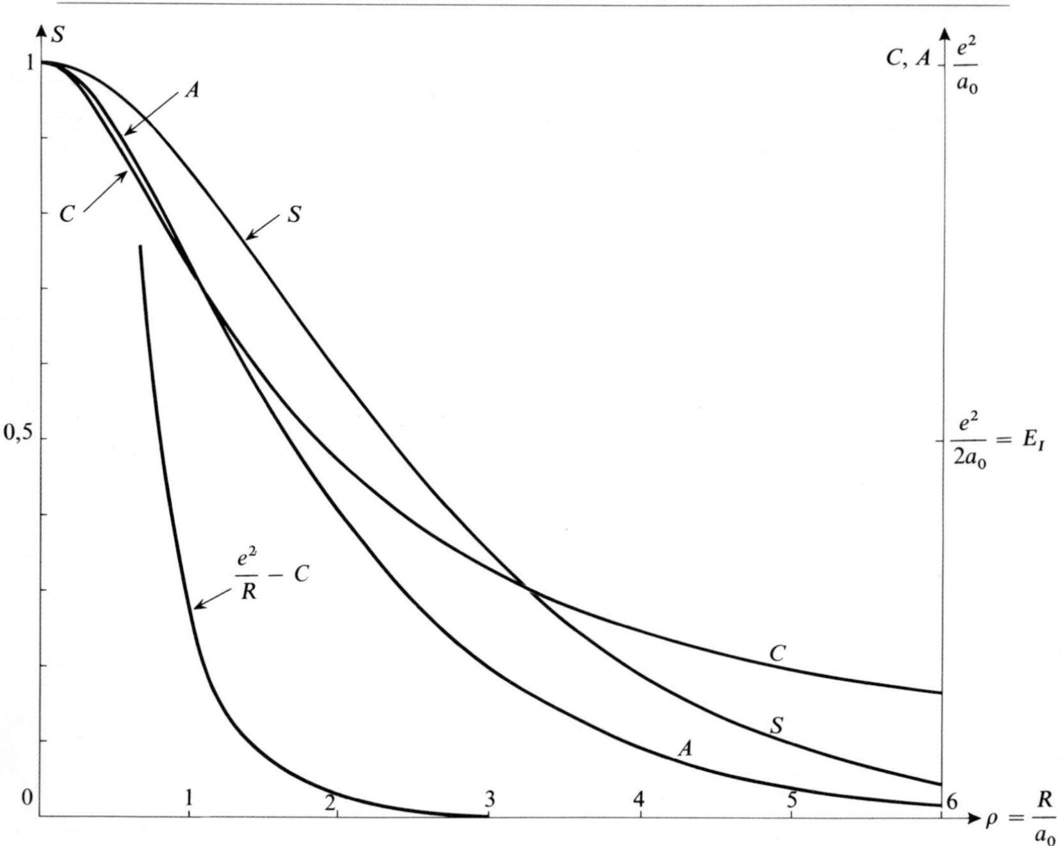

FIGURE 3

Variation of S (the overlap integral), C (the Coulomb integral) and A (the resonance integral) with respect to $\rho = R/a_0$. When $R \longrightarrow \infty$, S and A approach zero exponentially, while C decreases only with e^2/R (the "screened" interaction $\frac{e^2}{R} - C$ of the proton P_1 with the atom centered at P_2 also decreases exponentially, however).

d. BONDING AND ANTIBONDING STATES

α. *Calculation of the approximate energies*

We set:

$$\begin{aligned} E &= \varepsilon E_I \\ A &= \alpha E_I \\ C &= \gamma E_I \end{aligned} \tag{43}$$

Equation (21) can then be written:

$$\begin{vmatrix} -1 + \dfrac{2}{\rho} - \gamma - \varepsilon & \left(-1 + \dfrac{2}{\rho}\right)S - \alpha - \varepsilon S \\ \left(-1 + \dfrac{2}{\rho}\right)S - \alpha - \varepsilon S & -1 + \dfrac{2}{\rho} - \gamma - \varepsilon \end{vmatrix} = 0 \tag{44}$$

or:

$$\left[\gamma + \varepsilon + 1 - \frac{2}{\rho}\right]^2 = \left[\alpha + \left(\varepsilon + 1 - \frac{2}{\rho}\right)S\right]^2 \tag{45}$$

This gives the following two values for ε:

$$\varepsilon_+ = -1 + \frac{2}{\rho} + \frac{\alpha - \gamma}{1 - S} \tag{46-a}$$

$$\varepsilon_- = -1 + \frac{2}{\rho} - \frac{\alpha + \gamma}{1 + S} \tag{46-b}$$

ε_+ and ε_- both approach -1 when ρ approaches infinity. This means that the two approximate energies $E_\pm$ approach $-E_I$, the ground state energy of an isolated hydrogen atom, as expected (§2-a). Furthermore, it is convenient to choose this value as the energy origin, that is, to set:

$$\Delta E = E(\rho) - E(\infty) = E + E_I \tag{47}$$

Using (25), (32) and (40), the approximate energies ΔE_+ and ΔE_- can be written:

$$\Delta E_\pm = E_I \left\{ \frac{2}{\rho} \pm \frac{2\,\mathrm{e}^{-\rho}(1 + \rho) \mp \dfrac{2}{\rho}[1 - \mathrm{e}^{-2\rho}(1 + \rho)]}{1 \mp \mathrm{e}^{-\rho}(1 + \rho + \rho^2/3)} \right\} \tag{48}$$

The variation of $\Delta E_\pm/E_I$ with respect to ρ is shown in dashed lines in figure 2. We see that ΔE_- has a negative minimum for a certain value of the distance R between the two protons. Although this is an approximation (*cf.* fig. 2), it explains the existence of the chemical bond.

As we have already pointed out, the variation with respect to R of the diagonal elements H_{11} and H_{22} of determinant (21) has no minimum (dotted-line curve of figure 2). The minimum of ΔE_- therefore is due to the non-diagonal elements H_{12} and S_{12}. This shows that the phenomenon of the chemical bond appears only if the electronic orbitals of the two atoms participating in the bond overlap sufficiently.

β. *Eigenstates of H inside the subspace $\mathscr{F}$*

The eigenstate corresponding to E_- is called a *bonding state*, and the one corresponding to E_+, an *antibonding state*, since E_+ always remains greater than the energy $-E_I$ of the system formed by a hydrogen atom in the ground state and an infinitely distant proton.

According to (45):

$$\gamma + \varepsilon + 1 - \frac{2}{\rho} = \pm \left[\alpha + \left(\varepsilon + 1 - \frac{2}{\rho} \right) S \right] \tag{49}$$

System (20) then gives:

$$c_1 \pm c_2 = 0 \tag{50}$$

The bonding and antibonding states are therefore symmetrical and antisymmetrical linear combinations of the kets $| \varphi_1 \rangle$ and $| \varphi_2 \rangle$. To normalize them, it must be recalled that $| \varphi_1 \rangle$ and $| \varphi_2 \rangle$ are not orthogonal (their scalar product is equal to S). We therefore obtain:

$$| \psi_+ \rangle = \frac{1}{\sqrt{2(1 - S)}} [| \varphi_1 \rangle - | \varphi_2 \rangle] \tag{51-a}$$

$$| \psi_- \rangle = \frac{1}{\sqrt{2(1 + S)}} [| \varphi_1 \rangle + | \varphi_2 \rangle] \tag{51-b}$$

Note that the *bonding state* $| \psi_- \rangle$, associated with E_-, *is symmetrical* under exchange of $| \varphi_1 \rangle$ and $| \varphi_2 \rangle$, while the antibonding state is antisymmetrical.

COMMENT:

It could have been expected that the eigenstates of H inside the subspace $\mathscr{F}$ would be symmetrical and antisymmetrical combinations of $| \varphi_1 \rangle$ and $| \varphi_2 \rangle$: for given positions of the two protons, there is symmetry with respect to the bisecting plane of P_1P_2, and H remains unchanged if the roles of the two protons are exchanged.

The bonding and antibonding states are approximate stationary states of the system under study. We pointed out, furthermore, in complement E_{XI}, that the variational method can give a valid approximation for the energies but gives a more debatable result for the eigenfunctions. It is instructive, however, to have an idea of the mechanism of the chemical bond, to represent graphically the wave functions associated with the bonding and antibonding states, which are often called bonding and antibonding *molecular orbitals*. To do so, we can, for example, trace the surfaces of equal $|\psi|$ (the locus of points in space for which the modulus $|\psi|$ of the wave function has a given value). If ψ is real, we indicate by a + (or −) sign the regions in which it is positive (or negative). This is what is done in figure 4 for ψ_+ and ψ_- (the surfaces of equal $|\psi|$ are surfaces of revolution about the P_1P_2 axis, and figure 4 only shows their cross sections in a plane containing P_1P_2). The difference between the bonding orbital and the antibonding orbital is striking. In the first one, the electronic cloud "streches out" to include both protons, while in the second one, the position probability of the electron is zero in the bisecting plane of P_1P_2.

COMMENT:

We can calculate the mean value of the potential energy in the state $| \psi_- \rangle$, which, if we use (51-b), (31) and (39), is equal to:

$$\begin{aligned}\langle V \rangle &= \langle \psi_- | \left[\frac{e^2}{R} - \frac{e^2}{r_1} - \frac{e^2}{r_2} \right] | \psi_- \rangle \\ &= \frac{e^2}{R} - \frac{1}{1+S} \left[\langle \varphi_1 | \frac{e^2}{r_1} | \varphi_1 \rangle + \langle \varphi_1 | \frac{e^2}{r_1} | \varphi_2 \rangle \right. \\ &\qquad \left. + \langle \varphi_1 | \frac{e^2}{r_2} | \varphi_1 \rangle + \langle \varphi_1 | \frac{e^2}{r_2} | \varphi_2 \rangle \right] \\ &= E_I \left[\frac{2}{\rho} - \frac{1}{1+S} (2 + 2\alpha + \gamma) \right] \end{aligned} \tag{52}$$

Subtracting this from (46-b), we obtain the kinetic energy:

$$\begin{aligned}\langle T \rangle &= \langle \frac{\mathbf{P}^2}{2m} \rangle = \langle H - V \rangle \\ &= E_I \frac{1}{1+S} (1 - S + \alpha) \end{aligned} \tag{53}$$

We shall discuss later (§ 5) to what extent (52) and (53) give good approximations for the kinetic and potential energies.

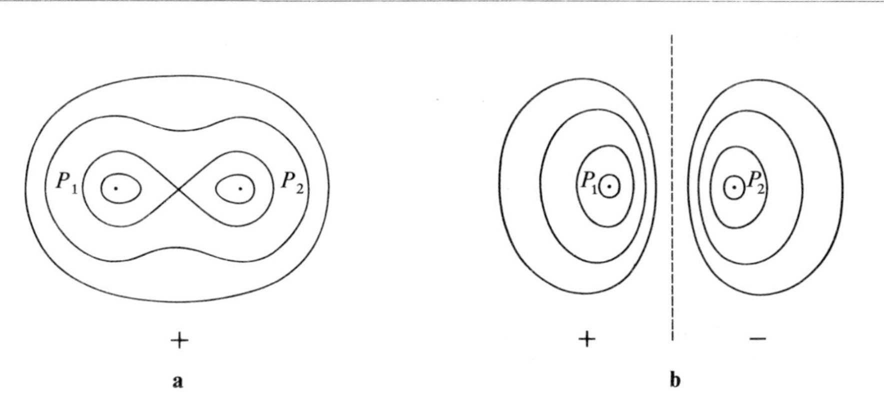

FIGURE 4

Schematic drawings of the bonding molecular orbital (fig. a) and the antibonding molecular orbital (fig. b) for the H_2^+ ion. We have shown the cross section in a plane containing P_1P_2 of a family of surfaces for which the modulus $|\psi|$ of the wave function has a constant given value. These are surfaces of revolution about P_1P_2 (we have shown 4 surfaces, corresponding to 4 different values of $|\psi|$). The + and − signs indicated in the figure are those of the wave function (which is real) in the corresponding regions. The dashed line is the trace of the bisecting plane of P_1P_2, which is a nodal plane for the antibonding orbit.

3. Critique of the preceding model. Possible improvements

a. RESULTS FOR SMALL R

What happens to the energy of the bonding state and the corresponding wave function when $R \longrightarrow 0$?

We see from figure 3 that S, A and C approach, respectively, 1, $2E_I$ and $2E_I$

when $\rho \longrightarrow 0$. If we subtract the repulsion term. e^2/R, of the two protons, to obtain the electronic energy, we find:

$$E_- - \frac{e^2}{R} \underset{R \to 0}{\longrightarrow} - 3E_I \tag{54}$$

In addition, since $| \varphi_1 \rangle$ approaches $| \varphi_2 \rangle$, $| \psi_- \rangle$ reduces to $| \varphi_1 \rangle$ (the ground state 1*s* of the hydrogen atom).

This result is obviously incorrect. When $R = 0$, we have the equivalent* of a helium ion He^+. The electronic energy of the ground state of H_2^+ must coincide, for $R = 0$, with that of the ground state of He^+. Since the helium nucleus is a $Z = 2$ nucleus, this energy is (*cf.* complement A_{VII}):

$$- Z^2 E_I = - 4E_I \tag{55}$$

and not $- 3E_I$. Furthermore, the wave function $\psi_-(\mathbf{r})$ should not approach $\varphi_1(\mathbf{r}) = (\pi a_0^3)^{-1/2} \text{e}^{-\rho_1}$, but rather $(\pi a_0^3/Z^3)^{-1/2} \text{e}^{-Z\rho_1}$ with $Z = 2$ (the Bohr orbit is twice as small). This enables us to understand why the disagreement between the exact result and that of § 2 above becomes important for small values of R (fig. 2): this calculation uses atomic orbitals which are too spread out when the two protons are too close to each other.

A possible improvement therefore consists of enlarging the family of trial kets because of these physical arguments and using kets of the form:

$$| \psi \rangle = c_1 \, | \varphi_1(Z) \rangle + c_2 \, | \varphi_2(Z) \rangle \tag{56}$$

where $| \varphi_1(Z) \rangle$ and $| \varphi_2(Z) \rangle$ are associated with 1*s* atomic orbitals of radius a_0/Z centered at P_1 and P_2. The ground state still corresponds, for reasons of symmetry, to $c_1 = c_2$. We consider Z like a variational parameter in seeking, for each value of R, the value of Z which minimizes the energy.

The calculation can be performed completely in elliptic coordinates. We find (*cf.* fig. 5) that the optimal value of Z decreases from $Z = 2$ for $R = 0$ to $Z = 1$ for $R \longrightarrow \infty$, as it should.

The curve obtained for ΔE_- is much closer to the exact curve (*cf.* fig. 2). Table I gives the values of the abscissa and ordinate of the minimum of ΔE_- obtained from the various models considered in this complement. It can be seen from this table that the energies found by the variational method are always greater than the exact energy of the ground state; in addition, we see that enlarging the family of trial kets improves the results for the energy.

b. RESULTS FOR LARGE R

When $R \longrightarrow \infty$, we see from (48) that E_+ and E_- exponentially approach the same value $- E_I$. Actually, this limit should not be obtained so rapidly. To see this, we shall use a perturbation approach, as in complement C_{XI} (Van der Waals forces) or E_{XII} (the Stark effect of the hydrogen atom). Let us evaluate the perturbation of the energy of a hydrogen atom (in the 1*s* state), situated at P_2, produced by the presence of a proton P_1 situated at a distance R much greater than $a_0 (\rho \gg 1)$. In the neighborhood of P_2, the proton P_1 creates an electric

* In addition to the two protons, the helium nucleus of course contains one or two neutrons.

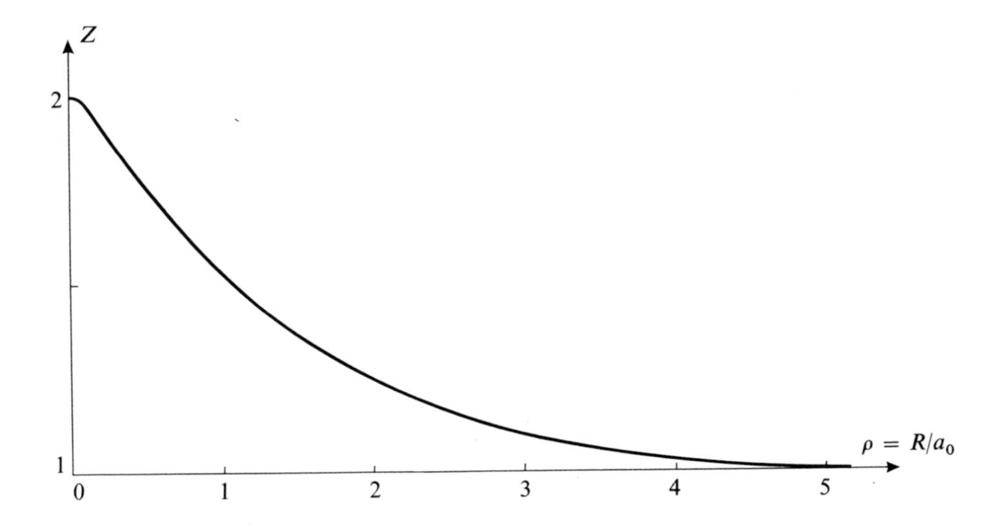

FIGURE 5

For each value of the internuclear distance, we have calculated the value of Z which minimizes the energy. For $R = 0$, we have the equivalent of a He^+ ion, and we indeed find $Z = 2$. For $R \gg a_0$, we have essentially an isolated hydrogen atom, which gives $Z = 1$. Between these two extremes, Z is a decreasing function of ρ. The corresponding optimal energies are represented by triangles in figure 2.

	Equilibrium distance between the two protons (the abscissa of the minimum of ΔE_-)	Depth of the minimum of ΔE_-
Variational method of §1 ($1s$ orbitals with $Z = 1$)	$2.50\ a_0$	1.76 eV
Variational method of §2-a ($1s$ orbitals with variable Z)..........	$2.00\ a_0$	2.35 eV
Variational method of §2-b (hybrid orbitals with variable Z, Z', σ).	$2.00\ a_0$	2.73 eV
Exact values	$2.00\ a_0$	2.79 eV

TABLE I

field **E**, which varies like $1/R^2$. This field polarizes the hydrogen atom and causes an electric dipole moment **D**, proportional to **E**, to appear. The electronic wave function is distorted, and the barycenter of the electronic charge distribution moves closer to P_1 (fig. 6). **E** and **D** are both proportional to $1/R^2$ and have the same sign. The electrostatic interaction between the proton P_1 and the atom situated at P_2 must therefore lower the energy by an amount which, like $-\mathbf{E}\cdot\mathbf{D}$, varies like $1/R^4$*. Consequently, the asymptotic behavior of ΔE_+ and ΔE_- must vary, not exponentially, but like $-a/R^4$ (where a is a positive constant).

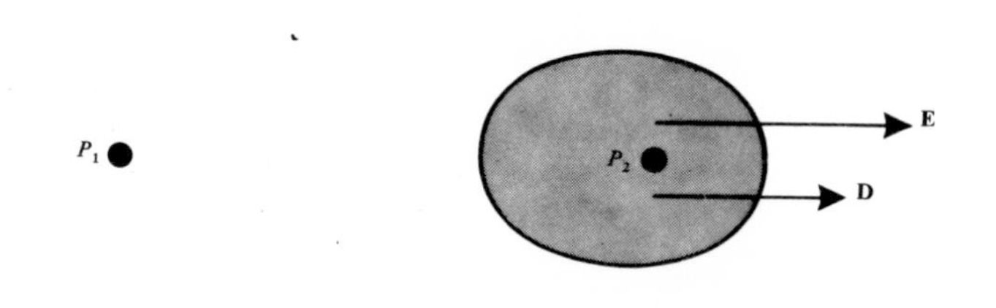

FIGURE 6

Under the effect of the electric field E created by the proton P_1, the electronic cloud of the hydrogen atom centered at P_2 becomes distorted, and this atom acquires an electric dipole moment D. An interaction energy results which decreases with $1/R^4$ when R increases.

It is actually possible to find this result by the variational method. Instead of linearly superposing $1s$ orbitals centered at P_1 and P_2, we shall superpose hybrid orbitals χ_1 and χ_2, which are not spherically symmetrical about P_1 and P_2. χ_2 is obtained, for example, by linearly superposing a $1s$ orbital and a $2p$ orbital, both centered at P_2**:

$$\chi_2(\mathbf{r}) = \varphi_{1s}^2(\mathbf{r}) + \sigma\,\varphi_{2p}^2(\mathbf{r}) \tag{57}$$

and has a form analogous to the one shown in figure 6. Now, consider determinant (21). The non-diagonal elements $H_{12} = \langle \chi_1 | H | \chi_2 \rangle$ and $S_{12} = \langle \chi_1 | \chi_2 \rangle$ still approach zero exponentially when $R \longrightarrow \infty$. This is because the product $\chi_1(\mathbf{r})\chi_2(\mathbf{r})$ appears in the corresponding integrals; even though distorted, the orbitals $\chi_1(\mathbf{r})$ and $\chi_2(\mathbf{r})$ still remain localized in the neighborhoods of P_1 and P_2 respectively, and their overlap goes to zero exponentially when $R \longrightarrow \infty$. The two eigenvalues E_+ and E_- therefore both approach $H_{11} = H_{22}$ when $R \longrightarrow \infty$, since determinant (21) becomes diagonal.

Now, what does H_{22} represent? As we have seen (*cf.* § 2-c), it is the energy of a hydrogen atom placed at P_2 and perturbed by the proton P_1. The calculation of § 2 neglected any polarization of the $1s$ electronic orbital due to the effect of the electric field created by P_1, and this is why we found an energy correction which decreased exponentially with R. However, if, as we are doing here, we take into account the polarization of the electronic orbital, we find a correction in $-a/R^4$. The fact that, in (57), we consider only the mixing with the $2p$ orbital causes the value of a given by the variational calculation to be approximate (whereas the perturbation calculation of the polarization involves all the excited states, *cf.* Complement E_{XII}).

* More precisely, the energy is lowered by $-\frac{1}{2}\,\mathbf{E}\cdot\mathbf{D}$ (*cf.* complement E_{XII}, § 1).
** The symmetry axis of the $2p$ orbital is chosen along the straight line joining the two protons.

The two curves ΔE_+ and ΔE_- therefore do approach each other exponentially, since the difference between E_+ and E_- involves only the non-diagonal elements H_{12} and S_{12}, and their common value for large R approaches zero like $-\ a/R^4$ (fig. 7).

The preceding discussion also suggests to us the use of polarized orbitals like the one in (57), not only for large R, but also for all other values of R as well. We would thus enlarge the family of trial kets and consequently improve the accuracy. In expression (57), we then

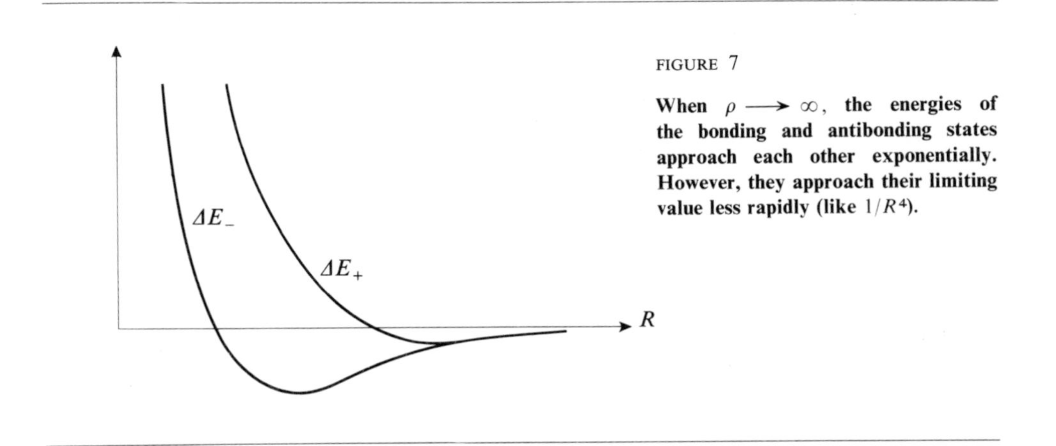

FIGURE 7

When $\rho \longrightarrow \infty$, the energies of the bonding and antibonding states approach each other exponentially. However, they approach their limiting value less rapidly (like $1/R^4$).

consider σ as a variation parameter, like the parameter Z which defines the Bohr radius a_0/Z associated with the $1s$ and $2p$ orbitals. To make the method even more flexible, we even choose different parameters Z and Z' for φ_{1s} and φ_{2p}. For each value of R, we then minimize the mean value of H in the state $|\chi_1\rangle + |\chi_2\rangle$ (which, for reasons of symmetry, is still the ground state), and we thus determine the optimal values of σ, Z, Z'. The agreement with the exact solution then becomes excellent (*cf.* table I).

4. Other molecular orbitals of the H_2^+ ion

In the preceding sections, we constructed, by the variational method, a bonding molecular orbital and an antibonding molecular orbital from the ground state $1s$ of each of the two hydrogen atoms which can be formed about the two protons. Of course, we chose the $1s$ state because it was clear that this would be the best choice for obtaining an approximation of the ground state of the system of two protons and one electron. We can obviously envisage, with the method of linear combination of atomic orbitals (§2-a), using excited states of the hydrogen atom to obtain other molecular orbitals of higher energies. The main interest of these excited orbitals here will be to give us an idea of the phenomena which can come into play in molecules which are more complex than the H_2^+ ion. For example, to understand the properties of a diatomic molecule containing several electrons, we can, in a first approximation, treat these electrons individually, as if they did not interact with each other. We thus determine the various possible stationary states for an isolated electron placed in the Coulomb field of the nuclei, and then place the electrons of the molecule in these states, taking the Pauli principle into account (chap. XIV, § D-1) and filling the lowest energy states first (this procedure is analogous to the one described for many-electron atoms in complement A_{XIV}). In this section, we shall indicate the principal properties of the excited molecular orbitals of the H_2^+ ion, while keeping in mind the possibilities of generalization to more complex molecules.

a. SYMMETRIES AND QUANTUM NUMBERS. SPECTROSCOPIC NOTATION

(*i*) The potential V created by the two protons is symmetric with respect to revolution about the P_1P_2 axis, which we shall choose as the Oz axis. This means that V and, consequently, the Hamiltonian H of the electron, do not depend on the angular variable φ which fixes the orientation about Oz of the MP_1P_2 plane containing the Oz axis and the point M. It follows that H commutes with the component L_z of the orbital angular momentum of the electron [in the $\{ | \mathbf{r} \rangle \}$ representation, L_z becomes the differential operator $\frac{\hbar}{i}\frac{\partial}{\partial\varphi}$, which commutes with any φ-independent operator]. We can then find a system of eigenstates of H which are also eigenstates of L_z, and class them according to the eigenvalues $m\hbar$ of L_z.

(*ii*) The potential V is also invariant under reflection through any plane containing P_1P_2, that is, the Oz axis. Under such a reflection, an eigenstate of L_z of eigenvalue $m\hbar$ is transformed into an eigenstate of L_z of eigenvalue $-m\hbar$ (the reflection changes the sense of revolution of the electron about Oz). Because of the invariance of V, the energy of a stationary state depends only on $|m|$.

In spectroscopic notation, we label each molecular orbital with a Greek letter indicating the value of $|m|$, as follows:

$$\begin{aligned} |m| &= 0 \longleftrightarrow \sigma \\ |m| &= 1 \longleftrightarrow \pi \\ |m| &= 2 \longleftrightarrow \delta \end{aligned} \tag{58}$$

(note the analogy with atomic spectroscopic notation : σ, π, δ recall s, p, d). For example, since the ground state $1s$ of the hydrogen atom has a zero orbital angular momentum, the two orbitals studied in the preceding sections are σ orbitals (it can be shown that this is also true for the exact stationary wave functions, and not only for the approximate states obtained by the variational method).

This notation does not use the fact that the two protons of the H_2^+ ion have equal charges. The σ, π, δ classification of molecular orbitals therefore remains valid for a heteropolar diatomic molecule.

(*iii*) In the H_2^+ ion (and, more generally, in homopolar diatomic molecules), the potential V is invariant under reflection through the middle O of P_1P_2. We can therefore choose eigenfunctions of the Hamiltonian H in such a way that they have a definite parity with respect to the point O. For an even orbital, we add to the Greek letter which characterizes $|m|$, an index g (from the German "gerade"); this index is u ("ungerade") for odd orbitals. Thus, the bonding orbital obtained above from the $1s$ atomic states is a σ_g orbital, while the corresponding antibonding orbital is σ_u.

(*iv*) Finally, we can use the invariance of H under reflection through the bisecting plane of P_1P_2 to choose stationary wave functions which have a definite parity in this operation, that is, a parity defined with respect to the change in sign of the single variable z. Functions which are odd under this reflection are labeled with an asterisk. They are necessarily zero at all points of the bisecting plane of P_1P_2, like the orbital shown in figure 4-b; these are antibonding orbitals.

COMMENT:

Reflection through the bisecting plane of P_1P_2 can be obtained by performing a reflection through O followed by a rotation of π about Oz. The parity (*iv*) is therefore not independent of the preceding symmetries (the "g" states will have an asterisk for odd $|m|$ and none for even $|m|$; the situation is reversed for the "u" states). However, it is convenient to consider this parity, since it enables us to determine the antibonding orbitals immediately.

b. MOLECULAR ORBITALS CONSTRUCTED FROM THE 2p ATOMIC ORBITALS

If we start with the excited state 2s of the hydrogen atom, arguments analogous to those of the preceding sections will give a bonding $\sigma_g(2s)$ orbital and an anti-bonding $\sigma^*(2s)$ orbital, with forms similar to those in figure 4. We shall therefore concern ourselves instead with molecular orbitals obtained from the excited atomic states 2p.

α. *Orbitals constructed from $2p_z$ states*

We shall denote by $| \varphi_{2p_z}^1 \rangle$ and $| \varphi_{2p_z}^2 \rangle$ the atomic states $2p_z$ (*cf.* complement E_{VII} § 2-b), centered at P_1 and P_2 respectively. The form of the corresponding orbitals is shown in figure 8 (note the choice of signs, indicated in the figure).

By a variational calculation analogous to the one in §2, we can construct, starting with these two atomic states, two approximate eigenstates of the Hamiltonian (13). The symmetries recalled in §4-a imply that, to within a normalization factor, these molecular states can be written :

$$| \varphi_{2p_z}^1 \rangle + | \varphi_{2p_z}^2 \rangle \tag{59-a}$$

$$| \varphi_{2p_z}^1 \rangle - | \varphi_{2p_z}^2 \rangle \tag{59-b}$$

The shape of the two molecular orbitals so obtained can easily be deduced from figure 8; they are shown in figure 9.

The two atomic states $2p_z$ are eigenstates of L_z with the eigenvalue zero; the same is therefore also true of the two states (59). The molecular orbital associated with (59-a) is even and is written $\sigma_g(2p_z)$; the one corresponding to (59-b) is odd under a reflection through O as well as under a reflection through the bisecting plane of P_1P_2, and we shall therefore denote it by $\sigma_u^*(2p_z)$.

β. *Orbitals constructed from $2p_x$ or $2p_y$ states*

We shall now start with the atomic states $| \varphi_{2p_x}^1 \rangle$ and $| \varphi_{2p_x}^2 \rangle$, with which are associated the real wave functions (*cf.* complement E_{VII}, § 2-b) shown in figure 10 (note that the surfaces of equal $|\psi|$ whose cross sections in the xOz plane are given in figure 10 are surfaces of revolution, not about Oz, but about axes parallel to Ox and passing through P_1 and P_2). Recall that the atomic orbital $2p_x$ is obtained by the linear combination of eigenstates of L_z corresponding to $m = 1$ and $m = -1$. The molecular orbitals constructed from these atomic orbitals therefore have $|m| = 1$; they are π orbitals.

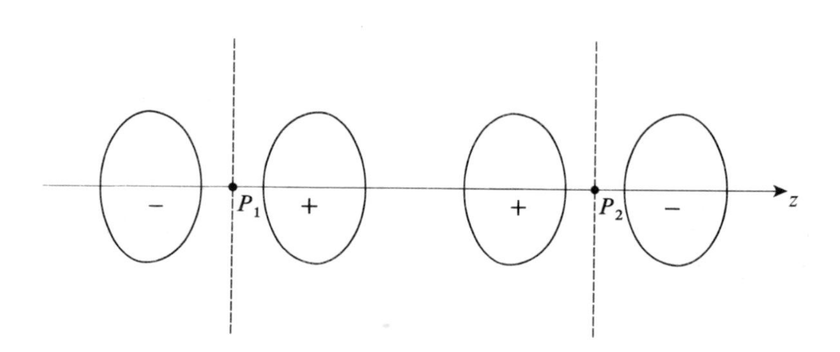

FIGURE 8

Schematic representation of the $2p_z$ atomic orbitals centered at P_1 and P_2 (the Oz axis is chosen along P_1P_2) and used as a basis for constructing the excited molecular orbitals $\sigma_g(2p_z)$ and $\sigma_u^*(2p_z)$ shown in figure 9 (note the sign convention chosen).

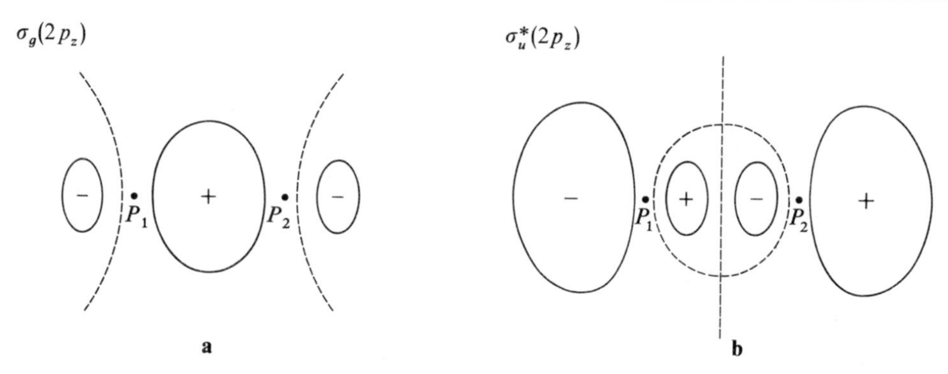

FIGURE 9

Schematic representation of the excited molecular orbitals : the bonding orbital $\sigma_g(2p_z)$ (fig. a) and the antibonding orbital $\sigma_u^*(2p_z)$ (fig. b). As in figure 8, we have drawn the cross section in a plane containing P_1P_2 of a constant modulus $|\psi|$ surface. This is a surface of revolution about P_1P_2. The sign shown is that of the (real) wave function. The dashed-line curves are the cross sections in the plane of the figure of the nodal surfaces ($|\psi| = 0$).

Here again, the approximate molecular states produced from the atomic states $2p_x$ are the symmetrical and antisymmetrical linear combinations:

$$|\varphi_{2p_x}^1\rangle + |\varphi_{2p_x}^2\rangle \tag{60-a}$$

$$|\varphi_{2p_x}^1\rangle - |\varphi_{2p_x}^2\rangle \tag{60-b}$$

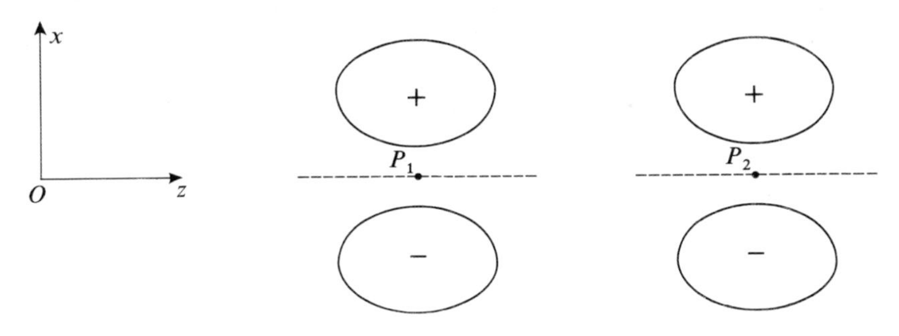

FIGURE 10

Schematic representation of the atomic orbitals $2p_x$ centered at P_1 and P_2 (the Oz axis is chosen along P_1P_2) and used as a basis for constructing the excited molecular orbitals $\pi_u(2p_x)$ and $\pi_g^*(2p_x)$ shown in figure 11. For each orbital, the surface of equal $|\psi|$, whose cross section in the xOz plane is shown, is a surface of revolution, no longer about Oz, but about a straight line parallel to Ox and passing either through P_1 or P_2.

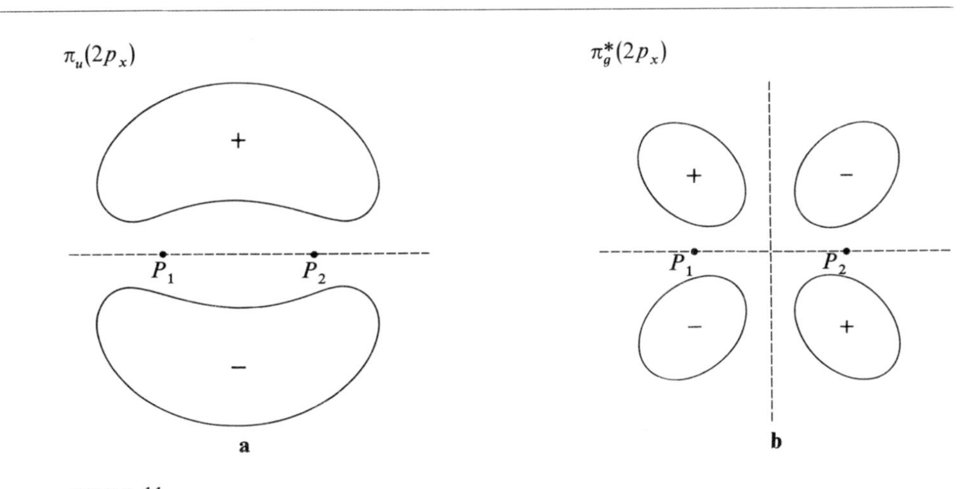

FIGURE 11

Schematic representation of the excited molecular orbitals: the bonding orbital $\pi_u(2p_x)$ (fig. a) and the antibonding orbital $\pi_g^*(2p_x)$ (fig. b). For each of these two orbitals, we have shown the cross section in the xOz plane of a surface on which $|\psi|$ has a given constant value. This surface is no longer a surface of revolution but is simply symmetrical with respect to the xOz plane. The meaning of the signs and the dashed lines is the same as in figures 4, 8, 9, 10.

The form of these molecular orbitals can easily be qualitatively deduced from figure 10. The surfaces of equal $|\psi|$ are not surfaces of revolution about Oz, but are simply symmetrical with respect to the xOz plane. Their cross sections in this plane are shown in figure 11. We see immediately in this figure that the orbital associated with state (60-a) is odd with respect to the middle O of P_1P_2 but even with respect to the bisecting plane of P_1P_2; it will therefore be denoted by $\pi_u(2p_x)$.

On the other hand, the orbital corresponding to (60-b) is even with respect to point O and odd with respect to the bisecting plane of P_1P_2 : it is an antibonding orbital, denoted by $\pi_g^*(2p_x)$. We stress the fact that these π orbitals have *planes of symmetry*, not axes of revolution like the σ orbitals.

Of course, the molecular orbitals produced by the atomic states $2p_y$ can be deduced from the preceding ones by a rotation of $\pi/2$ about P_1P_2.

π orbitals analogous to the preceding ones are involved in the double or triple bonds of atoms such as carbon (*cf.* complement E_{VII}, §§ 3-c and 4-c).

COMMENT:

We saw earlier (§2-d) that the energy separation of the bonding and antibonding levels is due to the overlap of the atomic wave functions. Now, for the same distance R, the overlap of the $\varphi^1_{2p_z}$ and $\varphi^2_{2p_z}$ orbitals, which point towards each other, is larger than that of $\varphi^1_{2p_x}$ and $\varphi^2_{2p_x}$, whose axes are parallel (fig. 8 and 10). We see that the energy difference between $\sigma_g\ (2p_z)$ and $\sigma_u^*(2p_z)$ is larger than that between $\pi_u(2p_x)$ and $\pi_g^*(2p_x)$ [or $\pi_u(2p_y)$ and $\pi_g^*(2p_y)$]. The hierarchy of the corresponding levels is indicated in figure 12.

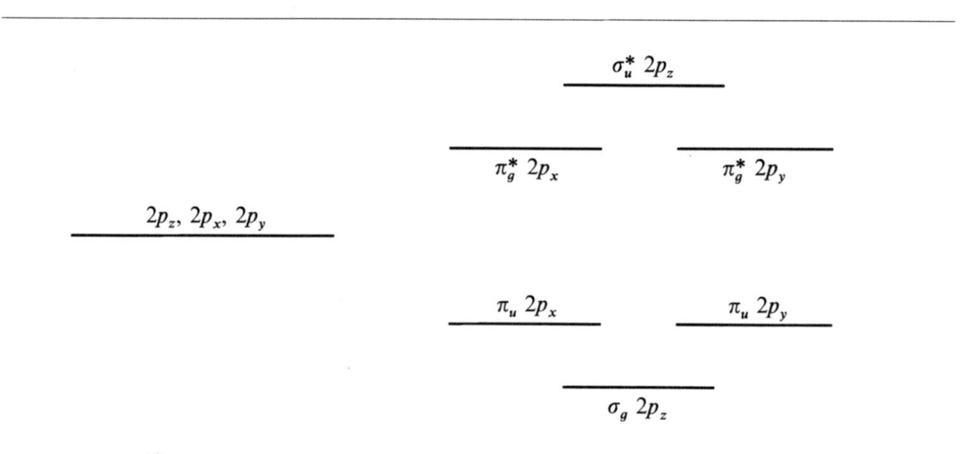

FIGURE 12

The energies of the various excited molecular orbitals constructed from the atomic orbitals $2p_z$, $2p_x$ and $2p_y$ centered at P_1 and P_2 (the Oz axis is chosen along P_1P_2). By symmetry, the molecular orbitals produced by the $2p_x$ atomic orbitals are degenerate with those produced by the $2p_y$ atomic orbitals. The difference between the bonding and antibonding molecular orbitals $\pi_u(2p_{x,y})$ and $\pi_g^*(2p_{x,y})$ is, however, smaller than the corresponding difference between the $\sigma_g(2p_z)$ and $\sigma_u^*(2p_z)$ molecular orbitals. This is due to the larger overlap of the two $2p_z$ atomic orbitals.

5. The origin of the chemical bond ; the virial theorem

a. STATEMENT OF THE PROBLEM

When the distance R between the protons decreases, their electrostatic repulsion e^2/R increases. The fact that the total energy $E_-(R)$ of the bonding state decreases (when R decreases from a very large value) and then passes through a minimum therefore means that the electronic energy begins by decreasing faster than e^2/R increases (of course, since this term diverges when $R \longrightarrow 0$, it is the repulsion between the protons which counts at short distances). We can then ask the following question : does the lowering of the electronic energy which makes the chemical bond possible arise from a lowering of the electronic potential energy or from a lowering of the kinetic energy or from both ?

We have already calculated, in (52) and (53), approximate expressions for the (total) potential and kinetic energies. We might then consider studying the variation of these expressions with respect to R. Such a method, however, would have to be used with caution, since, as we have already pointed out, the eigen-functions supplied by a variational calculation are much less precise than the energies. We shall discuss this point in greater detail in §5-d-β below.

Actually, it is possible to answer this question rigorously, thanks to the "virial theorem", which provides exact relations between $E(R)$ and the average kinetic and potential energies. Therefore, in this section, we shall prove this theorem and discuss its physical consequences. The results obtained, furthermore, are completely general and can be applied, not only to the molecular ion H_2^+, but also to all other molecules. Before considering the virial theorem itself, we shall begin by establishing some results which we shall need later.

b. SOME USEFUL THEOREMS

α. *Euler's theorem*

Recall that a function $f(x_1, x_2, \dots x_n)$ of several variables $x_1, x_2, \dots x_n$ is said to be homogeneous of degree s if it is multiplied by λ^s when all the variables are multiplied by λ:

$$f(\lambda x_1, \lambda x_2, \dots \lambda x_n) = \lambda^s f(x_1, x_2, \dots x_n) \tag{61}$$

For example, the potential of a three-dimensional harmonic oscillator:

$$V(x, y, z) = \frac{1}{2} m\omega^2(x^2 + y^2 + z^2) \tag{62}$$

is homogeneous of degree 2. Similarly, the electrostatic interaction energy of two particles:

$$\frac{e_a e_b}{r_{ab}} = \frac{e_a e_b}{\sqrt{(x_a - x_b)^2 + (y_a - y_b)^2 + (z_a - z_b)^2}} \tag{63}$$

is homogeneous of degree -1.

Euler's theorem indicates that any function f which is homogeneous of degree s satisfies the identity:

$$\sum_{i=1}^{n} x_i \frac{\partial f}{\partial x_i} = s f(x_1, ..., x_i, ..., x_n) \tag{64}$$

To prove this, we calculate the derivatives with respect to λ of both sides of (61). The left-hand side yields:

$$\sum_i \frac{\partial f}{\partial x_i}(\lambda x_1, ..., \lambda x_n) \times \frac{\partial}{\partial \lambda}(\lambda x_i) = \sum_i x_i \frac{\partial f}{\partial x_i}(\lambda x_1, ..., \lambda x_n) \tag{65}$$

and the right-hand side yields:

$$s \lambda^{s-1} f(x_1, ..., x_n) \tag{66}$$

If we set (65) equal to (66), with $\lambda = 1$, we obtain (64).

Euler's theorem can very easily be verified in examples (62) and (63).

β. *The Hellman-Feynman theorem*

Let $H(\lambda)$ be a Hermitian operator which depends on a real parameter λ, and $|\,\psi(\lambda)\,\rangle$, a normalized eigenvector of $H(\lambda)$ of eigenvalue $E(\lambda)$:

$$H(\lambda)\,|\,\psi(\lambda)\,\rangle = E(\lambda)\,|\,\psi(\lambda)\,\rangle \tag{67}$$

$$\langle\,\psi(\lambda)\,|\,\psi(\lambda)\,\rangle = 1 \tag{68}$$

The Hellmann-Feynman theorem indicates that:

$$\frac{\mathrm{d}}{\mathrm{d}\lambda} E(\lambda) = \langle\,\psi(\lambda)\,|\,\frac{\mathrm{d}}{\mathrm{d}\lambda} H(\lambda)\,|\,\psi(\lambda)\,\rangle \tag{69}$$

This relation can be proven as follows. According to (67) and (68), we have:

$$E(\lambda) = \langle\,\psi(\lambda)\,|\,H(\lambda)\,|\,\psi(\lambda)\,\rangle \tag{70}$$

If we differentiate this relation with respect to λ, we obtain:

$$\begin{aligned}\frac{\mathrm{d}}{\mathrm{d}\lambda} E(\lambda) &= \langle\,\psi(\lambda)\,|\,\frac{\mathrm{d}}{\mathrm{d}\lambda} H(\lambda)\,|\,\psi(\lambda)\,\rangle \\ &\quad + \left[\frac{\mathrm{d}}{\mathrm{d}\lambda}\langle\,\psi(\lambda)\,|\right] H(\lambda)\,|\,\psi(\lambda)\,\rangle + \langle\,\psi(\lambda)\,|\,H(\lambda)\left[\frac{\mathrm{d}}{\mathrm{d}\lambda}\,|\,\psi(\lambda)\,\rangle\right]\end{aligned} \tag{71}$$

that is, using (67) and the adjoint relation $[H(\lambda)$ is Hermitian, hence $E(\lambda)$ is real$]$:

$$\begin{aligned}\frac{\mathrm{d}}{\mathrm{d}\lambda} E(\lambda) &= \langle\,\psi(\lambda)\,|\,\frac{\mathrm{d}}{\mathrm{d}\lambda} H(\lambda)\,|\,\psi(\lambda)\,\rangle \\ &\quad + E(\lambda)\left\{\left[\frac{\mathrm{d}}{\mathrm{d}\lambda}\langle\,\psi(\lambda)\,|\right]|\,\psi(\lambda)\,\rangle + \langle\,\psi(\lambda)\,|\left[\frac{\mathrm{d}}{\mathrm{d}\lambda}\,|\,\psi(\lambda)\,\rangle\right]\right\}\end{aligned} \tag{72}$$

On the right-hand side, the expression inside curly brackets is the derivative of $\langle \psi(\lambda) \mid \psi(\lambda) \rangle$, which is zero since $\mid \psi(\lambda) \rangle$ is normalized; we therefore find (69).

γ. *Mean value of the commutator $[H, A]$ in an eigenstate of H*

Let $\mid \psi \rangle$ be a normalized eigenvector of the Hermitian operator H, of eigenvalue E. For any operator A:

$$\langle \psi \mid [H, A] \mid \psi \rangle = 0 \tag{73}$$

since, as $H \mid \psi \rangle = E \mid \psi \rangle$ and $\langle \psi \mid H = E \langle \psi \mid$:

$$\langle \psi \mid (HA - AH) \mid \psi \rangle = E \langle \psi \mid A \mid \psi \rangle - E \langle \psi \mid A \mid \psi \rangle = 0 \tag{74}$$

c. THE VIRIAL THEOREM APPLIED TO MOLECULES

α. *The potential energy of the system*

Consider an arbitrary molecule composed of N nuclei and Q electrons. We shall denote by $\mathbf{r}_k^n (k = 1, 2, ..., N)$ the classical positions of the nuclei, and by $\mathbf{r}_i^e$ and $\mathbf{p}_i^e$ $(i = 1, 2, ..., Q)$ the classical positions and momenta of the electrons. The components of these vectors will be written x_k^n, y_k^n, z_k^n, etc.

We shall use the Born-Oppenheimer approximation here, considering the $\mathbf{r}_k^n$ as given classical parameters. In the quantum mechanical calculation, only the $\mathbf{r}_i^e$ and $\mathbf{p}_i^e$ become operators, $\mathbf{R}_i^e$ and $\mathbf{P}_i^e$. We must therefore solve the eigenvalue equation:

$$H(\mathbf{r}_1^n, ..., \mathbf{r}_N^n) \mid \psi (\mathbf{r}_1^n, ..., \mathbf{r}_N^n) \rangle = E(\mathbf{r}_1^n, ..., \mathbf{r}_N^n) \mid \psi (\mathbf{r}_1^n, ..., \mathbf{r}_N^n) \rangle \tag{75}$$

of a Hamiltonian H which depends on the parameters $\mathbf{r}_1^n, ..., \mathbf{r}_N^n$ and which acts in the state space of the electrons. The expression for H can be written:

$$H = T_e + V(\mathbf{r}_1^n, ..., \mathbf{r}_N^n) \tag{76}$$

where T_e is the kinetic energy operator of the electrons:

$$T_e = \sum_{i=1}^{Q} \frac{1}{2m} (\mathbf{P}_i^e)^2 \tag{77}$$

and $V(\mathbf{r}_1^n, ..., \mathbf{r}_N^n)$ is the operator obtained by replacing the $\mathbf{r}_i^e$ by the operators $\mathbf{R}_i^e$ in the expression for the classical potential energy. The latter is the sum of the repulsion energy V_{ee} between the electrons, the attraction energy V_{en} between the electrons and the nuclei, and the repulsion energy V_{nn} between the nuclei, so that:

$$V(\mathbf{r}_1^n, ..., \mathbf{r}_N^n) = V_{ee} + V_{en}(\mathbf{r}_1^n, ..., \mathbf{r}_N^n) + V_{nn}(\mathbf{r}_1^n, ..., \mathbf{r}_N^n) \tag{78}$$

Actually, since V_{nn} depends only on the $\mathbf{r}_k^n$ and does not involve the $\mathbf{R}_i^e$, V_{nn} is a number and not an operator acting in the state space of the electrons. The only effect of V_{nn} is therefore to shift all the energies equally, since equation (75) is equivalent to:

$$H_e(\mathbf{r}_1^n, ..., \mathbf{r}_N^n) \mid \psi(\mathbf{r}_1^n, ..., \mathbf{r}_N^n) \rangle = E_e(\mathbf{r}_1^n, ..., \mathbf{r}_N^n) \mid \psi(\mathbf{r}_1^n, ..., \mathbf{r}_N^n) \rangle \tag{79}$$

where:

$$H_e(\mathbf{r}_1^n, ..., \mathbf{r}_N^n) = T_e + V_{ee} + V_{en}(\mathbf{r}_1^n, ..., \mathbf{r}_N^n) = H - V_{nn}(\mathbf{r}_1^n, ..., \mathbf{r}_N^n) \tag{80}$$

and where the electronic energy E_e is related to the total energy E by:

$$E_e(\mathbf{r}_1^n, ..., \mathbf{r}_N^n) = E(\mathbf{r}_1^n, ..., \mathbf{r}_N^n) - V_{nn}(\mathbf{r}_1^n, ..., \mathbf{r}_N^n) \tag{81}$$

We can apply Euler's theorem to the classical potential energy, since it is a homogeneous function of degree -1 of the *set* of electronic and nuclear coordinates. Since the operators $\mathbf{R}_i^e$ all commute with each other, we can find the relation between the quantum mechanical operators:

$$\sum_{k=1}^{N} \mathbf{r}_k^n . \boldsymbol{\nabla}_k^n V + \sum_{i=1}^{Q} \mathbf{R}_i^e . \boldsymbol{\nabla}_i^e V = - V \tag{82}$$

where $\boldsymbol{\nabla}_k^n$ and $\boldsymbol{\nabla}_i^e$ denote the operators obtained by substitution of the $\mathbf{R}_i^e$ for the $\mathbf{r}_i^e$ in the gradients with respect to $\mathbf{r}_k^n$ and $\mathbf{r}_i^e$ in the classical expression for the potential energy. Relation (82) will serve as the foundation of our proof of the virial theorem.

β. *Proof of the virial theorem*

We apply (73) to the special case in which:

$$A = \sum_{i=1}^{Q} \mathbf{R}_i^e . \mathbf{P}_i^e \tag{83}$$

To do so, we find the commutator of H with A:

$$\begin{aligned}\left[H, \sum_{i=1}^{Q} \mathbf{R}_i^e . \mathbf{P}_i^e\right] &= \sum_{i=1}^{Q} \sum_{x,y,z} \{ [H, X_i^e] P_{xi}^e + X_i^e [H, P_{xi}^e] \} \\ &= i\hbar \sum_{i=1}^{Q} \left\{ - \frac{(\mathbf{P}_i^e)^2}{m} + \mathbf{R}_i^e . \boldsymbol{\nabla}_i^e V \right\}\end{aligned} \tag{84}$$

(we have used the commutation relations of a function of the momentum with the position, or vice versa; *cf.* complement B_{II}, §4-c). The first term inside the curly brackets is proportional to the kinetic energy T_e. According to (82), the second term is equal to:

$$- V - \sum_{k=1}^{N} \mathbf{r}_k^n . \boldsymbol{\nabla}_k^n V \tag{85}$$

Consequently, relation (73) gives us:

$$2 \langle T_e \rangle + \langle V \rangle + \sum_{k=1}^{N} \mathbf{r}_k^n . \langle \boldsymbol{\nabla}_k^n V \rangle = 0 \tag{86}$$

that is, since the Hamiltonian H depends on the parameters $\mathbf{r}_k^n$ only through V:

$$2 \langle T_e \rangle + \langle V \rangle = - \sum_{k=1}^{N} \mathbf{r}_k^n . \langle \boldsymbol{\nabla}_k^n H \rangle \tag{87}$$

The components $\mathbf{r}_k^n$ here play a role analogous to that of the parameter λ in (69). Application of the Hellmann-Feynman theorem to the right-hand side of equation (87) then gives:

$$2\langle T_e \rangle + \langle V \rangle = -\sum_{k=1}^{N} \mathbf{r}_k^n \cdot \boldsymbol{\nabla}_k^n E(\mathbf{r}_1^n, ..., \mathbf{r}_k^n, ..., \mathbf{r}_N^n) \tag{88}$$

Furthermore, we obviously have:

$$\langle T_e \rangle + \langle V \rangle = E(\mathbf{r}_1^n, ..., \mathbf{r}_N^n) \tag{89}$$

We can then easily find from (88) and (89):

$$\boxed{\begin{aligned} \langle T_e \rangle &= -E - \sum_{k=1}^{N} \mathbf{r}_k^n \cdot \boldsymbol{\nabla}_k^n E \\ \langle V \rangle &= 2E + \sum_{k=1}^{N} \mathbf{r}_k^n \cdot \boldsymbol{\nabla}_k^n E \end{aligned}} \tag{90}$$

Thus, we obtain a very simple result: the virial theorem applied to molecules. It enables us to calculate the average kinetic and potential energies if we know the variation of the total energy with respect to the positions of the nuclei.

COMMENT:

The total electronic energy E_e and the electronic potential energy $\langle V_e \rangle$ are also related by:

$$\boxed{\langle V_e \rangle = 2E_e + \sum_{k=1}^{N} \mathbf{r}_k^n \cdot \boldsymbol{\nabla}_k^n E_e} \tag{91}$$

This relation can be proven by substituting (81) and the explicit expression for V_{nn} in terms of the $\mathbf{r}_k^n$ into the second relation of (90). However, it is simpler to note that the electronic potential energy $V_e = V_{ee} + V_{en}$, like the total potential energy V, is a homogeneous function of degree -1 of the coordinates of the system of particles. Consequently, the preceding arguments apply to H_e as well as to H, and we can simultaneously replace E by E_e and V by V_e in both relations (90).

γ. *A special case: the diatomic molecule*

When the number N of nuclei is equal to two, the energies depend only on the internuclear distance R. This further simplifies the expression for the virial theorem, which becomes:

$$\boxed{\begin{aligned} \langle T_e \rangle &= -E - R\frac{\mathrm{d}E}{\mathrm{d}R} \\ \langle V \rangle &= 2E + R\frac{\mathrm{d}E}{\mathrm{d}R} \end{aligned}} \tag{92}$$

Since E depends on the nuclear coordinates only through R, we have:

$$\frac{\partial E}{\partial x_k^n} = \frac{\mathrm{d}E}{\mathrm{d}R}\frac{\partial R}{\partial x_k^n} \tag{93}$$

and, consequently:

$$\sum_{k=1,2}\sum_{x,y,z} x_k^n \frac{\partial E}{\partial x_k^n} = \frac{\mathrm{d}E}{\mathrm{d}R}\sum_{k=1,2}\sum_{x,y,z} x_k^n \frac{\partial R}{\partial x_k^n} \tag{94}$$

Now, the distance R between the nuclei is a homogeneous function of degree 1 of the coordinates of the nuclei. Application of Euler's theorem to this function enables us to replace the double summation appearing on the right-hand side of (94) by R, and we finally obtain:

$$\sum_{k=1,2} \mathbf{r}_k^n . \boldsymbol{\nabla}_k^n E = R\frac{\mathrm{d}E}{\mathrm{d}R} \tag{95}$$

When this result is substituted into (90), it gives relations (92).

In (92) as in (90), we can replace E by E_e and V by V_e.

d. DISCUSSION

α. *The chemical bond is due to a lowering of the electronic potential energy*

Let E_∞ be the value of the total energy E of the system when the various nuclei are infinitely far apart. If it is possible to form a stable molecule by moving the nuclei closer together, there must exist a certain relative arrangement of these nuclei for which the total energy E passes through a minimum $E_0 < E_\infty$. For the corresponding values of $\mathbf{r}_k^n$, we then have:

$$\boldsymbol{\nabla}_k^n E = \mathbf{0} \tag{96}$$

Relations (90) then indicate that, for this equilibrium position, the kinetic and potential energies are equal to:

$$\begin{aligned} \langle T_e \rangle_0 &= -E_0 \\ \langle V \rangle_0 &= 2E_0 \end{aligned} \tag{97}$$

Furthermore, when the nuclei are infinitely far from each other, the system is composed of a certain number of atoms or ions without mutual interactions (the energy no longer depends on the $\mathbf{r}_k^n$). For each of these subsystems, the virial theorem indicates that $\langle T_e \rangle = -E$, $\langle V \rangle = 2E$, and, for the system as a whole, we must therefore also have:

$$\begin{aligned} \langle T_e \rangle_\infty &= -E_\infty \\ \langle V \rangle_\infty &= 2E_\infty \end{aligned} \tag{98}$$

Subtracting (98) from (97) then gives:

$$\begin{aligned} \langle T_e \rangle_0 - \langle T_e \rangle_\infty &= -(E_0 - E_\infty) > 0 \\ \langle V \rangle_0 - \langle V \rangle_\infty &= 2(E_0 - E_\infty) < 0 \end{aligned} \tag{99}$$

The formation of a stable molecule is therefore always accompanied by an increase in the kinetic energy of the electrons and a decrease in the total potential energy.

The electronic potential energy must, furthermore, decrease even more since the mean value $\langle V_{nn} \rangle$ (the repulsion between the nuclei), which is zero at infinity, is always positive. It is therefore a lowering of the potential energy of the electrons $\langle V_{ee} + V_{en} \rangle$ that is responsible for the chemical bond. At equilibrium, this lowering must outweigh the increase in $\langle T_e \rangle$ and $\langle V_{nn} \rangle$.

β. *The special case of the* H_2^+ *ion*

(*i*) Application of the virial theorem to the approximate variational energy.

We return to the study of the variation of $\langle T_e \rangle$ and $\langle V \rangle$ for the H_2^+ ion. We shall begin by examining the predictions of the variational model of § 2, which led to the approximate expressions (52) and (53). From the second of these relations, we deduce that:

$$\Delta T_e = \langle T_e \rangle - \langle T_e \rangle_\infty = \frac{1}{1 + S}(A - 2S\, E_I) \tag{100}$$

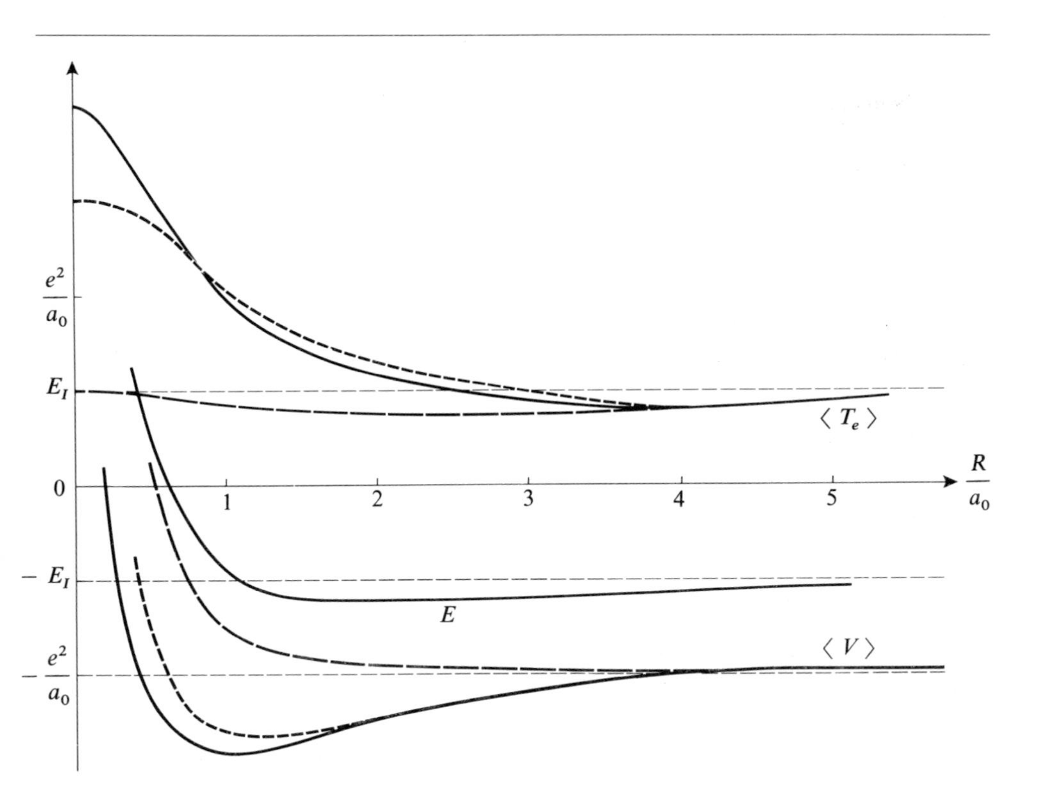

FIGURE 13

The electronic kinetic energy $\langle T_e \rangle$ and the potential energy $\langle V \rangle$ of the H_2^+ ion as functions of $\rho = R/a_0$ (for purposes of comparison, we have also shown the total energy $E = \langle T_e \rangle + \langle V \rangle$).
. solid lines: the exact values (the chemical bond is due to the fact that $\langle V \rangle$ decreases a little faster than $\langle T_e \rangle$ increases).
. long dashes: the mean values calculated from the bonding wave function given by the simple variational method of § 2.
. short dashes: the values obtained by the application of the virial theorem to the energy given by the same variational calculation.

Since S is always greater than $A/2E_I$ (*cf.* fig. 3), this calculation would tend to indicate that ΔT_e is always negative. This appears, moreover, in figure 13, where the dashed lines represent the variations of the approximate expressions (52) and (53). In particular, we see that, according to the variational calculation, ΔT_e is negative at equilibrium ($\rho \simeq 2.5$) and ΔV is positive. These results are both incorrect, according to (99). We see here the limits of a variational calculation, which gives an acceptable value for the total energy $\langle T_e + V \rangle$, but not for $\langle T_e \rangle$ and $\langle V \rangle$ separately. The mean values of the latter depend too strongly on the wave function.

The virial theorem enables us, without having to resort to the rigorous calculation mentioned in §1-c, to obtain a much better approximation for $\langle T_e \rangle$ and $\langle V \rangle$. All we need to do is apply the exact relations (92) to the energy E calculated by the variational method. We should expect to obtain an acceptable result, since the variational approximation is now used only to supply the total energy E. The values thus obtained for $\langle T_e \rangle$ and $\langle V \rangle$ are represented by short dashed lines in figure 13. For purposes of comparison, we have shown in solid lines the exact values of $\langle T_e \rangle$ and $\langle V \rangle$ (obtained by application of the virial theorem to the solid-line curve of figure 2). First of all, we see that for $\rho = 2.5$, the curve in short dashed lines indicates, as we should expect, that ΔT_e is positive and ΔV is negative. In addition, the general shape of these curves reproduces rather well that of the solid-line curves. As long as $\rho \gtrsim 1.5$, the virial theorem applied to the variational energy does give values which are very close to reality. This represents a considerable improvement over the direct calculation of the mean values in the approximate states.

(*ii*) Behavior of $\langle T \rangle$ and $\langle V \rangle$

The solid-line curves of figure 13 (the exact curves) show that $\langle T_e \rangle \longrightarrow 4E_I$ and $\langle V \rangle \longrightarrow +\infty$ when $R \longrightarrow 0$. Indeed, when $R = 0$, we have the equivalent of a He^+ ion for which the electronic kinetic energy is $4E_I$. The divergence of $\langle V \rangle$ is due to the term $\langle V_{nn} \rangle = e^2/R$, which becomes infinite when $R \longrightarrow 0$ (the electronic potential energy $\langle V_e \rangle = \langle V \rangle - e^2/R$ remains finite and approaches $-8E_I$, which is indeed its value in the He^+ ion).

The behavior for large R deserves a more detailed discussion. We have seen above (§ 3-b) that the energy E_- of the ground state behaves, for $R \gg a_0$, like:

$$E_- \simeq -E_I - \frac{a}{R^4} \tag{101}$$

where a is a constant which is proportional to the polarizability of the hydrogen atom. By substituting this result into formulas (92), we obtain:

$$\begin{aligned}\langle T_e \rangle &\simeq E_I - \frac{3a}{R^4} \\ \langle V \rangle &\simeq -2E_I + \frac{2a}{R^4}\end{aligned} \tag{102}$$

When R decreases from a very large value, $\langle T_e \rangle$ begins by decreasing with $1/R^4$ from its asymptotic value E_I, and $\langle V \rangle$ begins by increasing from $-2E_I$. These variations then change sign (this must be so since $\langle T_e \rangle_0$ is larger than $\langle T_e \rangle_\infty$ and $\langle V \rangle_0$ is smaller than $\langle V \rangle_\infty$): as R continues to decrease (*cf.* fig. 13), $\langle T_e \rangle$ passes through a minimum and then increases until it reaches its value $4E_I$ for $R = 0$. As for the potential energy $\langle V \rangle$, it passes through a maximum, then decreases, passes through a minimum, and then approaches infinity when $R \longrightarrow 0$. How can we interpret these variations?

As we have noted several times, the non-diagonal elements H_{12} and H_{21} of determinant (21) approach zero exponentially when $R \longrightarrow \infty$. We can therefore argue only in terms of H_{11} or H_{22} in discussing the variation of the energy of the H_2^+ ion at large internuclear distances.

The problem is then reduced to the study of the perturbation of a hydrogen atom centered at P_2 by the electric field of the proton P_1. This field tends to distort the electronic orbital by stretching it in the P_1 direction (*cf.* fig. 6). Consequently, the wave function extends into a larger volume. According to Heisenberg's uncertainty relations, this allows the kinetic energy to decrease; this can explain the behavior of $\langle T_e \rangle$ for large R.

Arguing in terms of H_{22}, we can also explain the asymptotic behavior of $\langle V \rangle$. The discussion of §3-b showed that, for $R \gg a_0$, the polarization of the hydrogen atom situated at P_2 makes its interaction energy $\langle -\frac{e^2}{r_1} + \frac{e^2}{R} \rangle$ with P_1 slightly negative (proportional to $-1/R^4$). If $\langle V \rangle$ is positive, it is because the potential energy $\langle -\frac{e^2}{r_2} \rangle$ of the atom at P_2 increases more rapidly, when P_1 is brought closer to P_2, than $\langle -\frac{e^2}{r_1} + \frac{e^2}{R} \rangle$ decreases. This increase in $\langle -\frac{e^2}{r_2} \rangle$ is due to the fact that the attraction of P_1 moves the electron slightly away from P_2 and carries it into regions of space in which the potential created by P_2 is less negative.

For $R \simeq R_0$ (the equilibrium position of the H_2^+ ion), the wave function of the bonding state is highly localized in the region between the two protons. The decrease in $\langle V \rangle$ (despite the increase in e^2/R) is due to the fact that the electron is in a region of space in which it benefits simultaneously from the attraction of both protons. This lowers its potential energy (*cf.* fig. 14). This combined attraction of the two protons also leads to a decrease in the spatial extension of the electronic wave function, which is concentrated in the intermediate region. This is why, for R close to R_0, $\langle T_e \rangle$ increases when R decreases.

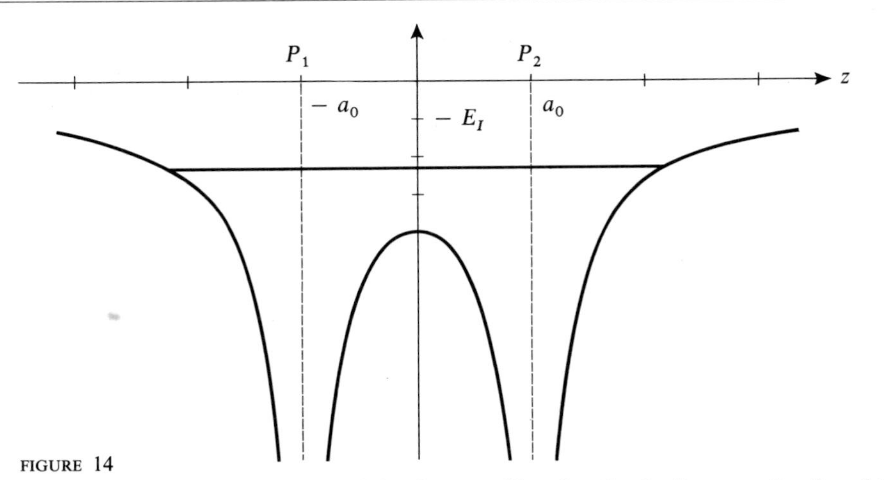

FIGURE 14

Variation of the potential energy V_e of the electron subjected to the simultaneous attraction of the two protons P_1 and P_2 as one moves along the line P_1P_2. In the bonding state, the wave function is concentrated in the region between P_1 and P_2, and the electron benefits simultaneously from the attraction of both protons.

References and suggestions for further reading (H_2^+ ion, H_2 molecule, nature of the chemical bond, etc.) :

Pauling (12.2); Pauling and Wilson (1.9), chaps. XII and XIII; Levine (12.3), chaps. 13 and 14; Karplus and Porter (12.1), chap. 5, § 6; Slater (1.6), chaps. 8 and 9; Eyring et al (12.5), chaps. XI and XII; Coulson (12.6), chap. IV; Wahl (12.13).

Complement H_{XI}

EXERCISES

1. A particle of mass m is placed in an infinite one-dimensional well of width a:

$$V(x) = 0 \qquad \text{for} \quad 0 \leqslant x \leqslant a$$
$$V(x) = +\infty \qquad \text{everywhere else}$$

It is subject to a perturbation W of the form:

$$W(x) = a w_0 \, \delta\left(x - \frac{a}{2}\right)$$

where w_0 is a real constant with the dimensions of an energy.

a. Calculate, to first order in w_0, the modifications induced by $W(x)$ in the energy levels of the particle.

b. Actually, the problem is exactly soluble. Setting $k = \sqrt{2mE/\hbar^2}$, show that the possible values of the energy are given by one of the two equations $\sin(ka/2) = 0$ or $\tan(ka/2) = -\hbar^2 k/maw_0$ (as in exercise 2 of complement L_I, watch out for the discontinuity of the derivative of the wave function at $x = a/2$).

Discuss the results obtained with respect to the sign and size of w_0. In the limit $w_0 \longrightarrow 0$, show that one obtains the results of the preceding question.

2. Consider a particle of mass m placed in an infinite two-dimensional potential well of width a (*cf.* complement G_{II}):

$$V(x, y) = 0 \qquad \text{if} \quad 0 \leqslant x \leqslant a \quad \text{and} \quad 0 \leqslant y \leqslant a$$
$$V(x, y) = +\infty \qquad \text{everywhere else}$$

This particle is also subject to a perturbation W described by the potential:

$$W(x, y) = w_0 \quad \text{for} \quad 0 \leqslant x \leqslant \frac{a}{2} \quad \text{and} \quad 0 \leqslant y \leqslant \frac{a}{2}$$

$$W(x, y) = 0 \qquad \text{everywhere else}$$

a. Calculate, to first order in w_0, the perturbed energy of the ground state.

b. Same question for the first excited state. Give the corresponding wave functions to zeroeth order in w_0.

3. A particle of mass m, constrained to move in the xOy plane, has a Hamiltonian:

$$H_0 = \frac{P_x^2}{2m} + \frac{P_y^2}{2m} + \frac{1}{2} m\omega^2 (X^2 + Y^2)$$

(a two-dimensional harmonic oscillator, of angular frequency ω). We want to study the effect on this particle of a perturbation W given by:

$$W = \lambda_1 W_1 + \lambda_2 W_2$$

where λ_1 and λ_2 are constants, and the expressions for W_1 and W_2 are:

$$W_1 = m\omega^2 XY$$

$$W_2 = \hbar\omega\left(\frac{L_z^2}{\hbar^2} - 2\right)$$

(L_z is the component along Oz of the orbital angular momentum of the particle).

In the perturbation calculations, consider only the corrections to first order for the energies and to zeroeth order for the state vectors.

a. Indicate without calculations the eigenvalues of H_0, their degrees of degeneracy and the associated eigenvectors.

In what follows, consider only the second excited state of H_0, of energy $3\hbar\omega$ and which is three-fold degenerate.

b. Calculate the matrices representing the restrictions of W_1 and W_2 to the eigensubspace of the eigenvalue $3\hbar\omega$ of H_0.

c. Assume $\lambda_2 = 0$ and $\lambda_1 \ll 1$.

Calculate, using perturbation theory, the effect of the term $\lambda_1 W_1$ on the second excited state of H_0.

d. Compare the results obtained in *c* with the limited expansion of the exact solution, to be found with the help of the methods described in complement H_V (normal vibrational modes of two coupled harmonic oscillators).

e. Assume $\lambda_2 \ll \lambda_1 \ll 1$. Considering the results of question *c* to be a new unperturbed situation, calculate the effect of the term $\lambda_2 W_2$.

f. Now assume that $\lambda_1 = 0$ and $\lambda_2 \ll 1$.

Using perturbation theory, find the effect of the term $\lambda_2 W_2$ on the second excited state of H_0.

g. Compare the results obtained in *f* with the exact solution, which can be found from the discussions of complement D_{VI}.

h. Finally, assume that $\lambda_1 \ll \lambda_2 \ll 1$. Considering the results of question *f* to be a new unperturbed situation, calculate the effect of the term $\lambda_1 W_1$.

4. Consider a particle P of mass μ, constrained to move in the xOy plane in a circle centered at O with fixed radius ρ (a two-dimensional rotator). The only variable of the system is the angle $\alpha = (Ox, OP)$, and the quantum state of the particle is defined by the wave function $\psi(\alpha)$ (which represents the probability amplitude of finding the particle at the point of the circle fixed by the angle α). At each point of the circle, $\psi(\alpha)$ can take on only one value, so that:

$$\psi(\alpha + 2\pi) = \psi(\alpha)$$

$\psi(\alpha)$ is normalized if:

$$\int_0^{2\pi} |\psi(\alpha)|^2 \, d\alpha = 1$$

a. Consider the operator $M = \frac{\hbar}{i}\frac{d}{d\alpha}$. Is M Hermitian? Calculate the eigenvalues and normalized eigenfunctions of M. What is the physical meaning of M?

b. The kinetic energy of the particle can be written:

$$H_0 = \frac{M^2}{2\mu\rho^2}$$

Calculate the eigenvalues and eigenfunctions of H_0. Are the energies degenerate?

c. At $t = 0$, the wave function of the particle is $N \cos^2 \alpha$ (where N is a normalization coefficient). Discuss the localization of the particle on the circle at a subsequent time t.

d. Assume that the particle has a charge q and that it interacts with a uniform electric field $\mathscr{E}$ parallel to Ox. We must therefore add to the Hamiltonian H_0 the perturbation:

$$W = - q\mathscr{E} \rho \cos \alpha$$

Calculate the new wave function of the ground state to first order in $\mathscr{E}$. Determine the proportionality coefficient χ (the linear suceptibility) between the electric dipole parallel to Ox acquired by the particle and the field $\mathscr{E}$.

e. Consider, for the ethane molecule CH_3-CH_3, a rotation of one CH_3 group relative to the other about the straight line joining the two carbon atoms.

To a first approximation, this rotation is free, and the Hamiltonian H_0 introduced in *b* describes the rotational kinetic energy of one of the CH_3 groups relative to the other ($2\mu\rho^2$ must, however, be replaced by λI, where I is the moment of inertia of the CH_3 group with respect to the rotational axis and λ is a constant). To take account of the electrostatic interaction energy between the two CH_3 groups, we add to H_0 a term of the form:

$$W = b \cos 3\alpha$$

where b is a real constant.

Give a physical justification for the α-dependence of W. Calculate the energy and wave function of the new ground state (to first order in b for the wave function and to second order for the energy). Give a physical interpretation of the result.

5. Consider a system of angular momentum $\mathbf{J}$. We confine ourselves in this exercise to a three-dimensional subspace, spanned by the three kets $| +1 \rangle$, $| 0 \rangle$, $| -1 \rangle$, common eigenstates of $\mathbf{J}^2$ (eigenvalue $2\hbar^2$) and J_z (eigenvalues $+\hbar, 0, -\hbar$). The Hamiltonian H_0 of the system is:

$$H_0 = aJ_z + \frac{b}{\hbar} J_z^2$$

where a and b are two positive constants, which have the dimensions of an angular frequency.

a. What are the energy levels of the system? For what value of the ratio b/a is there degeneracy?

b. A static field $\mathbf{B}_0$ is applied in a direction $\mathbf{u}$ with polar angles θ and φ. The interaction with $\mathbf{B}_0$ of the magnetic moment of the system:

$$\mathbf{M} = \gamma \mathbf{J}$$

(γ : the gyromagnetic ratio, assumed to be negative) is described by the Hamiltonian:

$$W = \omega_0 J_u$$

where $\omega_0 = -\gamma \, |\mathbf{B}_0|$ is the Larmor angular frequency in the field $\mathbf{B}_0$, and J_u is the component of $\mathbf{J}$ in the $\mathbf{u}$ direction:

$$J_u = J_z \cos\theta + J_x \sin\theta \cos\varphi + J_y \sin\theta \sin\varphi$$

Write the matrix which represents W in the basis of the three eigenstates of H_0.

c. Assume that $b = a$ and that the $\mathbf{u}$ direction is parallel to Ox. We also have $\omega_0 \ll a$.

Calculate the energies and eigenstates of the system, to first order in ω_0 for the energies and to zeroeth order for the eigenstates.

d. Assume that $b = 2a$ and that we again have $\omega_0 \ll a$, the direction of $\mathbf{u}$ now being arbitrary.

In the $\{ | +1 \rangle, | 0 \rangle, | -1 \rangle \}$ basis, what is the expansion of the ground state $| \psi_0 \rangle$ of $H_0 + W$, to first order in ω_0?

Calculate the mean value $\langle \mathbf{M} \rangle$ of the magnetic moment $\mathbf{M}$ of the system in the state $| \psi_0 \rangle$. Are $\langle \mathbf{M} \rangle$ and $\mathbf{B}_0$ parallel?

Show that one can write:

$$\langle M_i \rangle = \sum_j \chi_{ij} B_j$$

with $i, j = x, y, z$. Calculate the coefficients χ_{ij} (the components of the susceptibility tensor).

6. Consider a system formed by an electron spin $\mathbf{S}$ and two nuclear spins $\mathbf{I}_1$ and $\mathbf{I}_2$ ($\mathbf{S}$ is, for example, the spin of the unpaired electron of a paramagnetic diatomic molecule, and $\mathbf{I}_1$ and $\mathbf{I}_2$ are the spins of the two nuclei of this molecule).

Assume that $\mathbf{S}, \mathbf{I}_1, \mathbf{I}_2$ are all spin 1/2's. The state space of the three-spin system is spanned by the eight orthonormal kets $| \varepsilon_S, \varepsilon_1, \varepsilon_2 \rangle$, common eigenvectors of S_z, I_{1z}, I_{2z}, with respective eigenvalues $\varepsilon_S \hbar/2$, $\varepsilon_1 \hbar/2$, $\varepsilon_2 \hbar/2$ (with $\varepsilon_S = \pm, \varepsilon_1 = \pm, \varepsilon_2 = \pm$). For example, the ket $| +, -, + \rangle$ corresponds to the eigenvalues $+ \hbar/2$ for S_z, $- \hbar/2$ for I_{1z}, and $+ \hbar/2$ for I_{2z}.

a. We begin by neglecting any coupling of the three spins. We assume, however, that they are placed in a uniform magnetic field $\mathbf{B}$ parallel to Oz. Since the gyromagnetic ratios of $\mathbf{I}_1$ and $\mathbf{I}_2$ are equal, the Hamiltonian H_0 of the system can be written:

$$H_0 = \Omega S_z + \omega I_{1z} + \omega I_{2z}$$

where Ω and ω are real, positive constants, proportional to $|\mathbf{B}|$. Assume $\Omega > 2\omega$.

What are the possible energies of the three-spin system and their degrees of degeneracy? Draw the energy diagram.

b. We now take coupling of the spins into account by adding the Hamiltonian:

$$W = a\,\mathbf{S} \cdot \mathbf{I}_1 + a\,\mathbf{S} \cdot \mathbf{I}_2$$

where a is a real, positive constant (the direct coupling of $\mathbf{I}_1$ and $\mathbf{I}_2$ is negligible).

What conditions must be satisfied by $\varepsilon_S, \varepsilon_1, \varepsilon_2, \varepsilon_S', \varepsilon_1', \varepsilon_2'$ for $a\mathbf{S} \cdot \mathbf{I}_1$ to have a non-zero matrix element between $| \varepsilon_S, \varepsilon_1, \varepsilon_2 \rangle$ and $| \varepsilon_S', \varepsilon_1', \varepsilon_2' \rangle$? Same question for $a\,\mathbf{S} \cdot \mathbf{I}_2$.

c. Assume that:

$$a\hbar^2 \ll \hbar\Omega, \hbar\omega$$

so that W can be treated like a perturbation with respect to H_0. To first order in W, what are the eigenvalues of the total Hamiltonian $H = H_0 + W$? To zeroeth order in W, what are the eigenstates of H? Draw the energy diagram.

d. Using the approximation of the preceding question, determine the Bohr frequencies which can appear in the evolution of $\langle S_x \rangle$ when the coupling W of the spins is taken into account.

In an E.P.R. (Electronic Paramagnetic Resonance) experiment, the frequencies of the resonance lines observed are equal to the preceding Bohr frequencies. What is the shape of the E.P.R. spectrum observed for the three-spin system? How can the coupling constant a be determined from this spectrum?

e. Now assume that the magnetic field $\mathbf{B}$ is zero, so that $\Omega = \omega = 0$. The Hamiltonian then reduces to W.

α. Let $\mathbf{I} = \mathbf{I}_1 + \mathbf{I}_2$ be the total nuclear spin. What are the eigenvalues of $\mathbf{I}^2$ and their degrees of degeneracy? Show that W has no matrix elements between eigenstates of $\mathbf{I}^2$ of different eigenvalues.

β. Let $\mathbf{J} = \mathbf{S} + \mathbf{I}$ be the total spin. What are the eigenvalues of $\mathbf{J}^2$ and their degrees of degeneracy? Determine the energy eigenvalues of the three-spin system and their degrees of degeneracy. Does the set $\{ \mathbf{J}^2, J_z \}$ form a C.S.C.O.? Same question for $\{ \mathbf{I}^2, \mathbf{J}^2, J_z \}$.

7. Consider a nucleus of spin $I = 3/2$, whose state space is spanned by the four vectors $| m \rangle$ $(m = +3/2, +1/2, -1/2, -3/2)$, common eigenvectors of $\mathbf{I}^2$ (eigenvalue $15\hbar^2/4$) and I_z (eigenvalue $m\hbar$).

This nucleus is placed at the coordinate origin in a non-uniform electric field derived from a potential $U(x, y, z)$. The directions of the axes are chosen such that, at the origin:

$$\frac{\partial^2 U}{\partial x\, \partial y} = \frac{\partial^2 U}{\partial y\, \partial z} = \frac{\partial^2 U}{\partial z\, \partial x} = 0$$

Recall that U satisfies Laplace's equation:

$$\Delta U = 0$$

We shall assume that the interaction Hamiltonian between the electric field gradient at the origin and the electric quadrupole moment of the nucleus can be written:

$$H_0 = \frac{qQ}{2I(2I - 1)} \frac{1}{\hbar^2} \left[a_x I_x^2 + a_y I_y^2 + a_z I_z^2 \right]$$

where q is the electron charge, Q is a constant with the dimensions of a surface and proportional to the quadrupole moment of the nucleus, and:

$$a_x = \left(\frac{\partial^2 U}{\partial x^2} \right)_0 \quad ; \quad a_y = \left(\frac{\partial^2 U}{\partial y^2} \right)_0 \quad ; \quad a_z = \left(\frac{\partial^2 U}{\partial z^2} \right)_0$$

(the index 0 indicates that the derivatives are evaluated at the origin).

a. Show that, if U is symmetrical with respect to revolution about Oz, H_0 has the form:

$$H_0 = A[3I_z^2 - I(I+1)]$$

where A is a constant to be specified. What are the eigenvalues of H_0, their degrees of degeneracy and the corresponding eigenstates?

b. Show that, in the general case, H_0 can be written:

$$H_0 = A[3I_z^2 - I(I+1)] + B(I_+^2 + I_-^2)$$

where A and B are constants, to be expressed in terms of a_x and a_y.

What is the matrix which represents H_0 in the $\{ | m \rangle \}$ basis? Show that it can be broken down into two 2×2 submatrices. Determine the eigenvalues of H_0 and their degrees of degeneracy, as well as the corresponding eigenstates.

c. In addition to its quadrupole moment, the nucleus has a magnetic moment $\mathbf{M} = \gamma \mathbf{I}$ (γ: the gyromagnetic ratio). Onto the electrostatic field is superposed a magnetic field $\mathbf{B}_0$, of arbitrary direction $\mathbf{u}$. We set $\omega_0 = -\gamma |\mathbf{B}_0|$.

What term W must be added to H_0 in order to take into account the coupling between $\mathbf{M}$ and $\mathbf{B}_0$? Calculate the energies of the system to first order in B_0.

d. Assume $\mathbf{B}_0$ to be parallel to Oz and weak enough for the eigenstates found in *b* and the energies to first order in ω_0 found in *c* to be good approximations.

What are the Bohr frequencies which can appear in the evolution of $\langle I_x \rangle$? Deduce from them the shape of the nuclear magnetic resonance spectrum which can be observed with a radiofrequency field oscillating along Ox.

8. A particle of mass m is placed in an infinite one-dimensional potential well of width a:

$$\begin{cases} V(x) = 0 & \text{for } 0 \leqslant x \leqslant a \\ V(x) = +\infty & \text{elsewhere} \end{cases}$$

Assume that this particle, of charge $-q$, is subject to a uniform electric field $\mathscr{E}$, with the corresponding perturbation W being:

$$W = q\mathscr{E}\left(X - \frac{a}{2}\right)$$

a. Let ε_1 and ε_2 be the corrections to first- and second-order in $\mathscr{E}$ for the ground state energy.

Show that ε_1 is zero. Give the expression for ε_2 in the form of a series, whose terms are to be calculated in terms of q, $\mathscr{E}$, m, a, $\hbar$ (the integrals given at the end of the exercise can be used).

b. By finding upper bounds for the terms of the series for ε_2, give an upper bound for ε_2 (*cf.* §B-2-c of chapter XI). Similarly, give a lower bound ε_2, obtained by retaining only the principal term of the series.

With what accuracy do the two preceding bounds enable us to bracket the exact value of the shift ΔE in the ground state to second order in $\mathscr{E}$?

c. We now want to calculate the shift ΔE by using the variational method. Choose as a trial function:

$$\psi_\alpha(x) = \sqrt{\frac{2}{a}} \sin\left(\frac{\pi x}{a}\right)\left[1 + \alpha q \mathscr{E}\left(x - \frac{a}{2}\right)\right]$$

where α is the variational parameter. Explain this choice of trial functions.

Calculate the mean energy $\langle H \rangle(\alpha)$ of the ground state to second order in $\mathscr{E}$ [assuming the expansion of $\langle H \rangle(\alpha)$ to second order in $\mathscr{E}$ to be sufficient]. Determine the optimal value of α. Find the result ΔE_{var} given by the variational method for the shift in the ground state to second order in $\mathscr{E}$.

By comparing ΔE_{var} with the results of *b*, evaluate the accuracy of the variational method applied to this example.

We give the integrals:

$$\frac{2}{a}\int_0^a \left(x - \frac{a}{2}\right) \sin\left(\frac{\pi x}{a}\right) \sin\left(\frac{2n\pi x}{a}\right) \mathrm{d}x = -\frac{16na}{\pi^2}\frac{1}{(1 - 4n^2)^2}$$

$n = 1, 2, 3, \ldots$

$$\frac{2}{a}\int_0^a \left(x - \frac{a}{2}\right)^2 \sin^2\left(\frac{\pi x}{a}\right) \mathrm{d}x = \frac{a^2}{2}\left(\frac{1}{6} - \frac{1}{\pi^2}\right)$$

$$\frac{2}{a}\int_0^a \left(x - \frac{a}{2}\right) \sin\left(\frac{\pi x}{a}\right) \cos\left(\frac{\pi x}{a}\right) \mathrm{d}x = -\frac{a}{2\pi}$$

For all the numerical calculations, take $\pi^2 = 9.87$.

9. We want to calculate the ground state energy of the hydrogen atom by the variational method, choosing as trial functions the spherically symmetrical functions $\varphi_\alpha(\mathbf{r})$ whose r-dependence is given by:

$$\begin{cases} \varphi_\alpha(r) = C\left(1 - \dfrac{r}{\alpha}\right) & \text{for } r \leqslant \alpha \\ \varphi_\alpha(r) = 0 & \text{for } r > \alpha \end{cases}$$

C is a normalization constant and α is the variational parameter.

a. Calculate the mean value of the kinetic and potential energies of the electron in the state $|\varphi_\alpha\rangle$. Express the mean value of the kinetic energy in terms

of $\nabla\varphi$, so as to avoid the "delta functions" which appear in $\Delta\varphi$ (since $\nabla\varphi$ is discontinuous).

b. Find the optimal value α_0 of α. Compare α_0 with the Bohr radius a_0.

c. Compare the approximate value obtained for the ground state energy with the exact value $-E_I$.

10. We intend to apply the variational method to the determination of the energies of a particle of mass m in an infinite potential well:

$$V(x) = 0 \qquad -a \leqslant x \leqslant a$$

$$V(x) = \infty \qquad \text{everywhere else}$$

a. We begin by approximating, in the interval $[-a, +a]$, the wave function of the ground state by the simplest even polynomial which goes to zero at $x = \pm a$:

$$\psi(x) = a^2 - x^2 \qquad \text{for} \quad -a \leqslant x \leqslant a$$

$$\psi(x) = 0 \qquad \text{everywhere else}$$

(a variational family reduced to a single trial function).

Calculate the mean value of the Hamiltonian H in this state. Compare the result obtained with the true value.

b. Enlarge the family of trial functions by choosing an even fourth-degree polynomial which goes to zero at $x = \pm a$:

$$\psi_\alpha(x) = (a^2 - x^2)(a^2 - \alpha x^2) \qquad \text{for} \quad -a \leqslant x \leqslant a$$

$$\psi_\alpha(x) = 0 \qquad \text{everywhere else}$$

(a variational family depending on the real parameter α).

(α) Show that the mean value of H in the state $\psi_\alpha(x)$ is:

$$\langle H \rangle(\alpha) = \frac{\hbar^2}{2ma^2} \frac{33\alpha^2 - 42\alpha + 105}{2\alpha^2 - 12\alpha + 42}$$

(β) Show that the values of α which minimize or maximize $\langle H \rangle(\alpha)$ are given by the roots of the equation:

$$13\alpha^2 - 98\alpha + 21 = 0$$

(γ) Show that one of the roots of this equation gives, when substituted into $\langle H \rangle(\alpha)$, a value of the ground state energy which is much more precise than the one obtained in *a*.

(δ) What other eigenvalue is approximated when the second root of the equation obtained in *b*-β is used? Could this have been expected? Evaluate the precision of this determination.

c. Explain why the simplest polynomial which permits the approximation of the first excited state wave function is $x(a^2 - x^2)$.

What approximate value is then obtained for the energy of this state?

CHAPTER XII

An application of perturbation theory: the fine and hyperfine structure of the hydrogen atom

OUTLINE OF CHAPTER XII

E. *(continued)*

A. INTRODUCTION

The most important forces inside atoms are Coulomb electrostatic forces. We took them into account in chapter VII by choosing as the hydrogen atom Hamiltonian:

$$H_0 = \frac{\mathbf{P}^2}{2\mu} + V(R) \qquad \text{(A-1)}$$

The first term represents the kinetic energy of the atom in the center of mass frame (μ is the reduced mass). The second term:

$$V(R) = -\frac{q^2}{4\pi\varepsilon_0}\frac{1}{R} = -\frac{e^2}{R} \qquad \text{(A-2)}$$

represents the electrostatic interaction energy between the electron and the proton (q is the electron charge). In §C of chapter VII, we calculated in detail the eigenstates and eigenvalues of H_0.

Actually, expression (A-1) is only approximate: it does not take any relativistic effects into account. In particular, all the magnetic effects related to the electron spin are ignored. Moreover, we have not introduced the proton spin and the corresponding magnetic interactions. The error is, in reality, very small, since the hydrogen atom is a weakly relativistic system (recall that, in the Bohr model, the velocity v in the first orbit $n = 1$ satisfies $v/c = e^2/\hbar c = 1/137 \ll 1$). In addition, the magnetic moment of the proton is very small.

However, the considerable accuracy of spectroscopic experiments makes it possible to observe effects that cannot be explained in terms of the Hamiltonian (A-1). Therefore, we shall take into account the corrections we have just mentioned by writing the complete hydrogen atom Hamiltonian in the form:

$$H = H_0 + W \qquad \text{(A-3)}$$

where H_0 is given by (A-1) and where W represents all the terms neglected thus far. Since W is much smaller than H_0, it is possible to calculate its effects by using the perturbation theory presented in chapter XI. This is what we propose to do in this chapter. We shall show that W is responsible for a "fine structure", as well as for a "hyperfine structure" of the various energy levels calculated in chapter VII. Furthermore, these structures can be measured experimentally with very great accuracy (the hyperfine structure of the $1s$ ground state of the hydrogen atom is the physical quantity currently known to the largest number of significant figures). We shall also consider, in this chapter and its complements, the influence of an external static magnetic or electric field on the various levels of the hydrogen atom (the Zeeman effect and the Stark effect).

This chapter actually has two goals. On the one hand, we want to use a concrete and realistic case to illustrate the general stationary perturbation theory discussed in the preceding chapter. On the other hand, this study, which bears on one of the most fundamental systems of physics (the hydrogen atom), brings out certain concepts which are basic to atomic physics. For example, §B is devoted to a thorough discussion of various relativistic and magnetic corrections. This chapter, while not indispensable for the study of the last two chapters, presents concepts fundamental to atomic physics.

B. ADDITIONAL TERMS IN THE HAMILTONIAN

The first problem to be solved obviously consists of finding the expression for W.

1. The fine-structure Hamiltonian

a. THE DIRAC EQUATION IN THE WEAKLY RELATIVISTIC DOMAIN

In chapter IX, we mentioned that the spin appears naturally when we try to establish an equation for the electron which satisfies both the postulates of special relativity and those of quantum mechanics. Such an equation exists : it is the *Dirac equation*, which makes it possible to account for numerous phenomena (electron spin, the fine structure of hydrogen, etc.) and to predict the existence of positrons.

The most rigorous way of obtaining the expression for the relativistic corrections [appearing in the term W of (A-3)] therefore consists of first writing the Dirac equation for an electron placed in the potential $V(r)$ created by the proton (considered to be infinitely heavy and motionless at the coordinate origin). One then looks for its limiting form when the system is weakly relativistic, as is the case for the hydrogen atom. We then recognize that the description of the electron state must include a two-component spinor (*cf.* chap. IX, §C-1). The spin operators S_x, S_y, S_z, introduced in chapter IX then appear naturally. Finally, we obtain an expression such as (A-3) for the Hamiltonian H, in which W appears in the form of a power series expansion in v/c which we can evaluate.

It is out of the question here to study the Dirac equation, or to establish its form in the weakly relativistic domain. We shall confine ourselves to giving the first terms of the power series expansion in v/c of W and their interpretation.

$$H = \underbrace{m_e c^2 + \frac{\mathbf{P}^2}{2m_e} + V(R)}_{H_0} \underbrace{- \frac{\mathbf{P}^4}{8m_e^3c^2}}_{W_{mv}} \underbrace{+ \frac{1}{2m_e^2c^2}\frac{1}{R}\frac{\mathrm{d}V(R)}{\mathrm{d}R}\mathbf{L}\cdot\mathbf{S}}_{W_{SO}} \underbrace{+ \frac{\hbar^2}{8m_e^2c^2}\Delta V(R)}_{W_D} + \ldots \tag{B-1}$$

We recognize in (B-1) the rest-mass energy $m_e c^2$ of the electron (the first term) and the non-relativistic Hamiltonian H_0 (the second and third terms)★. The following terms are called fine structure terms.

COMMENT:

Note that it is possible to solve the Dirac equation exactly for an electron placed in a Coulomb potential. We thus obtain the energy levels of the hydrogen atom without having to make a limited power series expansion in v/c of the eigenstates and eigenvalues of H. The "perturbation" point of view which we are adopting here is, however, very useful in bringing out the form and physical meaning of the various interactions which exist inside an atom. This will later permit a generalization to the case of many-electron atoms (for which we do not know how to write the equivalent of the Dirac equation).

b. INTERPRETATION OF THE VARIOUS TERMS OF THE FINE-STRUCTURE HAMILTONIAN

α. *Variation of the mass with the velocity (W_{mv} term)*

(i) The physical origin

The physical origin of the W_{mv} term is very simple. If we start with the relativistic expression for the energy of a classical particle of rest-mass m_e and momentum $\mathbf{p}$:

$$E = c\sqrt{\mathbf{p}^2 + m_e^2 c^2} \tag{B-2}$$

and perform a limited expansion of E in powers of $|\mathbf{p}|/m_e c$, we obtain:

$$E = m_e c^2 + \frac{\mathbf{p}^2}{2m_e} - \frac{\mathbf{p}^4}{8m_e^3 c^2} + \dots \tag{B-3}$$

In addition to the rest-mass energy ($m_e c^2$) and the non-relativistic kinetic energy ($\mathbf{p}^2/2m_e$), we find the term $-\mathbf{p}^4/8m_e^3c^2$, which appears in (B-1). This term represents the first energy correction, due to the relativistic variation of the mass with the velocity.

(ii) Order of magnitude

To evaluate the size of this correction, we shall calculate the order of magnitude of the ratio W_{mv}/H_0:

$$\frac{W_{mv}}{H_0} \simeq \frac{\dfrac{\mathbf{p}^4}{8m_e^3c^2}}{\dfrac{\mathbf{p}^2}{2m_e}} = \frac{\mathbf{p}^2}{4m_e^2c^2} = \frac{1}{4}\left(\frac{v}{c}\right)^2 \simeq \alpha^2 \simeq \left(\frac{1}{137}\right)^2 \tag{B-4}$$

★ Expression (B-1) was obtained by assuming the proton to be infinitely heavy. This is why it is the mass m_e of the electron that appears, and not, as in (A-1), the reduced mass μ of the atom. As far as H_0 is concerned, the proton finite mass effect is taken into account by replacing m_e by μ. However, we shall neglect this effect in the subsequent terms of H, which are already corrections. It would, moreover, be difficult to evaluate, since the relativistic description of a system of two interacting particles poses serious problems [it is not sufficient to replace m_e by μ in the last terms of (B-1)].

since we have already mentioned that, for the hydrogen atom, $v/c \simeq \alpha$. Since $H_0 \simeq 10$ eV, we see that $W_{mv} \simeq 10^{-3}$ eV.

β. *Spin-orbit coupling (W_{SO} term)*

(i) The physical origin

The electron moves at a velocity $\mathbf{v} = \mathbf{p}/m_e$ in the electrostatic field $\mathbf{E}$ created by the proton. Special relativity indicates that there then appears, in the electron frame, a magnetic field $\mathbf{B}'$ given by:

$$\mathbf{B}' = -\frac{1}{c^2}\mathbf{v} \times \mathbf{E} \tag{B-5}$$

to first order in v/c. Since the electron possesses an intrinsic magnetic moment $\mathbf{M}_S = q\mathbf{S}/m_e$, it interacts with this field $\mathbf{B}'$. The corresponding interaction energy can be written:

$$W' = -\mathbf{M}_S \cdot \mathbf{B}' \tag{B-6}$$

Let us express W' more explicitly. The electrostatic field $\mathbf{E}$ appearing in (B-5) is equal to $-\frac{1}{q}\frac{\mathrm{d}V(r)}{\mathrm{d}r}\frac{\mathbf{r}}{r}$, where $V(r) = -\frac{e^2}{r}$ is the electrostatic energy of the electron. From this, we get:

$$\mathbf{B}' = \frac{1}{qc^2}\frac{1}{r}\frac{\mathrm{d}V(r)}{\mathrm{d}r}\frac{\mathbf{p}}{m_e} \times \mathbf{r} \tag{B-7}$$

In the corresponding quantum mechanical operator, there appears:

$$\mathbf{P} \times \mathbf{R} = -\mathbf{L} \tag{B-8}$$

Finally, we obtain:

$$W' = \frac{1}{m_e^2c^2}\frac{1}{R}\frac{\mathrm{d}V(R)}{\mathrm{d}R}\mathbf{L} \cdot \mathbf{S} = \frac{e^2}{m_e^2c^2}\frac{1}{R^3}\mathbf{L} \cdot \mathbf{S} \tag{B-9}$$

Thus we find, to within the factor 1/2*, the spin-orbit term W_{SO} which appears in (B-1). This term then represents the interaction of the magnetic moment of the electron spin with the magnetic field "seen" by the electron because of its motion in the electrostatic field of the proton.

(ii) Order of magnitude

Since $\mathbf{L}$ and $\mathbf{S}$ are of the order of $\hbar$, we have:

$$W_{SO} \simeq \frac{e^2}{m_e^2c^2}\frac{\hbar^2}{R^3} \tag{B-10}$$

* It can be shown that the factor 1/2 is due to the fact that the motion of the electron about the proton is not rectilinear. The electron spin therefore rotates with respect to the laboratory reference frame (Thomas precession; see Jackson (7.5) section 11-8, Omnès (16.13) chap. 4 §2, or Bacry (10.31) chap. 7 §5-d).

Let us compare W_{SO} with H_0, which is of the order of e^2/R:

$$\frac{W_{SO}}{H_0} \simeq \frac{\dfrac{e^2\hbar^2}{m_e^2c^2R^3}}{\dfrac{e^2}{R}} = \frac{\hbar^2}{m_e^2c^2R^2} \tag{B-11}$$

R is of the order of the Bohr radius, $a_0 = \hbar^2/m_e e^2$. Consequently:

$$\frac{W_{SO}}{H_0} \simeq \frac{e^4}{\hbar^2c^2} = \alpha^2 = \left(\frac{1}{137}\right)^2 \tag{B-12}$$

γ. *The Darwin term W_D*

(*i*) The physical origin

In the Dirac equation, the interaction between the electron and the Coulomb field of the nucleus is "local"; it only depends on the value of the field at the electron position **r**. However, the non-relativistic approximation (the series expansion in v/c) leads, for the two-component spinor which describes the electron state, to an equation in which the interaction between the electron and the field has become non-local. The electron is then affected by all the values taken on by the field in a domain centered at the point **r**, and whose size is of the order of the Compton wave length $\hbar/m_e c$ of the electron. This is the origin of the correction represented by the Darwin term.

To understand this more precisely, assume that the potential energy of the electron, instead of being equal to $V(\mathbf{r})$, is given by an expression of the form:

$$\int d^3\rho \; f(\boldsymbol{\rho}) \, V(\mathbf{r} + \boldsymbol{\rho}) \tag{B-13}$$

where $f(\boldsymbol{\rho})$ is a function whose integral is equal to 1, which only depends on $|\boldsymbol{\rho}|$, and which takes on significant values only inside a volume of the order of $(\hbar/m_e c)^3$, centered at $\boldsymbol{\rho} = \mathbf{0}$.

If we neglect the variation of $V(\mathbf{r})$ over a distance of the order of $\hbar/m_e c$, we can replace $V(\mathbf{r} + \boldsymbol{\rho})$ by $V(\mathbf{r})$ in (B-13) and take $V(\mathbf{r})$ outside the integral, which is then equal to 1. (B-13) reduces, in this case, to $V(\mathbf{r})$.

A better approximation consists of replacing, in (B-13), $V(\mathbf{r} + \boldsymbol{\rho})$ by its Taylor series expansion in the neighborhood of $\boldsymbol{\rho} = \mathbf{0}$. The zeroeth-order term gives $V(\mathbf{r})$. The first-order term is zero because of the spherical symmetry of $f(\boldsymbol{\rho})$. The second-order term involves the second derivatives of the potential energy $V(\mathbf{r})$ at the point **r** and quadratic functions of the components of $\boldsymbol{\rho}$, weighted by $f(\boldsymbol{\rho})$ and integrated over $d^3\rho$. This leads to a result of the order of

$$(\hbar/m_e c)^2 \Delta V(\mathbf{r})$$

It is therefore easy to accept the idea that this second-order term should be the Darwin term.

(*ii*) Order of magnitude

Replacing $V(R)$ by $-e^2/R$, we can write the Darwin term in the form :

$$-e^2\frac{\hbar^2}{8m_e^2c^2}\Delta\left(\frac{1}{R}\right)=\frac{\pi e^2\hbar^2}{2m_e^2c^2}\delta(\mathbf{R}) \tag{B-14}$$

(we have used the expression for the Laplacian of $1/R$ given by formula (61) of appendix II).

When we take the mean value of (B-14) in an atomic state, we find a contribution equal to:

$$\frac{\pi e^2\hbar^2}{2m_e^2c^2}|\psi(\mathbf{0})|^2$$

where $\psi(\mathbf{0})$ is the value of the wave function at the origin. The Darwin term therefore affects only the s electrons, which are the only ones for which $\psi(\mathbf{0}) \neq 0$ (*cf.* chap. VII, § C-4-c). The order of magnitude of $|\psi(\mathbf{0})|^2$ can be obtained by taking the integral of the square of the modulus of the wave function over a volume of the order of a_0^3 (where a_0 is the Bohr radius) to be equal to 1. Thus we obtain:

$$|\psi(\mathbf{0})|^2 \simeq \frac{1}{a_0^3}=\frac{m_e^3e^6}{\hbar^6} \tag{B-15}$$

which gives the order of magnitude of the Darwin term:

$$W_D \simeq \frac{\pi e^2\hbar^2}{2m_e^2c^2}|\psi(\mathbf{0})|^2 \simeq m_ec^2\frac{e^8}{\hbar^4c^4}=m_ec^2\alpha^4 \tag{B-16}$$

Since $H_0 \simeq m_ec^2\alpha^2$, we again see that:

$$\frac{W_D}{H_0}\simeq\alpha^2=\left(\frac{1}{137}\right)^2 \tag{B-17}$$

Thus, all the fine structure terms are about 10^4 times smaller than the non-relativistic Hamiltonian of chapter VII.

2. Magnetic interactions related to proton spin: the hyperfine Hamiltonian

a. PROTON SPIN AND MAGNETIC MOMENT

Thus far, we have considered the proton to be a physical point of masse M_p and charge $q_p = -q$. Actually, the proton, like the electron, is a spin 1/2 particle. We shall denote by $\mathbf{I}$ the corresponding spin observable.

With the spin $\mathbf{I}$ of the proton is associated a magnetic moment $\mathbf{M}_I$. However, the gyromagnetic ratio is different from that of the electron:

$$\mathbf{M}_I = g_p\mu_n\,\mathbf{I}/\hbar \tag{B-18}$$

where μ_n is the *nuclear Bohr magneton*:

$$\mu_n = \frac{q_p \hbar}{2M_p} \tag{B-19}$$

and the factor g_p, for the proton, is equal to: $g_p \simeq 5.585$. Because of the presence of M_p (the proton mass) in the denominator of (B-19), μ_n is close to 2 000 times smaller than the Bohr magneton μ_B (recall that $\mu_B = q\hbar/2m_e$). Although the angular momenta of the proton and the electron are the same, nuclear magnetism, because of the mass difference, is much less important than electronic magnetism. The magnetic interactions due to the proton spin **I** are therefore very weak.

b. THE MAGNETIC HYPERFINE HAMILTONIAN W_{hf}

The electron moves, therefore, not only in the electrostatic field of the proton, but also in the magnetic field created by $\mathbf{M}_I$. When we introduce the corresponding vector potential into the Schrödinger equation★, we find that we must add to the Hamiltonian (B-1) an additional series of terms for which the expression is (*cf.* complement A_{XII}):

$$W_{hf} = -\frac{\mu_0}{4\pi}\left\{\frac{q}{m_e R^3}\,\mathbf{L}\cdot\mathbf{M}_I + \frac{1}{R^3}\left[3(\mathbf{M}_S\cdot\mathbf{n})(\mathbf{M}_I\cdot\mathbf{n}) - \mathbf{M}_S\cdot\mathbf{M}_I\right] + \frac{8\pi}{3}\,\mathbf{M}_S\cdot\mathbf{M}_I\,\delta(\mathbf{R})\right\} \tag{B-20}$$

$\mathbf{M}_S$ is the spin magnetic moment of the electron, and **n** is the unit vector of the straight line joining the proton to the electron (fig. 1).

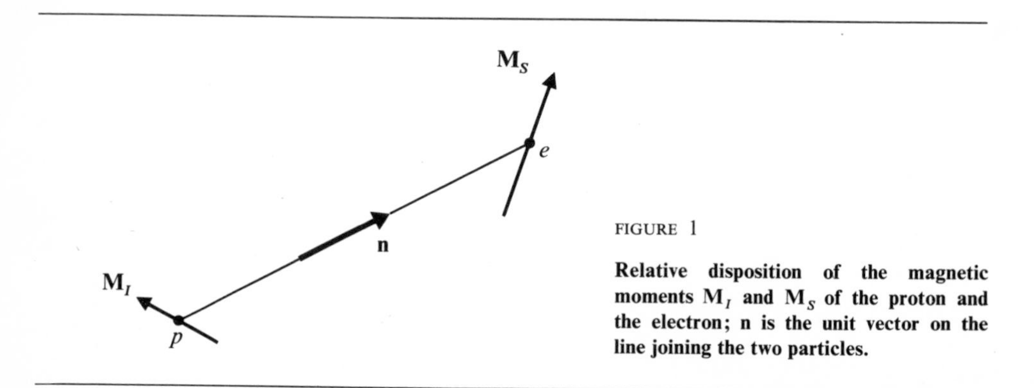

FIGURE 1

Relative disposition of the magnetic moments $\mathbf{M}_I$ and $\mathbf{M}_S$ of the proton and the electron; n is the unit vector on the line joining the two particles.

We shall see that W_{hf} introduces energy shifts which are small compared to those created by W_f. This is why W_{hf} is called the "hyperfine structure Hamiltonian".

★ Since the hyperfine interactions are very small corrective terms, they can be found using the non-relativistic Schrödinger equation.

c. INTERPRETATION OF THE VARIOUS TERMS OF W_{hf}

The first term of W_{hf} represents the interaction of the nuclear magnetic moment $\mathbf{M}_I$ with the magnetic field $(\mu_0/4\pi)q\mathbf{L}/m_e r^3$ created at the proton by the rotation of the electronic charge.

The second term represents the dipole-dipole interaction between the electronic and nuclear magnetic moments: the interaction of the magnetic moment of the electron spin with the magnetic field created by $\mathbf{M}_I$ (*cf.* complement B_{XI}) or vice versa.

Finally, the last term, also called Fermi's "contact term", arises from the singularity at $r = 0$ of the field created by the magnetic moment of the proton. In reality, the proton is not a point. It can be shown (*cf.* complement A_{XII}) that the magnetic field inside the proton does not have the same form as the one created outside by $\mathbf{M}_I$ (and which enters into the dipole-dipole interaction). The contact term describes the interaction of the magnetic moment of the electron spin with the magnetic field inside the proton (the "delta" function expresses the fact that this contact term exists, as its name indicates, only when the wave functions of the electron and proton overlap).

d. ORDERS OF MAGNITUDE

It can easily be shown that the order of magnitude of the first two terms of W_{hf} is:

$$\frac{q^2\hbar^2}{m_e M_p R^3}\frac{\mu_0}{4\pi} = \frac{e^2\hbar^2}{m_e M_p c^2}\frac{1}{R^3} \tag{B-21}$$

By using (B-10), we see that these terms are about 2 000 times smaller than W_{SO}.

As for the last term of (B-20), it is also 2 000 times smaller than the Darwin term, which also contains a $\delta(\mathbf{R})$ function.

C. THE FINE STRUCTURE OF THE $n = 2$ LEVEL

1. Statement of the problem

a. DEGENERACY OF THE $n = 2$ LEVEL

We saw in chapter VII that the energy of the hydrogen atom depends only on the quantum number n. The $2s$ ($n = 2$, $l = 0$) and $2p$ ($n = 2$, $l = 1$) states therefore have the same energy, equal to:

$$-\frac{E_I}{4} = -\frac{1}{8}\mu c^2\alpha^2$$

If the spins are ignored, the $2s$ subshell is composed of a single state, and the $2p$ subshell of three distinct states which differ by their eigenvalue $m_L\hbar$ of the component L_z of the orbital angular momentum $\mathbf{L}$ ($m_L = 1, 0, -1$). Because of the existence of electron and proton spins, the degeneracy of the $n = 2$ level is higher than the value calculated in chapter VII. The components S_z and I_z of the two spins can each take on two values : $m_S = \pm 1/2$, $m_I = \pm 1/2$. One possible orthonormal basis in the $n = 2$ level is then :

$$\left\{ | n = 2; l = 0; m_L = 0; m_S = \pm \frac{1}{2}; m_I = \pm \frac{1}{2} \rangle \right\} \tag{C-1}$$

(2*s* subshell, of dimension 4)

$$\left\{ | n = 2; l = 1; m_L = -1, 0, +1; m_S = \pm \frac{1}{2}; m_I = \pm \frac{1}{2} \rangle \right\} \tag{C-2}$$

(2*p* subshell, of dimension 12).

The $n = 2$ shell then has a total degeneracy equal to 16.

According to the results of chapter XI (§C), in order to calculate the effect of a perturbation W on the $n = 2$ level, it is necessary to diagonalize the 16×16 matrix representing the restriction of W to this level. The eigenvalues of this matrix are the first order corrections to the energy, and the corresponding eigenstates are the eigenstates of the Hamiltonian to zeroeth order.

b. THE PERTURBATION HAMILTONIAN

In all of this section, we shall assume that no external field is applied to the atom. The difference W between the exact Hamiltonian H and the Hamiltonian H_0 of chapter VII (§ C) contains fine structure terms, indicated in § B-1 above :

$$W_f = W_{mv} + W_{SO} + W_D \tag{C-3}$$

and hyperfine structure terms W_{hf}, introduced in § B-2. We thus have :

$$W = W_f + W_{hf} \tag{C-4}$$

Since W_f is close to 2 000 times larger than W_{hf} (*cf.* §B-2-d), we must obviously begin by studying the effect of W_f, before considering that of W_{hf}, on the $n = 2$ level. We shall see that the $n = 16$ degeneracy of this level is partially removed by W_f. The structure which appears in this way is called the "fine structure".

W_{hf} may then remove the remaining degeneracy of the fine structure levels and cause a "hyperfine structure" to appear inside each of these levels.

In this section (§C), we shall confine ourselves to the study of the fine structure of the $n = 1$ level. The calculations can easily be generalized to other levels.

2. Matrix representation of the fine-structure Hamiltonian W_f inside the $n = 2$ level

a. GENERAL PROPERTIES

The properties of W_f, as we shall see, enable us to show that the 16×16 matrix which represents it in the $n = 2$ level can be broken down into a series of square submatrices of smaller dimensions. This will considerably simplify the determination of the eigenvalues and eigenvectors of this matrix.

α. *W_f does not act on the spin variables of the proton*

We see from (B-1) that the fine structure terms do not depend on **I**. It follows that the proton spin can be ignored in the study of the fine structure (afterwards, we multiply by 2 all the degrees of degeneracy obtained). The dimension of the matrix to be diagonalized therefore falls from 16 to 8.

β. *W_f does not connect the 2s and 2p subshells*

Let us first prove that $\mathbf{L}^2$ commutes with W_f. The operator $\mathbf{L}^2$ commutes with the various components of **L**, with R ($\mathbf{L}^2$ acts only on the angular variables), with $\mathbf{P}^2$ [*cf.* formula (A-16) of chapter VII], and with **S** ($\mathbf{L}^2$ does not act on the spin variables). $\mathbf{L}^2$ therefore commutes with W_{mv} (which is proportional to $\mathbf{P}^4$), with W_{SO} (which depends only on R, **L**, **S**), and with W_D (which depends only on R).

The $2s$ and $2p$ states are eigenstates of $\mathbf{L}^2$ with different eigenvalues (0 and $2\hbar^2$). Therefore, W_f, which commutes with $\mathbf{L}^2$, has no matrix elements between a $2s$ state and a $2p$ state. The 8×8 matrix representing W_f inside the $n = 2$ level can be broken down, consequently, into a 2×2 matrix relative to the $2s$ state and a 6×6 matrix relative to the $2p$ state :

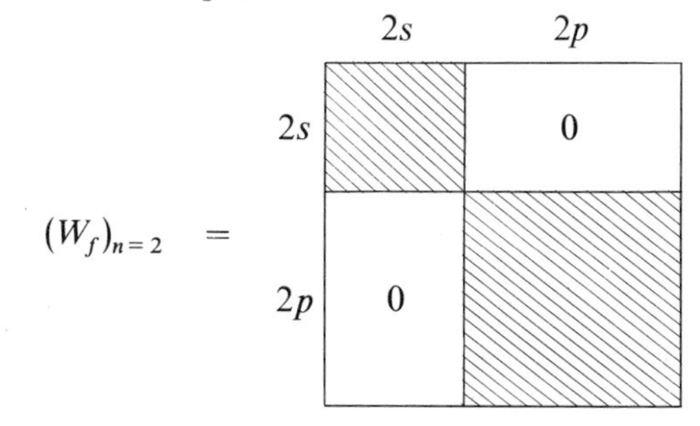

COMMENT:

The preceding property can also be considered to be a consequence of the fact that W_f is even. Under a reflection, **R** changes to $-$ **R** ($R = |\mathbf{R}|$ remains unchanged), **P** to $-$ **P**, **L** to **L**, and **S** to **S**. It is then easy to see that W_f remains invariant. W_f therefore has no matrix elements between the $2s$ and $2p$ states, which are of opposite parity (*cf.* complement F_{II}).

b. MATRIX REPRESENTATION OF W_f IN THE 2*s* SUBSHELL

The dimension 2 of the 2*s* subspace is the result of the two possible values $m_s = \pm 1/2$ of S_z (since we are ignoring I_z for the moment).

W_{mv} and W_D do not depend on **S**. The matrices which represent these two operators in the 2*s* subspace are therefore multiples of the unit matrix, with proportionality coefficients equal, respectively, to the purely orbital matrix elements:

$$\langle n = 2; l = 0; m_L = 0 \,|\, -\frac{\mathbf{P}^4}{8m_e^3c^2} \,|\, n = 2; l = 0; m_L = 0 \rangle$$

and:

$$\langle n = 2; l = 0; m_L = 0 \,|\, \frac{\hbar^2}{8m_e^2c^2}\, q\,\Delta V(R) \,|\, n = 2; l = 0; m_L = 0 \rangle$$

Since we know the eigenfunctions of H_0, the calculation of these matrix elements presents no theoretical difficulty. We find (*cf.* complement B_{XII}):

$$\langle W_{mv} \rangle_{2s} = -\frac{13}{128}\, m_e c^2 \alpha^4 \tag{C-5}$$

$$\langle W_D \rangle_{2s} = \frac{1}{16}\, m_e c^2 \alpha^4 \tag{C-6}$$

Finally, calculation of the matrix elements of W_{SO} involves "angular" matrix elements of the form $\langle l = 0, m_L = 0 \,|\, L_{x,y,z} \,|\, l = 0, m_L = 0 \rangle$, which are zero because of the value $l = 0$ of the quantum number l. Therefore:

$$\langle W_{SO} \rangle_{2s} = 0 \tag{C-7}$$

Thus, under the effect of the fine structure terms, the 2*s* subshell is shifted as a whole with respect to the position calculated in chapter VII by an amount equal to $-\,5m_e c^2\alpha^4/128$.

c. MATRIX REPRESENTATION OF W_f IN THE 2*p* SUBSHELL

α. W_{mv} *and* W_D *terms*

The W_{mv} and W_D terms commute with the various components of **L**, since **L** acts only on the angular variables and commutes with R and $\mathbf{P}^2$ (which depends on these variables only through $\mathbf{L}^2$; *cf.* chapter VII). **L** therefore commutes with W_{mv} and W_D. Consequently, W_{mv} and W_D are scalar operators with respect to the orbital variables (*cf.* complement B_{VI}, §5-b). Since W_{mv} and W_D do not act on the spin variables, it follows that the matrices which represent W_{mv} and W_D inside the 2*p* subspace are multiples of the unit matrix. The calculation of the proportionality coefficient is given in complement B_{XII} and leads to:

$$\langle W_{mv} \rangle_{2p} = -\frac{7}{384}\, m_e c^2 \alpha^4 \tag{C-8}$$

$$\langle W_D \rangle_{2p} = 0 \tag{C-9}$$

The result (C-9) is due to the fact that W_D is proportional to $\delta(\mathbf{R})$ and can therefore have a non-zero mean value only in an s state (for $l \geqslant 1$, the wave function is zero at the origin).

β. *W_{SO} term*

We must calculate the various matrix elements:

$$\langle\, n = 2\,; l = 1\,; s = \frac{1}{2}; m'_L\,; m'_S \mid \xi(R)\, \mathbf{L}\,.\,\mathbf{S} \mid n = 2\,; l = 1\,; s = \frac{1}{2}; m_L\,; m_S \,\rangle \tag{C-10}$$

with:

$$\xi(R) = \frac{e^2}{2m_e^2c^2}\,\frac{1}{R^3} \tag{C-11}$$

If we use the $\{\, |\, \mathbf{r}\,\rangle\,\}$ representation, we can separate the radial part of matrix element (C-10) from the angular and spin parts. Thus we obtain:

$$\xi_{2p}\,\langle\, l = 1\,; s = \frac{1}{2}; m'_L\,; m'_S \mid \mathbf{L}\,.\,\mathbf{S} \mid l = 1\,; s = \frac{1}{2}; m_L\,; m_S \,\rangle \tag{C-12}$$

where ξ_{2p} is a number, equal to the radial integral:

$$\xi_{2p} = \frac{e^2}{2m_e^2c^2}\int_0^\infty \frac{1}{r^3}\,|R_{21}(r)|^2\, r^2\, \mathrm{d}r \tag{C-13}$$

Since we know the radial function $R_{21}(r)$ of the $2p$ state, we can calculate ξ_{2p}. We find (*cf.* complement $\mathrm{B_{XII}}$):

$$\xi_{2p} = \frac{1}{48\hbar^2}\, m_e c^2 \alpha^4 \tag{C-14}$$

The radial variables have therefore disappeared. According to (C-12), the problem is reduced to the diagonalization of the operator $\xi_{2p}\mathbf{L}\,.\,\mathbf{S}$, which acts only on the angular and spin variables.

To represent the operator $\xi_{2p}\mathbf{L}\,.\,\mathbf{S}$ by a matrix, several different bases can be chosen :

– first of all, the basis:

$$\left\{\, |\, l = 1\,; s = \frac{1}{2}; m_L\,; m_S \,\rangle \,\right\} \tag{C-15}$$

which we have used thus far and which is constructed from common eigenstates of $\mathbf{L}^2$, $\mathbf{S}^2$, L_z, S_z;

– or, introducing the total angular momentum:

$$\mathbf{J} = \mathbf{L} + \mathbf{S} \tag{C-16}$$

the basis:

$$\left\{\, |\, l = 1\,; s = \frac{1}{2}; J\,; m_J \,\rangle \,\right\} \tag{C-17}$$

constructed from the eigenstates common to $\mathbf{L}^2$, $\mathbf{S}^2$, $\mathbf{J}^2$, J_z. According to the results of chapter X, since $l = 1$ and $s = 1/2$, J can take on two values : $J = 1 + 1/2 = 3/2$ and $J = 1 - 1/2 = 1/2$. Furthermore, we know how to go from one basis to the other, thanks to the Clebsch-Gordan coefficients [formulas (36) of complement A_X].

We shall now show that the second basis (C-17) is better adapted than the first one to the problem which interests us here, since $\xi_{2p}\mathbf{L} \cdot \mathbf{S}$ is diagonal in the basis (C-17). To see this, we square both sides of (C-16). We find ($\mathbf{L}$ and $\mathbf{S}$ commute):

$$\mathbf{J}^2 = (\mathbf{L} + \mathbf{S})^2 = \mathbf{L}^2 + \mathbf{S}^2 + 2\,\mathbf{L} \cdot \mathbf{S} \tag{C-18}$$

which gives :

$$\xi_{2p}\,\mathbf{L} \cdot \mathbf{S} = \frac{1}{2}\,\xi_{2p}\,(\mathbf{J}^2 - \mathbf{L}^2 - \mathbf{S}^2) \tag{C-19}$$

Each of the basis vectors (C-17) is an eigenstate of $\mathbf{L}^2$, $\mathbf{S}^2$, $\mathbf{J}^2$; we thus have :

$$\begin{aligned}\xi_{2p}\,\mathbf{L} \cdot \mathbf{S}\,|\, l = 1\,;\, s = \frac{1}{2}\,;\, J\,;\, m_J \rangle \\ = \frac{1}{2}\,\xi_{2p}\hbar^2\left[J(J+1) - 2 - \frac{3}{4}\right] |\, l = 1\,;\, s = \frac{1}{2}\,;\, J\,;\, m_J \rangle\end{aligned} \tag{C-20}$$

We see from (C-20) that the eigenvalues of $\xi_{2p}\mathbf{L} \cdot \mathbf{S}$ depend only on J and not on m_J; they are equal to:

$$\frac{1}{2}\,\xi_{2p}\left[\frac{3}{4} - 2 - \frac{3}{4}\right]\hbar^2 = -\,\xi_{2p}\,\hbar^2 = -\frac{1}{48}\,m_e c^2 \alpha^4 \tag{C-21}$$

for $J = 1/2$, and:

$$\frac{1}{2}\,\xi_{2p}\left[\frac{15}{4} - 2 - \frac{3}{4}\right]\hbar^2 = +\frac{1}{2}\,\xi_{2p}\,\hbar^2 = \frac{1}{96}\,m_e c^2 \alpha^4 \tag{C-22}$$

for $J = 3/2$.

The six-fold degeneracy of the $2p$ level is therefore partially removed by W_{SO}. We obtain a four-fold degenerate level corresponding to $J = 3/2$, and a two-fold degenerate level corresponding to $J = 1/2$. The $(2J + 1)$-fold degeneracy of each J state is an essential degeneracy related to the rotation invariance of W_f.

COMMENTS:

(*i*) In the $2s$ subspace ($l = 0$, $s = 1/2$), J can take on only one value, $J = 0 + 1/2 = 1/2$.

(*ii*) In the $2p$ subspace, W_{mv} and W_D are represented by multiples of the unit matrix. This property remains valid in any basis since the unit matrix is invariant under a change of basis. The choice of basis (C-17), required by the W_{SO} term, is therefore also adapted to the W_{mv} and W_D terms.

3. Results: the fine structure of the $n = 2$ level

a. SPECTROSCOPIC NOTATION

In addition to the quantum numbers n, l (and s), the preceding discussion introduced the quantum number J on which the energy correction due to the spin-orbit coupling term depends.

For the $2s$ level, $J = 1/2$; for the $2p$ level, $J = 1/2$ or $J = 3/2$. The level associated with a set of values, n, l, J is generally denoted by adding an index J to the symbol representing the (n, l) subshell in spectroscopic notation (*cf.* chap. VII, § C-4-b):

$$n\, l_J \tag{C-23}$$

where l stands for the letter s for $l = 0$, p for $l = 1$, d for $l = 2$, f for $l = 3$... Thus, the $n = 2$ level of the hydrogen atom gives rise to the $2s_{1/2}$, $2p_{1/2}$ and $2p_{3/2}$ levels.

b. POSITIONS OF THE $2s_{1/2}$, $2p_{1/2}$ AND $2p_{3/2}$ LEVELS

By regrouping the results of § 2, we can now calculate the positions of the $2s_{1/2}$, $2p_{1/2}$ and $2p_{3/2}$ levels with respect to the "unperturbed" energy of the $n = 2$ level calculated in chapter VII and equal to $-\mu c^2\alpha^2/8$.

According to the results of § 2-b, the $2s_{1/2}$ level is lowered by a quantity equal to:

$$-\frac{5}{128}\, m_e c^2 \alpha^4 \tag{C-24}$$

According to the results of § 2-c, the $2p_{1/2}$ level is lowered by a quantity equal to:

$$\left(-\frac{7}{384} - \frac{1}{48}\right) m_e c^2 \alpha^4 = -\frac{5}{128}\, m_e c^2 \alpha^4 \tag{C-25}$$

Thus we see that the $2s_{1/2}$ and $2p_{1/2}$ levels have the same energy. According to the theory presented here, this degeneracy must be considered to be accidental, as opposed to the essential $(2J + 1)$-fold degeneracy of each J level.

Finally, the $2p_{3/2}$ level is lowered by a quantity:

$$\left(-\frac{7}{384} + \frac{1}{96}\right) m_e c^2 \alpha^4 = -\frac{1}{128}\, m_e c^2 \alpha^4 \tag{C-26}$$

The preceding results are shown in figure 2.

COMMENTS:

(*i*) Only the spin-orbit coupling is responsible for the separation between the $2p_{1/2}$ and $2p_{3/2}$ levels, since W_{mv} and W_D shift the entire $2p$ level as a whole.

(*ii*) The hydrogen atom can go from the $2p$ state to the $1s$ state by emitting a Lyman α photon ($\lambda = 1\,216$ Å). The material presented in this chapter shows that, because of the spin-orbit coupling, the Lyman α line actually contains two neighboring lines*, $2p_{1/2} \longrightarrow 1s_{1/2}$ and $2p_{3/2} \longrightarrow 1s_{1/2}$, separated by an energy difference equal to:

$$\frac{4}{128}\, m_e c^2 \alpha^4 = \frac{1}{32}\, m_e c^2 \alpha^4$$

When they are observed with a sufficient resolution, the lines of the hydrogen spectrum therefore present a "fine structure".

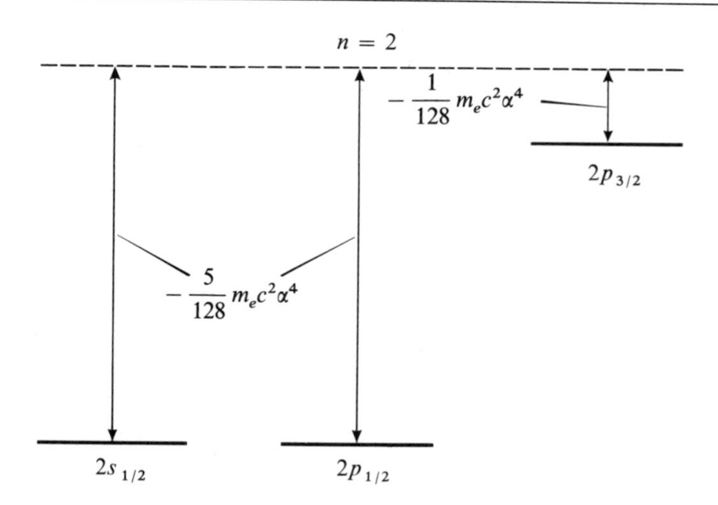

FIGURE 2

Fine structure of the $n = 2$ level of the hydrogen atom. Under the effect of the fine structure Hamiltonian W_f, the $n = 2$ level splits into three fine structure levels, written $2s_{1/2}$, $2p_{1/2}$ and $2p_{3/2}$. We have indicated the algebraic values of the shifts, calculated to first order in W_f. The shifts are the same for the $2s_{1/2}$ and $2p_{1/2}$ levels (a result which remains valid, moreover, to all orders in W_f). When we take into account the quantum mechanical nature of the electromagnetic field, we find that the degeneracy between the $2s_{1/2}$ and $2p_{1/2}$ levels is removed (the Lamb shift; see figure 4).

(*iii*) We see in figure 2 that the two levels with the same J have the same energy. This result is not merely true to first order in W_f: it remains valid to all orders. The exact solution of the Dirac equation gives, for the energy of a level characterized by the quantum numbers n, l, s, J, the value:

$$E_{n,J} = m_e c^2 \left[1 + \alpha^2 \left(n - J - \frac{1}{2} + \sqrt{(J + 1/2)^2 - \alpha^2} \right)^{-2} \right]^{-\frac{1}{2}} \qquad \text{(C-27)}$$

* In the ground state, $l = 0$ and $s = 1/2$, so J can take on only the value $J = 1/2$. W_f therefore does not remove the degeneracy of the $1s$ state, and there is only one fine structure level, the $1s_{1/2}$ level. This is a special case, since the ground state is the only one for which l is necessarily zero. This is why we have chosen here to study the excited $n = 2$ level.

We see that the energy depends only on n and J, and not on l.

If we make a limited expansion of formula (C-27) in powers of α, we obtain:

$$E_{n,J} = m_e c^2 - \frac{1}{2} m_e c^2 \alpha^2 \frac{1}{n^2} - \frac{m_e c^2}{2n^4}\left(\frac{n}{J + 1/2} - \frac{3}{4}\right)\alpha^4 + \dots \qquad \text{(C-28)}$$

The first term is the rest-mass-energy of the electron. The second term follows from the theory of chapter VII. The third term gives the correction to first order in W_f calculated in this chapter.

(*iv*) Even in the absence of an external field and incident photons, a fluctuating electromagnetic field must be considered to exist in space (*cf.* complement K_V, § 3-d-δ). This phenomenon is related to the quantum mechanical nature of the electromagnetic field, which we have not taken into consideration here. The coupling of the atom with these fluctuations of the electromagnetic field removes the degeneracy between the $2s_{1/2}$ and $2p_{1/2}$ levels. The $2s_{1/2}$ level is raised with respect to the $2p_{1/2}$ level by a quantity called the "Lamb shift" which is of the order of 1 060 MHz (fig. 4, page 1231).

The theoretical and experimental study of this phenomenon, which was discovered in 1949, has been the object of a great deal of research, leading to the development of modern quantum electrodynamics.

D. THE HYPERFINE STRUCTURE OF THE $n = 1$ LEVEL

It would now seem logical to study the effect of W_{hf} inside the fine structure levels $2s_{1/2}$, $2p_{1/2}$ and $2p_{3/2}$, in order to see if the interactions related to the proton spin **I** cause a hyperfine structure to appear in each of these levels. However, since W_f does not remove the degeneracy of the ground state $1s$, it is simpler to study the effect of W_{hf} on this state. The results obtained in this special case can easily be generalized to the $2s_{1/2}$, $2p_{1/2}$ and $2p_{3/2}$ levels.

1. Statement of the problem

a. THE DEGENERACY OF THE 1s LEVEL

For the $1s$ level, there is no orbital degeneracy ($l = 0$). On the other hand, the S_z and I_z components of **S** and **I** can still take on two values: $m_S = \pm 1/2$ and $m_I = \pm 1/2$. The degeneracy of the $1s$ level is therefore equal to 4, and a possible basis in this level is given by the vectors:

$$\left\{ | n = 1; l = 0; m_L = 0; m_S = \pm \frac{1}{2}; m_I = \pm \frac{1}{2} \rangle \right\} \qquad \text{(D-1)}$$

b. THE 1s LEVEL HAS NO FINE STRUCTURE

We shall show that the W_f term does not remove the degeneracy of the $1s$ level.

The W_{mv} and W_D terms do not act on m_S and m_I, and are represented in the $1s$ subspace by multiples of the unit matrix. We find (*cf.* complement B_{XII}):

$$\langle W_{mv} \rangle_{1s} = -\frac{5}{8} m_e c^2 \alpha^4 \tag{D-2}$$

$$\langle W_D \rangle_{1s} = \frac{1}{2} m_e c^2 \alpha^4 \tag{D-3}$$

Finally, calculation of the matrix elements of the W_{SO} term involves the "angular" matrix elements $\langle l = 0, m_L = 0 | L_{x,y,z} | l = 0, m_L = 0 \rangle$, which are obviously zero ($l = 0$); therefore:

$$\langle W_{SO} \rangle_{1s} = 0 \tag{D-4}$$

In conclusion, W_f merely shifts the $1s$ level as a whole by a quantity equal to:

$$\left(-\frac{5}{8} + \frac{1}{2} \right) m_e c^2 \alpha^4 = -\frac{1}{8} m_e c^2 \alpha^4 \tag{D-5}$$

without splitting the level. This result could have been foreseen : since $l = 0$ and $s = 1/2$, J can take on only one value, $J = 1/2$, and the $1s$ level therefore gives rise to only one fine structure level, $1s_{1/2}$.

Since the Hamiltonian W_f does not split the $1s$ level, we can now consider the effect of the W_{hf} term. To do so, we must first calculate the matrix which represents W_{hf} in the $1s$ level.

2. Matrix representation of W_{hf} in the $1s$ level

a. TERMS OTHER THAN THE CONTACT TERM

Let us show that the first two terms of W_{hf} [formula (B-20)] make no contribution.

Calculation of the contribution from the first term, $-\frac{\mu_0}{4\pi} \frac{q}{m_e R^3} \mathbf{L} \cdot \mathbf{M}_I$, leads to the "angular" matrix elements $\langle l = 0, m_L = 0 | \mathbf{L} | l = 0, m_L = 0 \rangle$, which are obviously zero ($l = 0$).

Similarly, it can be shown (*cf.* complement B_{XI}, § 3) that the matrix elements of the second term (the dipole-dipole interaction) are zero because of the spherical symmetry of the $1s$ state.

b. THE CONTACT TERM

The matrix elements of the last term of (B-20), that is, of the contact term, are of the form:

$$\langle n = 1 ; l = 0 ; m_L = 0 ; m'_S ; m'_I | \\ -\frac{2\mu_0}{3} \mathbf{M}_S \cdot \mathbf{M}_I \, \delta(\mathbf{R}) | n = 1 ; l = 0 ; m_L = 0 ; m_S ; m_I \rangle \tag{D-6}$$

If we go into the $\{ | \mathbf{r} \rangle \}$ representation, we can separate the orbital and spin parts of this matrix element and put it in the form:

$$\mathscr{A} \langle m'_S ; m'_I \mid \mathbf{I} \cdot \mathbf{S} \mid m_S ; m_I \rangle \tag{D-7}$$

where $\mathscr{A}$ is a number given by :

$$\begin{aligned} \mathscr{A} &= \frac{q^2}{3\varepsilon_0 c^2} \frac{g_p}{m_e M_p} \langle n = 1 ; l = 0 ; m_L = 0 \mid \delta(\mathbf{R}) \mid n = 1 ; l = 0 ; m_L = 0 \rangle \\ &= \frac{q^2}{3\varepsilon_0 c^2} \frac{g_p}{m_e M_p} \frac{1}{4\pi} |R_{10}(0)|^2 \\ &= \frac{4}{3} g_p \frac{m_e}{M_p} m_e c^2 \alpha^4 \left(1 + \frac{m_e}{M_p} \right)^{-3} \frac{1}{\hbar^2} \end{aligned} \tag{D-8}$$

We have used the expressions relating $\mathbf{M}_S$ and $\mathbf{M}_I$ to $\mathbf{S}$ and $\mathbf{I}$ [*cf.* (B-18)], as well as the expression for the radial function $R_{10}(r)$ given in §C-4-c of chapter VII*.

The orbital variables have therefore completely disappeared, and we are left with a problem of two spin 1/2's, $\mathbf{I}$ and $\mathbf{S}$, coupled by an interaction of the form:

$$\mathscr{A} \, \mathbf{I} \cdot \mathbf{S} \tag{D-9}$$

where $\mathscr{A}$ is a constant.

c. EIGENSTATES AND EIGENVALUES OF THE CONTACT TERM

To represent the operator $\mathscr{A}\mathbf{I} \cdot \mathbf{S}$, we have thus far considered only the basis:

$$\left\{ \left| s = \frac{1}{2} ; I = \frac{1}{2} ; m_S ; m_I \right\rangle \right\} \tag{D-10}$$

formed by the eigenvectors common to $\mathbf{S}^2$, $\mathbf{I}^2$, S_z, I_z. We can also, by introducing the total angular momentum**:

$$\mathbf{F} = \mathbf{S} + \mathbf{I} \tag{D-11}$$

use the basis:

$$\left\{ \left| s = \frac{1}{2} ; I = \frac{1}{2} ; F ; m_F \right\rangle \right\} \tag{D-12}$$

formed by the eigenstates common to $\mathbf{S}^2$, $\mathbf{I}^2$, $\mathbf{F}^2$, F_z. Since $s = I = 1/2$, F can take on only the two values $F = 0$ and $F = 1$. We can easily pass from one basis to the other by means of (B-22) and (B-23) of chapter X.

* The factor $(1 + m_e/M_p)^{-3}$ in (D-8) arises from the fact that it is the reduced mass μ which enters into $R_{10}(0)$. It so happens that, for the contact term, it is correct to take the nuclear finite mass effect into account in this way.

** The total angular momentum is actually $\mathbf{F} = \mathbf{L} + \mathbf{S} + \mathbf{I}$, that is, $\mathbf{F} = \mathbf{J} + \mathbf{I}$. However, for the ground state, the orbital angular momentum is zero, so $\mathbf{F}$ reduces to (D-11).

The $\{ | F, m_F \rangle \}$ basis is better adapted than the $\{ | m_S, m_I \rangle \}$ basis to the study of the operator $\mathscr{A}\mathbf{I}.\mathbf{S}$, as this operator is represented in the $\{ | F, m_F \rangle \}$ basis by a diagonal matrix (for the sake of simplicity, we do not explicitly write $s = 1/2$ and $I = 1/2$). This is true, since we obtain, from (D-11):

$$\mathscr{A}\mathbf{I}.\mathbf{S} = \frac{\mathscr{A}}{2}(\mathbf{F}^2 - \mathbf{I}^2 - \mathbf{S}^2) \tag{D-13}$$

It follows that the states $| F, m_F \rangle$ are eigenstates of $\mathscr{A}\mathbf{I}.\mathbf{S}$:

$$\mathscr{A}\mathbf{I}.\mathbf{S} | F, m_F \rangle = \frac{\mathscr{A}\hbar^2}{2}[F(F+1) - I(I+1) - S(S+1)] | F, m_F \rangle \tag{D-14}$$

We see from (D-14) that the eigenvalues depend only on F, and not on m_F. They are equal to:

$$\frac{\mathscr{A}\hbar^2}{2}\left[2 - \frac{3}{4} - \frac{3}{4}\right] = \frac{\mathscr{A}\hbar^2}{4} \tag{D-15}$$

for $F = 1$, and:

$$\frac{\mathscr{A}\hbar^2}{2}\left[0 - \frac{3}{4} - \frac{3}{4}\right] = -\frac{3\mathscr{A}\hbar^2}{4} \tag{D-16}$$

for $F = 0$.

The four-fold degeneracy of the $1s$ level is therefore partially removed by W_{hf}. We obtain a three-fold degenerate $F = 1$ level and a non-degenerate $F = 0$ level. The $(2F + 1)$-fold degeneracy of the $F = 1$ level is essential and is related to the invariance of W_{hf} under a rotation of the total system.

3. The hyperfine structure of the 1s level

a. POSITIONS OF THE LEVELS

Under the effect of W_f, the energy of the $1s$ level is lowered by a quantity $m_e c^2 \alpha^4/8$ with respect to the value $-\mu c^2 \alpha^2/2$ calculated in chapter VII. W_{hf} then splits the $1s_{1/2}$ level into two hyperfine levels, separated by an energy $\mathscr{A}\hbar^2$ (fig. 3). $\mathscr{A}\hbar^2$ is often called the "hyperfine structure of the ground state".

COMMENT:

It could be found, similarly, that W_{hf} splits each of the fine structure levels $2s_{1/2}$, $2p_{1/2}$ and $2p_{3/2}$ into a series of hyperfine levels, corresponding to all the values of F separated by one unit and included between $J + I$ and $|J - I|$. For the $2s_{1/2}$ and $2p_{1/2}$ levels, we have $J = 1/2$. Therefore, F takes on the two values $F = 1$ and $F = 0$. For the $2p_{3/2}$ level, $J = 3/2$, and, consequently, we have $F = 2$ and $F = 1$ (*cf.* fig. 4).

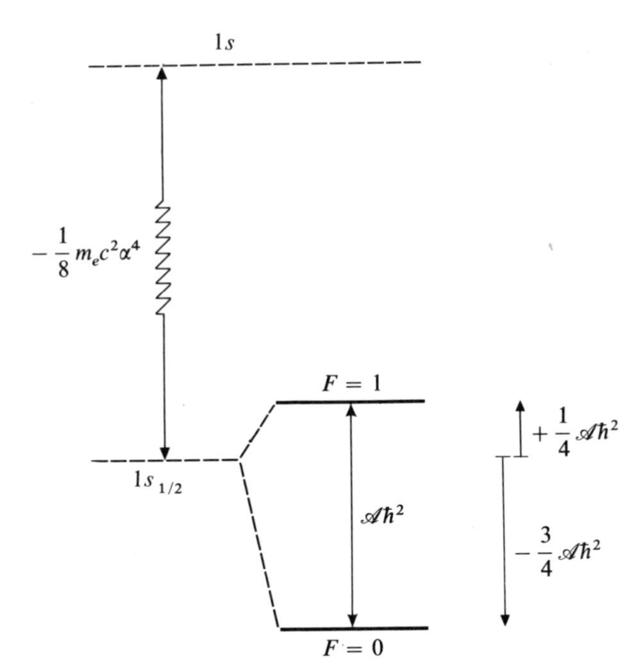

FIGURE 3

The hyperfine structure of the $n = 1$ level of the hydrogen atom. Under the effect of W_f, the $n = 1$ level undergoes a global shift equal to

$$- m_e c^2 \alpha^4 / 8;$$

J can take on only one value, $J = 1/2$. When the hyperfine coupling W_{hf} is taken into account, the $1s_{1/2}$ level splits into two hyperfine levels, $F = 1$ and $F = 0$. The hyperfine transition

$$F = 1 \longleftrightarrow F = 0$$

(the 21 cm line studied in radioastronomy) has a frequency which is known experimentally to twelve significant figures (thanks to the hydrogen maser).

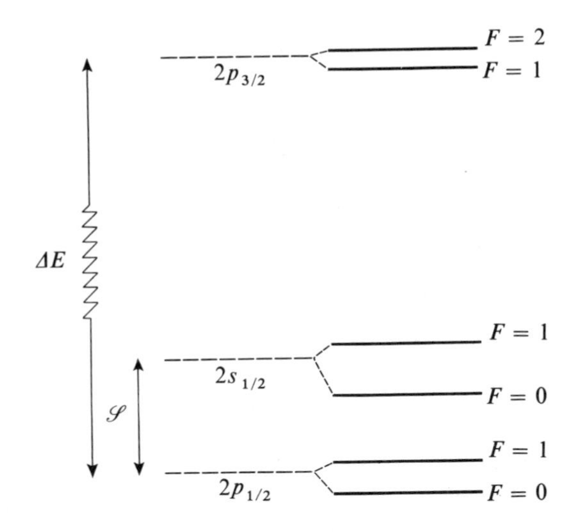

FIGURE 4

The hyperfine structure of the $n = 2$ level of the hydrogen atom. The separation $\mathscr{S}$ between the two levels $2s_{1/2}$ and $2p_{1/2}$ is the Lamb shift, which is about ten times smaller than the fine structure splitting ΔE separating the two levels $2p_{1/2}$ and $2p_{3/2}$ ($\mathscr{S} \simeq 1\,057.8$ MHz; $\Delta E \simeq 10\,969.1$ MHz). When the hyperfine coupling W_{hf} is taken into account, each level splits into two hyperfine sublevels (the corresponding value of the quantum number F is indicated on the right-hand side of the figure). The hyperfine splittings are equal to 23.7 MHz for the $2p_{3/2}$ level, 177.56 MHz for the $2s_{1/2}$ level and 59.19 MHz for the $2p_{1/2}$ level (for the sake of clarity, the figure is not drawn to scale).

b. IMPORTANCE OF THE HYPERFINE STRUCTURE OF THE 1*s* LEVEL

The hyperfine structure of the ground state of the hydrogen atom is currently the physical quantity which is known experimentally to the highest number of significant figures. Expressed in Hz, it is equal to★:

$$\frac{\mathscr{A}\hbar}{2\pi} = 1\,420\,405\,751.\,768 \pm 0.001 \text{ Hz} \tag{D-17}$$

Such a high degree of experimental accuracy was made possible by the development of the "hydrogen maser" in 1963. The principle of such a device is, very schematically, the following: hydrogen atoms, previously sorted (by a magnetic selection of the Stern-Gerlach type) so as to choose those in the upper hyperfine level $F = 1$, are stored in a glass cell (the arrangement is similar to the one shown in figure 6 of complement F_{IV}). This constitutes an amplifying medium for the hyperfine frequency $\dfrac{E(F = 1) - E(F = 0)}{h}$. If the cell is placed in a cavity tuned to the hyperfine frequency, and if the losses of the cavity are small enough for the gain to be greater than the losses, the system becomes unstable and can oscillate: we obtain an "atomic oscillator" (a maser). The frequency of the oscillator is very stable and of great spectral purity. Its measurement gives directly the value of the hyperfine splitting, expressed in Hz.

Note, finally, that hydrogen atoms in interstellar space are detected in radioastronomy by the radiation they emit spontaneously when they fall from the $F = 1$ hyperfine level to the $F = 0$ hyperfine level of the ground state (this transition corresponds to a wave length of 21 cm). Most of the information we possess about interstellar hydrogen clouds is supplied by the study of this 21 cm line.

E. THE ZEEMAN EFFECT OF THE 1*s* GROUND STATE HYPERFINE STRUCTURE

1. Statement of the problem

a. THE ZEEMAN HAMILTONIAN W_Z

We now assume the atom to be placed in a static uniform magnetic field $\mathbf{B}_0$ parallel to Oz. This field interacts with the various magnetic moments present in the atom : the orbital and spin magnetic moments of the electron, $\mathbf{M}_L = \dfrac{q}{2m_e}\mathbf{L}$ and $\mathbf{M}_S = \dfrac{q}{m_e}\mathbf{S}$, and the magnetic moment of the nucleus, $\mathbf{M}_I = -\dfrac{qg_p}{2M_p}\mathbf{I}$ [*cf.* expression (B-18)].

★ The calculations presented in this chapter are obviously completely incapable of predicting all these significant figures. Moreover, even the most advanced theories cannot, at the present time, explain more than the first five or six figures of (D-17).

The Zeeman Hamiltonian W_Z which describes the interaction energy of the atom with the field $\mathbf{B}_0$ can then be written:

$$\begin{aligned} W_Z &= -\mathbf{B}_0 \cdot (\mathbf{M}_L + \mathbf{M}_S + \mathbf{M}_I) \\ &= \omega_0(L_z + 2S_z) + \omega_n I_z \end{aligned} \quad \text{(E-1)}$$

where ω_0 (the Larmor angular frequency in the field $\mathbf{B}_0$) and ω_n are defined by:

$$\begin{cases} \omega_0 = -\dfrac{q}{2m_e} B_0 & \text{(E-2)} \\ \omega_n = \dfrac{q}{2M_p} g_p B_0 & \text{(E-3)} \end{cases}$$

Since $M_p \gg m_e$, we clearly have:

$$|\omega_0| \gg |\omega_n| \quad \text{(E-4)}$$

COMMENT:

Rigorously, W_Z contains another term, which is quadratic in B_0 (the diamagnetic term). This term does not act on the electronic and nuclear spin variables and merely shifts the $1s$ level as a whole, without modifying its Zeeman diagram, which we shall study later. Moreover, it is much smaller than (E-1). Recall that a detailed study of the effect of the diamagnetic term is presented in complement D_{VII}.

b. THE PERTURBATION "SEEN" BY THE 1s LEVEL

In this section, we propose to study the effect of W_Z on the $1s$ ground state of the hydrogen atom (the case of the $n = 2$ level is slightly more complicated since, in a zero magnetic field, this level possesses both a fine and a hyperfine structure, while the $n = 1$ level has only a hyperfine structure; the principle of the calculation is nevertheless the same). Even with the strongest magnetic fields that can be produced in the laboratory, W_Z is much smaller than the distance between the $1s$ level and the other levels; consequently, its effect can be treated by perturbation theory.

The effect of a magnetic field on an atomic energy level is called the "Zeeman effect". When B_0 is plotted on the x-axis and the energies of the various sublevels it creates are plotted on the y-axis, a *Zeeman diagram* is obtained.

If B_0 is sufficiently strong, the Zeeman Hamiltonian W_Z can be of the same order of magnitude as the hyperfine Hamiltonian W_{hf}★, or even larger. On the other hand, if B_0 is very weak, $W_Z \ll W_{hf}$. Therefore, in general it is not possible to establish the relative importance of W_Z and W_{hf}. To obtain the energies of the various sublevels, $(W_Z + W_{hf})$ must be diagonalized inside the $n = 1$ level.

★ Recall that W_f shifts the $1s$ level as a whole; it therefore also shifts the Zeeman diagram as a whole.

We showed in § D-2 that the restriction of W_{hf} to the $n = 1$ level could be put in the form $\mathscr{A}\,\mathbf{I}\,.\,\mathbf{S}$. Using expression (E-1) for W_Z, we see that we must also calculate matrix elements of the form:

$$\langle n = 1\,;\, l = 0\,;\, m_L = 0\,;\, m'_S\,;\, m'_I \,|\, \omega_0(L_z + 2S_z) + \omega_n I_z \,|\, n = 1\,;\, l = 0\,;\, m_L = 0\,;\, m_S\,;\, m_I \rangle \tag{E-5}$$

The contribution of $\omega_0 L_z$ is zero, since l and m_L are zero. Since $2\omega_0 S_z + \omega_n I_z$ acts only on the spin variables, we can, for this term, separate the orbital part of the matrix element:

$$\langle n = 1\,;\, l = 0\,;\, m_L = 0 \,|\, n = 1\,;\, l = 0\,;\, m_L = 0 \rangle = 1 \tag{E-6}$$

from the spin part.

In conclusion, therefore, we must, ignoring the quantum numbers n, l, m_L, diagonalize the operator:

$$\mathscr{A}\,\mathbf{I}\,.\,\mathbf{S} + 2\omega_0 S_z + \omega_n I_z \tag{E-7}$$

which acts only on the spin degrees of freedom. To do so, we can use either the $\{ \,|\, m_S, m_I \rangle \}$ basis or the $\{ \,|\, F, m_F \rangle \}$ basis.

According to (E-4), the last term of (E-7) is much smaller than the second one. To simplify the discussion, we shall neglect the term $\omega_n I_z$ from now on (it would be possible, however, to take it into account★). The perturbation "seen" by the $1s$ level can therefore be written, finally:

$$\mathscr{A}\,\mathbf{I}\,.\,\mathbf{S} + 2\omega_0 S_z \tag{E-8}$$

c. DIFFERENT DOMAINS OF FIELD STRENGTH

By varying B_0, we can continuously modify the magnitude of the Zeeman term $2\omega_0 S_z$. We shall consider three different field strengths, determined by the respective orders of magnitude of the hyperfine term and the Zeeman term:

(*i*) $\hbar\omega_0 \ll \mathscr{A}\hbar^2$: weak fields

(*ii*) $\hbar\omega_0 \gg \mathscr{A}\hbar^2$: strong fields

(*iii*) $\hbar\omega_0 \simeq \mathscr{A}\hbar^2$: intermediate fields

We shall later see that it is possible to diagonalize operator (E-8) exactly. However, in order to give a particularly simple example of perturbation theory, we shall use a slightly different method in cases (*i*) and (*ii*). In case (*i*), we shall treat $2\omega_0 S_z$ like a perturbation with respect to $\mathscr{A}\,\mathbf{I}\,.\,\mathbf{S}$. On the other hand, in case (*ii*), we shall treat $\mathscr{A}\,\mathbf{I}\,.\,\mathbf{S}$ like a perturbation with respect to $2\omega_0 S_z$. The exact diagonalization of the set of two operators, indispensable in case (*iii*), will allow us to check the preceding results.

2. The weak-field Zeeman effect

The eigenstates and eigenvalues of $\mathscr{A}\,\mathbf{I}\,.\,\mathbf{S}$ have already been determined (§ D-2). We therefore obtain two different levels : the three-fold degenerate level,

$$\{ \,|\, F = 1\,;\, m_F = -1, 0, +1 \rangle \},$$

★ This is what we do in complement C_{XII}, in which we study the hydrogen-like systems (muonium, positronium) for which it is not possible to neglect the magnetic moment of one of the two particles.

of energy $\mathscr{A}\hbar^2/4$, and the non-degenerate level, $\{ | F = 0; m_F = 0 \rangle \}$, of energy $- 3\mathscr{A}\hbar^2/4$. Since we are treating $2\omega_0 S_z$ like a perturbation with respect to $\mathscr{A}\mathbf{I}.\mathbf{S}$, we must now separately diagonalize the two matrices representing $2\omega_0 S_z$ in the two levels, $F = 1$ and $F = 0$, corresponding to two distinct eigenvalues of $\mathscr{A}\mathbf{I}.\mathbf{S}$.

a. MATRIX REPRESENTATION OF S_z IN THE $\{|F, m_F\rangle\}$ BASIS

Since we shall need it later, we shall begin by writing the matrix which represents S_z in the $\{ | F, m_F \rangle \}$ basis (for the problem which concerns us here, it would suffice to write the two submatrices corresponding to the $F = 1$ and $F = 0$ subspaces).

By using formulas (B-22) and (B-23) of chapter X, we easily obtain:

$$\begin{cases} S_z | F = 1; m_F = 1 \rangle = \frac{\hbar}{2} | F = 1; m_F = 1 \rangle \\ S_z | F = 1; m_F = 0 \rangle = \frac{\hbar}{2} | F = 0; m_F = 0 \rangle \\ S_z | F = 1; m_F = -1 \rangle = -\frac{\hbar}{2} | F = 1; m_F = -1 \rangle \\ S_z | F = 0; m_F = 0 \rangle = \frac{\hbar}{2} | F = 1; m_F = 0 \rangle \end{cases} \tag{E-9}$$

which gives the following expression for the matrix representing S_z in the $\{ | F, m_F \rangle \}$ basis (the basis vectors are arranged in the order $| 1, 1 \rangle$, $| 1, 0 \rangle$, $| 1, -1 \rangle$, $| 0, 0 \rangle$):

$$(S_z) = \frac{\hbar}{2} \times \begin{pmatrix} 1 & 0 & 0 & 0 \\ 0 & 0 & 0 & 1 \\ 0 & 0 & -1 & 0 \\ 0 & 1 & 0 & 0 \end{pmatrix} \tag{E-10}$$

COMMENT:

It is instructive to compare the preceding matrix with the one which represents F_z in the same basis:

$$(F_z) = \hbar \times \begin{pmatrix} 1 & 0 & 0 & 0 \\ 0 & 0 & 0 & 0 \\ 0 & 0 & -1 & 0 \\ 0 & 0 & 0 & 0 \end{pmatrix} \tag{E-11}$$

We see, first of all, that the two matrices are not proportional: the (F_z) matrix is diagonal, while the (S_z) one is not.

However, if we confine ourselves to the restrictions of the two matrices in the $F = 1$ subspace [limited by the darker line in expressions (E-10) and (E-11)], we see that they are proportional. Denoting by P_1 the projector onto the $F = 1$ subspace (*cf.* complement B_{II}), we have:

$$P_1 \, S_z \, P_1 = \frac{1}{2} P_1 \, F_z \, P_1 \tag{E-12}$$

It would be simple to show that the same relation exists between S_x and F_x on the one hand, and S_y and F_y, on the other.

We have thus found a special case of the Wigner-Eckart theorem (complement D_X), according to which, in a given eigensubspace of the total angular momentum, all the matrices which represent vector operators are proportional. It is clear from this example that this proportionality exists only for the restrictions of operators to a given eigensubspace of the total angular momentum, and not for the operators themselves.

Moreover, the proportionality coefficient 1/2 which appears in (E-12) can be obtained immediately from the projection theorem. According to formula (30) of complement E_X, this coefficient is equal to:

$$\frac{\langle \mathbf{S} \, . \, \mathbf{F} \rangle_{F=1}}{\langle \mathbf{F}^2 \rangle_{F=1}} = \frac{F(F+1) + s(s+1) - I(I+1)}{2F(F+1)} \tag{E-13}$$

Since $s = I = 1/2$, (E-13) is indeed equal to 1/2.

b. WEAK-FIELD EIGENSTATES AND EIGENVALUES

According to the results of §a, the matrix which represents $2\omega_0 S_z$ in the $F = 1$ level can be written:

$$\begin{pmatrix} \hbar\omega_0 & 0 & 0 \\ 0 & 0 & 0 \\ 0 & 0 & -\hbar\omega_0 \end{pmatrix} \tag{E-14}$$

In the $F = 0$ level, this matrix reduces to a number, equal to 0.

Since these two matrices are diagonal, we can immediately find the weak-field eigenstates (to zeroeth order in ω_0) and the eigenvalues (to first order in ω_0):

$$\begin{array}{lcl}
\textit{Eigenstates} & & \textit{Eigenvalues} \\
| F = 1 ; m_F = 1 \rangle & \longleftrightarrow & \dfrac{\mathscr{A}\hbar^2}{4} + \hbar\omega_0 \\
| F = 1 ; m_F = 0 \rangle & \longleftrightarrow & \dfrac{\mathscr{A}\hbar^2}{4} + 0 \\
| F = 1 ; m_F = -1 \rangle & \longleftrightarrow & \dfrac{\mathscr{A}\hbar^2}{4} - \hbar\omega_0 \\
| F = 0 ; m_F = 0 \rangle & \longleftrightarrow & -3\dfrac{\mathscr{A}\hbar^2}{4} + 0
\end{array} \tag{E-15}$$

In figure 5, we have plotted $\hbar\omega_0$ on the x-axis and the energies of the four Zeeman sublevels on the y-axis (Zeeman diagram). In a zero field, we have the two hyperfine levels, $F = 1$ and $F = 0$. When the field B_0 is turned on, the $| F = 0, m_F = 0 \rangle$ sublevel, which is not degenerate, starts horizontally; as for the $F = 1$ level, its three-fold degeneracy is completely removed : three equidistant sublevels are obtained, varying linearly with $\hbar\omega_0$, with slopes of $+1$, 0, -1 respectively.

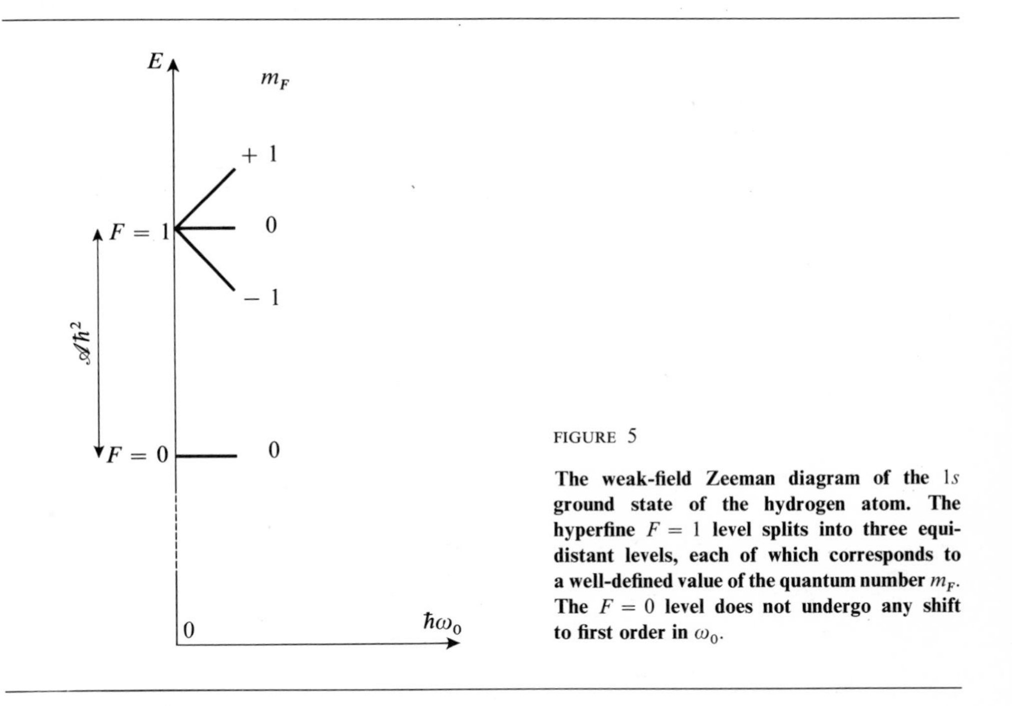

FIGURE 5

The weak-field Zeeman diagram of the $1s$ ground state of the hydrogen atom. The hyperfine $F = 1$ level splits into three equidistant levels, each of which corresponds to a well-defined value of the quantum number m_F. The $F = 0$ level does not undergo any shift to first order in ω_0.

The preceding treatment is valid as long as the difference $\hbar\omega_0$ between two adjacent Zeeman sublevels of the $F = 1$ level remains much smaller than the zero-field difference between the $F = 1$ and $F = 0$ levels (the hyperfine structure).

COMMENT:

The Wigner-Eckart theorem, mentioned above, makes it possible to show that, in a given level F of the total angular momentum, the Zeeman Hamiltonian $\omega_0(L_z + 2S_z)$ is represented by a matrix proportional to F_z. Thus, we can write, denoting the projector onto the F level by P_F:

$$P_F[\omega_0(L_z + 2S_z)]P_F = g_F\omega_0 P_F F_z P_F \tag{E-16}$$

g_F is called the *Landé factor* of the F state. In the case which concerns us here, $g_{F=1} = 1$.

c. THE BOHR FREQUENCIES INVOLVED IN THE EVOLUTION OF $\langle F \rangle$ AND $\langle S \rangle$. COMPARISON WITH THE VECTOR MODEL OF THE ATOM

In this section, we shall determine the different Bohr frequencies which appear in the evolution of $\langle \mathbf{F} \rangle$ and $\langle \mathbf{S} \rangle$, and show that certain aspects of the results obtained recall those found by using the vector model of the atom (*cf.* complement F_X).

First of all, we shall briefly review the predictions of the vector model of the atom (in which the various angular momenta are treated like classical vectors) as far as the hyperfine coupling between **I** and **S** is concerned. In a zero field, $\mathbf{F} = \mathbf{I} + \mathbf{S}$ is a constant of the motion. **I** and **S** precess about their resultant **F** with an angular velocity proportional to the coupling constant $\mathscr{A}$ between **I** and **S**. If the system is, in addition, placed in a weak static field $\mathbf{B}_0$ parallel to Oz, onto the rapid precessional motion of **I** and **S** about **F** is superposed a slow precessional motion of **F** about Oz (Larmor precession; fig. 6).

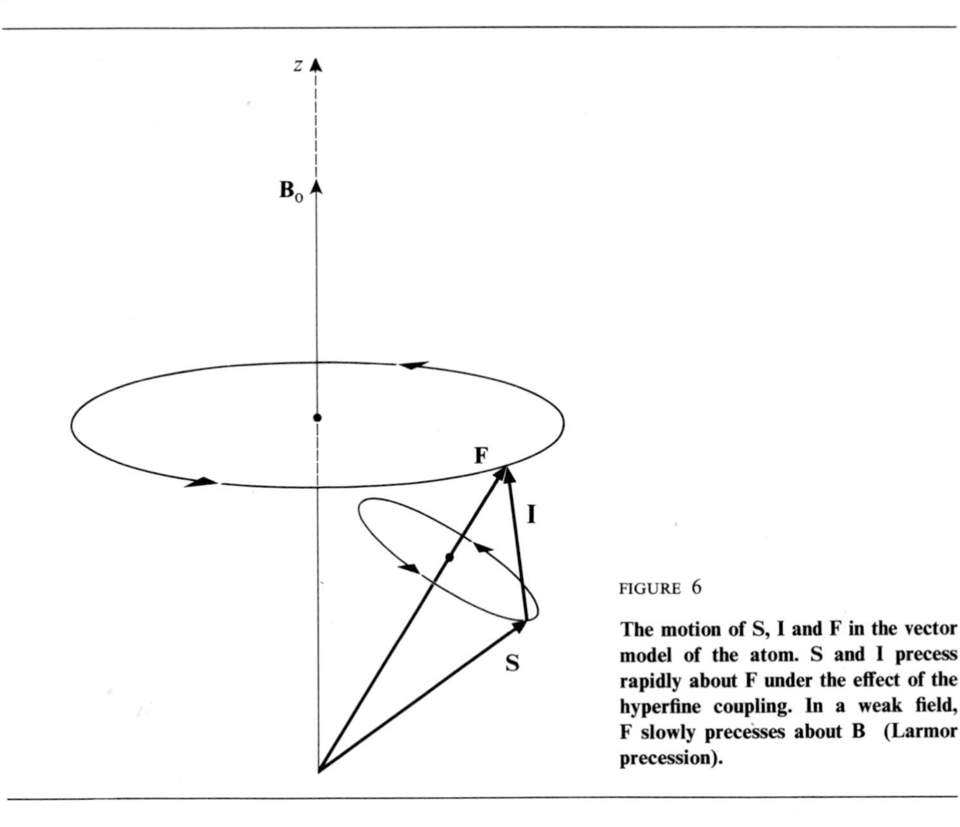

FIGURE 6

The motion of S, I and F in the vector model of the atom. S and I precess rapidly about F under the effect of the hyperfine coupling. In a weak field, F slowly precesses about B (Larmor precession).

F_z is therefore a constant of the motion, while S_z has a static part (the projection onto Oz of the component of **S** parallel to **F**), and a part which is modulated by the hyperfine precession frequency (the projection onto Oz of the component of **S** perpendicular to **F**, which precesses about **F**).

Let us compare these semi-classical results with those of the quantum theory presented earlier in this section. To do so, we must consider the time evolution of the mean values $\langle F_z \rangle$ and $\langle S_z \rangle$. According to the discussion of § D-2-d of chapter III, the mean value $\langle G \rangle(t)$ of a physical quantity G contains a series of components which oscillate at the various Bohr frequencies $(E - E')/h$ of the system. Also, a given Bohr frequency appears in $\langle G \rangle(t)$ only if the matrix element of G between the states corresponding to the two energies is different from zero. In the problem which concerns us here, the eigenstates of the weak-field Hamiltonian are the $| F, m_F \rangle$ states. Now consider the two matrices (E-10) and (E-11) which represent S_z and F_z in this basis. Since F_z has only diagonal matrix elements, no Bohr frequency different from zero can appear in $\langle F_z \rangle(t)$: $\langle F_z \rangle$ is therefore static. On the other hand, S_z has, not only diagonal matrix elements (with which is associated a static component of $\langle S_z \rangle$), but also a non-diagonal element between the $| F = 1, m_F = 0 \rangle$ and $| F = 0, m_F = 0 \rangle$ states, whose energy difference is $\mathscr{A}\hbar^2$, according to table (E-15) (or figure 5). It follows that $\langle S_z \rangle$ has, in addition to a static component, a component modulated at the angular frequency $\mathscr{A}\hbar$. This result recalls the one obtained using the vector model of the atom*.

COMMENT:

A relation can be established between perturbation theory and the vector model of the atom. The influence of a weak field B_0 on the $F = 1$ and $F = 0$ levels can be obtained by retaining in the Zeeman Hamiltonian $2\omega_0 S_z$ only the matrix elements in the $F = 1$ and $F = 0$ levels, "forgetting" the matrix element of S_z between $| F = 1; m_F = 0 \rangle$ and $| F = 0; m_F = 0 \rangle$. Proceeding in this way, we also "forget" the modulated component of $\langle S_z \rangle$, which is proportional to this matrix element. We therefore keep only the component of $\langle \mathbf{S} \rangle$ parallel to $\langle \mathbf{F} \rangle$.

Now, this is precisely what we do in the vector model of the atom when we want to evaluate the interaction energy with the field $\mathbf{B}_0$. In a weak field, $\mathbf{F}$ does precess about $\mathbf{B}_0$ much more slowly than $\mathbf{S}$ does about $\mathbf{F}$. The interaction of $\mathbf{B}_0$ with the component of $\mathbf{S}$ perpendicular to $\mathbf{F}$ therefore has no effect, on the average; only the projection of $\mathbf{S}$ onto $\mathbf{F}$ counts. This is how, for example, the Landé factor is calculated.

3. The strong-field Zeeman effect

We must now start by diagonalizing the Zeeman term.

a. EIGENSTATES AND EIGENVALUES OF THE ZEEMAN TERM

This term is diagonal in the $\{ | m_S, m_I \rangle \}$ basis:

$$2\omega_0 S_z | m_S, m_I \rangle = 2m_S\hbar\omega_0 | m_S, m_I \rangle \tag{E-17}$$

Since $m_S = \pm 1/2$, the eigenvalues are equal to $\pm \hbar\omega_0$. Each of them is therefore two-fold degenerate, because of the two possible values of m_I. We therefore have**:

$$\begin{cases} 2\omega_0 S_z | +, \pm \rangle = + \hbar\omega_0 | +, \pm \rangle \\ 2\omega_0 S_z | -, \pm \rangle = - \hbar\omega_0 | -, \pm \rangle \end{cases} \tag{E-18}$$

* A parallel could also be established between the evolution of $\langle F_x \rangle$, $\langle S_x \rangle$, $\langle F_y \rangle$, $\langle S_y \rangle$, and that of the projections of the vectors $\mathbf{F}$ and $\mathbf{S}$ of figure 6 onto Ox and Oy. However, the motion of $\langle \mathbf{F} \rangle$ and $\langle \mathbf{S} \rangle$ does not coincide perfectly with that of the classical angular momenta. In particular, the modulus of $\langle \mathbf{S} \rangle$ is not necessarily constant (in quantum mechanics, $\langle \mathbf{S}^2 \rangle \neq \langle \mathbf{S} \rangle^2$); see discussion of complement F_X.

** To simplify the notation, we shall often write $| \varepsilon_S, \varepsilon_I \rangle$ instead of $| m_S, m_I \rangle$, where ε_S and ε_I are equal to $+$ or $-$, depending on the signs of m_S and m_I.

b. EFFECTS OF THE HYPERFINE TERM CONSIDERED AS A PERTURBATION

The corrections to first order in $\mathscr{A}$ can be obtained by diagonalizing the restrictions of the operator $\mathscr{A}\,\mathbf{I}\,.\,\mathbf{S}$ to the two subspaces $\{\,|\,+,\pm\,\rangle\,\}$ and $\{\,|\,-,\pm\,\rangle\,\}$ corresponding to the two different eigenvalues of $2\omega_0 S_z$.

First of all, notice that, in each of these two subspaces, the two basis vectors $|\,+,+\,\rangle$ and $|\,+,-\,\rangle$ (or $|\,-,+\,\rangle$ and $|\,-,-\,\rangle$) are also eigenvectors of F_z, but do not correspond to the same value of $m_F = m_S + m_I$. Since the operator $\mathscr{A}\,\mathbf{I}\,.\,\mathbf{S} = \frac{\mathscr{A}}{2}(\mathbf{F}^2 - \mathbf{I}^2 - \mathbf{S}^2)$ commutes with F_z, it has no matrix elements between the two states $|\,+,+\,\rangle$ and $|\,+,-\,\rangle$, or $|\,-,+\,\rangle$ and $|\,-,-\,\rangle$. The two matrices representing $\mathscr{A}\,\mathbf{I}\,.\,\mathbf{S}$ in the two subspaces $\{\,|\,+,\pm\,\rangle\,\}$ and $\{\,|\,-,\pm\,\rangle\,\}$ are then diagonal, and their eigenvalues are simply the diagonal elements:

$$\langle\, m_S;\, m_I\,|\,\mathscr{A}\,\mathbf{I}\,.\,\mathbf{S}\,|\,m_S;\, m_I\,\rangle,$$

which can also be written, using the relation:

$$\mathbf{I}\,.\,\mathbf{S} = I_z S_z + \frac{1}{2}(I_+ S_- + I_- S_+) \tag{E-19}$$

in the form:

$$\langle\, m_S, m_I\,|\mathscr{A}\,\mathbf{I}\,.\,\mathbf{S}\,|\,m_S, m_I\,\rangle = \langle\, m_S, m_I\,|\,\mathscr{A} I_z S_z\,|\,m_S, m_I\,\rangle = \mathscr{A}\hbar^2 m_S m_I \tag{E-20}$$

Finally, in a strong field, the eigenstates (to zeroeth order in $\mathscr{A}$) and the eigenvalues (to first order in $\mathscr{A}$) are:

Eigenstates		*Eigenvalues*
$\|\,+,+\,\rangle$	$\longleftrightarrow$	$\hbar\omega_0 + \frac{\mathscr{A}\hbar^2}{4}$
$\|\,+,-\,\rangle$	$\longleftrightarrow$	$\hbar\omega_0 - \frac{\mathscr{A}\hbar^2}{4}$
$\|\,-,+\,\rangle$	$\longleftrightarrow$	$-\hbar\omega_0 - \frac{\mathscr{A}\hbar^2}{4}$
$\|\,-,-\,\rangle$	$\longleftrightarrow$	$-\hbar\omega_0 + \frac{\mathscr{A}\hbar^2}{4}$

(E-21)

In figure 7, the solid-line curves on the right-hand side (for $\hbar\omega_0 \gg \mathscr{A}\hbar^2$) represent the strong-field energy levels: we obtain two parallel straight lines of slope $+1$, separated by an energy $\mathscr{A}\hbar^2/2$, and two parallel straight lines of slope -1, also separated by $\mathscr{A}\hbar^2/2$. The perturbation treatments presented in this section and the preceding one therefore give the strong-field asymptotes and the tangents at the origin of the energy levels.

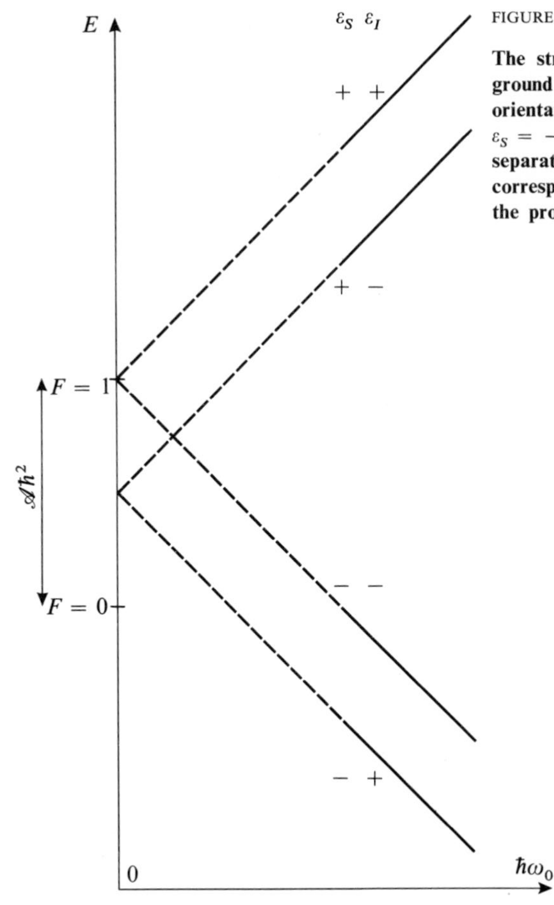

FIGURE 7

The strong-field Zeeman diagram of the $1s$ **ground state of the hydrogen atom. For each orientation of the electronic spin (**$\varepsilon_S = +$ **or** $\varepsilon_S = -$**), we obtain two parallel straight lines separated by an energy** $\mathscr{A}\hbar^2/2$**, each one corresponding to a different orientation of the proton spin (**$\varepsilon_I = +$ **or** $\varepsilon_I = -$**).**

COMMENT:

The strong-field splitting $\mathscr{A}\hbar^2/2$ of the two states, $| +, + \rangle$ and $| +, - \rangle$ or $| -, + \rangle$ and $| -, - \rangle$, can be interpreted in the following way. We have seen that only the term $I_z S_z$ of expression (E-19) for $\mathbf{I} \cdot \mathbf{S}$ is involved in a strong field, when the hyperfine coupling is treated like a perturbation of the Zeeman term. The total strong-field Hamiltonian (E-8) can therefore be written:

$$2\omega_0 S_z + \mathscr{A} I_z S_z = 2\left(\omega_0 + \frac{\mathscr{A}}{2} I_z\right) S_z \tag{E-22}$$

It is as if the electronic spin "saw", in addition to the external field $\mathbf{B}_0$, a smaller "internal field", arising from the hyperfine coupling between $\mathbf{I}$ and $\mathbf{S}$ and having two possible values, depending on whether the nuclear spin points up or down. This field adds to or substracts from $\mathbf{B}_0$ and is responsible for the energy difference between $| +, + \rangle$ and $| +, - \rangle$ or between $| -, + \rangle$ and $| -, - \rangle$.

c. THE BOHR FREQUENCIES INVOLVED IN THE EVOLUTION OF $\langle S_z \rangle$

In a strong field, the Zeeman coupling of **S** with $\mathbf{B}_0$ is more important than the hyperfine coupling of **S** with **I**. If we start by neglecting this hyperfine coupling, the vector model of the atom predicts that **S** will precess (very rapidly since $|\mathbf{B}_0|$ is large) about the Oz direction of $\mathbf{B}_0$ (**I** remains motionless, since we have assumed ω_n to be negligible).

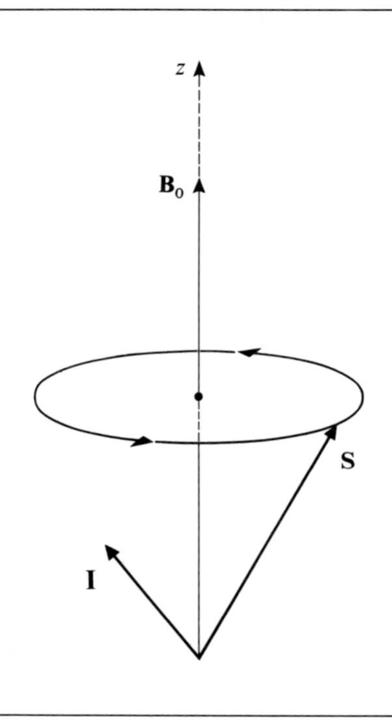

FIGURE 8

The motion of S in the vector model of the atom. In a strong field, S precesses rapidly about $\mathbf{B}_0$ (here we are neglecting both the Zeeman coupling between I and $\mathbf{B}_0$ and the hyperfine coupling between I and S, so that I remains motionless).

Expression (E-19) for the hyperfine coupling remains valid for classical vectors. Because of the very rapid precession of **S**, the terms S_+ and S_- oscillate very fast and have, on the average, no effect, so that only the term $I_z S_z$ counts. The effect of the hyperfine coupling is therefore to add a small field parallel to Oz and proportional to I_z (*cf.* comment of the preceding section), which accelerates or slows down the precession of **S** about Oz, depending on the sign of I_z. The vector model of the atom thus predicts that S_z will be static in a strong field.

We shall show that quantum theory gives an analogous result for the mean value $\langle S_z \rangle$ of the observable S_z. In a strong field, the well-defined energy states are, as we have seen, the states $| m_S, m_I \rangle$. Now, in this basis, the operator S_z has only diagonal matrix elements. No non-zero Bohr frequency can therefore appear in $\langle S_z \rangle$, which, consequently, is a static quantity*, unlike its weak-field counterpart (*cf.* § E-2-c).

* The study of $\langle S_x \rangle$ and $\langle S_y \rangle$ presents no difficulty. We find two Bohr angular frequencies : one, $\omega_0 + \mathcal{A}\hbar/2$, slightly larger than ω_0, and the other one, $\omega_0 - \mathcal{A}\hbar/2$, slightly smaller. They correspond to the two possible orientations of the "internal field", produced by I_z, which adds to the external field B_0.

Similarly, we find that **I** precesses about the "internal field" produced by S_z.

4. The intermediate-field Zeeman effect

a. THE MATRIX WHICH REPRESENTS THE TOTAL PERTURBATION IN THE $\{ | F, m_F \rangle \}$ BASIS

The $| F, m_F \rangle$ states are eigenstates of the operator $\mathscr{A} \mathbf{I} \cdot \mathbf{S}$. The matrix which represents this operator in the $\{ | F, m_F \rangle \}$ basis is therefore diagonal. The diagonal elements corresponding to $F = 1$ are equal to $\mathscr{A}\hbar^2/4$, and those corresponding to $F = 0$, to $- 3\mathscr{A}\hbar^2/4$. Furthermore, we have already written, in (E-10), the matrix representation of S_z in the same basis. It is now very simple to write the matrix which represents the total perturbation (E-8). Arranging the basis vectors in the order $| 1, 1 \rangle, | 1, -1 \rangle, | 1, 0 \rangle, | 0, 0 \rangle$, we thus obtain:

$$\begin{pmatrix} \frac{\mathscr{A}\hbar^2}{4} + \hbar\omega_0 & 0 & 0 & 0 \\ 0 & \frac{\mathscr{A}\hbar^2}{4} - \hbar\omega_0 & 0 & 0 \\ 0 & 0 & \frac{\mathscr{A}\hbar^2}{4} & \hbar\omega_0 \\ 0 & 0 & \hbar\omega_0 & -\frac{3\mathscr{A}\hbar^2}{4} \end{pmatrix} \tag{E-23}$$

COMMENT:

S_z and F_z commute; $2\omega_0 S_z$ can therefore have non-zero matrix elements only between two states with the same m_F. Thus, we could have predicted all the zeros of matrix (E-23).

b. ENERGY VALUES IN AN ARBITRARY FIELD

Matrix (E-23) can be broken into two 1×1 matrices and one 2×2 matrix. The two 1×1 matrices immediately yield two eigenvalues:

$$\begin{cases} E_1 = \dfrac{\mathscr{A}\hbar^2}{4} + \hbar\omega_0 \\ E_2 = \dfrac{\mathscr{A}\hbar^2}{4} - \hbar\omega_0 \end{cases} \tag{E-24}$$

corresponding respectively to the state $| 1, 1 \rangle$ (that is, the state $| +, + \rangle$) and to the state $|1, -1 \rangle$ (that is, the state $| -, - \rangle$). In figure 9, the two straight lines of slopes $+1$ and -1 passing through the point whose ordinate is $+ \mathscr{A}\hbar^2/4$ for a zero field (for which the perturbation theory treatment gave only the initial and asymptotic behavior) therefore represent, for any B_0, two of the Zeeman sublevels.

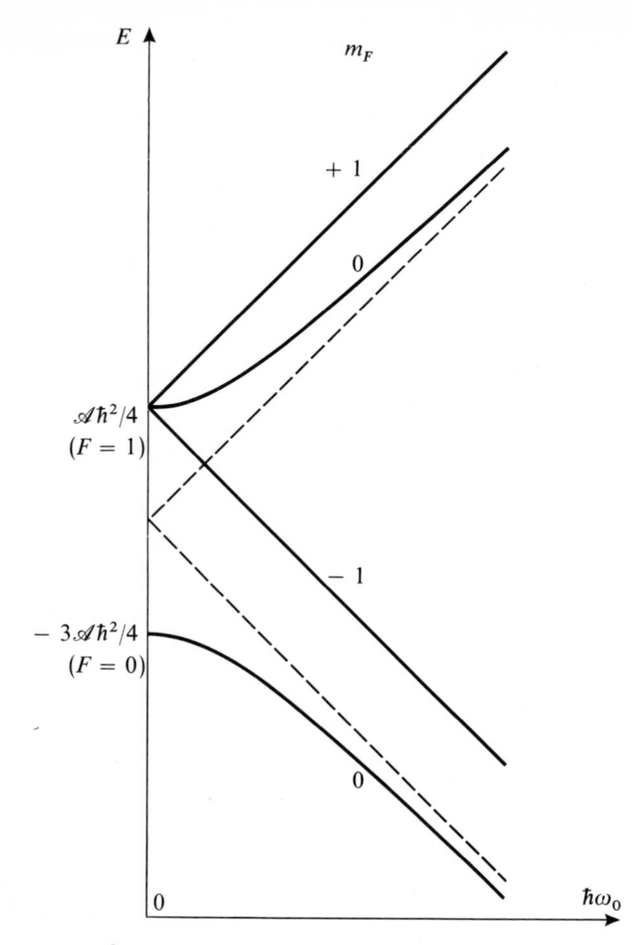

FIGURE 9

The Zeeman diagram (for an arbitrary field) of the 1s ground state of the hydrogen atom : m_F remains a good quantum number for any value of the field. We obtain two straight lines, of opposite slopes, corresponding to the values + 1 and − 1 of m_F, as well as a hyperbola whose two branches are associated with the two $m_F = 0$ levels. Figures 5 and 7 give, respectively, the tangents at the origin and the asymptotes of the levels shown in this diagram.

The eigenvalue equation of the remaining 2 × 2 matrix can be written:

$$\left(\frac{\mathscr{A}\hbar^2}{4} - E\right)\left(-\frac{3\mathscr{A}\hbar^2}{4} - E\right) - \hbar^2\omega_0^2 = 0 \tag{E-25}$$

The two roots of this equation can easily be found:

$$E_3 = -\frac{\mathscr{A}\hbar^2}{4} + \sqrt{\left(\frac{\mathscr{A}\hbar^2}{2}\right)^2 + \hbar^2\omega_0^2} \tag{E-26}$$

$$E_4 = -\frac{\mathscr{A}\hbar^2}{4} - \sqrt{\left(\frac{\mathscr{A}\hbar^2}{2}\right)^2 + \hbar^2\omega_0^2} \tag{E-27}$$

When $\hbar\omega_0$ varies, the two points of abscissas $\hbar\omega_0$ and ordinates E_3 and E_4 follow the two branches of a hyperbola (fig. 9). The asymptotes of this hyperbola are the two straight lines whose equation is $E = -(\mathscr{A}\hbar^2/4) \pm \hbar\omega_0$, obtained in §3 above. The two turning points of the hyperbola have abscissas of $\omega_0 = 0$ and ordinates of $-(\mathscr{A}\hbar^2/4) \pm \mathscr{A}\hbar^2/2$, that is, $\mathscr{A}\hbar^2/4$ and $-3\mathscr{A}\hbar^2/4$. The tangents at both these points are horizontal. This is in agreement with the results of §2 for the states $| F = 1 ; m_F = 0 \rangle$ and $| F = 0 ; m_F = 0 \rangle$.

The preceding results are summarized in figure 9, which is the Zeeman diagram of the $1s$ ground state.

c. PARTIAL HYPERFINE DECOUPLING

In a weak field, the well-defined energy states are the states $| F, m_F \rangle$; in a strong field, the states $| m_S, m_I \rangle$; in an intermediate field, the eigenstates of matrix (E-23), which are intermediate between the states $| F, m_F \rangle$ and the states $| m_S, m_I \rangle$.

One thus moves continuously from a strong coupling between **I** and **S** (coupled bases) to a total decoupling (uncoupled bases) via a partial coupling.

COMMENT:

An analogous phenomenon exists for the Zeeman fine structure effect. If, for simplicity, we neglect W_{hf}, we know (§ C) that, in a zero field, the eigenstates of the Hamiltonian H are the $| J, m_J \rangle$ states corresponding to a strong coupling between **L** and **S** (the spin-orbit coupling). This property remains valid as long as $W_Z \ll W_f$. If, on the other hand, B_0 is strong enough to make $W_Z \gg W_f$, we find that the eigenstates of H are the $| m_L, m_S \rangle$ states corresponding to a total decoupling of **L** and **S**. The intermediate zone ($W_Z \simeq W_f$) corresponds to a partial coupling of **L** and **S**. See, for example, complement D_{XII}, in which we study the Zeeman effect of the $2p$ level (without taking W_{hf} into account).

References and suggestions for further reading:

The hydrogen atom spectrum: Series (11.7), Bethe and Salpeter (11.10).

The Dirac equation : the subsection "Relativistic quantum mechanics" of section 2 of the bibliography and Messiah (1.17), chap. XX, especially §§ V and IV-27.

The fine structure of the $n = 2$ level and the Lamb shift: Lamb and Retherford (3.11); Frisch (3.13); Series (11.7), chaps. VI, VII and VIII.

The hyperfine structure of the ground state: Crampton et al (3.12).

The Zeeman effect and the vector model of the atom: Cagnac and Pebay-Peyroula (11.2), chap. XVII, §§3E and 4C; Born (11.4), chap. 6, §2.

Interstellar hydrogen: Roberts (11.17); Encrenaz (12.11), chap. IV.

A_{XII}: THE MAGNETIC HYPERFINE HAMILTONIAN	A_{XII}: derivation of the expression for the hyperfine Hamiltonian used in chapter XII. Gives the physical interpretation of the various terms appearing in this Hamiltonian – in particular, the contact term. Rather difficult.
B_{XII}: CALCULATION OF THE MEAN VALUES OF THE FINE-STRUCTURE HAMILTONIAN IN THE $1s$, $2s$ AND $2p$ STATES	B_{XII}: the detailed calculation of certain radial integrals appearing in the expressions obtained in chapter XII for the energy shifts. Not conceptually difficult.
C_{XII}: THE HYPERFINE STRUCTURE AND THE ZEEMAN EFFECT FOR MUONIUM AND POSITRONIUM	C_{XII}: extension of the study of §§ D and E of chapter XII to two important hydrogen-like systems, muonium and positronium, already presented in complement A_{VII}. Brief description of experimental methods for studying these two systems. Simple if the calculations of §§ D and E of chapter XII have been well understood.
D_{XII}: THE INFLUENCE OF THE ELECTRON SPIN ON THE ZEEMAN EFFECT OF THE HYDROGEN RESONANCE LINE	D_{XII}: study of the effect of the electronic spin on the frequencies and polarizations of the Zeeman components of the resonance line of hydrogen. Improves the results obtained in complement D_{VII}, in which the electron spin was ignored (uses certain results of that complement). Moderately difficult.
E_{XII}: THE STARK EFFECT FOR THE HYDROGEN ATOM	E_{XII}: study of the effect of a static electric field on the ground state ($n = 1$) and the first excited state ($n = 2$) of the hydrogen atom (the Stark effect). Shows the importance for the Stark effect of the existence of a degeneracy between two states of different parities. Rather simple.

Complement A_{XII}

THE MAGNETIC HYPERFINE HAMILTONIAN

The aim of this complement is to justify the expression for the hyperfine Hamiltonian given in chapter XII [relation (B-20)]. As in that chapter, we shall confine our reasoning to the hydrogen atom, which is composed of a single electron and a proton, although most of the ideas remain valid for any atom. We have already said that the origin of the hyperfine Hamiltonian is the coupling between the electron and the electromagnetic field created by the proton. We shall therefore call $\mathbf{A}_I(\mathbf{r})$ and $U_I(\mathbf{r})$ the vector and scalar potentials associated with this electromagnetic field. We shall begin by considering the Hamiltonian of an electron subjected to these potentials.

1. Interaction of the electron with the scalar and vector potentials created by the proton

Let $\mathbf{R}$ and $\mathbf{P}$ be the position and momentum of the electron, $\mathbf{S}$, its spin; m_e, its mass; and q, its charge; $\mu_B = q\hbar/2m_e$ is the Bohr magneton.

The Hamiltonian H of the electron in the field of the proton can be written:

$$H = \frac{1}{2m_e}[\mathbf{P} - q\mathbf{A}_I(\mathbf{R})]^2 + qU_I(\mathbf{R}) - 2\mu_B\left(\frac{\mathbf{S}}{\hbar}\right) \cdot \nabla \times \mathbf{A}_I(\mathbf{R}) \tag{1}$$

This operator is obtained by adding the coupling energy between the magnetic moment $2\mu_B\mathbf{S}/\hbar$ associated with the spin and the magnetic field $\nabla \times \mathbf{A}_I(\mathbf{r})$ to expression (B-46) of chapter III (the Hamiltonian of a spinless particle).

We shall begin by studying the terms which, in (1), arise from the scalar potential $U_I(\mathbf{r})$. According to complement E_X, this potential results from the superposition of several contributions, each of them associated with one of the electric multipole moments of the nucleus. For an arbitrary nucleus, we must consider:

(i) The total charge $-Zq$ of the nucleus (the moment of order $k = 0$), which yields a potential energy:

$$V_0(\mathbf{r}) = qU_0(\mathbf{r}) = -\frac{Zq^2}{4\pi\varepsilon_0 r} \tag{2}$$

(with, for the proton, $Z = 1$). Now, the Hamiltonian which we chose in chapter VII for the study of the hydrogen atom is precisely:

$$H_0 = \frac{\mathbf{P}^2}{2m_e} + V_0(\mathbf{R}) \tag{3}$$

$V_0(\mathbf{R})$ has therefore already been taken into account in the principal Hamiltonian H_0.

(ii) The electric quadrupole moment ($k = 2$) of the nucleus. The corresponding potential adds to the potential V_0 and yields a term of the hyperfine Hamiltonian, called the electric quadrupole term. The results of complement E_X enable us to write this term without difficulty. In the case of the hydrogen atom, it is zero, since the proton, which is a spin 1/2 particle, has no electric quadrupole moment (*cf.* §2-c-α of complement E_X).

(iii) The electric multipole moments of order $k = 4, 6$, etc... which are theoretically involved as long as $k \leqslant 2I$; for the proton, they are all zero.

Thus, for the hydrogen atom, potential (2) is really the potential seen by the electron★. There is no need to add any corrections to it (by hydrogen atom, we mean the electron-proton system, excluding isotopes such as deuterium : since the deuterium nucleus has a spin $I = 1$, we would have to take into account an electric quadrupole hyperfine Hamiltonian — see comment (i) at the end of this complement).

Now let us consider the terms arising from the vector potential $\mathbf{A}_I(\mathbf{r})$ in (1). We denote by $\mathbf{M}_I$ the magnetic dipole moment of the proton (which, for the same reason as above, cannot have magnetic multipole moments of order $k > 1$). We have:

$$\mathbf{A}_I(\mathbf{r}) = \frac{\mu_0}{4\pi}\frac{\mathbf{M}_I \times \mathbf{r}}{r^3} \tag{4}$$

The hyperfine Hamiltonian W_{hf} can now be obtained by retaining in (1) the terms which are linear in $\mathbf{A}_I$:

$$W_{hf} = -\frac{q}{2m_e}[\mathbf{P} \cdot \mathbf{A}_I(\mathbf{R}) + \mathbf{A}_I(\mathbf{R}) \cdot \mathbf{P}] - 2\mu_B\left(\frac{\mathbf{S}}{\hbar}\right) \cdot \boldsymbol{\nabla} \times \mathbf{A}_I(\mathbf{R}) \tag{5}$$

and by replacing $\mathbf{A}_I$ by expression (4) (since W_{hf} already makes a very small correction to the energy levels of H_0, it is perfectly legitimate to ignore the second-order term, in $\mathbf{A}_I^2$). This is what we shall do in the following section.

★ We are concerned here only with the potential outside the nucleus, where the multipole moment expansion is possible. Inside the nucleus, we know that the potential does not have form (2). This causes a shift in the atomic levels called the "volume effect". This effect was studied in complement D_{XI}, and we shall not take it into account here.

2. The detailed form of the hyperfine Hamiltonian

a. COUPLING OF THE MAGNETIC MOMENT OF THE PROTON WITH THE ORBITAL ANGULAR MOMENTUM OF THE ELECTRON

First of all, we shall calculate the first term of (5). Using (4), we have:

$$\mathbf{P} \cdot \mathbf{A}_I(\mathbf{R}) + \mathbf{A}_I(\mathbf{R}) \cdot \mathbf{P} = \frac{\mu_0}{4\pi} \left\{ \mathbf{P} \cdot (\mathbf{M}_I \times \mathbf{R}) \frac{1}{R^3} + \frac{1}{R^3} (\mathbf{M}_I \times \mathbf{R}) \cdot \mathbf{P} \right\} \tag{6}$$

We can apply the rules for a mixed vector product to vector operators as long as we do not change the order of two non-commuting operators. The components of $\mathbf{M}_I$ commute with $\mathbf{R}$ and $\mathbf{P}$, so we have:

$$(\mathbf{M}_I \times \mathbf{R}) \cdot \mathbf{P} = (\mathbf{R} \times \mathbf{P}) \cdot \mathbf{M}_I = \mathbf{L} \cdot \mathbf{M}_I \tag{7}$$

where:

$$\mathbf{L} = \mathbf{R} \times \mathbf{P} \tag{8}$$

is the orbital angular momentum of the electron. It can easily be shown that:

$$\left[\mathbf{L}, \frac{1}{R^3} \right] = \mathbf{0} \tag{9}$$

(any function of $|\mathbf{R}|$ is a scalar operator), so that:

$$\frac{1}{R^3} (\mathbf{M}_I \times \mathbf{R}) \cdot \mathbf{P} = \frac{\mathbf{L} \cdot \mathbf{M}_I}{R^3} \tag{10}$$

Similarly:

$$\mathbf{P} \cdot (\mathbf{M}_I \times \mathbf{R}) \frac{1}{R^3} = - \mathbf{M}_I \cdot (\mathbf{P} \times \mathbf{R}) \frac{1}{R^3} = \frac{\mathbf{M}_I \cdot \mathbf{L}}{R^3} \tag{11}$$

since:

$$- \mathbf{P} \times \mathbf{R} = \mathbf{L} \tag{12}$$

Thus, the first term of (5) makes a contribution W_{hf}^L to W_{hf} which is equal to:

$$W_{hf}^L = - \frac{\mu_0}{4\pi} \frac{q}{2m_e} 2 \frac{\mathbf{M}_I \cdot \mathbf{L}}{R^3} = - \frac{\mu_0}{4\pi} 2\mu_B \frac{\mathbf{M}_I \cdot (\mathbf{L}/\hbar)}{R^3} \tag{13}$$

This term corresponds to the coupling between the nuclear magnetic moment $\mathbf{M}_I$ and the magnetic field :

$$\mathbf{B}_L = \frac{\mu_0}{4\pi} \frac{q\mathbf{L}}{m_e r^3}$$

created by the current loop associated with the rotation of the electron (*cf.* fig. 1).

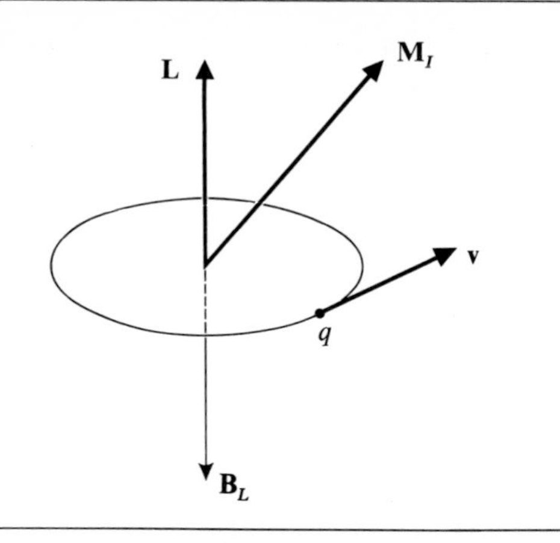

FIGURE 1

Relative disposition of the magnetic moment M_I of the proton and the field B_L created by the current loop associated with the motion of the electron of charge q and velocity v (B_L is anti-parallel to the orbital angular momentum L of the electron).

COMMENT:

The presence of the $1/R^3$ term in (13) might lead us to believe that there is a singularity at the origin, and that certain matrix elements of W_{hf}^L are infinite. Actually, this is not the case. Consider the matrix element $\langle \varphi_{k,l,m} | W_{hf}^L | \varphi_{k',l',m'} \rangle$, where $| \varphi_{k,l,m} \rangle$ and $| \varphi_{k',l',m'} \rangle$ are the stationary states of the hydrogen atom found in chapter VII. In the $\{ | \mathbf{r} \rangle \}$ representation, we have:

$$\langle \mathbf{r} | \varphi_{k,l,m} \rangle = \varphi_{k,l,m}(\mathbf{r}) = R_{k,l}(r) \, Y_l^m(\theta, \varphi) \tag{14}$$

with [*cf.* chap. VII, relation (A-28)]:

$$R_{k,l}(r) \underset{r \to 0}{\sim} C \, r^l \tag{15}$$

With the presence of the $r^2 \, \mathrm{d}r$ term in the integration volume element taken into account, the function to be integrated over r behaves at the origin like $r^{l+l'+2-3} = r^{l+l'-1}$. Furthermore, the presence of the Hermitian operator $\mathbf{L}$ in (13) means that the matrix element

$$\langle \varphi_{k,l,m} | W_{hf}^L | \varphi_{k',l',m'} \rangle$$

is zero when l or l' is zero. We then have $l + l' \geqslant 2$, and $r^{l+l'-1}$ remains finite at the origin.

b. COUPLING WITH THE ELECTRON SPIN

We shall see that, for the last term of (5), the problems related to the singularity at the origin of the vector potential (4) are important. This is why, in studying this term, we shall choose a proton of finite size, letting its radius approach zero at the end of the calculations. Furthermore, from a physical point of view, we now know that the proton does possess a certain spatial extension and that its "magnetism" is spread over a certain volume. However, the dimensions of the proton are much smaller than the Bohr radius a_0. This justifies our treating the proton as a point particle in the final stage of the calculation.

α. *The magnetic field associated with the proton*

Consider the proton to be a particle of radius ρ_0 (fig. 2), placed at the origin. The distribution of magnetism inside the proton creates, at a distant point, a field $\mathbf{B}$ which can be calculated by attributing to the proton a magnetic moment $\mathbf{M}_I$ which we shall choose parallel to Oz. For $r \gg \rho_0$, we obtain the components of $\mathbf{B}$ from the curl of the vector potential written in (4):

$$\begin{cases} B_x = \dfrac{\mu_0}{4\pi} 3M_I \dfrac{xz}{r^5} \\ B_y = \dfrac{\mu_0}{4\pi} 3M_I \dfrac{yz}{r^5} \\ B_z = \dfrac{\mu_0}{4\pi} M_I \dfrac{3z^2 - r^2}{r^5} \end{cases} \tag{16}$$

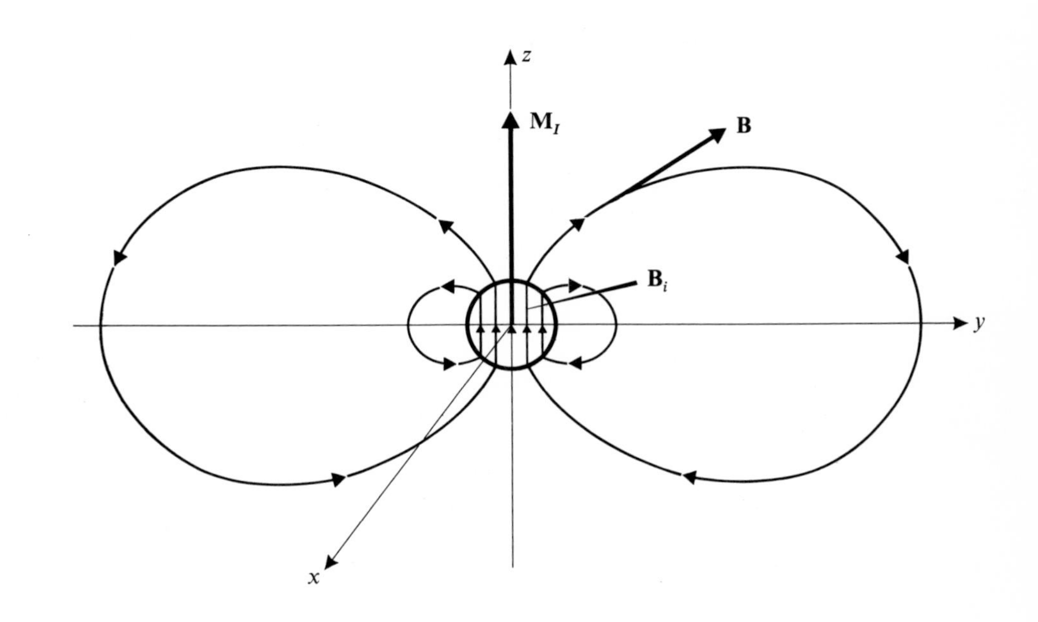

FIGURE 2

The magnetic field created by the proton. Outside the proton, the field is that of a dipole. Inside, the field depends on the exact distribution of the magnetism of the proton, but we can, in a first approximation, consider it to be uniform. The contact term corresponds to the interaction between the spin magnetic moment of the electron and this uniform field $\mathbf{B}_i$ inside the proton.

Expressions (16), moreover, remain valid even if r is not very large compared to ρ_0. We have already emphasized that the proton, since it is a spin 1/2 particle, has no magnetic multipole moments of order $k > 1$. The field outside the proton is therefore a pure dipole field.

Inside the proton, the magnetic field depends on the exact magnetic distribution. We shall assume this field $\mathbf{B}_i$ to be uniform (by symmetry, it must then be parallel to $\mathbf{M}_I$ and, therefore, to Oz)*.

To calculate the field $\mathbf{B}_i$ inside the proton, we shall write the equation stating that the flux of the magnetic field through a closed surface, bounded by the xOy plane and the upper hemisphere centered at O and of infinite radius, is zero. Since, as $r \longrightarrow \infty$, $|\mathbf{B}|$ decreases as $1/r^3$, the flux through this hemisphere is zero. Therefore, if $\Phi_i(\rho_0)$ denotes the flux through a disk centered at O of radius ρ_0 in the xOy plane, and $\Phi_e(\rho_0)$, the flux through the rest of the xOy plane, we have:

$$\Phi_i(\rho_0) + \Phi_e(\rho_0) = 0 \tag{17}$$

Relations (16) enable us to calculate $\Phi_e(\rho_0)$ easily, and we get:

$$\begin{aligned}\Phi_e(\rho_0) &= 2\pi \int_{\rho_0}^{+\infty} r\, \mathrm{d}r \left[-\frac{\mu_0}{4\pi} M_I \frac{1}{r^3} \right] \\ &= -\frac{\mu_0}{4\pi} M_I \frac{2\pi}{\rho_0}\end{aligned} \tag{18}$$

As for the flux $\Phi_i(\rho_0)$ of $\mathbf{B}_i$, it is equal to:

$$\Phi_i(\rho_0) = \pi \rho_0^2 B_i \tag{19}$$

so that (17) and (18) yield:

$$B_i = \frac{\mu_0}{4\pi} M_I \frac{2}{\rho_0^3} \tag{20}$$

Thus, we know the values of the field created by the proton at all points in space. We can now calculate the part of W_{hf} related to the electron spin $\mathbf{S}$.

β. *The magnetic dipole term*

If we substitute (16) into the term $-2\mu_B\left(\dfrac{\mathbf{S}}{\hbar}\right) . \nabla \times \mathbf{A}_I$, we obtain the operator:

$$W_{hf}^{\text{dip}} = -\frac{\mu_0}{4\pi}\frac{2\mu_B M_I}{\hbar} \left\{ 3Z \frac{XS_x + YS_y + ZS_z}{R^5} - \frac{S_z}{R^3} \right\} \tag{21}$$

that is, taking into account the fact that $\mathbf{M}_I$ is, by hypothesis, parallel to Oz:

$$W_{hf}^{\text{dip}} = \frac{\mu_0}{4\pi}\frac{2\mu_B}{\hbar}\frac{1}{R^3} \left\{ \mathbf{S} . \mathbf{M}_I - 3\frac{(\mathbf{S} . \mathbf{R})(\mathbf{M}_I . \mathbf{R})}{R^2} \right\} \tag{22}$$

* The following argument can be generalized to cases where $\mathbf{B}_i$ varies in a more complicated fashion (see comment (*iv*) at the end of this complement).

This is the expression for the Hamiltonian of the dipole-dipole interaction between two magnetic moments $\mathbf{M}_I$ and $\mathbf{M}_S = 2\mu_B\mathbf{S}/\hbar$ (*cf.* complement B_{XI}, § 1).

Actually, expression (16) for the magnetic field created by the proton is valid only for $r \geqslant \rho_0$, and (22) should be applied only to the part of the wave functions which satisfies this condition. However, when we let ρ_0 approach zero, expression (22) gives no singularity at the origin; it is therefore valid in all space.

Consider the matrix element:

$$\langle \varphi_{k,l,m,\varepsilon} | W_{hf}^{\text{dip}} | \varphi_{k',l',m',\varepsilon'} \rangle$$

(we are adding here the indices ε and ε' to the states $| \varphi_{k,l,m} \rangle$ considered above in order to label the eigenvalues $\varepsilon\hbar/2$ and $\varepsilon'\hbar/2$ of S_z) and, in particular, the radial integral which corresponds to it. At the origin, the function of r to be integrated behaves like $r^{l+l'+2-3} = r^{l+l'-1}$. Now, according to condition (8-c) of complement B_{XI}, the non-zero matrix elements are obtained for $l + l' \geqslant 2$. There is therefore no divergence at the origin. In the limit where $\rho_0 \longrightarrow 0$, the integral over r becomes an integral from 0 to infinity, and expression (22) is valid in all space.

γ. *The contact term*

We shall now substitute (20) into the last term of (5), so as to obtain the contribution of the internal field of the proton to W_{hf}. We then obtain an operator W_{hf}^c, which we shall call the "contact term", and whose matrix elements in the $\{|\varphi_{k,l,m,\varepsilon}\rangle\}$ representation are:

$$\langle \varphi_{k,l,m,\varepsilon} | W_{hf}^c | \varphi_{k',l',m',\varepsilon'} \rangle = -\frac{\mu_0}{4\pi}\frac{2\mu_B M_I}{\hbar} \langle \varepsilon | S_z | \varepsilon' \rangle \frac{2}{\rho^3} \iiint_{r \leqslant \rho_0} d^3r \; \varphi_{k,l,m}^*(\mathbf{r}) \, \varphi_{k',l',m'}(\mathbf{r}) \tag{23}$$

Let ρ_0 approach zero. The integration volume, $4\pi\rho_0^3/3$, also approaches zero, and the right-hand side of (23) becomes:

$$-\frac{\mu_0}{4\pi}\frac{2\mu_B M_I}{\hbar} \langle \varepsilon | S_z | \varepsilon' \rangle \frac{8\pi}{3} \varphi_{k,l,m}^*(\mathbf{r} = \mathbf{0}) \, \varphi_{k',l',m'}(\mathbf{r} = \mathbf{0}) \tag{24}$$

The contact term is therefore given by:

$$W_{hf}^c = -\frac{\mu_0}{4\pi}\frac{8\pi}{3} \mathbf{M}_I \cdot \left(\frac{2\mu_B \mathbf{S}}{\hbar}\right) \delta(\mathbf{R}) \tag{25}$$

Therefore, although the volume containing an internal magnetic field (20) approaches zero when $\rho_0 \longrightarrow 0$, the value of W_{hf}^c remains finite, since this internal field approaches infinity as $1/\rho_0^3$.

COMMENTS:

(*i*) In (25), the function $\delta(\mathbf{R})$ of the operator $\mathbf{R}$ is simply the projector:

$$\delta(\mathbf{R}) = | \mathbf{r} = \mathbf{0} \rangle \langle \mathbf{r} = \mathbf{0} | \tag{26}$$

(ii) The matrix element written in (25) is different from zero only if $l = l' = 0$. This is a necessary condition for $\varphi_{k,l,m}(\mathbf{r} = \mathbf{0})$ and $\varphi_{k',l',m'}(\mathbf{r} = \mathbf{0})$ to be non-zero (*cf.* chap. VII, §C-4-c-β). The contact term therefore exists only for the *s* states.

(iii) In order to study, in § 2-a, the coupling between $\mathbf{M}_I$ and the orbital angular momentum of the electron, we assumed expression (4) for $\mathbf{A}_I(\mathbf{r})$ to be valid in all space. This amounts to ignoring the fact that the field $\mathbf{B}$ actually has the form (20) inside the proton. We might wonder if this procedure is correct, or if there is not also an orbital contact term in W_{hf}^L.

Actually, this is not the case. The term in $\mathbf{P} \cdot \mathbf{A}_I + \mathbf{A}_I \cdot \mathbf{P}$ would lead, for the field $\mathbf{B}_i$, to an operator proportional to:

$$\mathbf{B}_i \cdot \mathbf{L} = \frac{\mu_0}{4\pi} M_I \frac{2}{\rho_0^3} L_z \tag{27}$$

Let us calculate the matrix element of such an operator in the $\{ | \varphi_{k,l,m} \rangle \}$ representation. The presence of the operator L_z requires, as above, $l, l' \geqslant 1$. The radial function to be integrated between 0 and ρ_0 then behaves at the origin like $r^{l+l'+2}$ and therefore goes to zero at least as rapidly as r^4. Despite the presence of the $1/\rho_0^3$ term in (27), the integral between $r = 0$ and $r = \rho_0$ therefore goes to zero in the limit where $\rho_0 \longrightarrow 0$.

3. Conclusion: the hyperfine-structure Hamiltonian

Now, let us take the sum of the operators W_{hf}^L, W_{hf}^{dip} and W_{hf}^c. We use the fact that the magnetic dipole moment $\mathbf{M}_I$ of the proton is proportional to its angular momentum $\mathbf{I}$:

$$\mathbf{M}_I = g_p \mu_n \left(\frac{\mathbf{I}}{\hbar} \right) \tag{28}$$

(*cf.* § B-2-a of chapter XII). We obtain:

$$W_{hf} = -\frac{\mu_0}{4\pi} \frac{2\mu_B \mu_n g_p}{\hbar^2} \left\{ \frac{\mathbf{I} \cdot \mathbf{L}}{R^3} + 3 \frac{(\mathbf{I} \cdot \mathbf{R})(\mathbf{S} \cdot \mathbf{R})}{R^5} - \frac{\mathbf{I} \cdot \mathbf{S}}{R^3} + \frac{8\pi}{3} \mathbf{I} \cdot \mathbf{S}\, \delta(\mathbf{R}) \right\} \tag{29}$$

This operator acts both in the state space of the electron and in the state space of the proton. It can be seen that this is indeed the operator introduced in chapter XII [*cf.* (B-20)].

COMMENTS:

(i) We will now discuss the generalization of formula (29) to the case of an atom having a nuclear spin $I > 1/2$.
First of all, if $I = 1$, we have already seen that the nucleus can have an electric quadrupole moment which adds a contribution to the potential $V_0(\mathbf{r})$ given by (2). An electric quadrupole hyperfine term is therefore present in the hyperfine Hamiltonian, in addition to the magnetic dipole term (29). Since

an electrical interaction does not directly affect the electron spin, this quadrupole term only acts on the orbital variables of the electrons.

If now $I > 1$, other nuclear electric or magnetic multipole moments can exist, increasing in number as I increases. The electric moments give rise to hyperfine terms acting only on the orbital electron variables, while the magnetic terms act on both the orbital and the spin variables. For elevated values of I, the hyperfine Hamiltonian has therefore a complex structure. In practice however, for the great majority of cases, one can limit the hyperfine Hamiltonian to magnetic dipole and electric quadrupole terms. This is due to the fact that the multipole nuclear moments of an order superior to 2 make extremely small contributions to the hyperfine atomic structures. These contributions are therefore difficult to observe experimentally. This arises essentially from the extremely small size of the nuclei compared to the spatial extent a_0 of the electronic wave functions.

(*ii*) The simplifying hypothesis which we have made concerning the field $\mathbf{B}(\mathbf{r})$ created by the proton (a perfectly uniform field within a sphere, a dipole field outside) is not essential. The form (25) of the magnetic dipole Hamiltonian remains valid whenever the nuclear magnetism has an arbitrary repartition, giving rise to more complicated internal fields $\mathbf{B}_i(\mathbf{r})$ (assuming however that the spatial extent of the nucleus is negligible compared to a_0; *cf.* the following comment). The argument is actually a direct generalization of that given in this complement. Consider a sphere S_ε centered at the origin, containing the nucleus and having a radius $\varepsilon \ll a_0$.

If $I = 1/2$, the field outside S_ε has the form (16) and, since ε is very small compared to a_0, its contribution leads to the terms (13) and (22). As for the contribution of the field $\mathbf{B}(\mathbf{r})$ inside S_ε, it depends only on the value at the origin of the electronic wave functions and on the integral of $\mathbf{B}(\mathbf{r})$ inside S_ε. Since the flux of $\mathbf{B}(\mathbf{r})$ across all closed surfaces is zero, the integral in S_ε of each component of $\mathbf{B}(\mathbf{r})$ can be transformed into an integral outside of S_ε, where $\mathbf{B}(\mathbf{r})$ has the form (16). A simple calculation will again give exactly expression (25) which is therefore independent of the simplifying hypothesis that we have made.

If $I > 1/2$, the nuclear contribution to the electromagnetic field outside of S_ε gives rise to the multipole hyperfine Hamiltonian which we have discussed in comment (*i*) above. On the other hand, one can easily show that the contribution of the field inside S_ε does not give rise to any new term : only the magnetic dipole posesses a contact term.

(*iii*) In all of the above, we have totally neglected the dimensions of the nucleus compared to those of the electronic wave functions (we have taken the limit $\rho_0/a_0 \longrightarrow 0$). This is obviously not always realistic, in particular for heavy atoms whose nuclei have a relatively large spatial extension. If one studies these "volume effects" (keeping for example several of the lower order terms in ρ_0/a_0), a series of new terms appears in the electron-nucleus interaction Hamiltonian. We have already encountered this type of effect in complement D_{XI} where we studied the effects of the radial distribution of the nuclear charge (nuclear multipole moments of order $k = 0$). Analogous phenomena occur concerning the spatial distribution of nuclear magnetism and lead to modifications of different terms of the hyperfine Hamiltonian (29). In particular, a new term must be added to the contact term (25) when the electronic wave functions vary significantly within the nucleus. This new term is neither simply proportional to $\delta(\mathbf{R})$, nor to the total magnetic moment of the nucleus. It depends on the spatial distribution of the nuclear magnetism. From a practical point of view, such a term is interesting since, using precise measurements of the hyperfine structure of heavy atoms, it permits obtaining information concerning the variation of the magnetism within the corresponding nuclei.

References and suggestions for further reading:

The hyperfine Hamiltonian including the electric quadrupole interaction term: Abragam (14.1), chap. VI; Kuhn (11.1), chap. VI, § B; Sobel'man (11.12), chap. 6.

Complement B_{XII}

CALCULATION OF THE MEAN VALUES OF THE FINE-STRUCTURE HAMILTONIAN IN THE 1s, 2s AND 2p STATES

For the hydrogen atom, the fine-structure Hamiltonian W_f is the sum of three terms:

$$W_f = W_{mv} + W_{SO} + W_D \tag{1}$$

studied in § B-1 of chapter XII.

The aim of this complement is to give the calculation of the mean values of these three operators for the $1s$, $2s$ and $2p$ states of the hydrogen atom, a calculation which was omitted in chapter XII for the sake of simplicity. We shall begin by calculating the mean values of $\langle 1/R \rangle$, $\langle 1/R^2 \rangle$ and $\langle 1/R^3 \rangle$ in these states.

1. Calculation of $\langle 1/R \rangle$, $\langle 1/R^2 \rangle$ and $\langle 1/R^3 \rangle$

The wave function associated with a stationary state of the hydrogen atom is (*cf.* chap. VII, § C):

$$\varphi_{n,l,m}(\mathbf{r}) = R_{n,l}(r)\, Y_l^m(\theta, \varphi) \tag{2}$$

$Y_l^m(\theta, \varphi)$ is a spherical harmonic. The expressions for the radial functions $R_{n,l}(r)$ corresponding to the $1s$, $2s$, $2p$ states are:

$$\begin{cases} R_{1,0}(r) = 2(a_0)^{-3/2}\, \mathrm{e}^{-r/a_0} \\ R_{2,0}(r) = 2(2a_0)^{-3/2}\left(1 - \dfrac{r}{2a_0}\right) \mathrm{e}^{-r/2a_0} \\ R_{2,1}(r) = (2a_0)^{-3/2}(3)^{-1/2}\, \dfrac{r}{a_0}\, \mathrm{e}^{-r/2a_0} \end{cases} \tag{3}$$

where a_0 is the Bohr radius:

$$a_0 = 4\pi\varepsilon_0 \frac{\hbar^2}{m_e q^2} = \frac{\hbar^2}{m_e e^2} \tag{4}$$

The Y_l^m are normalized with respect to θ and φ, so that the mean value $\langle R^q \rangle$ of the qth power (where q is a positive or negative integer) of the operator R associated with $r = |\mathbf{r}|$ in the state $| \varphi_{n,l,m} \rangle$ can be written★:

$$\langle R^q \rangle_{n,l,m} = \int_0^\infty r^{q+2} \, |R_{n,l}(r)|^2 \, \mathrm{d}r \tag{5}$$

It therefore does not depend on m. If (3) is substituted into (5), there appear integrals of the form:

$$I(k, p) = \int_0^\infty r^k \, \mathrm{e}^{-pr/a_0} \, \mathrm{d}r \tag{6}$$

where p and k are integers. We shall assume here that $k \geqslant 0$, that is, $q \geqslant -2$. An integration by parts then yields directly:

$$\begin{aligned} I(k, p) &= \left[-\frac{a_0}{p} \, \mathrm{e}^{-pr/a_0} \, r^k \right]_0^\infty + \frac{k a_0}{p} \int_0^\infty r^{k-1} \, \mathrm{e}^{-pr/a_0} \, \mathrm{d}r \\ &= \frac{k a_0}{p} \, I(k-1, p) \end{aligned} \tag{7}$$

Since, furthermore:

$$I(0, p) = \int_0^\infty \mathrm{e}^{-pr/a_0} \, \mathrm{dr} = \frac{a_0}{p} \tag{8}$$

we obtain, by recurrence:

$$I(k, p) = k! \left(\frac{a_0}{p} \right)^{k+1} \tag{9}$$

Now, let us apply this result to the mean values to be determined. We obtain:

$$\begin{aligned} \langle 1/R \rangle_{1s} &= \frac{4}{a_0^3} \int_0^\infty r \, \mathrm{e}^{-2r/a_0} \, \mathrm{d}r \\ &= \frac{4}{a_0^3} \, I(1, 2) = \frac{1}{a_0} \end{aligned} \tag{10-a}$$

$$\begin{aligned} \langle 1/R \rangle_{2s} &= \frac{4}{8a_0^3} \int_0^\infty r \left[1 - \frac{r}{2a_0} \right]^2 \mathrm{e}^{-r/a_0} \, \mathrm{d}r \\ &= \frac{1}{2a_0^3} \left[I(1, 1) - \frac{1}{a_0} \, I(2, 1) + \frac{1}{4a_0^2} \, I(3, 1) \right] \\ &= \frac{1}{4a_0} \end{aligned} \tag{10-b}$$

★ Of course, this mean value exists only for values of q which make integral (5) convergent.

$$\langle 1/R \rangle_{2p} = \frac{1}{8a_0^3}\frac{1}{3}\int_0^\infty r\left(\frac{r}{a_0}\right)^2 e^{-r/a_0}\, dr$$
$$= \frac{1}{24a_0^5} I(3, 1) = \frac{1}{4a_0} \tag{10-c}$$

Similarly:

$$\langle 1/R^2 \rangle_{1s} = \frac{4}{a_0^3} I(0, 2) = \frac{2}{a_0^2} \tag{11-a}$$

$$\langle 1/R^2 \rangle_{2s} = \frac{1}{2a_0^3}\left[I(0, 1) - \frac{1}{a_0} I(1, 1) + \frac{1}{4a_0^2} I(2, 1) \right] = \frac{1}{4a_0^2} \tag{11-b}$$

$$\langle 1/R^2 \rangle_{2p} = \frac{1}{24a_0^5} I(2, 1) = \frac{1}{12a_0^2} \tag{11-c}$$

It is clear that the expression for the mean value of $1/R^3$ is meaningless for the $1s$ and $2s$ states [since integral (5) is divergent]. For the $2p$ state, it is equal to :

$$\langle 1/R^3 \rangle_{2p} = \frac{1}{24a_0^5} I(1, 1) = \frac{1}{24a_0^3} \tag{12}$$

2. The mean values $\langle W_{mv} \rangle$

Let:

$$H_0 = \frac{\mathbf{P}^2}{2m_e} + V \tag{13}$$

be the Hamiltonian of the electron subjected to the Coulomb potential.
We have:

$$\mathbf{P}^4 = 4m_e^2[H_0 - V]^2 \tag{14-a}$$

with:

$$V = -\frac{e^2}{R} \tag{14-b}$$

so that:

$$W_{mv} = -\frac{1}{2m_e c^2}[H_0 - V]^2 \tag{15}$$

We shall take the mean values of both sides of this expression in a state $| \varphi_{n,l,m} \rangle$. Since H_0 and V are Hermitian operators, we obtain:

$$\langle W_{mv} \rangle_{n,l,m} = -\frac{1}{2m_e c^2}\left[(E_n)^2 + 2E_n\, e^2 \langle 1/R \rangle_{n,l} + e^4 \langle 1/R^2 \rangle_{n,l}\right] \tag{16}$$

In this expression, we have set:

$$E_n = -\frac{E_I}{n^2} = -\frac{1}{2n^2}\alpha^2 m_e c^2 \tag{17}$$

where:

$$\alpha = \frac{e^2}{\hbar c} \tag{18}$$

is the fine-structure constant.

If we apply relation (16) to the case of the 1s state, we obtain, using (10-a) and (11-a):

$$\langle W_{mv} \rangle_{1s} = -\frac{1}{2m_e c^2}\left[\frac{1}{4}\alpha^4 m_e^2 c^4 - \alpha^2 m_e c^2 \frac{e^2}{a_0} + 2\frac{e^4}{a_0^2}\right] \tag{19}$$

that is, since, according to (4) and (18), $e^2/a_0 = \alpha^2 m_e c^2$:

$$\langle W_{mv} \rangle_{1s} = -\frac{1}{2}\alpha^4 m_e c^2\left[\frac{1}{4} - 1 + 2\right] = -\frac{5}{8}\alpha^4 m_e c^2 \tag{20}$$

The same type of calculation, for the 2s state, leads to:

$$\langle W_{mv} \rangle_{2s} = -\frac{1}{2}\alpha^4 m_e c^2\left[\left(\frac{1}{8}\right)^2 - 2\frac{1}{8}\frac{1}{4} + \frac{1}{4}\right] = -\frac{13}{128}\alpha^4 m_e c^2 \tag{21}$$

and, for the 2p state, to:

$$\langle W_{mv} \rangle_{2p} = -\frac{1}{2}\alpha^4 m_e c^2\left[\left(\frac{1}{8}\right)^2 - 2\frac{1}{8}\frac{1}{4} + \frac{1}{12}\right] = -\frac{7}{384}\alpha^4 m_e c^2 \tag{22}$$

3. The mean values $\langle W_D \rangle$

With (14-b) and the fact that $\Delta(1/r) = -4\pi\delta(\mathbf{r})$ taken into account, the mean value of W_D in the $|\varphi_{n,l,m}\rangle$ state can be written [see also formula (B-14) of chapter XII]:

$$\langle W_D \rangle_{n,l,m} = \frac{\hbar^2}{8m_e^2c^2} 4\pi e^2 |\varphi_{n,l,m}(\mathbf{r} = \mathbf{0})|^2 \tag{23}$$

This expression goes to zero if $\varphi_{n,l,m}(\mathbf{r} = \mathbf{0}) = 0$, that is, if $l \neq 0$. Therefore:

$$\langle W_D \rangle_{2p} = 0 \tag{24-a}$$

For the 1s and 2s levels, we obtain, using (2), (23) and the fact that $Y_0^0 = 1/\sqrt{4\pi}$:

$$\langle W_D \rangle_{1s} = \frac{\hbar^2}{8m_e^2c^2} e^2 |R_{1,0}(0)|^2 = \frac{1}{2}\alpha^4 m_e c^2 \tag{24-b}$$

as well as:

$$\langle W_D \rangle_{2s} = \frac{\hbar^2}{8m_e^2c^2} e^2 |R_{2,0}(0)|^2 = \frac{1}{16}\alpha^4 m_e c^2 \tag{24-c}$$

4. Calculation of the coefficient ξ_{2p} associated with W_{SO} in the 2*p* level

In § C-2-c-β of chapter XII, we defined the coefficient:

$$\xi_{2p} = \frac{e^2}{2m_e^2c^2} \int_0^\infty \frac{|R_{2,1}(r)|^2}{r} \, \mathrm{d}r \tag{25}$$

According to (3):

$$\xi_{2p} = \frac{e^2}{2m_e^2c^2} \frac{1}{24a_0^5} I(1, 1) \tag{26}$$

Relation (9) then yields:

$$\xi_{2p} = \frac{e^2}{2m_e^2c^2} \frac{1}{24a_0^3} = \frac{1}{48\hbar^2} \alpha^4 m_e c^2 \tag{27}$$

References:

Several radial integrals for hydrogen-like atoms are given in Bethe and Salpeter (11.10).

Complement C_{XII}

THE HYPERFINE STRUCTURE AND THE ZEEMAN EFFECT FOR MUONIUM AND POSITRONIUM

In complement A_{VII}, we studied some hydrogen-like systems, composed, like the hydrogen atom, of two oppositely charged particles electrostatically attracted to each other. Of all these systems, two are particularly interesting : muonium (composed of an electron, e^-, and a positive muon, μ^+) and positronium (composed of an electron, e^-, and a positron, e^+). Their importance lies in the fact that the various particles which come into play (the electron, the positron and the muon) are not directly affected by strong interactions (while the proton is). The theoretical and experimental study of muonium and positronium therefore permits a very direct test of the validity of quantum electrodynamics.

Actually, the most precise information we now possess about these two systems comes from the study of the hyperfine structure of their 1*s* ground state [the optical lines joining the 1*s* ground state to the various excited states have just recently been observed for the positronium; *cf.* Ref. (11.25)]. This hyperfine structure is the result, as in the case of the hydrogen atom, of magnetic interactions between the spins of the two particles. We shall describe some interesting features of the hyperfine structure and the Zeeman effect for muonium and positronium in this complement.

1. The hyperfine structure of the 1*s* ground state

Let $\mathbf{S}_1$ be the electron spin and $\mathbf{S}_2$, the spin of the other particle (the muon or the positron, which are both spin 1/2 particles). The degeneracy of the 1*s* ground state is then, as for hydrogen, four-fold.

We can use stationary perturbation theory to study the effect on the 1*s* ground state of the magnetic interactions between $\mathbf{S}_1$ and $\mathbf{S}_2$. The calculation is analogous to the one in §D of chapter XII. We are left with a problem of two spin 1/2's coupled by an interaction of the form:

$$\mathscr{A}\mathbf{S}_1 \cdot \mathbf{S}_2 \tag{1}$$

where $\mathscr{A}$ is a constant which depends on the system under study. We shall denote by $\mathscr{A}_H$, $\mathscr{A}_M$, $\mathscr{A}_P$ the three values of $\mathscr{A}$ which correspond respectively to hydrogen, muonium and positronium.

It is easy to see that:

$$\mathscr{A}_H < \mathscr{A}_M < \mathscr{A}_P \tag{2}$$

since the smaller the mass of particle (2), the larger its magnetic moment. Now the positron is about 200 times lighter than the muon, which is close to 10 times lighter than the proton.

COMMENT:

The theory of chapter XII is insufficient for the extremely precise study of the hyperfine structure of hydrogen, muonium and positronium. In particular, the hyperfine Hamiltonian W_{hf} given in §B-2 of this chapter describes only part of the interactions between particles (1) and (2). For example, the fact that the electron and the positron are antiparticles of each other (they can annihilate to produce photons) is responsible for an additional coupling between them which has no equivalent for hydrogen and muonium. In addition, a series of corrections (relativistic, radiative, recoil effects, etc.) must be taken into account. These are complicated to calculate and must be treated by quantum electrodynamics. Finally, for hydrogen, nuclear corrections are also involved which are related to the structure and polarizability of the proton. However, it can be shown that the form (1) of the coupling between $\mathbf{S}_1$ and $\mathbf{S}_2$ remains valid, the constant $\mathscr{A}$ being given by an expression which is much more complicated than formula (D-8) of chapter XII. The hydrogen-like systems studied in this complement are important precisely because they enable us to compare the theoretical value of $\mathscr{A}$ with experimental results.

The eigenstates of $\mathscr{A}\mathbf{S}_1 \cdot \mathbf{S}_2$ are the $| F, m_F \rangle$ states, where F and m_F are the quantum numbers related to the total angular momentum:

$$\mathbf{F} = \mathbf{S}_1 + \mathbf{S}_2 \tag{3}$$

As in the case of the hydrogen atom, F can take on two values, $F = 1$ and $F = 0$. The two levels, $F = 1$ and $F = 0$, have energies equal to $\mathscr{A}\hbar^2/4$ and $-3\mathscr{A}\hbar^2/4$, respectively. Their separation $\mathscr{A}\hbar^2$ gives the hyperfine structure of the $1s$ ground state. Expressed in MHz, this interval is equal to:

$$\frac{\hbar}{2\pi}\mathscr{A}_M = 4\,463\,.\,317 \pm 0.021 \text{ MHz} \tag{4}$$

for muonium, and:

$$\frac{\hbar}{2\pi}\mathscr{A}_P = 203\,403 \pm 12 \text{ MHz} \tag{5}$$

for positronium.

2. The Zeeman effect in the 1*s* ground state

a. THE ZEEMAN HAMILTONIAN

If we apply a static field $\mathbf{B}_0$ parallel to Oz, we must add, to the hyperfine Hamiltonian (1), the Zeeman Hamiltonian which describes the coupling of $\mathbf{B}_0$ to the magnetic moments:

$$\mathbf{M}_1 = \gamma_1 \mathbf{S}_1 \tag{6}$$

and:

$$\mathbf{M}_2 = \gamma_2 \mathbf{S}_2 \tag{7}$$

of the two spins, with gyromagnetic ratios γ_1 and γ_2. If we set:

$$\omega_1 = -\gamma_1 B_0 \tag{8}$$

$$\omega_2 = -\gamma_2 B_0 \tag{9}$$

this Zeeman Hamiltonian can be written:

$$\omega_1 S_{1z} + \omega_2 S_{2z} \tag{10}$$

In the case of hydrogen, the magnetic moment of the proton is much smaller than that of the electron. We used this property in § E-1 of chapter XII to neglect the Zeeman coupling of the proton, compared to that of the electron★. Such an approximation is less justified for muonium, since the magnetic moment of the muon is larger than that of the proton. We shall therefore take both terms of (10) into account. For positronium, furthermore, they are equally important: the electron and positron have equal masses and opposite charges, so that:

$$\gamma_1 = -\gamma_2 \tag{11}$$

or:

$$\omega_1 = -\omega_2 \tag{12}$$

b. STATIONARY STATE ENERGIES

When B_0 is not zero, it is necessary, in order to find the stationary state energies, to diagonalize the matrix representing the total Hamiltonian:

$$\mathscr{A}\mathbf{S}_1 \cdot \mathbf{S}_2 + \omega_1 S_{1z} + \omega_2 S_{2z} \tag{13}$$

in an arbitrary orthonormal basis, for example, the $\{ | F, m_F \rangle \}$ basis. A calculation which is analogous to the one in §E-4 of chapter XII then leads to the following matrix (the four basis vectors are arranged in the order

$$\{ | 1, 1 \rangle, | 1, -1 \rangle, | 1, 0 \rangle, | 0, 0 \rangle \}): \tag{14}$$

★ Recall that the gyromagnetic ratio of the electron spin is $\gamma_1 = 2\mu_B/\hbar$ (μ_B : the Bohr magneton). Thus, if we set $\omega_0 = -\mu_B B_0/\hbar$ (the Larmor angular frequency), the constant ω_1 defined by (8) is equal to $2\omega_0$ (this is, furthermore, the notation used in §E of chapter XII; to obtain the results of that section, it therefore suffices, in this complement, to replace ω_1 by $2\omega_0$ and ω_2 by 0).

$\frac{\mathscr{A}\hbar^2}{4} + \frac{\hbar}{2}(\omega_1 + \omega_2)$	0	0	0
0	$\frac{\mathscr{A}\hbar^2}{4} - \frac{\hbar}{2}(\omega_1 + \omega_2)$	0	0
0	0	$\frac{\mathscr{A}\hbar^2}{4}$	$\frac{\hbar}{2}(\omega_1 - \omega_2)$
0	0	$\frac{\hbar}{2}(\omega_1 - \omega_2)$	$-\frac{3\mathscr{A}\hbar^2}{4}$

(14)

Matrix (14) can be broken down into two 1×1 submatrices and a 2×2 submatrix. Two eigenvalues are therefore obvious:

$$\begin{cases} E_1 = \frac{\mathscr{A}\hbar^2}{4} + \frac{\hbar}{2}(\omega_1 + \omega_2) & (15) \\ E_2 = \frac{\mathscr{A}\hbar^2}{4} - \frac{\hbar}{2}(\omega_1 + \omega_2) & (16) \end{cases}$$

They correspond, respectively, to the states $|\,1, 1\,\rangle$ and $|\,1, -1\,\rangle$, which, moreover, coincide with the states $|\,+, +\,\rangle$ and $|\,-, -\,\rangle$ of the $\{\,|\varepsilon_1, \varepsilon_2\,\rangle\,\}$ basis of common eigenstates of S_{1z} and S_{2z}. The other two eigenvalues can be obtained by diagonalizing the remaining 2×2 submatrix. They are equal to:

$$E_3 = -\frac{\mathscr{A}\hbar^2}{4} + \sqrt{\left(\frac{\mathscr{A}\hbar^2}{2}\right)^2 + \frac{\hbar^2}{4}(\omega_1 - \omega_2)^2} \tag{17}$$

$$E_4 = -\frac{\mathscr{A}\hbar^2}{4} - \sqrt{\left(\frac{\mathscr{A}\hbar^2}{2}\right)^2 + \frac{\hbar^2}{4}(\omega_1 - \omega_2)^2} \tag{18}$$

In a weak field, they correspond to the states $|\,1, 0\,\rangle$ and $|\,0, 0\,\rangle$, respectively, and, in a strong field, to the states $|\,+, -\,\rangle$ and $|\,-, +\,\rangle$.

c. THE ZEEMAN DIAGRAM FOR MUONIUM

The only differences with the results of § E-4 of chapter XII arise from the fact that here, we are taking the Zeeman coupling of particle (2) into account. These differences appear only in a sufficiently strong field.

Let us therefore consider the form taken on by the energies E_3 and E_4 when $\hbar(\omega_1 - \omega_2) \gg \mathscr{A}\hbar^2$. In this case:

$$E_3 \simeq -\frac{\mathscr{A}\hbar^2}{4} + \frac{\hbar}{2}(\omega_1 - \omega_2) \tag{19}$$

$$E_4 \simeq -\frac{\mathscr{A}\hbar^2}{4} - \frac{\hbar}{2}(\omega_1 - \omega_2) \tag{20}$$

Now, compare (19) with (15) and (20) with (16). We see that, in a strong field, the energy levels are no longer represented by pairs of parallel lines, as was the case in § E-3 of chapter XII. The slopes of the asymptotes of the E_1 and E_3 levels are, respectively, $-\frac{\hbar}{2}(\gamma_1 + \gamma_2)$ and $-\frac{\hbar}{2}(\gamma_1 - \gamma_2)$; those of the E_2 and E_4 levels, $\frac{\hbar}{2}(\gamma_1 + \gamma_2)$ and $\frac{\hbar}{2}(\gamma_1 - \gamma_2)$. Since the two particles (1) and (2) have opposite charges, γ_1 and γ_2 have opposite signs. Consequently, in a sufficiently strong field, the E_3 level (which then corresponds to the $| +, - \rangle$ state) moves above the E_1 level (the $| +, + \rangle$ state), since its slope, $-\frac{\hbar}{2}(\gamma_1 - \gamma_2)$ is greater than $-\frac{\hbar}{2}(\gamma_1 + \gamma_2)$.

The distance between the E_1 and E_3 levels therefore varies in the following way with respect to B_0 (*cf.* fig. 1): starting from 0, it increases to a maximum for the value of B_0 which makes the derivative of:

$$E_1 - E_3 = \frac{\mathscr{A}\hbar^2}{2} + f(B_0) \tag{21}$$

equal to zero, with:

$$f(B_0) = -\frac{\hbar}{2}(\gamma_1 + \gamma_2)B_0 - \sqrt{\left(\frac{\mathscr{A}\hbar^2}{2}\right)^2 + \frac{\hbar^2 B_0^2}{4}(\gamma_1 - \gamma_2)^2} \tag{22}$$

The distance then goes to zero again, and finally increases without bound. As for the distance between the E_2 and E_4 levels, it starts with the value $\mathscr{A}\hbar^2$, decreases to a minimum for the value of B_0 which makes the derivative of:

$$E_2 - E_4 = \frac{\mathscr{A}\hbar^2}{2} - f(B_0) \tag{23}$$

equal to zero and then increases without bound.

Since it is the same function $f(B_0)$ that appears in (21) and (23), we can show that, for the same value of B_0 [the one which makes the derivative of $f(B_0)$ go to zero], the distances between the E_1 and E_3 levels and that between the E_2 and E_4 levels are either maximal or minimal. This property was recently used to improve the accuracy of experimental determinations of the hyperfine structure of muonium.

By stopping polarized muons (for example, in the $| + \rangle$ state) in a rare gas target, one can prepare, in a strong field, muonium atoms which will be found preferentially in the $| +, + \rangle$ and $| -, + \rangle$ states. If we then apply simultaneously two radiofrequency fields whose frequencies are close to $(E_1 - E_3)/h$ and $(E_2 - E_4)/h$, we induce resonant transitions from $| +, + \rangle$ to $| +, - \rangle$ and from $| -, + \rangle$ to $| -, - \rangle$ (arrows in figure 1). It is these transitions which are detected experimentally, since they correspond to a flip of the muon spin which is revealed by a change in the anisotropy of the positrons emitted during the β-decay of the muons. If we are operating in a field B_0 such that the derivative of $f(B_0)$ is zero, the inhomogeneities of the static field, which may exist from one point to another of the cell containing the rare gas, are not troublesome, since the resonant frequencies of muonium, $(E_1 - E_3)/h$ and $(E_2 - E_4)/h$, are not affected, to first order, by a variation of B_0 [ref. (11.24)].

COMMENT:

For the ground state of the hydrogen atom, we obtain a Zeeman diagram analogous to the one in figure 1 when we take into account the Zeeman coupling between the proton spin and the field $\mathbf{B}_0$.

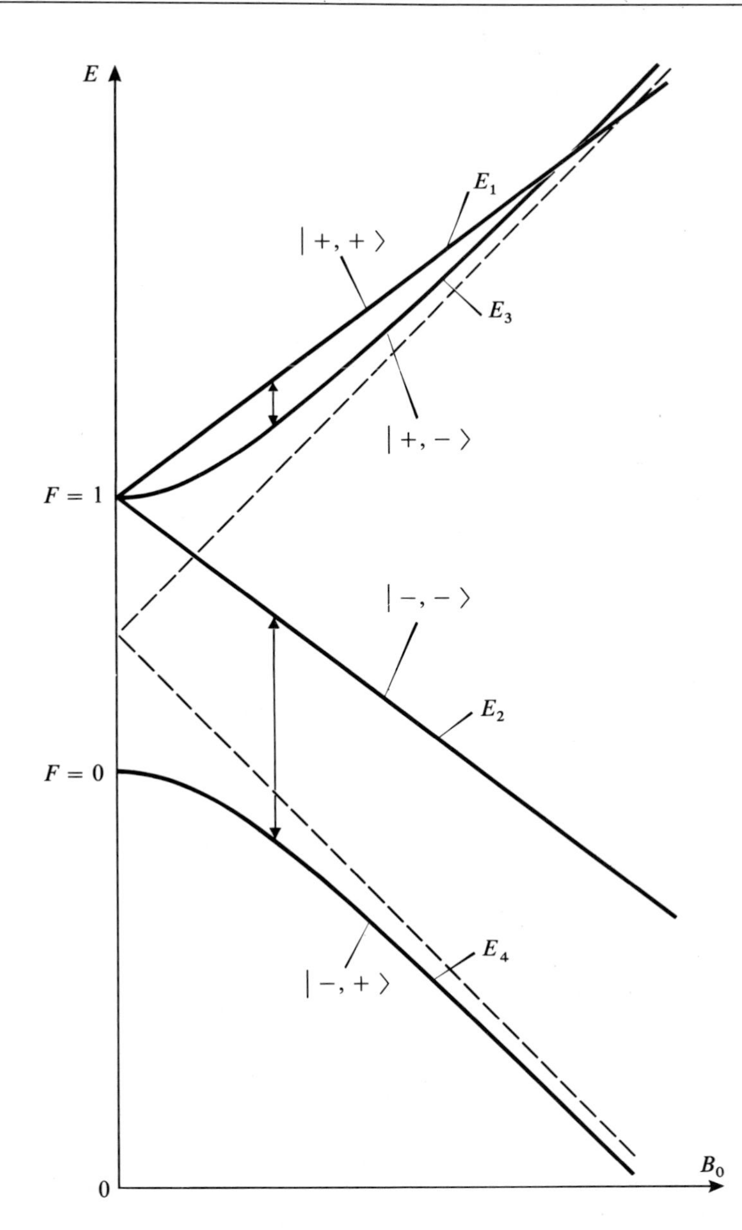

FIGURE 1

The Zeeman diagram for the $1s$ ground state of muonium. Since we are not neglecting here the Zeeman coupling between the magnetic moment of the muon and the static field $\mathbf{B}_0$, the two straight lines (which correspond, in a strong field, to the same electron spin orientation but different muon spin orientations) are no longer parallel, as was the case for hydrogen (in the Zeeman diagram of figure 9 of chapter XII, the Larmor angular frequency ω_n of the proton was neglected). For the same value of the static field B_0, the splitting between the E_1 and E_3 levels is maximal and that between the E_2 and E_4 levels is minimal. The arrows represent the transitions studied experimentally for this value of the field $\mathbf{B}_0$.

d. THE ZEEMAN DIAGRAM FOR POSITRONIUM

If we set $\omega_1 = -\omega_2$ (this property is a direct consequence of the fact that the positron is the antiparticle of the electron) in (15) and (16), we see that the E_1 and E_2 levels are independent of B_0:

$$E_1 = E_2 = \frac{\mathscr{A}\hbar^2}{4} \tag{24}$$

On the other hand, we obtain from (17) and (18):

$$E_3 = -\frac{\mathscr{A}\hbar^2}{4} + \sqrt{\left(\frac{\mathscr{A}\hbar^2}{2}\right)^2 + \hbar^2\gamma_1^2 B_0^2} \tag{25}$$

$$E_4 = -\frac{\mathscr{A}\hbar^2}{4} - \sqrt{\left(\frac{\mathscr{A}\hbar^2}{2}\right)^2 + \hbar^2\gamma_1^2 B_0^2} \tag{26}$$

The Zeeman diagram for positronium therefore has the form shown in figure 2. It is composed of two superposed straight lines parallel to the B_0 axis and one hyperbola.

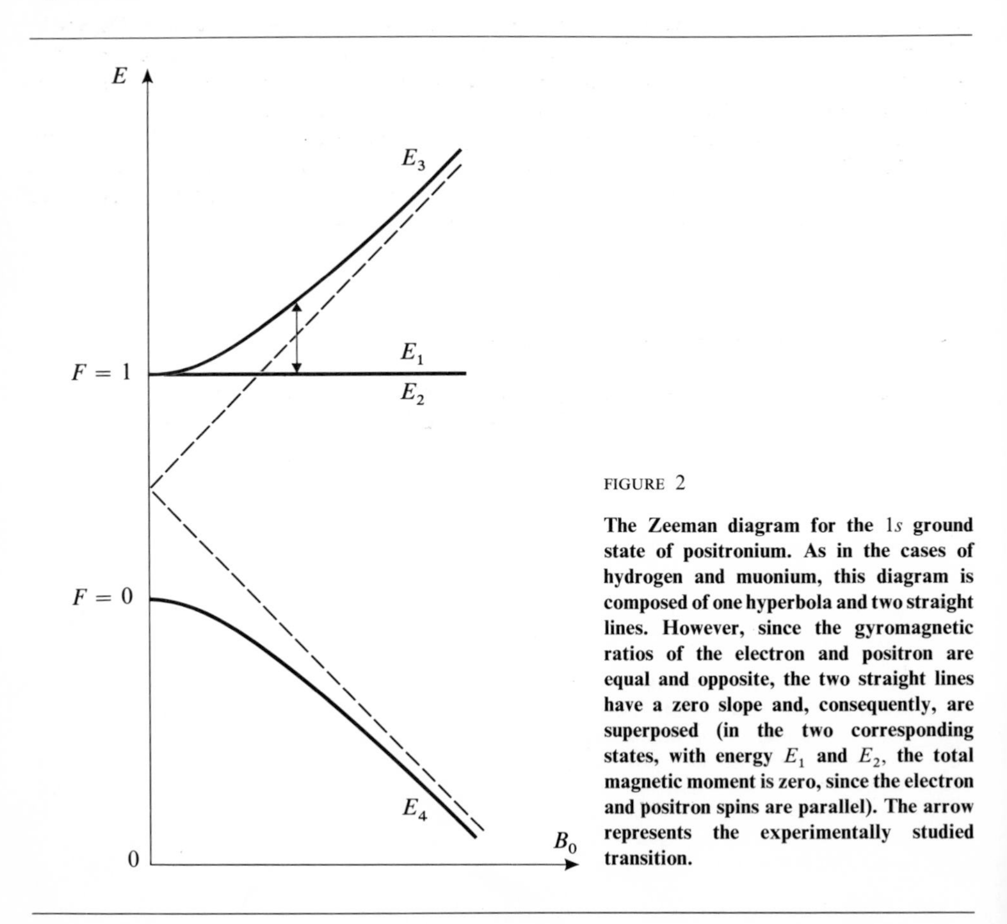

FIGURE 2

The Zeeman diagram for the 1*s* ground state of positronium. As in the cases of hydrogen and muonium, this diagram is composed of one hyperbola and two straight lines. However, since the gyromagnetic ratios of the electron and positron are equal and opposite, the two straight lines have a zero slope and, consequently, are superposed (in the two corresponding states, with energy E_1 and E_2, the total magnetic moment is zero, since the electron and positron spins are parallel). The arrow represents the experimentally studied transition.

Actually, positronium is not stable. It decays by emitting photons. In a zero field, it can be shown by symmetry considerations that the $F = 0$ state (the singlet spin state, or "para-positronium") decays by emitting two photons. Its half-life is of the order of $\tau_0 \simeq 1.25 \times 10^{-10}$ sec. On the other hand, the $F = 1$ state (the triplet spin state, or "orthopositronium") can decay only by emitting three photons (since the two-photon transition is forbidden). This process is much less probable, and the half-life of the triplet is much longer, on the order of $\tau_1 \simeq 1.4 \times 10^{-7}$ sec.

When a static field is applied, the E_1 and E_2 levels retain the same lifetimes since the corresponding eigenstates do not depend on B_0. On the other hand, the $| 1, 0 \rangle$ state is "mixed" with the $| 0, 0 \rangle$ state, and vice versa. Calculations analogous to those of complement H_{IV} show that the lifetime of the E_3 level is reduced relative to its zero-field value τ_1 (that of the E_4 level is increased relative to the value τ_0). The positronium atoms in the E_3 state then have a certain probability of decaying by emission of two photons.

This inequality of the lifetimes of the three states of energies E_1, E_2, E_3 when B_0 is non-zero is the basis of the methods for determining the hyperfine structure of positronium. Formation of positronium atoms by positron capture by an electron generally populates the four states of energies E_1, E_2, E_3, E_4 equally. In a non-zero field B_0, the two states E_1 and E_2 decay less rapidly than the E_3 state, so that in the stationary state, they are more populated. If we then apply a radiofrequency field oscillating at the frequency

$$(E_3 - E_1)/h = (E_3 - E_2)/h,$$

we induce resonant transitions from the E_1 and E_2 states to the E_3 state (the arrow of figure 2). This increases the decay rate via two-photon emission, which permits the detection of resonance when (with fixed B_0) we vary the frequency of the oscillating field. Determination of $E_3 - E_1$ for a given value of B_0 then allows us to find the constant $\mathscr{A}$ by using (24) and (25).

In a zero field, resonant transitions could also be induced between the unequally populated $F = 1$ and $F = 0$ levels. However, the corresponding resonant frequency, given by (5), is high and not easily produced experimentally. This is why one generally prefers to use the "low frequency" transition represented by the arrow of figure 2.

References and suggestions for further reading:

See the subsection "Exotic atoms" of section 11 of the bibliography.
The annihilation of positronium is discussed in Feynman III (1.2), §18-3.

Complement D_{XII}

THE INFLUENCE OF THE ELECTRONIC SPIN ON THE ZEEMAN EFFECT OF THE HYDROGEN RESONANCE LINE

1. Introduction

The conclusions of complement D_{VII} relative to the Zeeman effect for the resonance line of the hydrogen atom spectrum (the $1s \longleftrightarrow 2p$ transition) must be modified to take into account the electron spin and the associated magnetic interactions. This is what we shall do in this complement, using the results obtained in chapter XII.

To simplify the discussion, we shall neglect effects related to nuclear spin (which are much smaller than those related to the electron spin). Therefore, we shall not take the hyperfine coupling W_{hf} (chap. XII, § B-2) into account, choosing the Hamiltonian H in the form :

$$H = H_0 + W_f + W_Z \tag{1}$$

H_0 is the electrostatic Hamiltonian studied in chapter VII (§ C), W_f, the sum of the fine structure terms (chap. XII, §B-1):

$$W_f = W_{mv} + W_D + W_{SO} \tag{2}$$

and W_Z, the Zeeman Hamiltonian (chap. XII, §E-1) describing the interaction of the atom with a magnetic field $\mathbf{B}_0$ parallel to Oz:

$$W_Z = \omega_0(L_z + 2S_z) \tag{3}$$

where the Larmor angular frequency ω_0 is given by:

$$\omega_0 = -\frac{q}{2m_e} B_0 \tag{4}$$

[we shall neglect ω_n relative to ω_0; see formula (E-4) of chapter XII].

We shall determine the eigenvalues and eigenvectors of H by using a method analogous to that of §E of chapter XII : we shall treat W_f and W_Z like perturbations of H_0. Although they have the same unperturbed energy, the $2s$ and $2p$ levels can be studied separately since they are connected neither by W_f (chap. XII, §C-2-a-β) nor by W_Z. In this complement, the magnetic field $\mathbf{B}_0$ will be called weak or strong, depending on whether W_Z is small or large compared to W_f. Note that the magnetic fields considered here to be "weak" are those for which W_Z is small compared to W_f but large compared to W_{hf}, which we have neglected. These "weak fields" are therefore much stronger than those treated in § E of chapter XII.

Once the eigenstates and eigenvalues of H have been obtained, it is possible to study the evolution of the mean values of the three components of the electric dipole moment of the atom. Since an analogous calculation was performed in detail in complement D_{VII}, we shall not repeat it. We shall merely indicate, for weak fields and for strong fields, the frequencies and polarization states of the various Zeeman components of the resonance line of hydrogen (the Lyman α line).

2. The Zeeman diagrams of the 1*s* and 2*s* levels

We saw in § D-1-b of chapter XII that W_f shifts the $1s$ level as a whole and gives rise to only one fine-structure level, $1s_{1/2}$. The same is true for the $2s$ level, which becomes $2s_{1/2}$. In each of these two levels, we can choose a basis:

$$\left\{ | \, n \, ; l = 0 \, ; m_L = 0 \, ; m_S = \pm \frac{1}{2} \, ; m_I = \pm \frac{1}{2} \rangle \right\} \tag{5}$$

of eigenvectors common to H_0, $\mathbf{L}^2$, L_z, S_z, I_z (the notation is identical to that of chapter XII; since H does not act on the proton spin, we shall ignore m_I in all that follows).

The vectors (5) are obviously eigenvectors of W_Z with eigenvalues $2m_S\hbar\omega_0$. Thus, each $1s_{1/2}$ or $2s_{1/2}$ level splits, in a field B_0, into two Zeeman sublevels of energies:

$$E(n \, ; l = 0 \, ; m_L = 0 \, ; m_S) = E(ns_{1/2}) + 2m_S\hbar\omega_0 \tag{6}$$

where $E(ns_{1/2})$ is the zero-field energy of the $ns_{1/2}$ level, calculated in §§ C-2-b and D-1-b of chapter XII. The Zeeman diagram of the $1s_{1/2}$ level (as well as the one for the $2s_{1/2}$ level) is therefore composed of two straight lines of slopes $+1$ and -1 (fig. 1), corresponding, respectively, to the two possible orientations of the spin relative to $\mathbf{B}_0$ ($m_S = +1/2$ and $m_S = -1/2$).

Comparison of figure 1 and figure 9 of chapter XII shows that to neglect, as we are doing here, the effects related to nuclear spin amounts to considering fields $\mathbf{B}_0$ which are so large that $W_Z \gg W_{hf}$. We are then in the asymptotic region of the diagram of figure 9 of chapter XII, where we can ignore the splitting of the energy levels due to the proton spin and hyperfine coupling.

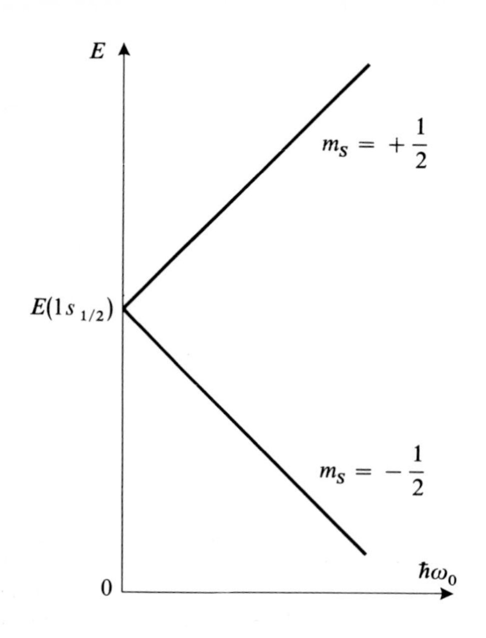

FIGURE 1

The Zeeman diagram of the $1s_{1/2}$ level when the hyperfine coupling W_{hf} is neglected. The ordinate of the point at which the two levels $m_S = \pm 1/2$ cross is the energy of the $1s_{1/2}$ level (i.e, the eigenvalue $- E_I$ of H_0, corrected for the global shift produced by the fine-structure Hamiltonian W_f). Figure 9 of chapter XII gives an idea of the modifications of this diagram produced by W_{hf}.

3. The Zeeman diagram of the 2p level

In the six-dimensional $2p$ subspace, we can choose one of the two bases:

$$\{ | n = 2; l = 1; m_L; m_S \rangle \} \tag{7}$$

or:

$$\{ | n = 2; l = 1; J; m_J \rangle \} \tag{8}$$

adapted, respectively, to the individual angular momenta **L** and **S** and to the total angular momentum $\mathbf{J} = \mathbf{L} + \mathbf{S}$ [*cf.* (36-a) and (36-b) of complement A_X].

The terms W_{mv} and W_D which appear in expression (2) for W_f shift the $2p$ level as a whole. Therefore, to study the Zeeman diagram of the $2p$ level, we simply diagonalize the 6×6 matrix which represents $W_{SO} + W_Z$ in either one of the two bases, (7) or (8). Actually, since W_Z and $W_{SO} = \xi_{2p}\mathbf{L} \cdot \mathbf{S}$ both commute with $J_z = L_z + S_z$, this 6×6 matrix can be broken down into as many submatrices as there are distinct values of m_J. Thus, there appear two one-dimensional submatrices (corresponding respectively to $m_J = +3/2$ and $m_J = -3/2$) and two two-dimensional submatrices (corresponding respectively to $m_J = +1/2$ and $m_J = -1/2$). The calculation of the eigenvalues and associated eigenvectors (which is very much like that of § E-4 of chapter XII) presents no difficulties and leads to the Zeeman diagram shown in figure 2. This diagram is composed of two straight lines and four hyperbolic branches.

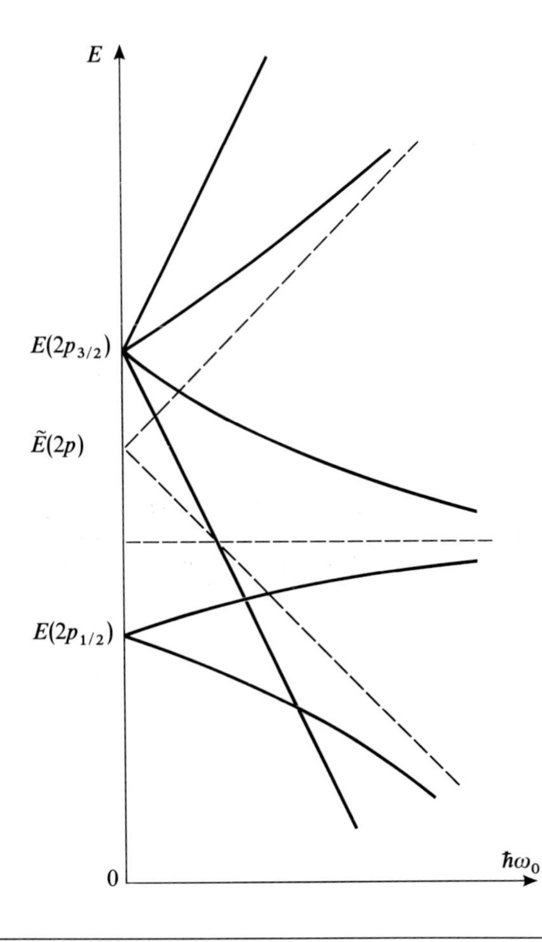

FIGURE 2

The Zeeman diagram of the $2p$ level when the hyperfine coupling W_{hf} is neglected. In a zero field, we find the fine-structure levels, $2p_{1/2}$ and $2p_{3/2}$. The Zeeman diagram is composed of two straight lines and two hyperbolas (for which the asymptotes are shown in dashed lines). The hyperfine coupling W_{hf} would significantly modify this diagram only in the neighborhood of $\omega_0 = 0$. $\tilde{E}(2p)$ is the $2p$ level energy (the eigenvalue $-E_I/4$ of H_0) corrected for the global shift produced by $W_{mv} + W_D$

In a zero field, the energies depend only on J. We obtain the two fine-structure levels, $2p_{3/2}$ and $2p_{1/2}$, already studied in § C of chapter XII, whose energies are equal to:

$$E(2p_{3/2}) = \tilde{E}(2p) + \frac{1}{2}\xi_{2p}\hbar^2 \tag{9}$$

$$E(2p_{1/2}) = \tilde{E}(2p) - \xi_{2p}\hbar^2 \tag{10}$$

$\tilde{E}(2p)$ is the $2p$ level energy $E(2p)$ corrected for the global shift due to W_{mv} and W_D [*cf.* expressions (C-8) and (C-9) of chapter XII]. ξ_{2p} is the constant which appears in the restriction $\xi_{2p}\mathbf{L}\cdot\mathbf{S}$ of W_{SO} to the $2p$ level [*cf.* expression (C-13) of chapter XII].

In weak fields ($W_Z \ll W_{SO}$), the slope of the energy levels can also be obtained by treating W_Z like a perturbation of W_f. It is then necessary to diagonalize the 4×4 and 2×2 matrices representing W_Z in the $2p_{3/2}$ and $2p_{1/2}$ levels. Calculations analogous to those of § E-2 of chapter XII show that these two submatrices are respectively proportional to those which represent $\omega_0 J_z$ in the same subspaces.

The proportionality coefficients, called "Landé factors" (*cf.* complement D_X, §3), are equal, respectively, to★:

$$g(2p_{3/2}) = \frac{4}{3} \tag{11}$$

$$g(2p_{1/2}) = \frac{2}{3} \tag{12}$$

In weak fields, each fine-structure level therefore splits into $2J + 1$ equidistant Zeeman sublevels. The eigenstates are the states of the "coupled" basis, (8), corresponding to the eigenvalues:

$$E(J, m_J) = E(2p_J) + m_J \, g(2p_J) \, \hbar\omega_0 \tag{13}$$

where the $E(2p_J)$ are given by expressions (9) and (10).

In strong fields ($W_Z \gg W_{SO}$), we can, on the other hand, treat $W_{SO} = \xi_{2p}\mathbf{L}\,.\,\mathbf{S}$ like a perturbation of W_Z, which is diagonal in basis (7). As in § E-3-b of chapter XII, it can easily be shown that only the diagonal elements of $\xi_{2p}\mathbf{L}\,.\,\mathbf{S}$ are involved when the corrections are calculated to first order in W_{SO}. Thus, we find that in strong fields, the eigenstates are the states of the "decoupled" basis, (7), and the corresponding eigenvalues are:

$$E(m_L, m_S) = \tilde{E}(2p) + (m_L + 2m_S)\hbar\omega_0 + m_L m_S \, \hbar^2 \xi_{2p} \tag{14}$$

Formula (14) gives the asymptotes of the diagram of figure 2.

As the magnetic field B_0 increases, we pass continuously from basis (8) to basis (7). The magnetic field gradually decouples the orbital angular momentum and the spin. This situation is the analogue of the one studied in §E of chapter XII, in which the angular momenta **S** and **I** were coupled or decoupled, depending on the relative importance of the hyperfine and Zeeman terms.

4. The Zeeman effect of the resonance line

a. STATEMENT OF THE PROBLEM

Arguments of the same type as those of §2-c of complement D_{VII} (see, in particular, the comment at the end of that complement) show that the optical transition between a $2p$ Zeeman sublevel and a $1s$ Zeeman sublevel is possible only if the matrix element of the electric dipole operator $q\mathbf{R}$ between these two states is different from zero★★. In addition, depending on whether it is the $q(X + iY)$, $q(X - iY)$ or qZ operator which has a non-zero matrix element between the two

★ These Landé factors can be calculated directly from formula (43) of complement D_X.

★★ The electric dipole, since it is an odd operator, has no matrix elements between the $1s$ and $2s$ states, which are both even. This is why we are ignoring the $2s$ states here.

Zeeman sublevels under consideration, the polarization state of the emitted light is σ^+, σ^- or π. Therefore, we use the previously determined eigenvectors and eigenvalues of H in order to obtain the frequencies of the various Zeeman components of the hydrogen resonance line and their polarization states.

COMMENT:

The $q(X + iY)$, $q(X - iY)$ and qZ operators act only on the orbital part of the wave function and cause m_L to vary, respectively, by $+1$, -1 and 0 (*cf.* complement D_{VII}, § 2-c); m_S is not affected. Since $m_J = m_L + m_S$ is a good quantum number (whatever the strength of the field B_0), $\Delta m_J = +1$ transitions have a σ^+ polarization; $\Delta m_J = -1$ transitions, a σ^- polarization; and $\Delta m_J = 0$ transitions, a π polarization.

b. THE WEAK-FIELD ZEEMAN COMPONENTS

Figure 3 shows the weak-field positions of the various Zeeman sublevels resulting from the $1s_{1/2}$, $2p_{1/2}$ and $2p_{3/2}$ levels, obtained from expressions (6), (13), (11) and (12). The vertical arrows indicate the various Zeeman components of the resonance line. The polarization is σ^+, σ^- or π, depending on whether $\Delta m_J = +1, -1$ or 0.

Figure 4 shows the positions of these various components on a frequency scale, relative to the zero-field positions of the lines. The result differs notably from that of complement D_{VII} (see figure 2 of that complement), where, observing in a direction perpendicular to $\mathbf{B}_0$, we had three equidistant components of polarization σ^+, π, σ^-, separated by a frequency difference $\omega_0/2\pi$.

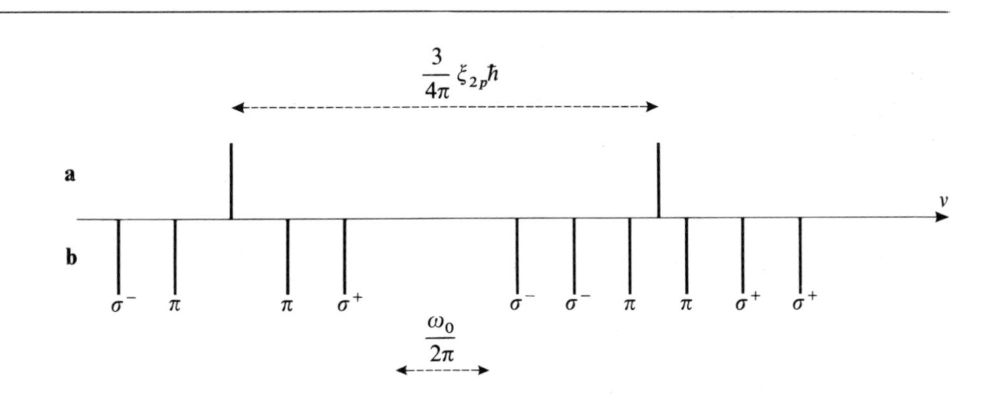

FIGURE 4

Frequencies of the various Zeeman components of the hydrogen resonance line.
a) In a zero field: two lines are observed, separated by the fine-structure splitting $3\xi_{2p}\hbar/4\pi$ (ξ_{2p} is the spin-orbit coupling constant of the $2p$ level) and corresponding respectively to the transitions $2p_{3/2} \longleftrightarrow 1s_{1/2}$ (the line on the right-hand side of the figure) and $2p_{1/2} \longleftrightarrow 1s_{1/2}$ (the line on the left-hand side).
b) In a weak field B_0: each line splits into a series of Zeeman components whose polarizations are indicated; $\omega_0/2\pi$ is the Larmor frequency in the field B_0.

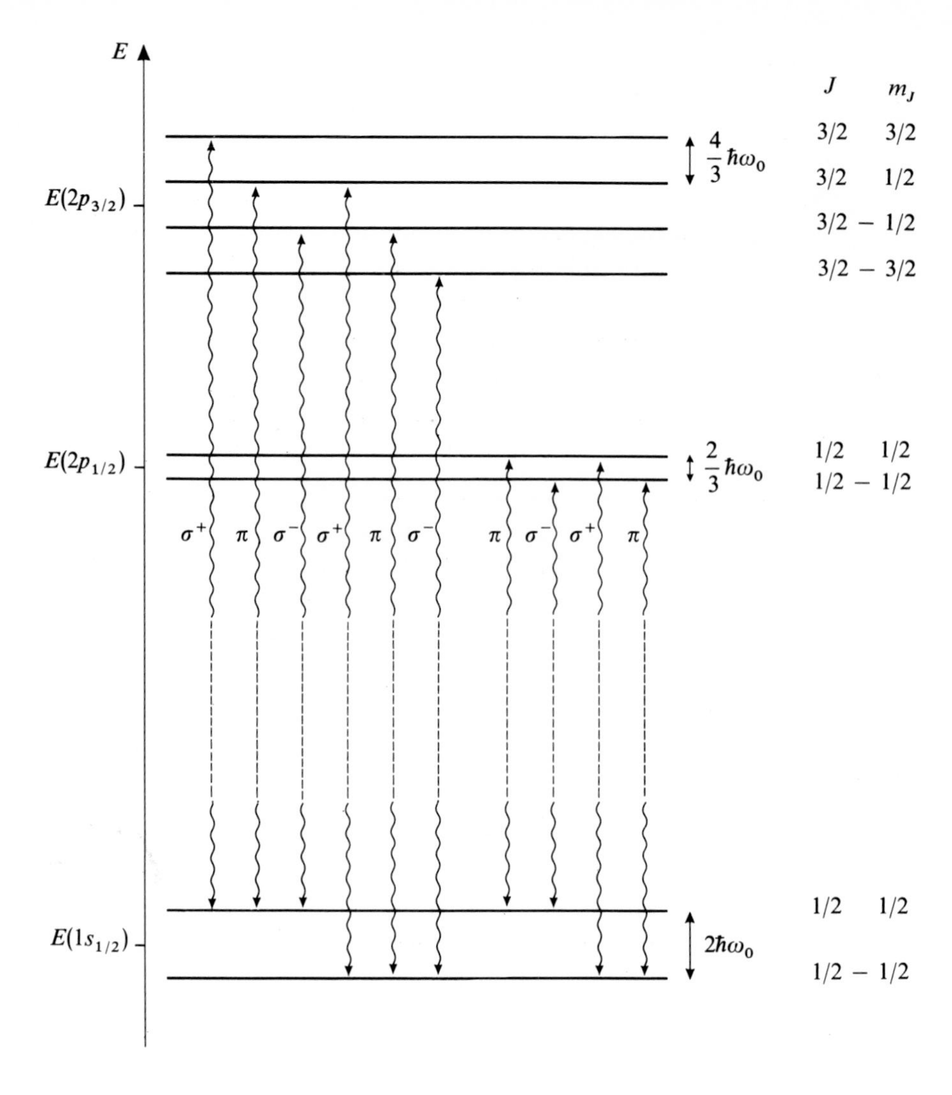

FIGURE 3

The disposition, in a weak field, of the Zeeman sublevels arising from the fine-structure levels, $1s_{1/2}$, $2p_{1/2}$, $2p_{3/2}$ (whose zero-field energies are marked on the vertical energy scale). On the right-hand side of the figure are indicated the splittings between adjacent Zeeman sublevels (for greater clarity, these splittings have been exaggerated with respect to the fine-structure splitting which separates the $2p_{1/2}$ and $2p_{3/2}$ levels), as well as the values of the quantum numbers J and m_J associated with each sublevel. The arrows indicate the Zeeman components of the resonance line, each of which has a well-defined polarization, σ^+, σ^- or π.

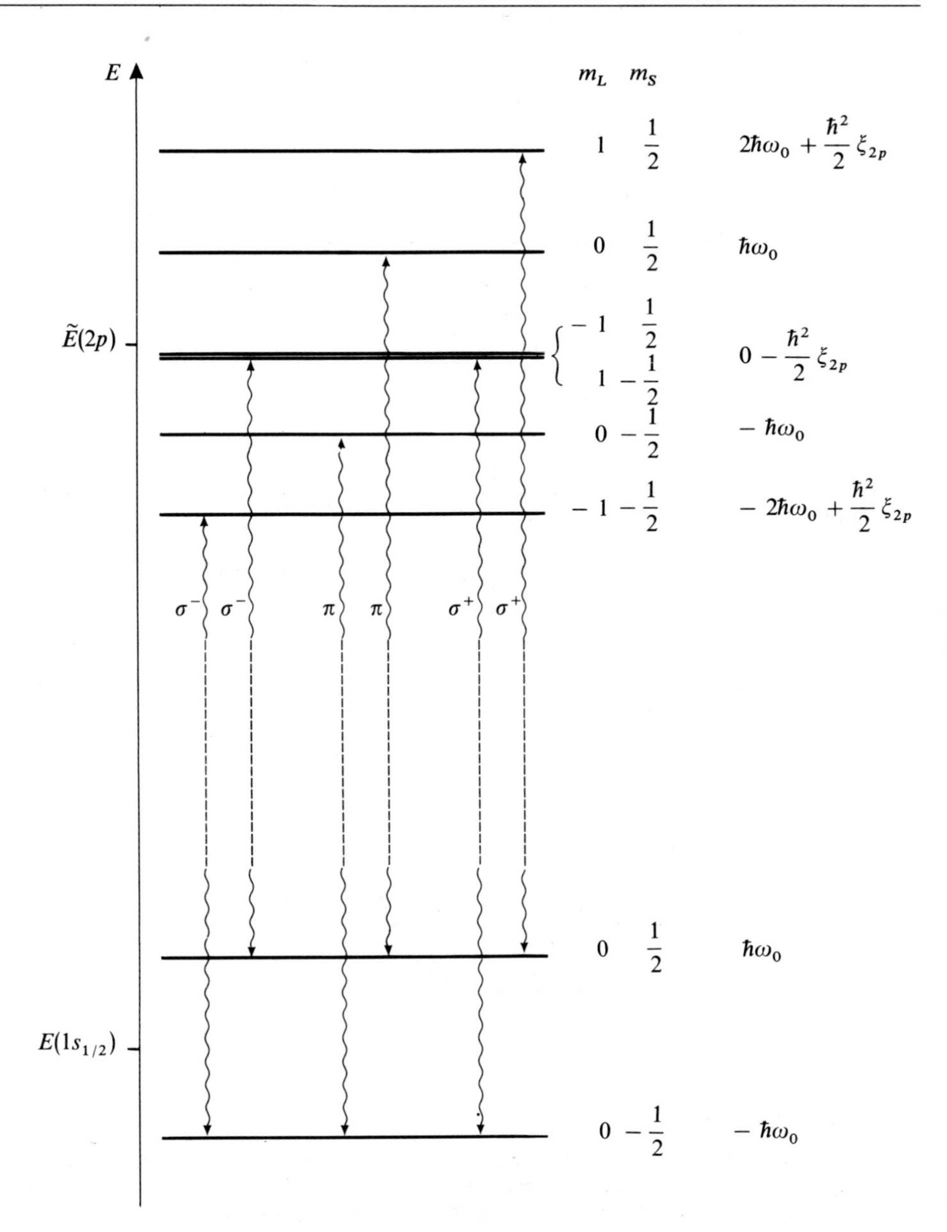

FIGURE 5

The disposition, in a strong field (decoupled fine structure), of the Zeeman sublevels arising from the $1s$ and $2p$ levels. On the right-hand side of the figure are indicated the values of the quantum numbers m_L and m_S associated with each Zeeman sublevel, as well as the corresponding energy, given relative to $E(1s_{1/2})$ or $\tilde{E}(2p)$. The vertical arrows indicate the Zeeman components of the resonance line.

c. THE STRONG-FIELD ZEEMAN COMPONENTS. THE PASCHEN-BACK EFFECT

Figure 5 shows the strong-field positions of the Zeeman sublevels arising from the $1s$ and $2p$ levels [see expressions (6) and (14)]. To first order in W_{SO}, the degeneracy between the states $| m_L = -1, m_S = 1/2 \rangle$ and $| m_L = 1, m_S = -1/2 \rangle$ is not removed. The vertical arrows indicate the Zeeman components of the resonance line. The polarization is σ^+, σ^- or π, depending on whether $\Delta m_L = +1$, -1 or 0 (recall that in an electric dipole transition, the quantum number m_S is not affected).

The corresponding optical spectrum is shown in figure 6. The two π transitions have the same frequency (*cf.* fig. 5). On the other hand, there is a small splitting, $\hbar\xi_{2p}/2\pi$, between the frequencies of the two σ^+ transitions and between those of the two σ^- transitions. The mean distance between the σ^+ doublet and the π line (or between the π line and the σ^- doublet) is equal to $\omega_0/2\pi$. The spectrum of figure 6 is therefore similar to that of figure 2 of complement D_{VII}. Furthermore,

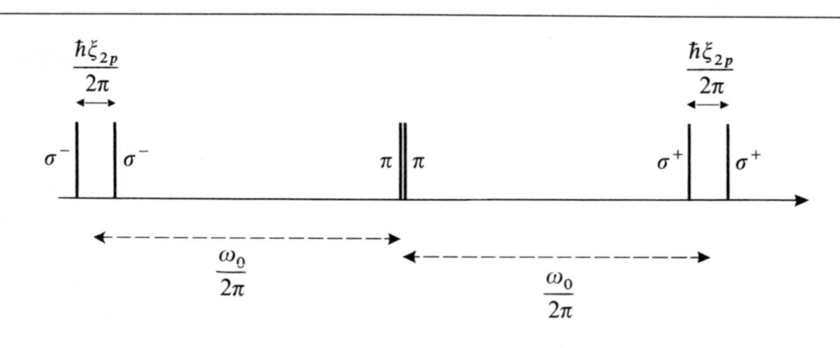

FIGURE 6

The strong-field positions of the Zeeman components of the hydrogen resonance line. Aside from the splitting of the σ^+ and σ^- lines, this spectrum is identical to the one obtained in complement D_{VII}, where the effects related to electron spin were ignored.

the splitting of the σ^+ and σ^- lines, due to the existence of the electron spin, is easy to understand. In strong fields, **L** and **S** are decoupled. Since the $1s \longleftrightarrow 2p$ transition is an electric dipole transition, only the orbital angular momentum **L** of the electron is affected by the optical transition. An argument analogous to the one in § E-3-b of chapter XII shows that the magnetic interactions related to the spin can be described by an "internal field" which adds to the external field $\mathbf{B}_0$ and whose sign changes, depending on whether the spin points up or down. It is this internal field that causes the splitting of the σ^+ and σ^- lines (the π line is not affected, since its quantum number m_L is zero).

References and suggestions for further reading:

Cagnac and Pebay-Peyroula (11.2), chaps. XI and XVII (especially §5-A of that chapter); White (11.5), chap. X; Kuhn (11.1), chap. III, §F; Sobel'man (11.12), chap. 8, § 29.

Complement E_{XII}

THE STARK EFFECT FOR THE HYDROGEN ATOM

Consider a hydrogen atom placed in a uniform static electric field $\boldsymbol{\mathscr{E}}$ parallel to Oz. To the Hamiltonian studied in chapter XII must be added the Stark Hamiltonian W_S, which describes the interaction energy of the electric dipole moment $q\mathbf{R}$ of the atom with the field $\boldsymbol{\mathscr{E}}$. W_S can be written:

$$W_S = - q\,\boldsymbol{\mathscr{E}}.\,\mathbf{R} = - q\mathscr{E}Z \tag{1}$$

Even for the strongest electric fields that can be produced in the laboratory, we always have $W_S \ll H_0$. On the other hand, if $\mathscr{E}$ is strong enough, W_S can have the same order of magnitude as W_f and W_{hf} or be even larger. To simplify the discussion, we shall assume throughout this complement that $\mathscr{E}$ is strong enough for the effect of W_S to be much larger than that of W_f or W_{hf}. We shall therefore calculate directly, using perturbation theory, the effect of W_S on the eigenstates of H_0 found in chapter VII (the next step, which we shall not consider here, would consist of evaluating the effect of W_f, and then of W_{hf}, on the eigenstates of $H_0 + W_S$).

Since both H_0 and W_S do not act on the spin variables, we shall ignore the quantum numbers m_S and m_I.

1. The Stark effect on the $n = 1$ level

a. THE SHIFT OF THE 1s STATE IS QUADRATIC IN $\mathscr{E}$

According to perturbation theory, the effect of the electric field can be obtained to first order by calculating the matrix element:

$$- q\mathscr{E} \langle\, n = 1, l = 0, m_L = 0 \,|\, Z \,|\, n = 1, l = 0, m_L = 0 \,\rangle.$$

Since the operator Z is odd, and since the ground state has a well-defined parity (it is even), the preceding matrix element is zero.

There is therefore no effect which is linear in $\mathscr{E}$, and we must go on to the next term of the perturbation series:

$$\varepsilon_2 = q^2\mathscr{E}^2 \sum_{\substack{n \neq 1 \\ l,m}} \frac{|\langle\, 1, 0, 0 \,|\, Z \,|\, n, l, m \,\rangle|^2}{E_1 - E_n} \tag{2}$$

where $E_n = -E_I/n^2$ is the eigenvalue of H_0 associated with the eigenstate $| n, l, m \rangle$ (*cf.* chap. VII, §C). The preceding sum is certainly not zero, since there exist states $| n, l, m \rangle$ whose parity is opposite to that of $| 1, 0, 0 \rangle$. We conclude that, to lowest order in $\mathscr{E}$, the Stark shift of the $1s$ ground state is quadratic. Since $E_1 - E_n$ is always negative, the ground state is lowered.

b. POLARIZABILITY OF THE 1*s* STATE

We have already mentioned that, for reasons of parity, the mean values of the components of the operator $q\mathbf{R}$ are zero in the state $| 1, 0, 0 \rangle$ (the unperturbed ground state).

In the presence of an electric field $\boldsymbol{\mathscr{E}}$ parallel to Oz, the ground state is no longer $| 1, 0, 0 \rangle$, but rather (according to the results of § B-1-b of chapter XI):

$$| \psi_0 \rangle = | 1, 0, 0 \rangle - q\mathscr{E} \sum_{\substack{n \neq 1 \\ l,m}} | n, l, m \rangle \frac{\langle n, l, m | Z | 1, 0, 0 \rangle}{E_1 - E_n} + \ldots \tag{3}$$

This shows that the mean value of the electric dipole moment $q\mathbf{R}$ in the perturbed ground state is, to first order in $\mathscr{E}$, $\langle \psi_0 | q\mathbf{R} | \psi_0 \rangle$. Using expression (3) for $| \psi_0 \rangle$, we then obtain:

$$\langle \psi_0 | q\mathbf{R} | \psi_0 \rangle = - q^2\mathscr{E} \sum_{\substack{n \neq 1 \\ l,m}} \frac{\langle 1,0,0 | \mathbf{R} | n,l,m \rangle \langle n,l,m | Z | 1,0,0 \rangle + \langle 1,0,0 | Z | n,l,m \rangle \langle n,l,m | \mathbf{R} | 1,0,0 \rangle}{E_1 - E_n} \tag{4}$$

Thus, we see that the electric field $\mathscr{E}$ causes an "induced" dipole moment to appear, proportional to $\mathscr{E}$. It can easily be shown, by using the spherical harmonic orthogonality relation*, that $\langle \psi_0 | qX | \psi_0 \rangle$ and $\langle \psi_0 | qY | \psi_0 \rangle$ are zero, and that the only non-zero mean value is:

$$\langle \psi_0 | qZ | \psi_0 \rangle = - 2q^2\mathscr{E} \sum_{\substack{n \neq 1 \\ l,m}} \frac{|\langle n, l, m | Z | 1, 0, 0 \rangle|^2}{E_1 - E_n} \tag{5}$$

In other words, the induced dipole moment is parallel to the applied field $\boldsymbol{\mathscr{E}}$. This is not surprising, given the spherical symmetry of the $1s$ state. The coefficient of proportionality χ between the induced dipole moment and the field is called the linear electric susceptibility. We see that quantum mechanics permits the calculation of this susceptibility for the $1s$ state:

$$\chi_{1s} = - 2q^2 \sum_{\substack{n \neq 1 \\ l,m}} \frac{|\langle n, l, m | Z | 1, 0, 0 \rangle|^2}{E_1 - E_n} \tag{6}$$

* This relation implies that $\langle 1, 0, 0 | Z | n, l, m \rangle$ is different from zero only if $l = 1$, $m = 0$ (the argument is the same as the one given for $\langle 2, 1, m | Z | 2, 0, 0 \rangle$ in the beginning of § 2 below). Consequently, in (2), (3), (4), (5), (6), the summation is actually carried out only over n (it includes, furthermore, the states of the positive energy continuum).

2. The Stark effect on the $n = 2$ level

The effect of W_S on the $n = 2$ level can be obtained to first order by diagonalizing the restriction of W_S to the subspace spanned by the four states of the $\{ | 2, 0, 0, \rangle ; | 2, 1, m \rangle, m = -1, 0, +1 \}$ basis.

The $| 2, 0, 0 \rangle$ state is even; the three $| 2, 1, m \rangle$ states are odd. Since W_S is odd, the matrix element $\langle 2, 0, 0 | W_S | 2, 0, 0 \rangle$ and the nine matrix elements $\langle 2, 1, m' | W_S | 2, 1, m \rangle$ are zero (*cf.* complement F_{II}). On the other hand, since the $| 2, 0, 0, \rangle$ and $| 2, 1, m \rangle$ states have opposite parities, $\langle 2, 1, m | W_S | 2, 0, 0 \rangle$ can be different from zero.

Let us show that actually only $\langle 2, 1, 0 | W_S | 2, 0, 0 \rangle$ is non-zero. W_S is proportional to $Z = R \cos \theta$ and, therefore, to $Y_1^0(\theta)$. The angular integral which enters into the matrix elements $\langle 2, 1, m | W_S | 2, 0, 0 \rangle$ is therefore of the form:

$$\int Y_1^{m*}(\Omega) \, Y_1^0(\Omega) \, Y_0^0(\Omega) \, d\Omega$$

Since Y_0^0 is a constant, this integral is proportional to the scalar product of Y_1^0 and Y_1^m and is therefore different from zero only if $m = 0$. Moreover, since Y_1^0, $R_{20}(r)$ and $R_{10}(r)$ are real, the corresponding matrix element of W_S is real. We shall set:

$$\langle 2, 1, 0 | W_S | 2, 0, 0 \rangle = \gamma \mathscr{E} \tag{7}$$

without concerning ourselves with the exact value of γ [which could be calculated without difficulty since we know the wave functions $\varphi_{2,1,0}(\mathbf{r})$ and $\varphi_{2,0,0}(\mathbf{r})$].

The matrix which represents W_S in the $n = 2$ level, therefore, has the following form (the basis vectors are arranged in the order $| 2, 1, 1 \rangle$, $| 2, 1, -1 \rangle$, $| 2, 1, 0 \rangle$, $| 2, 0, 0 \rangle$):

$$\begin{pmatrix} 0 & 0 & 0 & 0 \\ 0 & 0 & 0 & 0 \\ 0 & 0 & 0 & \gamma \mathscr{E} \\ 0 & 0 & \gamma \mathscr{E} & 0 \end{pmatrix} \tag{8}$$

We can immediately deduce the corrections to first order in $\mathscr{E}$ and the eigenstates to zeroeth order:

Eigenstates		*Corrections*
$\| 2, 1, 1 \rangle$	$\longleftrightarrow$	0
$\| 2, 1, -1 \rangle$	$\longleftrightarrow$	0
$\frac{1}{\sqrt{2}}(\| 2, 1, 0 \rangle + \| 2, 0, 0 \rangle)$	$\longleftrightarrow$	$\gamma \mathscr{E}$
$\frac{1}{\sqrt{2}}(\| 2, 1, 0 \rangle - \| 2, 0, 0 \rangle)$	$\longleftrightarrow$	$-\gamma \mathscr{E}$

(9)

Thus, we see that the degeneracy of the $n = 2$ level is partially removed and that the energy shifts are *linear*, and not quadratic, in $\mathscr{E}$. The appearance of a linear Stark effect is a typical result of the existence of two levels of opposite parities and the same energy, here the $2s$ and $2p$ levels. This situation exists only in the case of hydrogen (because of the l-fold degeneracy of the $n \neq 1$ shells).

COMMENT:

The states of the $n = 2$ level are not stable. Nevertheless, the lifetime of the $2s$ state is considerably longer than that of the $2p$ states, since the atom passes easily from $2p$ to $1s$ by spontaneous emission of a Lyman α photon (lifetime of the order of 10^{-9} sec), while decay from the $2s$ state requires the emission of two photons (lifetime of the order of a second). For this reason, the $2p$ states are said to be unstable and the $2s$ state, metastable.

Since the Stark Hamiltonian W_S has a non-zero matrix element between $2s$ and $2p$, any electric field (static or oscillating) "mixes" the metastable $2s$ state with the unstable $2p$ state, greatly reducing the $2s$ state's lifetime. This phenomenon is called "metastability quenching" (see also complement H_{IV}, in which we study the effect of a coupling between two states of different lifetimes).

References and suggestions for further reading:

The Stark effect in atoms: Kuhn (11.1), chap. III, §§A-6 and G. Ruark and Urey (11.9), chap. V, §§12 and 13; Sobel'man (11.12), chap. 8, §28.

The summation over the intermediate states which appears in (2) and (6) can be calculated exactly by the method of Dalgarno and Lewis; see Borowitz (1.7), §14-5; Schiff (1.18), §33. Original references: (2.34), (2.35), (2.36).

Quenching of metastability: see Lamb and Retherford (3.11), App. II; Sobel'man (11.12), chap. 8, § 28.5.

CHAPTER XIII

Aproximation methods for time-dependent problems

OUTLINE OF CHAPTER XIII

A. STATEMENT OF THE PROBLEM

Consider a physical system with Hamiltonian H_0. The eigenvalues and eigenvectors of H_0 will be denoted by E_n and $| \varphi_n \rangle$:

$$H_0 | \varphi_n \rangle = E_n | \varphi_n \rangle \tag{A-1}$$

For the sake of simplicity, we shall consider H_0 to be discrete and non-degenerate; the formulas obtained can easily be generalized (see, for example, §C-3). We assume that H_0 is not explicitly time-dependent, so that its eigenstates are stationary states.

At $t = 0$, a perturbation is applied to the system. Its Hamiltonian then becomes:

$$H(t) = H_0 + W(t) \tag{A-2}$$

with:

$$W(t) = \lambda \hat{W}(t) \tag{A-3}$$

where λ is a real dimensionless parameter much smaller than 1 and $\hat{W}(t)$ is an observable (which can be explicitly time-dependent) of the same order of magnitude as H_0 and zero for $t < 0$.

The system is assumed to be initially in the stationary state $| \varphi_i \rangle$, an eigenstate of H_0 of eigenvalue E_i. Starting at $t = 0$ when the perturbation is applied, the system evolves: the state $| \varphi_i \rangle$ is no longer, in general, an eigenstate of the perturbed Hamiltonian. We propose, in this chapter, to calculate the probability $\mathscr{P}_{if}(t)$ of finding the system in another eigenstate $| \varphi_f \rangle$ of H_0 at time t. In other words, we want to study the transitions which can be induced by the perturbation $W(t)$ between the stationary states of the unperturbed system.

The treatment is very simple. Between the times 0 and t, the system evolves in accordance with the Schrödinger equation:

$$i\hbar \frac{\mathrm{d}}{\mathrm{d}t} | \psi(t) \rangle = [H_0 + \lambda \hat{W}(t)] | \psi(t) \rangle \tag{A-4}$$

The solution $| \psi(t) \rangle$ of this first-order differential equation which corresponds to the initial condition:

$$| \psi(t = 0) \rangle = | \varphi_i \rangle \tag{A-5}$$

is unique. The desired probability $\mathscr{P}_{if}(t)$ can be written:

$$\mathscr{P}_{if}(t) = |\langle \varphi_f | \psi(t) \rangle|^2 \tag{A-6}$$

The whole problem therefore consists of finding the solution $| \psi(t) \rangle$ of (A-4) which corresponds to the initial condition (A-5). However, such a problem is not generally rigorously soluble. This is why we resort to approximation methods. We shall show in this chapter how, if λ is sufficiently small, the solution $| \psi(t) \rangle$ can be found in the form of a limited power series expansion in λ. Thus, we shall calculate $| \psi(t) \rangle$ explicitly to first order in λ, as well as the corresponding probability (§ B). The general formulas obtained will then be applied (§ C) to the study of an important special case, the one in which the perturbation is a sinusoidal function of time or a constant (the interaction of an atom with an electromagnetic wave, which falls into this category, is treated in detail in complement A_{XIII}). This is an example of the *resonance* phenomenon. Two situations will be considered : the one in which the spectrum of H_0 is discrete, and then the one in which the initial state $| \varphi_i \rangle$ is coupled to a continuum of final states. In the latter case, we shall prove an important formula known as "*Fermi's golden rule*".

COMMENT:

The situation treated in § C-3 of chapter IV can be considered to be a special case of the general problem discussed in this chapter. Recall that, in chapter IV, we discussed a two-level system (the states $| \varphi_1 \rangle$ and $| \varphi_2 \rangle$), initially in the state $| \varphi_1 \rangle$, subjected, beginning at time $t = 0$, to a constant perturbation W. The probability $\mathscr{P}_{12}(t)$ can then be calculated exactly, leading to *Rabi's formula*.

The problem we are taking up here is much more general. We shall consider a system with an arbitrary number of levels (sometimes, as in § C-3, with a continuum of states) and a perturbation $W(t)$ which is an arbitrary function of the time. This explains why, in general, we can obtain only an approximate solution.

B. APPROXIMATE SOLUTION OF THE SCHRÖDINGER EQUATION

1. The Schrödinger equation in the $\{ | \varphi_n \rangle \}$ representation

The probability $\mathscr{P}_{if}(t)$ explicitly involves the eigenstates $| \varphi_i \rangle$ and $| \varphi_f \rangle$ of H_0. It is therefore reasonable to choose the $\{ | \varphi_n \rangle \}$ representation.

a. THE SYSTEM OF DIFFERENTIAL EQUATIONS FOR THE COMPONENTS OF THE STATE VECTOR

Let $c_n(t)$ be the components of the ket $| \psi(t) \rangle$ in the $\{ | \varphi_n \rangle \}$ basis:

$$| \psi(t) \rangle = \sum_n c_n(t) | \varphi_n \rangle \tag{B-1}$$

with:

$$c_n(t) = \langle \varphi_n | \psi(t) \rangle \tag{B-2}$$

$\hat{W}_{nk}(t)$ denotes the matrix elements of the observable $\hat{W}(t)$ in the same basis:

$$\langle \varphi_n \mid \hat{W}(t) \mid \varphi_k \rangle = \hat{W}_{nk}(t) \tag{B-3}$$

Recall that H_0 is represented in the $\{ \mid \varphi_n \rangle \}$ basis by a diagonal matrix:

$$\langle \varphi_n \mid H_0 \mid \varphi_k \rangle = E_n \, \delta_{nk} \tag{B-4}$$

We shall project both sides of Schrödinger equation (A-4) onto $\mid \varphi_n \rangle$. To do so, we shall insert the closure relation:

$$\sum_k \mid \varphi_k \rangle \langle \varphi_k \mid = 1 \tag{B-5}$$

and use relations (B-2), (B-3) and (B-4). We obtain:

$$i\hbar \frac{\mathrm{d}}{\mathrm{d}t} c_n(t) = E_n \, c_n(t) + \sum_k \lambda \hat{W}_{nk}(t) \, c_k(t) \tag{B-6}$$

The set of equations (B-6), written for the various values of n, constitutes a system of coupled linear differential equations of first order in t, which enables us, in theory, to determine the components $c_n(t)$ of $\mid \psi(t) \rangle$. The coupling between these equations arises solely from the existence of the perturbation $\lambda \hat{W}(t)$, which, by its non-diagonal matrix elements $\lambda \hat{W}_{nk}(t)$, relates the evolution of $c_n(t)$ to that of all the other coefficients $c_k(t)$.

b. CHANGING FUNCTIONS

When $\lambda \hat{W}(t)$ is zero, equations (B-6) are no longer coupled, and their solution is very simple. It can be written:

$$c_n(t) = b_n \, \mathrm{e}^{-iE_n t/\hbar} \tag{B-7}$$

where b_n is a constant which depends on the initial conditions.

If, now, $\lambda \hat{W}(t)$ is not zero, while remaining much smaller than H_0 because of the condition $\lambda \ll 1$, we expect the solution $c_n(t)$ of equations (B-6) to be very close to solution (B-7). In other words, if we perform the change of functions:

$$c_n(t) = b_n(t) \, \mathrm{e}^{-iE_n t/\hbar} \tag{B-8}$$

we can predict that the $b_n(t)$ will be slowly varying functions of time.

We substitute (B-8) into equation (B-6); we obtain:

$$i\hbar \, \mathrm{e}^{-iE_n t/\hbar} \frac{\mathrm{d}}{\mathrm{d}t} b_n(t) + E_n \, b_n(t) \, \mathrm{e}^{-iE_n t/\hbar} = E_n \, b_n(t) \, \mathrm{e}^{-iE_n t/\hbar} + \sum_k \lambda \hat{W}_{nk} \, b_k(t) \, \mathrm{e}^{-iE_k t/\hbar} \tag{B-9}$$

We now multiply both sides of this relation by $\mathrm{e}^{+iE_n t/\hbar}$, and introduce the Bohr angular frequency:

$$\omega_{nk} = \frac{E_n - E_k}{\hbar} \tag{B-10}$$

related to the pair of states E_n and E_k. We obtain:

$$i\hbar \frac{\mathrm{d}}{\mathrm{d}t} b_n(t) = \lambda \sum_k \mathrm{e}^{i\omega_{nk}t} \hat{W}_{nk}(t)\, b_k(t) \tag{B-11}$$

2. Perturbation equations

The system of equations (B-11) is rigorously equivalent to Schrödinger equation (A-4). In general, we do not know how to find its exact solution. This is why we shall use the fact that λ is much smaller than 1 to try to determine this solution in the form of a power series expansion in λ (which we can hope to be rapidly convergent if λ is sufficiently small):

$$b_n(t) = b_n^{(0)}(t) + \lambda\, b_n^{(1)}(t) + \lambda^2 b_n^{(2)}(t) + \ldots \tag{B-12}$$

If we substitute this expansion into (B-11), and if we set equal the coefficients of λ^r on both sides of the equation, we find:

(i) for $r = 0$:

$$i\hbar \frac{\mathrm{d}}{\mathrm{d}t} b_n^{(0)}(t) = 0 \tag{B-13}$$

since the right-hand side of (B-11) has a common factor λ. Relation (B-13) expresses the fact that $b_n^{(0)}$ does not depend on t. Thus, if λ is zero, $b_n(t)$ reduces to a constant [*cf.* (B-7)].

(ii) for $r \neq 0$:

$$i\hbar \frac{\mathrm{d}}{\mathrm{d}t} b_n^{(r)}(t) = \sum_k \mathrm{e}^{i\omega_{nk}t} \hat{W}_{nk}(t)\, b_k^{(r-1)}(t) \tag{B-14}$$

Thus we see that, with the zeroeth-order solution determined by (B-13) and the initial conditions, recurrence relation (B-14) enables us to obtain the first-order solution. It then furnishes the second-order solution in terms of the first-order one, and, by recurrence, the solution to any order r in terms of the one to order $r - 1$.

3. Solution to first order in λ

a. THE STATE OF THE SYSTEM AT TIME t

For $t < 0$, the system is assumed to be in the state $| \varphi_i \rangle$. Of all the coefficients $b_n(t)$, only $b_i(t)$ is different from zero (and, furthermore, independent of t since $\lambda \hat{W}$ is then zero). At time $t = 0$, $\lambda \hat{W}(t)$ may become discontinuous in passing from a zero value to the value $\lambda \hat{W}(0)$. However, since $\lambda \hat{W}(t)$ remains finite, the solution of the Schrödinger equation is continuous at $t = 0$. It follows that:

$$b_n(t = 0) = \delta_{ni} \tag{B-15}$$

and this relation is valid for all λ. Consequently, the coefficients of expansion (B-12) must satisfy:

$$b_n^{(0)}(t = 0) = \delta_{ni} \tag{B-16}$$

$$b_n^{(r)}(t = 0) = 0 \qquad \text{if} \quad r \geqslant 1 \tag{B-17}$$

Equation (B-13) then immediately yields, for all positive t:

$$b_n^{(0)}(t) = \delta_{ni} \tag{B-18}$$

which completely determines the zeroeth-order solution.

This result now permits us to write (B-14), for $r = 1$, in the form:

$$\begin{aligned} i\hbar \frac{\mathrm{d}}{\mathrm{d}t} b_n^{(1)}(t) &= \sum_k \mathrm{e}^{i\omega_{nk}t}\, \hat{W}_{nk}(t)\, \delta_{ki} \\ &= \mathrm{e}^{i\omega_{ni}t}\, \hat{W}_{ni}(t) \end{aligned} \tag{B-19}$$

an equation which can be integrated without difficulty. Taking into account initial condition (B-17), we find:

$$b_n^{(1)}(t) = \frac{1}{i\hbar} \int_0^t \mathrm{e}^{i\omega_{ni}t'}\, \hat{W}_{ni}(t')\, \mathrm{d}t' \tag{B-20}$$

If we now substitute (B-18) and (B-20) into (B-8) and then into (B-1), we obtain the state $| \psi(t) \rangle$ of the system at time t, calculated to first order in λ.

b. THE TRANSITION PROBABILITY $\mathscr{P}_{if}(t)$

According to (A-6) and definition (B-2) of $c_f(t)$, the transition probability $\mathscr{P}_{if}(t)$ is equal to $|c_f(t)|^2$, that is, since $b_f(t)$ and $c_f(t)$ have the same modulus [*cf.* (B-8)]:

$$\mathscr{P}_{if}(t) = |b_f(t)|^2 \tag{B-21}$$

where:

$$b_f(t) = b_f^{(0)}(t) + \lambda\, b_f^{(1)}(t) + \ldots \tag{B-22}$$

can be calculated from the formulas established in the preceding section.

From now on, we shall assume that the states $| \varphi_i \rangle$ and $| \varphi_f \rangle$ are different. We shall therefore be concerned only with the transitions induced by $\lambda \hat{W}(t)$ between two distinct stationary states of H_0. We then have $b_f^{(0)}(t) = 0$, and, consequently:

$$\mathscr{P}_{if}(t) = \lambda^2\, |b_f^{(1)}(t)|^2 \tag{B-23}$$

Using (B-20) and replacing $\lambda \hat{W}(t)$ by $W(t)$ [*cf.* (A-3)], we finally obtain:

$$\mathscr{P}_{if}(t) = \frac{1}{\hbar^2} \left| \int_0^t e^{i\omega_{fi} t'} W_{fi}(t') \, dt' \right|^2 \tag{B-24}$$

Consider the function $\tilde{W}_{fi}(t')$, which is zero for $t' < 0$ and $t' > t$, and equal to $W_{fi}(t')$ for $0 \leqslant t' \leqslant t$ (*cf.* fig. 1). $\tilde{W}_{fi}(t')$ is the matrix element of the perturbation "seen" by the system between the time $t = 0$ and the measurement time t, when we try to determine if the system is in the state $| \varphi_f \rangle$. Result (B-24) shows that $\mathscr{P}_{if}(t)$ is proportional to the square of the modulus of the Fourier transform of the perturbation actually "seen", $\tilde{W}_{fi}(t')$. This Fourier transform is evaluated at an angular frequency equal to the Bohr angular frequency associated with the transition under consideration.

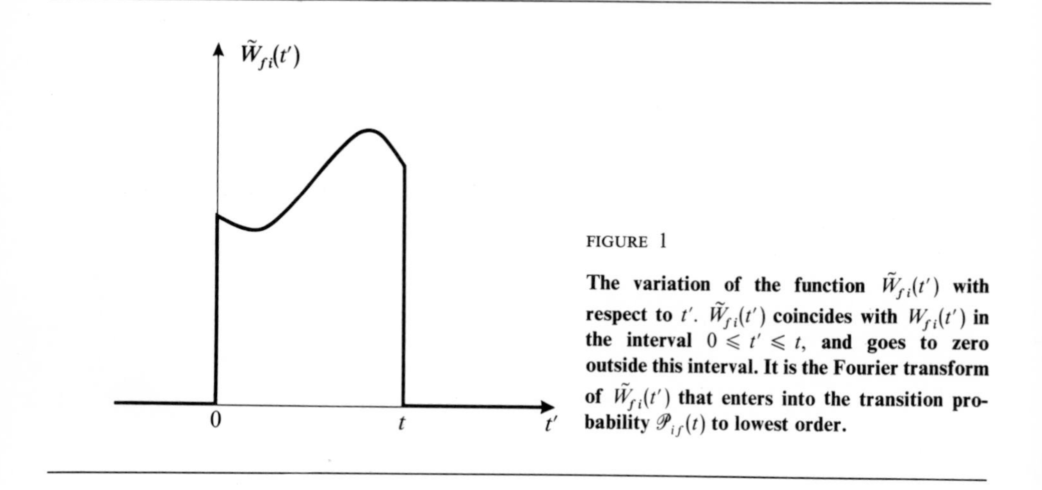

FIGURE 1

The variation of the function $\tilde{W}_{fi}(t')$ with respect to t'. $\tilde{W}_{fi}(t')$ coincides with $W_{fi}(t')$ in the interval $0 \leqslant t' \leqslant t$, and goes to zero outside this interval. It is the Fourier transform of $\tilde{W}_{fi}(t')$ that enters into the transition probability $\mathscr{P}_{if}(t)$ to lowest order.

Note also that the transition probability $\mathscr{P}_{if}(t)$ is zero to first order if the matrix element $W_{fi}(t)$ is zero for all t.

COMMENT:

We have not discussed the validity conditions of the approximation to first order in λ. Comparison of (B-11) with (B-19) shows that this approximation simply amounts to replacing, on the right-hand side of (B-11), the coefficients $b_k(t)$ by their values $b_k(0)$ at time $t = 0$. It is therefore clear that, so long as t remains small enough for $b_k(0)$ not to differ very much from $b_k(t)$, the approximation remains valid. On the other hand, when t becomes large, there is no reason why the corrections of order 2, 3, etc. in λ should be negligible.

C. AN IMPORTANT SPECIAL CASE: A SINUSOIDAL OR CONSTANT PERTURBATION

1. Application of the general equations

Now assume that $W(t)$ has one of the two simple forms:

$$\hat{W}(t) = \hat{W} \sin \omega t \tag{C-1-a}$$

$$\hat{W}(t) = \hat{W} \cos \omega t \tag{C-1-b}$$

where $\hat{W}$ is a time-independent observable and ω, a constant angular frequency. Such a situation is often encountered in physics. For example, in complements A_{XIII} and B_{XIII}, we consider the perturbation of a physical system by an electromagnetic wave of angular frequency ω; $\mathscr{P}_{if}(t)$ then represents the probability, induced by the incident monochromatic radiation, of a transition between the initial state $| \varphi_i \rangle$ and the final state $| \varphi_f \rangle$.

With the particular form (C-1-a) of $\hat{W}(t)$, the matrix elements $\hat{W}_{fi}(t)$ take on the form:

$$\hat{W}_{fi}(t) = \hat{W}_{fi} \sin \omega t = \frac{\hat{W}_{fi}}{2i} (e^{i\omega t} - e^{-i\omega t}) \tag{C-2}$$

where $\hat{W}_{fi}$ is a time-independent complex number. Let us now calculate the state vector of the system to first order in λ. If we substitute (C-2) into general formula (B-20), we obtain:

$$b_n^{(1)}(t) = -\frac{\hat{W}_{ni}}{2\hbar} \int_0^t [e^{i(\omega_{ni}+\omega)t'} - e^{i(\omega_{ni}-\omega)t'}] \, dt' \tag{C-3}$$

The integral which appears on the right-hand side of this relation can easily be calculated and yields:

$$b_n^{(1)}(t) = \frac{\hat{W}_{ni}}{2i\hbar} \left[\frac{1 - e^{i(\omega_{ni}+\omega)t}}{\omega_{ni} + \omega} - \frac{1 - e^{i(\omega_{ni}-\omega)t}}{\omega_{ni} - \omega} \right] \tag{C-4}$$

Therefore, in the special case we are treating, general equation (B-24) becomes:

$$\mathscr{P}_{if}(t;\omega) = \lambda^2 \, |b_f^{(1)}(t)|^2 = \frac{|W_{fi}|^2}{4\hbar^2} \left| \frac{1 - e^{i(\omega_{fi}+\omega)t}}{\omega_{fi} + \omega} - \frac{1 - e^{i(\omega_{fi}-\omega)t}}{\omega_{fi} - \omega} \right|^2 \tag{C-5-a}$$

(we have added the variable ω in the probability $\mathscr{P}_{if}$, since the latter depends on the frequency of the perturbation).

If we choose the special form (C-1-b) for $\hat{W}(t)$ instead of (C-1-a), a calculation analogous to the preceding one yields:

$$\mathscr{P}_{if}(t;\omega) = \frac{|W_{fi}|^2}{4\hbar^2} \left| \frac{1 - e^{i(\omega_{fi}+\omega)t}}{\omega_{fi} + \omega} + \frac{1 - e^{i(\omega_{fi}-\omega)t}}{\omega_{fi} - \omega} \right|^2 \tag{C-5-b}$$

$\hat{W}\cos\omega t$ becomes time-independent if we choose $\omega = 0$. The transition probability $\mathscr{P}_{if}(t)$ induced by a constant perturbation W can therefore be obtained by replacing ω by 0 in (C-5-b):

$$\begin{aligned}\mathscr{P}_{if}(t) &= \frac{|W_{fi}|^2}{\hbar^2\omega_{fi}^2}\,|1 - e^{i\omega_{fi}t}|^2 \\ &= \frac{|W_{fi}|^2}{\hbar^2}\,F(t, \omega_{fi})\end{aligned} \tag{C-6}$$

with:

$$F(t, \omega_{fi}) = \left[\frac{\sin(\omega_{fi}t/2)}{\omega_{fi}/2}\right]^2 \tag{C-7}$$

In order to study the physical content of equations (C-5) and (C-6), we shall first consider the case in which $|\,\varphi_i\,\rangle$ and $|\,\varphi_f\,\rangle$ are two discrete levels (§ 2), and then the one in which $|\,\varphi_f\,\rangle$ belongs to a continuum of final states (§ 3). In the first case, $\mathscr{P}_{if}(t\,;\omega)$ [or $\mathscr{P}_{if}(t)$] really represents a transition probability which can be measured, while, in the second case, we are actually dealing with a probability density (the truly measurable quantities then involve a summation over a set of final states). From a physical point of view, there is a distinct difference between these two cases. We shall see in complements C_{XIII} and D_{XIII} that, over a sufficiently long time interval, the system oscillates between the states $|\,\varphi_i\,\rangle$ and $|\,\varphi_f\,\rangle$ in the first case, while it leaves the state $|\,\varphi_i\,\rangle$ irreversibly in the second case.

In §2, in order to concentrate on the resonance phenomenon, we shall choose a sinusoidal perturbation, but the results obtained can easily be transposed to the case of a constant perturbation. Inversely, we shall use this latter case for the discussion of §3.

2. Sinusoidal perturbation which couples two discrete states: the resonance phenomenon

a. RESONANT NATURE OF THE TRANSITION PROBABILITY

When the time t is fixed, the transition probability $\mathscr{P}_{if}(t;\,\omega)$ is a function only of the variable ω. We shall see that this function has a maximum for:

$$\omega \simeq \omega_{fi} \tag{C-8-a}$$

or:

$$\omega \simeq -\,\omega_{fi} \tag{C-8-b}$$

A resonance phenomenon therefore occurs when the angular frequency of the perturbation coincides with the Bohr angular frequency associated with the pair of states $|\,\varphi_i\,\rangle$ and $|\,\varphi_f\,\rangle$. If we agree to choose $\omega \geqslant 0$, relations (C-8) give the resonance conditions corresponding respectively to the cases $\omega_{fi} > 0$ and $\omega_{fi} < 0$.

In the first case (*cf.* fig. 2-a), the system goes from the lower energy level E_i to the higher level E_f by the resonant absorption of an energy quantum $\hbar\omega$. In the second case (*cf.* fig. 2-b), the resonant perturbation stimulates the passage of the system

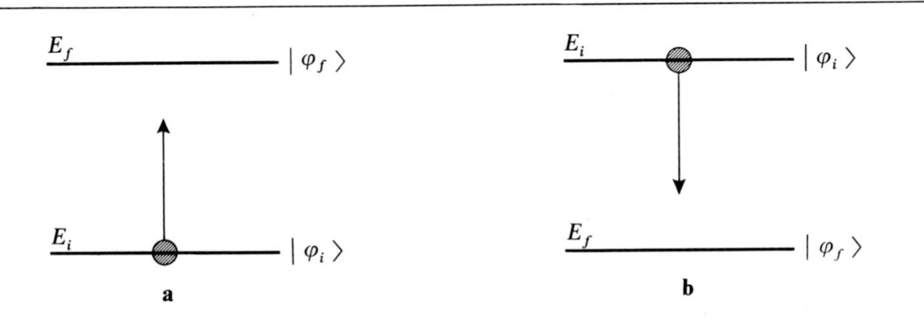

FIGURE 2

The relative disposition of the energies E_i and E_f associated with the states $|\varphi_i\rangle$ and $|\varphi_f\rangle$. If $E_i < E_f$ (fig. a), the $|\varphi_i\rangle \longrightarrow |\varphi_f\rangle$ transition occurs through absorption of an energy quantum $\hbar\omega$. If, on the other hand, $E_i > E_f$ (fig. b), the $|\varphi_i\rangle \longrightarrow |\varphi_f\rangle$ transition occurs through induced emission of an energy quantum $\hbar\omega$.

from the higher level E_i to the lower level E_f (accompanied by the induced emission of an energy quantum $\hbar\omega$). Throughout this section, we shall assume that ω_{fi} is positive (the situation of figure 2-a). The case in which ω_{fi} is negative could be treated analogously.

To reveal the resonant nature of the transition probability, we note that expressions (C-5-a) and (C-5-b) for $\mathscr{P}_{if}(t;\omega)$ involve the square of the modulus of a sum of two complex terms. The first of these terms is proportional to:

$$A_+ = \frac{1 - e^{i(\omega_{fi}+\omega)t}}{\omega_{fi}+\omega} = -\,i\,e^{i(\omega_{fi}-\omega)t/2}\,\frac{\sin\left[(\omega_{fi}+\omega)t/2\right]}{(\omega_{fi}+\omega)/2} \qquad \text{(C-9-a)}$$

and the second one, to:

$$A_- = \frac{1 - e^{i(\omega_{fi}-\omega)t}}{\omega_{fi}-\omega} = -\,i\,e^{i(\omega_{fi}-\omega)t/2}\,\frac{\sin\left[(\omega_{fi}-\omega)t/2\right]}{(\omega_{fi}-\omega)/2} \qquad \text{(C-9-b)}$$

The denominator of the A_- term goes to zero for $\omega = \omega_{fi}$, and that of the A_+ term, for $\omega = -\,\omega_{fi}$. Consequently, for ω close to ω_{fi}, we expect only the A_- term to be important; this is why it is called the "resonant term", while the A_+ term is called the "anti-resonant term" (A_+ would become resonant if, for negative ω_{fi}, ω were close to $-\,\omega_{fi}$).

Let us then consider the case in which:

$$|\omega - \omega_{fi}| \ll |\omega_{fi}| \qquad \text{(C-10)}$$

neglecting the anti-resonant term A_+ (the validity of this approximation will be discussed in § c below). Taking (C-9-b) into account, we then obtain:

$$\mathscr{P}_{if}(t\,;\omega) = \frac{|W_{fi}|^2}{4\hbar^2}\,F(t\,,\omega - \omega_{fi}) \tag{C-11}$$

with:

$$F(t,\omega - \omega_{fi}) = \left\{\frac{\sin\,[(\omega_{fi} - \omega)t/2]}{(\omega_{fi} - \omega)/2}\right\}^2 \tag{C-12}$$

Figure 3 represents the variation of $\mathscr{P}_{if}(t;\,\omega)$ with respect to ω, where t is fixed. It clearly shows the resonant nature of the transition probability. This probability presents a maximum for $\omega = \omega_{fi}$, when it is equal to $|W_{fi}|^2t^2/4\hbar^2$. As we move away from ω_{fi}, it decreases, going to zero for $|\omega - \omega_{fi}| = 2\pi/t$. When $|\omega - \omega_{fi}|$ continues to increase, it oscillates between the value $|W_{fi}|^2/\hbar^2(\omega - \omega_{fi})^2$ and zero ("diffraction pattern").

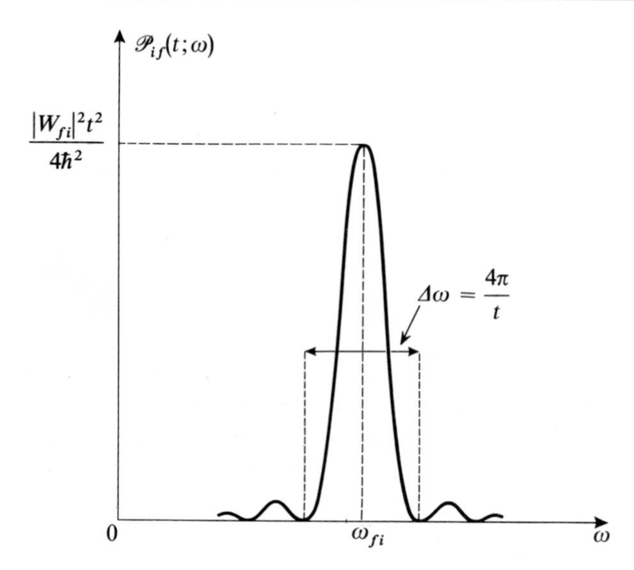

FIGURE 3

Variation, with respect to ω, of the first-order transition probability $\mathscr{P}_{if}(t;\,\omega)$ associated with a sinusoidal perturbation of angular frequency ω; t is fixed. When $\omega \simeq \omega_{fi}$, a resonance appears whose intensity is proportional to t^2 and whose width is inversely proportional to t.

b. THE RESONANCE WIDTH AND THE TIME-ENERGY UNCERTAINTY RELATION

The resonance width $\Delta\omega$ can be approximately defined as the distance between the first two zeros of $\mathscr{P}_{if}(t;\,\omega)$ about $\omega = \omega_{fi}$. It is inside this interval that the transition probability takes on its largest values [the first secondary maximum of $\mathscr{P}_{if}$, attained when $(\omega - \omega_{fi})t/2 = 3\pi/2$, is equal to $|W_{fi}|^2 t^2/9\pi^2\hbar^2$, that is, less than 5 % of the transition probability at resonance]. We then have:

$$\Delta\omega \simeq \frac{4\pi}{t} \qquad \text{(C-13)}$$

The larger the time t, the smaller this width.

Result (C-13) presents a certain analogy with the time-energy uncertainty relation (*cf.* chap. III, §D-2-e). Assume that we want to measure the energy difference $E_f - E_i = \hbar\omega_{fi}$ by applying a sinusoidal perturbation of angular frequency ω to the system and varying ω so as to detect the resonance. If the perturbation acts during a time t, the uncertainty ΔE on the value $E_f - E_i$ will be, according to (C-13), of the order of:

$$\Delta E = \hbar\,\Delta\omega \simeq \frac{\hbar}{t} \qquad \text{(C-14)}$$

Therefore, the product $t\Delta E$ cannot be smaller than $\hbar$. This recalls the time-uncertainty relation, although t here is not a time interval characteristic of the free evolution of the system, but is externally imposed.

c. VALIDITY OF THE PERTURBATION TREATMENT

Now let us examine the limits of validity of the calculations leading to result (C-11). We shall first discuss the resonant approximation, which consists of neglecting the anti-resonant term A_+, and then the first-order approximation in the perturbation expansion of the state vector.

α. *Discussion of the resonant approximation*

Using the hypothesis $\omega \simeq \omega_{fi}$, we have neglected A_+ relative to A_-. We shall therefore compare the moduli of A_+ and A_-.

The shape of the function $|A_-(\omega)|^2$ is shown in figure 3. Since $|A_+(\omega)|^2 = |A_-(-\omega)|^2$, $|A_+(\omega)|^2$ can be obtained by plotting the curve which is symmetric with respect to the preceding one relative to the vertical axis $\omega = 0$. If these two curves, of width $\Delta\omega$, are centered at points whose separation is much larger than $\Delta\omega$, it is clear that, in the neighborhood of $\omega = \omega_{fi}$, the modulus of A_+ is negligible compared to that of A_-. The resonant approximation is therefore justified on the condition* that:

$$2|\omega_{fi}| \gg \Delta\omega \qquad \text{(C-15)}$$

that is, using (C-13):

$$t \gg \frac{1}{|\omega_{fi}|} \simeq \frac{1}{\omega} \qquad \text{(C-16)}$$

Result (C-11) is therefore valid only if the sinusoidal perturbation acts during a time t which is large compared to $1/\omega$. The physical meaning of such a condition

* Note that if condition (C-15) is not satisfied, the resonant and anti-resonant terms interfere: it is not correct to simply add $|A_+|^2$ and $|A_-|^2$.

is clear: during the interval $[0, t]$, the perturbation must perform numerous oscillations to appear to the system as a sinusoidal perturbation. If, on the other hand, t were small compared to $1/\omega$, the perturbation would not have the time to oscillate and would be equivalent to a perturbation varying linearly in time [in the case (C-1-a)] or constant [in the case (C-1-b)].

COMMENT:

For a constant perturbation, condition (C-16) can never be satisfied, since ω is zero. However, it is not difficult to adapt the calculations of § b above to this case. We have already obtained [in (C-6)] the transition probability $\mathscr{P}_{if}(t)$ for a constant perturbation by directly setting $\omega = 0$ in (C-5-b). Note that the two terms A_+ and A_- are then equal, which shows that if (C-16) is not satisfied, the anti-resonant term is not negligible.

The variation of the probability $\mathscr{P}_{if}(t)$ with respect to the energy difference $\hbar\omega_{fi}$ (with the time t fixed) is shown in figure 4. This probability is maximal when $\omega_{fi} = 0$, which corresponds to what we found in §b above: if its angular frequency is zero, the perturbation is resonant when $\omega_{fi} = 0$ (degenerate levels). More generally, the considerations of § b concerning the features of the resonance can be transposed to this case.

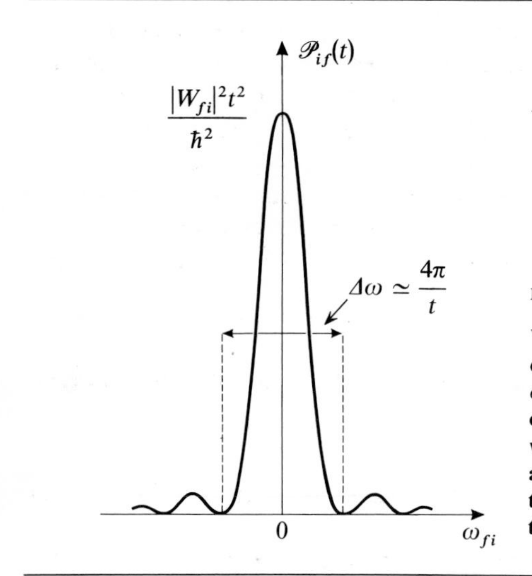

FIGURE 4

Variation of the transition probability $\mathscr{P}_{if}(t)$ associated with a constant perturbation with respect to $\omega_{fi} = (E_f - E_i)/\hbar$, for fixed t. A resonance appears, centered about $\omega_{fi} = 0$ (conservation of energy), with the same width as the resonance of figure 3, but an intensity four times greater (because of the constructive interference of the resonant and anti-resonant terms, which, for a constant perturbation, are equal).

β. *Limits of the first-order calculation*

We have already noted (*cf.* comment at the end of §B-3-b) that the first-order approximation can cease to be valid when the time t becomes too large. This can indeed be seen from expression (C-11), which, at resonance, can be written:

$$\mathscr{P}_{if}(t; \omega = \omega_{fi}) = \frac{|W_{fi}|^2}{4\hbar^2} t^2 \tag{C-17}$$

This function becomes infinite when $t \longrightarrow \infty$, which is absurd, since a probability can never be greater than 1.

In practice, for the first-order approximation to be valid at resonance, the probability in (C-17) must be much smaller than 1, that is★:

$$t \ll \frac{\hbar}{|W_{fi}|} \tag{C-18}$$

To show precisely why this inequality is related to the validity of the first-order approximation, it would be necessary to calculate the higher-order corrections from (B-14) and to examine under what conditions they are negligible. We would then see that although inequality (C-18) is necessary, it is not rigorously sufficient. For example, in the terms of second or higher order, there appear matrix elements $\hat{W}_{kn}$ of $\hat{W}$ other than $\hat{W}_{fi}$, on which certain conditions must be imposed for the corresponding corrections to be small.

Note that the problem of calculating the transition probability when t does not satisfy (C-18) is taken up in complement C_{XIII}, in which an approximation of a different type is used (the secular approximation).

3. Coupling with the states of the continuous spectrum

If the energy E_f belongs to a continuous part of the spectrum of H_0, that is, if the final states are labeled by continuous indices, we cannot measure the probability of finding the system in a *well-defined* state $| \varphi_f \rangle$ at time t. The postulates of chapter III indicate that in this case the quantity $|\langle \varphi_f | \psi(t) \rangle|^2$ which we found above (approximately) is a probability density. The physical predictions for a given measurement then involve an integration of this probability density over a certain group of final states (which depends on the measurement to be made). We shall consider what happens to the results of the preceding sections in this case.

a. INTEGRATION OVER A CONTINUUM OF FINAL STATES; DENSITY OF STATES

α. *Example*

To understand how this integration is performed over the final states, we shall first consider a concrete example.

We shall discuss the problem of the scattering of a spinless particle of mass m by a potential $W(\mathbf{r})$ (*cf.* chap. VIII). The state $| \psi(t) \rangle$ of the particle at time t can

★ For this theory to be meaningful, it is obviously necessary for conditions (C-16) and (C-18) to be compatible. That is, we must have:

$$\frac{1}{|\omega_{fi}|} \ll \frac{\hbar}{|W_{fi}|}$$

This inequality means that the energy difference $|E_f - E_i| = \hbar\,|\omega_{fi}|$ is much larger than the matrix element of $W(t)$ between $| \varphi_i \rangle$ and $| \varphi_f \rangle$.

be expanded on the states $| \mathbf{p} \rangle$ of well-defined momenta $\mathbf{p}$ and energies :

$$E = \frac{\mathbf{p}^2}{2m} \tag{C-19}$$

The corresponding wave functions are the plane waves:

$$\langle \mathbf{r} | \mathbf{p} \rangle = \left(\frac{1}{2\pi\hbar} \right)^{3/2} e^{i \mathbf{p} \cdot \mathbf{r}/\hbar} \tag{C-20}$$

The probability density associated with a measurement of the momentum is $|\langle \mathbf{p} | \psi(t) \rangle|^2$ [$| \psi(t) \rangle$ is assumed to be normalized].

The detector used in the experiment (see, for example, figure 2 of chapter VIII) gives a signal when the particle is scattered with the momentum $\mathbf{p}_f$. Of course, this detector always has a finite angular aperture, and its energy selectivity is not perfect : it emits a signal whenever the momentum $\mathbf{p}$ of the particle points within a solid angle $\delta\Omega_f$ about $\mathbf{p}_f$ and its energy is included in the interval δE_f centered at $E_f = \mathbf{p}_f^2/2m$. If D_f denotes the domain of $\mathbf{p}$-space defined by these conditions, the probability of obtaining a signal from the detector is therefore:

$$\delta\mathscr{P}(\mathbf{p}_f, t) = \int_{\mathbf{p} \in D_f} d^3p \, |\langle \mathbf{p} | \psi(t) \rangle|^2 \tag{C-21}$$

To use the results of the preceding sections, we shall have to perform a change of variables which results in an integral over the energies. This does not present any difficulties, since we can write :

$$d^3p = p^2 \, \mathrm{d}p \, \mathrm{d}\Omega \tag{C-22}$$

and replace the variable p by the energy E, to which it is related by (C-19). We thus obtain:

$$d^3p = \rho(E) \, \mathrm{d}E \, \mathrm{d}\Omega \tag{C-23}$$

where the function $\rho(E)$, called the *density of final states*, can be written, according to (C-19), (C-22) and (C-23):

$$\rho(E) = p^2 \frac{\mathrm{d}p}{\mathrm{d}E} = p^2 \frac{m}{p} = m\sqrt{2mE} \tag{C-24}$$

(C-21) then becomes:

$$\delta\mathscr{P}(\mathbf{p}_f, t) = \int_{\left\{ \begin{array}{l} \Omega \in \delta\Omega_f \\ E \in \delta E_f \end{array} \right.} \mathrm{d}\Omega \, \mathrm{d}E \, \rho(E) \, |\langle \mathbf{p} | \psi(t) \rangle|^2 \tag{C-25}$$

β. *The general case*

Assume that, in a particular problem, certain eigenstates of H_0 are labeled by a continuous set of indices, symbolized by α, such that the orthonormalization relation can be written:

$$\langle \alpha | \alpha' \rangle = \delta(\alpha - \alpha') \tag{C-26}$$

The system is described at time t by the normalized ket $| \psi(t) \rangle$. We want to calculate the probability $\delta\mathscr{P}(\alpha_f, t)$ of finding the system, in a measurement, in a given group of final states. We characterize this group of states by a domain D_f of values of the parameters α, centered at α_f, and we assume that their energies form a continuum. The postulates of quantum mechanics then yield:

$$\delta\mathscr{P}(\alpha_f, t) = \int_{\alpha \in D_f} \mathrm{d}\alpha \, |\langle \alpha | \psi(t) \rangle|^2 \tag{C-27}$$

As in the example of §α above, we shall change variables, and introduce the density of final states. Instead of characterizing these states by the parameters α, we shall use the energy E and a set of other parameters β (which are necessary when H_0 alone does not constitute a C.S.C.O.). We can then express $\mathrm{d}\alpha$ in terms of $\mathrm{d}E$ and $\mathrm{d}\beta$:

$$\mathrm{d}\alpha = \rho(\beta, E) \, \mathrm{d}\beta \, \mathrm{d}E \tag{C-28}$$

in which the density of final states $\rho(\beta, E)$★ appears. If we denote by $\delta\beta_f$ and δE_f the range of values of the parameters β and E defined by D_f, we obtain:

$$\delta\mathscr{P}(\alpha_f, t) = \int_{\substack{\beta \in \delta\beta_f \\ E \in \delta E_f}} \mathrm{d}\beta \, \mathrm{d}E \, \rho(\beta, E) \, |\langle \beta, E | \psi(t) \rangle|^2 \tag{C-29}$$

where the notation $| \alpha \rangle$ has been replaced by $| \beta, E \rangle$ in order to point up the E- and β-dependence of the probability density $|\langle \alpha | \psi(t) \rangle|^2$.

b. FERMI'S GOLDEN RULE

In expression (C-29), $| \psi(t) \rangle$ is the normalized state vector of the system at time t. As in §A of this chapter, we shall consider a system which is initially in an eigenstate $| \varphi_i \rangle$ of H_0 [$| \varphi_i \rangle$ therefore belongs to the discrete spectrum of H_0, since the initial state of the system must, like $| \psi(t) \rangle$, be normalizable]. In (C-29), we shall replace the notation $\delta\mathscr{P}(\alpha_f, t)$ by $\delta\mathscr{P}(\varphi_i, \alpha_f, t)$ in order to remember that the system starts from the state $| \varphi_i \rangle$.

The calculations of §B and their application to the case of a sinusoidal or constant perturbation (§§C-1 and C-2) remain valid when the final state of the system belongs to the continuous spectrum of H_0. If we assume W to be constant, we can therefore use (C-6) to find the probability density $|\langle \beta, E | \psi(t) \rangle|^2$ to first order in W. We then get:

$$|\langle \beta, E | \psi(t) \rangle|^2 = \frac{1}{\hbar^2} |\langle \beta, E | W | \varphi_i \rangle|^2 \, F\left(t, \frac{E - E_i}{\hbar}\right) \tag{C-30}$$

★ In the general case, the density of states ρ depends on both E and β. However, it often happens (*cf.* example of §α above) that ρ depends only on E.

where E and E_i are the energies of the states $| \beta, E \rangle$ and $| \varphi_i \rangle$ respectively, and F is the function defined by (C-7). We get for $\delta\mathscr{P}(\varphi_i, \alpha_f, t)$, finally:

$$\delta\mathscr{P}(\varphi_i, \alpha_f, t) = \frac{1}{\hbar^2} \int_{\substack{\beta \in \delta\beta_f \\ E \in \delta E_f}} d\beta \, dE \, \rho(\beta, E) \, |\langle \beta, E | W | \varphi_i \rangle|^2 \, F\left(t, \frac{E - E_i}{\hbar}\right) \tag{C-31}$$

The function $F\left(t, \frac{E - E_i}{\hbar}\right)$ varies rapidly about $E = E_i$ (*cf.* fig. 4). If t is sufficiently large, this function can be approximated, to within a constant factor, by the function $\delta(E - E_i)$, since, according to (11) and (20) of appendix II, we have:

$$\lim_{t \to \infty} F\left(t, \frac{E - E_i}{\hbar}\right) = \pi t \, \delta\left(\frac{E - E_i}{2\hbar}\right) = 2\pi\hbar t \, \delta(E - E_i) \tag{C-32}$$

On the other hand, the function $\rho(\beta, E) \, |\langle \beta, E | W | \varphi_i \rangle|^2$ generally varies much more slowly with E. We shall assume here that t is sufficiently large for the variation of this function over an energy interval of width $4\pi\hbar/t$ centered at $E = E_i$ to be negligible★. We can then in (C-31) replace $F\left(t, \frac{E - E_i}{\hbar}\right)$ by its limit (C-32). This enables us to perform the integral over E immediately. If, in addition, $\delta\beta_f$ is very small, integration over β is unnecessary, and we finally get:

– when the energy E_i belongs to the domain δE_f:

$$\delta\mathscr{P}(\varphi_i, \alpha_f, t) = \delta\beta_f \frac{2\pi}{\hbar} t \, | \langle \beta_f, E_f = E_i | W | \varphi_i \rangle |^2 \, \rho(\beta_f, E_f = E_i) \tag{C-33-a}$$

– when the energy E_i does not belong to this domain:

$$\delta\mathscr{P}(\varphi_i, \alpha_f, t) = 0 \tag{C-33-b}$$

As we saw in the comment of §C-2-c-α, a constant perturbation can induce transitions only between states of equal energies. The system must have the same energy (to within $2\pi\hbar/t$) in the initial and final states. This is why, if the domain δE_f excludes the energy E_i, the transition probability is zero.

The probability (C-33-a) increases linearly with time. Consequently, *the transition probability per unit time*, $\delta\mathscr{W}(\varphi_i, \alpha_f)$, defined by:

$$\delta\mathscr{W}(\varphi_i, \alpha_f) = \frac{d}{dt} \delta\mathscr{P}(\varphi_i, \alpha_f, t) \tag{C-34}$$

is time-independent. We introduce the transition probability density per unit time and per unit interval of the variable β_f:

$$w(\varphi_i, \alpha_f) = \frac{\delta\mathscr{W}(\varphi_i, \alpha_f)}{\delta\beta_f} \tag{C-35}$$

★ $\rho(\beta, E) \, |\langle \beta, E | W | \varphi_i \rangle|^2$ must vary slowly enough to enable the finding of values of t which satisfy the stated condition but remain small enough for the perturbation treatment of W to be valid. Here, we also assume that $\delta E_f \gg 4\pi\hbar/t$.

It is equal to:

$$w(\varphi_i, \alpha_f) = \frac{2\pi}{\hbar} |\langle \beta_f, E_f = E_i | W | \varphi_i \rangle|^2 \rho(\beta_f, E_f = E_i) \tag{C-36}$$

This important result is known as *Fermi's golden rule.*

COMMENTS:

(*i*) Assume that W is a sinusoidal perturbation of the form (C-1-a) or (C-1-b), which couples a state $| \varphi_i \rangle$ to a continuum of states $| \beta_f, E_f \rangle$ with energies E_f close to $E_i + \hbar\omega$. Starting with (C-11), we can carry out the same procedure as above, which yields:

$$w(\varphi_i, \alpha_f) = \frac{\pi}{2\hbar} |\langle \beta_f, E_f = E_i + \hbar\omega | W | \varphi_i \rangle|^2 \rho(\beta_f, E_f = E_i + \hbar\omega) \tag{C-37}$$

(*ii*) Let us return to the problem of the scattering of a particle by a potential W whose matrix elements in the $\{ | \mathbf{r} \rangle \}$ representation are given by:

$$\langle \mathbf{r} | W | \mathbf{r}' \rangle = W(\mathbf{r}) \, \delta(\mathbf{r} - \mathbf{r}') \tag{C-38}$$

Now assume that the initial state of the system is a well-defined momentum state:

$$| \psi(t = 0) \rangle = | \mathbf{p}_i \rangle \tag{C-39}$$

and we shall calculate the scattering probability of an incident particle of momentum $\mathbf{p}_i$ into the states of momentum $\mathbf{p}$ grouped about a given value $\mathbf{p}_f$ (with $|\mathbf{p}_f| = |\mathbf{p}_i|$). (C-36) gives the scattering probability $w(\mathbf{p}_i, \mathbf{p}_f)$ per unit time and per unit solid angle about $\mathbf{p} = \mathbf{p}_f$:

$$w(\mathbf{p}_i, \mathbf{p}_f) = \frac{2\pi}{\hbar} |\langle \mathbf{p}_f | W | \mathbf{p}_i \rangle|^2 \rho(E_f = E_i) \tag{C-40}$$

Taking into account (C-20), (C-38) and expression (C-24) for $\rho(E)$, we then get:

$$w(\mathbf{p}_i, \mathbf{p}_f) = \frac{2\pi}{\hbar} m \sqrt{2mE_i} \left(\frac{1}{2\pi\hbar}\right)^6 \left| \int d^3r \, e^{i(\mathbf{p}_i - \mathbf{p}_f).\mathbf{r}/\hbar} \, W(\mathbf{r}) \right|^2 \tag{C-41}$$

On the right-hand side of this relation, we recognize the Fourier transform of the potential $W(\mathbf{r})$, evaluated for the value of $\mathbf{p}$ equal to $\mathbf{p}_i - \mathbf{p}_f$.

Note that the initial state $| \mathbf{p}_i \rangle$ chosen here is not normalizable, and it cannot represent the physical state of a particle. However, although the norm of $| \mathbf{p}_i \rangle$ is infinite, the right-hand side of (C-41) maintains a finite value. Intuitively, we can therefore expect to obtain a correct physical result from this relation. If we divide the probability obtained by the probability current:

$$J_i = \left(\frac{1}{2\pi\hbar}\right)^3 \frac{\hbar k_i}{m} = \left(\frac{1}{2\pi\hbar}\right)^3 \sqrt{\frac{2E_i}{m}} \tag{C-42}$$

associated, according to (C-20), with the state $| \mathbf{p}_i \rangle$, we obtain:

$$\frac{w(\mathbf{p}_i, \mathbf{p}_f)}{J_i} = \frac{m^2}{4\pi^2\hbar^4} \left| \int d^3r \; \mathrm{e}^{i(\mathbf{p}_i - \mathbf{p}_f).\mathbf{r}/\hbar} \, W(\mathbf{r}) \right|^2 \tag{C-43}$$

which is the expression for the scattering cross section in the Born approximation (§B-4 of chap. VIII).

Although it is not rigorous, the preceding treatment enables us to show that the scattering cross sections of the Born approximation can also be obtained by a time-dependent approach, using Fermi's golden rule.

References and suggestions for further reading:

Perturbation series expansion of the evolution operator: Messiah (1.17), chap. XVII, §§1 and 2.

Sudden or adiabatic modification of the Hamiltonian: Messiah (1.17), chap. XVII, § II; Schiff (1.18), chap. 8, §35.

Diagramatic representation of a perturbation series (Feynman diagrams): Ziman (2.26), chap. 3; Mandl (2.9), chaps. 12 to 14; Bjorken and Drell (2.10), chaps. 16 and 17.

COMPLEMENTS OF CHAPTER XIII

A_{XIII}: INTERACTION OF AN ATOM WITH AN ELECTROMAGNETIC WAVE

A_{XIII}: illustration of the general considerations of §C-2 of chapter XIII, using the important example of an atom interacting with a sinusoidal electromagnetic wave. Introduces fundamental concepts such as: spectral line selection rules, absorption and induced emission of radiation, oscillator strength... Although moderately difficult, can be recommended for a first reading because of the importance of the concepts introduced.

B_{XIII}: LINEAR AND NON-LINEAR RESPONSE OF A TWO-LEVEL SYSTEM SUBJECTED TO A SINUSOIDAL PERTURBATION

B_{XIII}: a simple model for the study of some non-linear effects which appear in the interaction of an electromagnetic wave with an atomic system (saturation effects, multiple-quanta transitions,...). More difficult than A_{XIII} (graduate level); should therefore be reserved for a subsequent study.

C_{XIII}: OSCILLATIONS OF A SYSTEM BETWEEN TWO DISCRETE STATES UNDER THE EFFECT OF A RESONANT PERTURBATION

C_{XIII}: study of the behavior, over a long time interval, of a system which has discrete energy levels, subjected to a resonant perturbation. Completes, in greater detail, the results of §C-2 of chapter XIII, which are valid only for short times. Relatively simple.

D_{XIII}: DECAY OF A DISCRETE STATE RESONANTLY COUPLED TO A CONTINUUM OF FINAL STATES

D_{XIII}: study of the behavior, over a long time interval, of a discrete state resonantly coupled to a continuum of final states. Completes, in greater detail, the results obtained in §C-3 of chapter XIII (Fermi's golden rule), which were established only for short time intervals. Proves that the probability of finding the particle in the discrete level decreases exponentially and justifies the concept of lifetimes introduced phenomenologically in complement K_{III}. Important for its numerous physical applications; graduate level.

E_{XIII}: EXERCISES

E_{XIII}: exercise 10 can be done at the end of complement A_{XIII}; it is a step-by-step study of the effect of the external degrees of freedom of a quantum mechanical system on the frequencies of the electromagnetic radiation it can absorb (the Doppler effect, recoil energy, the Mössbauer effect).

Certain exercises (especially 8 and 9) are more difficult than those of other complements, but treat important physical phenomena.

Complement A_{XIII}

THE INTERACTION OF AN ATOM WITH AN ELECTROMAGNETIC WAVE

In §C of chapter XIII, we studied the special case of a sinusoidally time-dependent perturbation : $W(t) = W \sin \omega t$. We encountered the resonance phenomenon which occurs when ω is close to one of the Bohr angular frequencies $\omega_{fi} = (E_f - E_i)/\hbar$ of the physical system under consideration.

A particularly important application of this theory is the treatment of an atom interacting with a monochromatic wave. In this complement, we will use this example to illustrate the general considerations of chapter XIII and to point out certain fundamental concepts of atomic physics such as spectral line selection rules, induced absorption and emission of radiation, oscillator strength, etc...

As in chapter XIII, we shall confine ourselves to first-order perturbation calculations. Some higher-order effects in the interaction of an atom with an electromagnetic wave ("non-linear" effects) will be taken up in complement B_{XIII}.

We shall begin (§1) by analyzing the structure of the interaction Hamiltonian between an atom and the electromagnetic field. This will permit us to isolate the electric dipole, magnetic dipole and electric quadrupole terms, and to study the corresponding selection rules. Then we shall calculate the electric dipole moment induced by a non-resonant incident wave (§2) and compare the results obtained with those of the model of the elastically bound electron. Finally, we shall study (§3) the processes of induced absorption and emission of radiation which appear in the resonant excitation of an atom.

1. The interaction Hamiltonian. Selection rules

a. FIELDS AND POTENTIALS ASSOCIATED WITH A PLANE ELECTROMAGNETIC WAVE

Consider a plane electromagnetic wave★, of wave vector **k** (parallel to Oy) and angular frequency $\omega = ck$. The electric field of the wave is parallel to Oz and the magnetic field, to Ox (fig. 1).

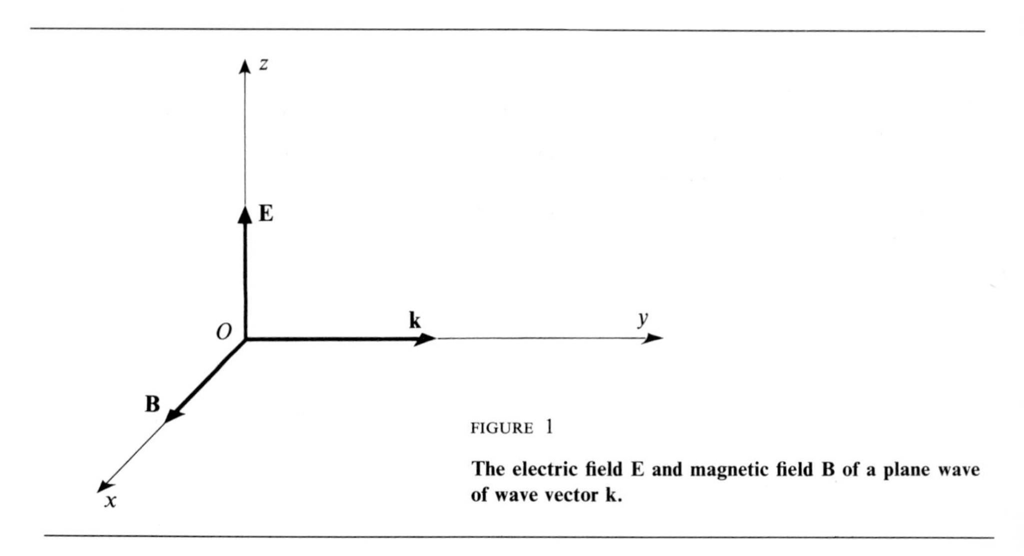

FIGURE 1

The electric field E and magnetic field B of a plane wave of wave vector k.

For such a wave, it is always possible, with a suitable choice of gauge (*cf.* appendix III, §4-b-α), to make the scalar potential $U(\mathbf{r}, t)$ zero. The vector potential $\mathbf{A}(\mathbf{r}, t)$ is then given by the real expression:

$$\mathbf{A}(\mathbf{r}, t) = \mathscr{A}_0 \, \mathbf{e}_z \, \mathrm{e}^{i(ky - \omega t)} + \mathscr{A}_0^* \, \mathbf{e}_z \, \mathrm{e}^{-i(ky - \omega t)} \tag{1}$$

where $\mathscr{A}_0$ is a complex constant whose argument depends on the choice of the time origin. We then have:

$$\mathbf{E}(\mathbf{r}, t) = -\frac{\partial}{\partial t}\mathbf{A}(\mathbf{r}, t) = i\omega\mathscr{A}_0 \, \mathbf{e}_z \, \mathrm{e}^{i(ky - \omega t)} - i\omega\mathscr{A}_0^* \, \mathbf{e}_z \, \mathrm{e}^{-i(ky - \omega t)} \tag{2}$$

$$\mathbf{B}(\mathbf{r}, t) = \boldsymbol{\nabla} \times \mathbf{A}(\mathbf{r}, t) = ik\mathscr{A}_0 \, \mathbf{e}_x \, \mathrm{e}^{i(ky - \omega t)} - ik\mathscr{A}_0^* \, \mathbf{e}_x \, \mathrm{e}^{-i(ky - \omega t)} \tag{3}$$

We shall choose the time origin such that the constant $\mathscr{A}_0$ is pure imaginary, and we set:

$$i\omega \mathscr{A}_0 = \frac{\mathscr{E}}{2} \tag{4-a}$$

$$ik \mathscr{A}_0 = \frac{\mathscr{B}}{2} \tag{4-b}$$

★ For the sake of simplicity, we shall confine ourselves here to the case of a plane wave. The results obtained in this complement, however, can be generalized to an arbitrary electromagnetic field.

where $\mathscr{E}$ and $\mathscr{B}$ are two real quantities such that:

$$\frac{\mathscr{E}}{\mathscr{B}} = \frac{\omega}{k} = c \tag{5}$$

We then obtain:

$$\mathbf{E}(\mathbf{r}, t) = \mathscr{E}\, \mathbf{e}_z \cos(ky - \omega t) \tag{6}$$
$$\mathbf{B}(\mathbf{r}, t) = \mathscr{B}\, \mathbf{e}_x \cos(ky - \omega t) \tag{7}$$

$\mathscr{E}$ and $\mathscr{B}$ are therefore the amplitudes of the electric and magnetic fields of the plane wave considered.

Finally, we shall calculate the Poynting vector* **G** associated with this plane wave:

$$\mathbf{G} = \varepsilon_0 c^2\, \mathbf{E} \times \mathbf{B} \tag{8}$$

Replacing **E** and **B** in (8) by their expressions (6) and (7), and taking the time-average value over a large number of periods, we obtain, using (5):

$$\overline{\mathbf{G}} = \varepsilon_0 c \frac{\mathscr{E}^2}{2}\, \mathbf{e}_y \tag{9}$$

b. THE INTERACTION HAMILTONIAN AT THE LOW-INTENSITY LIMIT

The preceding wave interacts with an atomic electron (of mass m and charge q) situated at a distance r from O and bound to this point O by a central potential $V(r)$ (created by a nucleus assumed to be motionless at O). The quantum mechanical Hamiltonian of this electron can be written:

$$H = \frac{1}{2m}[\mathbf{P} - q\mathbf{A}(\mathbf{R}, t)]^2 + V(R) - \frac{q}{m}\mathbf{S} \cdot \mathbf{B}(\mathbf{R}, t) \tag{10}$$

The last term of (10) represents the interaction of the spin magnetic moment of the electron with the oscillating magnetic field of the plane wave. $\mathbf{A}(\mathbf{R}, t)$ and $\mathbf{B}(\mathbf{R}, t)$ are the operators obtained by replacing, in the classical expressions (1) and (3), x, y, z by the observables X, Y, Z.

In expanding the square which appears on the right-hand side of (10), we should, in theory, remember that **P** does not generally commute with a function of **R**. Such a precaution is, however, unnecessary in the present case, since, as **A** is parallel to Oz [formula (1)], only the P_z component enters into the double product; now P_z commutes with the Y component of **R**, which is the only one to appear in expression (1) for $\mathbf{A}(\mathbf{R}, t)$. We can then take:

$$H = H_0 + W(t) \tag{11}$$

* Recall that the energy flux across a surface element dS perpendicular to the unit vector **n** is $\mathbf{G} \cdot \mathbf{n}\, \mathrm{d}S$.

where:

$$H_0 = \frac{\mathbf{P}^2}{2m} + V(R) \tag{12}$$

is the atomic Hamiltonian, and:

$$W(t) = -\frac{q}{m}\mathbf{P}\cdot\mathbf{A}(\mathbf{R}, t) - \frac{q}{m}\mathbf{S}\cdot\mathbf{B}(\mathbf{R}, t) + \frac{q^2}{2m}[\mathbf{A}(\mathbf{R}, t)]^2 \tag{13}$$

is the interaction Hamiltonian with the incident plane wave [the matrix elements of $W(t)$ approach zero when $\mathscr{A}_0$ approaches zero].

The first two terms of the right-hand side of (13) depend linearly on $\mathscr{A}_0$, and the third one depends on it quadratically. With ordinary light sources, the intensity is sufficiently low that the effect of the $\mathscr{A}_0^2$ term can be neglected compared to that of the $\mathscr{A}_0$ term. We shall therefore write:

$$W(t) \simeq W_I(t) + W_{II}(t) \tag{14}$$

with:

$$W_I(t) = -\frac{q}{m}\mathbf{P}\cdot\mathbf{A}(\mathbf{R}, t) \tag{15}$$

$$W_{II}(t) = -\frac{q}{m}\mathbf{S}\cdot\mathbf{B}(\mathbf{R}, t) \tag{16}$$

We shall evaluate the relative orders of magnitude of the matrix elements of $W_I(t)$ and $W_{II}(t)$ between two bound states of the electron. Those of $\mathbf{S}$ are of the order of $\hbar$, and $\mathbf{B}$ is of the order of $k\mathscr{A}_0$ [*cf.* formula (3)]. Thus:

$$\frac{W_{II}(t)}{W_I(t)} \simeq \frac{\frac{q}{m}\hbar k\mathscr{A}_0}{\frac{q}{m}p\mathscr{A}_0} = \frac{\hbar k}{p} \tag{17}$$

According to the uncertainty relations, $\hbar/p$ is, at most, of the order of atomic dimensions (characterized by the Bohr radius, $a_0 \simeq .5$ Å). k is equal to $2\pi/\lambda$, where λ is the wavelength associated with the incident wave. In the spectral domains used in atomic physics (the optical or Hertzian domains), λ is much greater than a_0, so that:

$$\frac{W_{II}(t)}{W_I(t)} \simeq \frac{a_0}{\lambda} \ll 1 \tag{18}$$

c. THE ELECTRIC DIPOLE HAMILTONIAN

α. *The electric dipole approximation. Interpretation*

Using expression (1) for $\mathbf{A}(\mathbf{R}, t)$, we can put $W_I(t)$ in the form:

$$W_I(t) = -\frac{q}{m}P_z[\mathscr{A}_0\,e^{ikY}\,e^{-i\omega t} + \mathscr{A}_0^*\,e^{-ikY}\,e^{i\omega t}] \tag{19}$$

Expanding the exponential $e^{\pm ikY}$ in powers of kY:

$$e^{\pm ikY} = 1 \pm ikY - \frac{1}{2}k^2Y^2 + ... \tag{20}$$

Since Y is of the order of atomic dimensions, we have, as above:

$$kY \simeq \frac{a_0}{\lambda} \ll 1 \tag{21}$$

We therefore obtain a good approximation for W_I by retaining only the first term of expansion (20). Let W_{DE} be the operator obtained by replacing $e^{\pm ikY}$ by 1 on the right-hand side of (19). Using (4-a), we get:

$$W_{DE}(t) = \frac{q\mathscr{E}}{m\omega} P_z \sin \omega t \tag{22}$$

$W_{DE}(t)$ is called the "electric dipole Hamiltonian". The electric dipole approximation, which is based on conditions (18) and (21), therefore consists of neglecting $W_{II}(t)$ relative to $W_I(t)$ and identifying $W_I(t)$ with $W_{DE}(t)$:

$$W(t) \simeq W_{DE}(t) \tag{23}$$

Let us show that, if we replace $W(t)$ by $W_{DE}(t)$, the electron oscillates as if it were subjected to a *uniform* sinusoidal electric field $\mathscr{E}\,\mathbf{e}_z \cos \omega t$, whose amplitude is that of the electric field of the incident plane wave evaluated at the point O. Physically, this means that the wave function of the bound electron is too localized about O for the electron to "feel" the spatial variation of the electric field of the incident plane wave. We shall therefore calculate the evolution of $\langle \mathbf{R} \rangle(t)$. Ehrenfest's theorem (*cf.* chap. III, § D-1-d) gives:

$$\begin{cases} \dfrac{d}{dt}\langle \mathbf{R} \rangle = \dfrac{1}{i\hbar}\langle [\mathbf{R}, H_0 + W_{DE}] \rangle = \dfrac{\langle \mathbf{P} \rangle}{m} + \dfrac{q\mathscr{E}}{m\omega}\mathbf{e}_z \sin \omega t \\ \dfrac{d}{dt}\langle \mathbf{P} \rangle = \dfrac{1}{i\hbar}\langle [\mathbf{P}, H_0 + W_{DE}] \rangle = -\langle \boldsymbol{\nabla} V(R) \rangle \end{cases} \tag{24}$$

Eliminating $\langle \mathbf{P} \rangle$ from these two equations, we obtain, after a simple calculation:

$$m\frac{d^2}{dt^2}\langle \mathbf{R} \rangle = -\langle \boldsymbol{\nabla} V(R) \rangle + q\mathscr{E}\,\mathbf{e}_z \cos \omega t \tag{25}$$

which is indeed the predicted result: the center of the wave packet associated with the electron moves like a particle of mass m and charge q which is subject to both the central force of the atomic bond [the first term on the right-hand side of (25)] and the influence of a uniform electric field [the second term of (25)].

COMMENT:

Expression (22) for the electric dipole interaction Hamiltonian seems rather unusual for a particle of charge q interacting with a uniform electric field $\mathbf{E} = \mathscr{E}\,\mathbf{e}_z \cos \omega t$. We would tend to write the interaction Hamiltonian in the form:

$$W'_{DE}(t) = -\mathbf{D}\cdot\mathbf{E} = -q\mathscr{E}Z \cos \omega t \tag{26}$$

where $\mathbf{D} = q\mathbf{R}$ is the electric dipole moment associated with the electron.

Actually, expressions (22) and (26) are equivalent. We shall show that we can go from one to the other by a gauge transformation (which does not modify the physical content of the quantum mechanics; *cf.* complement H_{III}). The gauge used to obtain (22) is:

$$\begin{cases} \mathbf{A}(\mathbf{r}, t) = \dfrac{\mathscr{E}}{\omega} \mathbf{e}_z \sin (ky - \omega t) & \text{(27-a)} \\ U(\mathbf{r}, t) = 0 & \text{(27-b)} \end{cases}$$

[to write (27-a), we have replace $\mathscr{A}_0$ by $\mathscr{E}/2i\omega$ in (1); *cf.* formula (4-a)]. Now consider the gauge transformation associated with the function:

$$\chi(\mathbf{r}, t) = z \frac{\mathscr{E}}{\omega} \sin \omega t \tag{28}$$

Thus, we introduce a new gauge $\{ \mathbf{A}', U' \}$ characterized by:

$$\begin{cases} \mathbf{A}' = \mathbf{A} + \nabla\chi = \mathbf{e}_z \dfrac{\mathscr{E}}{\omega} [\sin (ky - \omega t) + \sin \omega t] & \text{(29a)} \\ U' = U - \dfrac{\partial\chi}{\partial t} = - z\mathscr{E} \cos \omega t & \text{(29b)} \end{cases}$$

The electric dipole approximation amounts to replacing ky by 0 everywhere. We then see that in this approximation:

$$\mathbf{A}' \simeq \mathbf{e}_z \frac{\mathscr{E}}{\omega} [\sin (- \omega t) + \sin \omega t] = 0 \tag{30}$$

If, in addition, we neglect, as we did above, the magnetic interaction terms related to the spin, we obtain, for the system's Hamiltonian:

$$\begin{aligned} H' &= \frac{1}{2m} (\mathbf{P} - q\mathbf{A}')^2 + V(R) + qU'(\mathbf{R}, t) \\ &= \frac{\mathbf{P}^2}{2m} + V(R) + qU'(\mathbf{R}, t) \\ &= H_0 + W'(t) \end{aligned} \tag{31}$$

where H_0 is the atomic Hamiltonian given by (12), and:

$$W'(t) = qU'(\mathbf{R}, t) = - qZ \mathscr{E} \cos \omega t = W'_{DE}(t) \tag{32}$$

is the usual form (26) of the electric dipole interaction Hamiltonian.

Recall that the state of the system is no longer described by the same ket when we go from gauge (27) to gauge (29) (*cf.* complement H_{III}). The replacement of $W_{DE}(t)$ by $W'_{DE}(t)$ must therefore be accompanied by a change of state vector, the physical content, of course, remaining the same.

In the rest of this complement, we shall continue to use gauge (27).

β. *The matrix elements of the electric dipole Hamiltonian*

Later, we shall need the expressions for the matrix elements of W_{DE} between $| \varphi_i \rangle$ and $| \varphi_f \rangle$, eigenstates of H_0 of eigenvalues E_i and E_f. According to (22), these matrix elements can be written:

$$\langle \varphi_f | W_{DE}(t) | \varphi_i \rangle = \frac{q\mathscr{E}}{m\omega} \sin \omega t \langle \varphi_f | P_z | \varphi_i \rangle \tag{33}$$

It is simple to replace the matrix elements of P_z by those of Z on the right-hand side of (33). Insofar as we are neglecting all magnetic effects in the atomic Hamiltonian [*cf.* expression (12) for H_0], we can write:

$$[Z, H_0] = i\hbar \frac{\partial H_0}{\partial P_z} = i\hbar \frac{P_z}{m} \tag{34}$$

which yields:

$$\begin{aligned} \langle \varphi_f | [Z, H_0] | \varphi_i \rangle &= \langle \varphi_f | ZH_0 - H_0 Z | \varphi_i \rangle \\ &= -(E_f - E_i) \langle \varphi_f | Z | \varphi_i \rangle = \frac{i\hbar}{m} \langle \varphi_f | P_z | \varphi_i \rangle \end{aligned} \tag{35}$$

Introducing the Bohr angular frequency $\omega_{fi} = (E_f - E_i)/\hbar$,we then get :

$$\langle \varphi_f | P_z | \varphi_i \rangle = im\, \omega_{fi} \langle \varphi_f | Z | \varphi_i \rangle \tag{36}$$

and, consequently:

$$\langle \varphi_f | W_{DE}(t) | \varphi_i \rangle = iq \frac{\omega_{fi}}{\omega} \mathscr{E} \sin \omega t \langle \varphi_f | Z | \varphi_i \rangle \tag{37}$$

The matrix elements of $W_{DE}(t)$ are therefore proportional to those of Z.

COMMENT:

It is the matrix element of Z which appears in (37) because we have chosen the electric field $\mathbf{E}(\mathbf{r}, t)$ parallel to Oz. In practice, we may be led to choose a frame $Oxyz$ which is related, not to the light polarization, but to the symmetry of the states $| \varphi_i \rangle$ and $| \varphi_f \rangle$. For example, if the atoms are placed in a uniform magnetic field $\mathbf{B}_0$, the most convenient quantization axis for the study of their stationary states $| \varphi_n \rangle$ is obviously parallel to $\mathbf{B}_0$. The polarization of the electric field $\mathbf{E}(\mathbf{r}, t)$ can then be arbitrary relative to Oz. In this case, we must replace the matrix element of Z in (37) by that of a linear combination of X, Y and Z.

γ. *Electric dipole transition selection rules*

If the matrix element of W_{DE} between the states $| \varphi_i \rangle$ and $| \varphi_f \rangle$ is different from zero, that is, if $\langle \varphi_f | Z | \varphi_i \rangle$ is non-zero*, the transition $| \varphi_i \rangle \longrightarrow | \varphi_f \rangle$ is said to be an electric dipole transition. To study the transitions induced

* Actually, it suffices for one of the three numbers $\langle \varphi_f | Z | \varphi_i \rangle$, $\langle \varphi_f | X | \varphi_i \rangle$ or $\langle \varphi_f | Y | \varphi_i \rangle$ to be different from zero (*cf.* comment of § β above).

between $| \varphi_i \rangle$ and $| \varphi_f \rangle$ by the incident wave, we can then replace $W(t)$ by $W_{DE}(t)$. If, on the other hand, the matrix element of $W_{DE}(t)$ between $| \varphi_i \rangle$ and $| \varphi_f \rangle$ is zero, we must pursue the expansion of $W(t)$ further, and the corresponding transition is either a magnetic dipole transition or an electric quadrupole transition, etc...★ (see following sections). Since $W_{DE}(t)$ is much larger than the subsequent terms of the power series expansion of $W(t)$ in a_0/λ, electric dipole transitions will be, by far, the most intense. In fact, most optical lines emitted by atoms correspond to electric dipole transitions.

Let:

$$\begin{cases} \varphi_{n_i,l_i,m_i}(\mathbf{r}) = R_{n_i,l_i}(r)\, Y_{l_i}^{m_i}(\theta, \varphi) \\ \varphi_{n_f,l_f,m_f}(\mathbf{r}) = R_{n_f,l_f}(r)\, Y_{l_f}^{m_f}(\theta, \varphi) \end{cases} \tag{38}$$

be the wave functions associated with $| \varphi_i \rangle$ and $| \varphi_f \rangle$. Since:

$$z = r \cos \theta = \sqrt{\frac{4\pi}{3}}\, r\, Y_1^0(\theta) \tag{39}$$

the matrix element of Z between $| \varphi_i \rangle$ and $| \varphi_f \rangle$ is proportional to the angular integral:

$$\int \mathrm{d}\Omega\, Y_{l_f}^{m_f *}(\theta, \varphi)\, Y_1^0(\theta)\, Y_{l_i}^{m_i}(\theta, \varphi) \tag{40}$$

According to the results of complement C_X, this integral is different from zero only if:

$$l_f = l_i \pm 1 \tag{41}$$

and:

$$m_f = m_i \tag{42}$$

Actually, it would suffice to choose another polarization of the electric field (for example, parallel to Ox or Oy; see comment of § β) to have:

$$m_f = m_i \pm 1 \tag{43}$$

From (41), (42) and (43), we obtain the electric dipole transition selection rules:

$$\boxed{\begin{aligned} \Delta l &= l_f - l_i = \pm 1 \qquad &(44\text{-a}) \\ \Delta m &= m_f - m_i = -1, 0, +1 \qquad &(44\text{-b}) \end{aligned}}$$

★ It may happen that all the terms of the expansion have zero matrix elements. The transition is then said to be forbidden to all orders (it can be shown that this is always the case if $| \varphi_i \rangle$ and $| \varphi_f \rangle$ both have zero angular momenta).

COMMENTS:

(i) Z is an odd operator. It can connect only two states of different parities. Since the parities of $| \varphi_i \rangle$ and $| \varphi_f \rangle$ are those of l_i and l_f, $\Delta l = l_f - l_i$ must be odd, as is compatible with (44-a).

(ii) If there exists a spin-orbit coupling $\xi(r)\mathbf{L} \cdot \mathbf{S}$ between $\mathbf{L}$ and $\mathbf{S}$ (*cf.* chap. XII, § B-1-b-β), the stationary states of the electron are labeled by the quantum numbers l, s, J, m_J (with $\mathbf{J} = \mathbf{L} + \mathbf{S}$). The electric dipole transition selection rules can be obtained by looking for the non-zero matrix elements of $\mathbf{R}$ in the $\{ | l, s, J, m_J \rangle \}$ basis. By using the expansions of these basis vectors on the kets $| l, m \rangle | s, m_S \rangle$ (*cf.* complement A_X, § 2), we find, starting with (44-a) and (44-b), the selection rules:

$$\begin{cases} \Delta J = 0, \pm 1 & \text{(44-c)} \\ \Delta l = \pm 1 & \text{(44-d)} \\ \Delta m_J = 0, \pm 1 & \text{(44-e)} \end{cases}$$

Note that a $\Delta J = 0$ transition is not forbidden [unless $J_i = J_f = 0$; *cf.* note on the preceding page]. This is due to the fact that J is not related to the parity of the level.

Finally, we point out that selection rules (44-c, d, e) can be generalized to many-electron atoms.

d. THE MAGNETIC DIPOLE AND ELECTRIC QUADRUPOLE HAMILTONIANS

α. *Higher-order terms in the interaction Hamiltonian*

The interaction Hamiltonian given by (14) can be written in the form:

$$W(t) = W_I(t) + W_{II}(t) = W_{DE}(t) + [W_I(t) - W_{DE}(t)] + W_{II}(t) \qquad (45)$$

Thus far, we have studied $W_{DE}(t)$. As we have seen, the ratio of $W_I(t) - W_{DE}(t)$ and $W_{II}(t)$ to $W_{DE}(t)$ is of the order of a_0/λ.

To calculate $W_I(t) - W_{DE}(t)$, we simply replace $e^{\pm ikY}$ by $e^{\pm ikY} - 1 \simeq \pm ikY + ...$ in (19), which yields:

$$W_I(t) - W_{DE}(t) = -\frac{q}{m}[ik\mathscr{A}_0 e^{-i\omega t} - ik\mathscr{A}_0^* e^{i\omega t}]P_z Y + ... \qquad (46)$$

or, using (4-b):

$$W_I(t) - W_{DE}(t) = -\frac{q}{m}\mathscr{B}\cos\omega t\, P_z Y + ... \qquad (47)$$

If we write $P_z Y$ in the form:

$$P_z Y = \frac{1}{2}(P_z Y - ZP_y) + \frac{1}{2}(P_z Y + ZP_y) = \frac{1}{2}L_x + \frac{1}{2}(P_z Y + ZP_y) \qquad (48)$$

we obtain, finally:

$$W_I(t) - W_{DE}(t) = -\frac{q}{2m}L_x\mathscr{B}\cos\omega t - \frac{q}{2m}\mathscr{B}\cos\omega t[YP_z + ZP_y] + ... \qquad (49)$$

In the expression for $W_{II}(t)$ [formulas (16) and (3)], it is entirely justified to replace $e^{\pm ikY}$ by 1. We thus obtain a term of order a_0/λ relative to $W_I(t)$, that is, of the same order of magnitude as $W_I(t) - W_{DE}(t)$:

$$W_{II}(t) = -\frac{q}{m} S_x \mathscr{B} \cos \omega t + ... \tag{50}$$

Substituting (49) and (50) into (45) and grouping the terms differently, we obtain:

$$W(t) = W_{DE}(t) + W_{DM}(t) + W_{QE}(t) + ... \tag{51}$$

with:

$$W_{DM} = -\frac{q}{2m}(L_x + 2S_x) \mathscr{B} \cos \omega t \tag{52}$$

$$W_{QE} = -\frac{q}{2mc}(YP_z + ZP_y) \mathscr{E} \cos \omega t \tag{53}$$

[we have replaced $\mathscr{B}$ by $\mathscr{E}/c$ in (53)]. W_{DM} and W_{QE} (which are, *a priori*, of the same order of magnitude) are, respectively, the magnetic dipole and electric quadrupole Hamiltonians.

β. *Magnetic dipole transitions*

The transitions induced by W_{DM} are called magnetic dipole transitions. W_{DM} represents the interaction of the total magnetic moment of the electron with the oscillating magnetic field of the incident wave.

The magnetic dipole transition selection rules can be obtained by considering the conditions which must be met by $| \varphi_i \rangle$ and $| \varphi_f \rangle$ in order for W_{DM} to have a non-zero matrix element between these two states. Since neither L_x nor S_x changes the quantum number l, we must have, first of all, $\Delta l = 0$. L_x changes the eigenvalue m_L of L_z by ± 1, which gives $\Delta m_L = \pm 1$. S_x changes the eigenvalues m_S of S_z by ± 1, so that $\Delta m_S = \pm 1$. Note, furthermore, that if the magnetic field of the incident wave were parallel to Oz, we would have $\Delta m_L = 0$ and $\Delta m_S = 0$. Grouping these results, we obtain the magnetic dipole transition selection rules:

$$\begin{cases} \Delta l = 0 \\ \Delta m_L = \pm 1,0 \\ \Delta m_S = \pm 1,0 \end{cases} \tag{54}$$

COMMENT:

In the presence of a spin-orbit coupling, the eigenstates of H_0 are labeled by the quantum numbers l and J. Since L_x and S_x do not commute with $\mathbf{J}^2$, W_{DM} can connect states with the same l but different J. By using the addition formulas for an angular momentum l and an angular momentum 1/2 (*cf.* complement A_X, §2), it can easily be shown that selection rules (54) become:

$$\begin{cases} \Delta l = 0 \\ \Delta J = \pm 1, 0 \\ \Delta m_J = \pm 1, 0 \end{cases} \tag{55}$$

Note that the hyperfine transition $F = 0 \longleftrightarrow F = 1$ of the ground state of the hydrogen atom (*cf.* chap. XII, § D) is a magnetic dipole transition, since the components of **S** have non-zero matrix elements between the states of the $F = 1$ level and the $| F = 0, m_F = 0 \rangle$ state.

γ. *Electric quadrupole transitions*

Using (34), we can write:

$$\begin{aligned} YP_z + ZP_y = YP_z + P_yZ &= \frac{m}{i\hbar}\{ Y[Z, H_0] + [Y, H_0]Z \} \\ &= \frac{m}{i\hbar}(YZH_0 - H_0YZ) \end{aligned} \tag{56}$$

from which we obtain, as in (36):

$$\langle \varphi_f | W_{QE}(t) | \varphi_i \rangle = \frac{q}{2ic}\omega_{fi}\langle \varphi_f | YZ | \varphi_i \rangle \mathscr{E} \cos \omega t \tag{57}$$

The matrix element of $W_{QE}(t)$ is therefore proportional to that of YZ, which is a component of the electric quadrupole moment of the atom (*cf.* complement E_X). In addition, the following quantity appears in (57):

$$\frac{q\omega_{fi}}{c}\mathscr{E} = q\frac{\omega_{fi}}{\omega}\frac{\omega}{c}\mathscr{E} = q\frac{\omega_{fi}}{\omega}k\mathscr{E} \tag{58}$$

which, according to (2), is of the order of $q\partial\mathscr{E}_z/\partial y$. $W_{QE}(t)$ can therefore be interpreted as the interaction of the electric quadrupole moment of the atom with the gradient* of the electric field of the plane wave.

To obtain the electric quadrupole transition selection rules, we simply note that, in the $\{ | \mathbf{r} \rangle \}$ representation, YZ is a linear superposition of $r^2 Y_2^1(\theta, \varphi)$ and $r^2 Y_2^{-1}(\theta, \varphi)$. Therefore, in the matrix element $\langle \varphi_f | YZ | \varphi_i \rangle$ there appear angular integrals :

$$\int d\Omega\, Y_{l_f}^{m_f *}(\theta, \varphi)\, Y_2^{\pm 1}(\theta, \varphi)\, Y_{l_i}^{m_i}(\theta, \varphi) \tag{59}$$

which, according to the results of complement C_X, are different from zero only if $\Delta l = 0, \pm 2$ and $\Delta m = \pm 1$. This last relation becomes $\Delta m = \pm 2, \pm 1, 0$ when we consider an arbitrary polarization of the incident wave (*cf.* comment of §1-c-β), and the electric quadrupole transition selection rules can be written, finally:

$$\begin{cases} \Delta l = 0, \pm 2 \\ \Delta m = 0, \pm 1, \pm 2 \end{cases} \tag{60}$$

* It is normal for the electric field gradient to appear, since $W_{QE}(t)$ was obtained by expanding the potentials in a Taylor series in the neighborhood of O.

COMMENTS:

(*i*) W_{DM} and W_{QE} are even operators and can therefore connect only states of the same parity, which is compatible with (54) and (60). For a given transition, W_{DM} and W_{QE} are never in competition with W_{DE}. This facilitates the observation of magnetic dipole and electric quadrupole transitions.

Most of the transitions which occur in the microwave or radiofrequency domain – in particular, magnetic resonance transitions (*cf.* complement F_{IV}) – are magnetic dipole transitions.

(*ii*) For a $\Delta l = 0$, $\Delta m = 0$, ± 1 transition, the two operators W_{DM} and W_{QE} simultaneously have non-zero matrix elements. However, it is possible to find experimental conditions under which only magnetic dipole transitions are induced. All we need to do is place the atom, not in the path of a plane wave, but inside a cavity or radiofrequency loops, at a point where **B** is large but the gradient of **E** is negligible.

(*iii*) For a $\Delta l = 2$ transition, W_{DM} cannot be in competition with W_{QE}, and we have a pure quadrupole transition. As an example of a quadrupole transition, we can mention the green line of atomic oxygen (5 577 Å), which appears in the aurora borealis spectrum.

(*iv*) If we pursued the expansion of $e^{\pm ikY}$ further, we would find electric octupole and magnetic quadrupole terms, etc.

In the rest of this complement, we shall confine ourselves to electric dipole transitions. In the next complement, B_{XIII}, on the other hand, we shall consider a magnetic dipole transition.

2. Non-resonant excitation. Comparison with the elastically bound electron model

In this section, we shall assume that the atom, initially in the ground state $| \varphi_0 \rangle$, is excited by a non-resonant plane wave: ω coincides with none of the Bohr angular frequencies associated with transitions from $| \varphi_0 \rangle$.

Under the effect of this excitation, the atom acquires an electric dipole moment $\langle \mathbf{D} \rangle(t)$ which oscillates at the angular frequency ω (forced oscillation) and is proportional to $\mathscr{E}$ when $\mathscr{E}$ is small (linear response). We shall use perturbation theory to calculate this induced dipole moment, and we shall show that the results obtained are very close to those found with the classical model of the elastically bound electron.

This model has played a very important role in the study of the optical properties of materials. It enables us to calculate the polarization induced by the incident wave in a material. This polarization, which depends linearly on the field $\mathscr{E}$, behaves like a source term in Maxwell's equations. When we solve these equations, we find plane waves propagating in the material at a velocity different from c. This allows us to calculate the refractive index of the material in terms of various characteristics of elastically bound electrons (natural frequencies, number per unit volume, etc.). Thus, we see that it is very important to compare the predictions of this model (which we shall review in §a) with those of quantum mechanics.

a. CLASSICAL MODEL OF THE ELASTICALLY BOUND ELECTRON

α. *Equation of motion*

Consider an electron subject to a restoring force directed towards the point O and proportional to the displacement. In the classical Hamiltonian corresponding to (12), we then have:

$$V(r) = \frac{1}{2} m\omega_0^2 r^2 \tag{61}$$

where ω_0 is the electron's natural angular frequency.

If we make the same approximations, using the classical interaction Hamiltonian, as those which enabled us to obtain expression (22) for $W_{DE}(t)$ (the electric dipole approximation) in quantum mechanics, a calculation similar to that of §1-c-α [*cf.* equation (25)] yields the equation of motion:

$$\frac{\mathrm{d}^2}{\mathrm{d}t^2} z + \omega_0^2 z = \frac{q\mathscr{E}}{m} \cos \omega t \tag{62}$$

This is the equation of a harmonic oscillator subject to a sinusoidal force.

β. *General solution*

The general solution of (62) can be written:

$$z = A \cos (\omega_0 t - \varphi) + \frac{q\mathscr{E}}{m(\omega_0^2 - \omega^2)} \cos \omega t \tag{63}$$

where A and φ are real constants which depend on the initial conditions. The first term of (63), $A \cos (\omega_0 t - \varphi)$, represents the general solution of the homogeneous equation (the electron's free motion). The second term is a particular solution of the equation (forced motion of the electron).

We have not, thus far, taken damping into account. Without going into detail in the calculations, we shall cite the effects of weak damping: after a certain time τ, it causes the natural motion to disappear and very slightly modifies the forced motion (provided that we are sufficiently far from resonance: $|\omega - \omega_0| \gg 1/\tau$). We shall therefore retain only the second term of (63):

$$z = \frac{q\mathscr{E} \cos \omega t}{m(\omega_0^2 - \omega^2)} \tag{64}$$

COMMENT:

Far from resonance, the exact damping mechanism is of little importance, provided that it is weak. We shall not, therefore, take up the problem of the exact description of this damping, either in quantum or in classical mechanics. We shall merely use the fact that it exists to ignore the free motion of the electron.

It would be different for a resonant excitation: the induced dipole moment would then depend critically on the exact damping mechanism (spontaneous emission, thermal relaxation, etc.). This is why we shall not try to calculate $\langle \mathbf{D} \rangle(t)$ in § 3 (the case of a resonant excitation). We shall be concerned only with calculating the transition probabilities.

In complement B_{XIII}, we shall study a specific model of a system placed in an electromagnetic wave and at the same time subject to dissipative processes (Bloch equations of a system of spins). We shall then be able to calculate the induced dipole moment for any exciting frequency.

γ. *Susceptibility*

Let $\mathscr{D} = qz$ be the electric dipole moment of the system. According to (64), we have:

$$\mathscr{D} = qz = \frac{q^2}{m(\omega_0^2 - \omega^2)} \mathscr{E} \cos \omega t = \chi \, \mathscr{E} \cos \omega t \tag{65}$$

where the "susceptibility" χ is given by:

$$\chi = \frac{q^2}{m(\omega_0^2 - \omega^2)} \tag{66}$$

b. QUANTUM MECHANICAL CALCULATION OF THE INDUCED DIPOLE MOMENT

We shall begin by calculating, to first order in $\mathscr{E}$, the state vector $| \psi(t) \rangle$ of the atom at time t. We shall choose for the interaction Hamiltonian, the electric dipole Hamiltonian W_{DE} given by (22). In addition, we shall assume that:

$$| \psi(t = 0) \rangle = | \varphi_0 \rangle \tag{67}$$

We apply the results of § C-1 of chapter XIII, replacing W_{ni} by $\dfrac{q\mathscr{E}}{m\omega} \langle \varphi_n | P_z | \varphi_i \rangle$ and $| \varphi_i \rangle$ by $| \varphi_0 \rangle$. Thus we obtain*:

$$| \psi(t) \rangle = e^{-iE_0 t/\hbar} | \varphi_0 \rangle + \sum_{n \neq 0} \lambda \, b_n^{(1)}(t) \, e^{-iE_n t/\hbar} | \varphi_n \rangle \tag{68}$$

or, using (C-4) of chapter XIII and multiplying $| \psi(t) \rangle$ by the global phase factor $e^{iE_0 t/\hbar}$, which has no physical importance:

$$| \psi(t) \rangle = | \varphi_0 \rangle + \sum_{n \neq 0} \frac{q\mathscr{E}}{2im\hbar\omega} \langle \varphi_n | P_z | \varphi_0 \rangle \times \left\{ \frac{e^{-i\omega_{n0}t} - e^{i\omega t}}{\omega_{n0} + \omega} - \frac{e^{-i\omega_{n0}t} - e^{-i\omega t}}{\omega_{n0} - \omega} \right\} | \varphi_n \rangle \tag{69}$$

* Since W_{DE} is odd, $\langle \varphi_0 | W_{DE}(t) | \varphi_0 \rangle$ is zero, so $b_0^{(1)}(t) = 0$.

From this, we find $\langle \psi(t) |$ and $\langle D_z \rangle(t) = \langle \psi(t) | qZ | \psi(t) \rangle$. In the calculation of this mean value, we retain only the terms linear in $\mathscr{E}$, and we neglect all those which oscillate at angular frequencies $\pm \omega_{n0}$ (the natural motion, which would disappear if we took weak damping into account). Finally, after having replaced $\langle \varphi_n | P_z | \varphi_0 \rangle$ by its expression in terms of $\langle \varphi_n | Z | \varphi_0 \rangle$ [*cf.* equation (36)], we find:

$$\langle D_z \rangle(t) = \frac{2q^2}{\hbar} \mathscr{E} \cos \omega t \sum_n \frac{\omega_{n0} |\langle \varphi_n | Z | \varphi_0 \rangle|^2}{\omega_{n0}^2 - \omega^2} \tag{70}$$

c. DISCUSSION. OSCILLATOR STRENGTH

α. *The concept of oscillator strength*

We set:

$$f_{n0} = \frac{2m\, \omega_{n0} |\langle \varphi_n | Z | \varphi_0 \rangle|^2}{\hbar} \tag{71}$$

f_{n0} is a real dimensionless number, characteristic of the $|\varphi_0\rangle \longleftrightarrow |\varphi_n\rangle$ transition and called the oscillator strength★ of this transition. If $|\varphi_0\rangle$ is the ground state, f_{n0} is positive, like ω_{n0}.

Oscillator strengths satisfy the following sum rule (the Thomas-Reiche-Kuhn sum rule):

$$\sum_n f_{n0} = 1 \tag{72}$$

This can be shown as follows. Using (36), we can write:

$$f_{n0} = \frac{1}{i\hbar} \langle \varphi_0 | Z | \varphi_n \rangle \langle \varphi_n | P_z | \varphi_0 \rangle - \frac{1}{i\hbar} \langle \varphi_0 | P_z | \varphi_n \rangle \langle \varphi_n | Z | \varphi_0 \rangle \tag{73}$$

The summation over n can be performed by using the closure relation relative to the $\{ |\varphi_n\rangle \}$ basis, and we get:

$$\sum_n f_{n0} = \frac{1}{i\hbar} \langle \varphi_0 | (ZP_z - P_z Z) | \varphi_0 \rangle = \langle \varphi_0 | \varphi_0 \rangle = 1 \tag{74}$$

β. *The quantum mechanical justification for the elastically bound electron model*

We shall substitute definition (71) into (70) and multiply the expression so obtained by the number $\mathscr{N}$ of atoms contained in a volume whose linear dimensions are much smaller than the wavelength λ of the radiation. The total electric dipole moment induced in this volume can then be written:

$$\mathscr{N} \langle D_z \rangle(t) = \sum_n \mathscr{N} f_{n0} \frac{q^2}{m(\omega_{n0}^2 - \omega^2)} \mathscr{E} \cos \omega t \tag{75}$$

★ The operator Z enters into (71) because the incident wave is linearly polarized along Oz. It would, however, be possible to give a general definition of the oscillator strength, independent of the polarization of the incident wave.

Comparing (75) and (65), we see that it is like having $\mathcal{N}$ classical oscillators [since $\sum_n \mathcal{N} f_{n0} = \mathcal{N}$ according to (72)] whose natural angular frequencies are not all the same since they are equal to the various Bohr angular frequencies of the atom associated with the transitions from $| \varphi_0 \rangle$. According to (75), the proportion of oscillators with the angular frequency ω_{n0} is f_{n0}.

Thus, for a non-resonant wave, we have justified the classical model of the elastically bound electron. Quantum mechanics gives the frequencies of the various oscillators and the proportion of oscillators which have a given frequency. This result shows the importance of the concept of oscillator strength and enables us to understand *a posteriori* why the elastically bound electron model was so useful in the study of the optical properties of materials.

3. Resonant excitation. Induced absorption and emission

a. TRANSITION PROBABILITY ASSOCIATED WITH A MONOCHROMATIC WAVE

Consider an atom initially in the state $| \varphi_i \rangle$ placed in an electromagnetic wave whose angular frequency is close to a Bohr angular frequency ω_{fi}.

The results of §C-1 of chapter XIII (sinusoidal excitation) are directly applicable to the calculation of the transition probability $\mathscr{P}_{if}(t;\omega)$. We find, using expression (37) (thus making the electric dipole approximation):

$$\mathscr{P}_{if}(t;\omega) = \frac{q^2}{4\hbar^2}\left(\frac{\omega_{fi}}{\omega}\right)^2 |\langle \varphi_f | Z | \varphi_i \rangle|^2 \mathscr{E}^2 F(t, \omega - \omega_{fi}) \tag{76}$$

where:

$$F(t, \omega - \omega_{fi}) = \left\{ \frac{\sin\left[(\omega_{fi} - \omega)t/2\right]}{(\omega_{fi} - \omega)/2} \right\}^2 \tag{77}$$

We have already discussed the resonant nature of $\mathscr{P}_{if}(t;\omega)$ in chapter XIII. At resonance, $\mathscr{P}_{if}(t;\omega)$ is proportional to $\mathscr{E}^2$, that is, to the incident flux of electromagnetic energy [*cf.* formula (9)].

b. BROAD-LINE EXCITATION. TRANSITION PROBABILITY PER UNIT TIME

In practice, the radiation which strikes the atom is very often non-monochromatic. We shall denote by $\mathscr{I}(\omega)\,d\omega$ the incident flux of electromagnetic energy per unit surface within the interval $[\omega, \omega + d\omega]$. The variation of $\mathscr{I}(\omega)$ with respect to ω is shown in figure 2. Δ is the excitation line width. If Δ is infinite, we say that we are dealing with a "white spectrum".

The different monochromatic waves which constitute the incident radiation are generally incoherent : they have no well-defined phase relation. The total transition probability $\overline{\mathscr{P}}_{if}$ can therefore be obtained by summing the transition probabilities associated with each of these monochromatic waves. We must, consequently, replace $\mathscr{E}^2$ by $2\mathscr{I}(\omega)\,\mathrm{d}\omega/\varepsilon_0 c$ in (76) [formula (9)] and integrate over ω. This gives:

$$\overline{\mathscr{P}}_{if}(t) = \frac{q^2}{2\varepsilon_0 c\hbar^2} \left| \langle \varphi_f \mid Z \mid \varphi_i \rangle \right|^2 \times \int \mathrm{d}\omega \left(\frac{\omega_{fi}}{\omega} \right)^2 \mathscr{I}(\omega)\, F(t, \omega - \omega_{fi}) \tag{78}$$

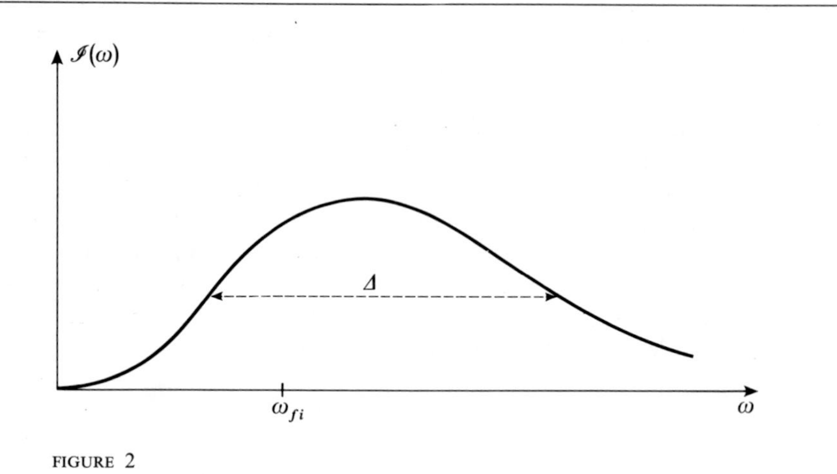

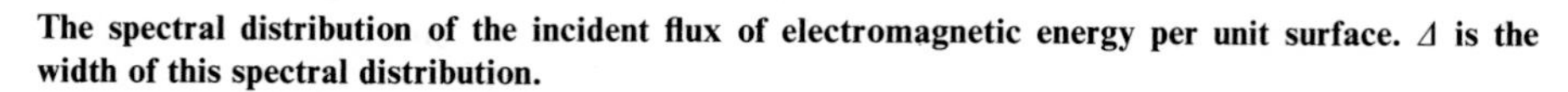

FIGURE 2

The spectral distribution of the incident flux of electromagnetic energy per unit surface. Δ is the width of this spectral distribution.

We can then proceed as in §C-3 of chapter XIII to evaluate the integral which appears in (78). Compared to a function of ω whose width is much larger than $4\pi/t$, the function $F(t, \omega - \omega_{fi})$ (see figure 3 of chapter XIII) behaves like $\delta(\omega - \omega_{fi})$. If t is large enough to make $4\pi/t \ll \Delta$ (Δ: excitation line width) while remaining small enough for the perturbation treatment to be valid, we can, in (78), assume that:

$$F(t, \omega - \omega_{fi}) \simeq 2\pi t\, \delta(\omega - \omega_{fi}) \tag{79}$$

which yields:

$$\overline{\mathscr{P}}_{if}(t) = \frac{\pi q^2}{\varepsilon_0 c\hbar^2} \left| \langle \varphi_f \mid Z \mid \varphi_i \rangle \right|^2 \mathscr{I}(\omega_{fi})\, t \tag{80}$$

We can write (80) in the form:

$$\overline{\mathscr{P}}_{if}(t) = C_{if}\, \mathscr{I}(\omega_{fi})\, t \tag{81}$$

where:

$$C_{if} = \frac{4\pi^2}{\hbar} \left| \langle \varphi_f \mid Z \mid \varphi_i \rangle \right|^2 \alpha \tag{82}$$

and α is the fine-structure constant:

$$\alpha = \frac{q^2}{4\pi\varepsilon_0}\frac{1}{\hbar c} = \frac{e^2}{\hbar c} \simeq \frac{1}{137} \tag{83}$$

This result shows that $\overline{\mathscr{P}}_{if}(t)$ increases linearly with time. The *transition probability per unit time* $\mathscr{W}_{if}$ is therefore equal to:

$$\mathscr{W}_{if} = C_{if}\,\mathscr{I}(\omega_{fi}) \tag{84}$$

$\mathscr{W}_{if}$ is proportional to the value of the incident intensity for the resonance frequency ω_{fi}, to the fine-structure constant α, and to the square of the modulus of the matrix element of Z, which is related [by (71)] to the oscillator strength of the $|\varphi_f\rangle \longleftrightarrow |\varphi_i\rangle$ transition.

In this complement, we have considered the case of radiation propagating along a given direction with a well-defined polarization state. By averaging the coefficients C_{if} over all propagation directions and over all possible polarization states, we could introduce coefficients B_{if}, analogous to the coefficients C_{if}, defining the transition probabilities per unit time for an atom placed in isotropic radiation. The coefficients B_{if} (and B_{fi}) are none other than the coefficients introduced by Einstein to describe the absorption (and induced emission). Thus, we see how quantum mechanics enables us to calculate these coefficients.

COMMENT:

A third coefficient, A_{fi}, was introduced by Einstein to describe the spontaneous emission of a photon which occurs when the atom falls back from the upper state $|\varphi_f\rangle$ to the lower state $|\varphi_i\rangle$. The theory presented in this complement does not explain spontaneous emission. In the absence of incident radiation, the interaction Hamiltonian is zero, and the eigenstates of H_0 (which is then the total Hamiltonian) are stationary states.

Actually, the preceding model is insufficient, since it treats asymmetrically the atomic system (which is quantized) and the electromagnetic field (which is considered classically). When we quantize both systems, we find, even in the absence of incident photons, that the coupling between the atom and the electromagnetic field continues to have observable effects (a simple interpretation of these effects is given in complement K_V). The eigenstates of H_0 are no longer stationary states, since H_0 is no longer the Hamiltonian of the total system, and we can indeed calculate the probability per unit time of spontaneous emission of a photon. Quantum mechanics therefore also enables us to obtain the Einstein coefficient A_{fi}.

References and suggestions for further reading:

See, for example: Schiff (1.18), chap. 11; Bethe and Jackiw (1.21), Part II, chaps. 10 and 11; Bohm (5.1), chap. 18, §§12 to 44.

For the elastically bound electron model: Berkeley 3 (7.1), supplementary topic 9; Feynman I (6.3), chap. 31 and Feynman II (7.2), chap. 32.

For Einstein coefficients: the original article (1.31), Cagnac and Pebay-Peyroula (11.2), chap. III and chap. XIX, § 4.

For the exact definition of oscillator strength: Sobel'man (11.12), chap. 9, §31.

For atomic multipole radiation and its selection rules: Sobel'man (11.12), chap. 9, § 32.

Complement B_{XIII}

LINEAR AND NON-LINEAR RESPONSES OF A TWO-LEVEL SYSTEM SUBJECT TO A SINUSOIDAL PERTURBATION

In the preceding complement, we applied first-order time-dependent perturbation theory to the treatment of some effects produced by the interaction of an atomic system and an electromagnetic wave: appearance of an induced dipole moment, induced emission and absorption processes, etc.

We shall now consider a simple example, in which it is possible to pursue the perturbation calculations to higher orders without too many complications. This will allow us to demonstrate some interesting "non-linear" effects: saturation effects, non-linear susceptibility, the absorption and induced emission of several photons, etc. In addition, the model which we shall describe takes into account (phenomenologically) the dissipative coupling of the atomic system with its surroundings (the relaxation process). This will enable us to complete the results related to the "linear response" obtained in the preceding complement. For example, we shall calculate the atom's induced dipole moment, not only far from resonance, but also at resonance.

Some of the effects we are going to describe are currently objects of a great deal of research. Their study necessitates very strong electromagnetic fields. Only recently (since the development of lasers) have we been able to produce such fields. New branches of research have thus appeared: quantum electronics, non-linear optics, etc. The calculation methods described in this complement (for a very simple model) are applicable to these problems.

1. Description of the model

a. BLOCH EQUATIONS FOR A SYSTEM OF SPIN 1/2'S INTERACTING WITH A RADIOFREQUENCY FIELD

We shall return to the system described in §4-a of complement F_{IV}: a system of spin 1/2's placed in a static field $\mathbf{B}_0$ parallel to Oz, interacting with an oscillating radiofrequency field and subject to "pumping" and "relaxation" processes.

If $\mathcal{M}(t)$ is the total magnetization of the spin system contained in the cell (fig. 6 of complement F_{IV}), we showed in complement F_{IV} that:

$$\frac{d}{dt}\mathcal{M}(t) = n\,\boldsymbol{\mu}_0 - \frac{1}{T_R}\mathcal{M}(t) + \gamma\mathcal{M}(t) \times \mathbf{B}(t) \tag{1}$$

The first term on the right-hand side describes the preparation, or the "pumping" of the system: n spins enter the cell per unit time, each one with an elementary magnetization $\boldsymbol{\mu}_0$ parallel to Oz. The second term arises from relaxation processes, characterized by the average time T_R required for a spin either to leave the cell or have its direction changed by collision with the walls. Finally, the last term of (1) corresponds to the precession of the spins about the total magnetic field :

$$\mathbf{B}(t) = B_0\,\mathbf{e}_z + \mathbf{B}_1(t) \tag{2}$$

$\mathbf{B}(t)$ is the sum of a static field $B_0\mathbf{e}_z$ parallel to Oz and a radiofrequency field $\mathbf{B}_1(t)$ of angular frequency ω.

COMMENTS:

(*i*) The transitions which we shall study in this complement (which connect the two states $|+\rangle$ and $|-\rangle$ of each spin 1/2) are magnetic dipole transitions.

(*ii*) One could question our using expression (1) relative to mean values rather than the Schrödinger equation. We do so because we are studying a statistical ensemble of spins coupled to a thermal reservoir (via collisions with the cell walls). We cannot describe this ensemble in terms of a state vector : we must use a density operator (see complement E_{III}). The equation of motion of this operator is called a "master equation" and we can show that it is exactly equivalent to (1) (see complement F_{IV}, §3 and 4, and complement E_{IV}, where we show that the mean value of the magnetization determines the density matrix of an ensemble of spin 1/2's).

Actually, the master equation satisfied by the density operator and the Schrödinger equation studied in §C-1 of chapter XIII have the same structure as (1): a linear differential equation, with constant or sinusoidally varying coefficients. The approximation methods we describe in this chapter are, therefore, applicable to any of these equations.

b. SOME EXACTLY AND APPROXIMATELY SOLUBLE CASES

If the radiofrequency field $\mathbf{B}_1(t)$ is rotating, that is, if:

$$\mathbf{B}_1(t) = B_1 \, (\mathbf{e}_x \cos \omega t + \mathbf{e}_y \sin \omega t) \tag{3}$$

equation (1) can be solved exactly [changing to the frame which is rotating with $\mathbf{B}_1$ transforms (1) into a time-independent linear differential system]. The exact solution of (1) corresponding to such a situation is given in §4-b of complement F_{IV}.

Here, we shall assume $\mathbf{B}_1$ to be linearly polarized along Ox:

$$\mathbf{B}_1(t) = B_1 \, \mathbf{e}_x \cos \omega t \tag{4}$$

In this case, it is not possible* to find an exact analytic solution of equation (1) (there is no transformation equivalent to changing to the rotating frame). We shall see, however, that a solution can be found in the form of a power series expansion in B_1.

COMMENT:

The calculations we shall present here for spin 1/2's can also be applied to other situations in which we can confine ourselves to two levels of the system and ignore all others. We know (*cf.* complement C_{IV}) that we can associate a fictitious spin 1/2 with any two-level system. The problem considered here is therefore that of an arbitrary two-level system subject to a sinusoidal perturbation.

c. RESPONSE OF AN ATOMIC SYSTEM

The set of terms which, in $\mathcal{M}_x$, $\mathcal{M}_y$, $\mathcal{M}_z$, depend on B_1 constitute the "response" of the atom to the electromagnetic perturbation. They represent the magnetic dipole moment induced in the spin system by the radiofrequency field. We shall see that such a dipole moment is not necessarily proportional to B_1; the terms in B_1 represent the linear response, and the others (terms in B_1^2, B_1^3, ...), the "non-linear response". In addition, we shall see that the induced dipole moment does not oscillate only at the angular frequency ω, but also at its various harmonics $p\omega$ ($p = 0, 2, 3, 4, ...$).

It is easy to see why we should be interested in calculating the response of an atomic system. Such a calculation constitutes an important part of the theory of the propagation of an electromagnetic wave in a material or of the theory of atomic oscillators, "masers" or "lasers".

* A linearly polarized field results from the superposition of a left and a right circular component. It would be possible to find an exact solution for each of these components taken separately. However, equation (1) is not linear, in the sense that a solution corresponding to (4) cannot be obtained by superposing two exact solutions, one of which corresponds to (3) and the other one to the field rotating in the opposite direction [in the term $\gamma \, \mathcal{M} \times \mathbf{B}$ which appears on the right-hand side of (1), $\mathcal{M}$ depends on $\mathbf{B}_1$].

Consider an electromagnetic field. Because of the coupling between this field and the atomic system, a polarization appears in the material, due to the atomic dipole moments (arrow directed towards the right in figure 1). This polarization acts like a source term in Maxwell's equations and contributes to the creation of the electromagnetic field (arrow directed towards

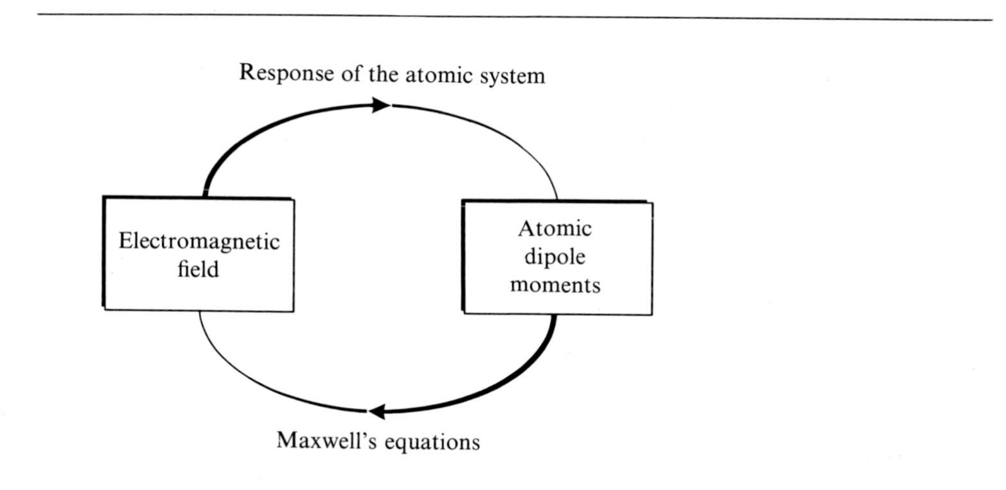

FIGURE 1

Schematic representation of the calculations to be performed in studying the propagation of an electromagnetic wave in a material (or the operation of an atomic oscillator, a laser or a maser). We begin by calculating the dipole moments induced in the material by a given electromagnetic field (the response of the atomic system). The corresponding polarization acts like a source term in Maxwell's equations and contributes to the creation of the electromagnetic field. We then take the field obtained to be equal to the one with which we started.

the left in figure 1). When we "close the loop", that is, when we take the field so created to be equal to the one with which we started, we obtain the wave propagation equations in the material (refractive index) or the oscillator equations (in the absence of external fields, an electromagnetic field may appear in the material, if there is sufficient amplification : the system becomes unstable and can oscillate spontaneously). In this complement, we shall be concerned only with the first step of the calculation (the atomic response).

2. The approximate solution of the Bloch equations of the system

a. PERTURBATION EQUATIONS

As in complement F_{IV}, we set:

$$\begin{cases} \omega_0 = -\gamma B_0 & (5) \\ \omega_1 = -\gamma B_1 & (6) \end{cases}$$

$\hbar\omega_0$ represents the energy difference of the spin states $| + \rangle$ and $| - \rangle$ (fig. 2). Substituting (4) into (2), and (2) into (1), we obtain, after a simple calculation :

$$\begin{cases} \dfrac{\mathrm{d}}{\mathrm{d}t}\mathcal{M}_z = n\mu_0 - \dfrac{\mathcal{M}_z}{T_R} + i\dfrac{\omega_1}{2}\cos\omega t\,(\mathcal{M}_- - \mathcal{M}_+) & \text{(7-a)} \\ \dfrac{\mathrm{d}}{\mathrm{d}t}\mathcal{M}_\pm = \qquad - \dfrac{\mathcal{M}_\pm}{T_R} \pm i\omega_0\mathcal{M}_\pm \mp i\omega_1\cos\omega t\,\mathcal{M}_z & \text{(7-b)} \end{cases}$$

with:

$$\mathcal{M}_\pm = \mathcal{M}_x \pm i\mathcal{M}_y \tag{8}$$

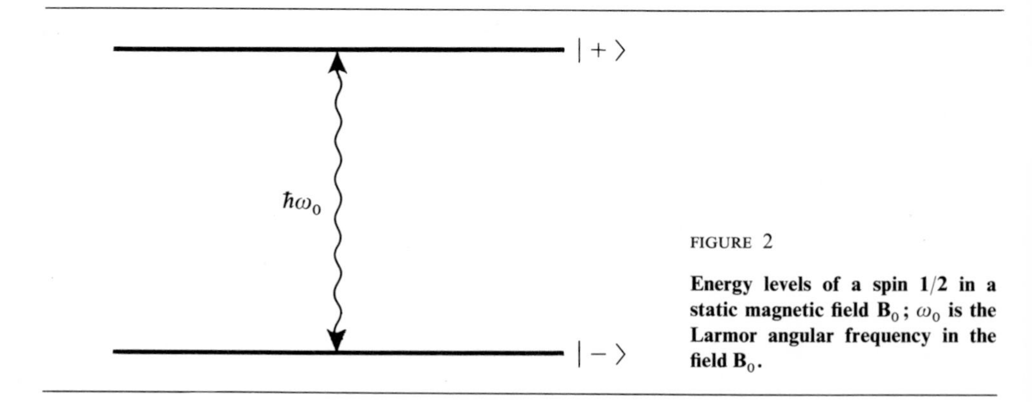

FIGURE 2

Energy levels of a spin 1/2 in a static magnetic field $\mathbf{B}_0$; ω_0 is the Larmor angular frequency in the field $\mathbf{B}_0$.

Note that the source term $n\mu_0$ exists only in the equation of motion of $\mathcal{M}_z$, since $\boldsymbol{\mu}_0$ is parallel to Oz, and the pumping is said to be longitudinal★. We also point out that the relaxation time can be different for the longitudinal components ($\mathcal{M}_z$) and the transverse components ($\mathcal{M}_\pm$) of the magnetization. For the sake of simplicity, we shall choose a single relaxation time here.

Equations (7-a) and (7-b), called the "Bloch equations", cannot be solved exactly. We shall therefore determine their solution in the form of a power series expansion in ω_1:

$$\begin{cases} \mathcal{M}_z = {}^{(0)}\mathcal{M}_z + \omega_1\,{}^{(1)}\mathcal{M}_z + \omega_1^2\,{}^{(2)}\mathcal{M}_z + \ldots + \omega_1^n\,{}^{(n)}\mathcal{M}_z + \ldots & \text{(9-a)} \\ \mathcal{M}_\pm = {}^{(0)}\mathcal{M}_\pm + \omega_1\,{}^{(1)}\mathcal{M}_\pm + \omega_1^2\,{}^{(2)}\mathcal{M}_\pm + \ldots + \omega_1^n\,{}^{(n)}\mathcal{M}_\pm + \ldots & \text{(9-b)} \end{cases}$$

Substituting (9-a) and (9-b) into (7-a) and (7-b), and setting equal the coefficients of terms in ω_1^n, we obtain the following perturbation equations:

$n = 0$

$$\begin{cases} \dfrac{\mathrm{d}}{\mathrm{d}t}\,{}^{(0)}\mathcal{M}_z = n\mu_0 - \dfrac{1}{T_R}\,{}^{(0)}\mathcal{M}_z & \text{(10-a)} \\ \dfrac{\mathrm{d}}{\mathrm{d}t}\,{}^{(0)}\mathcal{M}_\pm = -\dfrac{1}{T_R}\,{}^{(0)}\mathcal{M}_\pm \pm i\omega_0\,{}^{(0)}\mathcal{M}_\pm & \text{(10-b)} \end{cases}$$

★ In certain experiments, the pumping is "transversal" ($\boldsymbol{\mu}_0$ is perpendicular to $\mathbf{B}_0$). See exercise 1 at the end of the complement.

$n \neq 0$

$$\begin{cases} \dfrac{d}{dt}\, {}^{(n)}\mathcal{M}_z = -\dfrac{1}{T_R}\, {}^{(n)}\mathcal{M}_z + \dfrac{i}{2}\cos\omega t\,\left[{}^{(n-1)}\mathcal{M}_- - {}^{(n-1)}\mathcal{M}_+\right] & \text{(11-a)} \\ \dfrac{d}{dt}\, {}^{(n)}\mathcal{M}_\pm = -\dfrac{1}{T_R}\, {}^{(n)}\mathcal{M}_\pm \pm i\omega_0\, {}^{(n)}\mathcal{M}_\pm \mp i\cos\omega t\; {}^{(n-1)}\mathcal{M}_z & \text{(11-b)} \end{cases}$$

b. THE FOURIER SERIES EXPANSION OF THE SOLUTION

Since the only time-dependent terms on the right-hand side of (10) and (11) are sinusoidal, the steady-state solution of (10) and (11) is periodic, of period $2\pi/\omega$. We can expand it in a Fourier series:

$$\begin{cases} {}^{(n)}\mathcal{M}_z = \displaystyle\sum_{p=-\infty}^{+\infty} {}^{(n)}_{p}\mathcal{M}_z\, e^{ip\omega t} & \text{(12-a)} \\ {}^{(n)}\mathcal{M}_\pm = \displaystyle\sum_{p=-\infty}^{+\infty} {}^{(n)}_{p}\mathcal{M}_\pm\, e^{ip\omega t} & \text{(12-b)} \end{cases}$$

${}^{(n)}_{p}\mathcal{M}_z$ and ${}^{(n)}_{p}\mathcal{M}_\pm$ represent the $p\omega$ Fourier components of the nth-order solution.

By taking ${}^{(n)}\mathcal{M}_z$ real and ${}^{(n)}\mathcal{M}_+$ and ${}^{(n)}\mathcal{M}_-$ as complex conjugates of each other, we obtain the following reality conditions:

$$\begin{cases} {}^{(n)}_{p}\mathcal{M}_z = \left[{}^{(n)}_{-p}\mathcal{M}_z\right]^* & \text{(13-a)} \\ {}^{(n)}_{p}\mathcal{M}_\pm = \left[{}^{(n)}_{-p}\mathcal{M}_\mp\right]^* & \text{(13-b)} \end{cases}$$

Substituting (12-a) and (12-b) into (10) and (11), and setting equal to zero the coefficient of each exponential $e^{ip\omega t}$, we find:

$n = 0$

$$\begin{cases} {}^{(0)}_{0}\mathcal{M}_z = n\mu_0 T_R & \\ {}^{(0)}_{p}\mathcal{M}_z = 0 & \text{if } p \neq 0 \\ {}^{(0)}_{p}\mathcal{M}_\pm = 0 & \text{for all } p \end{cases} \qquad (14)$$

$n \neq 0$

$$\begin{cases} \left(ip\omega + \dfrac{1}{T_R}\right) {}^{(n)}_{p}\mathcal{M}_z = \dfrac{i}{4}\left[{}^{(n-1)}_{p+1}\mathcal{M}_- + {}^{(n-1)}_{p-1}\mathcal{M}_- - {}^{(n-1)}_{p+1}\mathcal{M}_+ - {}^{(n-1)}_{p-1}\mathcal{M}_+\right] & \text{(15-a)} \\ \left[i(p\omega \mp \omega_0) + \dfrac{1}{T_R}\right] {}^{(n)}_{p}\mathcal{M}_\pm = \mp \dfrac{i}{2}\left[{}^{(n-1)}_{p+1}\mathcal{M}_z + {}^{(n-1)}_{p-1}\mathcal{M}_z\right] & \text{(15-b)} \end{cases}$$

These algebraic equations can be solved immediately:

$$\begin{cases} {}^{(n)}_{p}\mathcal{M}_z = \dfrac{i}{4\left(ip\omega + \dfrac{1}{T_R}\right)}\left[{}^{(n-1)}_{p+1}\mathcal{M}_- + {}^{(n-1)}_{p-1}\mathcal{M}_- - {}^{(n-1)}_{p+1}\mathcal{M}_+ - {}^{(n-1)}_{p-1}\mathcal{M}_+\right] & \text{(16-a)} \\ {}^{(n)}_{p}\mathcal{M}_\pm = \mp \dfrac{i}{2\left[i(p\omega \mp \omega_0) + \dfrac{1}{T_R}\right]}\left[{}^{(n-1)}_{p+1}\mathcal{M}_z + {}^{(n-1)}_{p-1}\mathcal{M}_z\right] & \text{(16-b)} \end{cases}$$

Thus, expressions (16) give the nth-order solution explicitly in terms of the $(n - 1)$th-order solution. Since the zeroeth-order solution is known [*cf.* equations (14)], the problem is, in theory, entirely solved.

c. THE GENERAL STRUCTURE OF THE SOLUTION

It is possible to arrange the various terms of the expansion of the solution in a double-entry table in which the perturbation order n labels the columns and the degree p of the harmonic $p\omega$ being considered labels the row. To zeroeth-order, only ${}^{(0)}_{0}\mathcal{M}_z$ is different from zero. By iteration, using (16), we can deduce the other non-zero higher-order terms (table 1), thus obtaining a "tree-like structure". The following properties can be found directly by recurrence, using (16):

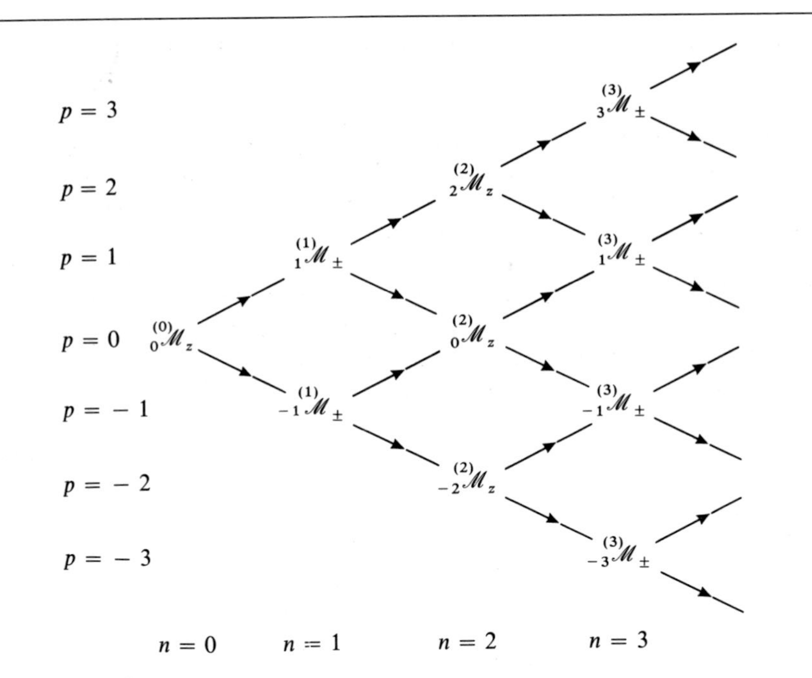

TABLE 1

Double-entry table indicating the $p\omega$ Fourier components of the magnetization which are non-zero to the nth perturbation order in ω_1.

(*i*) At even perturbation orders, only the longitudinal magnetization is modified; at odd orders, only the transverse magnetization.

(*ii*) At even perturbation orders, only the even harmonics are involved; at odd orders, only the odd harmonics.

(*iii*) For each value of n, the values of p to be retained are n, $n - 2$, ..., $-n + 2$, $-n$.

COMMENT:

This structure is valid only for a particular polarization of the radiofrequency field $\mathbf{B}_1(t)$ (perpendicular to $\mathbf{B}_0$). Analogous tables could be constructed for other radiofrequency polarizations.

3. Discussion

We shall now interpret the results of this calculation, through third order.

a. ZEROETH-ORDER SOLUTION: COMPETITION BETWEEN PUMPING AND RELAXATION

According to (14), the only non-zero zeroeth-order component is:

$$ {}^{(0)}_{0}\mathcal{M}_z = n\mu_0 T_R \tag{17}$$

In the absence of radiofrequency fields, there is therefore only a static longitudinal magnetization ($p = 0$). Since $\mathcal{M}_z$ is proportional to the population difference of the states $| + \rangle$ and $| - \rangle$ shown in figure 2 (*cf.* complement E_{IV}), it can also be said that the pumping populates these two states unequally.

The larger the number of particles entering the cell (the more efficient the pumping) and the longer T_R (the slower the relaxation), the larger ${}^{(0)}_{0}\mathcal{M}_z$. The zeroeth-order solution (17) therefore describes the dynamic equilibrium resulting from competition between the pumping and relaxation processes.

From now on, in order to simplify the notation, we shall set:

$$\mathcal{M}_0 = {}^{(0)}_{0}\mathcal{M}_z \tag{18-a}$$

$$\Gamma_R = \frac{1}{T_R} \tag{18-b}$$

b. FIRST-ORDER SOLUTION: THE LINEAR RESPONSE

To first order, only the transverse magnetization $\mathcal{M}_\perp$ is different from zero. Since $\mathcal{M}_+ = \mathcal{M}_-^*$, it suffices to study $\mathcal{M}_+$.

α. *Motion of the transverse magnetization*

According to table 1, for $n = 1$, we have $p = \pm 1$. Setting $n = 1$ and $p = \pm 1$ in (16-b), using (18), we get:

$$\begin{cases} {}^{(1)}_{1}\mathcal{M}_+ = \dfrac{\mathcal{M}_0}{2} \dfrac{1}{\omega_0 - \omega + i\Gamma_R} & \text{(19-a)} \\[2ex] {}^{(1)}_{-1}\mathcal{M}_+ = \dfrac{\mathcal{M}_0}{2} \dfrac{1}{\omega_0 + \omega + i\Gamma_R} & \text{(19-b)} \end{cases}$$

Substituting these expressions into (12-b) and then into (9-b), we obtain $\mathcal{M}_+$ to first order in ω_1:

$$\mathcal{M}_+ = \omega_1 \frac{\mathcal{M}_0}{2} \left[\frac{e^{i\omega t}}{\omega_0 - \omega + i\Gamma_R} + \frac{e^{-i\omega t}}{\omega_0 + \omega + i\Gamma_R} \right] \tag{20}$$

The point representing $\mathcal{M}_+$ describes the same motion in the complex plane as the projection $\boldsymbol{\mathcal{M}}_\perp$ of $\boldsymbol{\mathcal{M}}$ in the plane perpendicular to $\mathbf{B}_0$. According to (20), this motion results from the superposition of two circular motions with the same angular velocity, one of them right circular (the $e^{i\omega t}$ term) and the other left circular (the $e^{-i\omega t}$ term). The resulting motion, in the general case, is therefore elliptical.

β. *Existence of two resonances*

The right circular motion has a maximum amplitude when $\omega_0 = \omega$, and the left circular motion, when $\omega_0 = -\omega$. $\boldsymbol{\mathcal{M}}_\perp$ therefore presents two resonances (while for a rotating field, there was a single resonance; see complement F_{IV}). The interpretation of this phenomenon is as follows: the linear radiofrequency field can be broken down into a left and a right circular field, each of which induces a resonance; since the rotation directions are opposed, the static fields $\mathbf{B}_0$ for which these resonances appear are opposed.

γ. *Linear susceptibility*

Near a resonance ($\omega_0 \simeq \omega$, for example), we can neglect the non-resonant term in (20). We then get:

$$\mathcal{M}_+ \underset{\omega \simeq \omega_0}{\simeq} \omega_1 \frac{\mathcal{M}_0}{2} \frac{e^{i\omega t}}{\omega_0 - \omega + i\Gamma_R} \tag{21}$$

$\mathcal{M}_+$ is therefore proportional to the rotating radiofrequency field component in the direction corresponding to the resonance, $B_1 e^{i\omega t}/2$ in this case.

The ratio of $\mathcal{M}_+$ to this component is called the linear susceptibility $\chi(\omega)$:

$$\chi(\omega) = -\gamma \mathcal{M}_0 \frac{1}{\omega_0 - \omega + i\Gamma_R} \tag{22}$$

$\chi(\omega)$ is a complex susceptibility because of the existence of a phase difference between $\boldsymbol{\mathcal{M}}_\perp$ and the rotating component of the radiofrequency field responsible for the resonance.

The square of the modulus of $\chi(\omega)$ has the classical resonant form in the neighborhood of $\omega = \omega_0$ (fig. 3), over an interval of width:

$$\Delta\omega = 2\Gamma_R = \frac{2}{T_R} \tag{23}$$

The longer the relaxation time T_R, therefore, the sharper the resonance curve. From now on, we shall assume that the two resonances $\omega_0 = \omega$ and $\omega_0 = -\omega$ are completely separated, that is, that:

$$\omega/\Gamma_R = \omega T_R \gg 1 \tag{24}$$

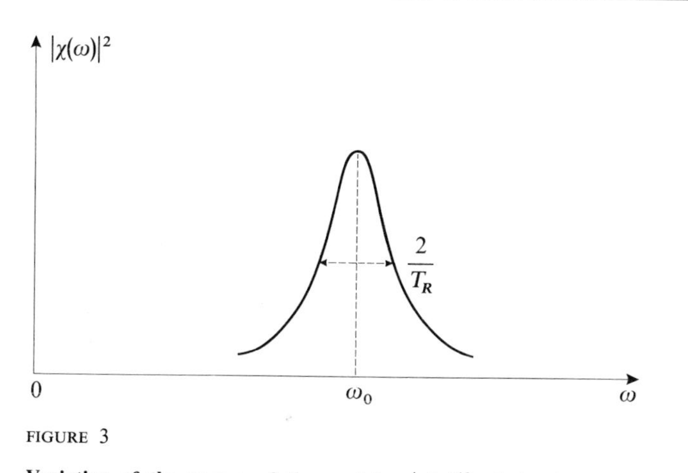

FIGURE 3

Variation of the square of the modulus $|\chi(\omega)|^2$ of the linear susceptibility of the spin system, with respect to ω. A resonance appears, of width $2/T_R$, centered at $\omega = \omega_0$.

The phase difference varies from 0 to $\pm\,\pi$ when we pass through resonance. It is equal to $\pm\,\pi/2$ at resonance: it is when $\boldsymbol{\mathcal{M}}_\perp$ and the rotating component are out of phase by $\pi/2$ that the work of the couple exerted by the field on $\boldsymbol{\mathcal{M}}$ is maximal. The sign of this work depends on the sign of $\mathcal{M}_0$, that is, on that of μ_0: it depends on whether the spin states of the entering particles are $|+\rangle$ or $|-\rangle$. In one case (spins entering in the lower level), the work is furnished by the field, and energy is transferred from the field to the spins (absorption). In the opposite case (particles entering in the higher level), the work is negative, and energy is transferred from the spins to the field (induced emission). The latter situation occurs in atomic amplifiers and oscillators (masers and lasers).

c. SECOND-ORDER SOLUTION: ABSORPTION AND INDUCED EMISSION

To second order, according to table 1, only ${}^{(2)}_{0}\mathcal{M}_z$ and ${}^{(2)}_{\pm 2}\mathcal{M}_z$ are non-zero. First, we shall study ${}^{(2)}_{0}\mathcal{M}_z$, that is, the static population difference of the states $|+\rangle$ and $|-\rangle$ to second order. We shall then consider ${}^{(2)}_{\pm 2}\mathcal{M}_z$, that is, the generation of the second harmonic.

α. *Variation of the population difference of the two states of the system*

${}^{(2)}_{0}\mathcal{M}_0$ corrects the zeroeth-order result obtained for ${}^{(0)}_{0}\mathcal{M}_0$. According to (16-a) and (13-b):

$$\begin{aligned}{}^{(2)}_{0}\mathcal{M}_z &= \frac{i}{4\Gamma_R}\left[\,{}^{(1)}_{1}\mathcal{M}_- + {}^{(1)}_{-1}\mathcal{M}_- - {}^{(1)}_{1}\mathcal{M}_+ - {}^{(1)}_{-1}\mathcal{M}_+\right] \\ &= \frac{i}{4\Gamma_R}\left[\,{}^{(1)}_{-1}\mathcal{M}^*_+ + {}^{(1)}_{1}\mathcal{M}^*_+ - {}^{(1)}_{1}\mathcal{M}_+ - {}^{(1)}_{-1}\mathcal{M}_+\right] \end{aligned} \tag{25}$$

which, according to first-order solutions (19-a) and (19-b), yields:

$$ {}^{(2)}_{\ 0}\mathcal{M}_z = -\frac{\mathcal{M}_0}{4}\left[\frac{1}{(\omega-\omega_0)^2+\Gamma_R^2}+\frac{1}{(\omega+\omega_0)^2+\Gamma_R^2}\right] \tag{26} $$

Grouping the static terms ($p = 0$) through second order in (9-a), we get:

$$ \mathcal{M}_z(\text{static}) = \mathcal{M}_0\left\{1-\frac{\omega_1^2}{4}\left[\frac{1}{(\omega-\omega_0)^2+\Gamma_R^2}+\frac{1}{(\omega+\omega_0)^2+\Gamma_R^2}\right]+\dots\right\} \tag{27} $$

Figure 4 represents this static longitudinal magnetization as a function of ω_0.

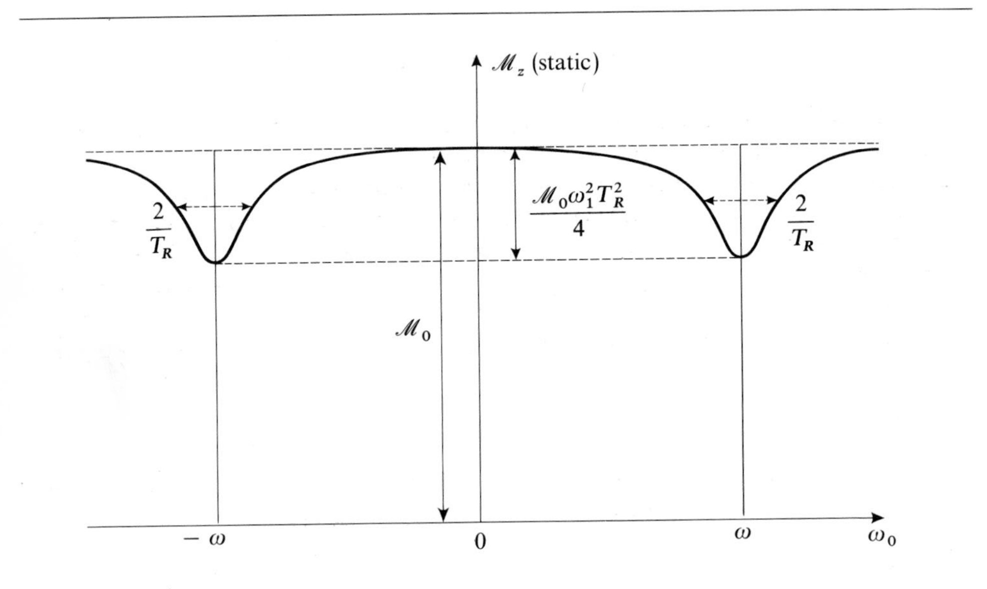

FIGURE 4

Variation of the static longitudinal magnetization with respect to ω_0. To second order in the perturbation treatment, there appear two resonances of width $2/T_R$, centered at $\omega_0 = \omega$ and $\omega_0 = -\omega$. The calculation is valid only if the relative intensity of the resonances is small, that is, if $\omega_1 T_R \ll 1$.

The population difference is therefore always *decreased*, to second order, relative to its value in the absence of radiofrequency, and the decrease is proportional to the *intensity* of the radiofrequency field. This is simple to understand : under the effect of the incident field, transitions are induced from $|+\rangle$ to $|-\rangle$ (induced emission) or from $|-\rangle$ to $|+\rangle$ (absorption); whatever the sign of the initial population difference, the transitions from the more populated state are the more numerous, so that they decrease the population difference.

COMMENT:

The maximum value of $\omega_1^2 \,|\, {}^{(2)}_{0}\mathcal{M}_z \,|$ is $\mathcal{M}_0 \omega_1^2 / 4\Gamma_R^2 = \mathcal{M}_0 \omega_1^2 T_R^2 / 4$ (the resonance amplitude which appears as a dip in figure 4). For the perturbation expansion to make sense, it is therefore necessary that:

$$\omega_1 T_R \ll 1 \tag{28}$$

β. *Generation of the second harmonic*

According to (16-a), (13-b), (19-a) and (19-b):

$$\begin{aligned} {}^{(2)}_{2}\mathcal{M}_z &= \frac{1}{4(2\omega - i\Gamma_R)} \left[{}^{(1)}_{-1}\mathcal{M}_+^* - {}^{(1)}_{1}\mathcal{M}_+ \right] \\ &= \frac{\mathcal{M}_0}{8(2\omega - i\Gamma_R)} \left[\frac{1}{\omega_0 + \omega - i\Gamma_R} - \frac{1}{\omega_0 - \omega + i\Gamma_R} \right] \end{aligned} \tag{29}$$

${}^{(2)}_{2}\mathcal{M}_z$ describes a vibration of the magnetic dipole along Oz at the angular frequency 2ω. The system can therefore radiate a wave of angular frequency 2ω, polarized (as far as the magnetic field is concerned) linearly along Oz.

Thus, we see that an atomic system is not generally a linear system. It can double the excitation frequency, triple it (as we shall see later), etc. The same type of phenomenon exists in optics for very high intensities ("non-linear optics"): a red laser beam (produced, for example, by a ruby laser) falling on a material such as a quartz crystal can give rise to an ultraviolet light beam (doubled frequency).

COMMENT:

It will prove useful to compare $|{}^{(2)}_{0}\mathcal{M}_z|$ and $|{}^{(2)}_{2}\mathcal{M}_z|$ in the neighborhood of $\omega_0 = \omega$. According to (29), for $\omega \simeq \omega_0$, we have:

$$|\, {}^{(2)}_{2}\mathcal{M}_z \,| \simeq \frac{\mathcal{M}_0}{16\omega_0 \Gamma_R} \tag{30}$$

Similarly, (26) indicates that:

$$|{}^{(2)}_{0}\mathcal{M}_z| \simeq \frac{\mathcal{M}_0}{4\Gamma_R^2} \tag{31}$$

Therefore, for $\omega \simeq \omega_0$:

$$\frac{|{}^{(2)}_{2}\mathcal{M}_z|}{|{}^{(2)}_{0}\mathcal{M}_z|} \simeq \frac{\Gamma_R}{4\omega_0} = \frac{1}{4\omega_0 T_R} \ll 1 \tag{32}$$

according to (24).

d. THIRD-ORDER SOLUTION: SATURATION EFFECTS AND MULTIPLE-QUANTA TRANSITIONS

To third order, table 1 shows that only ${}^{(3)}_{\pm 1}\mathcal{M}_{\pm}$ and ${}^{(3)}_{\pm 3}\mathcal{M}_{\pm}$ are non-zero; it suffices to study ${}^{(3)}\mathcal{M}_{+}$.

${}^{(3)}_{1}\mathcal{M}_{+}$ corrects to third order the right circular motion of $\boldsymbol{\mathcal{M}}_{\perp}$, found to first order and analyzed in §b above. We shall see that ${}^{(3)}_{1}\mathcal{M}_{+}$ corresponds to a saturation effect in the susceptibility of the system.

${}^{(3)}_{3}\mathcal{M}_{+}$ represents a new component of angular frequency 3ω of the motion of $\boldsymbol{\mathcal{M}}_{\perp}$ (generation of the third harmonic). Moreover, the resonant nature of ${}^{(3)}_{3}\mathcal{M}_{+}$ in the neighborhood of $\omega_0 = 3\omega$ can be interpreted as resulting from the simultaneous absorption of three radiofrequency photons, a process which conserves both the total energy and the total angular momentum.

α. *Saturation of the susceptibility of the system*

According to (16-b):

$$ {}^{(3)}_{1}\mathcal{M}_{+} = \frac{1}{2}\frac{1}{\omega_0 - \omega + i\Gamma_R}\left[{}^{(2)}_{2}\mathcal{M}_z + {}^{(2)}_{0}\mathcal{M}_z\right] \tag{33} $$

Since we are interested in the correction to the right circular motion discussed in §3-b, which is resonant at $\omega = \omega_0$, we shall place ourselves in the neighborhood of $\omega_0 = \omega$. We can then, according to the comment in the preceding section [*cf.* formula (32)], neglect ${}^{(2)}_{2}\mathcal{M}_z$ compared to ${}^{(2)}_{0}\mathcal{M}_z$. Thus we obtain, using expression (26) for ${}^{(2)}_{0}\mathcal{M}_z$ (neglecting the term whose resonance peak is at $\omega_0 = -\omega$):

$$ {}^{(3)}_{1}\mathcal{M}_{+} \simeq -\frac{\mathcal{M}_0}{8}\frac{1}{\omega_0 - \omega + i\Gamma_R}\frac{1}{(\omega - \omega_0)^2 + \Gamma_R^2} \tag{34} $$

If we regroup results (34) and (19-a), we find the expression for the right circular motion of $\mathcal{M}_{+}$ at the frequency $\omega/2\pi$, to third order in ω_1:

$\mathcal{M}_{+}$ (right circular) =

$$ \omega_1\frac{\mathcal{M}_0}{2}\frac{e^{i\omega t}}{\omega_0 - \omega + i\Gamma_R}\left[1 - \frac{\omega_1^2}{4}\frac{1}{(\omega - \omega_0)^2 + \Gamma_R^2}\right] \tag{35} $$

Comparing (35) and (21), we see that the susceptibility of the system goes from value (22) to the value:

$$ \chi(\omega) = -\gamma\mathcal{M}_0\frac{1}{\omega_0 - \omega + i\Gamma_R}\left[1 - \frac{\omega_1^2}{4}\frac{1}{(\omega_0 - \omega)^2 + \Gamma_R^2}\right] \tag{36} $$

It is therefore multiplied by a factor smaller than one; the greater the intensity of the radiofrequency field and the nearer we are to resonance, the smaller the factor. The system is then said to be "saturated". The ω_1^2 term of (36) is called the "non-linear susceptibility".

The physical meaning of this saturation is very clear. A weak electromagnetic field induces in the atomic system a dipole moment which is proportional to it. If the field amplitude is increased, the dipole cannot continue to increase proportionally to the field. The absorption and emission transitions induced by the field decrease the population difference of the atomic states involved. Consequently, the atomic system responds less and less to the field. Furthermore, we see that the term in brackets in (36) is none other than the term which expresses the decrease in the population difference to second order [*cf.* formula (27), in which the term resonant at $\omega_0 = -\omega$ was neglected].

COMMENT:

The saturation terms play a very important role in all maser or laser theories. Consider figure 1 again. If we keep only the linear response term in the first step of the calculation (arrow directed to the right), the induced dipole moment is proportional to the field. If the material amplifies (and if the losses of the electromagnetic cavity are sufficiently small), the reaction of the dipole on the field (arrow directed to the left) tends to increase the field by a quantity proportional to it. Thus, we obtain for the field a linear differential equation which leads to a solution which increases linearly with time.

It is the saturation terms that prevent this unlimited increase. They lead to an equation whose solution remains bounded and approaches a limit which is the steady-state laser field in the cavity. Physically, these saturation terms express the fact that the atomic system cannot furnish the field with an energy greater than that corresponding to the population difference initially introduced by the pumping.

β. *Three-photon transitions*

According to (16-b) and (29):

$$
\begin{aligned}
{}^{(3)}_{3}\mathcal{M}_+ &= \frac{1}{2}\,\frac{1}{\omega_0 - 3\omega + i\Gamma_R}\,{}^{(2)}_{2}\mathcal{M}_z \\
&= \frac{\mathcal{M}_0}{16}\,\frac{1}{\omega_0 - 3\omega + i\Gamma_R}\,\frac{1}{2\omega - i\Gamma_R}\left[\frac{1}{\omega_0 + \omega - i\Gamma_R} - \frac{1}{\omega_0 - \omega + i\Gamma_R}\right]
\end{aligned}
\tag{37}
$$

With respect to the term ${}^{(3)}_{3}\mathcal{M}_+$, we could make the same comment as for ${}^{(2)}_{2}\mathcal{M}_z$: the atomic system produces harmonics of the excitation frequency (here, the third harmonic).

The difference with the discussion of the preceding section relative to ${}^{(2)}_{2}\mathcal{M}_z$ is the appearance of a resonance centered at $\omega_0 = 3\omega$ [due to the first resonant denominator of (37)].

We can give a particle interpretation of the $\omega_0 = \omega$ resonance discussed in the preceding sections: the spin goes from the state $|-\rangle$ to the state $|+\rangle$ by absorbing a photon (or emitting it, depending on the relative positions of the $|+\rangle$ and $|-\rangle$ states). There is resonance when the energy $\hbar\omega$ of the photon is equal to the energy $\hbar\omega_0$ of the atomic transition. We could give an analogous particle interpretation of the $\omega_0 = 3\omega$ resonance. Since $\hbar\omega_0 = 3\hbar\omega$, the transition necessarily involves three photons, since the total energy must be conserved.

We may wonder why no resonance has appeared to second order for $\hbar\omega_0 = 2\hbar\omega$ (two-photon transition). The reason is that the total angular momentum must also be conserved during the transition. The linear radiofrequency field is, as we have already said, a superposition of two fields rotating in opposite directions. With each of these rotating fields are associated photons of a different type. For the right circular field, it is σ^+ photons, transporting an angular momentum $+\hbar$ relative to Oz. For the left circular field, it is σ^- photons, transporting an angular momentum $-\hbar$. To go from the $|-\rangle$ state to the $|+\rangle$ state, the spin must absorb an angular momentum $+\hbar$ relative to Oz (the difference between the two eigenvalues of S_z). It can do so by absorbing a σ^+ photon; if $\omega_0 = \omega$, there is also conservation of the total energy, which explains the appearance of the $\omega_0 = \omega$ resonance. The system can also acquire an angular momentum $+\hbar$ by absorbing three photons (fig. 5): two σ^+ photons and one σ^- photon. Therefore, if $\omega_0 \doteq 3\omega$, both energy and total angular momentum can be conserved, which explains the $\omega_0 = 3\omega$ resonance. On the other hand, two photons can never give the atom an angular momentum $+\hbar$: either both photons are σ^+ and they carry $2\hbar$, or they are both σ^- and they carry $-2\hbar$, or one is σ^+ and one is σ^- and they carry no total angular momentum.

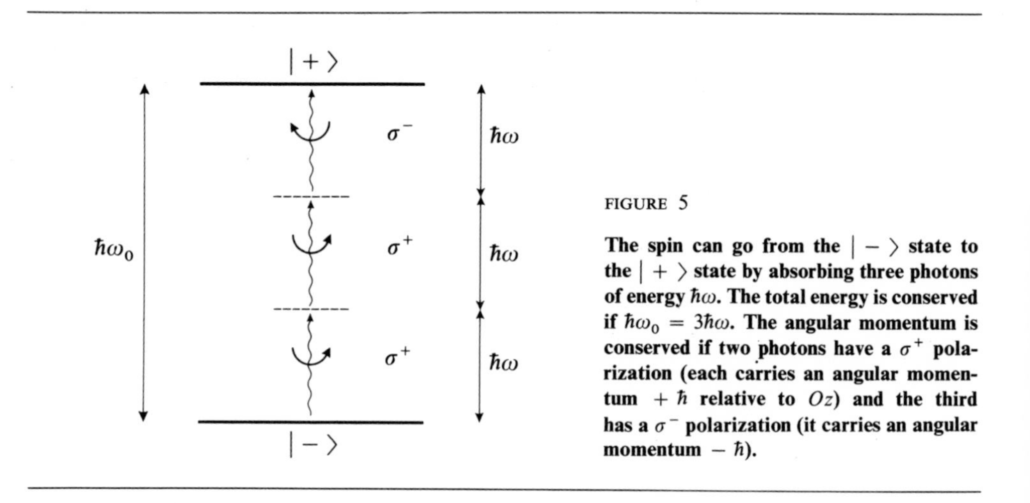

FIGURE 5

The spin can go from the $|-\rangle$ state to the $|+\rangle$ state by absorbing three photons of energy $\hbar\omega$. The total energy is conserved if $\hbar\omega_0 = 3\hbar\omega$. The angular momentum is conserved if two photons have a σ^+ polarization (each carries an angular momentum $+\hbar$ relative to Oz) and the third has a σ^- polarization (it carries an angular momentum $-\hbar$).

These arguments can easily be generalized and enable us to show that resonances appear when $\omega_0 = \omega, 3\omega, 5\omega, 7\omega, ..., (2n+1)\omega, ...$, corresponding to the absorption of an odd number of photons. Furthermore, we see from formula (16-b) that ${}^{(2n+1)}_{2n+1}\mathcal{M}_+$ gives rise to a resonance peak for $\omega_0 = (2n+1)\omega$. Nothing analogous occurs at even orders since, according to table 1, we must then use equation (16-a).

COMMENTS:

(*i*) If the field $\mathbf{B}_1$ is rotating, there is only one type of photon, σ^+ or σ^-. The same argument shows that a *single resonance* can then occur, at $\omega_0 = \omega$ if the photons are σ^+ and at $\omega_0 = -\omega$ if they are σ^-. This enables us to understand why the calculations are much simpler for a rotating field and lead to an exact solution. It is instructive to apply the method of this complement to the case of a rotating field and to show that the perturbation series can be summed to give the solution found directly in complement F_{IV}.

(*ii*) Consider a system having two levels of different parities, subject to the influence of an oscillating electric field. The interaction Hamiltonian then has the same structure as the one we are studying in this complement : S_x has only non-diagonal elements. Similarly, the electric dipole Hamiltonian, since it is odd, can have no diagonal elements. In the second case, the calculations are very similar to the preceding ones and lead to analogous conclusions : resonances are found for $\omega_0 = \omega, 3\omega, 5\omega, ...$ The interpretation of the "odd" nature of the spectrum is then as follows : the electric dipole photons have a negative parity, and the system must absorb an odd number of them in order to move from one level to another of different parity.

(*iii*) For the spin 1/2 case, assume that the linear radiofrequency field is neither parallel nor perpendicular to $\mathbf{B}_0$ (fig. 6). $\mathbf{B}_1$ can then be broken down into a component parallel to $\mathbf{B}_0$, $\mathbf{B}_{1//}$, with which are associated π photons (with zero angular momentum relative to Oz), and a component $\mathbf{B}_{1\perp}$, with which, as we have seen, σ^+ and σ^- photons are associated. In this case, the atom can increase its angular momentum relative to Oz by $+\hbar$, and move from $|-\rangle$

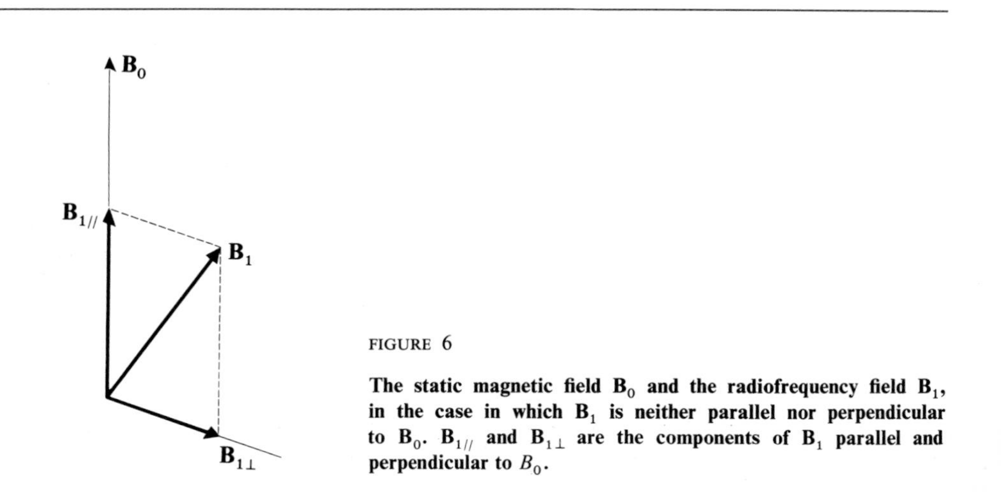

FIGURE 6

The static magnetic field $\mathbf{B}_0$ and the radiofrequency field $\mathbf{B}_1$, in the case in which $\mathbf{B}_1$ is neither parallel nor perpendicular to $\mathbf{B}_0$. $\mathbf{B}_{1//}$ and $\mathbf{B}_{1\perp}$ are the components of $\mathbf{B}_1$ parallel and perpendicular to B_0.

to $|+\rangle$, by absorbing two photons, one σ^+ and the other π. It can be shown, by applying the method of this complement, that for this polarization of the radiofrequency, a complete (even and odd) spectrum of resonances appears : $\omega_0 = \omega, 2\omega, 3\omega, 4\omega, ...$

4. Exercises: applications of this complement

EXERCISE 1

In equations (1), set $\omega_1 = 0$ (no radiofrequency) and choose $\boldsymbol{\mu}_0$ parallel to Ox (transverse pumping).

Calculate the steady-state values of $\mathcal{M}_x$, $\mathcal{M}_y$ and $\mathcal{M}_z$. Show that $\mathcal{M}_x$ and $\mathcal{M}_y$ undergo resonant variations when the static field is swept about zero (the Hanle effect). Give a physical interpretation of these resonances (pumping in competition with Larmor precession) and show that they permit the measurement of the product γT_R.

EXERCISE 2

Consider a spin system subject to the same static field $\mathbf{B}_0$ and to the same pumping and relaxation processes as in this complement. These spins are also subjected to two linear radiofrequency fields, the first one of angular frequency ω and amplitude B_1, parallel to Oz, and the second one of angular frequency ω' and amplitude B_1', parallel to Ox.

Using the general methods described in this complement, calculate the magnetization $\boldsymbol{\mathcal{M}}$ of the spin system to second order in $\omega_1 = -\gamma B_1$ and $\omega_1' = -\gamma B_1'$ (terms in ω_1^2, $\omega_1'^2$, $\omega_1\omega_1'$). We fix $\omega_0 = -\gamma B_0$ and ω_1. Assume $\omega_0 > \omega$, and let ω' vary. Show that, to this perturbation order, two resonances appear, one at $\omega' = \omega_0 - \omega$ and the other at $\omega' = \omega_0 + \omega$.

Give a physical interpretation of these two resonances (the first one corresponds to a two-photon absorption, and the second, to a Raman effect).

References and suggestions for further reading: see section 15

Semiclassical theories of masers and lasers: Lamb (15.4) and (15.2), Sargent et al. (15.5), chap. VIII, IX and X.

Non-linear optics: Baldwin (15.19), Bloembergen (15.21), Giordmaine (15.22).

Iterative solution of the master equation: Bloembergen (15.21), chap. 2, §§3, 4 and 5 and Appendix III.

Multiphoton processes in R. F. range, Hanle effect: Brossel's lectures in (15.2).

Complement C_{XIII}

OSCILLATIONS OF A SYSTEM BETWEEN TWO DISCRETE STATES UNDER THE EFFECT OF A SINUSOIDAL RESONANT PERTURBATION

The approximation method used to calculate the effect of a resonant perturbation in chapter XIII is not valid over long periods of time. We have seen [*cf.* condition (C-18) of this chapter] that t must satisfy:

$$t \ll \frac{\hbar}{|W_{fi}|} \tag{1}$$

Suppose we want to study the behavior of a system subjected to a resonant perturbation over a considerable time [for which condition (1) is not satisfied]. Since the first-order solution is then insufficient, we could try to calculate a certain number of higher-order terms to obtain a better expression for $\mathscr{P}_{if}(t;\omega)$:

$$\mathscr{P}_{if}(t;\omega) = |\lambda\, b_f^{(1)}(t) + \lambda^2 b_f^{(2)}(t) + \lambda^3 b_f^{(3)}(t) + \ldots|^2 \tag{2}$$

Such a method would lead to unnecessarily long calculations.

We shall see here that it is possible to solve the problem more elegantly and rapidly by fitting the approximation method to the resonant nature of the perturbation. The resonance condition $\omega \simeq \omega_{fi}$ implies that only the two discrete states $|\varphi_i\rangle$ and $|\varphi_f\rangle$ are effectively coupled by $W(t)$. Since the system, at the initial instant, is in the state $|\varphi_i\rangle$ [$b_i(0) = 1$], the probability amplitude $b_f(t)$ of finding it in the state $|\varphi_f\rangle$ at time t can be appreciable. On the other hand, all the coefficients $b_n(t)$ (with $n \neq i, f$) remain much smaller than 1 since they do not satisfy the resonance condition. This is the basis of the method we shall use.

1. The method: secular approximation

In chapter XIII, we replaced all the components $b_k(t)$ on the right-hand side of (B-11) by their values $b_k(0)$ at time $t = 0$. Here, we shall do the same thing for the components for which $k \neq i, f$. However, we shall explicitly keep $b_i(t)$

and $b_f(t)$. Thus, in order to determine $b_i(t)$ and $b_f(t)$, we are led to the system of equations [the perturbation having the form (C-1-a) of chap. XIII]:

$$i\hbar \frac{d}{dt} b_i(t) = \frac{1}{2i}\{ [e^{i\omega t} - e^{-i\omega t}]W_{ii}\, b_i(t) + [e^{i(\omega-\omega_{fi})t} - e^{-i(\omega+\omega_{fi})t}]W_{if}\, b_f(t) \}$$

$$i\hbar \frac{d}{dt} b_f(t) = \frac{1}{2i}\{ [e^{i(\omega+\omega_{fi})t} - e^{-i(\omega-\omega_{fi})t}]W_{fi}\, b_i(t) + [e^{i\omega t} - e^{-i\omega t}]W_{ff}\, b_f(t) \}$$

(3)

On the right-hand side of these equations, certain coefficients of $b_i(t)$ and $b_f(t)$ are proportional to $e^{\pm i(\omega-\omega_{fi})t}$, so they oscillate slowly in time when $\omega \simeq \omega_{fi}$. On the other hand, the coefficients proportional either to $e^{\pm i\omega t}$ or to $e^{\pm i(\omega+\omega_{fi})t}$ oscillate much more rapidly. Here, we shall use the secular approximation, which consists of neglecting the second type of terms. The remaining ones, called "secular terms", are then those whose coefficients reduce to constants for $\omega = \omega_{fi}$. When integrated over time, they make significant contributions to the variations of the components $b_i(t)$ and $b_f(t)$. On the other hand, the contribution of the other terms is negligible, since their variation is too rapid (the integration of $e^{i\Omega t}$ causes a factor $1/\Omega$ to appear, and the mean value of $e^{i\Omega t}$ over a large number of periods is practically zero).

COMMENT:

For the preceding argument to be valid, it is necessary for the temporal variation of a term $e^{i\omega t}b_{i,f}(t)$ to be due principally to the exponential, and not to the component $b_{i,f}(t)$. Since ω is very close to ω_{fi}, this means that $b_{i,f}(t)$ must not vary very much over a time interval of the order of $1/|\omega_{fi}|$. This is indeed true with the assumptions we have made, that is, with $W \ll H_0$. The variations of $b_i(t)$ and $b_f(t)$ (which are constants if $W = 0$) are due to the presence of the perturbation W, and are appreciable for times of the order of $\hbar/|W_{if}|$ [this can be verified directly from formulas (8), obtained below]. Since by hypothesis $|W_{if}| \ll \hbar\, |\omega_{fi}|$, this time is much greater than $1/|\omega_{fi}|$.

In conclusion, the secular approximation leads to the system of equations :

$$\frac{d}{dt} b_i(t) = -\frac{1}{2\hbar} e^{i(\omega-\omega_{fi})t}\, W_{if}\, b_f(t) \tag{4-a}$$

$$\frac{d}{dt} b_f(t) = \frac{1}{2\hbar} e^{-i(\omega-\omega_{fi})t}\, W_{fi}\, b_i(t) \tag{4-b}$$

whose solution, very close to that of system (3), is easier to calculate, as we shall see in the next section.

2. Solution of the system of equations

We shall begin by considering the case for which $\omega = \omega_{fi}$. Differentiating (4-a) and substituting (4-b) into the result, we obtain:

$$\frac{d^2}{dt^2} b_i(t) = -\frac{1}{4\hbar^2} |W_{fi}|^2\, b_i(t) \tag{5}$$

Since the system is in the state $| \varphi_i \rangle$ at time $t = 0$, the initial conditions are:

$$\begin{cases} b_i(0) = 1 & \text{(6-a)} \\ b_f(0) = 0 & \text{(6-b)} \end{cases}$$

which, according to (4), gives:

$$\begin{cases} \dfrac{\mathrm{d}b_i}{\mathrm{d}t}(0) = 0 & \text{(7-a)} \\ \dfrac{\mathrm{d}b_f}{\mathrm{d}t}(0) = \dfrac{W_{fi}}{2\hbar} & \text{(7-b)} \end{cases}$$

The solution of (5) which satisfies (6-a) and (7-a) can be written:

$$b_i(t) = \cos\left(\frac{|W_{fi}|\, t}{2\hbar}\right) \tag{8-a}$$

We can then calculate b_f from (4-a):

$$b_f(t) = \mathrm{e}^{i\alpha_{fi}} \sin\left(\frac{|W_{fi}|\, t}{2\hbar}\right) \tag{8-b}$$

where α_{fi} is the argument of W_{fi}. The probability $\mathscr{P}_{if}(t;\, \omega = \omega_{fi})$ of finding the system in the state $| \varphi_f \rangle$ at time t is therefore, in this case, equal to:

$$\mathscr{P}_{if}(t\,;\omega = \omega_{fi}) = \sin^2\left(\frac{|W_{fi}|\, t}{2\hbar}\right) \tag{9}$$

When ω is different from ω_{fi} (while remaining close to the resonance value), the differential system (4) is still exactly soluble. In fact, it is completely analogous to the one we obtained in complement $\mathrm{F_{IV}}$ [*cf.* equation (15)] in studying the magnetic resonance of a spin 1/2. The same type of calculation as in that complement leads to the analogue of relation (27) (Rabi's formula), which can be written here:

$$\mathscr{P}_{if}(t\,;\omega) = \frac{|W_{if}|^2}{|W_{if}|^2 + \hbar^2(\omega - \omega_{fi})^2} \sin^2\left[\sqrt{\frac{|W_{if}|^2}{\hbar^2} + (\omega - \omega_{fi})^2}\, \frac{t}{2}\right] \tag{10}$$

[when $\omega = \omega_{fi}$, this expression does reduce to (9)].

3. Discussion

The discussion of the result obtained in (10) is the same as that of the magnetic resonance of a spin 1/2 (*cf.* complement $\mathrm{F_{IV}}$, §2-c). The probability $\mathscr{P}_{if}(t;\, \omega)$ is an oscillating function of time; for certain values of t, $\mathscr{P}_{if}(t;\, \omega) = 0$, and the system has gone back into the initial state $| \varphi_i \rangle$.

Furthermore, equation (10) measures the magnitude of the resonance phenomenon. When $\omega = \omega_{fi}$, however small the perturbation is, it can cause the system to move completely from the state $| \varphi_i \rangle$ to the state $| \varphi_f \rangle$*. On the other hand, if the perturbation is not resonant, the probability $\mathscr{P}_{if}(t; \omega)$ always remains less than 1.

Finally, it is interesting to compare the result obtained in this complement with the one obtained using the first-order theory in chapter XIII. First of all, note that, for all values of t, the probability $\mathscr{P}_{if}(t; \omega)$ obtained in (10) is included between 0 and 1. The approximation method used here therefore enables us to avoid the difficulties encountered in chapter XIII (*cf.* § C-2-c-β). When we let t approach zero in (9), we get (C-17) of this chapter. Thus, first-order perturbation theory is indeed valid for t sufficiently small (*cf.* comment of § B-3-b). It amounts to replacing the sinusoid which represents $\mathscr{P}_{if}(t; \omega)$ as a function of time by a parabola.

* The magnitude of the perturbation, characterized by $|W_{fi}|$, enters, at resonance, only into the time taken by the system to move from $| \varphi_i \rangle$ to $| \varphi_f \rangle$. The smaller $| W_{fi}|$, the longer the time.

Complement D_{XIII}

DECAY OF A DISCRETE STATE RESONANTLY COUPLED TO A CONTINUUM OF FINAL STATES

1. Statement of the problem

In § C-3 of chapter XIII, we showed that the coupling induced by a constant perturbation between an initial discrete state of energy E_i and a continuum of final states (some of which have an energy equal to E_i) causes the system to go from the initial state to this continuum of final states. More precisely, the probability of finding the system in a well-defined group of states of the continuum at time t increases linearly with time. Consequently, the probability $\mathscr{P}_{ii}(t)$ of finding the system in the initial state $| \varphi_i \rangle$ at time t must decrease linearly over time from the value $\mathscr{P}_{ii}(0) = 1$. It is clear that this result is valid only over short times, since extrapolation of the linear decrease of $\mathscr{P}_{ii}(t)$ to long times would lead to negative values of $\mathscr{P}_{ii}(t)$, which would be absurd for a probability. This raises the problem of determining the behavior of the system after a long time.

We encountered an analogous problem when we studied the resonant transitions induced by a sinusoidal perturbation between two discrete states $| \varphi_i \rangle$ and $| \varphi_f \rangle$. First-order perturbation theory predicts a decrease proportional to t^2 of $\mathscr{P}_{ii}(t)$ from the initial value $\mathscr{P}_{ii}(0) = 1$. The method presented in complement C_{XIII} shows that the system actually oscillates between the states $| \varphi_i \rangle$ and $| \varphi_f \rangle$. The decrease with t^2 found in § C of chapter XIII merely represents the "beginning" of the corresponding sinusoid.

We might expect to find an analogous result in the problem with which we are concerned here (oscillations of the system between the discrete state and the continuum). We shall show that this is not the case : *the physical system leaves the state* $| \varphi_i \rangle$ *irreversibly*. We find an exponential decrease $e^{-\Gamma t}$ for $\mathscr{P}_{ii}(t)$ (for which the perturbation treatment gives only the short-time behavior $1 - \Gamma t$). Thus, the *continuous* nature of the set of final states causes the reversibility found in complement C_{XIII} to disappear; it is responsible for a *decay* of the initial state, which thus acquires a *finite lifetime* (unstable state; *cf.* complement K_{III}).

The situation envisaged in this complement is very frequently encountered in physics. For example, a system, initially in a discrete state, can split, under the effect of an internal coupling (described, consequently, by a time-independent Hamiltonian W), into two distinct parts whose energies (kinetic in the case of material particles and electromagnetic in the case of photons) can have, theoretically, any value; this gives the set of final states a continuous nature. Thus, in *α-decay*, a nucleus which is initially in a discrete state is transformed (via the tunnel effect) into a system composed of an α-particle and another nucleus. A many-electron atom A which is initially in a configuration (*cf.* complement A_{XIV} and B_{XIV}) in which several electrons are excited can, under the effect of electrostatic interactions between electrons, give rise to a system formed of an ion A^+ and a free electron (the energy of the initial configuration must, of course, be greater than the simple ionization limit of A) : this is the "*autoionization*" phenomenon. We can also cite the *spontaneous emission* of a photon by an excited atomic (or nuclear) state : the interaction of the atom with the quantized electromagnetic field couples the discrete initial state (the excited atom in the absence of photons) with a continuum of final states (the atom in a lower state in the presence of a photon of arbitrary direction, polarization and energy). Finally, we can mention the photoelectric effect, in which a perturbation, now sinusoidal, couples a discrete state of an atom A to a continuum of final states (the ion A^+ and the photoelectron e^-).

These few examples of unstable states taken from various domains of physics are sufficient to indicate the importance of the problem we are treating in this complement.

2. Description of the model

a. ASSUMPTIONS ABOUT THE UNPERTURBED HAMILTONIAN H_0

To simplify the calculations as much as possible, we shall make the following assumptions about the spectrum of the unperturbed Hamiltonian H_0. This spectrum includes:

(*i*) a discrete state $| \varphi_i \rangle$ of energy E_i (non-degenerate):

$$H_0 | \varphi_i \rangle = E_i | \varphi_i \rangle \tag{1}$$

(*ii*) a set of states $| \alpha \rangle$ which form a continuum:

$$H_0 | \alpha \rangle = E | \alpha \rangle \tag{2}$$

E can take on a continuous infinity of values, distributed over a portion of the real axis *including* E_i. We shall assume, for example, that E varies from 0 to $+\infty$:

$$E \geqslant 0 \tag{3}$$

Each state $|\,\alpha\,\rangle$ is characterized by its energy E and a set of other parameters which we shall denote by β (as in § C-3-a-β of chapter XIII). $|\,\alpha\,\rangle$ can therefore also be written in the form $|\,\beta, E\,\rangle$. We have [*cf.* formula (C-28) of chap. XIII]:

$$\mathrm{d}\alpha = \rho(\beta, E)\,\mathrm{d}\beta\,\mathrm{d}E \tag{4}$$

where $\rho(\beta, E)$ is the density of final states.

The eigenstates of H_0 satisfy the following relations (orthogonality and closure relations):

$$\begin{cases} \langle\,\varphi_i\,|\,\varphi_i\,\rangle = 1 & \text{(5-a)} \\ \langle\,\varphi_i\,|\,\alpha\,\rangle = 0 & \text{(5-b)} \\ \langle\,\alpha\,|\,\alpha'\,\rangle = \delta(\alpha - \alpha') & \text{(5-c)} \end{cases}$$

$$|\,\varphi_i\,\rangle\langle\,\varphi_i\,| + \int \mathrm{d}\alpha\,|\,\alpha\,\rangle\langle\,\alpha\,| = 1 \tag{6}$$

b. ASSUMPTIONS ABOUT THE COUPLING W

We shall assume that W is not explicitly time-dependent and has no diagonal elements:

$$\langle\,\varphi_i\,|\,W\,|\,\varphi_i\,\rangle = \langle\,\alpha\,|\,W\,|\,\alpha\,\rangle = 0 \tag{7}$$

(if these diagonal elements were not zero, we could always add them to those of H_0, which would simply amount to changing the unperturbed energies). Similarly, we shall assume that W cannot couple two states of the continuum:

$$\langle\,\alpha\,|\,W\,|\,\alpha'\,\rangle = 0 \tag{8}$$

The only non-zero matrix elements of W are then those which connect the discrete state $|\,\varphi_i\,\rangle$ with the states of the continuum. It is these matrix elements, $\langle\,\alpha\,|\,W\,|\,\varphi_i\,\rangle$, which are responsible for the decay of the state $|\,\varphi_i\,\rangle$.

The preceding assumptions are not too restrictive. In particular, condition (8) is very often satisfied in the physical problems alluded to at the end of §1. The advantage of this model is that it enables us to investigate the physics of the decay phenomenon without too many complicated calculations. The essential physical conclusions would not be modified by using a more elaborate model.

Before taking up the new method of solving the Schrödinger equation which we are describing in this complement, we shall indicate the results of the first-order perturbation theory of chapter XIII as they apply to this model.

c. RESULTS OF FIRST-ORDER PERTURBATION THEORY

The discussion of § C-3 of chapter XIII enables us to calculate [using, in particular, formula (C-36)] the probability of finding the physical system at time t (initially in the state $|\,\varphi_i\,\rangle$) in a final state of arbitrary energy belonging to a group of final states characterized by the interval $\delta\beta_f$ around the value β_f.

Here, we shall concern ourselves with the probability of finding the system in any of the final states $| \alpha \rangle$: neither E nor β is specified. We must therefore integrate expression (C-36) of chapter XIII with respect to β, which gives the probability density [the integration over the energy was already performed in (C-36)]. Thus, we introduce the constant:

$$\Gamma = \frac{2\pi}{\hbar} \int \mathrm{d}\beta \, |\langle \beta, E = E_i \,|\, W \,|\, \varphi_i \rangle|^2 \, \rho(\beta, E = E_i) \tag{9}$$

The desired probability is then equal to Γt. With the assumptions of § a, it represents the probability of the system's having left the state $| \varphi_i \rangle$ at time t. If we call $\mathscr{P}_{ii}(t)$ the probability that the system is still in this state at time t, we have:

$$\mathscr{P}_{ii}(t) = 1 - \Gamma t \tag{10}$$

In the discussion of the following sections, it is important to recall the validity conditions for (10):

(*i*) Expression (10) results from a first-order perturbation theory which is valid only if $\mathscr{P}_{ii}(t)$ differs only slightly from its initial value $\mathscr{P}_{ii}(0) = 1$. We then must have:

$$t \ll \frac{1}{\Gamma} \tag{11}$$

(*ii*) Furthermore, (10) is valid only for sufficiently long times t.

To state the second condition more precisely, and to see, in particular, if it is compatible with (11), we return to expression (C-31) of chapter XIII (β and E are no longer constrained to vary only inside the intervals $\delta\beta_f$ and δE_f). Instead of proceeding as we did in chapter XIII, we shall integrate the probability density appearing in (C-31), first over β and then over E. The following integral then appears:

$$\frac{1}{\hbar^2} \int_0^{\infty} \mathrm{d}E \, F\left(t, \frac{E - E_i}{\hbar}\right) K(E) \tag{12}$$

where $K(E)$, which results from the first integration over β, is given by:

$$K(E) = \int \mathrm{d}\beta \, |\langle \beta, E \,|\, W \,|\, \varphi_i \rangle|^2 \, \rho(\beta, E) \tag{13}$$

$F\left(t, \frac{E - E_i}{\hbar}\right)$ is the diffraction function defined by (C-7) of chapter XIII, centered at $E = E_i$ and of width $4\pi\hbar/t$.

Let $\hbar\Delta$ be the "width" of $K(E)$: $\hbar\Delta$ represents the order of magnitude of the E variation needed for $K(E)$ to change significantly (*cf.* fig. 1). As soon as t is sufficiently large that:

$$t \gg \frac{1}{\Delta} \tag{14}$$

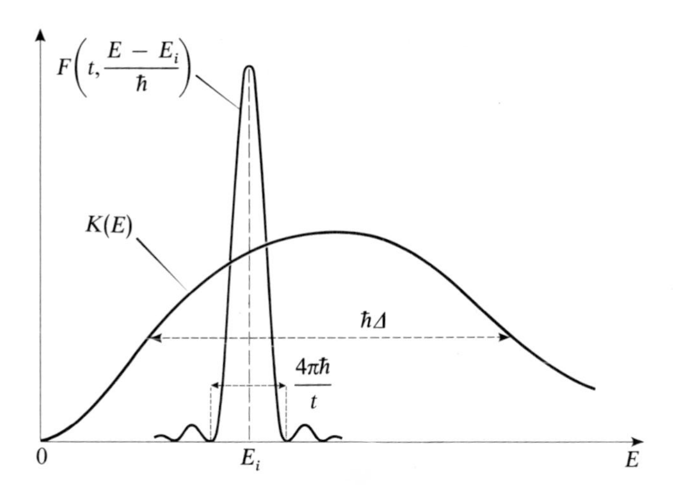

FIGURE 1

Variation of the functions $K(E)$ and $F\left(t, \frac{E - E_i}{\hbar}\right)$ with respect to E. The respective "widths" of the two curves are of the order of $\hbar\Delta$ and $4\pi\hbar/t$. For sufficiently large t, $F\left(t, \frac{E - E_i}{\hbar}\right)$ behaves like a "delta function" with respect to $K(E)$.

$F\left(t, \frac{E - E_i}{\hbar}\right)$ behaves like a "delta function" with respect to $K(E)$. Using relation (C-32) of chapter XIII, we can then write (12) in the form:

$$\frac{2\pi}{\hbar} t \int \mathrm{d}E\, \delta(E - E_i)\, K(E) = \frac{2\pi t}{\hbar} K(E = E_i) = \Gamma t \tag{15}$$

since by comparing (9) and (13), it can easily be seen that:

$$\frac{2\pi}{\hbar} K(E = E_i) = \Gamma \tag{16}$$

Again we find that the linear decrease appearing in (10) is valid only if t is large enough to satisfy (14).

Conditions (11) and (14), obviously, are compatible only if:

$$\Delta \gg \Gamma \tag{17}$$

We have thus given a quantitative form to the condition stated in the note on page 1300. In the rest of this complement, we shall assume that inequality (17) is satisfied.

d. INTEGRODIFFERENTIAL EQUATION EQUIVALENT TO THE SCHRÖDINGER EQUATION

It is easy to adapt expressions (B-11) of chapter XIII to the case we are studying here.

The state of the system at time t can be expanded on the $\{ | \varphi_i \rangle, | \alpha \rangle \}$ basis:

$$| \psi(t) \rangle = b_i(t) \, e^{-iE_i t/\hbar} | \varphi_i \rangle + \int d\alpha \, b(\alpha, t) \, e^{-iEt/\hbar} | \alpha \rangle \tag{18}$$

When we substitute state vector (18) into the Schrödinger equation, using the assumptions stated in §§ 2-a and 2-b, we obtain, after a calculation that is analogous to the one in § B-1 of chapter XIII, the following equations of motion:

$$\begin{cases} i\hbar \dfrac{d}{dt} b_i(t) = \displaystyle\int d\alpha \, e^{i(E_i - E)t/\hbar} \langle \varphi_i | W | \alpha \rangle \, b(\alpha, t) & (19) \\ i\hbar \dfrac{d}{dt} b(\alpha, t) = e^{i(E - E_i)t/\hbar} \langle \alpha | W | \varphi_i \rangle \, b_i(t) & (20) \end{cases}$$

The problem consists of using these rigorous equations to predict the behavior of the system after a long time, taking into account the initial conditions:

$$\begin{cases} b_i(0) = 1 & (21\text{-a}) \\ b(\alpha, 0) = 0 & (21\text{-b}) \end{cases}$$

The simplifying assumptions which we made for W imply that $\frac{d}{dt} b_i(t)$ depends only on $b(\alpha, t)$, and $\frac{d}{dt} b(\alpha, t)$, only on $b_i(t)$. Consequently, we can integrate equation (20), taking initial condition (21-b) into account. Substituting the value obtained in this way for $b(\alpha, t)$ into (19), we obtain the following equation describing the evolution of $b_i(t)$:

$$\frac{d}{dt} b_i(t) = -\frac{1}{\hbar^2} \int d\alpha \int_0^t dt' \, e^{i(E_i - E)(t - t')/\hbar} |\langle \alpha | W | \varphi_i \rangle|^2 \, b_i(t') \tag{22}$$

By using (4) and performing the integration over β, we obtain, using (13):

$$\frac{d}{dt} b_i(t) = -\frac{1}{\hbar^2} \int_0^\infty dE \int_0^t dt' \, K(E) \, e^{i(E_i - E)(t - t')/\hbar} \, b_i(t') \tag{23}$$

Thus, we have been able to obtain an equation involving only b_i. However, it must be noted that this equation is no longer a differential equation, but an integrodifferential equation: $\frac{d}{dt} b_i(t)$ depends on the entire "history of the system" between the times 0 and t.

Equation (23) is rigorously equivalent to the Schrödinger equation. We do not know how to solve it exactly. In the following sections, we shall describe two approximate methods for solving this equation. One of them (§ 3) is equivalent to the first-order theory of chapter XIII; the other one (§ 4) enables us to study the long-time behavior of the system more satisfactorily.

3. Short-time approximation. Relation to first-order perturbation theory

If t is not too large, that is, if $b_i(t)$ is not too different from $b_i(0) = 1$, we can replace $b_i(t')$ by $b_i(0) = 1$ on the right-hand side of (23). This right-hand side then reduces to a double integral, over E and t', whose integration presents no difficulties :

$$- \frac{1}{\hbar^2} \int_0^\infty \mathrm{d}E \int_0^t \mathrm{d}t' \; K(E) \, \mathrm{e}^{i(E_i - E)(t - t')/\hbar} \tag{24}$$

We shall perform this calculation explicitly, since it allows us to introduce two constants [one of which is Γ, defined by (9)] which play an important role in the more elaborate method described in § 4.

We shall begin by integrating over t' in (24). According to (47) of appendix II, the limit of this integral for $t \longrightarrow \infty$ is the Fourier transform of the Heaviside step function. More precisely:

$$\lim_{t \to \infty} \int_0^t \mathrm{e}^{i(E_i - E)\tau/\hbar} \, \mathrm{d}\tau = \hbar \left[\pi \delta(E_i - E) + i \mathcal{P} \left(\frac{1}{E_i - E} \right) \right] \tag{25}$$

(we have set $t - t' = \tau$).

Actually, it is not necessary to let t approach infinity in order to use (25) in the calculation of (24). It suffices for $\hbar/t$ to be very much smaller than the "width" $\hbar\Delta$ of $K(E)$, that is, for t to be very much greater than $1/\Delta$. We again find the validity condition (14). If this condition is satisfied, we can then use (25) to write (24) in the form:

$$- \frac{\pi}{\hbar} K(E = E_i) - \frac{i}{\hbar} \mathcal{P} \int_0^\infty \frac{K(E)}{E_i - E} \, \mathrm{d}E \tag{26}$$

The first term of (26) is, according to (16), simply $- \Gamma/2$. We shall set:

$$\delta E = \mathcal{P} \int_0^\infty \frac{K(E)}{E_i - E} \, \mathrm{d}E \tag{27}$$

Therefore, the double integral (24) is equal to:

$$- \frac{\Gamma}{2} - i \frac{\delta E}{\hbar} \tag{28}$$

When $b_i(t')$ is replaced by $b_i(0) = 1$ in (23), this equation then becomes [as soon as (14) is satisfied] :

$$\frac{d}{dt} b_i(t) = -\frac{\Gamma}{2} - i\frac{\delta E}{\hbar} \tag{29}$$

The solution of (29), using the initial condition (21-a), is very simple:

$$b_i(t) = 1 - \left(\frac{\Gamma}{2} + i\frac{\delta E}{\hbar}\right)t \tag{30}$$

Obviously, this result is valid only if $|b_i(t)|$ differs only slightly from 1, that is, if:

$$t \ll \frac{1}{\Gamma} \;,\; \frac{\hbar}{\delta E} \tag{31}$$

This is the other validity condition, (11), for first-order perturbation theory.

Using (30), we can easily calculate the probability $\mathscr{P}_{ii}(t) = |b_i(t)|^2$ that the system is still in the state $| \varphi_i \rangle$ at time t. If we neglect terms in Γ^2 and δE^2, we obtain:

$$\mathscr{P}_{ii}(t) = 1 - \Gamma t \tag{32}$$

All the results obtained in chapter XIII can then be deduced from equation (23) when $b_i(t')$ is replaced by $b_i(0)$. This equation has also enabled us to introduce the parameter δE, whose physical significance will be discussed later [note that δE does not appear in the treatment of chapter XIII because we were concerned only with the calculation of the probability $|b_i(t)|^2$, and not with that of the probability amplitude $b_i(t)$].

4. Another approximate method for solving the Schrödinger equation

A better approximation consists of replacing $b_i(t')$ by $b_i(t)$ rather than by $b_i(0)$ in (23). To see this, we shall begin by doing the integral over E which appears on the right-hand side of the rigorous equation, (23). We thus obtain a function of E_i and $t - t'$:

$$g(E_i, t - t') = -\frac{1}{\hbar^2}\int_0^\infty dE \;\; K(E)\, e^{i(E_i - E)(t - t')/\hbar} \tag{33}$$

which is clearly different from zero only if $t - t'$ is very small. In (33), we are integrating over E the product of $K(E)$, which varies slowly with E (*cf.* fig. 1), and an exponential whose period with respect to the variable E is $2\pi\hbar/(t - t')$. If we choose values of t and t' such that this period is very much smaller than the width $\hbar\Delta$ of $K(E)$, the product of these two functions undergoes numerous oscillations when E is varied, and its integral over E is negligible. Consequently, the modulus of $g(E_i, t - t')$ is large for $t - t' \simeq 0$ and becomes negligible as soon as $t - t' \gg 1/\Delta$. This property means that, for all t, the only values of $b_i(t')$ to enter significantly into the right-hand side of (23) are those which correspond to t' very close to $t (t - t' \lesssim 1/\Delta)$. Indeed, once the integration over E has been performed, this right-hand side becomes :

$$\int_0^t g(E_i, t - t')\; b_i(t')\,\mathrm{d}t' \tag{34}$$

and we see that the presence of $g(E_i, t - t')$ practically eliminates the contribution of $b_i(t')$ as soon as $t - t' \gg 1/\Delta$.

Thus, the derivative $\frac{\mathrm{d}}{\mathrm{d}t} b_i(t)$ has only a very short memory of the previous values of $b_i(t)$ between 0 and t. Actually, it depends only on the values of b_i at times immediately before t, and *this is true for all t*. This property enables us to transform the integrodifferential equation (23) into a differential equation. If $b_i(t)$ varies very little over a time interval of the order of $1/\Delta$, we make only a small error by replacing $b_i(t')$ by $b_i(t)$ in (34). This yields:

$$b_i(t) \int_0^t g(E_i, t - t')\,\mathrm{d}t' = -\left(\frac{\Gamma}{2} + i\frac{\delta E}{\hbar}\right) b_i(t) \tag{35}$$

[to write the right-hand side of (35), we used the fact that the integral over t' of $g(E_i, t - t')$ is simply, according to (33), the double integral (24) evaluated in § 3 above].

Now, according to the results of § 3 (and as we shall see later), the time scale characteristic of the evolution of $b_i(t)$ is of the order of $1/\Gamma$ or $\hbar/\delta E$. The validity condition for (35) is then:

$$\Gamma \;,\; \frac{\delta E}{\hbar} \ll \Delta \tag{36}$$

which we have already assumed to be fulfilled [*cf.* (17)].

To a good approximation, and for all t, equation (23) can therefore be written:

$$\frac{\mathrm{d}}{\mathrm{d}t} b_i(t) = -\left(\frac{\Gamma}{2} + i\frac{\delta E}{\hbar}\right) b_i(t) \tag{37}$$

whose solution, using (21-a), is obvious:

$$b_i(t) = \mathrm{e}^{-\Gamma t/2}\; \mathrm{e}^{-i\delta E t/\hbar} \tag{38}$$

It can easily be shown that the limited expansion of (38) gives (30) to first order in Γ and δE.

COMMENT:

No upper bound has been imposed on t. On the other hand, the integral $\int_0^t g(E_i, t - t')\,\mathrm{d}t'$ which appears in (35) is equal to $-(\Gamma/2 + i\delta E/\hbar)$ only if

$t \gg 1/\Delta$, as we saw in § 3 above. For very short times, the theory presented here suffers from the same limitations as perturbation theory; however, it has the great advantage of being valid for long times.

If we now substitute expression (38) for $b_i(t)$ into equation (20), we obtain a very simple equation which enables us to determine the probability amplitude $b(\alpha, t)$ associated with the state $| \alpha \rangle$:

$$b(\alpha, t) = \frac{1}{i\hbar} \langle \alpha | W | \varphi_i \rangle \int_0^t \mathrm{e}^{-\Gamma t'/2} \, \mathrm{e}^{i(E - E_i - \delta E)t'/\hbar} \, \mathrm{d}t' \tag{39}$$

that is:

$$b(\alpha, t) = \frac{\langle \alpha | W | \varphi_i \rangle}{\hbar} \frac{1 - \mathrm{e}^{-\Gamma t/2} \, \mathrm{e}^{i(E - E_i - \delta E)t/\hbar}}{\frac{1}{\hbar}(E - E_i - \delta E) + i\frac{\Gamma}{2}} \tag{40}$$

Equations (38) and (40), respectively, describe the decay of the initial state and the "filling" of the final states $| \alpha \rangle$. Now let us study in greater detail the physical content of these two equations.

5. Discussion

a. LIFETIME OF THE DISCRETE STATE

According to (38), we have:

$$\mathscr{P}_{ii}(t) = |b_i(t)|^2 = \mathrm{e}^{-\Gamma t} \tag{41}$$

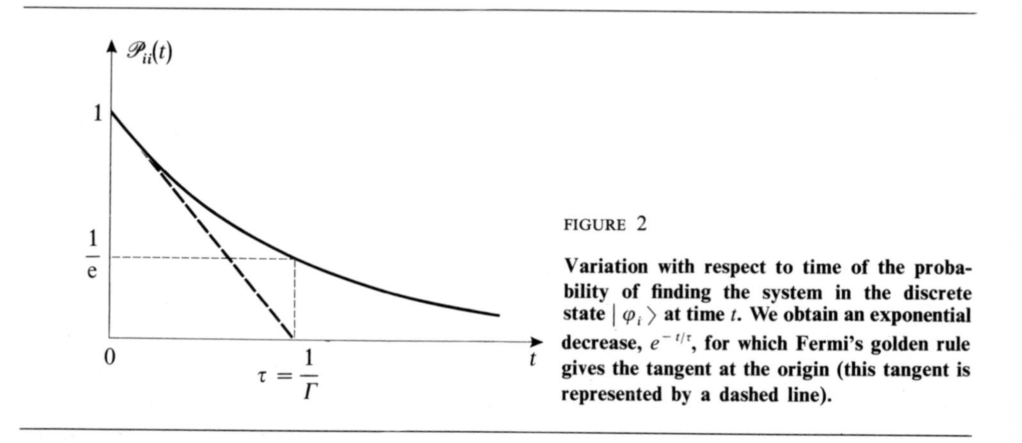

FIGURE 2

Variation with respect to time of the probability of finding the system in the discrete state $| \varphi_i \rangle$ at time t. We obtain an exponential decrease, $e^{-t/\tau}$, for which Fermi's golden rule gives the tangent at the origin (this tangent is represented by a dashed line).

$\mathscr{P}_{ii}(t)$ therefore decreases *irreversibly* from $\mathscr{P}_{ii}(0) = 1$ and approaches zero as $t \longrightarrow \infty$ (fig. 2). The discrete initial state is said to have a *finite lifetime* τ; τ is the time constant of the exponential of figure 2:

$$\tau = \frac{1}{\Gamma} \tag{42}$$

This irreversible behavior contrasts sharply with the oscillations of the system (Rabi's formula) between two discrete states when it is subject to a resonant perturbation coupling these two states.

b. SHIFT OF THE DISCRETE STATE DUE TO THE COUPLING WITH THE CONTINUUM

If we go from $b_i(t)$ to $c_i(t)$ [*cf.* formula (B-8) of chapter XIII], we obtain, from (38):

$$c_i(t) = e^{-\Gamma t/2} \, e^{-i(E_i + \delta E)t/\hbar} \tag{43}$$

Recall that, in the absence of the coupling W, we would have:

$$c_i(t) = e^{-iE_i t/\hbar} \tag{44}$$

In addition to the exponential decrease, $e^{-\Gamma t/2}$, the coupling with the continuum is therefore responsible for a shift in the discrete state energy, which goes from E_i to $E_i + \delta E$. This is the interpretation of the quantity δE introduced in § 3.

Let us analyze expression (27) for δE more closely. Substituting definition (13) of $K(E)$ into (27), we get:

$$\delta E = \mathscr{P} \int_0^\infty \frac{dE}{E_i - E} \int d\beta \; \rho(\beta, E) \, |\langle \beta, E \,|\, W \,|\, \varphi_i \rangle|^2 \tag{45}$$

or, if we use (4) and replace $\langle \beta, E \,|$ by $\langle \alpha \,|$:

$$\delta E = \mathscr{P} \int d\alpha \frac{|\langle \alpha \,|\, W \,|\, \varphi_i \rangle|^2}{E_i - E} \tag{46}$$

The contribution to this integral of a particular state $|\, \alpha \rangle$ of the continuum, for which $E \neq E_i$, is:

$$\frac{|\langle \alpha \,|\, W \,|\, \varphi_i \rangle|^2}{E_i - E} \tag{47}$$

We recognize (47) as a familiar expression in stationary perturbation theory [*cf.* formula (B-14) of chapter XI]. (47) represents the energy shift of the state $|\, \varphi_i \rangle$ due to the coupling with the state $|\alpha\rangle$, to second order in W. δE is simply the sum of the shifts due to the various states $|\, \alpha \rangle$ of the continuum. We might imagine that a problem would appear for the states $|\, \alpha \rangle$ for which $E = E_i$. Actually, the presence in (46) of the principal part $\mathscr{P}$ implies that the contribution of the states $|\, \alpha \rangle$ situated immediately above $|\, \varphi_i \rangle$ compensates that of the states situated immediately below.

Summing up:

(*i*) The coupling of $|\, \varphi_i \rangle$ with the states $|\, \alpha \rangle$ of the same energy is responsible for the finite lifetime of $|\, \varphi_i \rangle$ [the function $\delta(E_i - E)$ of formula (25) enters into the expression for Γ].

(*ii*) The coupling of $| \varphi_i \rangle$ with the states $| \alpha \rangle$ of different energies is responsible for an energy shift of the state $| \varphi_i \rangle$. This shift can be calculated by stationary perturbation theory (this was not obvious in advance).

COMMENT:

In the particular case of the spontaneous emission of a photon by an atom, δE represents the shift of the atomic level under study due to the coupling with the continuum of final states (an atom in another discrete state, in the presence of a photon). The difference between the shifts δE of the $2s_{1/2}$ and $2p_{1/2}$ states of the hydrogen atom is the "Lamb shift" [*cf.* complement K_V, § 3-d-δ and chapter XII, § C-3-b, comment (*iv*)].

c. ENERGY DISTRIBUTION OF THE FINAL STATES

Once the discrete state has decayed, that is, when $t \gg 1/\Gamma$, the final state of the system belongs to the continuum of states $| \alpha \rangle$. It is interesting to study the energy distribution of the possible final states. For example, in the spontaneous emission of a photon by an atom, this energy distribution is that of the photon emitted when the atom falls back from the excited level to a lower level (the natural width of spectral lines).

When $t \gg 1/\Gamma$, the exponential which appears in the numerator of (40) is practically zero. We then have:

$$|b(\alpha, t)|^2 \underset{t \gg \frac{1}{\Gamma}}{\sim} |\langle \alpha | W | \varphi_i \rangle|^2 \frac{1}{(E - E_i - \delta E)^2 + \hbar^2 \Gamma^2/4} \tag{48}$$

$|b(\alpha, t)|^2$ actually represents a probability density. The probability of finding the system, after the decay, in a group of final states characterized by the intervals $d\beta_f$ and dE_f about β_f and E_f can be calculated directly from (48):

$$d\mathscr{P}(\beta_f, E_f, t) = |\langle \beta_f, E_f | W | \varphi_i \rangle|^2 \rho(\beta_f, E_f) \frac{1}{(E_f - E_i - \delta E)^2 + \hbar^2 \Gamma^2/4} d\beta_f \, dE_f \tag{49}$$

Let us examine the E_f-dependence of the probability density:

$$\frac{d\mathscr{P}(\beta_f, E_f, t)}{d\beta_f \, dE_f}$$

Since $|\langle \beta_f, E_f | W | \varphi_i \rangle|^2 \rho(\beta_f, E_f)$ remains practically constant when E_f varies over an interval of the order of $\hbar\Gamma$, the variation of the probability density with respect to E_f is essentially determined by the function:

$$\frac{1}{(E_f - E_i - \delta E)^2 + \hbar^2 \Gamma^2/4} \tag{50}$$

and has, consequently, the form shown in figure 3. The energy distribution of the final states has a maximum for $E_f = E_i + \delta E$, that is, when the final state energy is equal to that of the initial state $| \varphi_i \rangle$, corrected by the shift δE.

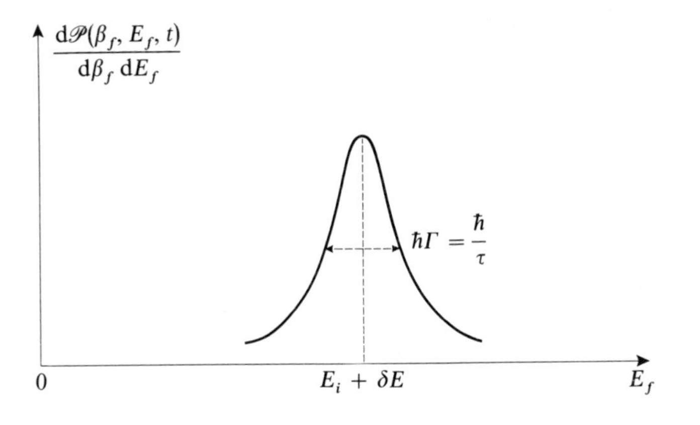

FIGURE 3

Form of the energy distribution of the final states attained by the system after the decay of the discrete state. We obtain a Lorentzian distribution centered at $E_i + \delta E$ (the energy of the discrete state corrected by the shift δE due to the coupling with the continuum). The shorter the lifetime τ of the discrete state, the wider the distribution (time-energy uncertainty relation).

The form of the distribution is that of a Lorentz curve of width $\hbar\Gamma$, called the "natural width" of the state $| \varphi_i \rangle$. An energy dispersion of the final states therefore appears. The larger $\hbar\Gamma$ (that is, the shorter the lifetime $\tau = 1/\Gamma$ of the discrete state), the greater the dispersion. More precisely:

$$\Delta E_f = \hbar\Gamma = \frac{\hbar}{\tau} \tag{51}$$

Note again the analogy between (51) and the time-energy uncertainty relation. In the presence of the coupling W, the state $| \varphi_i \rangle$ can be observed only during a finite time, of the order of its lifetime τ. When we want to determine its energy by measuring that of the final state of the system, the uncertainty ΔE of the result cannot be much less than $\hbar/\tau$.

References:

The original article: Weisskopf and Wigner (2.33).

Complement E_{XIII}

EXERCISES

1. Consider a one-dimensional harmonic oscillator of mass m, angular frequency ω_0 and charge q. Let $| \varphi_n \rangle$ and $E_n = (n + 1/2)\hbar\omega_0$ be the eigenstates and eigenvalues of its Hamiltonian H_0.

For $t < 0$, the oscillator is in the ground state $| \varphi_0 \rangle$. At $t = 0$, it is subjected to an electric field "pulse" of duration τ. The corresponding perturbation can be written :

$$W(t) = \begin{cases} - q\mathscr{E}X & \text{for } 0 \leqslant t \leqslant \tau \\ 0 & \text{for } t < 0 \text{ and } t > \tau \end{cases}$$

$\mathscr{E}$ is the field amplitude and X is the position observable. Let $\mathscr{P}_{0n}$ be the probability of finding the oscillator in the state $| \varphi_n \rangle$ after the pulse.

a. Calculate $\mathscr{P}_{01}$ by using first-order time-dependent perturbation theory. How does $\mathscr{P}_{01}$ vary with τ, for fixed ω_0?

b. Show that, to obtain $\mathscr{P}_{02}$, the time-dependent perturbation theory calculation must be pursued at least to second order. Calculate $\mathscr{P}_{02}$ to this perturbation order.

c. Give the exact expressions for $\mathscr{P}_{01}$ and $\mathscr{P}_{02}$ in which the translation operator used in complement F_V appears explicitly. By making a limited power series expansion in $\mathscr{E}$ of these expressions, find the results of the preceding questions.

2. Consider two spin 1/2's, $\mathbf{S}_1$ and $\mathbf{S}_2$, coupled by an interaction of the form $a(t)\mathbf{S}_1 \cdot \mathbf{S}_2$; $a(t)$ is a function of time which approaches zero when $|t|$ approaches infinity, and takes on non-negligible values (on the order of a_0) only inside an interval, whose width is of the order of τ, about $t = 0$.

a. At $t = -\infty$, the system is in the state $| +, - \rangle$ (an eigenstate of S_{1z} and S_{2z} with the eigenvalues $+\hbar/2$ and $-\hbar/2$). Calculate, without approximations, the state of the system at $t = +\infty$. Show that the probability $\mathscr{P}(+ - \longrightarrow - +)$ of finding, at $t = +\infty$, the system in the state $| -, + \rangle$ depends only on the integral $\int_{-\infty}^{+\infty} a(t)\, dt$.

b. Calculate $\mathscr{P}(+ - \longrightarrow - +)$ by using first-order time-dependent perturbation theory. Discuss the validity conditions for such an approximation by comparing the results obtained with those of the preceding question.

c. Now assume that the two spins are also interacting with a static magnetic field $\mathbf{B}_0$ parallel to Oz. The corresponding Zeeman Hamiltonian can be written :

$$H_0 = - B_0(\gamma_1 S_{1z} + \gamma_2 S_{2z})$$

where γ_1 and γ_2 are the gyromagnetic ratios of the two spins, assumed to be different.

Assume that $a(t) = a_0 \, \mathrm{e}^{-t^2/\tau^2}$. Calculate $\mathscr{P}(+ \; - \longrightarrow - \; +)$ by first-order time-dependent perturbation theory. With fixed a_0 and τ, discuss the variation of $\mathscr{P}(+ \; - \longrightarrow - \; +)$ with respect to B_0.

3. Two-photon transitions between non-equidistant levels

Consider an atomic level of angular momentum $J = 1$, subject to static electric and magnetic fields, both parallel to Oz. It can be shown that three non-equidistant energy levels are then obtained. The eigenstates $| \varphi_M \rangle$ of J_z ($M = -1, 0, +1$), of energies E_M correspond to them. We set $E_1 - E_0 = \hbar\omega_0$, $E_0 - E_{-1} = \hbar\omega_0'$ ($\omega_0 \neq \omega_0'$).

The atom is also subjected to a radiofrequency field rotating at the angular frequency ω in the xOy plane. The corresponding perturbation $W(t)$ can be written:

$$W(t) = \frac{\omega_1}{2} (J_+ \, \mathrm{e}^{-i\omega t} + J_- \, \mathrm{e}^{i\omega t})$$

where ω_1 is a constant proportional to the amplitude of the rotating field.

a. We set (notation identical to that of chapter XIII):

$$| \psi(t) \rangle = \sum_{M=-1}^{+1} b_M(t) \, \mathrm{e}^{-iE_M t/\hbar} \, | \varphi_M \rangle$$

Write the system of differential equations satisfied by the $b_M(t)$.

b. Assume that, at time $t = 0$, the system is in the state $| \varphi_{-1} \rangle$. Show that if we want to calculate $b_1(t)$ by time-dependent perturbation theory, the calculation must be pursued to second order. Calculate $b_1(t)$ to this perturbation order.

c. For fixed t, how does the probability $\mathscr{P}_{-1,+1}(t) = |b_1(t)|^2$ of finding the system in the state $| \varphi_1 \rangle$ at time t vary with respect to ω? Show that a resonance appears, not only for $\omega = \omega_0$ and $\omega = \omega_0'$, but also for $\omega = (\omega_0 + \omega_0')/2$. Give a particle interpretation of this resonance.

4. Returning to exercise 5 of complement H_{XI} and using its notation, assume that the field $\mathbf{B}_0$ is oscillating at angular frequency ω, and can be written $\mathbf{B}_0(t) = \mathbf{B}_0 \cos \omega t$. Assume that $b = 2a$ and that ω is not equal to any Bohr angular frequency of the system (non-resonant excitation).

Introduce the susceptibility tensor χ, of components $\chi_{ij}(\omega)$, defined by:

$$\langle M_i \rangle (t) = \sum_j \mathrm{Re} \, [\chi_{ij}(\omega) \, B_{0j} \, \mathrm{e}^{i\omega t}]$$

with $i, j = x, y, z$. Using a method analogous to the one in § 2 of complement A_{XIII}, calculate $\chi_{ij}(\omega)$. Setting $\omega = 0$, find the results of exercise 5 of complement H_{XI}.

5. The Autler-Townes effect

Consider a three-level system : $|\varphi_1\rangle$, $|\varphi_2\rangle$, and $|\varphi_3\rangle$, of energies E_1, E_2 and E_3. Assume $E_3 > E_2 > E_1$ and $E_3 - E_2 \ll E_2 - E_1$.

This system interacts with a magnetic field oscillating at the angular frequency ω. The states $|\varphi_2\rangle$ and $|\varphi_3\rangle$ are assumed to have the same parity, which is the opposite of that of $|\varphi_1\rangle$, so that the interaction Hamiltonian $W(t)$ with the oscillating magnetic field can connect $|\varphi_2\rangle$ and $|\varphi_3\rangle$ to $|\varphi_1\rangle$. Assume that, in the basis of the three states $|\varphi_1\rangle$, $|\varphi_2\rangle$, $|\varphi_3\rangle$, arranged in that order, $W(t)$ is represented by the matrix:

$$\begin{bmatrix} 0 & 0 & 0 \\ 0 & 0 & \omega_1 \sin \omega t \\ 0 & \omega_1 \sin \omega t & 0 \end{bmatrix}$$

where ω_1 is a constant proportional to the amplitude of the oscillating field.

a. Set (notation identical to that of chapter XIII):

$$|\psi(t)\rangle = \sum_{i=1}^{3} b_i(t)\, e^{-iE_it/\hbar} |\varphi_i\rangle$$

Write the system of differential equations satisfied by the $b_i(t)$.

b. Assume that ω is very close to $\omega_{32} = (E_3 - E_2)/\hbar$. Making approximations analogous to those used in complement C_{XIII}, integrate the preceding system, with the initial conditions:

$$b_1(0) = b_2(0) = \frac{1}{\sqrt{2}} \qquad b_3(0) = 0$$

(neglect, on the right-hand side of the differential equations, the terms whose coefficients, $e^{\pm i(\omega + \omega_{32})t}$, vary very rapidly, and keep only those whose coefficients are constant or vary very slowly, as $e^{\pm i(\omega - \omega_{32})t}$).

c. The component D_z along Oz of the electric dipole moment of the system is represented, in the basis of the three states $|\varphi_1\rangle$, $|\varphi_2\rangle$, $|\varphi_3\rangle$, arranged in that order, by the matrix:

$$\begin{bmatrix} 0 & d & 0 \\ d & 0 & 0 \\ 0 & 0 & 0 \end{bmatrix}$$

where d is a real constant (D_z is an odd operator and can connect only states of different parities).

Calculate $\langle D_z \rangle(t) = \langle \psi(t) | D_z | \psi(t) \rangle$, using the vector $|\psi(t)\rangle$ calculated in *b*.

Show that the time evolution of $\langle D_z \rangle(t)$ is given by a superposition of sinusoidal terms. Determine the frequencies ν_k and relative intensities π_k of these terms.

These are the frequencies that can be absorbed by the atom when it is placed in an oscillating electric field parallel to Oz. Describe the modifications of this absorption spectrum when, for ω fixed and equal to ω_{32}, ω_1 is increased from zero. Show that the presence of the magnetic field oscillating at the frequency $\omega_{32}/2\pi$ splits the electric dipole absorption line at the frequency $\omega_{21}/2\pi$, and that the separation of the two components of the doublet is proportional to the oscillating magnetic field amplitude (the Autler-Townes doublet).

What happens when, for ω_1 fixed, $\omega - \omega_{32}$ is varied?

6. Elastic scattering by a particle in a bound state. Form factor

Consider a particle (a) in a bound state $| \varphi_0 \rangle$ described by the wave function $\varphi_0(\mathbf{r}_a)$ localized about a point O. Towards this particle (a) is directed a beam of particles (b), of mass m, momentum $\hbar\mathbf{k}_i$, energy $E_i = \hbar^2\mathbf{k}_i^2/2m$ and wave function $\frac{1}{(2\pi)^{3/2}} e^{i\mathbf{k}_i.\mathbf{r}_b}$. Each particle ($b$) of the beam interacts with particle (a). The corresponding potential energy, W, depends only on the relative position $\mathbf{r}_b - \mathbf{r}_a$ of the two particles.

a. Calculate the matrix element:

$$\langle a : \varphi_0 ; b : \mathbf{k}_f | W(\mathbf{R}_b - \mathbf{R}_a) | a : \varphi_0 ; b : \mathbf{k}_i \rangle$$

of $W(\mathbf{R}_b - \mathbf{R}_a)$ between two states in which particle (a) is in the same state $| \varphi_0 \rangle$ and particle (b) goes from the state $| \mathbf{k}_i \rangle$ to the state $| \mathbf{k}_f \rangle$. The expression for this matrix element should include the Fourier transform $\overline{W}(\mathbf{k})$ of the potential $W(\mathbf{r}_b - \mathbf{r}_a)$:

$$W(\mathbf{r}_b - \mathbf{r}_a) = \frac{1}{(2\pi)^{3/2}} \int \overline{W}(\mathbf{k}) \, e^{i\mathbf{k}.(\mathbf{r}_b - \mathbf{r}_a)} \, d^3k$$

b. Consider the scattering processes in which, under the effect of the interaction W, particle (b) is scattered in a certain direction, with particle (a) remaining in the same quantum state $| \varphi_0 \rangle$ after the scattering process (elastic scattering).

Using a method analogous to the one in chapter XIII [*cf.* comment (*ii*) of § C-3-b], calculate, in the Born approximation, the elastic scattering cross section of particle (b) by particle (a) in the state $| \varphi_0 \rangle$.

Show that this cross section can be obtained by multiplying the cross section for scattering by the potential $W(\mathbf{r})$ (in the Born approximation) by a factor which characterizes the state $| \varphi_0 \rangle$, called the "form factor".

Show that, if the Fourier transform $\overline{W}(\mathbf{k})$ of $W(\mathbf{r})$ is known, studying the variation of the cross section with the scattering angle enables one to determine experimentally the probability density $|\varphi_0(\mathbf{r}_a)|^2$ associated with the state $| \varphi_0 \rangle$.

7. A simple model of the photoelectric effect

Consider, in a one-dimensional problem, a particle of mass m, placed in a potential of the form $V(x) = -\alpha\delta(x)$, where α is a real positive constant.

Recall (*cf.* exercises 2 and 3 of complement K_I) that, in such a potential, there is a single bound state, of negative energy $E_0 = -m\alpha^2/2\hbar^2$, associated with a normalized wave function $\varphi_0(x) = \sqrt{m\alpha/\hbar^2}\, e^{-\frac{m\alpha}{\hbar^2}|x|}$. For each positive value of the energy $E = \hbar^2k^2/2m$, on the other hand, there are two stationary wave functions, corresponding, respectively, to an incident particle coming from the left or from the right. The expression for the first eigenfunction, for example, is:

$$\chi_k(x) = \begin{cases} \dfrac{1}{\sqrt{2\pi}}\left[e^{ikx} - \dfrac{1}{1 + i\hbar^2k/m\alpha}\, e^{-ikx}\right] & \text{for } x < 0 \\[2ex] \dfrac{1}{\sqrt{2\pi}}\, \dfrac{i\hbar^2k/m\alpha}{1 + i\hbar^2k/m\alpha}\, e^{ikx} & \text{for } x > 0 \end{cases}$$

a. Show that the $\chi_k(x)$ satisfy the orthonormalization relation (in the extended sense):

$$\langle \chi_k \mid \chi_{k'} \rangle = \delta(k - k')$$

The following relation [*cf.* formula (47) of appendix II] can be used:

$$\int_{-\infty}^{0} e^{iqx}\,dx = \int_{0}^{\infty} e^{-iqx}\,dx = \lim_{\varepsilon \to 0} \frac{1}{\varepsilon + iq}$$

$$= \pi\,\delta(q) - i\,\mathscr{P}\left(\frac{1}{q}\right)$$

Calculate the density of states $\rho(E)$ for a positive energy E.

b. Calculate the matrix element $\langle \chi_k \mid X \mid \varphi_0 \rangle$ of the position observable X between the bound state $\mid \varphi_0 \rangle$ and the positive energy state $\mid \chi_k \rangle$ whose wave function was given above.

c. The particle, assumed to be charged (charge q) interacts with an electric field oscillating at the angular frequency ω. The corresponding perturbation is:

$$W(t) = -q\mathscr{E}X \sin \omega t$$

where $\mathscr{E}$ is a constant.

The particle is initially in the bound state $\mid \varphi_0 \rangle$. Assume that $\hbar\omega > -E_0$. Calculate, using the results of § C of chapter XIII [see, in particular, formula (C-37)], the transition probability w per unit time to an arbitrary positive energy state (the photoelectric or photoionization effect). How does w vary with ω and $\mathscr{E}$?

8. Disorientation of an atomic level due to collisions with rare gas atoms

Consider a motionless atom A at the origin of a coordinate frame $Oxyz$ (see figure). This atom A is in a level of angular momentum $J = 1$, to which correspond the three orthonormal kets $| M \rangle$ ($M = -1, 0, +1$), eigenstates of J_z of eigenvalues $M\hbar$.

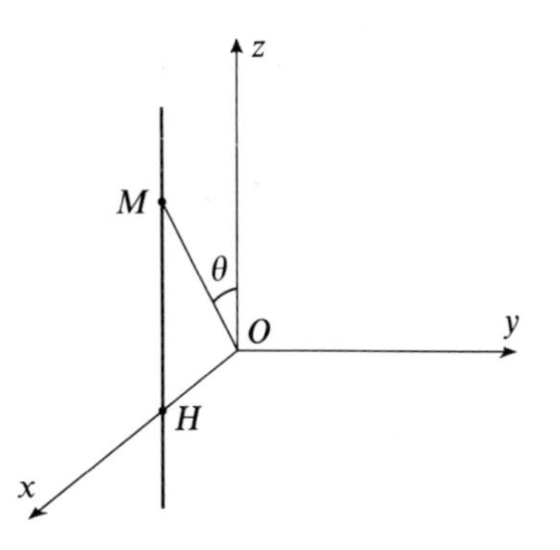

A second atom B, in a level of zero angular momentum, is in uniform rectilinear motion in the xOz plane: it is travelling at the velocity v along a straight line parallel to Oz and situated at a distance b from this axis (b is the "impact parameter"). The time origin is chosen at the time when B arrives at point H of the Ox axis ($OH = b$). At time t, atom B is therefore at point M, where $HM = vt$. Call θ the angle between Oz and OM.

The preceding model, which treats the external degrees of freedom of the two atoms classically, permits the simple calculation of the effect on the internal degrees of freedom of atom A (which are treated quantum mechanically) of a collision with atom B (which is, for example, a rare gas atom in the ground state). It can be shown that, because of the Van der Waals forces (*cf.* complement C_{XI}) between the two atoms, atom A is subject to a perturbation W acting on its internal degrees of freedom, and given by:

$$W = \frac{C}{r^6} J_u^2$$

where C is a constant, r is the distance between the two atoms, and J_u is the component of the angular momentum $\mathbf{J}$ of atom A on the OM axis joining the two atoms.

a. Express W in terms of C, b, v, t, J_z, $J_\pm = J_x \pm iJ_y$. Introduce the dimensionless parameter $\tau = vt/b$.

b. Assume that there is no external magnetic field, so that the three states $| +1 \rangle$, $| 0 \rangle$, $| -1 \rangle$ of atom A have the same energy.

Before the collision, that is, at $t = -\infty$, atom A is in the state $| -1 \rangle$. Using first-order time-dependent perturbation theory, calculate the probability $\mathscr{P}_{-1,+1}$ of finding, after the collision (that is, at $t = +\infty$), atom A in the state $| +1 \rangle$. Discuss the variation of $\mathscr{P}_{-1,+1}$ with respect to b and v. Similarly, calculate $\mathscr{P}_{-1,0}$.

c. Now assume that there is a static field $\mathbf{B}_0$ parallel to Oz, so that the three states $| M \rangle$ have an additional energy $M\hbar\omega_0$ (the Zeeman effect), where ω_0 is the Larmor angular frequency in the field $\mathbf{B}_0$.

α. With ordinary magnetic fields ($B_0 \sim 10^2$ gauss), $\omega_0 \simeq 10^9$ rad sec^{-1}; b is of the order of 5 Å, and v, of the order of 5×10^2 m . sec^{-1}. Show that, under these conditions, the results of question b remain valid.

β. Without going into detailed calculations, explain what happens for much higher values of B_0. Starting with what value of ω_0 (where b and v have the values indicated in α) will the results of b no longer be valid?

d. Without going into detailed calculations, explain how to calculate the disorientation probabilities $\mathscr{P}_{-1,+1}$ and $\mathscr{P}_{-1,0}$ for an atom A placed in a gas of atoms B in thermodynamic equilibrium at the temperature T, containing a number n of atoms per unit volume sufficiently small that only binary collisions need be considered.

N.B. We give : $\displaystyle\int_{-\infty}^{+\infty} \frac{\mathrm{d}\tau}{(1 + \tau^2)^4} = \frac{5\pi}{16}$

9. Transition probability per unit time under the effect of a random perturbation. Simple relaxation model

A physical system, subject to a perturbation $W(t)$, is at time $t = 0$ in the eigenstate $| \varphi_i \rangle$ of its Hamiltonian H_0. Let $\mathscr{P}_{if}(t)$ be the probability of finding the system at time t in another eigenstate of H_0, $| \varphi_f \rangle$. The transition probability per unit time $w_{if}(t)$ is defined by $w_{if}(t) = \dfrac{\mathrm{d}}{\mathrm{d}t} \mathscr{P}_{if}(t)$.

a. Show that, to first order in perturbation theory, we have:

$$w_{if}(t) = \frac{1}{\hbar^2} \int_0^t \mathrm{e}^{i\omega_{fi}\tau}\, W_{fi}(t) W_{fi}^*(t - \tau)\, \mathrm{d}\tau + \text{c.c.} \tag{1}$$

with $\hbar\omega_{fi} = E_f - E_i$ (notation identical to that of chapter XIII).

b. Consider a very large number $\mathscr{N}$ of systems (k), which are identical and without mutual interactions ($k = 1, 2, ..., \mathscr{N}$). Each of them has a different microscopic environment and, consequently, "sees" a different perturbation $W^{(k)}(t)$. It is, of course, impossible to know with certainty each of the individual perturbations $W^{(k)}(t)$; we can specify only statistical averages such as:

$$\overline{W_{fi}(t)} = \lim_{\mathscr{N} \to \infty} \frac{1}{\mathscr{N}} \sum_{k=1}^{\mathscr{N}} W_{fi}^{(k)}(t)$$

$$\overline{W_{fi}(t)\, W_{fi}^*(t - \tau)} = \lim_{\mathscr{N} \to \infty} \frac{1}{\mathscr{N}} \sum_{k=1}^{\mathscr{N}} W_{fi}^{(k)}(t)\, W_{fi}^{(k)*}(t - \tau) \tag{2}$$

This perturbation is said to be "random".

This random perturbation is called stationary if the preceding averages do not depend on the time t. The unperturbed Hamiltonian H_0 is then redefined so as to make all the $\overline{W_{fi}}$ zero, and we set:

$$g_{fi}(\tau) = \overline{W_{fi}(t)\, W^*_{fi}(t-\tau)} \tag{3}$$

$g_{fi}(\tau)$ is called the "correlation function" of the perturbation (for the pair of states $|\varphi_i\rangle$, $|\varphi_f\rangle$). $g_{fi}(\tau)$ generally goes to zero for $\tau \gg \tau_c$, is a characteristic time, called the "correlation time" of the perturbation. The perturbation has a "memory" which extends into the past only over an interval of the order of τ_c.

α. The $\mathcal{N}$ systems are all in the state $|\varphi_i\rangle$ at time $t = 0$ and are subject to a stationary random perturbation, whose correlation function is $g_{fi}(\tau)$ and whose correlation time is τ_c ($\mathcal{N}$ can be considered to be infinite in the calculations).

Calculate the proportion $\pi_{if}(t)$ of systems which go into the state $|\varphi_f\rangle$ per unit time. Show that after a certain value t_1 of t, to be specified, $\pi_{if}(t)$ no longer depends on t.

β. For fixed τ_c, how does π_{if} vary with ω_{fi}? Consider the case for which $g_{fi}(\tau) = |v_{fi}|^2\, e^{-\tau/\tau_c}$, with v_{fi} constant.

γ. The preceding theory is rigorously valid only for $t \ll t_2$ [since formula (1) results from a perturbation theory]. What is the order of magnitude of t_2? Taking $t_2 \gg t_1$, find the condition for introducing a transition probability per unit time which is independent of t [use the form of $g_{fi}(\tau)$ given in the preceding question]. Would it be possible to extend the preceding theory beyond $t = t_2$?

c. Application to a simple system.

The $\mathcal{N}$ systems under consideration are $\mathcal{N}$ spin 1/2 particles, with gyromagnetic ratio γ, placed in a static field $\mathbf{B}_0$ (set $\omega_0 = -\gamma B_0$). These particles are enclosed in a spherical cell of radius R. Each of them bounces constantly back and forth between the walls. The mean time between two collisions of the same particle with the wall is called the "flight time" τ_v. During this time, the particle "sees" only the field $\mathbf{B}_0$. In a collision with the wall, each particle remains adsorbed on the surface during a mean time τ_a ($\tau_a \ll \tau_v$), during which it "sees", in addition to $\mathbf{B}_0$, a constant microscopic magnetic field $\mathbf{b}$, due to the paramagnetic impurities contained in the wall. The direction of $\mathbf{b}$ varies randomly from one collision to another; the mean amplitude of $\mathbf{b}$ is denoted by b_0.

α. What is the correlation time of the perturbation seen by the spins? Give the physical justification for the following form, to be chosen for the correlation function of the components of the microscopic field $\mathbf{b}$:

$$\overline{b_x(t) b_x(t-\tau)} = \frac{1}{3}\, b_0^2\, \frac{\tau_a}{\tau_v}\, e^{-\tau/\tau_a} \tag{4}$$

and analogous expressions for the components along Oy and Oz, all the cross terms $\overline{b_x(t) b_y(t-\tau)}$... being zero.

β. Let $\mathcal{M}_z$ be the component along the Oz axis defined by the field $\mathbf{B}_0$ of the macroscopic magnetization of the $\mathcal{N}$ particles. Show that, under the effect of the collisions with the wall, $\mathcal{M}_z$ "relaxes", with a time constant T_1:

$$\frac{d\mathcal{M}_z}{dt} = -\frac{\mathcal{M}_z}{T_1}$$

(T_1 is called the longitudinal relaxation time). Calculate T_1 in terms of γ, B_0, τ_a, τ_v, b_0.

γ. Show that studying the variation of T_1 with B_0 permits the experimental determination of the mean adsorption time τ_a.

δ. We have at our disposition several cells, of different radii R, constructed from the same material. By measuring T_1, how can we determine experimentally the mean amplitude b_0 of the microscopic field at the wall?

10. Absorption of radiation by a many-particle system forming a bound state. The Doppler effect. Recoil energy. The Mössbauer effect

In complement A_{XIII}, we consider the absorption of radiation by a charged particle attracted by a fixed center O (the hydrogen atom model for which the nucleus is infinitely heavy). In this exercise, we treat a more realistic situation, in which the incident radiation is absorbed by a system of many particles of finite masses interacting with each other and forming a bound state. Thus, we are studying the effect on the absorption phenomenon of the degrees of freedom of the center of mass of the system.

I. ABSORPTION OF RADIATION BY A FREE HYDROGEN ATOM. THE DOPPLER EFFECT. RECOIL ENERGY

Let $\mathbf{R}_1$ and $\mathbf{P}_1$, $\mathbf{R}_2$ and $\mathbf{P}_2$ be the position and momentum observables of two particles, (1) and (2), of masses m_1 and m_2 and opposite charges q_1 and q_2 (a hydrogen atom). Let $\mathbf{R}$ and $\mathbf{P}$, $\mathbf{R}_G$ and $\mathbf{P}_G$ be the position and momentum observables of the relative particle and the center of mass (*cf.* chap. VII, § B). $M = m_1 + m_2$ is the total mass, and $m = m_1 m_2/(m_1 + m_2)$ is the reduced mass. The Hamiltonian H_0 of the system can be written:

$$H_0 = H_e + H_i \tag{1}$$

where:

$$H_e = \frac{1}{2M}\mathbf{P}_G^2 \tag{2}$$

is the translational kinetic energy of the atom, assumed to be free ("external" degrees of freedom), and where H_i (which depends only on $\mathbf{R}$ and $\mathbf{P}$) describes the internal energy of the atom ("internal" degrees of freedom). We denote by $| \mathbf{K} \rangle$ the eigenstates of H_e, with eigenvalues $\hbar^2\mathbf{K}^2/2M$. We concern ourselves with only two eigenstates of H_i, $| \chi_a \rangle$ and $| \chi_b \rangle$, of energies E_a and E_b ($E_b > E_a$). We set:

$$E_b - E_a = \hbar\omega_0 \tag{3}$$

a. What energy must be furnished to the atom to move it from the state $| \mathbf{K}; \chi_a \rangle$ (the atom in the state $| \chi_a \rangle$ with a total momentum $\hbar\mathbf{K}$) to the state $| \mathbf{K}'; \chi_b \rangle$?

b. This atom interacts with a plane electromagnetic wave of wave vector $\mathbf{k}$ and angular frequency $\omega = ck$, polarized along the unit vector $\mathbf{e}$ perpendicular to $\mathbf{k}$. The corresponding vector potential $\mathbf{A}(\mathbf{r}, t)$ is:

$$\mathbf{A}(\mathbf{r}, t) = \mathscr{A}_0 \, \mathbf{e} \, e^{i(\mathbf{k}\cdot\mathbf{r} - \omega t)} + c.c. \tag{4}$$

where $\mathscr{A}_0$ is a constant. The principal term of the interaction Hamiltonian between this plane wave and the two-particle system can be written (*cf.* complement A_{XIII}, §1-b):

$$W(t) = - \sum_{i=1}^{2} \frac{q_i}{m_i} \mathbf{P}_i \, . \, \mathbf{A}(\mathbf{R}_i, t) \tag{5}$$

Express $W(t)$ in terms of $\mathbf{R}$, $\mathbf{P}$, $\mathbf{R}_G$, $\mathbf{P}_G$, m, M and q (setting $q_1 = - q_2 = q$), and show that, in the electric dipole approximation which consists of neglecting $\mathbf{k} \, . \, \mathbf{R}$ (but not $\mathbf{k} \, . \, \mathbf{R}_G$) compared to 1, we have:

$$W(t) = W e^{-i\omega t} + W^\dagger e^{i\omega t} \tag{6}$$

where:

$$W = - \frac{q\mathscr{A}_0}{m} \mathbf{e} \, . \, \mathbf{P} \, e^{i\mathbf{k}.\mathbf{R}_G} \tag{7}$$

c. Show that the matrix element of W between the state $| \mathbf{K}; \chi_a \rangle$ and the state $| \mathbf{K}'; \chi_b \rangle$ is different from zero only if there exists a certain relation between $\mathbf{K}$, $\mathbf{k}$, $\mathbf{K}'$ (to be specified). Interpret this relation in terms of the total momentum conservation during the absorption of an incident photon by the atom.

d. Show from this that if the atom in the state $| \mathbf{K}; \chi_a \rangle$ is placed in the plane wave (4), resonance occurs when the energy $\hbar\omega$ of the photons associated with the incident wave differs from the energy $\hbar\omega_0$ of the atomic transition $| \chi_a \rangle \longrightarrow | \chi_b \rangle$ by a quantity δ which is to be expressed in terms of $\hbar$, ω_0, $\mathbf{K}$, $\mathbf{k}$, M, c (since δ is a corrective term, we can replace ω by ω_0 in the expression for δ). Show that δ is the sum of two terms, one of which, δ_1, depends on $\mathbf{K}$ and on the angle between $\mathbf{K}$ and $\mathbf{k}$ (the Doppler effect), and the other, δ_2, is independent of $\mathbf{K}$. Give a physical interpretation of δ_1 and δ_2 (showing that δ_2 is the recoil kinetic energy of the atom when, having been initially motionless, it absorbs a resonant photon).

Show that δ_2 is negligible compared to δ_1 when $\hbar\omega_0$ is of the order of 10 eV (the domain of atomic physics). Choose, for M, a mass of the order of that of the proton ($Mc^2 \simeq 10^9$ eV), and, for $|\mathbf{K}|$, a value corresponding to a thermal velocity at $T = 300$ °K. Would this still be true if $\hbar\omega_0$ were of the order of 10^5 eV (the domain of nuclear physics)?

II. RECOILLESS ABSORPTION OF RADIATION BY A NUCLEUS VIBRATING ABOUT ITS EQUILIBRIUM POSITION IN A CRYSTAL. THE MÖSSBAUER EFFECT

The system under consideration is now a nucleus of mass M vibrating at the angular frequency Ω about its equilibrium position in a crystalline lattice (the Einstein model; *cf.* complement A_V, § 2). We again denote by $\mathbf{R}_G$ and $\mathbf{P}_G$ the position and momentum of the center of mass of this nucleus. The vibrational energy of the nucleus is described by the Hamiltonian:

$$H_e = \frac{1}{2M}\mathbf{P}_G^2 + \frac{1}{2}M\Omega^2(X_G^2 + Y_G^2 + Z_G^2) \tag{8}$$

which is that of a three-dimensional isotropic harmonic oscillator. Denote by $|\psi_{n_x,n_y,n_z}\rangle$ the eigenstate of H_e of eigenvalue $(n_x + n_y + n_z + 3/2)\hbar\Omega$. In addition to these external degrees of freedom, the nucleus possesses internal degrees of freedom with which are associated observables which all commute with $\mathbf{R}_G$ and $\mathbf{P}_G$. Let H_i be the Hamiltonian which describes the internal energy of the nucleus. As above, we concern ourselves with two eigenstates of H_i, $|\chi_a\rangle$ and $|\chi_b\rangle$, of energies E_a and E_b, and we set $\hbar\omega_0 = E_b - E_a$. Since $\hbar\omega_0$ falls into the γ-ray domain, we have, of course:

$$\omega_0 \gg \Omega \tag{9}$$

e. What energy must be furnished to the nucleus to allow it to go from the state $|\psi_{0,0,0};\chi_a\rangle$ (the nucleus in the vibrational state defined by the quantum numbers $n_x = 0$, $n_y = 0$, $n_z = 0$ and the internal state $|\chi_a\rangle$) to the state $|\psi_{n,0,0};\chi_b\rangle$?

f. This nucleus is placed in an electromagnetic wave of the type defined by (4), whose wave vector $\mathbf{k}$ is parallel to Ox. It can be shown that, in the electric dipole approximation, the interaction Hamiltonian of the nucleus with this plane wave (responsible for the absorption of the γ-rays) can be written as in (6), with:

$$W = \mathscr{A}_0 S_i(k)\, e^{ikX_G} \tag{10}$$

where $S_i(k)$ is an operator which acts on the internal degrees of freedom and consequently commutes with $\mathbf{R}_G$ and $\mathbf{P}_G$. Set $s(k) = \langle\chi_b|S_i(k)|\chi_a\rangle$.

The nucleus is initially in the state $|\psi_{0,0,0};\chi_a\rangle$. Show that, under the influence of the incident plane wave, a resonance appears whenever $\hbar\omega$ coincides with one of the energies calculated in *e*, with the intensity of the corresponding resonance proportional to $|s(k)|^2\,|\langle\psi_{n,0,0}|e^{ikX_G}|\psi_{0,0,0}\rangle|^2$, where the value of k is to be specified. Show, furthermore, that condition (9) allows us to replace k by $k_0 = \omega_0/c$ in the expression for the intensity of the resonance.

g. We set:

$$\pi_n(k_0) = |\langle\varphi_n|e^{ik_0X_G}|\varphi_0\rangle|^2 \tag{11}$$

where the states $|\varphi_n\rangle$ are the eigenstates of a one-dimensional harmonic oscillator of position X_G, mass M and angular frequency Ω.

α. Calculate $\pi_n(k_0)$ in terms of $\hbar$, M, Ω, k_0, n (see also exercise 7 of complement M_V). Set $\xi = \dfrac{\hbar^2 k_0^2}{2M} \Big/ \hbar\Omega$. Hint: establish a recurrence relation between $\langle \varphi_n \mid e^{ik_0 X_G} \mid \varphi_0 \rangle$ and $\langle \varphi_{n-1} \mid e^{ik_0 X_G} \mid \varphi_0 \rangle$, and express all the $\pi_n(k_0)$ in terms of $\pi_0(k_0)$, which is to be calculated directly from the wave function of the harmonic oscillator ground state. Show that the $\pi_n(k_0)$ are given by a Poisson distribution.

β. Verify that $\sum\limits_{n=0}^{\infty} \pi_n(k_0) = 1$.

γ. Show that $\sum\limits_{n=0}^{\infty} n\hbar\Omega\, \pi_n(k_0) = \hbar^2\omega_0^2/2Mc^2$.

h. Assume that $\hbar\Omega \gg \hbar^2\omega_0^2/2Mc^2$, that is, that the vibrational energy of the nucleus is much greater than the recoil energy (very rigid crystalline bonds). Show that the absorption spectrum of the nucleus is essentially composed of a single line of angular frequency ω_0. This line is called the recoilless absorption line. Justify this name. Why does the Doppler effect disappear?

i. Now assume that $\hbar\Omega \ll \hbar^2\omega_0^2/2Mc^2$ (very weak crystalline bonds). Show that the absorption spectrum of the nucleus is composed of a very large number of equidistant lines whose barycenter (obtained by weighting the abscissa of each line by its relative intensity) coincides with the position of the absorption line of the free and initially motionless nucleus. What is the order of magnitude of the width of this spectrum (the dispersion of the lines about their barycenter)? Show that one obtains the results of the first part in the limit $\Omega \longrightarrow 0$.

Exercise 3 :

References : see Brossel's lectures in (15.2).

Exercise 5 :

References : see Townes and Schawlow (12.10), chap. 10, § 9.

Exercise 6 :

References : see Wilson (16.34).

Exercise 9 :

References : see Abragam (14.1), chap. VIII; Slichter (14.2), chap. 5.

Exercise 10 :

References : see De Benedetti (16.23); Valentin (16.1), annexe XV.

CHAPTER XIV

Systems of identical particles

OUTLINE OF CHAPTER XIV

In chapter III, we stated the postulates of non-relativistic quantum mechanics, and in chapter IX, we concentrated on those which concern spin degrees of freedom. Here, we shall see (§A) that, in reality, these postulates are not sufficient when we are dealing with systems containing many identical particles since, in this case, their application leads to ambiguities in the physical predictions. To eliminate these ambiguities, it is necessary to introduce a new postulate, concerning the quantum mechanical description of systems of identical particles. We shall state this postulate in §C and discuss its physical implications in §D. Before we do so, however, we shall (in §B) define and study permutation operators, which considerably facilitate the reasoning and the calculations.

A. STATEMENT OF THE PROBLEM

1. Identical particles: definition

Two particles are said to be identical if all their intrinsic properties (mass, spin, charge, etc.) are exactly the same: no experiment can distinguish one from the other. Thus, all the electrons in the universe are identical, as are all the protons and all the hydrogen atoms. On the other hand, an electron and a positron are not identical, since, although they have the same mass and the same spin, they have different electrical charges.

An important consequence can be deduced from this definition: when a physical system contains two identical particles, there is no change in its properties or its evolution if the roles of these two particles are exchanged.

COMMENT:

Note that this definition is independent of the experimental conditions. Even if, in a given experiment, the charges of the particles are not measured, an electron and a positron can never be treated like identical particles.

2. Identical particles in classical mechanics

In classical mechanics, the presence of identical particles in a system poses no particular problems. This special case is treated just like the general case. Each particle moves along a well-defined trajectory, which enables us to distinguish it from the others and "follow" it throughout the evolution of the system.

To treat this point in greater detail, we shall consider a system of two identical particles. At the initial time t_0, the physical state of the system is defined by specifying the position and velocity of each of the two particles; we denote this initial data by $\{ \mathbf{r}_0, \mathbf{v}_0 \}$ and $\{ \mathbf{r}'_0, \mathbf{v}'_0 \}$. To describe this physical state and calculate its evolution, we number the two particles: $\mathbf{r}_1(t)$ and $\mathbf{v}_1(t)$ denote the position and velocity of particle (1) at time t, and $\mathbf{r}_2(t)$ and $\mathbf{v}_2(t)$, those of particle (2). This numbering has no physical foundation, as it would if we were dealing with two particles having different natures. It follows that the initial physical state which we have just defined may, in theory, be described by two different "mathematical states" as we can set, either:

$$\begin{array}{ll} \mathbf{r}_1(t_0) = \mathbf{r}_0 & \mathbf{r}_2(t_0) = \mathbf{r}'_0 \\ \mathbf{v}_1(t_0) = \mathbf{v}_0 & \mathbf{v}_2(t_0) = \mathbf{v}'_0 \end{array} \tag{A-1}$$

or:

$$\begin{array}{ll} \mathbf{r}_1(t_0) = \mathbf{r}'_0 & \mathbf{r}_2(t_0) = \mathbf{r}_0 \\ \mathbf{v}_1(t_0) = \mathbf{v}'_0 & \mathbf{v}_2(t_0) = \mathbf{v}_0 \end{array} \tag{A-2}$$

Now, let us consider the evolution of the system. Suppose that the solution of the equations of motion defined by initial conditions (A-1) can be written:

$$\mathbf{r}_1(t) = \mathbf{r}(t) \qquad \mathbf{r}_2(t) = \mathbf{r}'(t) \tag{A-3}$$

where $\mathbf{r}(t)$ and $\mathbf{r}'(t)$ are two vector functions. The fact that the two particles are identical implies that the system is not changed if they exchange roles. Consequently, the Lagrangian $\mathscr{L}(\mathbf{r}_1, \mathbf{v}_1; \mathbf{r}_2, \mathbf{v}_2)$ and the classical Hamiltonian $\mathscr{H}(\mathbf{r}_1, \mathbf{p}_1; \mathbf{r}_2, \mathbf{p}_2)$ are invariant under exchange of indices 1 and 2. It follows that the solution of the equations of motion corresponding to the initial state (A-2) is:

$$\mathbf{r}_1(t) = \mathbf{r}'(t) \qquad \mathbf{r}_2(t) = \mathbf{r}(t) \tag{A-4}$$

where the functions $\mathbf{r}(t)$ and $\mathbf{r}'(t)$ are the same as in (A-3).

The two possible mathematical descriptions of the physical state under consideration are therefore perfectly equivalent, since they lead to the same physical predictions. The particle which started from $\{ \mathbf{r}_0, \mathbf{v}_0 \}$ at t_0 is at $\mathbf{r}(t)$ with the velocity $\mathbf{v}(t) = \mathrm{d}\mathbf{r}/\mathrm{d}t$ at time t, and the one which started from $\{ \mathbf{r}'_0, \mathbf{v}'_0 \}$ is at $\mathbf{r}'(t)$ with the velocity $\mathbf{v}'(t) = \mathrm{d}\mathbf{r}'/\mathrm{d}t$ (fig. 1). Under these conditions, all we need to do is choose, at the initial time, either one of the two possible "mathematical states" and ignore the existence of the other one. Thus, we treat the system as if the two particles were actually of different natures. The numbers (1) and (2), with which we label them arbitrarily at t_0, then act like intrinsic properties to distinguish the two particles. Since we can follow each particle step-by-step along its trajectory (arrows in figure 1), we can determine the locations of the particle numbered (1) and the one numbered (2) at any time.

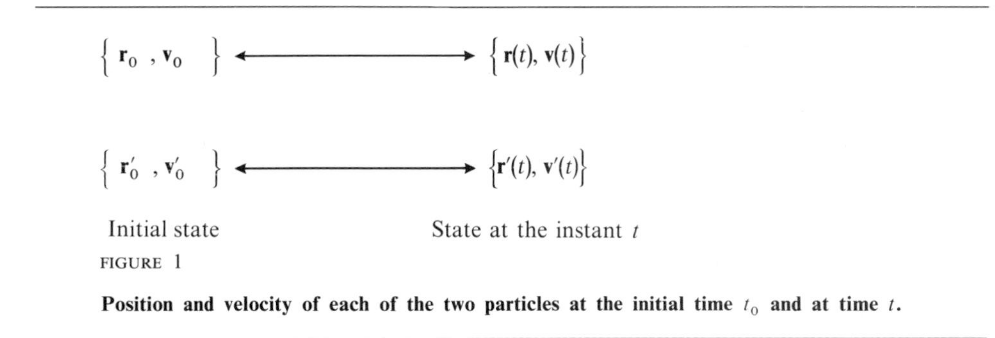

FIGURE 1

Position and velocity of each of the two particles at the initial time t_0 and at time t.

3. Identical particles in quantum mechanics: the difficulties of applying the general postulates

a. QUALITATIVE DISCUSSION OF A FIRST SIMPLE EXAMPLE

It is immediately apparent that the situation is radically different in quantum mechanics, since the particles no longer have definite trajectories. Even if, at t_0, the wave packets associated with two identical particles are completely separated in space, their subsequent evolution may mix them. We then "lose track" of the particles; when we detect one particle in a region of space in which both of them have a non-zero position probability, we have no way of knowing if the particle detected is the one numbered (1) or the one numbered (2). Except in special cases – for example, when the two wave packets never overlap – the numbering of the two particles becomes ambiguous when their positions are measured, since, as we shall see, there exist several distinct "paths" taking the system from its initial state to the state found in the measurement.

To investigate this point in greater detail, consider a concrete example: a collision between two identical particles in their center of mass frame (fig. 2). Before the collision, we have two completely separate wave packets, directed towards each other (fig. 2-a). We can agree, for example, to denote by (1) the particle on the left and by (2), the one on the right. During the collision (fig. 2-b), the two wave packets overlap. After the collision, the region of space in which the probability density of the two particles is non-zero* looks like a spherical shell whose radius increases over time (fig. 2-c). Suppose that a detector placed in the direction which makes an angle θ with the initial velocity of wave packet (1) detects a particle. It is then certain (because momentum is conserved in the collision) that the other particle is moving away in the opposite direction. However, it is impossible to know if the particle detected at D is the one initially numbered (1) or the one numbered (2). Thus, there

* The two-particle wave function depends on six variables (the components of the two particles coordinates $\mathbf{r}$ and $\mathbf{r}'$) and is not easily represented in 3 dimensions. Figure 2 is therefore very schematic: the grey regions are those to which both $\mathbf{r}$ and $\mathbf{r}'$ must belong for the wave function to take on significant values.

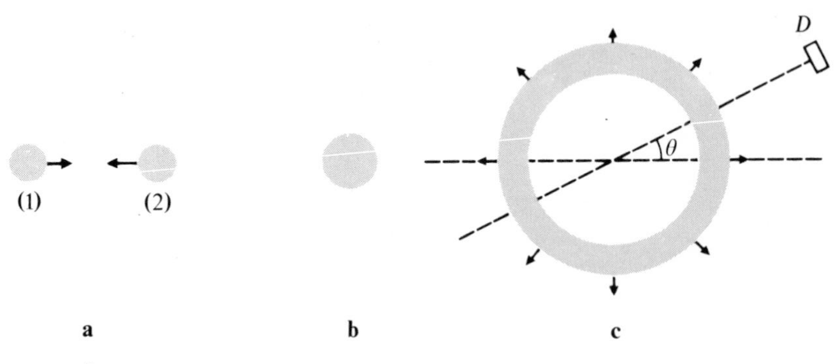

FIGURE 2

Collision between two identical particles in the center of mass frame : schematic representation of the probability density of the two particles.
Before the collision (fig. a), the two wave packets are clearly separated and can be labeled. During the collision (fig. b), the two wave packets overlap. After the collision (fig. c), the probability density is non-zero in a region shaped like a spherical shell whose radius increases over time. Because the two particles are identical, it is impossible, when a particle is detected at *D*, to know with which wave packet, (1) or (2), it was associated before the collision.

are two different "paths" that could have led the system from the initial state shown in figure 2-a to the final state found in the measurement. These two paths are represented schematically in figures 3-a and 3-b. Nothing enables us to determine which one was actually followed.

A fundamental difficulty then arises in quantum mechanics when using the postulates of chapter III. In order to calculate the probability of a given measurement result it is necessary to know the final state vectors associated with this result. Here, there are two, which correspond respectively to figures 3a and 3b. These two kets are distinct (and, furtheremore, orthogonal). Nevertheless, they are

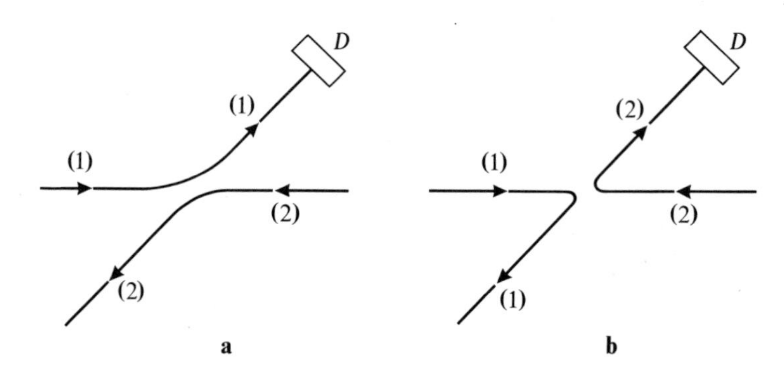

FIGURE 3

Schematic representation of two types of "paths" which the system could have followed in going from the initial state to the state found in the measurement. Because the two particles are identical, we cannot determine the path that was actually followed.

associated with a single physical state since it is impossible to imagine a more complete measurement that would permit distinguishing between them. Under these conditions, should one calculate the probability using path 3a, path 3b or both? in the latter case, should one take the sum of the probabilities associated with each path, or the sum of their probability amplitudes (and in this case, with what sign)? These different possibilities lead, as we shall verify later, to different predictions.

The answer to the preceding questions will be given in §D after we have stated the symmetrization postulate. Before going on, we shall study another example which will aid us in understanding the difficulties related to the indistinguishability of two particles.

b. ORIGIN OF THE DIFFICULTIES: EXCHANGE DEGENERACY

In the preceding example, we considered two wave packets which, initially, did not overlap, which enabled us to label each of them arbitrarily with a number, (1) or (2). Ambiguities appeared, however, when we tried to determine the mathematical state (or ket) associated with a given result of a position measurement. Actually, the same difficulty arises in the choice of the mathematical ket used to describe the initial physical state. This type of difficulty is related to the concept of "exchange degeneracy" which we shall introduce in this section. To simplify the reasoning, we shall first consider a different example, so as to confine ourselves to a finite-dimensional space. Then, we shall generalize the concept of exchange degeneracy, showing that it can be generalized to all quantum mechanical systems containing identical particles.

α. *Exchange degeneracy for a system of two spin 1/2 particles*

Let us consider a system composed of two identical spin 1/2 particles, confining ourselves to the study of its spin degrees of freedom. As in §A-2, we shall distinguish between the physical state of the system and its mathematical description (a ket in state space).

It would seem natural to suppose that, if we made a complete measurement of each of the two spins, we would then know the physical state of the total system perfectly. Here, we shall assume that the component along Oz of one of them is equal to $+\hbar/2$ and that of the other one, $-\hbar/2$ (this is the equivalent for the two spins of the specification of $\{ \mathbf{r}_0, \mathbf{v}_0 \}$ and $\{ \mathbf{r}'_0, \mathbf{v}'_0 \}$ in §A-2).

To describe the system mathematically, we number the particles: $\mathbf{S}_1$ and $\mathbf{S}_2$ denote the two spin observables, and $\{ | \varepsilon_1, \varepsilon_2 \rangle \}$ (where ε_1 and ε_2 can be equal to $+$ or $-$) is the orthonormal basis of the state space formed by the common eigenkets of S_{1z} (eigenvalue $\varepsilon_1\hbar/2$) and S_{2z} (eigenvalue $\varepsilon_2\hbar/2$).

Just as in classical mechanics, two different "mathematical states" could be associated with the same physical state. Either one of the two orthogonal kets:

$$| \varepsilon_1 = +, \varepsilon_2 = - \rangle \qquad \text{(A-5-a)}$$

$$| \varepsilon_1 = -, \varepsilon_2 = + \rangle \qquad \text{(A-5-b)}$$

can, *a priori*, describe the physical state considered here.

These two kets span a two-dimensional subspace whose normalized vectors are of the form:

$$\alpha | +, - \rangle + \beta | -, + \rangle \qquad \text{(A-6)}$$

with:

$$|\alpha|^2 + |\beta|^2 = 1 \tag{A-7}$$

By the superposition principle, all mathematical kets (A-6) can represent the same physical state as (A-5-a) or (A-5-b) (one spin pointing up and the other one pointing down). This is called "*exchange degeneracy*".

Exchange degeneracy creates fundamental difficulties, since application of the postulates of chapter III to the various kets (A-6) can lead to physical predictions which depend on the ket chosen. Let us determine, for example, the probability of finding the components of the two spins along Ox both equal to $+\hbar/2$. With this measurement result is associated a single ket of the state space. According to formula (A-20) of chapter IV, this ket can be written:

$$\begin{aligned}\frac{1}{\sqrt{2}}[|\varepsilon_1 = +\rangle + |\varepsilon_1 = -\rangle] \otimes \frac{1}{\sqrt{2}}[|\varepsilon_2 = +\rangle + |\varepsilon_2 = -\rangle] \\ = \frac{1}{2}[|+,+\rangle + |-,+\rangle + |+,-\rangle + |-,-\rangle]\end{aligned} \tag{A-8}$$

Consequently, the desired probability, for the vector (A-6), is equal to:

$$\left|\frac{1}{2}(\alpha + \beta)\right|^2 \tag{A-9}$$

This probability does depend on the coefficients α and β. It is not possible, therefore, to describe the physical state under consideration by the set of kets (A-6) or by any one of them chosen arbitrarily. The exchange degeneracy must be removed. That is, we must indicate unambiguously which of the kets (A-6) is to be used.

COMMENT:

In this example, exchange degeneracy appears only in the initial state, since we chose the same value for the components of the two spins in the final state. In the general case (for example, if the measurement result corresponds to two different eigenvalues of S_x), exchange degeneracy appears in both the initial and the final state.

β. *Generalization*

The difficulties related to exchange degeneracy arise in the study of all systems containing an arbitrary number N of identical particles ($N > 1$).

Consider, for example, a three-particle system. With each of the three particles, taken separately, are associated a state space and observables acting in this space. Thus, we are led to number the particles: $\mathscr{E}(1)$, $\mathscr{E}(2)$ and $\mathscr{E}(3)$ will denote the three one-particle state spaces, and the corresponding observables will be labeled by the same indices. The state space of the three-particle system is the tensor product:

$$\mathscr{E} = \mathscr{E}(1) \otimes \mathscr{E}(2) \otimes \mathscr{E}(3) \tag{A-10}$$

Now, consider an observable $B(1)$, initially defined in $\mathscr{E}(1)$. We shall assume that $B(1)$ alone constitutes a C.S.C.O. in $\mathscr{E}(1)$ [or that $B(1)$ actually denotes several observables which form a C.S.C.O.]. The fact that the three particles are identical implies that the observables $B(2)$ and $B(3)$ exist and that they constitute C.S.C.O.'s in $\mathscr{E}(2)$ and $\mathscr{E}(3)$ respectively. $B(1)$, $B(2)$ and $B(3)$ have the same spectrum, $\{ b_n ; n = 1, 2, ... \}$. Using the bases which define these three observables in $\mathscr{E}(1)$, $\mathscr{E}(2)$ and $\mathscr{E}(3)$, we can construct, by taking the tensor product, an orthonormal basis of $\mathscr{E}$, which we shall denote by:

$$\{ | 1 : b_i ; 2 : b_j ; 3 : b_k \rangle ; i, j, k = 1, 2, ... \} \tag{A-11}$$

The kets $| 1 : b_i ; 2 : b_j ; 3 : b_k \rangle$ are common eigenvectors of the extensions of $B(1)$, $B(2)$ and $B(3)$ in $\mathscr{E}$, with respective eigenvalues b_i, b_j and b_k.

Since the three particles are identical, we cannot measure $B(1)$ or $B(2)$ or $B(3)$, since the numbering has no physical significance. However, we can measure the physical quantity B for each of the three particles. Suppose that such a measurement has resulted in three different eigenvalues, b_n, b_p and b_q. Exchange degeneracy then appears, since the state of the system after this measurement can, *a priori*, be represented by any one of the kets of the subspace of $\mathscr{E}$ spanned by the six basis vectors:

$$\begin{array}{lll} | 1 : b_n ; 2 : b_p ; 3 : b_q \rangle, & | 1 : b_q ; 2 : b_n ; 3 : b_p \rangle, & | 1 : b_p ; 2 : b_q ; 3 : b_n \rangle, \\ | 1 : b_n ; 2 : b_q ; 3 : b_p \rangle, & | 1 : b_p ; 2 : b_n ; 3 : b_q \rangle, & | 1 : b_q ; 2 : b_p ; 3 : b_n \rangle \end{array} \tag{A-12}$$

Therefore, *a complete measurement on each of the particles does not permit the determination of a unique ket* of the state space of the system.

COMMENT:

The indeterminacy due to exchange degeneracy is, of course, less important if two of the eigenvalues found in the measurement are equal. This indeterminacy disappears in the special case in which the three results are identical.

B. PERMUTATION OPERATORS

Before stating the additional postulate which enables us to remove the indeterminacy related to exchange degeneracy, we shall study certain operators, defined in the total state space of the system under consideration, which actually permute the various particles of the system. The use of these permutation operators will simplify the calculations and reasoning in §§ C and D.

1. Two-particle systems

a. DEFINITION OF THE PERMUTATION OPERATOR P_{21}

Consider a system composed of two particles with the same spin s. Here it is not necessary for these two particles to be identical; it is sufficient that their individual state spaces be isomorphic. Therefore, to avoid the problems which arise when the two particles are identical, we shall assume that they are not: the numbers (1) and (2) with which they are labeled indicate their natures. For example, (1) will denote a proton and (2), an electron.

We choose a basis, $\{ | u_i \rangle \}$, in the state space $\mathscr{E}(1)$ of particle (1). Since the two particles have the same spin, $\mathscr{E}(2)$ is isomorphic to $\mathscr{E}(1)$, and it can be spanned by the same basis. By taking the tensor product, we construct, in the state space $\mathscr{E}$ of the system, the basis:

$$\{ | 1 : u_i ; 2 : u_j \rangle \} \tag{B-1}$$

Since the order of the vectors is of no importance in a tensor product, we have:

$$| 2 : u_j ; 1 : u_i \rangle \equiv | 1 : u_i ; 2 : u_j \rangle \tag{B-2}$$

However, note that:

$$| 1 : u_j ; 2 : u_i \rangle \neq | 1 : u_i ; 2 : u_j \rangle \qquad \text{if} \quad i \neq j \tag{B-3}$$

The permutation operator P_{21} is then defined as the linear operator whose action on the basis vectors is given by:

$$P_{21} | 1 : u_i ; 2 : u_j \rangle = | 2 : u_i ; 1 : u_j \rangle = | 1 : u_j ; 2 : u_i \rangle \tag{B-4}$$

Its action on any ket of $\mathscr{E}$ can easily be obtained by expanding this ket on the basis (B-1)*.

COMMENT:

If we choose the basis formed by the common eigenstates of the position observable $\mathbf{R}$ and the spin component S_z, (B-4) can be written:

$$P_{21} | 1 : \mathbf{r}, \varepsilon ; 2 : \mathbf{r}', \varepsilon' \rangle = | 1 : \mathbf{r}', \varepsilon' ; 2 : \mathbf{r}, \varepsilon \rangle \tag{B-5}$$

Any ket $| \psi \rangle$ of the state space $\mathscr{E}$ can be represented by a set of $(2s + 1)^2$ functions of six variables:

$$| \psi \rangle = \sum_{\varepsilon, \varepsilon'} \int d^3r \, d^3r' \, \psi_{\varepsilon, \varepsilon'}(\mathbf{r}, \mathbf{r}') \, | 1 : \mathbf{r}, \varepsilon ; 2 : \mathbf{r}', \varepsilon' \rangle \tag{B-6}$$

with:

$$\psi_{\varepsilon, \varepsilon'}(\mathbf{r}, \mathbf{r}') = \langle 1 : \mathbf{r}, \varepsilon ; 2 : \mathbf{r}', \varepsilon' | \psi \rangle \tag{B-7}$$

* It can easily be shown that the operator P_{21} so defined does not depend on the $\{ | u_i \rangle \}$ basis chosen.

We then have:

$$P_{21} | \psi \rangle = \sum_{\varepsilon,\varepsilon'} \int d^3r \, d^3r' \, \psi_{\varepsilon,\varepsilon'}(\mathbf{r}, \mathbf{r}') \, | 1 : \mathbf{r}', \varepsilon' ; 2 : \mathbf{r}, \varepsilon \rangle \tag{B-8}$$

By changing the names of the dummy variables:

$$\begin{aligned} &\varepsilon \longleftrightarrow \varepsilon' \\ &\mathbf{r} \longleftrightarrow \mathbf{r}' \end{aligned} \tag{B-9}$$

we transform formula (B-8) into:

$$P_{21} | \psi \rangle = \sum_{\varepsilon,\varepsilon'} \int d^3r \, d^3r' \, \psi_{\varepsilon',\varepsilon}(\mathbf{r}', \mathbf{r}) \, | 1 : \mathbf{r}, \varepsilon ; 2 : \mathbf{r}', \varepsilon' \rangle \tag{B-10}$$

Consequently, the functions:

$$\psi'_{\varepsilon,\varepsilon'}(\mathbf{r}, \mathbf{r}') = \langle 1 : \mathbf{r}, \varepsilon ; 2 : \mathbf{r}', \varepsilon' | P_{21} | \psi \rangle \tag{B-11}$$

which represent the ket $| \psi' \rangle = P_{21} | \psi \rangle$ can be obtained from the functions (B-7) which represent the ket $| \psi \rangle$ by inverting $(\mathbf{r}, \varepsilon)$ and $(\mathbf{r}', \varepsilon')$:

$$\psi'_{\varepsilon,\varepsilon'}(\mathbf{r}, \mathbf{r}') = \psi_{\varepsilon',\varepsilon}(\mathbf{r}', \mathbf{r}) \tag{B-12}$$

b. PROPERTIES OF P_{21}

We see directly from definition (B-4) that:

$$(P_{21})^2 = 1 \tag{B-13}$$

The operator P_{21} is its own inverse.

It can easily be shown that P_{21} *is Hermitian*:

$$P_{21}^\dagger = P_{21} \tag{B-14}$$

The matrix elements of P_{21} in the $\{ | 1 : u_i ; 2 : u_j \rangle \}$ basis are:

$$\begin{aligned} \langle 1 : u_{i'} ; 2 : u_{j'} | P_{21} | 1 : u_i ; 2 : u_j \rangle &= \langle 1 : u_{i'} ; 2 : u_{j'} | 1 : u_j ; 2 : u_i \rangle \\ &= \delta_{i'j} \, \delta_{j'i} \end{aligned} \tag{B-15}$$

Those of $P_{21}^\dagger$ are, by definition:

$$\begin{aligned} \langle 1 : u_{i'} ; 2 : u_{j'} | P_{21}^\dagger | 1 : u_i ; 2 : u_j \rangle &= (\langle 1 : u_i ; 2 : u_j | P_{21} | 1 : u_{i'} ; 2 : u_{j'} \rangle)^* \\ &= (\langle 1 : u_i ; 2 : u_j | 1 : u_{j'} ; 2 : u_{i'} \rangle)^* \\ &= \delta_{ij'} \, \delta_{ji'} \end{aligned} \tag{B-16}$$

Each of the matrix elements of $P_{21}^\dagger$ is therefore equal to the corresponding matrix element of P_{21}. This leads to relation (B-14).

It follows from (B-13) and (B-14) that P_{21} is also *unitary*:

$$P_{21}^\dagger P_{21} = P_{21} P_{21}^\dagger = 1 \tag{B-17}$$

c. SYMMETRIC AND ANTISYMMETRIC KETS. SYMMETRIZER AND ANTISYMMETRIZER

According to (B-14), the eigenvalues of P_{21} must be real. Since, according to (B-13), their squares are equal to 1, these eigenvalues are simply $+1$ and -1. The eigenvectors of P_{21} associated with the eigenvalue $+1$ are called *symmetric*, those corresponding to the eigenvalue -1, *antisymmetric*:

$$\begin{array}{ll} P_{21} | \psi_S \rangle = | \psi_S \rangle & \Longrightarrow | \psi_S \rangle \text{ symmetric} \\ P_{21} | \psi_A \rangle = - | \psi_A \rangle & \Longrightarrow | \psi_A \rangle \text{ antisymmetric} \end{array} \tag{B-18}$$

Now consider the two operators:

$$S = \frac{1}{2}(1 + P_{21}) \tag{B-19-a}$$

$$A = \frac{1}{2}(1 - P_{21}) \tag{B-19-b}$$

These operators are *projectors*, since (B-13) implies that:

$$S^2 = S \tag{B-20-a}$$

$$A^2 = A \tag{B-20-b}$$

and, in addition, (B-14) enables us to show that:

$$S^\dagger = S \tag{B-21-a}$$

$$A^\dagger = A \tag{B-21-b}$$

S and A are projectors onto orthogonal subspaces, since, according to (B-13):

$$SA = AS = 0 \tag{B-22}$$

These subspaces are supplementary, since definitions (B-19) yield:

$$S + A = 1 \tag{B-23}$$

If $| \psi \rangle$ is an arbitrary ket of the state space $\mathscr{E}$, $S | \psi \rangle$ is a symmetric ket and $A | \psi \rangle$, an antisymmetric ket, as it is easy to see, using (B-13) again, that:

$$\begin{array}{l} P_{21} S | \psi \rangle = S | \psi \rangle \\ P_{21} A | \psi \rangle = - A | \psi \rangle \end{array} \tag{B-24}$$

For this reason, S and A are called, respectively, a *symmetrizer* and an *antisymmetrizer*.

COMMENT:

The same symmetric ket is obtained by applying S to $P_{21} | \psi \rangle$ or to $| \psi \rangle$ itself:

$$SP_{21} | \psi \rangle = S | \psi \rangle \tag{B-25}$$

For the antisymmetrizer, we have, similarly:

$$AP_{21} | \psi \rangle = - A | \psi \rangle \tag{B-26}$$

d. TRANSFORMATION OF OBSERVABLES BY PERMUTATION

Consider an observable $B(1)$, initially defined in $\mathscr{E}(1)$ and then extended into $\mathscr{E}$. It is always possible to construct the $\{ | u_i \rangle \}$ basis in $\mathscr{E}(1)$ from eigenvectors of $B(1)$ (the corresponding eigenvalues will be written b_i). Let us now calculate the action of the operator $P_{21}B(1)P_{21}^{\dagger}$ on an arbitrary basis ket of $\mathscr{E}$:

$$\begin{aligned} P_{21}B(1)P_{21}^{\dagger} \,|\, 1 : u_i ; 2 : u_j \rangle &= P_{21}B(1) \,|\, 1 : u_j ; 2 : u_i \rangle \\ &= b_j P_{21} \,|\, 1 : u_j ; 2 : u_i \rangle \\ &= b_j \,|\, 1 : u_i ; 2 : u_j \rangle \end{aligned} \tag{B-27}$$

We would obtain the same result by applying the observable $B(2)$ directly to the basis ket chosen. Consequently:

$$P_{21}B(1)P_{21}^{\dagger} = B(2) \tag{B-28}$$

The same reasoning shows that:

$$P_{21}B(2)P_{21}^{\dagger} = B(1) \tag{B-29}$$

In $\mathscr{E}$, there are also observables, such as $B(1) + C(2)$ or $B(1)C(2)$, which involve both indices simultaneously. We obviously have:

$$P_{21}[B(1) + C(2)]P_{21}^{\dagger} = B(2) + C(1) \tag{B-30}$$

Similarly, using (B-17), we find:

$$\begin{aligned} P_{21}B(1)C(2)P_{21}^{\dagger} &= P_{21}B(1)P_{21}^{\dagger}P_{21}C(2)P_{21}^{\dagger} \\ &= B(2)C(1) \end{aligned} \tag{B-31}$$

These results can be generalized to all observables in $\mathscr{E}$ which can be expressed in terms of observables of the type of $B(1)$ and $C(2)$, to be denoted by $\mathcal{O}(1, 2)$:

$$P_{21}\mathcal{O}(1, 2)P_{21}^{\dagger} = \mathcal{O}(2, 1) \tag{B-32}$$

$\mathcal{O}(2, 1)$ is the observable obtained from $\mathcal{O}(1, 2)$ by exchanging indices 1 and 2 throughout.

An observable $\mathcal{O}_S(1, 2)$ is said to be *symmetric* if:

$$\mathcal{O}_S(2, 1) = \mathcal{O}_S(1, 2) \tag{B-33}$$

According to (B-32), all symmetric observables satisfy:

$$P_{21}\mathcal{O}_S(1, 2) = \mathcal{O}_S(1, 2)P_{21} \tag{B-34}$$

that is :

$$[\mathcal{O}_S(1, 2), P_{21}] = 0 \tag{B-35}$$

Symmetric observables commute with the permutation operator.

2. Systems containing an arbitrary number of particles

In the state space of a system composed of N particles with the same spin (temporarily assumed to be of different natures), $N!$ permutation operators can be defined (one of which is the identity operator). If N is greater than 2, the properties of these operators are more complex than those of P_{21}. To have an idea of the changes involved when N is greater than 2, we shall briefly study the case in which $N = 3$.

a. DEFINITION OF THE PERMUTATION OPERATORS

Consider, therefore, a system of three particles which are not necessarily identical but have the same spin. As in § B-1-a, we construct a basis of the state space of the system by taking a tensor product:

$$\{ | 1 : u_i ; 2 : u_j ; 3 : u_k \rangle \} \tag{B-36}$$

In this case, there exist six permutation operators, which we shall denote by:

$$P_{123}, P_{312}, P_{231}, P_{132}, P_{213}, P_{321} \tag{B-37}$$

By definition, the operator P_{npq} (where n, p, q is an arbitrary permutation of the numbers 1, 2, 3) is the linear operator whose action on the basis vectors obeys:

$$P_{npq} | 1 : u_i ; 2 : u_j ; 3 : u_k \rangle = | n : u_i ; p : u_j ; q : u_k \rangle \tag{B-38}$$

For example:

$$\begin{aligned} P_{231} | 1 : u_i ; 2 : u_j ; 3 : u_k \rangle &= | 2 : u_i ; 3 : u_j ; 1 : u_k \rangle \\ &= | 1 : u_k ; 2 : u_i ; 3 : u_j \rangle \end{aligned} \tag{B-39}$$

P_{123} therefore coincides with the identity operator. The action of P_{npq} on any ket of the state space can easily be obtained by expanding this ket on the basis (B-36).

The $N!$ permutation operators associated with a system of N particles with the same spin could be defined analogously.

b. PROPERTIES

α. *The set of permutation operators constitutes a group*

This can easily be shown for the operators (B-37):

(*i*) P_{123} is the identity operator.

(*ii*) The product of two permutation operators is also a permutation operator. We can show, for example, that:

$$P_{312} P_{132} = P_{321} \tag{B-40}$$

To do so, we apply the left-hand side to an arbitrary basis ket:

$$\begin{aligned} P_{312} P_{132} | 1 : u_i ; 2 : u_j ; 3 : u_k \rangle &= P_{312} | 1 : u_i ; 3 : u_j ; 2 : u_k \rangle \\ &= P_{312} | 1 : u_i ; 2 : u_k ; 3 : u_j \rangle \\ &= | 3 : u_i ; 1 : u_k ; 2 : u_j \rangle \\ &= | 1 : u_k ; 2 : u_j ; 3 : u_i \rangle \end{aligned} \tag{B-41}$$

The action of P_{321} effectively leads to the same result:

$$\begin{aligned} P_{321} \mid 1 : u_i ; 2 : u_j ; 3 : u_k \rangle &= \mid 3 : u_i ; 2 : u_j ; 1 : u_k \rangle \\ &= \mid 1 : u_k ; 2 : u_j ; 3 : u_i \rangle \end{aligned} \tag{B-42}$$

(*iii*) Each permutation operator has an inverse, which is also a permutation operator. Reasoning as in (*ii*), we can easily show that:

$$\begin{aligned} P_{123}^{-1} &= P_{123} ; \quad P_{312}^{-1} = P_{231} ; \quad P_{231}^{-1} = P_{312} \\ P_{132}^{-1} &= P_{132} ; \quad P_{213}^{-1} = P_{213} ; \quad P_{321}^{-1} = P_{321} \end{aligned} \tag{B-43}$$

Note that *the permutation operators do not commute with each other.*

For example:

$$P_{132} P_{312} = P_{213} \tag{B-44}$$

which, compared to (B-40), shows that the commutator of P_{132} and P_{312} is not zero.

β. *Transpositions. Parity of a permutation operator*

A *transposition* is a permutation which simply exchanges the roles of two of the particles, without touching the others. Of the operators (B-37), the last three are transposition operators★. Transposition operators are Hermitian, and each of them is the same as its inverse, so that they are also unitary [the proofs of these properties are identical to those for (B-14), (B-13) and (B-17)].

Any permutation operator can be broken down into a product of transposition operators. For example, the second operator (B-37) can be written:

$$P_{312} = P_{132} P_{213} = P_{321} P_{132} = P_{213} P_{321} = P_{132} P_{213} (P_{132})^2 = \ldots \tag{B-45}$$

This decomposition is not unique. However, for a given permutation, it can be shown that the parity of the number of transpositions into which it can be broken down is always the same : it is called the *parity of the permutation.* Thus, the first three operators (B-37) are even, and the last three, odd. For any N, there are always as many even permutations as odd ones.

γ. *Permutation operators are unitary*

Permutation operators, which are products of transposition operators, all of which are unitary, are therefore also unitary. However, they are not necessarily Hermitian, since transposition operators do not generally commute with each other.

Finally, note that the adjoint of a given permutation operator has the same parity as that of the operator, since it is equal to the product of the same transposition operators, taken in the opposite order.

★ Of course, for $N = 2$, the only permutation possible is transposition.

c. COMPLETELY SYMMETRIC OR ANTISYMMETRIC KETS. SYMMETRIZER AND ANTISYMMETRIZER

Since the permutation operators do not commute for $N > 2$, it is not possible to construct a basis formed by common eigenvectors of these operators. Nevertheless, we shall see that there exist certain kets which are simultaneously eigenvectors of all the permutation operators.

We shall denote by P_α an arbitrary permutation operator associated with a system of N particles with the same spin; α represents an arbitrary permutation of the first N integers. A ket $| \psi_S \rangle$ such that:

$$P_\alpha | \psi_S \rangle = | \psi_S \rangle \tag{B-46}$$

for any permutation P_α, is said to be *completely symmetric*. Similarly, a *completely antisymmetric* ket $| \psi_A \rangle$ satisfies, by definition*:

$$P_\alpha | \psi_A \rangle = \varepsilon_\alpha | \psi_A \rangle \tag{B-47}$$

where:

$$\begin{aligned} \varepsilon_\alpha &= +1 \quad \text{if } P_\alpha \text{ is an even permutation} \\ \varepsilon_\alpha &= -1 \quad \text{if } P_\alpha \text{ is an odd permutation} \end{aligned} \tag{B-48}$$

The set of completely symmetric kets constitutes a vector subspace $\mathscr{E}_S$ of the state space $\mathscr{E}$; the set of completely antisymmetric kets, a subspace $\mathscr{E}_A$.

Now consider the two operators:

$$S = \frac{1}{N!} \sum_\alpha P_\alpha \tag{B-49}$$

$$A = \frac{1}{N!} \sum_\alpha \varepsilon_\alpha P_\alpha \tag{B-50}$$

where the summations are performed over the $N!$ permutations of the first N integers, and ε_α is defined by (B-48). We shall show that S and A are the projectors onto $\mathscr{E}_S$ and $\mathscr{E}_A$ respectively. For this reason, they are called a *symmetrizer* and an *antisymmetrizer*.

S and A are Hermitian operators:

$$S^\dagger = S \tag{B-51}$$

$$A^\dagger = A \tag{B-52}$$

The adjoint $P_\alpha^\dagger$ of a given permutation operator is, as we saw above (*cf.* §B-2-b-γ), another permutation operator, of the same parity (which coincides, furthermore, with P_α^{-1}). Taking the adjoints of the right-hand sides of the definitions of S and A therefore amounts simply to changing the order of the terms in the summations (since the set of the P_α^{-1} is again the permutation group).

* According to the property stated in § B-2-b-β, this definition can also be based solely on the transposition operators : any transposition operator leaves a completely symmetric ket invariant and transforms a completely antisymmetric ket into its opposite.

Also, if P_{α_0} is an arbitrary permutation operator, we have:

$$P_{\alpha_0}S = SP_{\alpha_0} = S \tag{B-53-a}$$

$$P_{\alpha_0}A = AP_{\alpha_0} = \varepsilon_{\alpha_0}A \tag{B-53-b}$$

This is due to the fact that $P_{\alpha_0}P_\alpha$ is also a permutation operator:

$$P_{\alpha_0}P_\alpha = P_\beta \tag{B-54}$$

such that:

$$\varepsilon_\beta = \varepsilon_{\alpha_0}\varepsilon_\alpha \tag{B-55}$$

If, for P_{α_0} fixed, we choose successively for P_α all the permutations of the group, we see that the P_β are each identical to one and only one of these permutations (in, of course, a different order). Consequently:

$$P_{\alpha_0}S = \frac{1}{N!}\sum_\alpha P_{\alpha_0}P_\alpha = \frac{1}{N!}\sum_\beta P_\beta = S \tag{B-56-a}$$

$$P_{\alpha_0}A = \frac{1}{N!}\sum_\alpha \varepsilon_\alpha P_{\alpha_0}P_\alpha = \frac{1}{N!}\varepsilon_{\alpha_0}\sum_\beta \varepsilon_\beta P_\beta = \varepsilon_{\alpha_0}A \tag{B-56-b}$$

Similarly, we could prove analogous relations in which S and A are multiplied by P_{α_0} from the right.

From (B-53), we see that:

$$\begin{aligned} S^2 &= S \\ A^2 &= A \end{aligned} \tag{B-57}$$

and, moreover:

$$AS = SA = 0 \tag{B-58}$$

This is because:

$$\begin{aligned} S^2 &= \frac{1}{N!}\sum_\alpha P_\alpha S = \frac{1}{N!}\sum_\alpha S = S \\ A^2 &= \frac{1}{N!}\sum_\alpha \varepsilon_\alpha P_\alpha A = \frac{1}{N!}\sum_\alpha \varepsilon_\alpha^2 A = A \end{aligned} \tag{B-59}$$

as each summation includes $N!$ terms; furthermore:

$$AS = \frac{1}{N!}\sum_\alpha \varepsilon_\alpha P_\alpha S = \frac{1}{N!} S\sum_\alpha \varepsilon_\alpha = 0 \tag{B-60}$$

since half the ε_α are equal to $+1$ and half equal to -1 (*cf.* § B-2-b-β).

S and A are therefore *projectors*. They project respectively onto $\mathscr{E}_S$ and $\mathscr{E}_A$ since, according to (B-53), their action on any ket $|\psi\rangle$ of the state space yields a completely symmetric or completely antisymmetric ket:

$$P_{\alpha_0}\, S\,|\psi\rangle = S\,|\psi\rangle \tag{B-61-a}$$

$$P_{\alpha_0}\, A\,|\psi\rangle = \varepsilon_{\alpha_0}A\,|\psi\rangle \tag{B-61-b}$$

COMMENTS:

(*i*) The completely symmetric ket constructed by the action of S on $P_\alpha | \psi \rangle$, where P_α is an arbitrary permutation, is the same as that obtained from $| \psi \rangle$, since expressions (B-53) indicate that:

$$SP_\alpha | \psi \rangle = S | \psi \rangle \tag{B-62}$$

As for the corresponding completely antisymmetric kets, they differ at most by their signs:

$$AP_\alpha | \psi \rangle = \varepsilon_\alpha A | \psi \rangle \tag{B-63}$$

(*ii*) For $N > 2$, the symmetrizer and antisymmetrizer are not projectors onto supplementary subspaces. For example, when $N = 3$, it is easy to obtain [by using the fact that the first three permutations (B-37) are even and the others odd] the relation:

$$S + A = \frac{1}{3}(P_{123} + P_{231} + P_{312}) \neq 1 \tag{B-64}$$

In other words, the state space is not the direct sum of the subspace $\mathscr{E}_S$ of completely symmetric kets and the subspace $\mathscr{E}_A$ of completely antisymmetric kets.

d. TRANSFORMATION OF OBSERVABLES BY PERMUTATION

We have indicated (§ B-2-b-β) that any permutation operator of an N-particle system can be broken down into a product of transposition operators analogous to the operator P_{21} studied in § B-1. For these transposition operators, we can use the arguments of §B-1-d to determine the behavior of the various observables of the system when they are multiplied from the left by an arbitrary permutation operator P_α and from the right by $P_\alpha^\dagger$.

In particular, the observables $\mathscr{O}_S(1, 2, ..., N)$, which are completely symmetric under exchange of the indices 1, 2, ..., N, commute with all the transposition operators, and, therefore, with all the permutation operators:

$$[\mathscr{O}_S(1, 2, ..., N), P_\alpha] = 0 \tag{B-65}$$

C. THE SYMMETRIZATION POSTULATE

1. Statement of the postulate

When a system includes several identical particles, only certain kets of its state space can describe its physical states. Physical kets are, depending on the nature of the identical particles, either completely symmetric or completely antisymmetric with respect to permutation of these particles. Those particles for which the physical kets are symmetric are called *bosons*, and those for which they are antisymmetric, *fermions*.

The symmetrization postulate thus limits the state space for a system of identical particles. This space is no longer, as it was in the case of particles of different natures, the tensor product $\mathscr{E}$ of the individual state spaces of the particles constituting the system. It is only a subspace of $\mathscr{E}$, $\mathscr{E}_S$ or $\mathscr{E}_A$, depending on whether the particles are bosons or fermions.

From the point of view of this postulate, particles existing in nature are divided into two categories. All currently known particles obey the following *empirical rule*★: particles of half-integral spin (electrons, positrons, protons, neutrons, muons, etc.) are fermions, and particles of integral spin (photons, mesons, etc.) are bosons.

COMMENT:

Once this rule has been verified for the particles which are called "elementary", it holds for all other particles as well, inasmuch as they are composed of these elementary particles. Consider a system of many identical composite particles. Permuting two of them is equivalent to simultaneously permuting all the particles composing the first one with the corresponding particles (necessarily identical to the aforementioned ones) of the second one. This permutation must leave the ket describing the state of the system unchanged if the composite particles being studied are formed only of elementary bosons or if each of them contains an even number of fermions (no sign change, or an even number of sign changes); in this case, the particles are bosons. On the other hand, composite particles containing an odd number of fermions are themselves fermions (an odd number of sign changes in the permutation). Now, the spin of these composite particles is necessarily integral in the first case and half-integral in the second one (chap. X, § C-3-c). They therefore obey the rule just stated. For example, atomic nuclei are known to be composed of neutrons and protons, which are fermions (spin 1/2). Consequently, nuclei whose mass number A (the total number of nucleons) is even are bosons, and those whose mass number is odd are fermions. Thus, the nucleus of the ^{3}He isotope of helium is a fermion, and that of the ^{4}He isotope, a boson.

2. Removal of exchange degeneracy

We shall begin by examining how this new postulate removes the exchange degeneracy and the corresponding difficulties.

The discussion of § A can be summarized in the following way. Let $|u\rangle$ be a ket which can mathematically describe a well-defined physical state of a system containing N identical particles. For any permutation operator P_α, $P_\alpha|u\rangle$ can describe this physical state as well as $|u\rangle$. The same is true for any ket belonging to the subspace $\mathscr{E}_u$ spanned by $|u\rangle$ and all its permutations $P_\alpha|u\rangle$. Depending on the ket $|u\rangle$ chosen, the dimension of $\mathscr{E}_u$ can vary between 1 and $N!$. If this dimension is greater than 1, several mathematical kets correspond to the same physical state: there is then an exchange degeneracy.

★ The "spin-statistics theorem", proven in quantum field theory, makes it possible to consider this rule to be a consequence of very general hypotheses. However, these hypotheses may not all be correct, and discovery of a boson of half-integral spin or a fermion of integral spin remains possible. It is not inconceivable that, for certain particles, the physical kets might have more complex symmetry properties than those envisaged here.

The new postulate which we have introduced considerably restricts the class of mathematical kets able to describe a physical state : these kets must belong to $\mathscr{E}_S$ for bosons and to $\mathscr{E}_A$ for fermions. We shall be able to say that the difficulties related to exchange degeneracy are eliminated if we can show that $\mathscr{E}_u$ contains *a single* ket of $\mathscr{E}_S$ or *a single* ket of $\mathscr{E}_A$.

To do so, we shall use the relations $S = SP_\alpha$ or $A = \varepsilon_\alpha AP_\alpha$, proven in (B-53). We obtain:

$$S \mid u \rangle = SP_\alpha \mid u \rangle \tag{C-1-a}$$

$$A \mid u \rangle = \varepsilon_\alpha AP_\alpha \mid u \rangle \tag{C-1-b}$$

These relations express the fact that the projections onto $\mathscr{E}_S$ and $\mathscr{E}_A$ of the various kets which span $\mathscr{E}_u$ and, consequently, of all the kets of $\mathscr{E}_u$, are collinear. The symmetrization postulate thus unambiguously indicates (to within a constant factor) *the* ket of $\mathscr{E}_u$ which must be associated with the physical state considered : $S \mid u \rangle$ for bosons and $A \mid u \rangle$ for fermions. This ket is called the *physical ket*.

COMMENT:

It is possible for all the kets of $\mathscr{E}_u$ to have a zero projection onto $\mathscr{E}_A$ (or $\mathscr{E}_S$). In this case, the symmetrization postulate excludes the corresponding physical state. Later (§§3-b and 3-c), we shall see examples of such a situation when dealing with fermions.

3. Construction of physical kets

a. THE CONSTRUCTION RULE

The discussion of the preceding section leads directly to the following rule for the construction of the *unique* ket (the physical ket) corresponding to a given physical state of a system of N identical particles:

(*i*) Number the particles arbitrarily, and construct the ket $\mid u \rangle$ corresponding to the physical state considered and to the numbers given to the particles.

(*ii*) Apply S or A to $\mid u \rangle$, depending on whether the identical particles are bosons or fermions.

(*iii*) Normalize the ket so obtained.

We shall describe some simple examples which illustrate this rule.

b. APPLICATION TO SYSTEMS OF TWO IDENTICAL PARTICLES

Consider a system composed of two identical particles. Suppose that one of them is known to be in the individual state characterized by the normalized ket $\mid \varphi \rangle$, and the other one, in the individual state characterized by the normalized ket $\mid \chi \rangle$.

First of all, we shall envisage the case in which the two kets, $\mid \varphi \rangle$ and $\mid \chi \rangle$, are distinct. The preceding rule is applied in the following way:

(*i*) We label with the number 1, for example, the particle in the state $| \varphi \rangle$, and with the number 2, the one in the state $| \chi \rangle$. This gives:

$$| u \rangle = | 1 : \varphi ; 2 : \chi \rangle \tag{C-2}$$

(*ii*) We symmetrize $| u \rangle$ if the particles are bosons:

$$S | u \rangle = \frac{1}{2} [| 1 : \varphi ; 2 : \chi \rangle + | 1 : \chi ; 2 : \varphi \rangle] \tag{C-3-a}$$

We antisymmetrize $| u \rangle$ if the particles are fermions:

$$A | u \rangle = \frac{1}{2} [| 1 : \varphi ; 2 : \chi \rangle - | 1 : \chi ; 2 : \varphi \rangle] \tag{C-3-b}$$

(*iii*) The kets (C-3-a) and (C-3-b), in general, are not normalized. If we assume $| \varphi \rangle$ and $| \chi \rangle$ to be orthogonal, the normalization constant is very simple to calculate. All we have to do to normalize $S | u \rangle$ or $A | u \rangle$ is replace the factor 1/2 appearing in formulas (C-3) by $1/\sqrt{2}$. The normalized physical ket, in this case, can therefore be written:

$$| \varphi ; \chi \rangle = \frac{1}{\sqrt{2}} [| 1 : \varphi ; 2 : \chi \rangle + \varepsilon | 1 : \chi ; 2 : \varphi \rangle] \tag{C-4}$$

with $\varepsilon = +1$ for bosons and -1 for fermions.

We shall now assume that the two individual states, $| \varphi \rangle$ and $| \chi \rangle$, are identical:

$$| \varphi \rangle = | \chi \rangle \tag{C-5}$$

(C-2) then becomes:

$$| u \rangle = | 1 : \varphi ; 2 : \varphi \rangle \tag{C-6}$$

$| u \rangle$ is already symmetric. If the two particles are bosons, (C-6) is then the physical ket associated with the state in which the two bosons are in the same individual state $| \varphi \rangle$. If, on the other hand, the two particles are fermions, we see that:

$$A | u \rangle = \frac{1}{2} [| 1 : \varphi ; 2 : \varphi \rangle - | 1 : \varphi ; 2 : \varphi \rangle] = 0 \tag{C-7}$$

Consequently, there exists no ket of $\mathscr{E}_A$ able to describe the physical state in which two fermions are in the same individual state $| \varphi \rangle$. Such a physical state is therefore excluded by the symmetrization postulate. We have thus established, for a special case, a fundamental result known as "*Pauli's exclusion principle*": *two identical fermions cannot be in the same individual state*. This result has some very important physical consequences which we shall discuss in § D-1.

c. GENERALIZATION TO AN ARBITRARY NUMBER OF PARTICLES

These ideas can be generalized to an arbitrary number N of particles. To see how this can be done, we shall first treat the case $N = 3$.

Consider a physical state of the system defined by specifying the three individual normalized states $|\varphi\rangle$, $|\chi\rangle$ and $|\omega\rangle$. The state $|u\rangle$ which enters into the rule of § a can be chosen in the form:

$$|u\rangle = |1:\varphi; 2:\chi; 3:\omega\rangle \tag{C-8}$$

We shall discuss the cases of bosons and fermions separately.

α. *The case of bosons*

The application of S to $|u\rangle$ gives:

$$\begin{aligned} S|u\rangle &= \frac{1}{3!}\sum_\alpha P_\alpha |u\rangle \\ &= \frac{1}{6}[|1:\varphi;2:\chi;3:\omega\rangle + |1:\omega;2:\varphi;3:\chi\rangle + |1:\chi;2:\omega;3:\varphi\rangle \\ &+ |1:\varphi;2:\omega;3:\chi\rangle + |1:\chi;2:\varphi;3:\omega\rangle + |1:\omega;2:\chi;3:\varphi\rangle] \end{aligned} \tag{C-9}$$

It then suffices to normalize the ket (C-9).

First of all, let us assume that the three kets $|\varphi\rangle$, $|\chi\rangle$ and $|\omega\rangle$ are orthogonal. The six kets appearing on the right-hand side of (C-9) are then also orthogonal. To normalize (C-9), all we must do is replace the factor $1/6$ by $1/\sqrt{6}$.

If the two states $|\varphi\rangle$ and $|\chi\rangle$ coincide, while remaining orthogonal to $|\omega\rangle$, only three distinct kets now appear on the right-hand side of (C-9). It can easily be shown that the normalized physical ket can then be written:

$$\begin{aligned} |\varphi;\varphi;\omega\rangle = \frac{1}{\sqrt{3}}[&|1:\varphi;2:\varphi;3:\omega\rangle \\ &+ |1:\varphi;2:\omega;3:\varphi\rangle + |1:\omega;2:\varphi;3:\varphi\rangle] \end{aligned} \tag{C-10}$$

Finally, if the three states $|\varphi\rangle$, $|\chi\rangle$, $|\omega\rangle$ are the same, the ket:

$$|u\rangle = |1:\varphi;2:\varphi;3:\varphi\rangle \tag{C-11}$$

is already symmetric and normalized.

β. *The case of fermions*

The application of A to $|u\rangle$ gives:

$$A|u\rangle = \frac{1}{3!}\sum_\alpha \varepsilon_\alpha P_\alpha |1:\varphi;2:\chi;3:\omega\rangle \tag{C-12}$$

The signs of the various terms of the sum (C-12) are determined by the same rule as those of a 3 × 3 determinant. This is why it is convenient to write $A|u\rangle$ in the form of a *Slater determinant:*

$$A|u\rangle = \frac{1}{3!}\begin{vmatrix} |1:\varphi\rangle & |1:\chi\rangle & |1:\omega\rangle \\ |2:\varphi\rangle & |2:\chi\rangle & |2:\omega\rangle \\ |3:\varphi\rangle & |3:\chi\rangle & |3:\omega\rangle \end{vmatrix} \tag{C-13}$$

$A \mid u \rangle$ is zero if two of the individual states $\mid \varphi \rangle$, $\mid \chi \rangle$ or $\mid \omega \rangle$ coincide, since the determinant (C-13) then has two identical columns. We obtain Pauli's exclusion principle, already mentioned in §C-3-b: the same quantum mechanical state cannot be simultaneously occupied by several identical fermions.

Finally, note that if the three states $\mid \varphi \rangle$, $\mid \chi \rangle$, $\mid \omega \rangle$ are orthogonal, the six kets appearing on the right-hand side of (C-12) are orthogonal. All we must then do to normalize $A \mid u \rangle$ is replace the factor $1/3!$ appearing in (C-12) or (C-13) by $1/\sqrt{3!}$.

If, now, the system being considered contains more than three identical particles, the situation actually remains similar to the one just described. It can be shown that, for N identical bosons, it is always possible to construct the physical state $S \mid u \rangle$ from arbitrary individual states $\mid \varphi \rangle$, $\mid \chi \rangle$, ... On the other hand, for fermions, the physical ket $A \mid u \rangle$ can be written in the form of an $N \times N$ Slater determinant; this excludes the case in which two individual states coincide (the ket $A \mid u \rangle$ is then zero). This shows, and we shall return to this in detail in §D, how different the consequences of the new postulate can be for fermion and boson systems.

d. CONSTRUCTION OF A BASIS IN THE PHYSICAL STATE SPACE

Consider a system of N identical particles. Starting with a basis, $\{ \mid u_i \rangle \}$, in the state space of a single particle, we can construct the basis:

$$\{ \mid 1 : u_i ; 2 : u_j ; \ldots ; N : u_p \rangle \}$$

in the tensor product space $\mathscr{E}$. However, since the physical state space of the system is not $\mathscr{E}$, but, rather, one of the subspaces, $\mathscr{E}_S$ or $\mathscr{E}_A$, the problem arises of how to determine a basis in this physical state space.

By application of S (or A) to the various kets of the basis:

$$\{ \mid 1 : u_i ; 2 : u_j ; \ldots ; N : u_p \rangle \}$$

we can obtain a set of vectors spanning $\mathscr{E}_S$ (or $\mathscr{E}_A$). Let $\mid \varphi \rangle$ be an arbitrary ket of $\mathscr{E}_S$, for example (the case in which $\mid \varphi \rangle$ belongs to $\mathscr{E}_A$ can be treated in the same way). $\mid \varphi \rangle$, which belongs to $\mathscr{E}$, can be expanded in the form:

$$\mid \varphi \rangle = \sum_{i,j,\ldots,p} a_{i,j,\ldots,p} \mid 1 : u_i ; 2 : u_j ; \ldots N : u_p \rangle \tag{C-14}$$

Since $\mid \varphi \rangle$, by hypothesis, belongs to $\mathscr{E}_S$, we have $S \mid \varphi \rangle = \mid \varphi \rangle$, and we simply apply the operator S to both sides of (C-14) to show that $\mid \varphi \rangle$ can be expressed in the form of a linear combination of the various kets $S \mid 1 : u_i ; 2 : u_j ; \ldots ; N : u_p \rangle$.

However, it must be noted that the various kets $S \mid 1 : u_i ; 2 : u_j ; \ldots ; N : u_p \rangle$ are not independent. Let us permute the roles of the various particles in one of the kets $\mid 1 : u_i ; 2 : u_j ; \ldots ; N : u_p \rangle$ of the initial basis (before symmetrization). On this new ket, application of S or A leads, according to (B-62) and (B-63), to the same ket of $\mathscr{E}_S$ or $\mathscr{E}_A$ (possibly with a change of sign).

Thus, we are led to introduce the concept of an *occupation number*: by definition, for the ket $\mid 1 : u_i ; 2 : u_j ; \ldots ; N : u_p \rangle$, the occupation number n_k of the individual state $\mid u_k \rangle$ is equal to the number of times that the state $\mid u_k \rangle$ appears in the sequence $\{ | u_i \rangle, | u_j \rangle \dots | u_p \rangle \}$, that is, the number of particles in the state $| u_k \rangle$ $\left(\text{we have, obviously, } \sum_k n_k = N \right)$. Two different kets $| 1 : u_i ; 2 : u_j ; \dots ; N : u_p \rangle$ for which the occupation numbers are equal can be obtained from each other by the action of a permutation operator. Consequently, after the action of the symmetrizer S (or the antisymmetrizer A), they give the same physical state, which we shall denote by $| n_1, n_2, \dots, n_k, \dots \rangle$:

$$| n_1, n_2, \dots, n_k, \dots \rangle$$
$$= c\, S \, | \underbrace{1 : u_1 ; 2 : u_1 ; \dots n_1 : u_1}_{\substack{n_1 \text{ particles} \\ \text{in the state } | u_1 \rangle}} ; \underbrace{n_1 + 1 : u_2 ; \dots ; n_1 + n_2 : u_2}_{\substack{n_2 \text{ particles} \\ \text{in the state } | u_2 \rangle}} ; \dots \rangle \tag{C-15}$$

For fermions, S would be replaced by A in (C-15) (c is a factor which permits the normalization of the state obtained in this way*). We shall not study the states $| n_1, n_2, \dots, n_k, \dots \rangle$ in detail here; we shall confine ourselves to giving some of their important properties:

(*i*) The scalar product of two kets $| n_1, n_2, \dots, n_k, \dots \rangle$ and $| n'_1, n'_2, \dots, n'_k, \dots \rangle$ is different from zero only if all the occupation numbers are equal ($n_k = n'_k$ for all k).

By using (C-15) and definitions (B-49) and (B-50) of S and A, we can obtain the expansion of the two kets under consideration on the orthonormal basis, $\{ | 1 : u_i ; 2 : u_j ; \dots ; N : u_p \rangle \}$. It is then easy to see that, if the occupation numbers are not all equal, these two kets cannot simultaneously have non-zero components on the same basis vector.

(*ii*) If the particles under study are bosons, the kets $| n_1, n_2, \dots, n_k, \dots \rangle$, in which the various occupation numbers n_k are arbitrary $\left(\text{with, of course } \sum_k n_k = N \right)$, form an orthonormal basis of the physical state space.

Let us show that, for bosons, the kets $| n_1, n_2, \dots, n_k, \dots \rangle$ defined by (C-15) are never zero. To do so, we replace S by its definition (B-49). There then appear, on the right-hand side of (C-15), various orthogonal kets $| 1 : u_i ; 2 : u_j ; \dots ; N : u_p \rangle$, all with positive coefficients. $| n_1, n_2, \dots, n_k, \dots \rangle$ cannot, therefore, be zero.

The $| n_1, n_2, \dots, n_k, \dots \rangle$ form a basis in $\mathscr{E}_S$ since these kets span $\mathscr{E}_S$, are all non-zero, and are orthogonal to each other.

(*iii*) If the particles under study are fermions, a basis of the physical state space $\mathscr{E}_A$ is obtained by choosing the set of kets $| n_1, n_2, \dots, n_k, \dots \rangle$ in which all the occupation numbers are equal either to 1 or to 0 $\left(\text{again with } \sum_k n_k = N \right)$.

* A simple calculation yields : $c = \sqrt{N!/n_1!\, n_2! \dots}$ for bosons and $\sqrt{N!}$ for fermions.

The preceding proof is not applicable to fermions because of the minus signs which appear before the odd permutations in definition (B-50) of A. Furthermore, we saw in §c that two identical fermions cannot occupy the same individual quantum state : if any one of the occupation numbers is greater than 1, the vector defined by (C-15) is zero. On the other hand, it is never zero if all the occupation numbers are equal to one or zero, since two particles are then never in the same individual quantum state, so that the kets $| 1 : u_i; 2 : u_j; ...; N : u_p \rangle$ and $P_\alpha | 1 : u_i; 2 : u_j; ...; N : u_p \rangle$ are always distinct and orthogonal. Relation (C-15) therefore defines a non-zero physical ket in this case. The rest of the proof is the same as for bosons.

4. Application of the other postulates

It remains for us to show how the general postulates of chapter III can be applied in light of the symmetrization postulate introduced in § C-1, and to verify that no contradictions arise. More precisely, we shall see how measurement processes can be described with kets belonging only to either $\mathscr{E}_S$ or $\mathscr{E}_A$, and we shall show that the time evolution process does not take the ket $| \psi(t) \rangle$ associated with the state of the system out of this subspace. Thus, all the quantum mechanical formalism can be applied inside either $\mathscr{E}_S$ or $\mathscr{E}_A$.

a. MEASUREMENT POSTULATES

α. *Probability of finding the system in a given physical state*

Consider a measurement performed on a system of identical particles. The ket $| \psi(t) \rangle$ describing the quantum state of the system before the measurement must, according to the symmetrization postulate, belong to $\mathscr{E}_S$ or to $\mathscr{E}_A$, depending on whether the system is formed of bosons or fermions. To apply the postulates of chapter III concerning measurements, we must take the scalar product of $| \psi(t) \rangle$ with the ket $| u \rangle$ corresponding to the physical state of the system after the measurement. This ket $| u \rangle$ is to be constructed by applying the rule given in §C-3-a. The probability amplitude $\langle u | \psi(t) \rangle$ can therefore be expressed in terms of two vectors, both belonging either to $\mathscr{E}_S$ or to $\mathscr{E}_A$. In §D-2, we shall discuss a certain number of examples of such calculations.

If the measurement envisaged is a "complete" measurement (yielding, for example, the positions and spin components S_z for all the particles), the physical ket $| u \rangle$ is unique (to within a constant factor). On the other hand, if the measurement is "incomplete" (for example, a measurement of the spins only, or a measurement bearing on a single particle), several orthogonal physical kets are obtained, and the corresponding probabilities must then be summed.

β. *Physical observables; invariance of $\mathscr{E}_S$ and $\mathscr{E}_A$*

In certain cases, it is possible to specify the measurement performed on the system of identical particles by giving the explicit expression of the corresponding observable in terms of $\mathbf{R}_1$, $\mathbf{P}_1$, $\mathbf{S}_1$, $\mathbf{R}_2$, $\mathbf{P}_2$, $\mathbf{S}_2$, etc.

We shall give some concrete examples of observables which can be measured in a three-particle system:

– Position of the center of mass $\mathbf{R}_G$, total momentum $\mathbf{P}$ and total angular momentum $\mathbf{L}$:

$$\mathbf{R}_G = \frac{1}{3}(\mathbf{R}_1 + \mathbf{R}_2 + \mathbf{R}_3) \tag{C-16}$$

$$\mathbf{P} = \mathbf{P}_1 + \mathbf{P}_2 + \mathbf{P}_3 \tag{C-17}$$

$$\mathbf{L} = \mathbf{L}_1 + \mathbf{L}_2 + \mathbf{L}_3 \tag{C-18}$$

– Electrostatic repulsion energy:

$$W = \frac{q^2}{4\pi\varepsilon_0}\left(\frac{1}{|\mathbf{R}_1 - \mathbf{R}_2|} + \frac{1}{|\mathbf{R}_2 - \mathbf{R}_3|} + \frac{1}{|\mathbf{R}_3 - \mathbf{R}_1|}\right) \tag{C-19}$$

– Total spin:

$$\mathbf{S} = \mathbf{S}_1 + \mathbf{S}_2 + \mathbf{S}_3 \tag{C-20}$$

etc.

It is clear from these expressions that the observables associated with the physical quantities considered involve the various particles symmetrically. This important property follows directly from the fact that the particles are identical. In (C-16), for example, $\mathbf{R}_1$, $\mathbf{R}_2$ and $\mathbf{R}_3$ have the same coefficient, since the three particles have the same mass. It is the equality of the charges which is at the basis of the symmetrical form of (C-19). In general, since no physical properties are modified when the roles of the N identical particles are permuted, these N particles must play a symmetrical role* in any actually measurable observable. Mathematically, the corresponding observable G, which we shall call a *physical* observable, must be invariant under all permutations of the N identical particles. It must therefore commute with all the permutation operators P_α of the N particles (*cf.* § B-2-d):

$$[G, P_\alpha] = 0 \qquad \text{for all} \quad P_\alpha \tag{C-21}$$

For a system of two identical particles, for example, the observable $\mathbf{R}_1 - \mathbf{R}_2$ (the vector difference of the positions of the two particles), which is not invariant under the effect of the permutation P_{21} ($\mathbf{R}_1 - \mathbf{R}_2$ changes signs) is not a physical observable; indeed, a measurement of $\mathbf{R}_1 - \mathbf{R}_2$ assumes that particle (1) can be distinguished from particle (2). On the other hand, we can measure the distance between the two particles, that is, $\sqrt{(\mathbf{R}_1 - \mathbf{R}_2)^2}$, which is symmetrical.

Relation (C-21) implies that $\mathscr{E}_S$ and $\mathscr{E}_A$ are both invariant under the action of a physical observable G. Let us show that, if $|\psi\rangle$ belongs to $\mathscr{E}_A$, $G|\psi\rangle$ also belongs to $\mathscr{E}_A$ (the same proof also applies, of course, to $\mathscr{E}_S$). The fact that $|\psi\rangle$ belongs to $\mathscr{E}_A$ means that:

$$P_\alpha|\psi\rangle = \varepsilon_\alpha|\psi\rangle \tag{C-22}$$

Now let us calculate $P_\alpha G|\psi\rangle$. According to (C-21) and (C-22), we have:

$$P_\alpha G|\psi\rangle = GP_\alpha|\psi\rangle = \varepsilon_\alpha G|\psi\rangle \tag{C-23}$$

* Note that this reasoning is valid for fermions as well as for bosons.

Since the permutation P_α is arbitrary, (C-23) expresses the fact that $G | \psi \rangle$ is completely antisymmetric and therefore belongs to $\mathscr{E}_A$.

All operations normally performed on an observable – in particular, the determination of eigenvalues and eigenvectors – can therefore be applied to G entirely within one of the subspaces, $\mathscr{E}_S$ or $\mathscr{E}_A$. Only the eigenkets of G belonging to the physical subspace, and the corresponding eigenvalues, are retained.

COMMENTS:

(*i*) All the eigenvalues of G which exist in the total space $\mathscr{E}$ are not necessarily found if we restrict ourselves to the subspace $\mathscr{E}_S$ (or $\mathscr{E}_A$). The effect of the symmetrization postulate on the spectrum of a symmetric observable G may therefore be to abolish certain eigenvalues. On the other hand, it adds no new eigenvalues to this spectrum, since, because of the global invariance of $\mathscr{E}_S$ (or $\mathscr{E}_A$) under the action of G, any eigenvector of G in $\mathscr{E}_S$ (or $\mathscr{E}_A$) is also an eigenvector of G in $\mathscr{E}$ with the same eigenvalue.

(*ii*) Consider the problem of writing mathematically, in terms of the observables $\mathbf{R}_1$, $\mathbf{P}_1$, $\mathbf{S}_1$, etc., the observables corresponding to the different types of measurement envisaged in § α. This problem is not always simple. For example, for a system of three identical particles, we shall try to write the observables corresponding to the simultaneous measurement of the three positions in terms of $\mathbf{R}_1$, $\mathbf{R}_2$ and $\mathbf{R}_3$. We can resolve this problem by considering several physical observables chosen such that we can, using the results obtained by measuring them, unambiguously deduce the position of each particle (without, of course, being able to associate a numbered particle with each position). For example, we can choose the set

$$X_1 + X_2 + X_3,\; X_1X_2 + X_2X_3 + X_3X_1,\; X_1X_2X_3$$

(and the corresponding observables for the Y and Z coordinates). However, this point of view is rather formal. Rather than trying to write the expressions for the observables in all cases, it is simpler to follow the method used in § α, in which we confined ourselves to using the physical eigenkets of the measurement.

b. TIME-EVOLUTION POSTULATES

The Hamiltonian of a system of identical particles must be a physical observable. We shall write, for example, the Hamiltonian describing the motion of the two electrons of the helium atom about the nucleus, assumed to be motionless*:

$$H(1, 2) = \frac{\mathbf{P}_1^2}{2m_e} + \frac{\mathbf{P}_2^2}{2m_e} - \frac{2e^2}{R_1} - \frac{2e^2}{R_2} + \frac{e^2}{|\mathbf{R}_1 - \mathbf{R}_2|} \tag{C-24}$$

The first two terms represent the kinetic energy of the system; they are symmetrical because the two masses are equal. The next two terms are due to the attraction of the nucleus (whose charge is twice that of the proton). The electrons are obviously equally affected by this attraction. Finally, the last term describes

* Here, we shall consider only the most important terms of this Hamiltonian. See complement B_{XIV} for a more detailed study of the helium atom.

the mutual interaction of the electrons. It is also symmetrical, since neither of the two electrons is in a privileged position. It is clear that this argument can be generalized to any system of identical particles. Consequently, all the permutation operators commute with the Hamiltonian of the system:

$$[H, P_\alpha] = 0 \tag{C-25}$$

Under these conditions, if the ket $| \psi(t_0) \rangle$ describing the state of the system at a given time t_0 is a physical ket, the same must be true of the ket $| \psi(t) \rangle$ obtained from $| \psi(t_0) \rangle$ by solving the Schrödinger equation. According to this equation:

$$| \psi(t + \mathrm{d}t) \rangle = \left(1 + \frac{\mathrm{d}t}{i\hbar} H\right) | \psi(t) \rangle \tag{C-26}$$

Now, applying P_α and using relation (C-25):

$$P_\alpha | \psi(t + \mathrm{d}t) \rangle = \left(1 + \frac{\mathrm{d}t}{i\hbar} H\right) P_\alpha | \psi(t) \rangle \tag{C-27}$$

If $| \psi(t) \rangle$ is an eigenvector of P_α, $| \psi(t + \mathrm{d}t) \rangle$ is also an eigenvector of P_α, with the same eigenvalue. Since $| \psi(t_0) \rangle$, by hypothesis, is a completely symmetric or completely antisymmetric ket, this property is conserved over time.

The symmetrization postulate is therefore also compatible with the postulate which gives the time evolution of physical systems : the Schrödinger equation does not remove the ket $| \psi(t) \rangle$ from $\mathscr{E}_S$ or $\mathscr{E}_A$.

D. DISCUSSION

In this final section, we shall examine the consequences of the symmetrization postulate on the physical properties of systems of identical particles. First of all, we shall indicate the fundamental differences introduced by Pauli's exclusion principle between systems of identical fermions and systems of identical bosons. Then, we shall discuss the implications of the symmetrization postulate concerning the calculation of the probabilities associated with the various physical processes.

1. Differences between bosons and fermions. Pauli's exclusion principle

In the statement of the symmetrization postulate, the difference between bosons and fermions may appear insignificant. Actually, this simple sign difference in the symmetry of the physical ket has extremely important consequences. As we saw in §C-3 the symmetrization postulate does not restrict the individual states accessible to a system of identical bosons. On the other hand, it requires fermions to obey Pauli's exclusion principle : two identical fermions cannot occupy the same quantum mechanical state.

The exclusion principle was formulated initially in order to explain the properties of many-electron atoms (§D-1-a below and complement A_{XIV}). It can now be seen to be more than a principle applicable only to electrons: it is a consequence of the symmetrization postulate, valid for all systems of identical fermions. Predictions based on this principle, which are often spectacular, have always been confirmed experimentally. We shall give some examples of them.

a. GROUND STATE OF A SYSTEM OF INDEPENDENT IDENTICAL PARTICLES

The Hamiltonian of a system of identical particles (bosons or fermions) is always symmetrical with respect to permutations of these particles (§ C-4). Consider such a system in which the various particles are independent, that is, do not interact with each other (at least in a first approximation). The corresponding Hamiltonian is then a sum of one-particle operators of the form:

$$H(1, 2, ..., N) = h(1) + h(2) + ... + h(N) \tag{D-1}$$

$h(1)$ is a function only of the observables associated with the particle numbered (1); the fact that the particles are identical [which implies a symmetrical Hamiltonian $H(1, 2, ..., N)$] requires this function h to be the same in the N terms of expression (D-1). In order to determine the eigenstates and eigenvalues of the total Hamiltonian $H(1, 2, ..., N)$, we simply calculate those of the individual Hamiltonian $h(j)$ in the state space $\mathscr{E}(j)$ of one of the particles:

$$h(j) \mid \varphi_n \rangle = \mathrm{e}_n \mid \varphi_n \rangle ; \quad \mid \varphi_n \rangle \in \mathscr{E}(j) \tag{D-2}$$

For the sake of simplicity, we shall assume that the spectrum of $h(j)$ is discrete and non-degenerate.

If we are considering a system of identical bosons, the physical eigenvectors of the Hamiltonian $H(1, 2, ..., N)$ can be obtained by symmetrizing the tensor products of N arbitrary individual states $\mid \varphi_n \rangle$:

$$\mid \Phi^{(S)}_{n_1,n_2,...,n_N} \rangle = c \sum_{\alpha} P_\alpha \mid 1 : \varphi_{n_1} ; 2 : \varphi_{n_2} ; ... ; N : \varphi_{n_N} \rangle \tag{D-3}$$

where the corresponding energy is the sum of the N individual energies:

$$E_{n_1,n_2,...,n_N} = \mathrm{e}_{n_1} + \mathrm{e}_{n_2} + ... + \mathrm{e}_{n_N} \tag{D-4}$$

[it can easily be shown that each of the kets appearing on the right-hand side of (D-3) is an eigenket of H with the eigenvalue (D-4); this is also true of their sum]. In particular, if e_1 is the smallest eigenvalue of $h(j)$, and $\mid \varphi_1 \rangle$ is the associated eigenstate, the ground state of the system is obtained when the N identical bosons are all in the state $\mid \varphi_1 \rangle$. The energy of this ground state is therefore:

$$E_{1,1,...,1} = N \mathrm{e}_1 \tag{D-5}$$

and its state vector is:

$$\mid \varphi^{(S)}_{1,1,...,1} \rangle = \mid 1 : \varphi_1 ; 2 : \varphi_1 ; ... ; N : \varphi_1 \rangle \tag{D-6}$$

Now, suppose that the N identical particles considered are fermions. It is no longer possible for these N particles all to be in the individual state $\mid \varphi_1 \rangle$. To obtain the ground state of the system, Pauli's exclusion principle must be taken into account. If the individual energies e_n are arranged in increasing order:

$$e_1 < e_2 < \dots < e_{n-1} < e_n < e_{n+1} < \dots, \tag{D-7}$$

the ground state of the system of N identical fermions has an energy of:

$$E_{1,2,\dots,N} = e_1 + e_2 + \dots + e_N \tag{D-8}$$

and it is described by the normalized physical ket:

$$| \Phi^{(A)}_{1,2,\dots,N} \rangle = \frac{1}{\sqrt{N!}} \begin{vmatrix} |1:\varphi_1\rangle & |1:\varphi_2\rangle \dots & |1:\varphi_N\rangle \\ |2:\varphi_1\rangle & |2:\varphi_2\rangle \dots & |2:\varphi_N\rangle \\ \vdots & & \\ |N:\varphi_1\rangle & |N:\varphi_2\rangle \dots & |N:\varphi_N\rangle \end{vmatrix} \tag{D-9}$$

The highest individual energy e_N found in the ground state is called the *Fermi energy* of the system.

Pauli's exclusion principle thus plays a role of primary importance in all domains of physics in which many-electron systems are involved, such as atomic and molecular physics (*cf.* complements A_{XIV} and B_{XIV}) and solid state physics (*cf.* complement C_{XIV}), and in all those in which many-proton and many-neutron systems are involved, such as nuclear physics★.

COMMENT:

In most cases, the individual energies e_n are actually degenerate. Each of them can then enter into a sum such as (D-8) a number of times equal to its degree of degeneracy.

b. QUANTUM STATISTICS

The object of statistical mechanics is to study systems composed of a very large number of particles (in numerous cases, the mutual interactions between these particles are weak enough to be neglected in a first approximation). Since we do not know the microscopic state of the system exactly, we content ourselves with describing it globally by its macroscopic properties (pressure, temperature, density, etc.). A particular macroscopic state corresponds to a whole set of microscopic states. We then use probabilities: the statistical weight of a macroscopic state is proportional to the number of distinct microscopic states which correspond to it, and the system, at thermodynamic equilibrium, is in its most probable macroscopic state (with any constraints that may be imposed taken into account). To study the macroscopic properties of the system, it is therefore essential to

★ The ket representing the state of a nucleus must be antisymmetric both with respect to the set of protons and with respect to the set of neutrons.

determine how many different microscopic states possess certain characteristics and, in particular, a given energy.

In classical statistical mechanics (Maxwell-Boltzmann statistics), the N particles of the system are treated as if they were of different natures, even if they are actually identical. Such a microscopic state is defined by specifying the individual state of each of the N particles. Two microscopic states are considered to be distinct when these N individual states are the same but the permutation of the particles is different.

In quantum statistical mechanics, the symmetrization postulate must be taken into account. A microscopic state of a system of identical particles is characterized by the enumeration of the N individual states which form it, the order of these states being of no importance since their tensor product must be symmetrized or antisymmetrized. The numbering of the microscopic states therefore does not lead to the same result as in classical statistical mechanics. In addition, Pauli's principle radically differentiates systems of identical bosons and systems of identical fermions: the number of particles occupying a given individual state cannot exceed one for fermions, while it can take on any value for bosons (*cf.* § C-3). Different statistical properties result: bosons obey *Bose-Einstein statistics* and fermions, *Fermi-Dirac statistics*. This is the origin of the terms "bosons" and "fermions".

The physical properties of systems of identical fermions and systems of identical bosons are very different. These differences can be observed, for example, at low temperatures. The particles then tend to accumulate in the lowest-energy individual states, as is possible for identical bosons (this phenomenon is called *Bose condensation*), while identical fermions are subject to the restrictions of Pauli's principle. Bose condensation is at the origin of the remarkable properties (superfluidity) of the ^{4}He isotope of helium, while the ^{3}He isotope, which is a fermion (*cf.* comment of § C-1), does not possess the same properties.

2. The consequences of particle indistinguishability on the calculation of physical predictions

In quantum mechanics, all the predictions concerning the properties of a system are expressed in terms of probability amplitudes (scalar products of two state vectors) or matrix elements of an operator. It is then not surprising that the symmetrization or antisymmetrization of state vectors causes special *interference effects* to appear in systems of identical particles. First, we shall specify these effects, and then we shall see how they disappear under certain conditions (the particles of the system, although identical, then behave as if they were of different natures). To simplify the discussion, we shall confine ourselves to systems containing only two identical particles.

a. INTERFERENCES BETWEEN DIRECT AND EXCHANGE PROCESSES

α. *Predictions concerning a measurement on a system of identical particles: the direct term and the exchange term*

Consider a system of two identical particles, one of which is known to be in the individual state $| \varphi \rangle$ and the other, in the individual state $| \chi \rangle$. We shall assume $| \varphi \rangle$ and $| \chi \rangle$ to be orthogonal, so that the state of the system is described by the normalized physical ket [*cf.* formula (C-4)]:

$$| \varphi ; \chi \rangle = \frac{1}{\sqrt{2}} [1 + \varepsilon P_{21}] | 1 : \varphi ; 2 : \chi \rangle \tag{D-10}$$

where:

$$\begin{aligned} \varepsilon &= +1 \quad \text{if the particles are bosons} \\ \varepsilon &= -1 \quad \text{if the particles are fermions} \end{aligned} \tag{D-11}$$

With the system in this state, suppose that we want to measure on each of the two particles the same physical quantity B with which the observables $B(1)$ and $B(2)$ are associated. For the sake of simplicity, we shall assume that the spectrum of B is entirely discrete and non-degenerate :

$$B | u_i \rangle = b_i | u_i \rangle \tag{D-12}$$

What is the probability of finding certain given values in this measurement (b_n for one of the particles and $b_{n'}$ for the other one)? We shall begin by assuming b_n and $b_{n'}$ to be different, so that the corresponding eigenvectors $| u_n \rangle$ and $| u_{n'} \rangle$ are orthogonal. Under these conditions, the normalized physical ket defined by the result of this measurement can be written:

$$| u_n ; u_{n'} \rangle = \frac{1}{\sqrt{2}} [1 + \varepsilon P_{21}] | 1 : u_n ; 2 : u_{n'} \rangle \tag{D-13}$$

which gives the probability amplitude associated with this result:

$$\langle u_n ; u_{n'} | \varphi ; \chi \rangle = \frac{1}{2} \langle 1 : u_n ; 2 : u_{n'} | (1 + \varepsilon P_{21}^{\dagger})(1 + \varepsilon P_{21}) | 1 : \varphi ; 2 : \chi \rangle \tag{D-14}$$

Using properties (B-13) and (B-14) of the operator P_{21}, we can write:

$$\frac{1}{2}(1 + \varepsilon P_{21}^{\dagger})(1 + \varepsilon P_{21}) = 1 + \varepsilon P_{21} \tag{D-15}$$

(D-14) then becomes:

$$\langle u_n ; u_{n'} | \varphi ; \chi \rangle = \langle 1 : u_n ; 2 : u_{n'} |(1 + \varepsilon P_{21})| 1 : \varphi ; 2 : \chi \rangle \tag{D-16}$$

Letting $1 + \varepsilon P_{21}$ act on the bra, we obtain:

$$\begin{aligned} \langle u_n ; u_{n'} | \varphi ; \chi \rangle &= \langle 1 : u_n ; 2 : u_{n'} | 1 : \varphi ; 2 : \chi \rangle \\ &\qquad + \varepsilon \langle 1 : u_{n'} ; 2 : u_n | 1 : \varphi ; 2 : \chi \rangle \\ &= \langle 1 : u_n | 1 : \varphi \rangle \langle 2 : u_{n'} | 2 : \chi \rangle \\ &\qquad + \varepsilon \langle 1 : u_{n'} | 1 : \varphi \rangle \langle 2 : u_n | 2 : \chi \rangle \\ &= \langle u_n | \varphi \rangle \langle u_{n'} | \chi \rangle + \varepsilon \langle u_{n'} | \varphi \rangle \langle u_n | \chi \rangle \end{aligned} \tag{D-17}$$

The numbering has disappeared from the probability amplitude, which is now expressed directly in terms of the scalar products $\langle u_n | \varphi \rangle \ldots \langle u_n | \chi \rangle$. Also, the probability amplitude appears either as a sum (for bosons) or a difference (for fermions) of two terms, with which we can associate the diagrams of figures 4-a and 4-b.

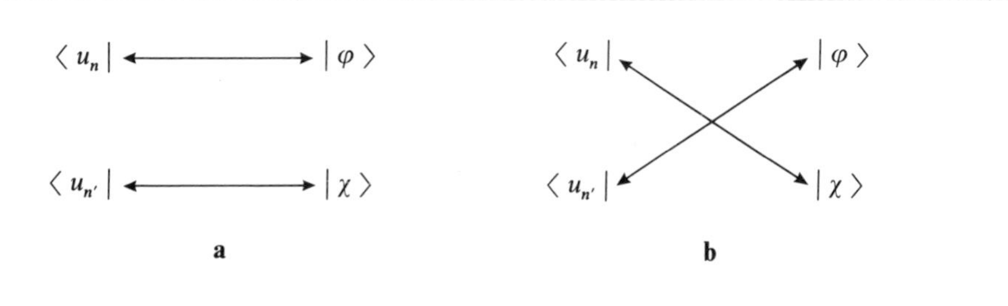

FIGURE 4

Schematic representation of the direct term and the exchange term associated with a measurement performed on a system of two identical particles. Before the measurement, one of the particles is known to be in the state $| \varphi \rangle$ and the other one, in the state $| \chi \rangle$. The measurement result obtained corresponds to a situation in which one particle is in the state $| u_n \rangle$ and the other one, in the state $| u_{n'} \rangle$. Two probability amplitudes are associated with such a measurement; they are represented schematically by figures a and b. These amplitudes interfere with a + sign for bosons and with a − sign for fermions.

We can interpret result (D-17) in the following way. The two kets $| \varphi \rangle$ and $| \chi \rangle$ associated with the initial state can be connected to the two bras $\langle u_n |$ and $\langle u_{n'} |$ associated with the final state by two different "paths", represented schematically by figures 4-a and 4-b. With each of these paths is associated a probability amplitude, $\langle u_n | \varphi \rangle \langle u_{n'} | \chi \rangle$ or $\langle u_{n'} | \varphi \rangle \langle u_n | \chi \rangle$, and *these two amplitudes interfere with a + sign for bosons and a − sign for fermions.* Thus, we obtain the answer to the question posed in § A-3-a above: the desired probability $\mathscr{P}(b_n; b_{n'})$ is equal to the square of the modulus of (D-17):

$$\mathscr{P}(b_n; b_{n'}) = |\langle u_n | \varphi \rangle \langle u_{n'} | \chi \rangle + \varepsilon \langle u_{n'} | \varphi \rangle \langle u_n | \chi \rangle|^2 \tag{D-18}$$

One of the two terms on the right-hand side of (D-17), the one which corresponds, for example, to path 4-a, is often called the *direct term*. The other term is called the *exchange term*.

COMMENT:

Let us examine what happens if the two particles, instead of being identical, are of different natures. We shall then choose as the initial state of the system the tensor product ket:

$$| \psi \rangle = | 1 : \varphi ; 2 : \chi \rangle \tag{D-19}$$

Now, consider a measurement instrument which, although the two particles, (1) and (2), are not identical, is not able to distinguish between them. If it yields the results b_n and $b_{n'}$, we do not know if b_n is associated with particle (1) or particle (2) (for example, for a system composed of a muon μ^- and an electron e^-, the measurement device may be sensitive only to the charge of the particles, giving no information about their masses). The two eigenstates $| 1 : u_n ; 2 : u_{n'} \rangle$ and $| 1 : u_{n'} ; 2 : u_n \rangle$ (which, in this case, represent different physical states) then correspond to the same measurement result. Since they are orthogonal, we must add the corresponding probabilities, which gives:

$$\begin{aligned} \mathscr{P}'(b_n ; b_{n'}) &= |\langle 1 : u_n ; 2 : u_{n'} | 1 : \varphi ; 2 : \chi \rangle|^2 \\ &\qquad + |\langle 1 : u_{n'} ; 2 : u_n | 1 : \varphi ; 2 : \chi \rangle|^2 \\ &= |\langle u_n | \varphi \rangle|^2 \, |\langle u_{n'} | \chi \rangle|^2 + |\langle u_{n'} | \varphi \rangle|^2 \, |\langle u_n | \chi \rangle|^2 \end{aligned} \tag{D-20}$$

Comparison of (D-18) with (D-20) clearly reveals the significant difference in the physical predictions of quantum mechanics depending on whether the particles under consideration are identical or not.

Now consider the case in which the two states $| u_n \rangle$ and $| u_{n'} \rangle$ are the same. When the two particles are fermions, the corresponding physical state is excluded by Pauli's principle, and the probability $\mathscr{P}(b_n ; b_n)$ is zero. On the other hand, if the two particles are bosons, we have:

$$| u_n ; u_n \rangle = | 1 : u_n ; 2 : u_n \rangle \tag{D-21}$$

and, consequently:

$$\begin{aligned} \langle u_n ; u_n | \varphi ; \chi \rangle &= \frac{1}{\sqrt{2}} \langle 1 : u_n ; 2 : u_n | (1 + P_{21}) | 1 : \varphi ; 2 : \chi \rangle \\ &= \sqrt{2} \, \langle u_n | \varphi \rangle \langle u_n | \chi \rangle \end{aligned} \tag{D-22}$$

which gives:

$$\mathscr{P}(b_n ; b_n) = 2 \, |\langle u_n | \varphi \rangle \langle u_n | \chi \rangle|^2 \tag{D-23}$$

COMMENTS:

(*i*) Let us compare this result with the one which would be obtained in the case, already considered above, in which the two particles are different. We must then replace $| \varphi ; \chi \rangle$ by $| 1 : \varphi ; 2 : \chi \rangle$ and $| u_n ; u_n \rangle$ by $| 1 : u_n ; 2 : u_n \rangle$, which gives the value for the probability amplitude:

$$\langle u_n | \varphi \rangle \langle u_n | \chi \rangle \tag{D-24}$$

and, consequently:

$$\mathscr{P}'(b_n ; b_n) = |\langle u_n | \varphi \rangle \langle u_n | \chi \rangle|^2 \tag{D-25}$$

(*ii*) For a system containing N identical particles, there are, in general, $N!$ distinct exchange terms which add (or subtract) in the probability amplitude. For example, consider a system of three identical particles in the individual states $| \varphi \rangle$, $| \chi \rangle$ and $| \omega \rangle$, and the probability of finding, in a measurement, the results b_n, $b_{n'}$ and $b_{n''}$. The possible "paths" are then shown in figure 5. There are six such paths (all different if the three eigenvalues,

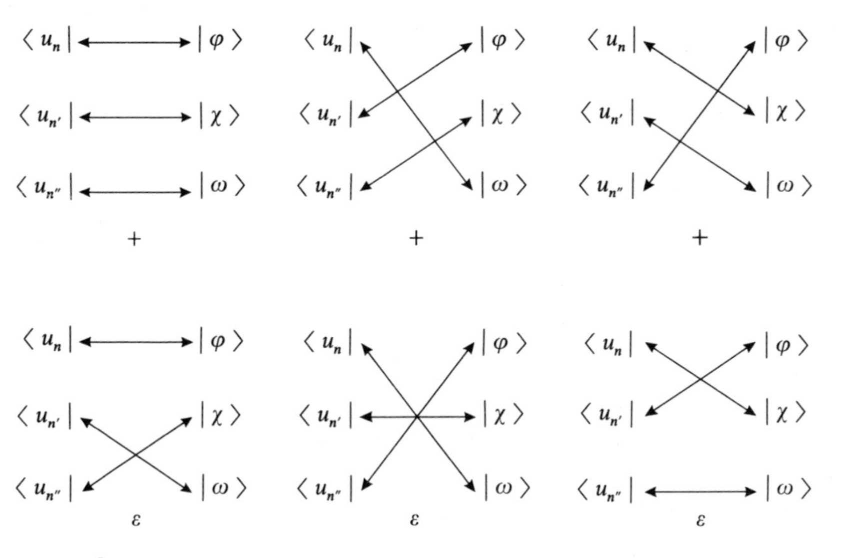

FIGURE 5

Schematic representation of the six probability amplitudes associated with a system of three identical particles. Before the measurement, one particle is known to be in the state $| \varphi \rangle$, another, in the state $| \chi \rangle$, and the last one, in the state $| \omega \rangle$. The result obtained corresponds to a situation in which one particle is in the state $| u_n \rangle$, another, in the state $| u_{n'} \rangle$, and the last one, in the state $| u_{n''} \rangle$. The six amplitudes interfere with a sign which is shown beneath each one ($\varepsilon = +1$ for bosons, -1 for fermions).

b_n, $b_{n'}$ and $b_{n''}$, are different). Some always contribute to the probability amplitude with a + sign, others with an ε sign (+ for bosons and − for fermions).

β. *Example : elastic collision of two identical particles*

To understand the physical meaning of the exchange term, let us examine a concrete example (already alluded to in §A-3-a): that of the elastic collision of two identical particles in their center of mass frame★. Unlike the situation in §α above, here we must take into account the evolution of the system between the initial time when it is in the state $| \psi_i \rangle$ and the time t when the measurement is performed. However, as we shall see, this evolution does not change the problem radically, and the exchange term enters the problem as before.

★ We shall give a simplified treatment of this problem, intended only to illustrate the relation between the direct term and the exchange term. In particular, we ignore the spin of the two particles. However, the calculations of this section remain valid in the case in which the interactions are not spin-dependent and the two particles are initially in the same spin state.

In the initial state of the system (fig. 6-a), the two particles are moving towards each other with opposite momenta. We choose the Oz axis along the direction of these momenta, and we denote their modulus by p. One of the particles thus possesses the momentum $p\mathbf{e}_z$, and the other one, the momentum $-p\mathbf{e}_z$ (where $\mathbf{e}_z$ is the unit vector of the Oz axis). We shall write the physical ket $| \psi_i \rangle$ representing this initial state in the form:

$$| \psi_i \rangle = \frac{1}{\sqrt{2}}(1 + \varepsilon P_{21}) | 1 : p\mathbf{e}_z ; 2 : - p\mathbf{e}_z \rangle \tag{D-26}$$

$| \psi_i \rangle$ describes the state of the system at t_0 before the collision.

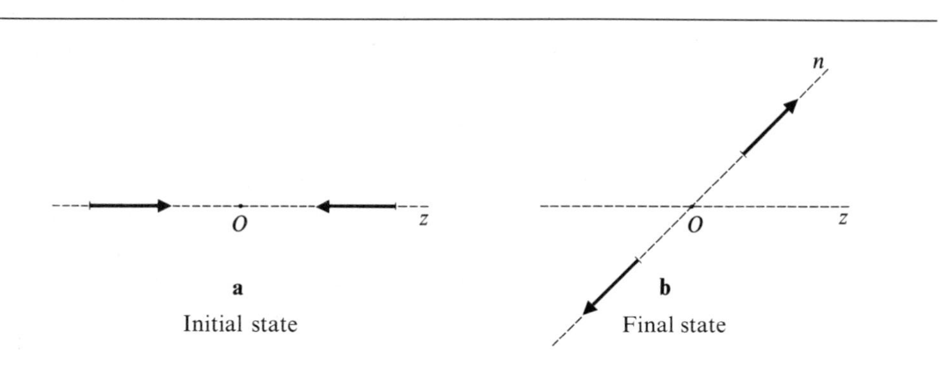

FIGURE 6

Collision between two identical particles in the center of mass frame : the momenta of the two particles in the initial state (fig. a) and in the final state found in the measurement (fig. b) are represented. For the sake of simplicity, we ignore the spin of the particles.

The Schrödinger equation which governs the time evolution of the system is linear. Consequently, there exists a linear operator $U(t, t')$, which is a function of the Hamiltonian H, such that the state vector at time t is given by:

$$| \psi(t) \rangle = U(t, t_0) | \psi_i \rangle \tag{D-27}$$

(complement F_{III}). In particular, after the collision, the state of the system at time t_1 is represented by the physical ket:

$$| \psi(t_1) \rangle = U(t_1, t_0) | \psi_i \rangle \tag{D-28}$$

Note that, since the Hamiltonian H is symmetric, the evolution operator U commutes with the permutation operator:

$$[U(t, t'), P_{21}] = 0 \tag{D-29}$$

Now, let us calculate the probability amplitude of the result envisaged in § A-3-a, in which the particles are detected in the two opposite directions of the On axis, of unit vector $\mathbf{n}$ (fig. 6-b). We denote the physical ket associated with this final state by:

$$| \psi_f \rangle = \frac{1}{\sqrt{2}}(1 + \varepsilon P_{21}) \,|\, 1 : p\mathbf{n} ; 2 : - p\mathbf{n} \rangle \tag{D-30}$$

The desired probability amplitude can therefore be written:

$$\begin{aligned} \langle \psi_f \,|\, \psi(t_1) \rangle &= \langle \psi_f \,|\, U(t_1, t_0) \,|\, \psi_i \rangle \\ &= \frac{1}{2} \langle 1 : p\mathbf{n} ; 2 : - p\mathbf{n} \,|\, (1 + \varepsilon P_{21}^\dagger) U(t_1, t_0)(1 + \varepsilon P_{21}) \,|\, 1 : p\mathbf{e}_z ; 2 : - p\mathbf{e}_z \rangle \end{aligned} \tag{D-31}$$

According to relation (D-29) and the properties of the operator P_{21}, we finally obtain:

$$\begin{aligned} &\langle \psi_f \,|\, U(t_1, t_0) \,|\, \psi_i \rangle \\ &\quad = \langle 1 : p\mathbf{n} ; 2 : - p\mathbf{n} \,|\, (1 + \varepsilon P_{21}^\dagger) U(t_1, t_0) \,|\, 1 : p\mathbf{e}_z ; 2 : - p\mathbf{e}_z \rangle \\ &\quad = \langle 1 : p\mathbf{n} ; 2 : - p\mathbf{n} \,|\, U(t_1, t_0) \,|\, 1 : p\mathbf{e}_z ; 2 : - p\mathbf{e}_z \rangle \\ &\qquad\qquad + \varepsilon \langle 1 : - p\mathbf{n} ; 2 : p\mathbf{n} \,|\, U(t_1, t_0) \,|\, 1 : p\mathbf{e}_z ; 2 : - p\mathbf{e}_z \rangle \end{aligned} \tag{D-32}$$

The direct term corresponds, for example, to the process shown in figure 7-a, and the exchange term is then represented by figure 7-b. Again, the probability amplitudes associated with these two processes must be added or subtracted. This causes an

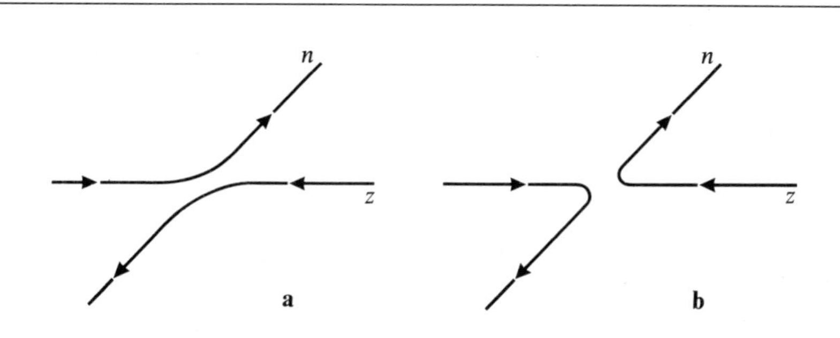

FIGURE 7

Collision between two identical particles in the center of mass frame : schematic representation of the physical processes corresponding to the direct term and the exchange term. The scattering amplitudes associated with these two processes interfere with a + sign for bosons and a − sign for fermions.

interference term to appear when the square of the modulus of expression (D-32) is taken. Note also that this expression is simply multiplied by ε if $\mathbf{n}$ is changed to $-\mathbf{n}$, so that the corresponding probability is invariant under this change.

b. SITUATIONS IN WHICH THE SYMMETRIZATION POSTULATE CAN BE IGNORED

If application of the symmetrization postulate were always indispensable, it would be impossible to study the properties of a system containing a restricted number of particles, because it would be necessary to take into account all the particles in the universe which are identical to those in the system. We shall see in this section that this is not the case. In fact, under certain special conditions, identical particles behave as if they were actually different, and it is not necessary to take the symmetrization postulate into account in order to obtain correct physical predictions. It seems natural to expect, considering the results of §D-2-a, that such a situation would arise whenever the exchange terms introduced by the symmetrization postulate are zero. We shall give two examples.

α. *Identical particles situated in two distinct regions of space*

Consider two identical particles, one of which is in the individual state $| \varphi \rangle$ and the other, in the state $| \chi \rangle$. To simplify the notation, we shall ignore their spin. Suppose that the domain of the wave functions representing the kets $| \varphi \rangle$ and $| \chi \rangle$ are well separated in space:

$$\begin{cases} \varphi(\mathbf{r}) = \langle \mathbf{r} | \varphi \rangle = 0 \quad \text{if} \quad \mathbf{r} \notin D \\ \chi(\mathbf{r}) = \langle \mathbf{r} | \chi \rangle = 0 \quad \text{if} \quad \mathbf{r} \notin \Delta \end{cases} \tag{D-33}$$

where the domains D and Δ do not overlap. The situation is analogous to the classical mechanical one (§A-2): as long as the domains D and Δ do not overlap, each of the particles can be "followed"; we therefore expect application of the symmetrization postulate to be unnecessary.

In this case, we can envisage measuring an observable related to one of the two particles. All we need is a measurement device placed so that it cannot record what happens in the domain D, or in the domain Δ. If it is D which is excluded in this way, the measurement will only concern the particle in Δ, an vice versa.

Now, imagine a measurement concerning the two particles simultaneously, but performed with two distinct measurement devices, one of which is not sensitive to phenomena occurring in Δ, and the other, to those in D. How can the probability of obtaining a given result be calculated? Let $| u \rangle$ and $| v \rangle$ be the individual states associated respectively with the results of the two measurement devices. Since the two particles are identical, the symmetrization postulate must, in theory, be taken into account. In the probability amplitude associated with the measurement result, the direct term is then $\langle u | \varphi \rangle \langle v | \chi \rangle$, and the exchange term is $\langle u | \chi \rangle \langle v | \varphi \rangle$. Now, the spatial disposition of the measurement devices implies that:

$$\begin{aligned} u(\mathbf{r}) &= \langle \mathbf{r} | u \rangle = 0 \quad \text{if} \quad \mathbf{r} \in \Delta \\ v(\mathbf{r}) &= \langle \mathbf{r} | v \rangle = 0 \quad \text{if} \quad \mathbf{r} \in D \end{aligned} \tag{D-34}$$

According to (D-33) and (D-34), the wave functions $u(\mathbf{r})$ and $\chi(\mathbf{r})$ do not overlap; neither do $v(\mathbf{r})$ and $\varphi(\mathbf{r})$, so that:

$$\langle u | \chi \rangle = \langle v | \varphi \rangle = 0 \tag{D-35}$$

The exchange term is therefore zero. Consequently, it is unnecessary, in this situation, to use the symmetrization postulate. We obtain the desired result directly by reasoning as if the particles were of different natures, labeling, for example, the one in the domain D with the number 1, and the one situated in Δ with the number 2. Before the measurement, the state of the system is then described by the ket $|\, 1 : \varphi ; 2 : \chi \,\rangle$, and with the measurement result envisaged is associated the ket $|\, 1 : u ; 2 : v \,\rangle$. Their scalar product gives the probability amplitude $\langle \, u \,|\, \varphi \,\rangle \langle \, v \,|\, \chi \,\rangle$.

This argument shows that the existence of identical particles does not prevent the separate study of restricted systems, composed of a small number of particles.

COMMENT:

In the initial state chosen, the two particles are situated in two distinct regions of space. In addition, we have defined the state of the system by specifying two individual states. We might wonder if, after the system has evolved, it is still possible to study one of the two particles and ignore the other one. For this to be the case, it is necessary, not only that the two particles remain in two distinct regions of space, but also that they do not interact. Whether the particles are identical or not, an interaction always introduces correlations between them, and it is no longer possible to describe each of them by a state vector.

β. *Particles which can be identified by the direction of their spins*

Consider an elastic collision between two identical spin 1/2 particles (electrons, for example), assuming that spin-dependent interactions can be neglected, so that the spin states of the two particles are conserved during the collision. If these spin states are initially orthogonal, they enable us to distinguish between the two particles at all times, as if they were not identical; consequently, the symmetrization postulate should again have no effect here.

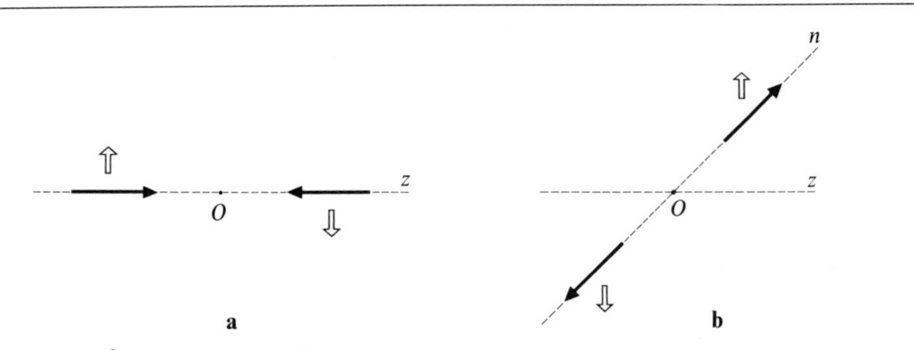

FIGURE 8

Collision between two identical spin 1/2 particles in the center of mass frame : a schematic representation of the momenta and spins of the two particles in the initial state (fig. a) and in the final state found in the measurement (fig. b). If the interactions between the two particles are spin-independent, the orientation of the spins does not change during the collision. When the two particles are not in the same spin state before the collision (the case of the figure), it is possible to determine the "path" followed by the system in arriving at a given final state. For example, the only scattering process which leads to the final state of figure b and which has a non-zero amplitude is of the type shown in figure 7-a.

We can show this, using the calculation of §D-2-a-β. The initial physical ket will be, for example (fig. 8-a):

$$| \psi_i \rangle = \frac{1}{\sqrt{2}} (1 - P_{21}) | 1 : p\mathbf{e}_z , + ; 2 : - p\mathbf{e}_z , - \rangle \tag{D-36}$$

(where the symbol + or − added after each momentum indicates the sign of the spin component along a particular axis). The final state which we are considering (fig. 8-b) will be described by:

$$| \psi_f \rangle = \frac{1}{\sqrt{2}} (1 - P_{21}) | 1 : p\mathbf{n} , + ; 2 : - p\mathbf{n} , - \rangle \tag{D-37}$$

Under these conditions, only the first term of (D-32) is different from zero, since the second one can be written:

$$\langle 1 : - p\mathbf{n}, - ; 2 : p\mathbf{n}, + \mid U(t_1, t_0) \mid 1 : p\mathbf{e}_z, + ; 2 : - p\mathbf{e}_z, - \rangle \tag{D-38}$$

This is the matrix element of a spin-independent operator (by hypothesis) between two kets whose spin states are orthogonal; it is therefore zero. Consequently, we would obtain the same result if we treated the two particles directly as if they were different, that is, if we did not antisymmetrize the initial and final kets and if we associated index 1 with the spin state $| + \rangle$ and index 2 with the spin state $| - \rangle$. Of course, this is no longer possible if the evolution operator U, that is, the Hamiltonian H of the system, is spin-dependent.

References and suggestions for further reading:

The importance of interference between direct and exchange terms is stressed in Feynman III (1.2), § 3.4 and chap. 4.

Quantum statistics : Reif (8.4), Kittel (8.2).

Permutation groups : Messiah (1.17), app. D, § IV; Wigner (2.23), chap. 13; Bacry (10.31), §§ 41 and 42.

The effect of the symmetrization postulate on molecular spectra : Herzberg (12.4), Vol. I, chap. III, § 2f.

An article giving a popularized version: Gamow (1.27).

COMPLEMENTS OF CHAPTER XIV

A_{XIV}: MANY-ELECTRON ATOMS. ELECTRONIC CONFIGURATIONS

A_{XIV}: simple study of many-electron atoms in the central-field approximation. Discusses the consequences of Pauli's exclusion principle and introduces the concept of a configuration. Remains qualitative.

B_{XIV}: ENERGY LEVELS OF THE HELIUM ATOM : CONFIGURATIONS, TERMS, MULTIPLETS

B_{XIV}: study, in the case of the helium atom, of the effect of the electrostatic repulsion between electrons and of the magnetic interactions. Introduces the concepts of terms and multiplets. Can be reserved for later study.

C_{XIV}: PHYSICAL PROPERTIES OF AN ELECTRON GAS. APPLICATION TO SOLIDS

C_{XIV}: study of the ground state of a gas of free electrons enclosed in a "box". Introduces the concept of a Fermi energy and periodic boundary conditions. Generalization to electrons in solids and qualitative discussion of the relation between electrical conductivity and the position of the Fermi level. Moderately difficult. The physical discussions are emphasized. Can be considered to be a sequel of F_{XI}.

D_{XIV}: EXERCISES

Complement A_{XIV}

MANY-ELECTRON ATOMS. ELECTRONIC CONFIGURATIONS

The energy levels of the hydrogen atom were studied in detail in chapter VII. Such a study is considerably simplified by the fact that the hydrogen atom possesses a single electron, so that Pauli's principle is not relevant. In addition, by using the center of mass frame, we can reduce the problem to the calculation of the energy levels of a single particle (the relative particle) subject to a central potential.

In this complement, we shall consider many-electron atoms, for which these simplifications cannot be made. In the center-of-mass frame, we must solve a problem involving several non-independent particles. We shall see that this is a complex problem and give only an approximate solution, using the central-field approximation (which will be outlined, without going into details of the calculations). In addition, Pauli's principle, as we shall show, plays an important role.

1. The central-field approximation

Consider a Z-electron atom. Since the mass of its nucleus is much larger (several thousand times) than that of the electrons, the center-of-mass of the atom practically coincides with the nucleus, which we shall therefore assume to be motionless at the coordinate origin*. The Hamiltonian describing the motion of the electrons, neglecting relativistic corrections and, in particular, spin-dependent terms, can be written :

$$H = \sum_{i=1}^{Z} \frac{\mathbf{P}_i^2}{2m_e} - \sum_{i=1}^{Z} \frac{Ze^2}{R_i} + \sum_{i<j} \frac{e^2}{|\mathbf{R}_i - \mathbf{R}_j|} \tag{1}$$

We have numbered the electrons arbitrarily from 1 to Z, and we have set:

$$e^2 = \frac{q^2}{4\pi\varepsilon_0} \tag{2}$$

* Making this approximation amounts to neglecting the nuclear finite mass effect.

where q is the electron charge. The first term of the Hamiltonian (1) represents the total kinetic energy of the system of Z electrons. The second one arises from the attraction exerted on each of them by the nucleus, which bears a positive charge equal to $-Zq$. The last one describes the mutual repulsion of the electrons [note that the summation is carried out here over the $Z(Z-1)/2$ different ways of pairing the Z-electrons].

The Hamiltonian (1) is too complicated for us to solve its eigenvalue equation exactly, even in the simplest case, that of helium ($Z = 2$).

a. DIFFICULTIES RELATED TO ELECTRON INTERACTIONS

In the absence of the mutual interaction term $\sum_{i<j} \frac{e^2}{|\mathbf{R}_i - \mathbf{R}_j|}$ in H, the electrons would be independent. It would then be easy to determine the energies of the atom. We would simply sum the energies of the Z electrons placed individually in the Coulomb potential $-Ze^2/r$, and the theory presented in chapter VII would yield the result immediately. As for the eigenstates of the atom, they could be obtained by antisymmetrizing the tensor product of the stationary states of the various electrons.

It is then the presence of the mutual interaction term that makes it difficult to solve the problem exactly. We might try to treat this term by perturbation theory. However, a rough evaluation of its relative magnitude shows that this would not yield a good approximation. We expect the distance $|\mathbf{R}_i - \mathbf{R}_j|$ between two electrons to be, on the average, roughly the distance R_i of an electron from the nucleus. The ratio ρ of the third term of formula (1) to the second one is therefore approximately equal to:

$$\rho \simeq \frac{\frac{1}{2} Z(Z-1)}{Z^2} \tag{3}$$

ρ varies between 1/4 for $Z = 2$ and 1/2 for Z much larger than 1. Consequently, the perturbation treatment of the mutual interaction term would yield, at most, more or less satisfactory results for helium ($Z = 2$), but it is out of the question to apply it to other atoms (ρ is already equal to 1/3 for $Z = 3$). A more elaborate approximation method must therefore be found.

b. PRINCIPLE OF THE METHOD

To understand the concept of a central field, we shall use a semi-classical argument. Consider a particular electron (i). In a first approximation, the existence of the $Z - 1$ other electrons affects it only because their charge distribution partially compensates the electrostatic attraction of the nucleus. In this approximation, the electron (i) can be considered to move in a potential which depends only on its position $\mathbf{r}_i$ and takes into account the average effect of the repulsion of the other electrons. We choose a potential $V_c(r_i)$ which depends only on the modulus or $\mathbf{r}_i$ and call it the "central potential" of the atom under consideration. Of course, this can only be an approximation : since the motion of the electron (i) actually influences that of the ($Z - 1$) other electrons, it is not possible to ignore the correlations which exist between them. Moreover, when the electron (i) is in the immediate vicinity of another electron (j), the repulsion exerted by the latter becomes preponderant, and the corresponding force is not central. However, the idea of an average potential appears more valid in quantum mechanics, where we consider the delocalization of the electrons as distributing their charges throughout an extended region of space.

These considerations thus lead us to write the Hamiltonian (1) in the form:

$$H = \sum_{i=1}^{Z} \left[\frac{\mathbf{P}_i^2}{2m_e} + V_c(R_i) \right] + W \tag{4}$$

with:

$$W = - \sum_{i=1}^{Z} \frac{Ze^2}{R_i} + \sum_{i<j} \frac{e^2}{|\mathbf{R}_i - \mathbf{R}_j|} - \sum_{i=1}^{Z} V_c(R_i) \tag{5}$$

If the central potential $V_c(r_i)$ is suitably chosen, W should play the role of a small correction in the Hamiltonian H. The central-field approximation then consists of neglecting this correction, that is of choosing the approximate Hamiltonian:

$$H_0 = \sum_{i=1}^{Z} \left[\frac{\mathbf{P}_i^2}{2m_e} + V_c(R_i) \right] \tag{6}$$

W will then be treated like a perturbation of H_0 (*cf.* complement B_{XIV}, §2). The diagonalization of H_0 leads to a problem of *independent particles* : to obtain the eigenstates of H_0, we simply determine those of the one-electron Hamiltonian:

$$\frac{\mathbf{P}^2}{2m_e} + V_c(R) \tag{7}$$

Definitions (4) and (5) do not, of course, fix the central potential $V_c(r)$, since we always have $H = H_0 + W$, for all $V_c(r)$. However, in order to treat W like a perturbation, $V_c(r)$ must be wisely chosen. We shall not take up the problem of the existence and determination of such an optimal potential here. This is a complex problem. The potential $V_c(r)$ to which a given electron is subject depends on the spatial distribution of the $(Z - 1)$ other electrons, and this distribution, in turn, depends on the potential $V_c(r)$, since the wave functions of the $(Z - 1)$ electrons must also be calculated from $V_c(r)$. We must therefore arrive at a coherent solution (one generally says "self-consistent"), for which the wave functions determined from $V_c(r)$ give a charge distribution which reconstitutes this same potential $V_c(r)$.

c. ENERGY LEVELS OF THE ATOM

While the exact determination of the potential $V_c(r)$ requires rather long calculations, the short- and long-distance behavior of this potential is simple to predict. We expect, for small r, the electron (i) under consideration to be inside the charge distribution created by the other electrons, so that it "sees" only the attractive potential of the nucleus. On the other hand, for large r, that is, outside the "cloud" formed by the $(Z - 1)$ electrons treated globally, it is as if we had a single point charge situated at the coordinate origin and equal to the sum of the charges of the nucleus and the "cloud" [the $(Z - 1)$ electrons screen the field of the nucleus]. Consequently (fig. 1):

$$\begin{aligned} V_c(r) &\simeq -\frac{e^2}{r} \quad \text{for large } r \\ V_c(r) &\simeq -\frac{Ze^2}{r} \quad \text{for small } r \end{aligned} \tag{8}$$

For intermediate values of r, the variation of $V_c(r)$ can be more or less complicated, depending on the atom under consideration.

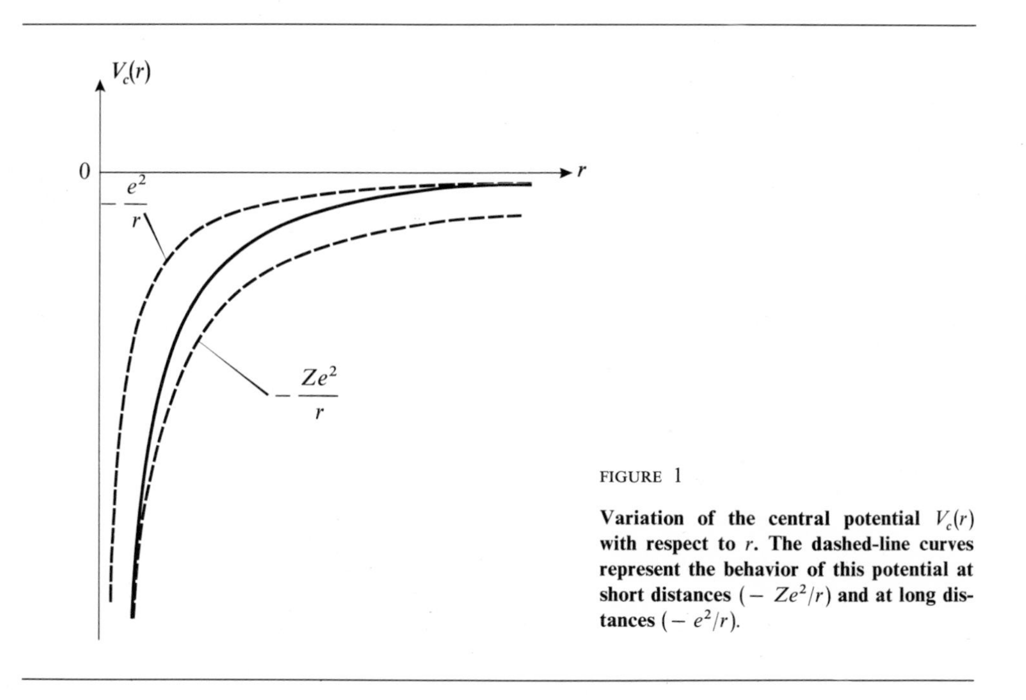

FIGURE 1

Variation of the central potential $V_c(r)$ with respect to r. The dashed-line curves represent the behavior of this potential at short distances ($-Ze^2/r$) and at long distances ($-e^2/r$).

Although these considerations are qualitative, they give an idea of the spectrum of the one-electron Hamiltonian (7). Since $V_c(r)$ is not simply proportional to $1/r$, the accidental degeneracy found for the hydrogen atom (chap. VII, §C-4-b) is no longer observed. The eigenvalues of the Hamiltonian (7) depend on the two quantum numbers n and l [however, they remain independent of m, since $V_c(r)$ is central]. l, of course, characterizes the eigenvalue of the operator $\mathbf{L}^2$, and n is, by definition (as for the hydrogen atom), the sum of the azimuthal quantum number l, and the radial quantum number k introduced in solving the radial equation corresponding to l; n and l are therefore integral and satisfy:

$$0 \leqslant l \leqslant n - 1 \tag{9}$$

Obviously, for a given value of l, the energies $E_{n,l}$ increase with n:

$$E_{n,l} > E_{n',l} \quad \text{if} \quad n > n' \tag{10}$$

For fixed n, the energy is lower when the corresponding eigenstate is more "penetrating", that is, when the probability density of the electron in the vicinity of the nucleus is larger [according to (8), the screening effect is then smaller]. The energies $E_{n,l}$ associated with the same value of n can therefore be arranged in order of increasing angular momenta:

$$E_{n,0} < E_{n,l} < \dots < E_{n,n-1} \tag{11}$$

It so happens that the hierarchy of states is approximately the same for all atoms, although the absolute values of the corresponding energies obviously vary with Z. Figure 2 indicates this hierarchy, as well as the $2(2l + 1)$-fold degeneracy of each state (the factor 2 comes from the electron spin). The various states are represented in spectroscopic notation (*cf.* chap. VII, §C-4-b). Those shown inside the same brackets are very close to each other, and may even, in certain atoms, practically coincide (we stress the fact that figure 2 is simply a schematic representation intended to situate the eigenvalues $E_{n,l}$ with respect to each other; no attempt is made to establish an even moderately realistic energy scale).

Note the great difference between the energy spectrum shown and that of the hydrogen atom (*cf.* chap. VII, fig. 4). As we have already pointed out, the energy depends here on the orbital quantum number l, and, in addition, the order of the states is different. For example, figure 2 indicates that the $4s$ shell has a slightly lower energy than that of the $3d$ shell. This is explained, as we have seen, by the fact that the $4s$ wave function is more penetrating. Analogous inversions occur for the $n = 4$ and $n = 5$ shells, etc. This demonstrates the importance of inter-electron repulsion.

2. Electron configurations of various elements

In the central-field approximation, the eigenstates of the total Hamiltonian H_0 of the atom are Slater determinants, constructed from the individual electron states associated with the energy states $E_{n,l}$ which we have just described. This is therefore the situation envisaged in § D-1-a of chapter XIV: the ground state of the atom is obtained when the Z electrons occupy the lowest individual states compatible with Pauli's principle. The maximum number of electrons which can have a given energy $E_{n,l}$ is equal to the $2(2l + 1)$-fold degeneracy of this energy level. The set of individual states associated with the same energy $E_{n,l}$ is called a *shell*. The list of occupied shells with the number of electrons found in each is called the *electronic configuration*. The notation used will be specified below in a certain number of examples. The concept of a configuration also plays an important role in the chemical properties of atoms. Knowledge of the wave functions of the various electrons and of the corresponding energies makes it possible to interpret the number, stability, and geometry of the chemical bonds which can be formed by this atom (*cf.* complement E_{VII}).

To determine the electronic configuration of a given atom in its ground state, we simply "fill" the various shells successively, in the order indicated in figure 2 (starting, of course, with the $1s$ level) until the Z electrons are exhausted. This is what we shall do, in a rapid review of Mendeleev's table.

In the ground state of the hydrogen atom, the single electron of this atom

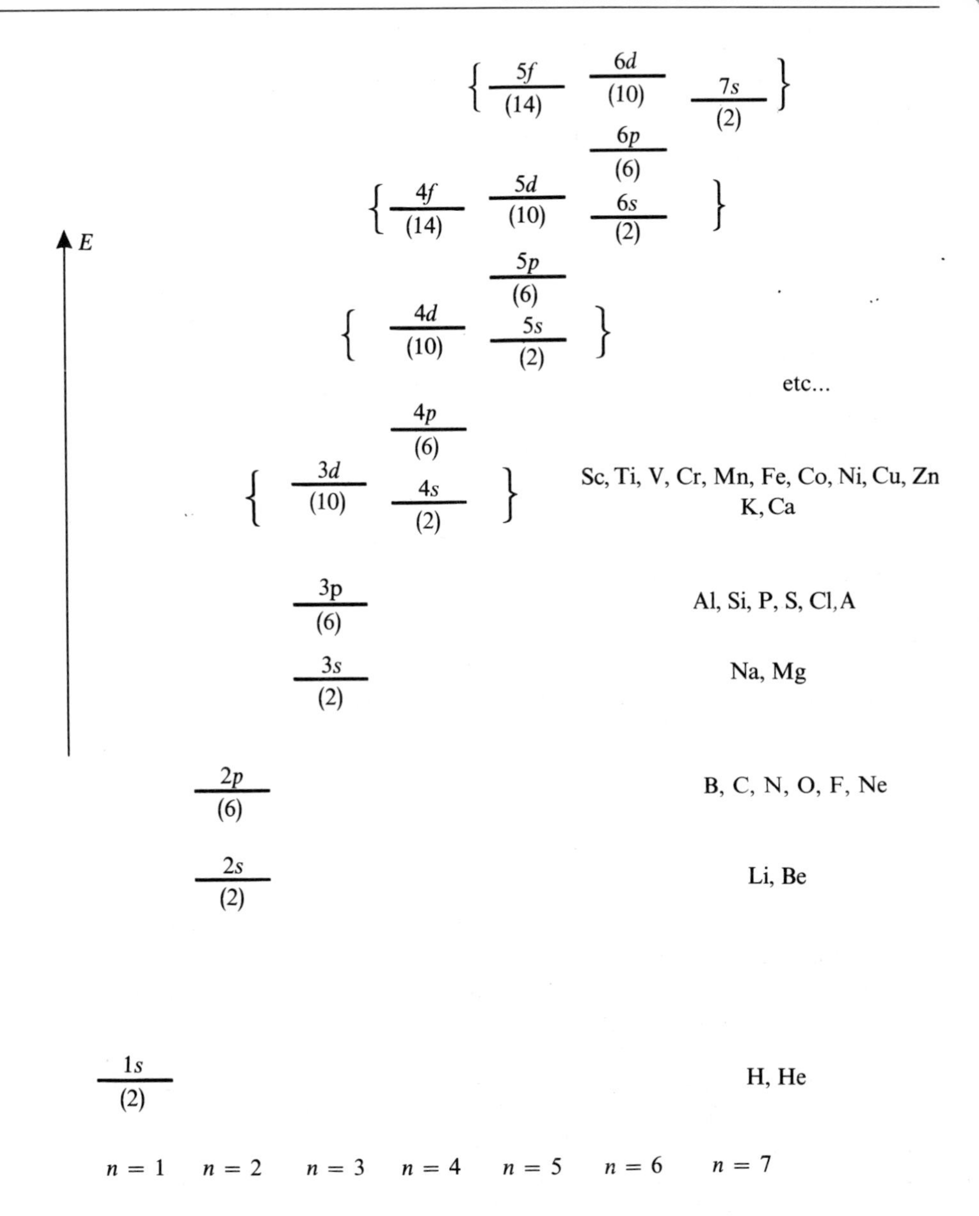

FIGURE 2

Schematic representation of the hierarchy of energy levels (electronic shells) in a central potential of the type shown in figure 1. For each value of n, the energy increases with l. The degeneracy of each level is indicated in parentheses. The levels which appear inside the same bracket are very close to each other, and their relative disposition can vary from one atom to another.
On the right-hand side of the figure, we have indicated the chemical symbols of the atoms for which the electronic shell appearing on the same line is the outermost shell occupied in the ground state configuration.

occupies the $1s$ level. The electronic configuration of the next element (helium, $Z = 2$) is:

$$\text{He}: 1s^2 \tag{12}$$

which means that the two electrons occupy the two orthogonal states of the $1s$ shell (same spatial wave function, orthogonal spin states). Then comes lithium ($Z = 3$), whose electronic configuration is:

$$\text{Li}: 1s^2, 2s \tag{13}$$

The $1s$ shell can accept only two electrons, so the third one must go into the level which is directly above it, that is, according to figure 2, into the $2s$ shell. This shell can accept a second electron, which gives beryllium ($Z = 4$) the electronic configuration:

$$\text{Be}: 1s^2, 2s^2 \tag{14}$$

For $Z > 4$, the $2p$ shell (*cf.* fig. 2) is the first to be gradually filled, and so on. As the number Z of electrons increases, higher and higher electronic shells are brought in (on the right-hand side of figure 2, we have shown, opposite each of the lowest shells, the symbols of the atoms for which this shell is the outermost). Thus, we obtain the configurations of the ground state for all the atoms. This explains Mendeleev's classification. However, it must be noted that levels which are very close to each other (those grouped in brackets in figure 2) may be filled in a very irregular fashion. For example, although figure 2 gives the $4s$ shell a lower energy than that of the $3d$ shell, chromium ($Z = 24$) has five $3d$ electrons although the $4s$ shell is incomplete. Similar irregularities arise for copper ($Z = 29$), niobium ($Z = 41$), etc.

COMMENTS:

(*i*) The electronic configurations which we have analyzed characterize the ground state of various atoms in the central-field approximation. The lowest excited states of the Hamiltonian H_0 are obtained when one of the electrons moves to an individual energy level which is higher than the last shell occupied in the ground state. We shall see, for example, in complement B_{XIV}, that the first excited configuration of the helium atom is:

$$1s, 2s \tag{15}$$

(*ii*) A single non-zero Slater determinant is associated with an electronic configuration ending with a complete shell, since there are then as many orthogonal individual states as there are electrons. Thus, the ground state of the rare gases (... ns^2, np^6) is non-degenerate, as is that of the alkaline-earths (..., ns^2). On the other hand, when the number of external electrons is smaller than the degree of degeneracy of the outermost shell, the ground state of the atom is degenerate. For the alkalines (..., ns), the degree of degeneracy is equal to 2; for carbon ($1s^2$, $2s^2$, $2p^2$), it is equal to $C_6^2 = 15$, since two individual states can be chosen arbitrarily from the six orthogonal states constituting the $2p$ shell.

(*iii*) It can be shown that, for a complete shell, the total angular momentum is zero, as are the total orbital angular momentum and the total spin (the sums, respectively, of the orbital angular momenta and the spins of the electrons occupying this shell). Consequently, the angular momentum of an atom* is due only to its outer electrons. Thus, the total angular momentum of a helium atom in its ground state is zero, and that of an alkali metal is equal to 1/2 (a single external electron of zero orbital angular momentum and spin 1/2).

References and suggestions for further reading:

Pauling and Wilson (1.9), chap. IX; Levine (12.3), chap. 11, §§1, 2 and 3; Kuhn (11.1), chap. IV, §§A and B; Schiff (1.18), §47; Slater (1.6), chap. 6; Landau and Lifshitz (1.19), §§ 68, 69 and 70. See also references of chap. XI (Hartree and Hartree-Fock methods).

The shell model in nuclear physics : Valentin (16.1), chap. VI; Preston (16.4), chap. 7; Deshalit and Feshbach (16.6), chap. IV and V. See also articles by Mayer (16.20), Peierls (16.21) and Baranger (16.22).

* The angular momentum being discussed here is that of the electronic cloud of the atom. The nucleus also possesses an angular momentum which should be added to this one.

Complement B_{XIV}

ENERGY LEVELS OF THE HELIUM ATOM: CONFIGURATIONS, TERMS, MULTIPLETS

In the preceding complement, we studied many-electron atoms in the central-field approximation in which the electrons are independent. This enabled us to introduce the concept of a configuration. We shall evaluate the corrections which must be made to this approximation, taking into account the inter-electron electrostatic repulsion more precisely. In order to simplify the reasoning, we shall confine ourselves to the simplest many-electron atom, the helium atom. We shall show that, under the effect of the inter-electron electrostatic repulsion, the configurations of this atom (§1) split into spectral terms (§2), which give rise to fine-structure multiplets (§3) when smaller terms in the atomic Hamiltonian (magnetic interactions) are taken into account. The concepts we shall bring out in this treatment can be generalized to more complex atoms.

1. The central-field approximation. Configurations

a. THE ELECTROSTATIC HAMILTONIAN

As in the preceding complement, we shall take into account only the electrostatic forces at first, writing the Hamiltonian of the helium atom [formula (C-24) of chapter XIV] in the form:

$$H = H_0 + W \tag{1}$$

where:

$$H_0 = \frac{\mathbf{P}_1^2}{2m_e} + \frac{\mathbf{P}_2^2}{2m_e} + V_c(R_1) + V_c(R_2) \tag{2}$$

and:

$$W = -\frac{2e^2}{R_1} - \frac{2e^2}{R_2} + \frac{e^2}{|\mathbf{R}_1 - \mathbf{R}_2|} - V_c(R_1) - V_c(R_2) \tag{3}$$

The central potential $V_c(r)$ is chosen so as to make W a small correction of H_0.

When W is neglected, the electrons can be considered to be independent (although their average electrostatic repulsion is partially taken into account by the potential V_c). The energy levels of H_0 then define the electronic configurations which we shall study in this section. We shall then examine the effect of W by using stationary perturbation theory in §2.

b. THE GROUND STATE CONFIGURATION AND FIRST EXCITED CONFIGURATIONS

According to the discussion of complement A_{XIV} (§ 2), the configurations of the helium atom are specified by the quantum numbers n, l and n', l' of the two electrons (placed in the central potential V_c). The corresponding energy E_c can be written:

$$E_c = E_{n,l} + E_{n',l'} \tag{4}$$

Thus (fig. 1), the ground state configuration, written $1s^2$, is obtained when the two electrons are in the $1s$ shell; the first excited configuration $1s$, $2s$, when one electron is in the $1s$ shell and the other one is in the $2s$ shell. Similarly, the second excited configuration is the $1s$, $2p$ configuration.

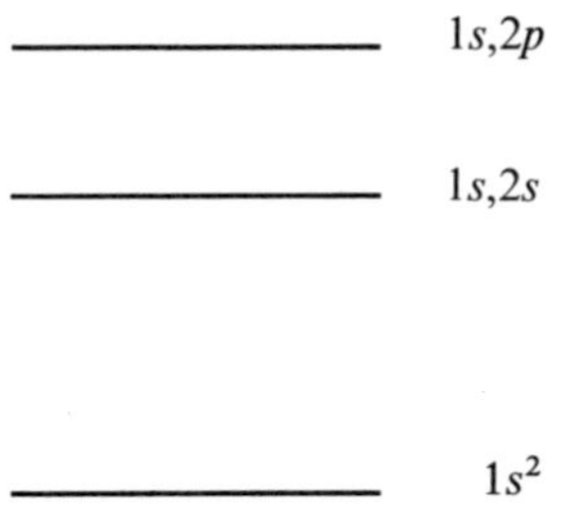

$1s,2p$

$1s,2s$

$1s^2$

FIGURE 1

The ground state configuration and first excited configurations of the helium atom (the energies are not shown to scale).

The excited configurations of the helium atom are of the form $1s$, $n'l'$. Actually, there also exist "doubly excited" configurations of the type nl, $n'l'$ (with $n, n' > 1$). But, for helium, their energy is greater than the ionization energy E_I of the atom (the limit of the energy of the configuration $1s$, $n'l'$ when $n' \longrightarrow \infty$). Most of the corresponding states, therefore, are very unstable : they tend to dissociate rapidly into an ion and an electron and are called "autoionizing states". However, there exist levels belonging to doubly excited configurations which are not autoionizing, but which decay by emitting photons. Some of the corresponding spectral lines have been observed experimentally.

c. DEGENERACY OF THE CONFIGURATIONS

Since V_c is central and not spin-dependent, the energy of a configuration does not depend on the magnetic quantum numbers m and m' $(-l \leqslant m \leqslant l, -l' \leqslant m' \leqslant l')$ or on the spin quantum numbers ε and ε' $(\varepsilon = \pm, \varepsilon' = \pm)$ associated with the two electrons. Most of the configurations, therefore, are degenerate; it is this degeneracy which we shall now calculate.

A state belonging to a configuration is defined by specifying the four quantum numbers (n, l, m, ε) and $(n', l', m', \varepsilon')$ of each electron. Since the electrons are identical particles, the symmetrization postulate must be taken into account. The physical ket associated with this state can, according to the results of §C-3-b of chapter XIV, be written in the form:

$$| n, l, m, \varepsilon ; n', l', m', \varepsilon' \rangle = \frac{1}{\sqrt{2}}(1 - P_{21}) \,|\, 1 : n, l, m, \varepsilon ; 2 : n', l', m', \varepsilon' \rangle \tag{5}$$

Pauli's principle excludes the states of the system for which the two electrons would be in the same individual quantum state $(n = n', l = l', m = m', \varepsilon = \varepsilon')$. According to the discussion of § C-3-b of chapter XIV, the set of physical kets (5) for which n, l, n', l' are fixed and which are not null (that is, not excluded by Pauli's principle) constitute an orthonormal basis in the subspace $\mathscr{E}(n, l; n', l')$ of $\mathscr{E}_A$ associated with the configuration $nl, n'l'$.

To evaluate the degeneracy of a configuration $nl, n'l'$, we shall distinguish between two cases:

(*i*) The two electrons are not in the same shell (we do not have $n = n'$ and $l = l'$).

The individual states of the two electrons can never coincide, and $m, m', \varepsilon, \varepsilon'$ can independently take on any value. The degeneracy of the configuration, consequently, is equal to:

$$2(2l + 1) \times 2(2l' + 1) = 4(2l + 1)(2l' + 1) \tag{6}$$

The $1s$, $2s$ and $1s$, $2p$ configurations enter into this category; their degeneracies are equal to 4 and 12 respectively.

(*ii*) The two electrons are in the same shell $(n = n'$ and $l = l')$.

In this case, the states for which $m = m'$ and $\varepsilon = \varepsilon'$ must be excluded. Since the number of distinct individual quantum states is equal to $2(2l + 1)$, the degree of degeneracy of the nl^2 configuration is equal to the number of pairs that can be formed from these individual states (*cf.* § C-3-b of chapter XIV), that is:

$$C^2_{2(2l+1)} = (2l + 1)(4l + 1) \tag{7}$$

Thus, the $1s^2$ configuration, which enters into this category, is not degenerate. It is useful to expand the Slater determinant corresponding to this configuration. If, in (5), we set $n = n' = 1$, $l = l' = m = m' = 0$, $\varepsilon = +$, $\varepsilon' = -$, we obtain, writing the spatial part as a common factor:

$$| 1s^2 \rangle = | 1 : 1, 0, 0; 2 : 1, 0, 0 \rangle \otimes \frac{1}{\sqrt{2}}(| 1 : + ; 2 : - \rangle - | 1 : - ; 2 : + \rangle) \tag{8}$$

In the spin part of (8), we recognize the expression for the singlet state $| S = 0, M_S = 0 \rangle$, where S and M_S are the quantum numbers related to the total spin $\mathbf{S} = \mathbf{S}_1 + \mathbf{S}_2$ (*cf.* chap. X, §B-4). Thus, although the Hamiltonian H_0 does not depend on the spins, the constraints introduced by the symmetrization postulate require the total spin of the ground state to have the value $S = 0$.

2. The effect of the inter-electron electrostatic repulsion: exchange energy, spectral terms

We shall now study the effect of W by using stationary perturbation theory. To do so, we must diagonalize the restriction of W inside the subspace $\mathscr{E}(n, l; n', l')$ associated with the nl, $n'l'$ configuration. The eigenvalues of the corresponding matrix give the corrections of the configuration energy E_c to first order in W; the associated eigenstates are the zero-order eigenstates.

To calculate the matrix which represents W inside $\mathscr{E}(n, l; n', l')$, we can choose any basis, in particular, the basis of kets (5). Actually, it is to our advantage to use a basis which is well adapted to the symmetries of W. We shall see that we can choose a basis in which the restriction of W is already diagonal.

a. CHOICE OF A BASIS OF $\mathscr{E}(n, l; n', l')$ ADAPTED TO THE SYMMETRIES OF W

α. *Total orbital momentum* **L** *and total spin* **S**

W does not commute with the individual orbital angular momenta $\mathbf{L}_1$ and $\mathbf{L}_2$ of each electron. However, we have already shown (*cf.* chap. X, §A-2) that, if **L** denotes the total orbital angular momentum:

$$\mathbf{L} = \mathbf{L}_1 + \mathbf{L}_2 \tag{9}$$

we have:

$$[W, \mathbf{L}] = \left[\frac{e^2}{R_{12}}, \mathbf{L}\right] = \mathbf{0} \tag{10}$$

Therefore, **L** is a constant of the motion★. Moreover, since W does not act in the spin state space, this is also true for the total spin **S**:

$$[W, \mathbf{S}] = \mathbf{0} \tag{11}$$

Now, consider the set of four operators, $\mathbf{L}^2$, $\mathbf{S}^2$, L_z, S_z. They commute with each other and with W. We shall show that they constitute a C.S.C.O. in the subspace $\mathscr{E}(n, l; n', l')$ of $\mathscr{E}_A$. This will enable us in § b to find directly the eigenvalues of the restriction of W in this subspace.

★ This result is related to the fact that, under a rotation involving both electrons, the distance between them, R_{12}, is invariant. However, it changes if only one of the two electrons is rotated. This is why W commutes with neither $\mathbf{L}_1$ nor $\mathbf{L}_2$.

To do this, we shall return to the space $\mathscr{E}$, the tensor product of the state spaces $\mathscr{E}(1)$ and $\mathscr{E}(2)$ relative to the two electrons, numbered arbitrarily. The subspace $\mathscr{E}(n, l; n', l')$ of $\mathscr{E}_A$ associated with the nl, $n'l'$ configuration can be obtained by antisymmetrizing the various kets of the subspace $\mathscr{E}_{n,l}(1) \otimes \mathscr{E}_{n',l'}(2)$ of $\mathscr{E}$★. If we choose the basis $| 1 : n, l, m, \varepsilon \rangle \otimes | 2 : n', l', m', \varepsilon' \rangle$ in this subspace, we obtain the basis of physical kets (5) by antisymmetrization.

However, we know from the results of chapter X that we can also choose in $\mathscr{E}_{n,l}(1) \otimes \mathscr{E}_{n',l'}(2)$ another basis composed of common eigenvectors of $\mathbf{L}^2$, L_z, $\mathbf{S}^2$, S_z and entirely defined by the specification of the corresponding eigenvalues. We shall write this basis:

$$\{ | 1 : n, l; 2 : n', l'; L, M_L \rangle \otimes | S, M_S \rangle \} \tag{12}$$

with:

$$\begin{cases} L = l + l', l + l' - 1, ..., |l - l'| \\ S = 1, 0 \end{cases} \tag{13}$$

Since $\mathbf{L}^2$, L_z, $\mathbf{S}^2$, S_z are all symmetric operators (they commute with P_{21}), the vectors (12) remain, after antisymmetrization, eigenvectors of $\mathbf{L}^2$, L_z, $\mathbf{S}^2$, S_z with the same eigenvalues (some of them may, of course, have a zero projection onto $\mathscr{E}_A$, in which case the corresponding physical states are excluded by Pauli's principle; see § β below). The non-zero kets obtained by antisymmetrization of (12) are therefore orthogonal, since they correspond to different eigenvalues of at least one of the four observables under consideration. Since they span $\mathscr{E}(n, l; n', l')$, they constitute an orthonormal basis of this subspace, which we shall write:

$$\{ | n, l; n', l'; L, M_L; S, M_S \rangle \} \tag{14}$$

with:

$$| n, l; n', l'; L, M_L; S, M_S \rangle = c(1 - P_{21}) \{ | 1 : n, l; 2 : n', l'; L, M_L \rangle \otimes | S, M_S \rangle \} \tag{15}$$

where c is a normalization constant. $\mathbf{L}^2$, L_z, $\mathbf{S}^2$, S_z therefore form a C.S.C.O. inside $\mathscr{E}(n, l; n', l')$.

Now, we shall introduce the permutation operator $P_{21}^{(S)}$ in the spin state space:

$$P_{21}^{(S)} | 1 : \varepsilon; 2 : \varepsilon' \rangle = | 1 : \varepsilon'; 2 : \varepsilon \rangle \tag{16}$$

We showed in §B-4 of chapter X [*cf.* comment (*ii*)] that:

$$P_{21}^{(S)} | S, M_S \rangle = (-1)^{S+1} | S, M_S \rangle \tag{17}$$

★ We could also start with the subspace $\mathscr{E}_{n',l'}(1) \otimes \mathscr{E}_{n,l}(2)$ [*cf.* comment (*i*) of §B-2-c of chapter XIV, p. 1386].

Furthermore, if $P_{21}^{(0)}$ is the permutation operator in the state space of the orbital variables, we have:

$$P_{21} = P_{21}^{(0)} \otimes P_{21}^{(S)} \tag{18}$$

Using (17) and (18), we can, finally, put (15) in the form:

$$\begin{aligned} &| n, l; n', l'; L, M_L; S, M_S \rangle \\ &= c \left\{ \left[1 - (-1)^{S+1} P_{21}^{(0)}\right] | 1 : n, l; 2 : n', l'; L, M_L \rangle \right\} \otimes | S, M_S \rangle \end{aligned} \tag{19}$$

β. *Constraints imposed by the symmetrization postulate*

We have seen that the dimension of $\mathscr{E}(n, l; n', l')$ is not always equal to $4(2l + 1)(2l' + 1)$, that is, to the dimension of $\mathscr{E}_{n,l}(1) \otimes \mathscr{E}_{n',l'}(2)$. Certain kets of $\mathscr{E}_{n,l}(1) \otimes \mathscr{E}_{n',l'}(2)$ can therefore have a zero projection onto $\mathscr{E}(n, l; n', l')$. It is interesting to study the consequences for the basis (14) of this constraint imposed by the symmetrization postulate.

First of all, assume that the two electrons do not occupy the same shell. It is then easy to see that the orbital part of (19) is a sum or a difference of two orthogonal kets and, consequently, is never zero*. Since the same is true of $| S, M_S \rangle$, we see that all the possible values of L and S [*cf.* formula (13)] are allowed. For example, for the $1s$, $2s$ configuration, we can have $S = 0$, $L = 0$ and $S = 1$, $L = 0$; for the $1s$, $2p$ configuration, we can have $S = 0$, $L = 1$ and $S = 1$, $L = 1$, etc.

If we now assume that the two electrons occupy the same shell, we have $n = n'$ and $l = l'$, and certain of the kets (19) can be zero. Let us write $| 1 : n, l; 2 : n', l'; L, M_L \rangle$ in the form:

$$\begin{aligned} &| 1 : n, l; 2 : n', l'; L, M_L \rangle \\ &\qquad = \sum_m \sum_{m'} \langle l, l'; m, m' | L, M_L \rangle | 1 : n, l, m; 2 : n', l', m' \rangle \end{aligned} \tag{20}$$

According to relation (25) of complement B_X:

$$\langle l, l; m, m' | L, M_L \rangle = (-1)^L \langle l, l; m', m | L, M_L \rangle \tag{21}$$

By using (20), we then obtain:

$$P_{21}^{(0)} | 1 : n, l; 2 : n, l; L, M_L \rangle = (-1)^L | 1 : n, l; 2 : n, l; L, M_L \rangle \tag{22}$$

Substituting this result into (19), we obtain**:

$$| n, l; n, l; L, M_L; S, M_S \rangle = \begin{cases} 0 & \text{if } L + S \text{ is odd} \\ | 1 : n, l; 2 : n, l; L, M_L \rangle \otimes | S, M_S \rangle & \text{if } L + S \text{ is even} \end{cases} \tag{23}$$

Therefore, L and S cannot be arbitrary : $L + S$ must be even. In particular, for the $1s^2$ configuration, we must have $L = 0$, so $S = 1$ is excluded. This is a result found previously.

* The normalization constant c is then equal to $\frac{1}{\sqrt{2}}$.

** The normalization constant is then 1/2.

Finally, note that the symmetrization postulate introduces a close correlation between the symmetry of the orbital part and that of the spin part of the physical ket (19). Since the total ket must be antisymmetric, and the spin part, depending on the value of S, is symmetric ($S = 1$) or antisymmetric ($S = 0$), the orbital part must be antisymmetric when $S = 1$ and symmetric when $S = 0$. We shall see the importance of this point later.

b. SPECTRAL TERMS. SPECTROSCOPIC NOTATION

W commutes with the four observables $\mathbf{L}^2$, L_z, $\mathbf{S}^2$, S_z, which form a C.S.C.O. inside $\mathscr{E}(n, l; n', l')$. It follows that the restriction of W inside $\mathscr{E}(n, l; n', l')$ is diagonal in the basis:

$$\{ \mid n, l; n', l'; L, M_L; S, M_S \rangle \}$$

and has eigenvalues of:

$$\delta(L, S) = \langle n, l; n', l'; L, M_L; S, M_S \mid W \mid n, l; n', l'; L, M_L; S, M_S \rangle \qquad (24)$$

This energy depends neither on M_L nor on M_S, since relations (10) and (11) imply that W commutes not only with L_z and S_z but also with $L_\pm$ and $S_\pm$: W is therefore a scalar operator in both the orbital state space and the spin state space (*cf.* complement B_{VI}, §§5-b and 6-c).

Inside each nl, $n'l'$ configuration, we thus obtain energy levels $E_c(n, l; n', l') + \delta(L, S)$, labeled by their values of L and S. Each of them is $(2L + 1)(2S + 1)$-fold degenerate. Such levels are called *spectral terms* and denoted in the following way. With each value of L is associated, in spectroscopic notation (chap. VII, §C-4-b) a letter of the alphabet; we write the corresponding capital letter and add, at the upper left, a number equal to $2S + 1$. For example, the $1s^2$ configuration leads to a single spectral term, written ^1S (the ^3S, as we have seen, is forbidden by Pauli's principle). The $1s, 2s$ configuration produces two terms, ^1S (non-degenerate) and ^3S (three-fold degenerate) : the $1s, 2p$ configuration, two terms, ^1P (degeneracy 3) and ^3P (degeneracy 9). For a more complicated configuration such as, for example, $2p^2$, we obtain (*cf.* § 2-a-β) the spectral terms ^1S, ^1D and ^3P ($L + S$ must be even), etc.

Under the effect of the electrostatic repulsion, the degeneracy of each configuration is therefore partially removed (the $1s^2$ configuration, which is non-degenerate, is simply shifted). We shall study this effect in greater detail in the simple example of the $1s, 2s$ configuration. We shall try to understand why the two terms ^1S and ^3S resulting from this configuration, and whose total spin values are different, have different energies although the original Hamiltonian is purely electrostatic.

c. DISCUSSION

α. *Energies of the spectral terms arising from the $1s, 2s$ configuration*

In the $1s, 2s$ configuration, $l = l' = L = 0$. It is then easy to obtain from (20):

$$\begin{aligned} &| 1 : n = 1, l = 0; 2 : n' = 2, l' = 0; L = M_L = 0 \rangle = \\ &| 1 : n = 1, l = m = 0; 2 : n' = 2, l' = m' = 0 \rangle \end{aligned} \qquad (25)$$

a vector which we shall write, more simply, $| 1 : 1s ; 2 : 2s \rangle$. If $| {}^3S, M_S \rangle$ and $| {}^1S, 0 \rangle$ denote the states corresponding to the two spectral terms ${}^3S$ and ${}^1S$ arising from the 1s, 2s configuration, we obtain, substituting (25) into (19):

$$\begin{cases} | {}^3S, M_S \rangle = \dfrac{1}{\sqrt{2}} [(1 - P_{21}^{(0)}) | 1 : 1s ; 2 : 2s \rangle] \otimes | S = 1, M_S \rangle & \text{(26-a)} \\ | {}^1S, 0 \rangle = \dfrac{1}{\sqrt{2}} [(1 + P_{21}^{(0)}) | 1 : 1s ; 2 : 2s \rangle] \otimes | S = 0, M_S = 0 \rangle & \text{(26-b)} \end{cases}$$

Since W does not act on the spin variables, the eigenvalues given by (24) can be written:

$$\delta({}^3S) = \frac{1}{2} \langle 1 : 1s ; 2 : 2s | (1 - P_{21}^{(0)}) \, W \, (1 - P_{21}^{(0)}) | 1 : 1s ; 2 : 2s \rangle \qquad \text{(27-a)}$$

$$\delta({}^1S) = \frac{1}{2} \langle 1 : 1s ; 2 : 2s | (1 + P_{21}^{(0)}) \, W \, (1 + P_{21}^{(0)}) | 1 : 1s ; 2 : 2s \rangle \qquad \text{(27-b)}$$

(we have used the fact that $P_{21}^{(0)}$ is Hermitian). Moreover, $P_{21}^{(0)}$ commutes with W, and the square of $P_{21}^{(0)}$ is the identity operator. Therefore:

$$(1 \pm P_{21}^{(0)}) \, W \, (1 \pm P_{21}^{(0)}) = (1 \pm P_{21}^{(0)})^2 \, W = 2(1 \pm P_{21}^{(0)}) \, W \qquad (28)$$

Finally, we obtain:

$$\begin{cases} \delta({}^3S) = K - J & \text{(29-a)} \\ \delta({}^1S) = K + J & \text{(29-b)} \end{cases}$$

with:

$$K = \langle 1 : 1s ; 2 : 2s | W | 1 : 1s ; 2 : 2s \rangle \qquad (30)$$

$$\begin{aligned} J &= \langle 1 : 1s ; 2 : 2s | P_{21}^{(0)} W | 1 : 1s ; 2 : 2s \rangle \\ &= \langle 1 : 2s ; 2 : 1s | W | 1 : 1s ; 2 : 2s \rangle \end{aligned} \qquad (31)$$

K therefore represents an overall shift of the energy of the two terms and does not contribute to their separation. J is more interesting, as it introduces an energy difference between the ${}^3S$ and ${}^1S$ terms (*cf.* fig. 2). We shall therefore study it in a little more detail.

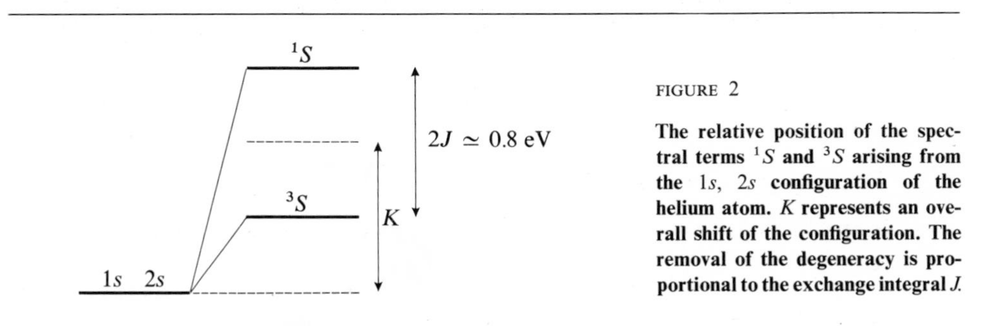

FIGURE 2

The relative position of the spectral terms ${}^1S$ and ${}^3S$ arising from the 1s, 2s configuration of the helium atom. K represents an overall shift of the configuration. The removal of the degeneracy is proportional to the exchange integral J.

β. *The exchange integral*

When we substitute expression (3) for W into (31), there appear terms of the form:

$$\langle 1 : 2s ; 2 : 1s \mid V_c(R_1) \mid 1 : 1s ; 2 : 2s \rangle = \langle 1 : 2s \mid V_c(R_1) \mid 1 : 1s \rangle \langle 2 : 1s \mid 2 : 2s \rangle \tag{32}$$

Now, the scalar product of the two orthogonal states, $\mid 2 : 1s \rangle$ and $\mid 2 : 2s \rangle$ is zero. Expression (32) is then equal to zero. The same type of reasoning shows that the terms which arise from the operators $V_c(R_2)$, $-\ 2e^2/R_1$, $-\ 2e^2/R_2$ are also zero, since each of these operators acts only in the single-electron spaces while the state of the two electrons is different in the ket and bra of (31). Finally, there remains:

$$J = \langle 1 : 2s ; 2 : 1s \mid \frac{e^2}{|\mathbf{R}_1 - \mathbf{R}_2|} \mid 1 : 1s ; 2 : 2s \rangle \tag{33}$$

J therefore involves only the electrostatic repulsion between the electrons.

Let $\varphi_{n,l,m}(\mathbf{r})$ be the wave functions associated with the states $\mid n, l, m \rangle$ (the stationary states of an electron in the central potential V_c):

$$\varphi_{n,l,m}(\mathbf{r}) = \langle \mathbf{r} \mid n, l, m \rangle \tag{34}$$

In the $\{ \mid \mathbf{r} \rangle \}$ representation, the calculation of J from (33) yields:

$$J = \int d^3r_1 \int d^3r_2 \; \varphi^*_{2,0,0}(\mathbf{r}_1)\, \varphi^*_{1,0,0}(\mathbf{r}_2) \frac{e^2}{|\mathbf{r}_1 - \mathbf{r}_2|} \varphi_{1,0,0}(\mathbf{r}_1)\, \varphi_{2,0,0}(\mathbf{r}_2) \tag{35}$$

This integral is called the "exchange integral". We shall not calculate it explicitly here; we point out, however, that it is positive.

γ. *The physical origin of the energy difference between the two spectral terms*

We see from expressions (26) and (27) that the origin of the energy separation of the ^3S and ^1S terms lies in the symmetry differences of the orbital parts of these terms. As we emphasized at the end of § 2-a, a triplet term ($S = 1$) must have an orbital part which is antisymmetric under exchange of the two electrons; hence the $-$ sign before $P_{21}^{(0)}$ in (26-a) and (27-a). On the other hand, a singlet term ($S = 0$) must have a symmetric orbital part [+ sign in (26-b) and (27-b)].

This explains the relative position of the ^3S and ^1S terms shown in figure 2. For the singlet term, the orbital wave function is symmetric with respect to exchange of the two electrons, which then have a non-zero probability of being at the same point in space. This is why the electrostatic repulsion, which gives an energy of e^2/r_{12} which is large when the electrons are near each other, significantly increases the singlet state energy. On the other hand, for the triplet state, the orbital function is antisymmetric with respect to exchange of the two electrons, which then have a zero probability of being at the same point in space. The mean value of the electrostatic repulsion is then smaller. Therefore, the energy difference between the singlet and triplet states arises from the fact that the correlations between the orbital variables of the two electrons depend, because of the symmetrization postulate, on the value of the total spin.

δ. *Analysis of the role played by the symmetrization postulate*

At this point in the discussion, it might be thought that the degeneracy of a configuration is removed by the symmetrization postulate. We shall show* that this is not the case. This postulate merely fixes the value of the total spin of the terms arising from a given configuration (because of the inter-electron electrostatic repulsion).

To see this, imagine for a moment that we do not need to apply the symmetrization postulate. Suppose, for example, that the two electrons are replaced by two particles (fictitious, of course) of the same mass, the same charge and the same spin as the electrons but with another intrinsic property which permits us to distinguish between them (without, however, changing the Hamiltonian H of the problem, which is still given by formula (1)]. Since H is not spin-dependent and we do not have to apply the symmetrization postulate, we can ignore the spins completely until the end of the calculations, and then multiply the degeneracies obtained by 4. The energy level of H_0 corresponding to the $1s$, $2s$ configuration is two-fold degenerate from the orbital point of view because two orthogonal states, $| 1 : 1s ; 2 : 2s \rangle$ and $| 1 : 2s ; 2 : 1s \rangle$ correspond to it (they are different physical states since the two particles are of different natures). To study the effect of W, we must diagonalize W in the two-dimensional space spanned by these two kets. The corresponding matrix can be written :

$$\begin{pmatrix} K & J \\ J & K \end{pmatrix} \tag{36}$$

where J and K are given by (30) and (31) [the two diagonal elements of (36) are equal because W is invariant under permutation of the two particles]. Matrix (36) can be diagonalized immediately. The eigenvalues found are $K + J$ and $K - J$, associated respectively with the symmetric and antisymmetric linear combinations of the two kets $| 1 : 1s ; 2 : 2s \rangle$ and $| 1 : 2s ; 2 : 1s \rangle$. The fact that these orbital eigenstates have well-defined symmetries relative to exchange of the two particles has nothing to do with Pauli's principle. It arises only from the fact that W commutes with $P_{21}^{(0)}$ (common eigenstates of W and $P_{21}^{(0)}$ can therefore be found).

When the two particles are not identical, we obtain the same arrangement of levels and the same orbital symmetry as before. On the other hand, the degeneracy of the levels is obviously different: the lower level, with energy $K - J$, can have a total spin of either $S = 0$ or $S = 1$, as can the upper level.

If we return to the real helium atom, we now see very clearly the role played by Pauli's principle. It is not responsible for the splitting of the initial level $1s$, $2s$ into the two energy levels $K + J$ and $K - J$, since this splitting would also appear for two particles of different natures. Similarly, the symmetric or antisymmetric character of the orbital part of the eigenvectors is related to the invariance of the electrostatic interaction under permuation of the two electrons. Pauli's principle merely forbids the lower state to have a total spin $S = 0$ and the upper state to have a total spin $S = 1$, since the corresponding states would be globally symmetric, which is unacceptable for fermions.

* See also comment (*i*) of § C-4-a-β of chapter XIV, p. 1395.

ε. *The effective spin-dependent Hamiltonian*

We replace W by the operator:

$$\tilde{W} = \alpha + \beta \, \mathbf{S}_1 \, . \, \mathbf{S}_2 \tag{37}$$

where $\mathbf{S}_1$ and $\mathbf{S}_2$ denote the two electron spins. We also have:

$$\tilde{W} = \alpha - \frac{3\beta\hbar^2}{4} + \frac{\beta}{2}\mathbf{S}^2 \tag{38}$$

so that the eigenstates of $\tilde{W}$ are the triplet states, with the eigenvalue $\alpha + \beta\hbar^2/4$, and the singlet state, with the eigenvalue $\alpha - 3\beta\hbar^2/4$. Therefore, if we set :

$$\begin{cases} \alpha = K - \dfrac{J}{2} \\ \beta = -\dfrac{2J}{\hbar^2} \end{cases} \tag{39}$$

we obtain, by diagonalizing $\tilde{W}$, the same eigenstates and eigenvalues we found above*. We can then consider that it is as if the perturbation responsible for the appearance of the terms were $\tilde{W}$ (the "effective" Hamiltonian), which is of the same form as the magnetic interaction between two spins. However, one should not conclude that the coupling energy between the electrons, which is responsible for the appearance of the two terms, is of magnetic origin : two magnetic moments equal to that of the electron and placed at a distance of the order of 1 Å from each other would have an interaction energy much smaller than J. However, because of the very simple form of $\tilde{W}$, this effective Hamiltonian is often used instead of W.

An analogous situation arises in the study of ferromagnetic materials. In these substances, the electron spins tend to align themselves parallel to each other. Since the spin state is then completely symmetric, Pauli's principle requires the orbital state to be completely antisymmetric. For the same reasons as for the helium atom, the electronic repulsion energy is then minimal. When we study such phenomena, we often use effective Hamiltonians of the same type as (37). However, it must be noted that the physical interaction which is at the origin of the coupling is again electrostatic and not magnetic.

COMMENTS:

(*i*) The $1s$, $2p$ configuration can be treated in the same way. We then have $L = 1$, so that $M_L = +1, 0$ or -1. As for the $1s$, $2s$ configuration, the shells occupied by the two electrons are different, so that the two terms ^3P and ^1P exist simultaneously. The first one is nine-fold degenerate, and the second, three-fold. It can be shown, as above, that the ^3P term has an energy lower than that of the ^1P term, and the difference between the two energies is proportional to an exchange integral which is analogous to the one written

* We must, obviously, keep only the eigenvectors of $\tilde{W}$ which belong to $\mathscr{E}_A$.

in (35). We would proceed in the same way for all other configurations of the type $1s, n'l'$.

(*ii*) We have treated W like a perturbation of H_0. For this approach to be coherent, the energy shifts associated with W [for example, the exchange integral written in (35)] must be much smaller than the energy differences between configurations. Now, this is not the case; for the $1s,2s$ and $1s,2p$ configurations, for example, while the energy difference $\Delta E(^1S - {}^3S)$ in the $1s,2s$ configuration is of the order of 0.8 eV, the minimum distance between levels is $\Delta E[(1s,2p)^3P - (1s,2s)^1S] \simeq 0.35$ eV. We might therefore believe that it is not valid to treat W like a perturbation of H_0.

However, the approach we have given is correct. This is due to the fact that, for all configurations of the type $1s,n'l'$, we have $L = l'$. Therefore, W, which, according to (10), commutes with **L**, has zero matrix elements between the states of the $1s,2s$ configuration and those of the $1s, 2p$ configuration, which correspond to different values of L. W couples a $1s,n'l'$ configuration only to configurations with distinctly higher energies, of the $1s,n''l''$ type with $l'' = l'$ (only the values of n are different) or of the $nl,n''l''$ type, with n and n'' different from 1 (the angular momenta l and l'' can be added to give l').

3. Fine-structure levels; multiplets

Thus far, we have taken into account in the Hamiltonian only interactions of purely electrostatic origin: we have neglected all effects of relativistic and magnetic origin. Actually, such effects exist, and we have already studied them in the case of the hydrogen atom (*cf.* chap. XII, § B-1), where they arise from the variation of the electron mass with the velocity, from the $\mathbf{L} \cdot \mathbf{S}$ spin-orbit coupling, and from the Darwin term. For helium, the situation is more complicated because of the simultaneous presence of two electrons. For example, there is a spin-spin magnetic coupling term in the Hamiltonian (*cf.* complement B_{XI}) which acts in both the spin state space and the orbital state space of the two electrons*. Nevertheless, a great simplification arises from the fact that the energy differences associated with these couplings of relativistic and magnetic origin are much weaker than those which exist between two different spectral terms. This enables us to treat the corresponding Hamiltonian (the fine-structure Hamiltonian) like a perturbation.

The detailed study of the fine structure levels of helium falls outside the domain of this complement. We shall confine ourselves here to describing the symmetries of the problem and indicating how to distinguish between the different energy levels. We shall use the fact that the fine-structure Hamiltonian H_{SF} is invariant under a simultaneous rotation of all the orbital and spin variables. This means (*cf.* complement B_{VI}, § 6) that, if **J** denotes the total angular momentum of the electrons:

$$\mathbf{J} = \mathbf{L} + \mathbf{S} \tag{40}$$

we have:

$$[H_{SF}, \mathbf{J}] = \mathbf{0} \tag{41}$$

* See for example § 19.6 in Sobel'man (11.12) for an explicit expression of the different terms of the fine structure Hamiltonian (Breit Hamiltonian).

On the other hand, the fine-structure Hamiltonian changes if the rotation acts only on the orbital variables or only on the spins :

$$[H_{SF}, \mathbf{L}] = - [H_{SF}, \mathbf{S}] \neq \mathbf{0} \tag{42}$$

These properties can easily be seen for the operators $\sum_i \xi(r_i)\mathbf{L}_i \cdot \mathbf{S}_i$, for example, or for the dipole-dipole magnetic interaction Hamiltonian (*cf.* complement B_{XI}).

The state space associated with a term is spanned by the states $| n, l; n', l'; L, M_L; S, M_S \rangle$ written in (19), where L and S are fixed, and where:

$$\begin{cases} - L \leqslant M_L \leqslant + L \\ - S \leqslant M_S \leqslant + S \end{cases} \tag{43}$$

In this subspace, it can be shown that $\mathbf{J}^2$ and J_z form a C.S.C.O. which, according to (41), commutes with H_{SF}. The eigenvectors $| J, M_J \rangle$ common to $\mathbf{J}^2$ [eigenvalue $J(J+1)\hbar^2$] and J_z (eigenvalue $M_J\hbar$) are therefore necessarily eigenvectors of H_{SF}, with an eigenvalue which depends on J but not on M_J (this last property arises from the fact that H_{SF} commutes with J_+ and J_-). According to the general theory of addition of angular momenta, the possible values of J are:

$$J = L + S, L + S - 1, L + S - 2, ..., |L - S| \tag{44}$$

The effect of H_{SF} is therefore a partial removal of the degeneracy. For each "term", there appear as many distinct levels as there are different values of J, according to relation (44). Each of these levels is $(2J+1)$- fold degenerate and is called a "multiplet". The usual spectroscopic notation consists of denoting a multiplet by adding a right lower index equal to the value of J to the symbol representing the term from which it arises. For example, the ground state of the helium atom gives a single multiplet, ${}^1S_0$. Similarly, each of the terms ${}^1S$ and ${}^3S$ of the $1s$, $2s$ configuration leads to a single multiplet : ${}^1S_0$ and ${}^3S_1$, respectively. On the other hand, the ${}^3P$ term arising from $1s$, $2p$ yields three multiplets, ${}^3P_2$, ${}^3P_1$ and ${}^3P_0$ (*cf.* fig. 3), and so on. We point out that the measurement and theoretical calculation of the fine structure of the ${}^3P$ level of the $1s$, $2p$ configuration is of great fundamental interest, since it can lead to the very precise knowledge of the "fine structure constant", $\alpha = e^2/\hbar c$.

COMMENTS:

(*i*) For many atoms, the fine-structure Hamiltonian is essentially given by:

$$H_{SF} \simeq \sum_{i=1}^{N} \xi(R_i)\, \mathbf{L}_i \cdot \mathbf{S}_i \tag{45}$$

where $\mathbf{R}_i$, $\mathbf{L}_i$ and $\mathbf{S}_i$ denote the positions, angular momenta and spins of each of the N electrons. It can then be shown, using the Wigner-Eckart theorem (*cf.* complement D_X), that the energy of the J multiplet is proportional to $J(J+1) - L(L+1) - S(S+1)$. This result is sometimes called the "Landé interval rule".

For helium, the ${}^3P_1$ and ${}^3P_2$ levels arising from the $1s,2p$ configuration are much closer than would be predicted by this rule. This arises from the importance of the dipole-dipole magnetic coupling of the spins of the two electrons.

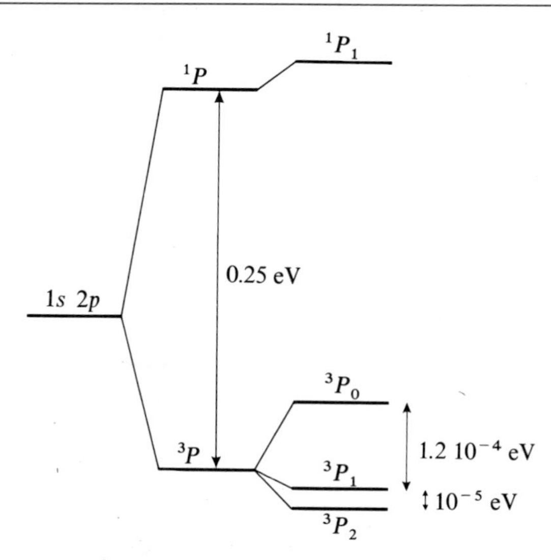

FIGURE 3

The relative position of the spectral terms and multiplets arising from the $1s,2p$ configuration of the helium atom (the splitting of the three multiplets $^3P_0$, $^3P_1$, $^3P_2$ has been greatly exaggerated in order to make the figure clearer).

(*ii*) In this complement, we have neglected the "hyperfine effects" related to nuclear spin (*cf.* chap. XII, §B-2). Such effects actually exist only for the ^{3}He isotope, whose nucleus has a spin $I = 1/2$ (the nucleus of the ^{4}He isotope has a zero spin). Each multiplet of electronic angular momentum J splits, in the case of ^{3}He, into two hyperfine levels of total angular momentum $F = J \pm 1/2$, $(2F + 1)$-fold degenerate (unless, of course, $J = 0$).

References and suggestions for further reading:

Kuhn (11.1), chap. III-B; Slater (11.8), chap. 18; Bethe and Salpeter (11.10).

Multiplet theory and the Pauli principle: Landau and Lifshitz (1.19), §§ 64 and 65; Slater (1.6), chap. 7 and (11.8), chap. 13; Kuhn (11.1), chap. V, § A; Sobel'man (11.12), chap. 2, § 5.3.

Complement C_{XIV}

PHYSICAL PROPERTIES OF AN ELECTRON GAS. APPLICATION TO SOLIDS

In complements A_{XIV} and B_{XIV}, we studied, taking the symmetrization postulate into account, the energy levels of a small number of independent electrons placed in a central potential (the shell model of many-electron atoms). Now, we shall consider systems composed of a much larger number of electrons, and we shall show that Pauli's exclusion principle has an equally spectacular effect on their behavior.

To simplify the discussion, we shall neglect interactions between electrons. Moreover, we shall assume, at first (§ 1), that they are subject to no external potential other than the one that restricts them to a given volume and which exists only in the immediate vicinity of the boundary (a free electrons gas enclosed in a "box"). We shall introduce the important concept of the *Fermi energy* E_F, which depends only on the number of electrons per unit volume. We shall also show that the physical properties of the electron gas (specific heat, magnetic susceptibility, ...) are essentially determined by the electrons whose energy is close to E_F.

A free-electron model describes the principal properties of certain metals rather well. However the electrons of a solid are actually subject to the periodic potential created by the ions of the crystal. We know that the energy levels of each electron are then grouped into allowed energy bands, separated by forbidden bands (*cf.* complements F_{XI} and O_{III}). We shall show qualitatively in § 2 that the electric conductivity of a solid is essentially determined by the position of the Fermi level of the electron system relative to the allowed energy bands. Depending on this position, the solid is an insulator or a conductor.

1. Free electrons enclosed in a box

a. GROUND STATE OF AN ELECTRON GAS; FERMI ENERGY E_F

Consider a system of N electrons, whose mutual interactions we shall neglect, and which, furthermore, are subjected to no external potential. These N electrons, however, are enclosed in a box, which, for simplicity, we shall choose to be a cube with edges of length L.

If the electrons cannot pass through the walls of the box, it is because the walls constitute practically infinite potential barriers. Since the potential energy of the electrons is zero inside the box, the problem is reduced to that of the three-dimensional infinite square well (*cf.* complements G_{II} and H_I). The stationary states of a particle in such a well are described by the wave functions:

$$\varphi_{n_x,n_y,n_z}(\mathbf{r}) = \left(\frac{2}{L}\right)^{3/2} \sin\left(n_x \frac{\pi x}{L}\right) \sin\left(n_y \frac{\pi y}{L}\right) \sin\left(n_z \frac{\pi z}{L}\right) \tag{1-a}$$

$$n_x, n_y, n_z = 1, 2, 3... \tag{1-b}$$

[expression (1-a) is valid for $0 \leqslant x, y, z \leqslant L$, since the wave function is zero outside this region]. The energy associated with φ_{n_x,n_y,n_z} is equal to:

$$E_{n_x,n_y,n_z} = \frac{\pi^2\hbar^2}{2m_e L^2}(n_x^2 + n_y^2 + n_z^2) \tag{2}$$

Of course, the electron spin must be taken into account: each of the wave functions (1) describes the spatial part of two distinct stationary states which differ by their spin orientation; these two states correspond to the same energy, since the Hamiltonian of the problem is spin-independent.

The set of these stationary states constitutes a discrete basis, enabling us to construct any state of an electron enclosed in this box (that is, whose wave function goes to zero at the walls). Note that, by increasing the dimensions of the box, we can make the interval between two consecutive individual energies as small as we wish, since this interval is inversely proportional to L^2. If L is sufficiently large, therefore, we cannot, in practice, distinguish between the discrete spectrum (2) and a continuous spectrum containing all the positive values of the energy.

The ground state of the system of the N independent electrons can be obtained by antisymmetrizing the tensor product of the N individual states associated with the lowest energies compatible with Pauli's principle. If N is small, it is thus simple to fill the first individual levels (2) and to find the ground state of the system, as well as its degree of degeneracy and the antisymmetrized kets which correspond to it. However, when N is much larger than 1 (in a macroscopic solid, N is of the order of 10^{23}), this method cannot be used in practice, and we must use a more global reasoning.

We shall begin by evaluating the number $n(E)$ of individual stationary states whose energies are lower than a given value E. To do so, we shall write expression (2) for the possible energies in the form:

$$E_{n_x,n_y,n_z} = \frac{\hbar^2}{2m_e}\mathbf{k}^2_{n_x,n_y,n_z} \tag{3}$$

with:

$$\left(\mathbf{k}_{n_x,n_y,n_z}\right)_x = n_x \frac{\pi}{L}$$

$$\left(\mathbf{k}_{n_x,n_y,n_z}\right)_y = n_y \frac{\pi}{L}$$

$$\left(\mathbf{k}_{n_x,n_y,n_z}\right)_z = n_z \frac{\pi}{L} \tag{4}$$

According to (1), a vector $\mathbf{k}_{n_x,n_y,n_z}$ corresponds to each function $\varphi_{n_x,n_y,n_z}(\mathbf{r})$. Conversely, to each of these vectors, there corresponds one and only one function φ_{n_x,n_y,n_z}. The number of states $n(E)$ can then be obtained by multiplying by 2 the number of vectors $\mathbf{k}_{n_x,n_y,n_z}$ whose modulus is smaller than $\sqrt{2m_eE/\hbar^2}$ (the factor 2 arises, of course, from the existence of electron spin). The tips of the vectors $\mathbf{k}_{n_x,n_y,n_z}$ divide $\mathbf{k}$-space into elementary cubes of edge π/L (see figure 1, in which, for simplicity, a two-dimensional rather than a three-dimensional space is

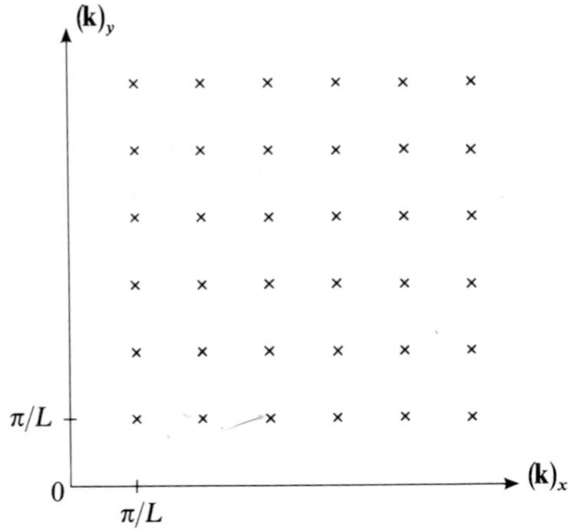

FIGURE 1

Tips of the vectors $\mathbf{k}_{n_x,n_y}$ characterizing the stationary wave functions in a two-dimensional infinite square well.

shown). Each of these tips is common to eight neighboring cubes, and each cube has eight corners. Consequently, if the elementary cubes are sufficiently small (that is, if L is sufficiently large), there can be considered to be one vector $\mathbf{k}_{n_x,n_y,n_z}$ per volume element $(\pi/L)^3$ of $\mathbf{k}$-space.

The value E of the energy which we have chosen defines, in $\mathbf{k}$-space, a sphere centered at the origin, of radius $\sqrt{2m_eE/\hbar^2}$. Only one-eigth of the volume of this sphere is involved, since the components of $\mathbf{k}$ are positive [*cf.* (1-b) and (4)]. If we divide it by the volume element $(\pi/L)^3$ associated with each stationary state, and if we take into account the factor 2 due to the spin, we obtain:

$$n(E) = 2\frac{1}{8}\frac{4}{3}\pi\left(\frac{2m_e}{\hbar^2}E\right)^{3/2}\frac{1}{(\pi/L)^3} = \frac{L^3}{3\pi^2}\left(\frac{2m_e}{\hbar^2}E\right)^{3/2} \tag{5}$$

This result enables us to calculate immediately the maximal individual energy of an electron in the ground state of the system, that is, the *Fermi energy* E_F of the electron gas. This energy E_F satisfies:

$$n(E_F) = N \tag{6}$$

which gives:

$$E_F = \frac{\hbar^2}{2m_e}\left(3\pi^2 \frac{N}{L^3}\right)^{2/3} \tag{7}$$

Note that, as might be expected, the Fermi energy depends only on the number N/L^3 of electrons per unit volume. At absolute zero, all the individual states of energy less than E_F are occupied, and all those whose energies are greater than E_F are empty. We shall see in § 1-b what happens at non-zero temperatures.

We can also deduce the *density of states* $\rho(E)$ from (5). $\rho(E)\,\mathrm{d}E$ is, by definition, the number of states whose energies are included between E and $E + \mathrm{d}E$. This state density, as we shall see later, is of considerable physical importance. It can be obtained simply by differentiating $n(E)$ with respect to E:

$$\rho(E) = \frac{\mathrm{d}n(E)}{\mathrm{d}E} = \frac{L^3}{2\pi^2}\left(\frac{2m_e}{\hbar^2}\right)^{3/2} E^{1/2} \tag{8}$$

$\rho(E)$ therefore varies like $\sqrt{E}$. At absolute zero, the number of electrons with a given energy between E and $E + \mathrm{d}E$ (less than E_F, of course) is equal to $\rho(E)\,\mathrm{d}E$. By using the value (7) of the Fermi energy E_F, we can put $\rho(E)$ in the form:

$$\rho(E) = \frac{3}{2} N \frac{E^{1/2}}{E_F^{3/2}} \tag{9}$$

COMMENT:

It can be seen from (5) that the dimensions of the box are involved only through the intermediary of the volume element $(\pi/L)^3$ associated, in **k**-space, with each stationary state. If, instead of choosing a cubic box of edge L, we had considered a parallelepiped of edges L_1, L_2, L_3, we would have obtained a volume element of $\pi^3/L_1L_2L_3$: only the volume $L_1L_2L_3$ of the box, therefore, enters into the density of states. This result can be shown to remain valid, whatever the exact form of the box, provided it is sufficiently large.

b. IMPORTANCE OF THE ELECTRONS WITH ENERGIES CLOSE TO E_F

The results obtained in the preceding section make it possible to understand the physical properties of a free electron gas. We shall give two simple examples here, that of the specific heat and that of the magnetic susceptibility of the system. We shall confine ourselves, however, to semi-quantitative arguments which simply illustrate the fundamental importance of Pauli's exclusion principle.

α. *Specific heat*

At absolute zero, the electron gas is in its ground state: all the individual levels of energy less than E_F are occupied, and all the others are empty. Taking into account the form (8) of the density of states $\rho(E)$, we can represent the situation schematically as in figure 2-a: the number $\nu(E)\,\mathrm{d}E$ of electrons with an energy between E and $E + \mathrm{d}E$ is $\rho(E)\,\mathrm{d}E$ for $E < E_F$ and zero for $E > E_F$. What happens if the temperature T is low but not strictly zero?

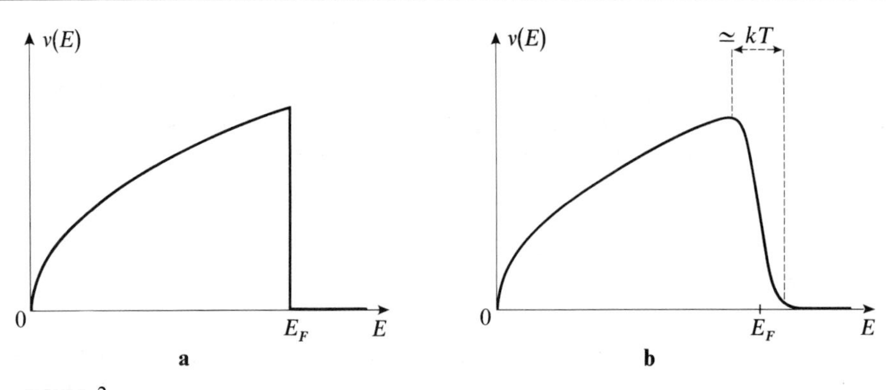

FIGURE 2

Variation of $\nu(E)$ with respect to E [$\nu(E)\,\mathrm{d}E$ is the number of electrons with energy between E and $E + \mathrm{d}E$]. At absolute zero, all the levels whose energies are less than the Fermi energy E_F are occupied (fig. a). At a slightly higher temperature T, the transition between empty and occupied levels occurs over an energy interval of a few kT (fig. b).

If the electrons obeyed classical mechanics, each of them, in going from absolute zero to the temperature T, would gain an energy of the order of kT (where k is the Boltzmann constant). The total energy per unit volume of the electron gas would then be approximately:

$$U_{cl}(T) \simeq \frac{N}{L^3} kT \tag{10}$$

which would give a constant volume specific heat $\partial U_{cl}/\partial T$ which is independent of the temperature.

In reality, the physical phenomena are totally different, since Pauli's principle prevents most of the electrons from gaining energy. For an electron whose initial energy E is much less than E_F (more precisely, if $E_F - E \gg kT$), the states to which it could go if its energy increased by kT are already occupied and are therefore forbidden to it. Only electrons having an initial energy E close to E_F ($E_F - E \simeq kT$) can "heat up", as shown by figure 2-b. The number of these electrons is approximately:

$$\Delta N \simeq \rho(E_F)\, kT = \frac{3}{2} N \frac{kT}{E_F} \tag{11}$$

[according to (9)]. Since the energy of each one increases by about kT, the total energy per unit volume can be written:

$$U(T) \simeq \frac{N}{L^3} \frac{kT}{E_F} kT \tag{12}$$

instead of the classical expression, (10). Consequently, the constant volume specific heat is proportional to the absolute temperature T:

$$c_V = \frac{\partial U}{\partial T} \simeq \frac{Nk}{L^3} \frac{kT}{E_F} \tag{13}$$

For a metal, to which the free-electron model can be applied, E_F is typically on the order of a few eV. Since kT is about 0.03 eV at ordinary temperatures, we see that in this case the factor kT/E_F introduced by Pauli's principle is of the order of 1/100.

COMMENTS:

(*i*) In order to calculate the specific heat of the electron gas quantitatively, we must know the probability $f(E, T)$ for an individual state of energy E to be occupied when the system is at thermodynamic equilibrium at the temperature T. The number $\nu(E)\,\mathrm{d}E$ of electrons whose energies are included between E and $E + \mathrm{d}E$ is then:

$$\nu(E)\,\mathrm{d}E = f(E, T)\,\rho(E)\,\mathrm{d}E \tag{14}$$

It is shown in statistical mechanics that, for fermions, the function $f(E, T)$ can be written:

$$f(E, T) = \frac{1}{\mathrm{e}^{(E-\mu)/kT} + 1} \tag{15}$$

where μ is the *chemical potential*, also called the *Fermi level* of the system. This is the *Fermi-Dirac distribution*. The Fermi level is determined by the condition that the total number of electrons must be equal to N:

$$\int_0^{+\infty} \frac{\rho(E)\,\mathrm{d}E}{\mathrm{e}^{(E-\mu)/kT} + 1} = N \tag{16}$$

μ depends on the temperature, but it can be shown that it varies very slowly for small T. The shape of the function $f(E, T)$ is shown in figure 3. At absolute zero, $f(E, 0)$ is equal to 1 for $E < \mu$ and to 0 for $E > \mu$ ("step" function). At non-zero temperatures, $f(E, T)$ has the form of a rounded "step" (the energy interval over which it varies is of the order of a few kT as long as $kT \ll \mu$).

For a free electron gas, it is clear that the Fermi level μ at absolute zero coincides with the Fermi energy E_F calculated in §1-a. According to (14) and the form that $f(E, T)$ takes for $T = 0$ (fig. 3), μ then characterizes, like E_F, the highest individual energy.

On the other hand, for a system with a discrete spectrum of energy levels ($E_1, E_2, ..., E_i, ...$), the Fermi level μ obtained from formula (16) does not coincide with the highest individual energy E_m in the ground state at absolute zero. In this case, the density of states is composed of a series of "delta functions" centered at E_1, E_2, ..., E_i, ...;

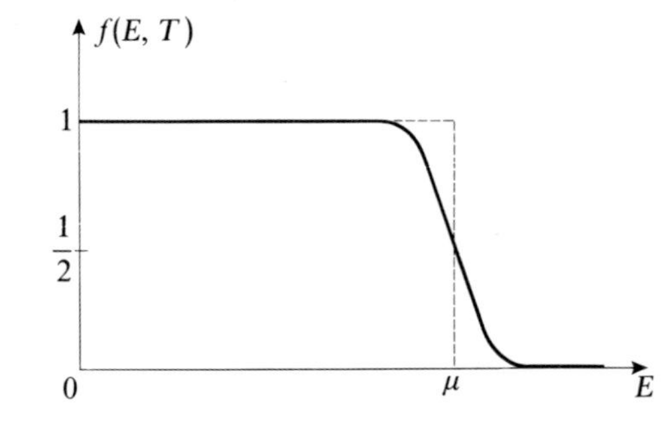

FIGURE 3

Shape of the Fermi-Dirac distribution at absolute zero (dashed line) and at low temperatures (solid line).
For an electron gas at absolute zero, the Fermi level μ coincides with the Fermi energy E_F. The curves in figure 2 can be obtained by multiplying the density of states $\rho(E)$ by $f(E, T)$.

consequently, at absolute zero, μ can take on any value between E_m and E_{m+1}, since, according to (14), all these possibilities lead to the same value of $\nu(E)$. We choose to define μ at absolute zero as the limit of $\mu(T)$ as T approaches zero. Since at non-zero temperatures the level E_m empties a little, and E_{m+1} begins to fill, the limit of $\mu(T)$ is found to be a value between E_m and E_{m+1} (halfway between these two values if the two states E_m and E_{m+1} have the same degree of degeneracy).

Similarly, for a system containing a series of allowed energy bands separated by forbidden bands (electrons of a solid; *cf.* complement F_{XI}), the Fermi level μ is in a forbidden band when the highest individual energy at absolute zero coincides with the upper limit of an allowed band. On the other hand, the Fermi level μ is equal to E_F when E_F falls in the middle of an allowed band.

(*ii*) The preceding results explain the behavior of the specific heat of metals at very low temperatures. At ordinary temperatures, the specific heat is essentially due to vibrations of the ionic lattice (*cf.* complement L_V), since that of the electron gas is practically negligible. However, the specific heat of the lattice approaches zero as T^3 for small T. Therefore, that of the electron gas becomes preponderant at low temperatures (around 1 °K), where, for metals, a decrease which is linear with respect to T is actually observed.

β. *Magnetic susceptibility*

Now suppose that a free electron gas is placed in a uniform magnetic field **B** parallel to Oz. The energy of an individual stationary state then depends on the corresponding spin state, since the Hamiltonian contains a paramagnetic spin term (*cf.* chap. IX, §A-2):

$$W = -2\frac{\mu_B}{\hbar} B\, S_z \tag{17}$$

where μ_B is the Bohr magneton:

$$\mu_B = \frac{q\hbar}{2m_e} \tag{18}$$

and **S** is the electron spin operator. For the sake of simplicity, we shall treat (17) as the only additional term in the Hamiltonian (the behavior of the spatial wave functions was studied in detail in complement E_{VI}). Under these conditions, the stationary states remain the same as in the absence of a magnetic field, and the corresponding energy is increased or decreased by $\mu_B B$ depending on the spin state. The densities of states $\rho_+(E)$ and $\rho_-(E)$ corresponding respectively to the spin states $| + \rangle$ and $| - \rangle$ can therefore be obtained very simply from the density $\rho(E)$ calculated in §1-a:

$$\rho_\pm(E) = \frac{1}{2}\,\rho(E \pm \mu_B B) \tag{19}$$

Thus, at absolute zero, we arrive at the situation shown in figure 4.

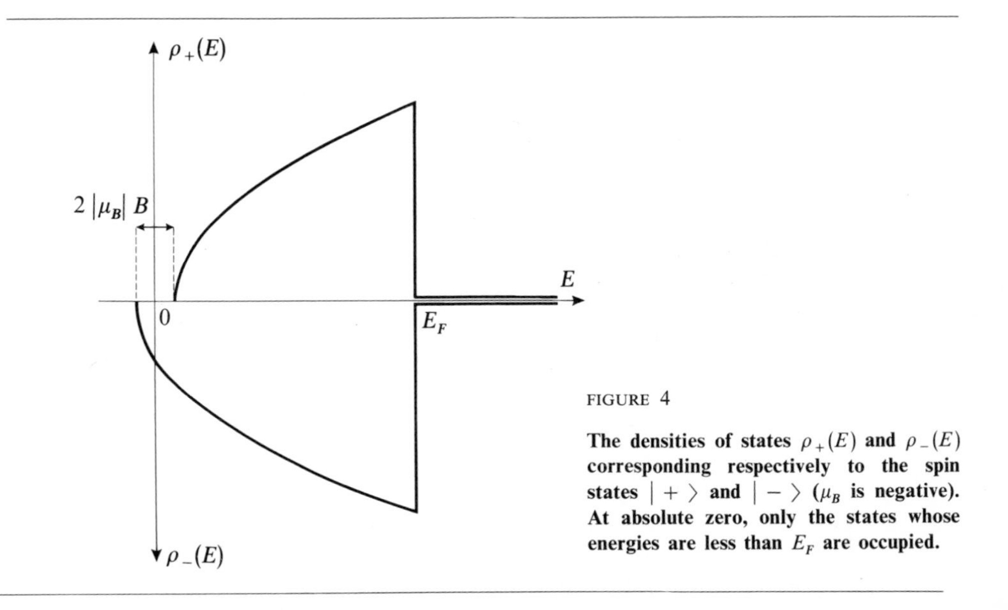

FIGURE 4

The densities of states $\rho_+(E)$ and $\rho_-(E)$ corresponding respectively to the spin states $| + \rangle$ and $| - \rangle$ (μ_B is negative). At absolute zero, only the states whose energies are less than E_F are occupied.

Since the magnetic energy $|\mu_B| B$ is much smaller than E_F, the difference between the number of electrons whose spins are antiparallel to the magnetic field and the number whose spins are parallel to **B** is practically, at absolute zero:

$$N_- - N_+ \simeq \frac{1}{2}\,\rho(E_F)\,2\,|\mu_B|\,B \tag{20}$$

The magnetic moment M per unit volume can therefore be written:

$$\begin{aligned} M &= |\mu_B| \frac{1}{L^3}(N_- - N_+) \\ &= \mu_B^2\, B \frac{1}{L^3}\,\rho(E_F) \end{aligned} \tag{21}$$

This magnetic moment is proportional to the applied field, so that the magnetic susceptibility per unit volume is equal to:

$$\chi = \frac{M}{B} = \mu_B^2 \frac{1}{L^3}\,\rho(E_F) \tag{22}$$

or, using expression (9) for $\rho(E)$:

$$\chi = \frac{3}{2} \frac{N}{L^3} \frac{\mu_B^2}{E_F} \tag{23}$$

COMMENTS:

(*i*) We have assumed the system to be at absolute zero, but result (23) remains valid at low temperatures, since the modifications of the number of occupied states (fig. 2-b) are practically the same for both spin orientations. We therefore find a temperature-independent magnetic susceptibility. This is indeed what is observed for metals.

(*ii*) As in the preceding section, we see that the system behavior in the presence of a magnetic field is essentially determined by the electrons whose energies are close to E_F. This is another manifestation of Pauli's principle. When the magnetic field is applied, the electrons in the $| + \rangle$ spin state tend to go into the $| - \rangle$ state, which is energetically more favorable. But most of them are prevented from doing so by the exclusion principle, since all the neighbouring $| - \rangle$ states are already occupied.

c. PERIODIC BOUNDARY CONDITIONS

α. *Introduction*

The functions φ_{n_x,n_y,n_z} given by formula (1-a) have a completely different structure from that of the plane waves $e^{i\mathbf{k}.\mathbf{r}}$ which usually describe the stationary states of free electrons. This difference arises solely from the boundary conditions imposed by the walls of the box, since, inside the box, the plane waves satisfy the same equation as the φ_{n_x,n_y,n_z}:

$$-\frac{\hbar^2}{2m_e} \Delta\varphi(\mathbf{r}) = E\,\varphi(\mathbf{r}) \tag{24}$$

The functions (1-a) are less convenient to handle than plane waves; this is why the latter are preferably used. To do so, we impose on the solutions of equation (24) new, artificial, boundary conditions which do not exclude plane waves. Of course, since these conditions are different from those actually created by the walls of the box, this changes the physical problem. However, we shall show in this section that we can find the most important physical properties of the initial system in this way. For this to be true, it is necessary for the new boundary conditions to lead to a discrete set of possible values of $\mathbf{k}$ such that:

(*i*) The system of plane waves corresponding to these values of $\mathbf{k}$ constitutes a basis on which can be expanded any function whose domain is inside the box.

(*ii*) The density of states $\rho'(E)$ associated with this set of values of $\mathbf{k}$ is identical to the density of states $\rho(E)$ calculated in § 1-a from the true stationary states.

Of course, the fact that the new boundary conditions are different from the real conditions means that the plane waves cannot correctly describe what happens near the walls (surface effects). However, it is clear that they can, because of condition (*ii*), lead to a very simple explanation of the volume effects, which, according to what we have seen in §1-b, depend only on the density of states $\rho(E)$. Moreover, because of condition (*i*), the motion of any wave packet far from the walls can be correctly described by superposing plane waves, since, between two collisions with the walls, the wave packet propagates freely.

β. *The Born-Von Karman conditions*

We shall no longer require the individual wave functions to go to zero at the walls of the box, but, rather, to be periodic with a period L:

$$\varphi(x + L, y, z) = \varphi(x, y, z) \tag{25}$$

with analogous relations in y and z. Wave functions of the form $e^{i\mathbf{k}\cdot\mathbf{r}}$ satisfy these conditions if the components of the vector $\mathbf{k}$ satisfy:

$$\begin{cases} k_x = n'_x \dfrac{2\pi}{L} \\ k_y = n'_y \dfrac{2\pi}{L} \\ k_z = n'_z \dfrac{2\pi}{L} \end{cases} \tag{26}$$

where, now, n'_x, n'_y *and* n'_z *are positive or negative integers or zero*. We therefore introduce a new system of wave functions :

$$\varphi'_{n'_x,n'_y,n'_z}(\mathbf{r}) = \frac{1}{L^{3/2}}\, e^{i\frac{2\pi}{L}(n'_x x + n'_y y + n'_z z)} \tag{27}$$

which are normalized inside the volume of the box. The corresponding energy, according to (24), can be written:

$$E_{n'_x,n'_y,n'_z} = \frac{\hbar^2}{2m_e}\frac{4\pi^2}{L^2}\left(n'^2_x + n'^2_y + n'^2_z\right) \tag{28}$$

Any wave function defined inside the box can be extended into a periodic function in x, y, z, of period L. Since this periodic function can always be expanded in a Fourier series (*cf.* appendix I, § 1-b), the $\{ \varphi'_{n'_x,n'_y,n'_z}(\mathbf{r}) \}$ system constitutes a basis for wave functions with a domain inside the box. To each vector $\mathbf{k}_{n'_x,n'_y,n'_z}$, whose components are given by (26), there corresponds a well-defined value of the energy $E_{n'_x,n'_y,n'_z}$, given by (28). Note, however, that the vectors $\mathbf{k}_{n'_x,n'_y,n'_z}$ can now have positive, negative or zero components, and that their tips divide space into elementary cubes whose edges are twice that found in § 1-a.

In order to show that boundary conditions (25) lead to the same physical results (as far as the volume effects are concerned) as those of § 1-a, it suffices to calculate the number $n'(E)$ of stationary states of energy less than E, and find the value (5) [the Fermi energy E_F and the density of states $\rho(E)$ can be derived directly from $n(E)$]. We evaluate $n'(E)$ in the same way as in §1-a, taking into account the new characteristics of the vectors $\mathbf{k}_{n'_x,n'_y,n'_z}$. Since the components of $\mathbf{k}$ can now have arbitrary signs, the volume of the sphere of radius $\sqrt{2m_eE/\hbar^2}$ must no longer be divided by 8. However, this modification is compensated by the fact that the volume element $(2\pi/L)^3$ associated with each of the states (27) is eight times larger than the one corresponding to the boundary conditions of §1-a. Consequently, $n'(E)$ is the same as expression (5) for $n(E)$.

The periodic boundary conditions (25) therefore permit us to meet conditions (i) and (ii) of the preceding section. They are usually called the Born-Von Karman conditions ("B.V.K. conditions").

COMMENT:

Consider a truly free electron (not enclosed in a box). The eigenfunctions of the three components of the momentum $\mathbf{P}$ (and, consequently, those of the Hamiltonian $H = \mathbf{P}^2/2m_e$) form a "continuous basis":

$$\left\{\left(\frac{1}{2\pi\hbar}\right)^{3/2} e^{i\mathbf{p}.\mathbf{r}/\hbar}\right\} \tag{29}$$

We have already indicated several times that the states for which the form (29) is valid in all space are not physical states, but, can be used as mathematical intermediaries in studying the physical states, which are wave packets.

We sometimes prefer to use the discrete basis (27) rather than the continuous basis (29). To do so, we consider the electron to be enclosed in a fictitious box of edge L, much larger than any dimension involved in the problem, and we impose the B.V.K. conditions. Any wave packet, which will always be inside the box for sufficiently large L, can be as well expanded on the discrete basis (27) as on the continuous basis (29). The states (27) can therefore, like the states (29), be considered to be intermediaries of the calculation; however, they present the advantage of being normalized inside the box. We must, of course, check, at the end of the calculations, that the various physical quantities obtained (transition probabilities, cross sections,...) do not depend on L, provided that L is sufficiently large.

Obviously, for a truly free electron, L has no physical meaning and can be arbitrary, as long as it is sufficiently large for the states (27) to form a basis on which the wave packets involved in the problem can be expanded [condition (i) of § 1-c-α]. On the other hand, in the physical problem which we are studying here, L^3 is the volume inside which the N electrons are actually confined and has, consequently, a definite value.

2. Electrons in solids

a. ALLOWED BANDS

The model of free electron gas enclosed in a box can be applied rather well to the conduction electrons of a metal. These electrons can be considered to move freely inside the metal, the electrostatic attraction of the crystalline lattice preventing them from escaping when they approach the surface of the metal. However, this model does not explain why some solids are good electrical conductors while others are insulators. This is a remarkable experimental fact : the electric properties of crystals are due to the electrons of the atoms of which they are composed; yet, the intrinsic conductivity can vary by a factor of 10^{30} between a good insulator and a pure metal. We shall see, in a very qualitative way, how this can be explained by Pauli's principle and by the existence of energy bands arising from the periodic nature of the potential created by the ions (*cf.* complements O_{III} and F_{XI}).

We showed in complement F_{XI} that if, in a first approximation, we consider the electrons of a solid to be independent, their possible individual energies are grouped into *allowed bands*, separated by *forbidden bands*. Assuming that each electron is subject to the influence of a linear chain of regularly spaced positive ions, we found, in the strong-bond approximation, a series of bands, each one containing $2\mathcal{N}$ levels, where $\mathcal{N}$ is the number of ions (the factor 2 arises from the spin).

The situation, of course, is more complex in a real crystal, in which the positive ions occupy the nodes of a three-dimensional lattice. The theoretical understanding of the properties of a solid requires a detailed study of the energy bands, a study which is based on the spatial characteristics of the crystalline lattice. We shall not treat in detail these specific problems of solid state physics. We shall content ourselves with a qualitative discussion of the phenomena.

b. POSITION OF THE FERMI LEVEL AND ELECTRIC CONDUCTIVITY

Knowing the band structure and the number of states per band, we obtain the ground state of the electron system of a solid by successively "filling" the individual states of the various allowed bands, beginning, of course, with the lowest energies. The electron system is really in the ground state only at absolute zero. However, as we pointed out in §1-b-α, the characteristics of this ground state permit the semi-quantitative understanding of the behavior of the system at non-zero temperatures – often, up to ordinary temperatures. Like the thermal and magnetic properties (*cf.* §1-b), the electrical properties of the system are principally determined by the electrons whose individual energies are very close to the highest value E_F. If we place the solid in an electric field, an electron whose initial energy is much lower than E_F cannot gain energy by being accelerated, since the states it would reach in this way are already occupied. It is therefore essential to know the position of E_F relative to the allowed energy bands.

First of all, we shall assume (fig. 5-a) that E_F falls in the middle of an allowed band. The Fermi level μ is then equal to E_F [*cf.* comment (*i*) of §1-b-α]. The electrons whose energies are close to E_F can easily be accelerated, in this case, since the slightly higher energy states are empty and accessible. Consequently, *a solid for which the Fermi level falls in the middle of an allowed band is a conductor*. The electrons with the highest energies then behave approximately like free particles.

Consider, on the other hand, a solid for which the ground state is composed of entirely occupied allowed bands (fig. 5-b). E_F is then equal to the upper limit of an allowed band, and the Fermi level μ falls inside the adjacent forbidden band [*cf.* comment (*i*) of §1-b-α]. In this case, no electrons can be accelerated, since the energy states immediately above theirs are forbidden. Therefore, *a solid for which the Fermi level falls inside a forbidden band is an insulator.* The larger the interval ΔE between the last occupied band and the first empty allowed band, the better the insulator. We shall return to this point later.

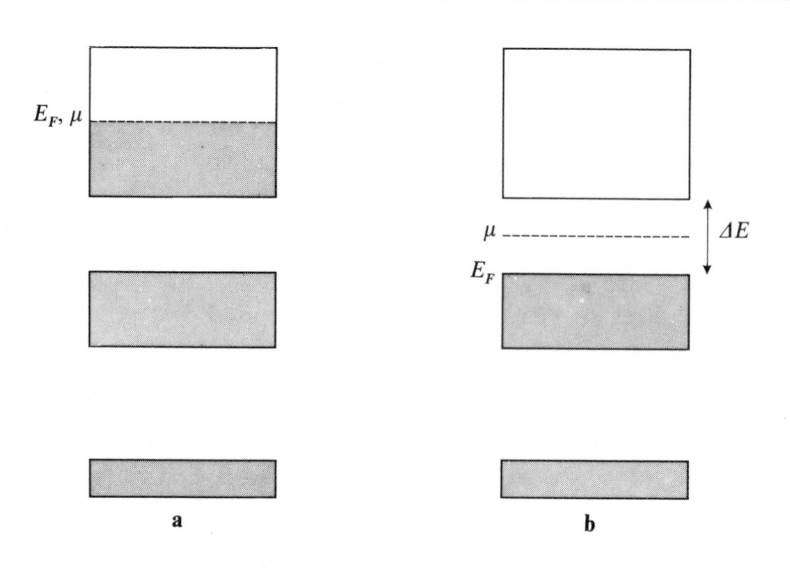

FIGURE 5

Schematic representation of the individual levels occupied by the electrons at absolute zero (in grey). E_F is the highest individual energy.
In a conductor (fig. a), E_F (which then coincides with the Fermi level μ) falls inside an allowed band, called the "conduction band". The electrons whose energies are near E_F can then be accelerated easily, since the slightly higher energy states are accessible to them.
In an insulator (fig. b), E_F falls on the upper boundary of an allowed band called the "valence band" (the Fermi level μ is then situated in the adjacent forbidden band). The electrons can be excited only by crossing the forbidden band. This requires an energy at least equal to the width ΔE of this band.

The deep allowed bands, completely occupied by electrons and, consequently, inert from an electrical and thermal point of view. are called *valence bands.* They are generally narrow. In a "strong-bond" model (*cf.* complement F_{XI}, §2), these bands arise from the atomic levels of lowest energies, which are only slightly affected by the presence of the other atoms in the crystal. On the other hand, the higher bands are wider; a partially occupied band is called a *conduction band.*

For a solid to be a good insulator, the last occupied band must not only be entirely full in the ground state, but also, separated from the immediately higher allowed band by a sufficiently wide forbidden band. As we have indicated (§1-b-α), at non-zero temperatures, some states of energy lower than E_F can empty, while some higher energy states fill (fig. 2-b). For the solid to remain an insulator at the temperature T, the width ΔE of the forbidden band which prevents this excitation of electrons must be much larger than kT. If ΔE is less than or of the order of kT, a certain number of electrons leave the last valence band to occupy states of the immediately higher allowed band (which would be completely empty at absolute zero). The crystal then possesses conduction electrons, but in restricted numbers: it is a *semiconductor* (such a semiconductor is called *intrinsic*; see comment below). For example, diamond, for which ΔE is close to 5 eV, remains an insulator at ordinary temperatures, while silicon and germanium, although quite similar to diamond, are semiconductors: their forbidden bands have a width ΔE less than 1 eV. These considerations, while very qualitative, enable us to understand why the electrical conductivity of a semiconductor increases very rapidly with the temperature; with more quantitative arguments, we indeed find a dependence of the form $e^{-\Delta E/2kT}$.

The properties of semiconductors also reveal an apparently paradoxical phenomenon. It is as if, in addition to the electrons which have crossed the forbidden band ΔE at a temperature T, there existed in the crystal an equal number of particles with a positive charge. These particles also contribute to the electric current, but their contribution to the Hall effect*, for example, is opposite in sign to what would be expected for electrons. This can be explained very well by band theory, and constitutes a spectacular demonstration of Pauli's principle. To understand this qualitatively, we must recall that the last valence band, when it is completely full in the vicinity of absolute zero, does not conduct any current (Pauli's principle forbids the corresponding electrons from being accelerated). When, by thermal excitation, certain electrons move into the conduction band, they free the states they had occupied in the valence band. These empty states in an almost full band are called "holes". Holes behave like particles of charge opposite to that of the electron. If an electric field is applied to the system, the electrons remaining in the valence band can move, without leaving this band, and occupy the empty states. In this way, they "fill holes" but also "leave new holes behind them". Holes therefore move in the direction opposite to that of the electrons, that is, as if they had a positive charge. This very rough argument can be made more precise, and it can indeed be shown that holes are in every way equivalent to positive charge carriers.

COMMENT:

We have been speaking only of chemically pure and geometrically perfect crystals. However, in practice, all solids have imperfections and impurities, which often play an important role, particularly in semiconductors. Consider, for example, a quadrivalent silicon or germanium crystal, in which certain atoms are replaced by pentavalent impurity atoms, such as phosphorus, arsenic or antimony (this often happens, without any important change in the crystal structure). An atom of such an impurity possesses one too many outer electrons relative to the neighboring silicon or germanium atoms; it is called an electron *donor*. The binding energy ΔE_d of the additional electron is considerably lower in the crystal than in the free atom (it is of the order of a few hundredths of an eV); this is due essentially to the large dielectric constant of the crystal, which reduces the Coulomb force (*cf.* complement A_{VII}, § 1-a-δ). The result is that the excess electrons brought in by the donor atoms move more easily into the conduction band than

* Recall what the Hall effect is : in a sample carrying a current and placed in a magnetic field perpendicular to this current, the moving charges are subject to the Lorentz force. In the steady state, this causes a transverse electric field to appear (perpendicular to the current and to the magnetic field).

do the "normal" electrons which occupy the valence band (fig. 6-a). The crystal thus becomes a conductor at a temperature much lower than would pure silicon or germanium. This conductivity due to impurities is called *extrinsic*. Analogously, a trivalent impurity (like boron, aluminium or gallium) behaves in silicon or germanium like an electron *acceptor*: it can easily capture a valence band electron (fig. 6-b), leaving a hole which can conduct the current. In a pure (intrinsic) semiconductor, the number of conduction electrons is always equal to the number of holes in the valence band. An extrinsic semiconductor, on the other hand, can, depending on the relative proportion of donor and acceptor atoms, contain more conduction electrons than holes (it is then said to be of the

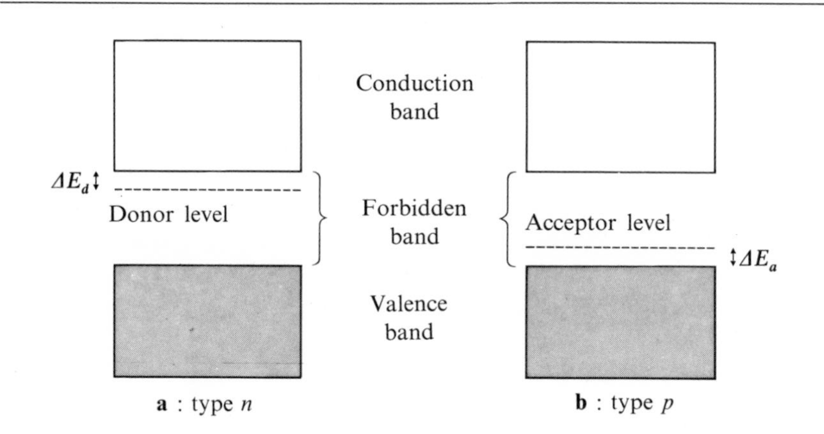

FIGURE 6

Extrinsic semiconductors: donor atoms (fig. a) bring in electrons which move easily into the conduction band, since their ground states are separated from it only by an energy interval ΔE_d which is much smaller than the width of the forbidden band. Acceptor atoms (fig. b) easily capture valence band electrons, since, for this to happen, these electrons need only an excitation energy ΔE_a which is much smaller than that needed to reach the conduction band. This process creates, in the valence band, holes which can conduct current.

n-type, since the majority of charge carriers are negative), or more holes than conduction electrons (*p-type* semiconductors with a majority of positive charge carriers). These properties serve as the foundation of numerous technological applications (transistors, rectifiers, photoelectric cells, etc.). This is why impurities are often intentionally added to a semiconductor to modify its characteristics : this is called "doping".

References and suggestions for further reading:

See section 8 of the bibliography, especially Kittel (8.2) and Reif (8.4).

For the solid state physics part, see Feynman III (1.2), chap. 14 and section 13 of the bibliography.

Complement D_{XIV}

EXERCISES

1. Let h_0 be the Hamiltonian of a particle. Assume that the operator h_0 acts only on the orbital variables and has three equidistant levels of energies 0, $\hbar\omega_0$, $2\hbar\omega_0$ (where ω_0 is a real positive constant) which are non-degenerate in the orbital state space $\mathscr{E}_{\mathbf{r}}$ (in the total state space, the degeneracy of each of these levels is equal to $2s + 1$, where s is the spin of the particle). From the point of view of the orbital variables, we are concerned only with the subspace of $\mathscr{E}_{\mathbf{r}}$ spanned by the three corresponding eigenstates of h_0.

a. Consider a system of three independent electrons whose Hamiltonian can be written:

$$H = h_0(1) + h_0(2) + h_0(3)$$

Find the energy levels of H and their degrees of degeneracy.

b. Same question for a system of three identical bosons of spin 0.

2. Consider a system of two identical bosons of spin $s = 1$ placed in the same central potential $V(r)$. What are the spectral terms (*cf.* complement B_{XIV}, §2-b) corresponding to the $1s^2$, $1s2p$, $2p^2$ configurations?

3. Consider the state space of an electron, spanned by the two vectors $|\varphi_{p_x}\rangle$ and $|\varphi_{p_y}\rangle$ which represent two atomic orbitals, p_x and p_y, of wave functions $\varphi_{p_x}(\mathbf{r})$ and $\varphi_{p_y}(\mathbf{r})$ (*cf.* complement E_{VII}, §2-b):

$$\varphi_{p_x}(\mathbf{r}) = xf(r) = \sin\theta\cos\varphi\, rf(r)$$
$$\varphi_{p_y}(\mathbf{r}) = yf(r) = \sin\theta\sin\varphi\, rf(r)$$

a. Write, in terms of $|\varphi_{p_x}\rangle$ and $|\varphi_{p_y}\rangle$, the state $|\varphi_{p_\alpha}\rangle$ which represents the p_α orbital pointing in the direction of the xOy plane which makes an angle α with Ox.

b. Consider two electrons whose spins are both in the $|+\rangle$ state, the eigenstate of S_z of eigenvalue $+\hbar/2$.

Write the normalized state vector $|\psi\rangle$ which represents the system of two electrons, one of which is in the state $|\varphi_{p_x}\rangle$ and the other, in the state $|\varphi_{p_y}\rangle$.

c. Same question, with one of the electrons in the state $|\varphi_{p_\alpha}\rangle$ and the other one in the state $|\varphi_{p_\beta}\rangle$, where α and β are two arbitrary angles. Show that the state vector $|\psi\rangle$ obtained is the same.

d. The system is in the state $|\psi\rangle$ of question *b*. Calculate the probability density $\mathscr{P}(r, \theta, \varphi; r', \theta', \varphi')$ of finding one electron at (r, θ, φ) and the other one at (r', θ', φ'). Show that the electronic density $\rho(r, \theta, \varphi)$, [the probability density of finding any electron at (r, θ, φ)] is symmetrical with respect to revolution about the Oz axis. Determine the probability density of having $\varphi - \varphi' = \varphi_0$, where φ_0 is given. Discuss the variation of this probability density with respect to φ_0.

4. Collision between two identical particles

The notation used is that of §D-2-a-β of chapter XIV.

a. Consider two particles, (1) and (2), with the same mass m, assumed for the moment to have no spin and to be distinguishable. These two particles interact through a potential $V(r)$ which depends only on the distance between them, r. At the initial time t_0, the system is in the state $| 1 : p\mathbf{e}_z ; 2 : -p\mathbf{e}_z \rangle$. Let $U(t, t_0)$ be the evolution operator of the system. The probability amplitude of finding it in the state $| 1 : p\mathbf{n} ; 2 : -p\mathbf{n} \rangle$ at time t_1 is:

$$F(\mathbf{n}) = \langle 1 : p\mathbf{n} ; 2 : -p\mathbf{n} | U(t_1, t_0) | 1 : p\mathbf{e}_z ; 2 : -p\mathbf{e}_z \rangle$$

Let θ and φ be the polar angles of the unit vector $\mathbf{n}$ in a system of orthonormal axes $Oxyz$. Show that $F(\mathbf{n})$ does not depend on φ. Calculate in terms of $F(\mathbf{n})$ the probability of finding any one of the particles (without specifying which one) with the momentum $p\mathbf{n}$ and the other one with the momentum $-p\mathbf{n}$. What happens to this probability if θ is changed to $\pi - \theta$?

b. Consider the same problem [with the same spin-independent interaction potential $V(r)$], but now with two identical particles, one of which is initially in the state $| p\mathbf{e}_z, m_s \rangle$, and the other, in the state $| -p\mathbf{e}_z, m'_s \rangle$ (the quantum numbers m_s and m'_s refer to the eigenvalues $m_s\hbar$ and $m'_s\hbar$ of the spin component along Oz). Assume that $m_s \neq m'_s$. Express in terms of $F(\mathbf{n})$ the probability of finding, at time t_1, one particle with momentum $p\mathbf{n}$ and spin m_s and the other one with momentum $-p\mathbf{n}$ and spin m'_s. If the spins are not measured, what is the probability of finding one particle with momentum $p\mathbf{n}$ and the other one with momentum $-p\mathbf{n}$? What happens to these probabilities when θ is changed to $\pi - \theta$?

c. Treat problem *b* for the case $m_s = m'_s$. In particular, examine the $\theta = \pi/2$ direction, distinguishing between two possibilities, depending on whether the particles are bosons or fermions. Show that, again, the scattering probability is the same in the θ and $\pi - \theta$ directions.

5. Collision between two identical unpolarized particles

Consider two identical particles, of spin s, which collide. Assume that their initial spin states are not known: each of the two particles has the same probability of being in the $2s + 1$ possible orthogonal spin states. Show that, with the notation of the preceding exercise, the probability of observing scattering in the $\mathbf{n}$ direction is:

$$|F(\mathbf{n})|^2 + |F(-\mathbf{n})|^2 + \frac{\varepsilon}{2s+1}[F^*(\mathbf{n})F(-\mathbf{n}) + \text{c.c.}]$$

($\varepsilon = +1$ for bosons, -1 for fermions).

6. Possible values of the relative angular momentum of two identical particles

Consider a system of two identical particles interacting by means of a potential which depends only on their relative distance, so that the Hamiltonian of the system can be written:

$$H = \frac{\mathbf{P}_1^2}{2m} + \frac{\mathbf{P}_2^2}{2m} + V(|\mathbf{R}_1 - \mathbf{R}_2|)$$

As in § B of chapter VII, we set:

$$\mathbf{R}_G = \frac{1}{2}(\mathbf{R}_1 + \mathbf{R}_2) \qquad \mathbf{P}_G = \mathbf{P}_1 + \mathbf{P}_2$$

$$\mathbf{R} = \mathbf{R}_1 - \mathbf{R}_2 \qquad \mathbf{P} = \frac{1}{2}(\mathbf{P}_1 - \mathbf{P}_2)$$

H then becomes:

$$H = H_G + H_r$$

with:

$$H_G = \frac{\mathbf{P}_G^2}{4m}$$

$$H_r = \frac{\mathbf{P}^2}{m} + V(R)$$

a. First, we assume that the two particles are identical bosons of zero spin (π mesons, for example).

α. We use the $\{ | \mathbf{r}_G, \mathbf{r} \rangle \}$ basis of the state space $\mathscr{E}$ of the system, composed of common eigenvectors of the observables $\mathbf{R}_G$ and $\mathbf{R}$. Show that, if P_{21} is the permutation operator of the two particles:

$$P_{21} | \mathbf{r}_G, \mathbf{r} \rangle = | \mathbf{r}_G, -\mathbf{r} \rangle$$

β. We now go to the $\{ | \mathbf{p}_G ; E_n, l, m \rangle \}$ basis of common eigenvectors of $\mathbf{P}_G$, H_r, $\mathbf{L}^2$ and L_z ($\mathbf{L} = \mathbf{R} \times \mathbf{P}$ is the relative angular momentum of the two particles). Show that these new basis vectors are given by expressions of the form:

$$| \mathbf{p}_G ; E_n, l, m \rangle = \frac{1}{(2\pi\hbar)^{3/2}} \int d^3 r_G \, e^{i \mathbf{p}_G \cdot \mathbf{r}_G / \hbar} \times \int d^3 r \, R_{n,l}(r) \, Y_l^m(\theta, \varphi) \, | \mathbf{r}_G, \mathbf{r} \rangle$$

Show that:

$$P_{21} | \mathbf{p}_G ; E_n, l, m \rangle = (-1)^l \, | \mathbf{p}_G ; E_n, l, m \rangle$$

γ. What values of l are allowed by the symmetrization postulate ?

b. The two particles under consideration are now identical fermions of spin 1/2 (electrons or protons).

α. In the state space of the system, we first use the $\{ | \mathbf{r}_G, \mathbf{r}; S, M \rangle \}$ basis of common eigenstates of $\mathbf{R}_G$, $\mathbf{R}$, $\mathbf{S}^2$ and S_z, where $\mathbf{S} = \mathbf{S}_1 + \mathbf{S}_2$ is the total spin of the system (the kets $| S, M \rangle$ of the spin state space were determined in §B of chapter X). Show that:

$$P_{21} | \mathbf{r}_G, \mathbf{r}; S, M \rangle = (-1)^{S+1} | \mathbf{r}_G, -\mathbf{r}; S, M \rangle$$

β. We now go to the $\{ | \mathbf{p}_G; E_n, l, m; S, M \rangle \}$ basis of common eigenstates of $\mathbf{P}_G$, H_r, $\mathbf{L}^2$, L_z, $\mathbf{S}^2$ and S_z.

As in question *a* - *β*, show that:

$$P_{21} | \mathbf{p}_G; E_n, l, m; S, M \rangle = (-1)^{S+1}(-1)^l | \mathbf{p}_G; E_n, l, m; S, M \rangle$$

γ. Derive the values of l allowed by the symmetrization postulate for each of the values of S (triplet and singlet).

c. *(more difficult)*

Recall that the total scattering cross section in the center of mass system of two distinguishable particles interacting through the potential $V(r)$ can be written:

$$\sigma = \frac{4\pi}{k^2} \sum_{l=0}^{\infty} (2l + 1) \sin^2 \delta_l$$

where the δ_l are the phase shifts associated with $V(r)$ [*cf.* chap. VIII, formula (C-58)].

α. What happens to this cross section if the measurement device is equally sensitive to both particles (the two particles have the same mass)?

β. Show that, in the case envisaged in question *a*, the expression for σ becomes:

$$\sigma = \frac{16\pi}{k^2} \sum_{l \text{ even}} (2l + 1) \sin^2 \delta_l$$

γ. For two unpolarized identical fermions of spin 1/2 (the case of question *b*), prove that:

$$\sigma = \frac{4\pi}{k^2} \left\{ \sum_{l \text{ even}} (2l + 1) \sin^2 \delta_l + 3 \sum_{l \text{ odd}} (2l + 1) \sin^2 \delta_l \right\}$$

7. Position probability densities for a system of two identical particles

Let $| \varphi \rangle$ and $| \chi \rangle$ be two normalized orthogonal states belonging to the orbital state space $\mathscr{E}_\mathbf{r}$ of an electron, and let $| + \rangle$ and $| - \rangle$ be the two eigenvectors, in the spin state space $\mathscr{E}_s$, of the S_z component of its spin.

a. Consider a system of two electrons, one in the state $| \varphi, + \rangle$ and the other, in the state $| \chi, - \rangle$. Let $\rho_{II}(\mathbf{r}, \mathbf{r}')d^3r d^3r'$ be the probability of finding one of them in a volume d^3r centered at point $\mathbf{r}$, and the other in a volume d^3r' centered at $\mathbf{r}'$ (two-particle density function). Similarly, let $\rho_I(\mathbf{r})d^3r$ be the probability of finding one of the electrons in a volume d^3r centered at point $\mathbf{r}$ (one-particule density function). Show that:

$$\begin{aligned}\rho_{II}(\mathbf{r}, \mathbf{r}') &= |\varphi(\mathbf{r})|^2 \, |\chi(\mathbf{r}')|^2 + |\varphi(\mathbf{r}')|^2 \, |\chi(\mathbf{r})|^2 \\ \rho_I(\mathbf{r}) &= |\varphi(\mathbf{r})|^2 + |\chi(\mathbf{r})|^2\end{aligned}$$

Show that these expressions remain valid even if $| \varphi \rangle$ and $| \chi \rangle$ are not orthogonal in $\mathscr{E}_\mathbf{r}$.

Calculate the integrals over all space of $\rho_I(\mathbf{r})$ and $\rho_{II}(\mathbf{r}, \mathbf{r}')$. Are they equal to 1 ?

Compare these results with those which would be obtained for a system of two distinguishable particles (both spin 1/2), one in the state $| \varphi, + \rangle$ and the other in the state $| \chi, - \rangle$; the device which measures their positions is assumed to be unable to distinguish between the two particles.

b. Now assume that one electron is in the state $| \varphi, + \rangle$ and the other one, in the state $| \chi, + \rangle$. Show that we then have:

$$\begin{aligned}\rho_{II}(\mathbf{r}, \mathbf{r}') &= |\varphi(\mathbf{r})\chi(\mathbf{r}') - \varphi(\mathbf{r}')\chi(\mathbf{r})|^2 \\ \rho_I(\mathbf{r}) &= |\varphi(\mathbf{r})|^2 + |\chi(\mathbf{r})|^2\end{aligned}$$

Calculate the integrals over all space of $\rho_I(\mathbf{r})$ and $\rho_{II}(\mathbf{r}, \mathbf{r}')$.

What happens to ρ_I and ρ_{II} if $| \varphi \rangle$ and $| \chi \rangle$ are no longer orthogonal in $\mathscr{E}_\mathbf{r}$?

c. Same questions for two identical bosons, either in the same spin state or in two orthogonal spin states.

8. The aim of this exercise is to demonstrate the following point : once the state vector of a system of N identical bosons (or fermions) has been suitably symmetrized (or antisymmetrized), it is not indispensable, in order to calculate the probability of any measurement result, to perform another symmetrization (or antisymmetrization) of the kets associated with the measurement. More precisely, provided that the state vector belongs to $\mathscr{E}_S$ (or $\mathscr{E}_A$), the physical predictions can be calculated as if we were confronted with a system of distinguishable particles studied by imperfect measurement devices unable to distinguish between them.

Let $| \psi \rangle$ be the state vector of a system of N identical bosons (all of the following reasoning is equally valid for fermions). We have:

$$S \, | \psi \rangle = | \psi \rangle \tag{1}$$

I.

a. Let $| \chi \rangle$ be the normalized physical ket associated with a measurement in which the N bosons are found to be in the different and orthonormal individual states $| u_\alpha \rangle, | u_\beta \rangle, \ldots, | u_\nu \rangle$. Show that:

$$| \chi \rangle = \sqrt{N!} \, S \, | 1 : u_\alpha ; 2 : u_\beta ; \ldots ; N : u_\nu \rangle \tag{2}$$

b. Show that, because of the symmetry properties of $| \psi \rangle$:

$$|\langle 1 : u_\alpha ; 2 : u_\beta ; ... ; N : u_\nu | \psi \rangle|^2 = |\langle i : u_\alpha ; j : u_\beta ; ... ; l : u_\nu | \psi \rangle|^2$$

where $i, j, ..., l$ is an arbitrary permutation of the numbers 1, 2, ..., N.

c. Show that the probability of finding the system in the state $| \chi \rangle$ can be written:

$$\begin{aligned} |\langle \chi | \psi \rangle|^2 &= N! \, |\langle 1 : u_\alpha ; 2 : u_\beta ; ... ; N : u_\nu | \psi \rangle|^2 \\ &= \sum_{\{i,j,...,l\}} |\langle i : u_\alpha ; j : u_\beta ; ... ; l : u_\nu | \psi \rangle|^2 \end{aligned} \tag{3}$$

where the summation is performed over all permutations of the numbers 1, 2, ..., N.

d. Now assume that the particles are distinguishable, and that their state is described by the ket $| \psi \rangle$. What would be the probability of finding any one of them in the state $| u_\alpha \rangle$, another one in the state $| u_\beta \rangle$, ..., and the last one in the state $| u_\nu \rangle$?

Conclude, by comparison with the results of *c*, that, for identical particles, it is sufficient to apply the symmetrization postulate to the state vector $| \psi \rangle$ of the system.

e. How would the preceding argument be modified if several of the individual states constituting the state $| \chi \rangle$ were identical? (For the sake of simplicity, consider only the case where $N = 3$).

II. *(more difficult)*

Now, consider the general case, in which the measurement result being considered is not necessarily defined by the specification of individual states, since the measurement may no longer be complete. According to the postulates of chapter XIV, we must proceed in the following way in order to calculate the corresponding probability:

– first of all, we treat the particles as distinguishable, and we number them; their state space is then $\mathscr{E}$. Then let $\mathscr{E}_m$ be the subspace of $\mathscr{E}$ associated with the measurement result envisaged, with the measurement being performed with devices incapable of distinguishing between the particles;

– with $| \psi_m \rangle$ denoting an arbitrary ket of $\mathscr{E}_m$, we construct the set of kets $S | \psi_m \rangle$ which constitutes a vector space $\mathscr{E}_m^S$ ($\mathscr{E}_m^S$ is the projection of $\mathscr{E}_m$ onto $\mathscr{E}_S$); if the dimension of $\mathscr{E}_m^S$ is greater than 1, the measurement is not complete;

– the desired probability is then equal to the square of the norm of the orthogonal projection of the ket $| \psi \rangle$ onto $\mathscr{E}_m^S$ which describes the state of the N identical particles.

a. If P_α is an arbitrary permutation operator of the N particles, show that, by construction of $\mathscr{E}_m$:

$$P_\alpha | \psi_m \rangle \in \mathscr{E}_m$$

Show that $\mathscr{E}_m$ is globally invariant under the action of S and that $\mathscr{E}_m^S$ is simply the intersection of $\mathscr{E}_S$ and $\mathscr{E}_m$.

b. We construct an orthonormal basis in $\mathscr{E}_m$:

$$\{ | \varphi_m^1 \rangle, | \varphi_m^2 \rangle, ..., | \varphi_m^k \rangle, | \varphi_m^{k+1} \rangle, ..., | \varphi_m^p \rangle \}$$

the first k vectors of which constitute a basis of $\mathscr{E}_m^S$. Show that the kets $S | \varphi_m^n \rangle$, where $k + 1 \leqslant n \leqslant p$, must be linear combinations of the first k vectors of this basis. Show, by taking their scalar products with the bras $\langle \varphi_m^1 |, \langle \varphi_m^2 |, ..., \langle \varphi_m^k |$, that these kets $S | \varphi_m^n \rangle$ (with $n \geqslant k + 1$) are necessarily zero.

c. Show from the preceding results that the symmetric nature of $| \psi \rangle$ implies that:

$$\sum_{n=1}^{p} |\langle \varphi_m^n | \psi \rangle|^2 = \sum_{n=1}^{k} |\langle \varphi_m^n | \psi \rangle|^2$$

that is:

$$\langle \psi | P_m^S | \psi \rangle = \langle \psi | P_m | \psi \rangle$$

where P_m^S and P_m denote respectively the projectors onto $\mathscr{E}_m^S$ and $\mathscr{E}_m$.

Conclusion: The probabilities of the measurement results can be calculated from the projection of the ket $| \psi \rangle$ (belonging to $\mathscr{E}_S$) onto an eigensubspace $\mathscr{E}_m$ whose kets do not all belong to $\mathscr{E}_S$, but in which all the particles play equivalent roles.

9. One- and two-particle density functions in a electron gas at absolute zero

I.

a. Consider a system of N particles 1, 2, ..., i, ..., N with the same spin s. First of all, assume that they are not identical. In the state space $\mathscr{E}(i)$ of particle (i), the ket $| i : \mathbf{r}_0, m \rangle$ represents a state in which particle (i) is localized at the point $\mathbf{r}_0$ in the spin state $| m \rangle$ ($m\hbar$: the eigenvalue of S_z).

Consider the operator:

$$F_m(\mathbf{r}_0) = \sum_{i=1}^{N} \left\{ | i : \mathbf{r}_0, m \rangle \langle i : \mathbf{r}_0, m | \otimes \prod_{j \neq i} I(j) \right\}$$

where $I(j)$ is the identity operator in the space $\mathscr{E}(j)$.

Let $| \psi \rangle$ be the state of the N-particle system. Show that $\langle \psi | F_m(\mathbf{r}_0) | \psi \rangle \, d\tau$ represents the probability of finding any one of the particles in the infinitesimal volume element $d\tau$ centered at $\mathbf{r}_0$, the component of its spin being equal to $m\hbar$.

b. Consider the operator:

$$G_{mm'}(\mathbf{r}_0, \mathbf{r}'_0) = \sum_{i=1}^{N} \sum_{j \neq i} \left\{ | i : \mathbf{r}_0, m; j : \mathbf{r}'_0, m' \rangle \langle i : \mathbf{r}_0, m; j : \mathbf{r}'_0, m' | \otimes \prod_{k \neq i,j} I(k) \right\}$$

What is the physical meaning of the quantity $\langle \psi | G_{mm'}(\mathbf{r}_0, \mathbf{r}'_0) | \psi \rangle \, d\tau \, d\tau'$, where $d\tau$ and $d\tau'$ are infinitesimal volumes?

The mean values $\langle \psi \mid F_m(\mathbf{r}_0) \mid \psi \rangle$ and $\langle \psi \mid G_{mm'}(\mathbf{r}_0, \mathbf{r}'_0) \mid \psi \rangle$ will be written, respectively, $\rho^{\text{I}}_m(\mathbf{r}_0)$ and $\rho^{\text{II}}_{mm'}(\mathbf{r}_0, \mathbf{r}'_0)$ and will be called the one- and two-particle density functions of the N-particle system.

The preceding expressions remain valid when the particles are identical, provided that $\mid \psi \rangle$ is the suitably symmetrized or antisymmetrized state vector of the system (*cf.* preceding exercise).

II.

Consider a system of N particles in the normalized and orthogonal individual states $\mid u_1 \rangle, \mid u_2 \rangle, ..., \mid u_N \rangle$.

The normalized state vector of the system is:

$$\mid \psi \rangle = \sqrt{N!}\, T \mid 1 : u_1 ; 2 : u_2 ; ...; N : u_N \rangle$$

where T is the symmetrizer for bosons and the antisymmetrizer for fermions. In this part, we want to calculate the mean values in the state $\mid \psi \rangle$ of symmetric one-particle operators of the type:

$$F = \sum_{i=1}^{N} \left\{ f(i) \otimes \prod_{j \neq i} I(j) \right\}$$

or of symmetric two-particle operators of the type:

$$G = \sum_{i=1}^{N} \sum_{j \neq i} \left\{ g(i, j) \otimes \sum_{k \neq i,j} I(k) \right\}$$

a. Show that:

$$\langle \psi \mid F \mid \psi \rangle = \langle 1 : u_1 ; 2 : u_2 ; ...; N : u_N \mid \left[\sum_\alpha \varepsilon_\alpha P_\alpha \right] F \mid 1 : u_1 ; 2 : u_2 ; ...; N : u_N \rangle$$

where $\varepsilon_\alpha = +1$ for bosons, and $+1$ or -1 for fermions, depending on whether the permutation P_α is even or odd.

Show that the same expression is valid for the operator G.

b. Derive the relations:

$$\langle \psi \mid F \mid \psi \rangle = \sum_{i=1}^{N} \langle i : u_i \mid f(i) \mid i : u_i \rangle$$

$$\langle \psi \mid G \mid \psi \rangle = \sum_{i=1}^{N} \sum_{j \neq i} \{ \langle i : u_i ; j : u_j \mid g(i, j) \mid i : u_i ; j : u_j \rangle + \varepsilon \langle i : u_j ; j : u_i \mid g(i, j) \mid i : u_i ; j : u_j \rangle \}$$

with $\varepsilon = +1$ for bosons, $\varepsilon = -1$ for fermions.

III.

We now want to apply the results of part II to the operators $F_m(\mathbf{r}_0)$ and $G_{mm'}(\mathbf{r}_0, \mathbf{r}'_0)$ introduced in part I. The physical system under study is a gas of N free electrons enclosed in a cubic box of edge L at absolute zero (complement C_{XIV}, § 1). By applying periodic boundary conditions, we obtain individual states of the form $| \varphi_{\mathbf{k}} \rangle | \pm \rangle$, where the wave function associated with $| \varphi_{\mathbf{k}} \rangle$ is a plane wave $\frac{1}{L^{3/2}} e^{i\mathbf{k}.\mathbf{r}}$, and the components of $\mathbf{k}$ satisfy relations (26) of complement C_{XIV}. We shall call $E_F = \hbar^2 k_F^2/2m$ the Fermi energy of the system and $\lambda_F = 2\pi/k_F$, the Fermi wavelength.

a. Show that the two one-particle density functions $\rho_+^{\text{I}}(\mathbf{r}_0)$ and $\rho_-^{\text{I}}(\mathbf{r}_0)$ are both equal to:

$$\rho_+^{\text{I}}(\mathbf{r}_0) = \rho_-^{\text{I}}(\mathbf{r}_0) = \sum_{\mathbf{k}} | \varphi_{\mathbf{k}}(\mathbf{r}_0)|^2$$

where the summation over $\mathbf{k}$ is performed over all values of $\mathbf{k}$ of modulus less then k_F, satisfying the periodic boundary conditions. By using § 1 of complement C_{XIV}, show that $\rho_+^{\text{I}}(\mathbf{r}_0) = \rho_-^{\text{I}}(\mathbf{r}_0) = k_F^3/6\pi^2 = N/2L^3$. Could this result have been predicted simply ?

b. Show that the two two-particle density functions $\rho_{+-}^{\text{II}}(\mathbf{r}_0, \mathbf{r}'_0)$ and $\rho_{-+}^{\text{II}}(\mathbf{r}_0, \mathbf{r}'_0)$ are both equal to :

$$\sum_{\mathbf{k}} \sum_{\mathbf{k}'} | \varphi_{\mathbf{k}}(\mathbf{r}_0)\, \varphi_{\mathbf{k}'}(\mathbf{r}'_0)|^2 = N^2/4L^6$$

where the summations over $\mathbf{k}$ and $\mathbf{k}'$ are defined as above. Give a physical interpretation.

c. Finally, consider the two two-particle density functions $\rho_{++}^{\text{II}}(\mathbf{r}_0, \mathbf{r}'_0)$ and $\rho_{--}^{\text{II}}(\mathbf{r}_0, \mathbf{r}'_0)$. Prove that they are both equal to:

$$\sum_{\mathbf{k}} \sum_{\mathbf{k}' \neq \mathbf{k}} \left\{ |\varphi_{\mathbf{k}}(\mathbf{r}_0)\, \varphi_{\mathbf{k}'}(\mathbf{r}'_0)|^2 - \varphi_{\mathbf{k}}^*(\mathbf{r}'_0)\, \varphi_{\mathbf{k}'}^*(\mathbf{r}_0)\, \varphi_{\mathbf{k}}(\mathbf{r}_0)\, \varphi_{\mathbf{k}'}(\mathbf{r}'_0) \right\}$$

Show that the restriction $\mathbf{k}' \neq \mathbf{k}$ can be omitted, and show that the two two-particle density functions are equal to:

$$\frac{N^2}{4L^6} - \left| \sum_{\mathbf{k}} \varphi_{\mathbf{k}}^*(\mathbf{r}_0)\, \varphi_{\mathbf{k}}(\mathbf{r}'_0) \right|^2 = \frac{N^2}{4L^6} [1 - C^2(k_F d)]$$

with $d = |\mathbf{r}_0 - \mathbf{r}'_0|$, where the function $C(x)$ is defined by:

$$C(x) = \frac{3}{x^3} [\sin x - x \cos x]$$

$\left(\sum_{\mathbf{k}} \text{ can be replaced by an integral over } \mathbf{k} \right)$.

How do the two-particle density functions $\rho_{++}^{\text{II}}(\mathbf{r}_0, \mathbf{r}'_0)$ and $\rho_{--}^{\text{II}}(\mathbf{r}_0, \mathbf{r}'_0)$ vary with respect to the distance d between $\mathbf{r}_0$ and $\mathbf{r}'_0$? Show that it is practically impossible to find two electrons with the same spin separated by a distance much smaller than λ_F.

APPENDIX I

Fourier series and Fourier transforms

Appendix I

FOURIER SERIES AND FOURIER TRANSFORMS

In this appendix, we shall review a certain number of definitions, formulas and properties which are useful in quantum mechanics. We do not intend to enter into the details of the derivations, nor shall we give rigorous proofs of the mathematical theorems.

1. Fourier series

a. PERIODIC FUNCTIONS

A function $f(x)$ of a variable is said to be *periodic* if there exists a real non-zero number L such that, for all x:

$$\boxed{f(x + L) = f(x)} \tag{1}$$

L is called the *period* of the function $f(x)$.

If $f(x)$ is periodic with a period of L, all numbers nL, where n is a positive or negative integer, are also periods of $f(x)$. The *fundamental period* L_0 of such a function is defined as being its smallest positive period (the term "period" is often used in physics to denote what is actually the fundamental period of a function).

COMMENT:

We can take a function $f(x)$ defined only on a finite interval $[a, b]$ of the real axis and construct a function $f_p(x)$ which is equal to $f(x)$ inside $[a, b]$ and is periodic, with a period $(b - a)$. $f_p(x)$ is continuous if $f(x)$ is and if:

$$f(b) = f(a) \tag{2}$$

We know that the *trigonometric functions* are periodic. In particular:

$$\cos 2\pi \frac{x}{L} \quad \text{and} \quad \sin 2\pi \frac{x}{L} \tag{3}$$

have fundamental periods equal to L.

Other particularly important examples of periodic functions are the *periodic exponentials*. For an exponential $e^{\alpha x}$ to have a period of L, it is necessary and sufficient, according to definition (1), that:

$$e^{\alpha L} = 1 \tag{4}$$

that is:

$$\alpha L = 2in\pi \tag{5}$$

where n is an integer. There are therefore two exponentials of fundamental period L:

$$e^{\pm 2i\pi \frac{x}{L}} \tag{6}$$

which are, furthermore, related to the trigonometric functions (3) which have the same period:

$$e^{\pm 2i\pi \frac{x}{L}} = \cos 2\pi \frac{x}{L} \pm i \sin 2\pi \frac{x}{L} \tag{7}$$

The exponential $e^{2in\pi \frac{x}{L}}$ also has a period of L, but its fundamental period is L/n.

b. EXPANSION OF A PERIODIC FUNCTION IN A FOURIER SERIES

Let $f(x)$ be a periodic function with a fundamental period of L. If it satisfies certain mathematical conditions (as is practically always the case in physics), it can be expanded in a series of imaginary exponentials or trigonometric functions.

α. *Series of imaginary exponentials*

We can write $f(x)$ in the form:

$$\boxed{f(x) = \sum_{n=-\infty}^{+\infty} c_n \, e^{ik_n x}} \tag{8}$$

with:

$$k_n = n \frac{2\pi}{L} \tag{9}$$

The coefficients c_n of the Fourier series (8) are given by the formula:

$$\boxed{c_n = \frac{1}{L} \int_{x_0}^{x_0+L} dx \, e^{-ik_n x} f(x)} \tag{10}$$

where x_0 is an arbitrary real number.

To prove (10), we multiply (8) by $e^{-ik_p x}$ and integrate between x_0 and $x_0 + L$:

$$\int_{x_0}^{x_0+L} dx\, e^{-ik_p x} f(x) = \sum_{n=-\infty}^{+\infty} c_n \int_{x_0}^{x_0+L} dx\, e^{i(k_n - k_p)x} \tag{11}$$

The integral of the right-hand side is zero for $n \neq p$ and equal to L for $n = p$. Hence formula (10). It can easily be shown that the value obtained for c_n is independent of the number x_0 chosen.

The set of values $|c_n|$ is called the *Fourier spectrum* of $f(x)$. Note that $f(x)$ is real if and only if:

$$c_{-n} = c_n^* \tag{12}$$

β. *Cosine and sine series*

If, in the series (8), we group the terms corresponding to opposite values of n, we obtain:

$$f(x) = c_0 + \sum_{n=1}^{\infty} (c_n e^{ik_n x} + c_{-n} e^{-ik_n x}) \tag{13}$$

that is, according to (7):

$$f(x) = a_0 + \sum_{n=1}^{\infty} (a_n \cos k_n x + b_n \sin k_n x) \tag{14}$$

with:

$$\begin{aligned} a_0 &= c_0 \\ a_n &= c_n + c_{-n} \\ b_n &= i(c_n - c_{-n}) \end{aligned} \Bigg\} n > 0 \tag{15}$$

The formulas giving the coefficients a_n and b_n can therefore be derived from (10):

$$\begin{aligned} a_0 &= \frac{1}{L}\int_{x_0}^{x_0+L} dx\, f(x) \\ a_n &= \frac{2}{L}\int_{x_0}^{x_0+L} dx\, f(x) \cos k_n x \\ b_n &= \frac{2}{L}\int_{x_0}^{x_0+L} dx\, f(x) \sin k_n x \end{aligned} \tag{16}$$

If $f(x)$ has a definite parity, expansion (14) is particularly convenient, since:

$$\begin{aligned} b_n &= 0 \quad \text{if} \quad f(x) \text{ is even} \\ a_n &= 0 \quad \text{if} \quad f(x) \text{ is odd} \end{aligned} \tag{17}$$

Moreover, if $f(x)$ is real, the coefficients a_n and b_n are real.

c. THE BESSEL-PARSEVAL RELATION

It can easily be shown from the Fourier series (8) that:

$$\boxed{\frac{1}{L}\int_{x_0}^{x_0+L} \mathrm{d}x \, |f(x)|^2 = \sum_{n=-\infty}^{+\infty} |c_n|^2} \tag{18}$$

This can be shown using equation (8):

$$\frac{1}{L}\int_{x_0}^{x_0+L} \mathrm{d}x \, |f(x)|^2 = \sum_{n,p} c_p^* c_n \frac{1}{L}\int_{x_0}^{x_0+L} \mathrm{d}x \, \mathrm{e}^{i(k_n-k_p)x} \tag{19}$$

As in (11), the integral of the right-hand side is equal to $L\delta_{np}$. This proves (18).

When expansion (14) is used, the Bessel-Parseval relation (18) can also be written:

$$\frac{1}{L}\int_{x_0}^{x_0+L} \mathrm{d}x \, |f(x)|^2 = |a_0|^2 + \frac{1}{2}\sum_{n=1}^{\infty} [|a_n|^2 + |b_n|^2] \tag{20}$$

If we have two functions, $f(x)$ and $g(x)$, with the same period L, whose Fourier coefficients are, respectively, c_n and d_n, we can generalize relation (18) to the form:

$$\frac{1}{L}\int_{x_0}^{x_0+L} \mathrm{d}x \, g^*(x) f(x) = \sum_{n=-\infty}^{+\infty} d_n^* c_n \tag{21}$$

2. Fourier transforms

a. DEFINITIONS

α. *The Fourier integral as the limit of a Fourier series*

Now, consider a function $f(x)$ which is not necessarily periodic. We define $f_L(x)$ to be the periodic function of period L which is equal to $f(x)$ inside the interval $[-L/2, L/2]$. $f_L(x)$ can be expanded in a Fourier series:

$$f_L(x) = \sum_{n=-\infty}^{+\infty} c_n \, \mathrm{e}^{ik_n x} \tag{22}$$

where k_n is defined by formula (9), and:

$$c_n = \frac{1}{L}\int_{x_0}^{x_0+L} \mathrm{d}x \, \mathrm{e}^{-ik_n x} f_L(x) = \frac{1}{L}\int_{-\frac{L}{2}}^{+\frac{L}{2}} \mathrm{d}x \, \mathrm{e}^{-ik_n x} f(x) \tag{23}$$

When L approaches infinity, $f_L(x)$ becomes the same as $f(x)$. We shall therefore let L approach infinity in the expressions above.

Definition (9) of k_n then yields:

$$k_{n+1} - k_n = \frac{2\pi}{L} \tag{24}$$

We shall now replace $1/L$ by its expression in terms of $(k_{n+1} - k_n)$ in (23) and substitute this value of c_n into the series (22):

$$f_L(x) = \sum_{n=-\infty}^{+\infty} \frac{k_{n+1} - k_n}{2\pi} e^{ik_n x} \int_{-\frac{L}{2}}^{+\frac{L}{2}} d\xi \, e^{-ik_n \xi} f(\xi) \tag{25}$$

When $L \longrightarrow \infty$, $k_{n+1} - k_n$ approaches zero [*cf.* (24)], so that the sum over n is transformed into a definite integral. $f_L(x)$ approaches $f(x)$. The integral appearing in (25) becomes a function of the continuous variable k. If we set:

$$\tilde{f}(k) = \frac{1}{\sqrt{2\pi}} \int_{-\infty}^{+\infty} dx \, e^{-ikx} f(x) \tag{26}$$

relation (25) can be written in the limit of infinite L:

$$f(x) = \frac{1}{\sqrt{2\pi}} \int_{-\infty}^{+\infty} dk \, e^{ikx} \tilde{f}(k) \tag{27}$$

$f(x)$ and $\tilde{f}(k)$ are called *Fourier transforms* of each other.

β. *Fourier transforms in quantum mechanics*

In quantum mechanics, we actually use a slightly different convention. If $\psi(x)$ is a (one-dimensional) wave function, its Fourier transform $\overline{\psi}(p)$ is defined by:

$$\boxed{\overline{\psi}(p) = \frac{1}{\sqrt{2\pi\hbar}} \int_{-\infty}^{+\infty} dx \, e^{-ipx/\hbar} \psi(x)} \tag{28}$$

and the inverse formula is:

$$\boxed{\psi(x) = \frac{1}{\sqrt{2\pi\hbar}} \int_{-\infty}^{+\infty} dp \, e^{ipx/\hbar} \overline{\psi}(p)} \tag{29}$$

To go from (26) and (27) to (28) and (29), we set:

$$p = \hbar k \tag{30}$$

(p has the dimensions of a momentum if x is a length), and:

$$\overline{\psi}(p) = \frac{1}{\sqrt{\hbar}} \tilde{\psi}(k) = \frac{1}{\sqrt{\hbar}} \tilde{\psi}\left(\frac{p}{\hbar}\right) \tag{31}$$

In this appendix, as is usual in quantum mechanics, we shall use definition (28) of the Fourier transform instead of the traditional definition, (26). To return to the latter definition, furthermore, all we need to do is replace $\hbar$ by 1 and p by k in all the following expressions.

b. SIMPLE PROPERTIES

We shall state (28) and (29) in the condensed notation:

$$\bar{\psi}(p) = \mathscr{F}[\psi(x)] \tag{32-a}$$

$$\psi(x) = \bar{\mathscr{F}}[\bar{\psi}(p)] \tag{32-b}$$

The following properties can easily be demonstrated:

(*i*)

$$\bar{\psi}(p) = \mathscr{F}[\psi(x)] \Longrightarrow \bar{\psi}(p - p_0) = \mathscr{F}[e^{ip_0x/\hbar}\,\psi(x)] \tag{33}$$

$$e^{-ipx_0/\hbar}\bar{\psi}(p) = \mathscr{F}[\psi(x - x_0)]$$

This follows directly from definition (28).

(*ii*)

$$\bar{\psi}(p) = \mathscr{F}[\psi(x)] \Longrightarrow \mathscr{F}[\psi(cx)] = \frac{1}{|c|}\,\bar{\psi}\left(\frac{p}{c}\right) \tag{34}$$

To see this, all we need to do is change the integration variable:

$$u = cx \tag{35}$$

In particular:

$$\mathscr{F}[\psi(-x)] = \bar{\psi}(-p) \tag{36}$$

Therefore, if the function $\psi(x)$ has a definite parity, its Fourier transform has the same parity.

(*iii*)

$$\psi(x) \text{ real} \longleftrightarrow [\bar{\psi}(p)]^* = \bar{\psi}(-p) \tag{37-a}$$

$$\psi(x) \text{ pure imaginary} \longleftrightarrow [\bar{\psi}(p)]^* = -\bar{\psi}(-p) \tag{37-b}$$

The same expressions are valid if the functions ψ and $\bar{\psi}$ are inverted.

(*iv*) If $f^{(n)}$ denotes the nth derivative of the function f, successive differentiations inside the summation yield, according to (28) and (29):

$$\mathscr{F}[\psi^{(n)}(x)] = \left(\frac{ip}{\hbar}\right)^n \bar{\psi}(p) \tag{38-a}$$

$$\bar{\psi}^{(n)}(p) = \mathscr{F}\left[\left(-\frac{ix}{\hbar}\right)^n \psi(x)\right] \tag{38-b}$$

(*v*) The *convolution* of two functions $\psi_1(x)$ and $\psi_2(x)$ is, by definition, the function $\psi(x)$ equal to:

$$\psi(x) = \int_{-\infty}^{+\infty} dy\, \psi_1(y)\, \psi_2(x - y) \tag{39}$$

Its Fourier transform is proportional to the ordinary product of the transforms of $\psi_1(x)$ and $\psi_2(x)$:

$$\bar{\psi}(p) = \sqrt{2\pi\hbar}\, \bar{\psi}_1(p)\, \bar{\psi}_2(p) \tag{40}$$

This can be shown as follows.
We take the Fourier transform of expression (39):

$$\bar{\psi}(p) = \frac{1}{\sqrt{2\pi\hbar}} \int_{-\infty}^{+\infty} \mathrm{d}x\, \mathrm{e}^{-ipx/\hbar} \int_{-\infty}^{+\infty} \mathrm{d}y\, \psi_1(y)\, \psi_2(x-y) \tag{41}$$

and perform the change of integration variables:

$$\{ x, y \} \Longrightarrow \{ u = x - y, y \} \tag{42}$$

If we multiply and divide by $\mathrm{e}^{ipy/\hbar}$ we obtain:

$$\bar{\psi}(p) = \frac{1}{\sqrt{2\pi\hbar}} \int_{-\infty}^{+\infty} \mathrm{d}y\, \mathrm{e}^{-ipy/\hbar} \psi_1(y) \int_{-\infty}^{+\infty} \mathrm{d}u\, \mathrm{e}^{-ipu/\hbar} \psi_2(u) \tag{43}$$

which proves (40).

(*vi*) When $\psi(x)$ is a peaked function of width Δx, the width Δp of $\bar{\psi}(p)$ satisfies:

$$\Delta x \,.\, \Delta p \gtrsim \hbar \tag{44}$$

(see § C-2 of chapter I, where this inequality is analyzed, and complement C_{III}).

c. THE PARSEVAL-PLANCHEREL FORMULA

A function and its Fourier transform have the same norm:

$$\boxed{\int_{-\infty}^{+\infty} \mathrm{d}x\, |\psi(x)|^2 = \int_{-\infty}^{+\infty} \mathrm{d}p\, |\bar{\psi}(p)|^2} \tag{45}$$

To prove this, all we need to do is use (28) and (29) in the following way:

$$\begin{aligned}\int_{-\infty}^{+\infty} \mathrm{d}x\, |\psi(x)|^2 &= \int_{-\infty}^{+\infty} \mathrm{d}x\, \psi^*(x) \frac{1}{\sqrt{2\pi\hbar}} \int_{-\infty}^{+\infty} \mathrm{d}p\, \mathrm{e}^{ipx/\hbar}\, \bar{\psi}(p) \\ &= \int_{-\infty}^{+\infty} \mathrm{d}p\, \bar{\psi}(p) \frac{1}{\sqrt{2\pi\hbar}} \int_{-\infty}^{+\infty} \mathrm{d}x\, \mathrm{e}^{ipx/\hbar}\, \psi^*(x) \\ &= \int_{-\infty}^{+\infty} \mathrm{d}p\, \bar{\psi}^*(p)\, \bar{\psi}(p)\end{aligned} \tag{46}$$

As in § 1-c, the Parseval-Plancherel formula can be generalized:

$$\boxed{\int_{-\infty}^{+\infty} \mathrm{d}x\, \varphi^*(x)\, \psi(x) = \int_{-\infty}^{+\infty} \mathrm{d}p\, \bar{\varphi}^*(p)\, \bar{\psi}(p)} \tag{47}$$

d. EXAMPLES

We shall confine ourselves to three examples of Fourier transforms, for which the calculations are straightforward.

(*i*) Square function

$$\left.\begin{aligned} \bar{\psi}(x) &= \frac{1}{a} \quad \text{for} \quad -\frac{a}{2} < x < \frac{a}{2} \\ &= 0 \quad \text{for} \quad |x| > \frac{a}{2} \end{aligned}\right\} \Longleftrightarrow \quad \bar{\psi}(p) = \frac{1}{\sqrt{2\pi\hbar}} \frac{\sin(pa/2\hbar)}{pa/2\hbar} \tag{48}$$

(*ii*) Decreasing exponential

$$\psi(x) = e^{-|x|/a} \Longleftrightarrow \bar{\psi}(p) = \sqrt{\frac{2}{\pi\hbar}} \frac{1/a}{(p^2/\hbar^2) + (1/a^2)} \tag{49}$$

(*iii*) Gaussian function

$$\psi(x) = e^{-x^2/a^2} \Longleftrightarrow \bar{\psi}(p) = \frac{a}{\sqrt{2\hbar}} e^{-p^2a^2/4\hbar^2} \tag{50}$$

(note the remarkable fact that the Gaussian form is conserved by the Fourier transform).

COMMENT:

In each of these three cases, the widths Δx and Δp can be defined for $\psi(x)$ and $\bar{\psi}(p)$ respectively, and they verify inequality (44).

e. FOURIER TRANSFORMS IN THREE-DIMENSIONAL SPACE

For wave functions $\psi(\mathbf{r})$ which depend on the three spatial variables x, y, z, (28) and (29) are replaced by:

$$\bar{\psi}(\mathbf{p}) = \frac{1}{(2\pi\hbar)^{3/2}} \int d^3r \, e^{-i\mathbf{p}.\mathbf{r}/\hbar} \, \psi(\mathbf{r}) \tag{51-a}$$

$$\psi(\mathbf{r}) = \frac{1}{(2\pi\hbar)^{3/2}} \int d^3p \, e^{i\mathbf{p}.\mathbf{r}/\hbar} \, \bar{\psi}(\mathbf{p}) \tag{51-b}$$

The properties stated above (§§2-b and 2-c) can easily be generalized to three dimensions.

If ψ depends only on the modulus r of the radius-vector $\mathbf{r}$, $\bar{\psi}$ depends only on the modulus p of the momentum $\mathbf{p}$ and can be calculated from the expression:

$$\bar{\psi}(p) = \frac{1}{\sqrt{2\pi\hbar}} \frac{2}{p} \int_0^\infty r \, dr \sin\frac{pr}{\hbar} \, \psi(r) \tag{52}$$

First, we shall find using (51-a) the value of $\overline{\psi}$ for a vector $\mathbf{p}'$ obtained from $\mathbf{p}$ by an arbitrary rotation $\mathscr{R}$:

$$\mathbf{p}' = \mathscr{R}\, \mathbf{p} \tag{53}$$

$$\overline{\psi}(\mathbf{p}') = \frac{1}{(2\pi\hbar)^{3/2}} \int d^3r \, \mathrm{e}^{-i\mathbf{p}'.\mathbf{r}/\hbar} \, \psi(r) \tag{54}$$

In this integral, we replace the variable $\mathbf{r}$ by $\mathbf{r}'$ and set:

$$\mathbf{r}' = \mathscr{R}\, \mathbf{r} \tag{55}$$

Since the volume element is conserved under rotation:

$$d^3r' = d^3r \tag{56}$$

In addition, the function ψ is unchanged, since the modulus of $\mathbf{r}'$ remains equal to r; finally:

$$\mathbf{p}' . \mathbf{r}' = \mathbf{p} . \mathbf{r} \tag{57}$$

since the scalar product is rotation-invariant. We thus find:

$$\overline{\psi}(\mathbf{p}') = \overline{\psi}(\mathbf{p}) \tag{58}$$

that is, $\overline{\psi}$ depends only on the modulus of $\mathbf{p}$ and not on its direction.

We can then choose $\mathbf{p}$ along Oz to evaluate $\overline{\psi}(p)$:

$$\begin{aligned}
\overline{\psi}(p) &= \frac{1}{(2\pi\hbar)^{3/2}} \int d^3r \, \mathrm{e}^{-ipz/\hbar} \, \psi(r) \\
&= \frac{1}{(2\pi\hbar)^{3/2}} \int_0^\infty r^2 \, \mathrm{d}r \, \psi(r) \int_0^{2\pi} \mathrm{d}\varphi \int_0^\pi \mathrm{d}\theta \sin\theta \, \mathrm{e}^{-\frac{ipr\cos\theta}{\hbar}} \\
&= \frac{1}{(2\pi\hbar)^{3/2}} \int_0^\infty r^2 \, \mathrm{d}r \, \psi(r) \, 2\pi \frac{2\hbar}{pr} \sin\frac{pr}{\hbar} \\
&= \frac{1}{\sqrt{2\pi\hbar}} \frac{2}{p} \int_0^\infty r \, \mathrm{d}r \, \psi(r) \sin\frac{pr}{\hbar}
\end{aligned} \tag{59}$$

This proves (52).

References and suggestions for further reading:

See, for example, Arfken (10.4), chaps. 14 and 15, or Butkov (10.8), chaps. 4 and 7; Bass (10.1), vol. I, chaps. XVIII through XX; section 10 of the bibliography, especially the subsection "Fourier transforms; distributions".

APPENDIX II

The Dirac δ- "function"

Appendix II

THE DIRAC δ-"FUNCTION"

The δ-"function" is actually a distribution. However, like most physicists, we shall treat it like an ordinary function. This approach, although not mathematically rigorous, is sufficient for quantum mechanical applications.

1. Introduction; principal properties

a. INTRODUCTION OF THE δ-"FUNCTION"

Consider the function $\delta^{(\varepsilon)}(x)$ given by (*cf.* fig. 1):

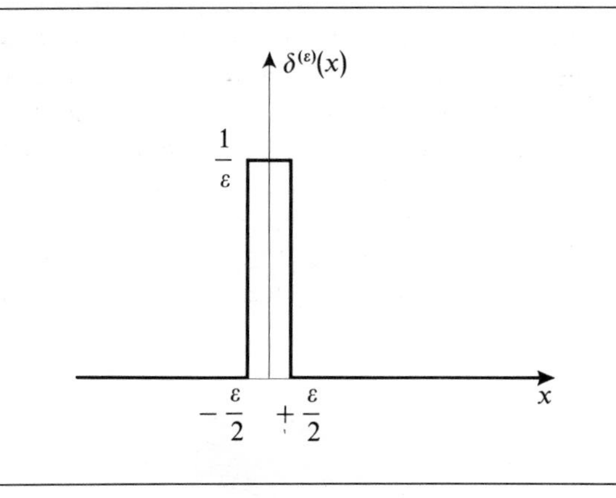

FIGURE 1

The function $\delta^{(\varepsilon)}(x)$: a square function of width ε and height $1/\varepsilon$, centered at $x = 0$.

$$\begin{aligned} \delta^{(\varepsilon)}(x) &= \frac{1}{\varepsilon} \quad \text{for} \quad -\frac{\varepsilon}{2} < x < \frac{\varepsilon}{2} \\ &= 0 \quad \text{for} \quad |x| > \frac{\varepsilon}{2} \end{aligned} \tag{1}$$

where ε is a positive number, We shall evaluate the integral:

$$\int_{-\infty}^{+\infty} dx\, \delta^{(\varepsilon)}(x)\, f(x) \tag{2}$$

where $f(x)$ is an arbitrary function, well-defined for $x = 0$. If ε is sufficiently small, the variation of $f(x)$ over the effective integration interval $[-\varepsilon/2, \varepsilon/2]$ is negligible, and $f(x)$ remains practically equal to $f(0)$. Therefore:

$$\int_{-\infty}^{+\infty} dx\, \delta^{(\varepsilon)}(x)\, f(x) \simeq f(0) \int_{-\infty}^{+\infty} dx\, \delta^{(\varepsilon)}(x) = f(0) \tag{3}$$

The smaller ε, the better the approximation. We therefore examine the limit $\varepsilon = 0$ and define the δ-"function" by the relation:

$$\int_{-\infty}^{+\infty} dx\, \delta(x)\, f(x) = f(0) \tag{4}$$

which is valid for any function $f(x)$ defined at the origin. More generally, $\delta(x - x_0)$ is defined by:

$$\boxed{\int_{-\infty}^{+\infty} dx\, \delta(x - x_0)\, f(x) = f(x_0)} \tag{5}$$

COMMENTS:

(*i*) Actually, the integral notation in (5) is not mathematically justified. δ is defined rigorously not as a function but as a distribution. Physically, this distinction is not an essential one as it becomes impossible to distinguish between $\delta^{(\varepsilon)}(x)$ and $\delta(x)$ as soon as ε becomes negligible compared to all the distances involved in a given physical problem*: any function $f(x)$ which we might have to consider does not vary significantly over an interval of length ε. Whenever a mathematical difficulty might arise, all we need to do is assume that $\delta(x)$ is actually $\delta^{(\varepsilon)}(x)$ [or an analogous but more regular function, for example, one of those given in (7), (8), (9), (10), (11)], with ε extremely small but not strictly zero.

(*ii*) For arbitrary integration limits a and b, we have:

$$\begin{aligned} \int_a^b dx\, \delta(x)\, f(x) &= f(0) \quad \text{if} \quad 0 \in [a, b] \\ &= 0 \quad \text{if} \quad 0 \notin [a, b] \end{aligned} \tag{6}$$

* The accuracy of present-day physical measurements does not, in any case, allow us to investigate phenomena on a scale of less than a fraction of a Fermi (1 Fermi $= 10^{-15}$ m).

b. FUNCTIONS WHICH APPROACH δ

It can easily be shown that, in addition to $\delta^{(\varepsilon)}(x)$ defined by (1), the following functions approach $\delta(x)$, that is, satisfy (5), when the parameter ε approaches zero from the positive side:

$$(i) \quad \frac{1}{2\varepsilon}\,e^{-|x|/\varepsilon} \tag{7}$$

$$(ii) \quad \frac{1}{\pi}\,\frac{\varepsilon}{x^2+\varepsilon^2} \tag{8}$$

$$(iii) \quad \frac{1}{\varepsilon\sqrt{\pi}}\,e^{-x^2/\varepsilon^2} \tag{9}$$

$$(iv) \quad \frac{1}{\pi}\,\frac{\sin(x/\varepsilon)}{x} \tag{10}$$

$$(v) \quad \frac{\varepsilon}{\pi}\,\frac{\sin^2(x/\varepsilon)}{x^2} \tag{11}$$

We shall also mention an identity which is often useful in quantum mechanics (particularly in collision theory):

$$\lim_{\varepsilon\to 0_+} \frac{1}{x \pm i\varepsilon} = \mathscr{P}\frac{1}{x} \mp i\pi\delta(x) \tag{12}$$

where $\mathscr{P}$ denotes the Cauchy principal part, defined by★ [$f(x)$ in a function which is regular at $x = 0$]:

$$\mathscr{P}\int_{-A}^{+B} \frac{dx}{x}\,f(x) = \lim_{\eta\to 0_+}\left[\int_{-A}^{-\eta} + \int_{+\eta}^{+B}\right]\frac{dx}{x}\,f(x)\ ; \qquad A, B > 0 \tag{13}$$

To prove (12), we separate the real and imaginary parts of $1/(x \pm i\varepsilon)$:

$$\frac{1}{x \pm i\varepsilon} = \frac{x \mp i\varepsilon}{x^2+\varepsilon^2} \tag{14}$$

Since the imaginary part is proportional to the function (8), we have :

$$\lim_{\varepsilon\to 0_+} \mp i\frac{\varepsilon}{x^2+\varepsilon^2} = \mp i\pi\,\delta(x) \tag{15}$$

As for the real part, we shall multiply it by a function $f(x)$ which is regular at the origin and integrate over x:

$$\lim_{\varepsilon\to 0_+}\int_{-\infty}^{+\infty} \frac{x\,dx}{x^2+\varepsilon^2}\,f(x) = \lim_{\varepsilon\to 0_+}\lim_{\eta\to 0_+}\left[\int_{-\infty}^{-\eta} + \int_{-\eta}^{+\eta} + \int_{+\eta}^{+\infty}\right]\frac{x\,dx}{x^2+\varepsilon^2}\,f(x) \tag{16}$$

★ One often uses one of the following relations :

$$\mathscr{P}\int_{-A}^{+B}\frac{dx}{x}\,f(x) = \int_{-B}^{+B} dx\,\frac{f_-(x)}{x} + \int_{-A}^{-B} dx\,\frac{f(x)}{x}$$

$$= \int_{-A}^{+B} dx\,\frac{f(x)-f(0)}{x} + f(0)\,\mathrm{Log}\,\frac{B}{A}$$

where $f_-(x) = [f(x) - f(-x)]/2$ is the odd part of $f(x)$. These formulas allow us to explicitely eliminate the divergence at the origin.

The second integral is zero:

$$\lim_{\eta \to 0_+} \int_{-\eta}^{+\eta} \frac{x \, \mathrm{d}x}{x^2 + \varepsilon^2} f(x) = f(0) \lim_{\eta \to 0_+} \frac{1}{2} \left[\mathrm{Log}\,(x^2 + \varepsilon^2) \right]_{-\eta}^{+\eta} = 0 \tag{17}$$

If we now reverse the order of the evaluation of the limits in (16), the $\varepsilon \longrightarrow 0$ limit presents no difficulties in the two other integrals. Thus :

$$\lim_{\varepsilon \to 0_+} \int_{-\infty}^{+\infty} \frac{x \, \mathrm{d}x}{x^2 + \varepsilon^2} f(x) = \lim_{\eta \to 0_+} \left[\int_{-\infty}^{-\eta} + \int_{+\eta}^{+\infty} \right] \frac{\mathrm{d}x}{x} f(x) \tag{18}$$

This establishes identity (12).

c. PROPERTIES OF δ

The properties we shall now state can be demonstrated using (5). Multiplying both sides of the equations below by a function $f(x)$ and integrating, we see that the results obtained are indeed equal.

(*i*) $\quad \delta(-x) = \delta(x) \qquad (19)$

(*ii*) $\quad \delta(cx) = \dfrac{1}{|c|} \delta(x) \qquad (20)$

and, more generally:

$$\delta[g(x)] = \sum_j \frac{1}{|g'(x_j)|} \delta(x - x_j) \tag{21}$$

where $g'(x)$ is the derivative of $g(x)$ and the x_j are the simple zeros of the function $g(x)$:

$$\begin{aligned} g(x_j) &= 0 \\ g'(x_j) &\neq 0 \end{aligned} \tag{22}$$

The summation is performed over all the simple zeros of $g(x)$. If $g(x)$ has multiple zeros [that is, for which $g'(x_j)$ is zero], the expression $\delta[g(x)]$ makes no sense.

(*iii*) $\quad x\,\delta(x - x_0) = x_0\,\delta(x - x_0) \qquad (23)$

and, in particular:

$$x\,\delta(x) = 0 \tag{24}$$

The converse is also true and it can be shown that the equation :

$$x\,u(x) = 0 \tag{25}$$

has the general solution:

$$u(x) = c\,\delta(x) \tag{26}$$

where c is an arbitrary constant.

More generally:

$$g(x)\,\delta(x - x_0) = g(x_0)\,\delta(x - x_0) \tag{27}$$

$$(iv) \quad \int_{-\infty}^{+\infty} \mathrm{d}x \, \delta(x - y) \, \delta(x - z) = \delta(y - z) \tag{28}$$

Equation (28) can be understood by examining functions $\delta^{(\varepsilon)}(x)$ like the one shown in figure 1. The integral:

$$F^{(\varepsilon)}(y, z) = \int_{-\infty}^{+\infty} \mathrm{d}x \; \delta^{(\varepsilon)}(x - y) \, \delta^{(\varepsilon)}(x - z) \tag{29}$$

is zero as long as $|y - z| < \varepsilon$, that is, as long as the two square functions do not overlap (fig. 2).

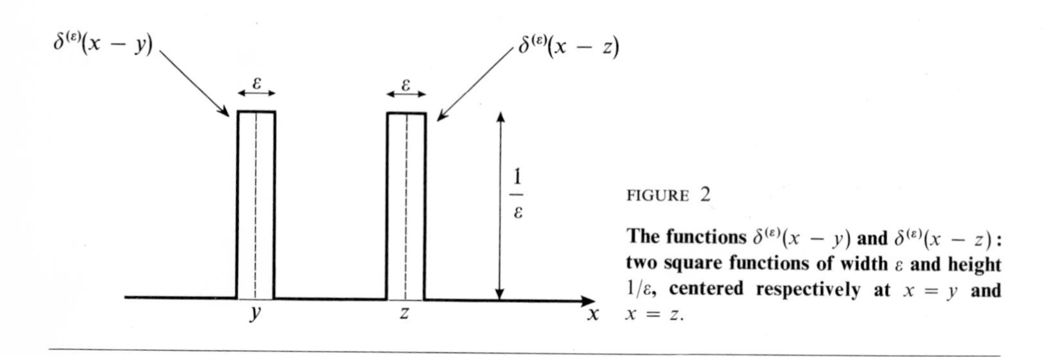

FIGURE 2

The functions $\delta^{(\varepsilon)}(x - y)$ and $\delta^{(\varepsilon)}(x - z)$: two square functions of width ε and height $1/\varepsilon$, centered respectively at $x = y$ and $x = z$.

The maximum value of the integral, obtained for $y = z$, is equal to $1/\varepsilon$. Between this maximum value and 0, the variation of $F^{(\varepsilon)}(y, z)$ with respect to $y - z$ is linear (fig. 3). We see immediately that $F^{(\varepsilon)}(y, z)$ approaches $\delta(y - z)$ when $\varepsilon \longrightarrow 0$.

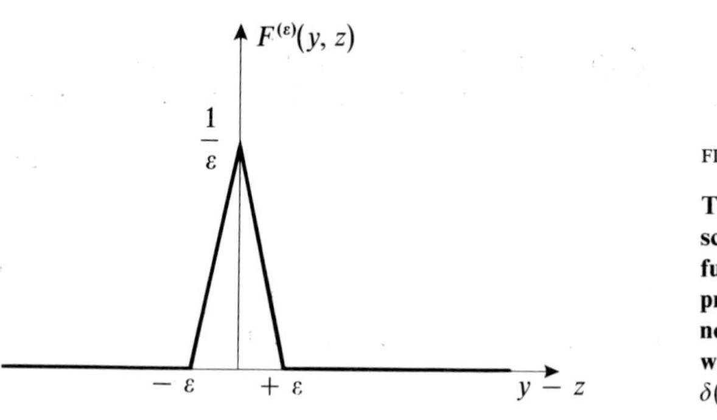

FIGURE 3

The variation with respect to $y - z$ of the scalar product $F^{(\varepsilon)}(y, z)$ of the two square functions shown in figure 2. This scalar product is zero when the two functions do not overlap ($|y - z| \geqslant \varepsilon$), and maximal when they coincide. $F^{(\varepsilon)}(y, z)$ approaches $\delta(y - z)$ when $\varepsilon \longrightarrow 0$.

COMMENT:

A sum of regularly spaced δ-functions:

$$\sum_{q=-\infty}^{+\infty} \delta(x - qL) \tag{30}$$

can be considered to be a periodic "function" of period L. By applying formulas (8), (9) and (10) of appendix I, we can write it in the form:

$$\sum_{q=-\infty}^{+\infty} \delta(x - qL) = \frac{1}{L} \sum_{n=-\infty}^{+\infty} e^{2i\pi \frac{nx}{L}} \tag{31}$$

2. The δ-"function" and the Fourier transform

a. THE FOURIER TRANSFORM OF δ

Definition (28) of appendix I and equation (5) enable us to calculate directly the Fourier transform $\bar{\delta}_{x_0}(p)$ of $\delta(x - x_0)$:

$$\bar{\delta}_{x_0}(p) = \frac{1}{\sqrt{2\pi\hbar}} \int_{-\infty}^{+\infty} \mathrm{d}x\, e^{-ipx/\hbar}\, \delta(x - x_0) = \frac{1}{\sqrt{2\pi\hbar}}\, e^{-ipx_0/\hbar} \tag{32}$$

In particular, that of $\delta(x)$ is a constant:

$$\bar{\delta}_0(p) = \frac{1}{\sqrt{2\pi\hbar}} \tag{33}$$

The inverse Fourier transform [formula (29) of appendix I] then yields:

$$\boxed{\delta(x - x_0) = \frac{1}{2\pi\hbar} \int_{-\infty}^{+\infty} \mathrm{d}p\, e^{ip(x-x_0)/\hbar} = \frac{1}{2\pi} \int_{-\infty}^{+\infty} \mathrm{d}k\, e^{ik(x-x_0)}} \tag{34}$$

This result can also be found by using the function $\delta^{(\varepsilon)}(x)$ defined by (1) or any of the functions given in § 1-b. For example, (48) of appendix I enables us to write:

$$\delta^{(\varepsilon)}(x) = \frac{1}{2\pi\hbar} \int_{-\infty}^{+\infty} \mathrm{d}p\, e^{ipx/\hbar} \frac{\sin(p\varepsilon/2\hbar)}{p\varepsilon/2\hbar} \tag{35}$$

If we let ε approach zero, we indeed obtain (34).

b. APPLIÇATIONS

Expression (34) for the δ-function is often very convenient. We shall show, for example, how it simplifies finding the inverse Fourier transform and the Parseval-Plancherel relation [formulas (29) and (45) of appendix I].

Starting with:

$$\bar{\psi}(p) = \frac{1}{\sqrt{2\pi\hbar}} \int_{-\infty}^{+\infty} \mathrm{d}x\, e^{-ipx/\hbar}\, \psi(x) \tag{36}$$

we calculate:

$$\frac{1}{\sqrt{2\pi\hbar}}\int_{-\infty}^{+\infty} \mathrm{d}p\ \mathrm{e}^{ipx/\hbar}\ \bar{\psi}(p) = \frac{1}{2\pi\hbar}\int_{-\infty}^{+\infty} \mathrm{d}\xi\ \psi(\xi)\int_{-\infty}^{+\infty} \mathrm{d}p\ \mathrm{e}^{ip(x-\xi)/\hbar} \tag{37}$$

In the second integral, we recognize $\delta(x - \xi)$, so that:

$$\frac{1}{\sqrt{2\pi\hbar}}\int_{-\infty}^{+\infty} \mathrm{d}p\ \mathrm{e}^{ipx/\hbar}\ \bar{\psi}(p) = \int_{-\infty}^{+\infty} \mathrm{d}\xi\ \psi(\xi)\ \delta(x - \xi) = \psi(x) \tag{38}$$

which is the inversion formula of the Fourier transform.

Similarly:

$$|\bar{\psi}(p)|^2 = \frac{1}{2\pi\hbar}\int_{-\infty}^{+\infty} \mathrm{d}x\ \mathrm{e}^{ipx/\hbar}\ \psi^*(x)\int_{-\infty}^{+\infty} \mathrm{d}x'\ \mathrm{e}^{-ipx'/\hbar}\ \psi(x') \tag{39}$$

If we integrate this expression over p, we find:

$$\int_{-\infty}^{+\infty} \mathrm{d}p\ |\bar{\psi}(p)|^2 = \frac{1}{2\pi\hbar}\int_{-\infty}^{+\infty} \mathrm{d}x\ \psi^*(x)\int_{-\infty}^{+\infty} \mathrm{d}x'\ \psi(x')\int_{-\infty}^{+\infty} \mathrm{d}p\ \mathrm{e}^{ip(x-x')/\hbar} \tag{40}$$

that is, according to (34):

$$\int_{-\infty}^{+\infty} \mathrm{d}p\ |\bar{\psi}(p)|^2 = \int_{-\infty}^{+\infty} \mathrm{d}x\ \psi^*(x)\int_{-\infty}^{+\infty} \mathrm{d}x'\psi(x')\ \delta(x - x') = \int_{-\infty}^{+\infty} \mathrm{d}x\ |\psi(x)|^2 \tag{41}$$

which is none other than the Parseval-Plancherel formula.

We can obtain the Fourier transform of a convolution product in an analogous way [*cf.* formulas (39) and (40) of appendix I].

3. Integral and derivatives of the δ-"function"

a. δ IS THE DERIVATIVE OF THE "UNIT STEP-FUNCTION"

We shall evaluate the integral:

$$\theta^{(\varepsilon)}(x) = \int_{-\infty}^{x} \delta^{(\varepsilon)}(x')\ \mathrm{d}x' \tag{42}$$

where the function $\delta^{(\varepsilon)}(x)$ is defined in (1). It can easily be seen that $\theta^{(\varepsilon)}(x)$ is equal to 0 for $x \leqslant -\frac{\varepsilon}{2}$, to 1 for $x \geqslant \frac{\varepsilon}{2}$, and to $\frac{1}{\varepsilon}\left(x + \frac{\varepsilon}{2}\right)$ for $-\frac{\varepsilon}{2} \leqslant x \leqslant \frac{\varepsilon}{2}$. The variation of $\theta^{(\varepsilon)}(x)$ with respect to x is shown in figure 4. When $\varepsilon \longrightarrow 0$, $\theta^{(\varepsilon)}(x)$ approaches the *Heaviside "step-function"* $\theta(x)$, which, by definition, is equal to:

$$\begin{aligned} \theta(x) &= 1 \quad \text{if} \quad x > 0 \\ \theta(x) &= 0 \quad \text{if} \quad x < 0 \end{aligned} \tag{43}$$

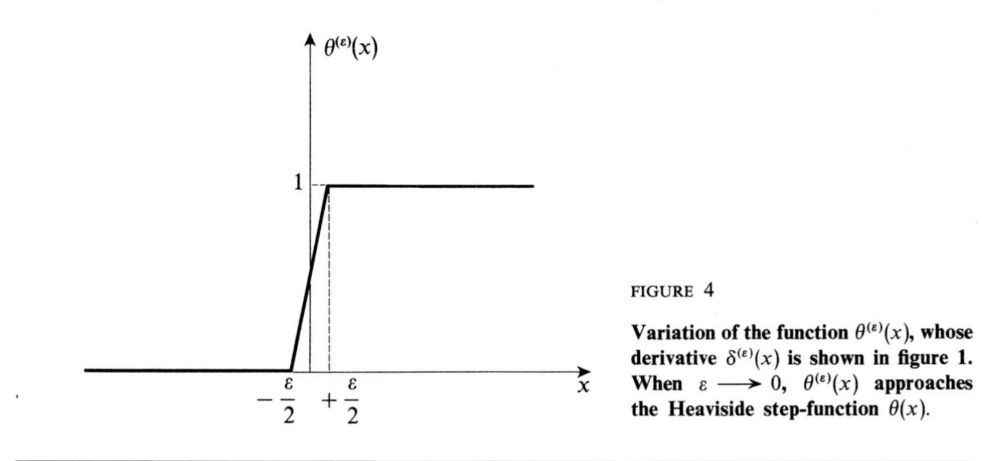

FIGURE 4

Variation of the function $\theta^{(\varepsilon)}(x)$, whose derivative $\delta^{(\varepsilon)}(x)$ is shown in figure 1. When $\varepsilon \longrightarrow 0$, $\theta^{(\varepsilon)}(x)$ approaches the Heaviside step-function $\theta(x)$.

$\delta^{(\varepsilon)}(x)$ is the derivative of $\theta^{(\varepsilon)}(x)$. By considering the limit $\varepsilon \longrightarrow 0$, we see that $\delta(x)$ is the derivative of $\theta(x)$:

$$\frac{\mathrm{d}}{\mathrm{d}x}\,\theta(x) = \delta(x) \tag{44}$$

Now, consider a function $g(x)$ which has a discontinuity σ_0 at $x = 0$:

$$\operatorname*{Lim}_{x \to 0_+} g(x) - \operatorname*{Lim}_{x \to 0_-} g(x) = \sigma_0 \tag{45}$$

Such a function can be written in the form $g(x) = g_1(x)\theta(x) + g_2(x)\theta(-x)$, where $g_1(x)$ and $g_2(x)$ are continuous functions which satisfy $g_1(0) - g_2(0) = \sigma_0$. If we differentiate this expression, using (44), we obtain:

$$\begin{aligned} g'(x) &= g_1'(x)\,\theta(x) + g_2'(x)\,\theta(-x) + g_1(x)\,\delta(x) - g_2(x)\,\delta(-x) \\ &= g_1'(x)\,\theta(x) + g_2'(x)\,\theta(-x) + \sigma_0\,\delta(x) \end{aligned} \tag{46}$$

according to properties (19) and (27) of δ. For a discontinuous function, there is then added to the ordinary derivative [the first two terms of (46)] a term proportional to the δ-function, the proportionality coefficient being the magnitude of the function's discontinuity★.

COMMENT:

The Fourier transform of the step-function $\theta(k)$ can be found simply from (12). We get:

$$\int_{-\infty}^{+\infty} \theta(k)\,\mathrm{e}^{ikx}\,\mathrm{d}k = \operatorname*{Lim}_{\varepsilon \to 0_+} \int_0^{\infty} \mathrm{d}k\,\mathrm{e}^{ik(x+i\varepsilon)} = \operatorname*{Lim}_{\varepsilon \to 0_+} \frac{i}{x + i\varepsilon} = i\mathscr{P}\frac{1}{x} + \pi\delta(x) \tag{47}$$

★ Of course, if the function is discontinuous at $x = x_0$, the additional term is of the form : $[g_1(x_0) - g_2(x_0)]\delta(x - x_0)$.

b. DERIVATIVES OF δ

By analogy with the expression for integration by parts, the derivative $\delta'(x)$ of the δ-function is defined by the relation★:

$$\int_{-\infty}^{+\infty} \mathrm{d}x \, \delta'(x) \, f(x) = - \int_{-\infty}^{+\infty} \mathrm{d}x \, \delta(x) \, f'(x) = - f'(0) \tag{48}$$

From this definition, we get immediately:

$$\delta'(-x) = -\delta'(x) \tag{49}$$

and:

$$x\,\delta'(x) = -\delta(x) \tag{50}$$

Conversely it can be shown that the general solution of the equation:

$$x\,u(x) = \delta(x) \tag{51}$$

can be written:

$$u(x) = -\delta'(x) + c\,\delta(x) \tag{52}$$

where the second term arises from the homogeneous equation [*cf.* formulas (25) and (26)].

Equation (34) allows us to write $\delta'(x)$ in the form:

$$\delta'(x) = \frac{1}{2\pi\hbar} \int_{-\infty}^{+\infty} \mathrm{d}p \left(\frac{ip}{\hbar}\right) \mathrm{e}^{ipx/\hbar} = \frac{i}{2\pi} \int_{-\infty}^{+\infty} k \, \mathrm{d}k \, \mathrm{e}^{ikx} \tag{53}$$

The nth-order derivative $\delta^{(n)}(x)$ can be defined in the same way:

$$\int_{-\infty}^{+\infty} \mathrm{d}x \, \delta^{(n)}(x) \, f(x) = (-1)^n f^{(n)}(0) \tag{54}$$

Relations (49) and (50) can then be generalized to the forms:

$$\delta^{(n)}(-x) = (-1)^n \, \delta^{(n)}(x) \tag{55}$$

and:

$$x\,\delta^{(n)}(x) = -n\,\delta^{(n-1)}(x) \tag{56}$$

4. The δ-"function" in three-dimensional space

The δ-"function" in three-dimensional space, which we shall write simply as $\delta(\mathbf{r})$, is defined by an expression analogous to (4):

$$\int d^3r \, \delta(\mathbf{r}) \, f(\mathbf{r}) = f(\mathbf{0}) \tag{57}$$

★ $\delta'(x)$ can be considered to be the limit, for $\varepsilon \longrightarrow 0$, of the derivative of one of the functions given in § 1-b.

and, more generally:

$$\int d^3r \, \delta(\mathbf{r} - \mathbf{r}_0) \, f(\mathbf{r}) = f(\mathbf{r}_0) \tag{58}$$

$\delta(\mathbf{r} - \mathbf{r}_0)$ can be broken down into a product of three one-dimensional functions:

$$\delta(\mathbf{r} - \mathbf{r}_0) = \delta(x - x_0) \, \delta(y - y_0) \, \delta(z - z_0) \tag{59}$$

or, if we use polar coordinates:

$$\begin{aligned} \delta(\mathbf{r} - \mathbf{r}_0) &= \frac{1}{r^2 \sin \theta} \delta(r - r_0) \delta(\theta - \theta_0) \, \delta(\varphi - \varphi_0) \\ &= \frac{1}{r^2} \delta(r - r_0) \, \delta(\cos \theta - \cos \theta_0) \, \delta(\varphi - \varphi_0) \end{aligned} \tag{60}$$

The properties stated above for $\delta(x)$ can therefore easily be generalized to $\delta(\mathbf{r})$. We shall mention, in addition, the important relation:

$$\Delta \left(\frac{1}{r} \right) = - 4\pi \, \delta(\mathbf{r}) \tag{61}$$

where Δ is the Laplacian operator.

Equation (61) can easily be understood if it is recalled that in electrostatics, an electrical point charge q placed at the origin can be described by a volume density $\rho(\mathbf{r})$ equal to:

$$\rho(\mathbf{r}) = q \, \delta(\mathbf{r}) \tag{62}$$

We know that the expression for the electrostatic potential produced by this charge is:

$$U(\mathbf{r}) = \frac{q}{4\pi\varepsilon_0} \frac{1}{r} \tag{63}$$

Equation (61) is thus simply the Poisson equation for this special case:

$$\Delta U(\mathbf{r}) = - \frac{1}{\varepsilon_0} \rho(\mathbf{r}) \tag{64}$$

To prove (61) rigorously, it is necessary to use mathematical distribution theory. We shall confine ourselves here to an elementary "proof".

First of all, note that the Laplacian of $1/r$ is everywhere zero, except, perhaps, at the origin, which is a singular point:

$$\left(\frac{\mathrm{d}^2}{\mathrm{d}r^2} + \frac{2}{r} \frac{\mathrm{d}}{\mathrm{d}r} \right) \frac{1}{r} = 0 \qquad \text{for } r \neq 0 \tag{65}$$

Let $g_\varepsilon(\mathbf{r})$ be a function equal to $1/r$ when $\mathbf{r}$ is outside the sphere S_ε, centered at O and of a radius ε, and which takes on values (of the order of $1/\varepsilon$) inside this sphere such that $g_\varepsilon(\mathbf{r})$

is sufficiently regular (continuous, differentiable, etc.). Let $f(\mathbf{r})$ be an arbitrary function of $\mathbf{r}$ which is also regular at all points in space. We now find the limit of the integral:

$$I(\varepsilon) = \int d^3r\, f(\mathbf{r})\, \Delta g_\varepsilon(\mathbf{r}) \tag{66}$$

for $\varepsilon \longrightarrow 0$. According to (65), this integral can receive contributions only from inside the sphere S_ε, and:

$$I(\varepsilon) = \int_{r \leqslant \varepsilon} d^3r\, f(\mathbf{r})\, \Delta g_\varepsilon(\mathbf{r}) \tag{67}$$

We choose ε small enough for the variation of $f(\mathbf{r})$ inside S_ε to be negligible. Then:

$$I(\varepsilon) \simeq f(\mathbf{0}) \int_{r \leqslant \varepsilon} d^3r\, \Delta g_\varepsilon(\mathbf{r}) \tag{68}$$

Transforming the integral so obtained into an integral over the surface $\mathscr{S}_\varepsilon$ of S_ε, we obtain:

$$I(\varepsilon) \simeq f(\mathbf{0}) \int_{\mathscr{S}_\varepsilon} \boldsymbol{\nabla} g_\varepsilon(\mathbf{r}) \,.\, \mathrm{d}\mathbf{n} \tag{69}$$

Now, since $g_\varepsilon(\mathbf{r})$ is continuous on the surface $\mathscr{S}_\varepsilon$, we get:

$$[\boldsymbol{\nabla} g_\varepsilon(\mathbf{r})]_{r=\varepsilon} = \left[-\frac{1}{r^2} \right]_{r=\varepsilon} \mathbf{e}_r = -\frac{1}{\varepsilon^2}\mathbf{e}_r \tag{70}$$

(where $\mathbf{e}_r$ is the unit vector $\mathbf{r}/r$). This yields:

$$\begin{aligned} I(\varepsilon) &\simeq f(\mathbf{0}) \times 4\pi\varepsilon^2 \times \left[-\frac{1}{\varepsilon^2} \right] \\ &\simeq -4\pi f(\mathbf{0}) \end{aligned} \tag{71}$$

that is:

$$\lim_{\varepsilon \to 0} \int d^3r\, \Delta g_\varepsilon(\mathbf{r})\, f(\mathbf{r}) = -4\pi f(\mathbf{0}) \tag{72}$$

According to definition (57), this is simply (61).

Equation (61) can be used, for example, to derive an expression which is useful in collision theory (*cf.* chap. VIII):

$$(\Delta + k^2)\frac{\mathrm{e}^{\pm ikr}}{r} = -4\pi\, \delta(\mathbf{r}) \tag{73}$$

To do so, it is sufficient to consider $\mathrm{e}^{\pm ikr}/r$ as a product:

$$\Delta\left[\frac{\mathrm{e}^{\pm ikr}}{r}\right] = \frac{1}{r}\Delta(\mathrm{e}^{\pm ikr}) + \mathrm{e}^{\pm ikr}\Delta\left(\frac{1}{r}\right) + 2\boldsymbol{\nabla}\left(\frac{1}{r}\right) . \boldsymbol{\nabla}(\mathrm{e}^{\pm ikr}) \tag{74}$$

Now:

$$\nabla(e^{\pm ikr}) = \pm\, ik\, e^{\pm ikr}\frac{\mathbf{r}}{r}$$

$$\Delta(e^{\pm ikr}) = -\,k^2\, e^{\pm ikr} \pm \frac{2ik}{r}\, e^{\pm ikr} \tag{75}$$

We therefore find, finally:

$$\begin{aligned}(\Delta + k^2)\frac{e^{\pm ikr}}{r} &= \left[-\frac{k^2}{r} \pm \frac{2ik}{r^2} - 4\pi\,\delta(\mathbf{r}) - \frac{2}{r^2}\times(\pm\, ik) + \frac{k^2}{r}\right]e^{\pm ikr} \\ &= -\,4\pi\, e^{\pm ikr}\,\delta(\mathbf{r}) \\ &= -\,4\pi\,\delta(\mathbf{r})\end{aligned} \tag{76}$$

according to (27).

Equation (61) can, furthermore, be generalized : the Laplacian of the function $Y_l^m(\theta, \varphi)/r^{l+1}$ involves lth-order derivatives of $\delta(\mathbf{r})$. Consider, for example $\cos\theta/r^2$. We know that the expression for the electrostatic potential created at a distant point by an electric dipole of moment $\mathbf{D}$ directed along Oz is $\frac{D}{4\pi\varepsilon_0}\frac{\cos\theta}{r^2}$. If q is the absolute value of each of the two charges which make up the dipole and a is the distance between them, the modulus D of the dipole moment is the product qa, and the corresponding charge density can be written:

$$\rho(\mathbf{r}) = q\,\delta\left(\mathbf{r} - \frac{a}{2}\mathbf{e}_z\right) - q\,\delta\left(\mathbf{r} + \frac{a}{2}\mathbf{e}_z\right) \tag{77}$$

(where $\mathbf{e}_z$ denotes the unit vector of the Oz axis). If we let a approach zero, while maintaining $D = qa$ finite, this charge density becomes:

$$\rho(\mathbf{r}) \underset{a\to 0}{\longrightarrow} D\frac{\partial}{\partial z}\,\delta(\mathbf{r}) \tag{78}$$

Therefore in the limit where $a \longrightarrow 0$, the Poisson equation, (64), yields:

$$\Delta\left(\frac{\cos\theta}{r^2}\right) = -\,4\pi\frac{\partial}{\partial z}\,\delta(\mathbf{r}) \tag{79}$$

Of course, this formula could be justified as (61) was above, or proven by distribution theory. Analogous reasoning could be applied to the function $Y_l^m(\theta, \varphi)/r^{l+1}$ which gives the potential created by an electric multipole moment $\mathcal{Q}_l^m$ located at the origin (complement E_X).

References and suggestions for further reading:

See Dirac (1.13) § 15, and, for example, Butkov (10.8), chap. 6, or Bass (10.1), vol. I, §§ 21.7 and 21.8; section 10 of the bibliography, especially the subsection "Fourier transforms; distributions".

APPENDIX III

Lagrangian and Hamiltonian in classical mechanics

Appendix III

LAGRANGIAN AND HAMILTONIAN IN CLASSICAL MECHANICS

We shall review the definition and principal properties of the Lagrangian and the Hamiltonian in classical mechanics. This appendix is not meant to be a course in analytical mechanics. Its goal is simply to indicate the classical basis for applying the quantization rules (*cf.* chap. III) to a physical system. In particular, we shall concern ourselves essentially with systems of point particles.

1. Review of Newton's laws

a. DYNAMICS OF A POINT PARTICLE

Non-relativistic classical mechanics is based on the hypothesis that there exists at least one geometrical frame, called the *Galilean* or *inertial frame*, in which the following law is valid:

The fundamental law of dynamics : a point particle has, at all times, an acceleration γ which is proportional to the resultant $\mathbf{F}$ of the forces acting on it:

$$\mathbf{F} = m\gamma \tag{1}$$

The constant m is an intrinsic property of the particle, called its *inertial mass*.

It can easily be shown that if a Galilean frame exists, all frames which are in uniform translational motion with respect to it are also Galilean frames. This leads us to the *Galilean relativity principle*: there is no absolute frame; there is no experiment which can give one inertial frame a privileged role with respect to all others.

b. SYSTEMS OF POINT PARTICLES

If we are dealing with a system composed of n point particles, we apply the fundamental law to each of them★:

$$m_i \ddot{\mathbf{r}}_i = \mathbf{F}_i \quad ; \quad i = 1, 2, \ldots n \tag{2}$$

The forces which act on the particles can be classed in two categories : *internal forces* represent the interactions between the particles of the system, and *external forces* originate outside the system. The internal forces are postulated to satisfy the *principle of action and reaction*: the force exerted by particle (i) on particle (j) is equal and opposite to the one exerted by (j) on (i). This principle is true for gravitational forces (Newton's law) and electrostatic forces, but not for magnetic forces (whose origin is relativistic).

If all the forces can be derived from a potential, the equations of motion, (2), can be written:

$$m_i \ddot{\mathbf{r}}_i = - \boldsymbol{\nabla}_i V \tag{3}$$

where $\boldsymbol{\nabla}_i$ denotes the gradient with respect to the $\mathbf{r}_i$ coordinates, and the potential energy V is of the form:

$$V = \sum_{i=1}^{n} V_i(\mathbf{r}_i) + \sum_{i<j} V_{ij}(\mathbf{r}_i - \mathbf{r}_j) \tag{4}$$

(the first term in this expression corresponds to the external forces and the second one, to the internal forces). *In cartesian coordinates*, the motion of the system is therefore described by the $3n$ differential equations:

$$\left.\begin{aligned} m_i \ddot{x}_i &= - \frac{\partial V}{\partial x_i} \\ m_i \ddot{y}_i &= - \frac{\partial V}{\partial y_i} \\ m_i \ddot{z}_i &= - \frac{\partial V}{\partial z_i} \end{aligned}\right\} i = 1, 2, \ldots n \tag{5}$$

★ In mechanics, a simplified notation is generally used for the time-derivatives; by definition, $\dot{u} = \frac{\mathrm{d}u}{\mathrm{d}t}$, $\ddot{u} = \frac{\mathrm{d}^2u}{\mathrm{d}t^2}$, etc...

c. FUNDAMENTAL THEOREMS

We shall first review a few definitions. The *center of mass* or *center of gravity* of a system is the point G whose coordinates are:

$$\mathbf{r}_G = \frac{\sum_{i=1}^{n} m_i \, \mathbf{r}_i}{\sum_{i=1}^{n} m_i} \tag{6}$$

The total *kinetic energy* of the system is equal to:

$$T = \sum_{i=1}^{n} \frac{1}{2} m_i \, \dot{\mathbf{r}}_i^2 \tag{7}$$

where $\dot{\mathbf{r}}_i$ is the velocity of particle (i). The *angular momentum* with respect to the origin is the vector:

$$\mathscr{L} = \sum_{i=1}^{n} \mathbf{r}_i \times m_i \, \dot{\mathbf{r}}_i \tag{8}$$

The following theorems can then be easily proven:

(i) The center of mass of a system moves like a point particle with a mass equal to the total mass of the system, subject to a force equal to the resultant of all the forces involved in the system:

$$\left[\sum_{i=1}^{n} m_i \right] \ddot{\mathbf{r}}_G = \sum_{i=1}^{n} \mathbf{F}_i \tag{9}$$

(ii) The time-derivative of the angular momentum evaluated at a fixed point is equal to the moment of the forces with respect to this point:

$$\dot{\mathscr{L}} = \sum_{i=1}^{n} \mathbf{r}_i \times \mathbf{F}_i \tag{10}$$

(iii) The variation of the kinetic energy between time t_1 and t_2 is equal to the work performed by all the forces during the motion between these two times:

$$T(t_2) - T(t_1) = \int_{t_1}^{t_2} \sum_{i=1}^{n} \mathbf{F}_i \cdot \dot{\mathbf{r}}_i \, \mathrm{d}t \tag{11}$$

If the internal forces satisfy the principle of action and reaction, and if they are directed along the straight lines joining the interacting particles, their contribution to the resultant [equation (9)] and to the moment with respect to the origin [equation (10)] is zero. If, in addition, the system is isolated (that is, if it is not subject to any external forces), the total angular momentum $\mathscr{L}$ is constant, and the center of mass is in uniform rectilinear motion. This means that the total mechanical momentum:

$$\sum_{i=1}^{n} m_i \dot{\mathbf{r}}_i \tag{12}$$

is also a constant of the motion.

2. The Lagrangian and Lagrange's equations

Consider a system of n particles in which the forces are derived from a potential energy [*cf.* formula (4)], which we shall write simply $V(\mathbf{r}_i)$. The *Lagrangian*, or *Lagrange's function*, of this system is the function of $6n$ variables

$$\{ x_i, y_i, z_i; \dot{x}_i, \dot{y}_i, \dot{z}_i; i = 1, 2, ..., n \}$$

given by:

$$\begin{aligned} \mathscr{L}(\mathbf{r}_i, \dot{\mathbf{r}}_i) &= T - V \\ &= \frac{1}{2} \sum_{i=1}^{n} m_i \dot{\mathbf{r}}_i^2 - V(\mathbf{r}_i) \end{aligned} \tag{13}$$

It can immediately be shown that the equations of motion written in (5) are identical to *Lagrange's equations*:

$$\begin{aligned} &\frac{\mathrm{d}}{\mathrm{d}t}\frac{\partial \mathscr{L}}{\partial \dot{x}_i} - \frac{\partial \mathscr{L}}{\partial x_i} = 0 \\ &\frac{\mathrm{d}}{\mathrm{d}t}\frac{\partial \mathscr{L}}{\partial \dot{y}_i} - \frac{\partial \mathscr{L}}{\partial y_i} = 0 \\ &\frac{\mathrm{d}}{\mathrm{d}t}\frac{\partial \mathscr{L}}{\partial \dot{z}_i} - \frac{\partial \mathscr{L}}{\partial z_i} = 0 \end{aligned} \tag{14}$$

A very interesting feature of Lagrange's equations is that they always have the same form, independent of the type of coordinates used (whether they are cartesian or not). In addition, they can be applied to systems which are more general than particle systems. Many physical systems (including for example one or several solid bodies) can be described at a given time t by a set of N independent parameters q_i ($i = 1, 2, ..., N$), called *generalized coordinates*. Knowledge of the q_i permits the calculation of the position in space of any point of the system. The motion of this system is therefore characterized by specifying the N functions of time $q_i(t)$. The time-derivatives $\dot{q}_i(t)$ are called the *generalized velocities*. The state of the system at a given instant t_0 is therefore defined by the set of $q_i(t_0)$ and $\dot{q}_i(t_0)$. If the forces acting on the system can be derived from a potential energy $V(q_1, q_2, ..., q_N)$, the Lagrangian $\mathscr{L}(q_1, q_2, ..., q_N; \dot{q}_1, \dot{q}_2, ..., \dot{q}_N)$ is again the difference between the total kinetic energy T and the potential energy V. It can be shown that, for any choice of the coordinates q_i, the equations of motion can always be written:

$$\boxed{\frac{\mathrm{d}}{\mathrm{d}t}\frac{\partial \mathscr{L}}{\partial \dot{q}_i} - \frac{\partial \mathscr{L}}{\partial q_i} = 0} \tag{15}$$

where $\frac{d}{dt}$ denotes the *total time-derivative*

$$\frac{d}{dt} = \frac{\partial}{\partial t} + \sum_{i=1}^{N} \dot{q}_i \frac{\partial}{\partial q_i} + \sum_{i=1}^{N} \ddot{q}_i \frac{\partial}{\partial \dot{q}_i} \tag{16}$$

Furthermore, it is not really necessary for the forces to be derived from a potential for us to be able to define a Lagrangian and use Lagrange's equations (we shall see an example of this situation in § 4-b). In the general case, the Lagrangian is a function of the coordinates q_i and the velocities $\dot{q}_i$, and can also be explicitly time-dependent*. We shall then write it:

$$\mathscr{L}(q_i, \dot{q}_i ; t) \tag{17}$$

Lagrange's equations are important in classical mechanics for several reasons. For one thing, as we have just indicated, they always have the same form, independent of the coordinates which are used. Furthermore, they are more convenient than Newton's equations when the system is complex. Finally, they are of considerable theoretical interest, since they form the foundation of the Hamiltonian formalism (*cf.* § 3 below), and since they can be derived from a variational principle (§ 5). The first two points are secondary as far as quantum mechanics is concerned, since quantum mechanics treats particle systems almost exclusively and since the quantization rules are stated in cartesian coordinates (*cf.* chap. III, § B-5). However, the last point is an essential one, since the Hamiltonian formalism constitutes the point of departure for the quantization of physical systems.

3. The classical Hamiltonian and the canonical equations

For a physical system described by N generalized coordinates, Lagrange's equations, (15), constitute a system of N coupled second-order differential equations with N unknown functions, the $q_i(t)$. We shall see that this system can be replaced by a system of $2N$ first-order equations with $2N$ unknown functions.

a. THE CONJUGATE MOMENTA OF THE COORDINATES

The conjugate momentum p_i of the generalized coordinate q_i is defined as :

$$\boxed{p_i = \frac{\partial \mathscr{L}}{\partial \dot{q}_i}} \tag{18}$$

p_i is also called the *generalized momentum*. In the case of a particle system for which the forces are derived from a potential energy, the conjugate momenta of the position variables $\mathbf{r}_i(x_i, y_i, z_i)$ are simply [see (13)] the mechanical momenta:

$$\mathbf{p}_i = m_i \, \dot{\mathbf{r}}_i \tag{19}$$

However, we shall see in § 4-b-γ that this is no longer true in the presence of a magnetic field.

* The Lagrangian is not unique : two functions $\mathscr{L}(q_i, \dot{q}_i ; t)$ and $\mathscr{L}'(q_i, \dot{q}_i ; t)$ may lead, using (15), to the same equations of motion. This is true, in particular, if the difference between $\mathscr{L}$ and $\mathscr{L}'$ is the total derivative with respect to time of a function $F(q_i ; t)$:

$$\mathscr{L}' - \mathscr{L} = \frac{d}{dt} F(q_i ; t) \equiv \frac{\partial F}{\partial t} + \sum_i \dot{q}_i \frac{\partial F}{\partial q_i}$$

Instead of defining the state of the system at a given time t by the N coordinates $q_i(t)$ and the N velocities $\dot{q}_i(t)$, we shall henceforth characterize it by the $2N$ variables:

$$\{ q_i(t), p_i(t); i = 1, 2, ..., N \} \tag{20}$$

This amounts to assuming that from the $2N$ parameters $q_i(t)$ and $p_i(t)$, we can determine the $\dot{q}_i(t)$ uniquely.

b. THE HAMILTON-JACOBI CANONICAL EQUATIONS

The *classical Hamiltonian*, or *Hamilton's function*, of the system is, by definition:

$$\mathscr{H} = \sum_{i=1}^{N} p_i \dot{q}_i - \mathscr{L} \tag{21}$$

In accordance with convention (20), we eliminate the $\dot{q}_i$ and consider the Hamiltonian to be a function of the coordinates and their conjugate momenta. Like $\mathscr{L}$, $\mathscr{H}$ may be explicitly time-dependent:

$$\mathscr{H}(q_i, p_i; t) \tag{22}$$

The total differential of the function $\mathscr{H}$:

$$\mathrm{d}\mathscr{H} = \sum_i \frac{\partial \mathscr{H}}{\partial q_i} \mathrm{d}q_i + \sum_i \frac{\partial \mathscr{H}}{\partial p_i} \mathrm{d}p_i + \frac{\partial \mathscr{H}}{\partial t} \mathrm{d}t \tag{23}$$

is equal to, using definitions (21) and (18):

$$\begin{aligned} \mathrm{d}\mathscr{H} &= \sum_i [p_i \, \mathrm{d}\dot{q}_i + \dot{q}_i \, \mathrm{d}p_i] - \sum_i \frac{\partial \mathscr{L}}{\partial q_i} \mathrm{d}q_i - \sum_i \frac{\partial \mathscr{L}}{\partial \dot{q}_i} \mathrm{d}\dot{q}_i - \frac{\partial \mathscr{L}}{\partial t} \mathrm{d}t \\ &= \sum_i \dot{q}_i \, \mathrm{d}p_i - \sum_i \frac{\partial \mathscr{L}}{\partial q_i} \mathrm{d}q_i - \frac{\partial \mathscr{L}}{\partial t} \mathrm{d}t \end{aligned} \tag{24}$$

Setting (23) and (24) equal, we see that the change from the $\{ q_i, \dot{q}_i \}$ variables to the $\{ q_i, p_i \}$ variables leads to:

$$\frac{\partial \mathscr{H}}{\partial q_i} = - \frac{\partial \mathscr{L}}{\partial q_i} \tag{25-a}$$

$$\frac{\partial \mathscr{H}}{\partial p_i} = \dot{q}_i \tag{25-b}$$

$$\frac{\partial \mathscr{H}}{\partial t} = - \frac{\partial \mathscr{L}}{\partial t} \tag{25-c}$$

Furthermore, using (18) and (25-a), we can write Lagrange's equations (15) in the form:

$$\frac{\mathrm{d}}{\mathrm{d}t} p_i = - \frac{\partial \mathscr{H}}{\partial q_i} \tag{26}$$

By grouping terms in (25-b) and (26), we obtain the equations of motion:

$$\boxed{\begin{aligned} \frac{dq_i}{dt} &= \frac{\partial \mathcal{H}}{\partial p_i} \\ \frac{dp_i}{dt} &= -\frac{\partial \mathcal{H}}{\partial q_i} \end{aligned}} \tag{27}$$

which are called the Hamilton-Jacobi canonical equations. As we said, (27) is a system of $2N$ first-order differential equations for $2N$ unknown functions, the $q_i(t)$ and $p_i(t)$.

For an n-particle system whose potential energy is $V(\mathbf{r}_i)$, we have, according to (13):

$$\begin{aligned} \mathcal{H} &= \sum_{i=1}^{n} \mathbf{p}_i \cdot \dot{\mathbf{r}}_i - \mathcal{L} \\ &= \sum_{i=1}^{n} \mathbf{p}_i \cdot \dot{\mathbf{r}}_i - \frac{1}{2} \sum_{i=1}^{n} m_i \, \dot{\mathbf{r}}_i^2 + V(\mathbf{r}_i) \end{aligned} \tag{28}$$

To express the Hamiltonian in terms of the variables $\mathbf{r}_i$ and $\mathbf{p}_i$, we use (19). This yields:

$$\mathcal{H}(\mathbf{r}_i, \mathbf{p}_i) = \sum_{i=1}^{n} \frac{\mathbf{p}_i^2}{2m_i} + V(\mathbf{r}_i) \tag{29}$$

Note that the Hamiltonian is thus equal to the *total energy* of the system. The canonical equations:

$$\begin{aligned} \frac{d\mathbf{r}_i}{dt} &= \frac{\mathbf{p}_i}{m_i} \\ \frac{d\mathbf{p}_i}{dt} &= - \boldsymbol{\nabla}_i V \end{aligned} \tag{30}$$

are equivalent to Newton's equations, (3).

4. Applications of the Hamiltonian formalism

a. A PARTICLE IN A CENTRAL POTENTIAL

Consider a system composed of a single particle of mass m whose potential energy $V(r)$ depends only on its distance from the origin. In polar coordinates (r, θ, φ), the components of the particle's velocity on the local axes (fig. 1) are:

$$\begin{aligned} v_r &= \dot{r} \\ v_\theta &= r\,\dot{\theta} \\ v_\varphi &= r \sin\theta \, \dot{\varphi} \end{aligned} \tag{31}$$

so that the Lagrangian, (13), can be written:

$$\mathscr{L}(r, \theta, \varphi; \dot{r}, \dot{\theta}, \dot{\varphi}) = \frac{1}{2} m \left[\dot{r}^2 + r^2\dot{\theta}^2 + r^2 \sin^2\theta \, \dot{\varphi}^2\right] - V(r) \tag{32}$$

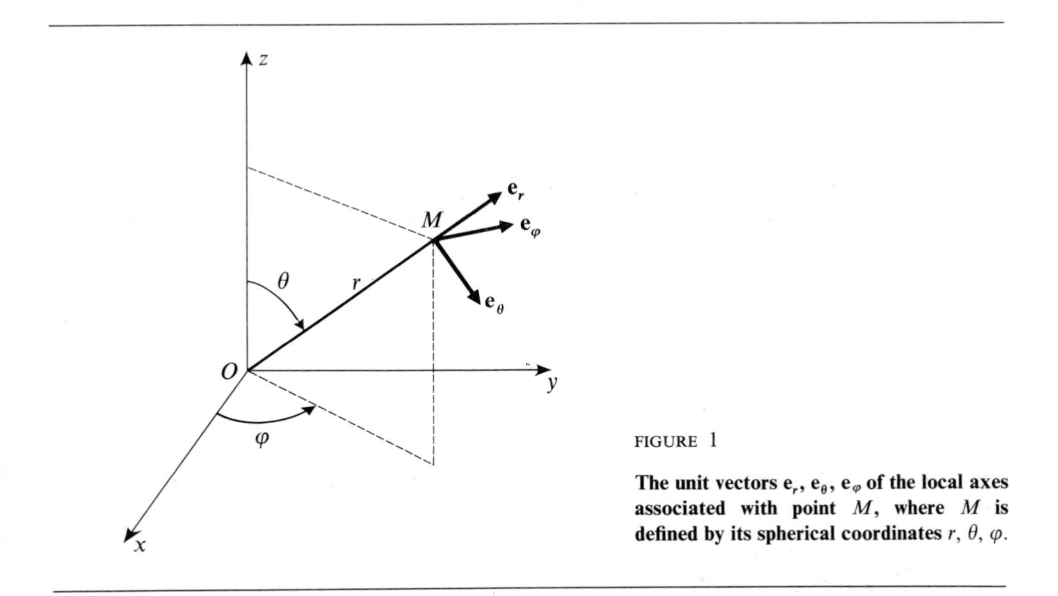

FIGURE 1

The unit vectors $\mathbf{e}_r$, $\mathbf{e}_\theta$, $\mathbf{e}_\varphi$ of the local axes associated with point M, where M is defined by its spherical coordinates r, θ, φ.

The conjugate momenta of the three variables r, θ, φ can then be calculated:

$$p_r = \frac{\partial \mathscr{L}}{\partial \dot{r}} = m\dot{r} \tag{33-a}$$

$$p_\theta = \frac{\partial \mathscr{L}}{\partial \dot{\theta}} = mr^2\dot{\theta} \tag{33-b}$$

$$p_\varphi = \frac{\partial \mathscr{L}}{\partial \dot{\varphi}} = mr^2 \sin^2\theta \, \dot{\varphi} \tag{33-c}$$

To obtain the Hamiltonian of the particle, we use definition (21). This amounts to adding $V(r)$ to the kinetic energy, expressed in terms of r, θ, φ and p_r, p_θ, p_φ. We find:

$$\mathscr{H}(r, \theta, \varphi; p_r, p_\theta, p_\varphi) = \frac{p_r^2}{2m} + \frac{1}{2mr^2}\left(p_\theta^2 + \frac{p_\varphi^2}{\sin^2\theta}\right) + V(r) \tag{34}$$

The system of canonical equations [formulas (27)] can be written here:

$$\frac{\mathrm{d}r}{\mathrm{d}t} = \frac{\partial \mathscr{H}}{\partial p_r} = \frac{p_r}{m} \tag{35-a}$$

$$\frac{\mathrm{d}\theta}{\mathrm{d}t} = \frac{\partial \mathscr{H}}{\partial p_\theta} = \frac{p_\theta}{mr^2} \tag{35-b}$$

$$\frac{\mathrm{d}\varphi}{\mathrm{d}t} = \frac{\partial \mathscr{H}}{\partial p_\varphi} = \frac{p_\varphi}{mr^2 \sin^2\theta} \tag{35-c}$$

$$\frac{\mathrm{d}p_r}{\mathrm{d}t} = -\frac{\partial \mathscr{H}}{\partial r} = \frac{1}{mr^3}\left(p_\theta^2 + \frac{p_\varphi^2}{\sin^2\theta}\right) - \frac{\partial V}{\partial r} \tag{35-d}$$

$$\frac{\mathrm{d}p_\theta}{\mathrm{d}t} = -\frac{\partial \mathscr{H}}{\partial \theta} = \frac{p_\varphi^2 \cos\theta}{mr^2 \sin^3\theta} \tag{35-e}$$

$$\frac{\mathrm{d}p_\varphi}{\mathrm{d}t} = -\frac{\partial \mathscr{H}}{\partial \varphi} = 0 \tag{35-f}$$

The first three of these equations simply give (33); the last three are the real equations of motion.

Now, consider the angular momentum of the particle with respect to the origin:

$$\boldsymbol{\mathscr{L}} = \mathbf{r} \times m\,\mathbf{v} \tag{36}$$

Its local components can easily be calculated from (31):

$$\begin{aligned}
\mathscr{L}_r &= 0 \\
\mathscr{L}_\theta &= -\,mrv_\varphi = -\,mr^2 \sin\theta\,\dot{\varphi} = -\,\frac{p_\varphi}{\sin\theta} \\
\mathscr{L}_\varphi &= mrv_\theta = mr^2\dot{\theta} = p_\theta
\end{aligned} \tag{37}$$

so that:

$$\boldsymbol{\mathscr{L}}^2 = p_\theta^2 + \frac{p_\varphi^2}{\sin^2\theta} \tag{38}$$

From the angular momentum theorem [formula (10)], we know that $\boldsymbol{\mathscr{L}}$ is a vector which is constant over time, since the force derived from the potential $V(r)$ is central, that is, collinear at each instant with the vector $\mathbf{r}$★.

By comparing (34) and (38), we see that the Hamiltonian $\mathscr{H}$ depends on the angular variables and their conjugate momenta only through the intermediary of $\boldsymbol{\mathscr{L}}^2$:

$$\mathscr{H}(r, \theta, \varphi\,; p_r, p_\theta, p_\varphi) = \frac{p_r^2}{2m} + \frac{1}{2mr^2}\,\boldsymbol{\mathscr{L}}^2(\theta, p_\theta, p_\varphi) + V(r) \tag{39}$$

Now, assume that the initial angular momentum of the particle is $\boldsymbol{\mathscr{L}}_0$. Since the angular momentum remains constant, the Hamiltonian (39) and the equation of motion (35-d) are the same as they would be for a particle of mass m, in a one-dimensional problem, placed in the effective potential:

$$V_{\text{eff}}(r) = V(r) + \frac{\boldsymbol{\mathscr{L}}_0^2}{2mr^2} \tag{40}$$

★ This conclusion can also be derived from (35-e) and (35-f) by calculating the time-derivatives of the components of $\boldsymbol{\mathscr{L}}$ on the fixed axes Ox, Oy, Oz.

b. A CHARGED PARTICLE PLACED IN AN ELECTROMAGNETIC FIELD

Now, consider a particle of mass m and charge q placed in an electromagnetic field characterized by the electric field vector $\mathbf{E}(\mathbf{r}, t)$ and the magnetic field vector $\mathbf{B}(\mathbf{r}, t)$.

α. *Description of the electromagnetic field. Gauges*

$\mathbf{E}(\mathbf{r}, t)$ and $\mathbf{B}(\mathbf{r}, t)$ satisfy Maxwell's equations:

$$\boldsymbol{\nabla} . \mathbf{E} = \frac{\rho}{\varepsilon_0} \tag{41-a}$$

$$\boldsymbol{\nabla} \times \mathbf{E} = -\frac{\partial \mathbf{B}}{\partial t} \tag{41-b}$$

$$\boldsymbol{\nabla} . \mathbf{B} = 0 \tag{41-c}$$

$$\boldsymbol{\nabla} \times \mathbf{B} = \mu_0 \mathbf{j} + \varepsilon_0 \mu_0 \frac{\partial \mathbf{E}}{\partial t} \tag{41-d}$$

where $\rho(\mathbf{r}, t)$ and $\mathbf{j}(\mathbf{r}, t)$ are the volume charge density and the current density producing the electromagnetic field. The fields $\mathbf{E}$ and $\mathbf{B}$ can be described by a scalar potential $U(\mathbf{r}, t)$ and a vector potential $\mathbf{A}(\mathbf{r}, t)$, since equation (41-c) implies that there exists a vector field $\mathbf{A}(\mathbf{r}, t)$ such that:

$$\mathbf{B} = \boldsymbol{\nabla} \times \mathbf{A}(\mathbf{r}, t) \tag{42}$$

(41-b) can thus be written:

$$\boldsymbol{\nabla} \times \left[\mathbf{E} + \frac{\partial \mathbf{A}}{\partial t} \right] = \mathbf{0} \tag{43}$$

Consequently, there exists a scalar function $U(\mathbf{r}, t)$ such that:

$$\mathbf{E} + \frac{\partial \mathbf{A}}{\partial t} = -\boldsymbol{\nabla} U(\mathbf{r}, t) \tag{44}$$

The set of the two potentials $\mathbf{A}(\mathbf{r}, t)$ and $U(\mathbf{r}, t)$ constitutes what is called a *gauge* for describing the electromagnetic field. The electric and magnetic fields can be calculated from the $\{ \mathbf{A}, U \}$ gauge by:

$$\mathbf{B}(\mathbf{r}, t) = \boldsymbol{\nabla} \times \mathbf{A}(\mathbf{r}, t) \tag{45-a}$$

$$\mathbf{E}(\mathbf{r}, t) = -\boldsymbol{\nabla} U(\mathbf{r}, t) - \frac{\partial}{\partial t} \mathbf{A}(\mathbf{r}, t) \tag{45-b}$$

A given electromagnetic field, that is, a pair of fields $\mathbf{E}(\mathbf{r}, t)$ and $\mathbf{B}(\mathbf{r}, t)$, can be described by an infinite number of gauges, which, for this reason, are said to be equivalent. If we know one gauge, $\{ \mathbf{A}, U \}$, which yields the fields $\mathbf{E}$ and $\mathbf{B}$, all the equivalent gauges, $\{ \mathbf{A}', U' \}$, can be found from the *gauge transformation formulas*:

$$\mathbf{A}'(\mathbf{r}, t) = \mathbf{A}(\mathbf{r}, t) + \boldsymbol{\nabla} \chi(\mathbf{r}, t) \tag{46-a}$$

$$U'(\mathbf{r}, t) = U(\mathbf{r}, t) - \frac{\partial}{\partial t} \chi(\mathbf{r}, t) \tag{46-b}$$

where $\chi(\mathbf{r}, t)$ is any scalar function.

First of all, it is easy to show from (46) that:

$$\begin{cases} \boldsymbol{\nabla} \times \mathbf{A}'(\mathbf{r}, t) = \boldsymbol{\nabla} \times \mathbf{A}(\mathbf{r}, t) \\ -\boldsymbol{\nabla} U'(\mathbf{r}, t) - \dfrac{\partial}{\partial t}\mathbf{A}'(\mathbf{r}, t) = -\boldsymbol{\nabla} U(\mathbf{r}, t) - \dfrac{\partial}{\partial t}\mathbf{A}(\mathbf{r}, t) \end{cases} \tag{47}$$

Any gauge, $\{ \mathbf{A}', U' \}$, which satisfies (46) therefore yields the same electric and magnetic fields as $\{ \mathbf{A}, U \}$.

Conversely we shall show that if two gauges, $\{ \mathbf{A}, U \}$ and $\{ \mathbf{A}', U' \}$, are equivalent, there must exist a function $\chi(\mathbf{r}, t)$ which establishes relations (46) between them. Since, by hypothesis:

$$\mathbf{B}(\mathbf{r}, t) = \boldsymbol{\nabla} \times \mathbf{A}(\mathbf{r}, t) = \boldsymbol{\nabla} \times \mathbf{A}'(\mathbf{r}, t) \tag{48}$$

we have:

$$\boldsymbol{\nabla} \times (\mathbf{A}' - \mathbf{A}) = \mathbf{0} \tag{49}$$

This implies that $\mathbf{A}' - \mathbf{A}$ is the gradient of a scalar function:

$$\mathbf{A}' - \mathbf{A} = \boldsymbol{\nabla} \chi(\mathbf{r}, t) \tag{50}$$

$\chi(\mathbf{r}, t)$ is, for the moment, determined only to within an arbitrary function of t, $f(t)$. Furthermore, the fact that the two gauges are equivalent means that:

$$\mathbf{E}(\mathbf{r}, t) = -\boldsymbol{\nabla} U(\mathbf{r}, t) - \frac{\partial}{\partial t}\mathbf{A}(\mathbf{r}, t) = -\boldsymbol{\nabla} U'(\mathbf{r}, t) - \frac{\partial}{\partial t}\mathbf{A}'(\mathbf{r}, t) \tag{51}$$

that is:

$$\boldsymbol{\nabla}(U' - U) + \frac{\partial}{\partial t}(\mathbf{A}' - \mathbf{A}) = \mathbf{0} \tag{52}$$

According to (50), we must have:

$$\boldsymbol{\nabla}(U' - U) = -\boldsymbol{\nabla}\frac{\partial}{\partial t}\chi(\mathbf{r}, t) \tag{53}$$

Consequently, the functions $U' - U$ and $-\dfrac{\partial}{\partial t}\chi(\mathbf{r}, t)$ can differ only by a function of t; thus, we can choose $f(t)$ so as to make them equal:

$$U' - U = -\frac{\partial}{\partial t}\chi(\mathbf{r}, t) \tag{54}$$

This completes the determination of the function $\chi(\mathbf{r}, t)$ (to within an additive constant). Two equivalent gauges must therefore satisfy relations of the form (46).

β. *Equations of motion and the Lagrangian*

In the electromagnetic field, the charged particle is subject to the *Lorentz force*:

$$\mathbf{F} = q\,[\mathbf{E} + \mathbf{v} \times \mathbf{B}] \tag{55}$$

(where $\mathbf{v}$ is the velocity of the particle at the time t). Newton's law therefore gives the equations of motion in the form:

$$m\,\ddot{\mathbf{r}} = q\left[\mathbf{E}(\mathbf{r}, t) + \dot{\mathbf{r}} \times \mathbf{B}(\mathbf{r}, t)\right] \tag{56}$$

Projecting this equation onto Ox and using (45), we obtain:

$$\begin{aligned} m\,\ddot{x} &= q[E_x + \dot{y}B_z - \dot{z}B_y] \\ &= q\left[-\frac{\partial U}{\partial x} - \frac{\partial A_x}{\partial t} + \dot{y}\left(\frac{\partial A_y}{\partial x} - \frac{\partial A_x}{\partial y}\right) - \dot{z}\left(\frac{\partial A_x}{\partial z} - \frac{\partial A_z}{\partial x}\right)\right] \end{aligned} \tag{57}$$

It can easily be shown that these equations can be derived from the Lagrangian by using (15):

$$\mathscr{L}(\mathbf{r}, \dot{\mathbf{r}}, t) = \frac{1}{2}m\,\dot{\mathbf{r}}^2 + q\,\dot{\mathbf{r}}\,.\,\mathbf{A}(\mathbf{r}, t) - qU(\mathbf{r}, t) \tag{58}$$

Therefore, although the Lorentz force is not derived from a potential energy, we can find a Lagragian for the problem.

Let us show that Lagrange's equations (15) do yield the equations of motion (56), using the Lagrangian (58). To do so, we shall first calculate :

$$\begin{aligned} \frac{\partial \mathscr{L}}{\partial \dot{x}} &= m\dot{x} + qA_x(\mathbf{r}, t) \\ \frac{\partial \mathscr{L}}{\partial x} &= q\,\dot{\mathbf{r}}\,.\,\frac{\partial}{\partial x}\mathbf{A}(\mathbf{r}, t) - q\frac{\partial}{\partial x}U(\mathbf{r}, t) \end{aligned} \tag{59}$$

Lagrange's equation for the x-coordinate can therefore be written:

$$\frac{\mathrm{d}}{\mathrm{d}t}\left[m\dot{x} + qA_x(\mathbf{r}, t)\right] - q\,\dot{\mathbf{r}}\,.\,\frac{\partial}{\partial x}\mathbf{A}(\mathbf{r}, t) + q\frac{\partial}{\partial x}U(\mathbf{r}, t) = 0 \tag{60}$$

Writing this equation explicitly and using (16), we again get (57):

$$m\ddot{x} + q\left[\frac{\partial A_x}{\partial t} + \dot{x}\frac{\partial A_x}{\partial x} + \dot{y}\frac{\partial A_x}{\partial y} + \dot{z}\frac{\partial A_x}{\partial z}\right] - q\left[\dot{x}\frac{\partial A_x}{\partial x} + \dot{y}\frac{\partial A_y}{\partial x} + \dot{z}\frac{\partial A_z}{\partial x}\right] + q\frac{\partial U}{\partial x} = 0 \tag{61}$$

that is:

$$m\ddot{x} = q\left[-\frac{\partial U}{\partial x} - \frac{\partial A_x}{\partial t} + \dot{y}\left(\frac{\partial A_y}{\partial x} - \frac{\partial A_x}{\partial y}\right) - \dot{z}\left(\frac{\partial A_x}{\partial z} - \frac{\partial A_z}{\partial x}\right)\right] \tag{62}$$

γ. *Momentum. The classical Hamiltonian*

The Lagrangian (58) enables us to calculate the conjugate momenta of the cartesian coordinates x, y, z of the particle. For example:

$$p_x = \frac{\partial \mathscr{L}}{\partial \dot{x}} = m\dot{x} + qA_x(\mathbf{r}, t) \tag{63}$$

The *momentum of the particle*, which is, by definition, the vector whose components are (p_x, p_y, p_z), *is no longer equal*, as in (19), to the mechanical momentum $m\dot{\mathbf{r}}$:

$$\mathbf{p} = m\,\dot{\mathbf{r}} + q\mathbf{A}(\mathbf{r}, t) \tag{64}$$

Finally, we shall write the classical Hamiltonian:

$$\begin{aligned}\mathscr{H}(\mathbf{r}, \mathbf{p}; t) &= \mathbf{p} \, . \, \dot{\mathbf{r}} - \mathscr{L} \\ &= \mathbf{p} \, . \, \frac{1}{m}(\mathbf{p} - q\mathbf{A}) - \frac{1}{2m}(\mathbf{p} - q\mathbf{A})^2 - \frac{q}{m}(\mathbf{p} - q\mathbf{A}) \, . \, \mathbf{A} + qU\end{aligned} \tag{65}$$

that is:

$$\mathscr{H}(\mathbf{r}, \mathbf{p}; t) = \frac{1}{2m}\left[\mathbf{p} - q\mathbf{A}(\mathbf{r}, t)\right]^2 + qU(\mathbf{r}, t) \tag{66}$$

COMMENT:

Hamiltonian formalism therefore uses the potentials $\mathbf{A}$ and U, and not the fields $\mathbf{E}$ and $\mathbf{B}$ directly. The result is that the description of the particle depends on the gauge chosen. It is reasonable to expect, however, since the Lorentz force is expressed in terms of the fields, that predictions concerning the physical behavior of the particle must be the same for two equivalent gauges. The physical consequences of the Hamiltonian formalism are said to be *gauge-invariant*. The concept of gauge invariance is analyzed in detail in complement $\mathrm{H_{III}}$.

5. The principle of least action

Classical mechanics can be based on a variational principle, the principle of least action. In addition to its theoretical importance, the concept of action serves as the foundation of the *Lagrangian formulation of quantum mechanics* (*cf.* complement $\mathrm{J_{III}}$). This is why we shall now briefly discuss the principle of least action and show how it leads to Lagrange's equations.

a. GEOMETRICAL REPRESENTATION OF THE MOTION OF A SYSTEM

First of all, consider a particle constrained to move along the Ox axis. Its motion can be represented by tracing, in the (x, t) plane, the curve defined by the law of motion which yields $x(t)$.

More generally, let us study a physical system described by N generalized coordinates q_i (for an n-particle system in three-dimensional space, $N = 3n$). It is convenient to interpret the q_i to be the coordinates of a point Q in an N-dimensional Euclidean space R_N. There is then a one-to-one correspondance between the positions of the system and the points of R_N. With each motion of the system is associated a motion of point Q in R_N, characterized by the N-dimensional vector function $Q(t)$ whose components are the $q_i(t)$. As in the simple case of a single particle moving in one dimension, the motion of point Q, that is, the motion of the system, can be represented by the graph of $Q(t)$, which is a curve in an $(N + 1)$-dimensional space-time (the time axis is added to the N dimensions of R_N). This curve characterizes the motion being studied.

b. THE PRINCIPLE OF LEAST ACTION

The $q_i(t)$ can be fixed arbitrarily; this gives point Q and the system an arbitrary motion. But their real behavior is defined by the initial conditions and the equations of motion. Suppose that we know that, in the course of the real motion, Q is at Q_1 at time t_1 and at Q_2 at a subsequent time t_2 (as is shown schematically by figure 2):

$$Q(t_1) = Q_1$$
$$Q(t_2) = Q_2 \tag{67}$$

There is an infinite number of *a priori* possible motions which satisfy conditions (67). They are represented by all the curves★, or *paths in space time*, which connect the points (Q_1, t_1) and (Q_2, t_2) (*cf.* fig. 2).

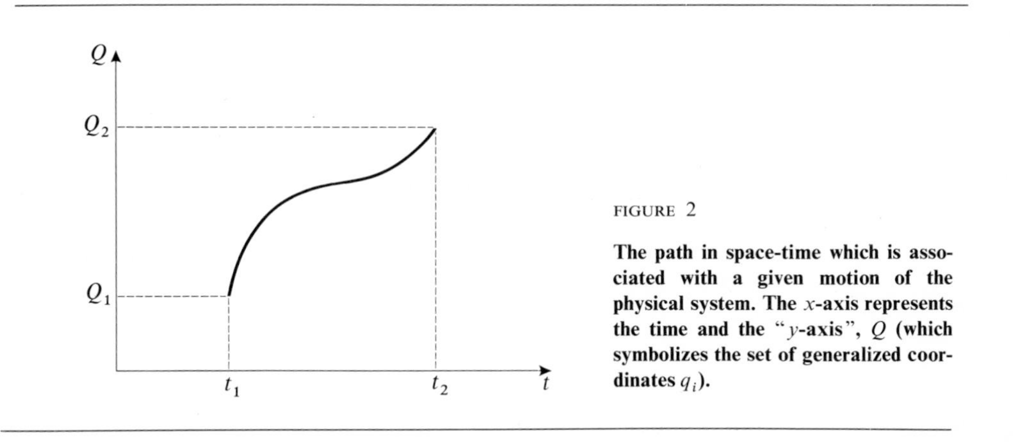

FIGURE 2

The path in space-time which is associated with a given motion of the physical system. The x-axis represents the time and the "y-axis", Q (which symbolizes the set of generalized coordinates q_i).

Consider such a path in space-time, Γ, characterized by the vector function $Q(t)$ which satisfies (67). If:

$$\mathscr{L}(q_1, q_2, ..., q_N; \dot{q}_1, \dot{q}_2, ..., \dot{q}_N; t) \equiv \mathscr{L}(Q, \dot{Q}; t) \tag{68}$$

is the Lagrangian of the system, the *action* S_Γ which corresponds to the path Γ is, by definition:

$$S_\Gamma = \int_{t_1}^{t_2} \mathrm{d}t \; \mathscr{L}[Q_\Gamma(t), \dot{Q}_\Gamma(t); t] \tag{69}$$

[the function to be integrated depends only on t; it is obtained by replacing the q_i and $\dot{q}_i$ by the time-dependent coordinates of $Q_\Gamma(t)$ and $\dot{Q}_\Gamma(t)$ in the Lagrangian (68)].

The *principle of least action* can then be stated in the following way : of all the paths in space-time connecting (Q_1, t_1) with (Q_2, t_2), the one which is actually followed (that is, the one which characterizes the real motion of the system) is the one for which the action is minimal. In other words, when we go from the path which is actually followed to one infinitely close to it, the action does not vary to first order. Note the analogy with other variational principles, such as Fermat's principle in optics.

★ Excluding, of course, the curves which "go backward", that is, which would give two distinct positions of Q for the same time t.

c. LAGRANGE'S EQUATIONS AS A CONSEQUENCE OF THE PRINCIPLE OF LEAST ACTION

In conclusion, we shall show how Lagrange's equations can be deduced from the principle of least action.

Suppose that the real motion of the system under study is characterized by the N functions of time $q_i(t)$, that is by the path in space-time Γ connecting the points (Q_1, t_1) and (Q_2, t_2). Now consider an infinitely close path, Γ' (fig. 3), for which the generalized coordinates are equal to:

$$q_i'(t) = q_i(t) + \delta q_i(t) \tag{70}$$

where the $\delta q_i(t)$ are infinitesimally small and satisfy conditions (67), that is:

$$\delta q_i(t_1) = \delta q_i(t_2) = 0 \tag{71}$$

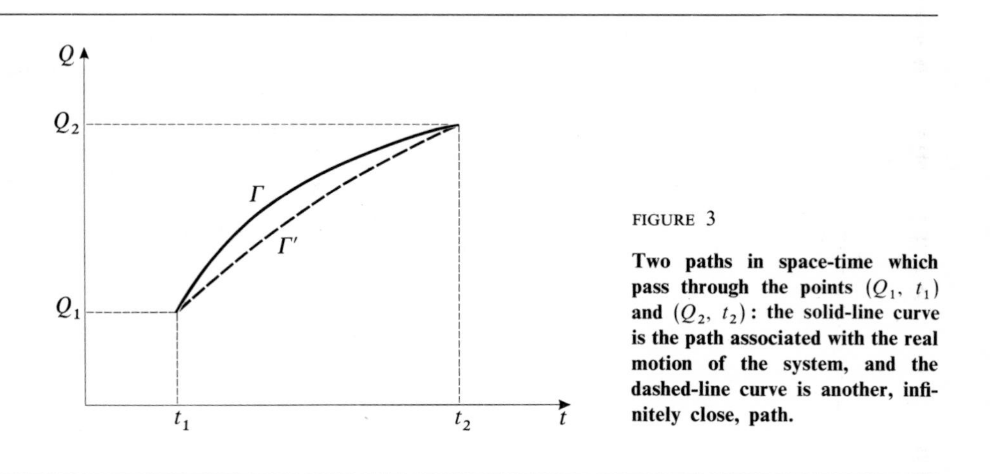

FIGURE 3

Two paths in space-time which pass through the points (Q_1, t_1) and (Q_2, t_2): the solid-line curve is the path associated with the real motion of the system, and the dashed-line curve is another, infinitely close, path.

The generalized velocities $\dot{q}_i'(t)$ corresponding to Γ' can be obtained by differentiating relations (70):

$$\dot{q}_i'(t) = \dot{q}_i(t) + \frac{\mathrm{d}}{\mathrm{d}t}\,\delta q_i(t) \tag{72}$$

Thus, their increments $\delta\dot{q}_i(t)$ are simply:

$$\delta\dot{q}_i(t) = \frac{\mathrm{d}}{\mathrm{d}t}\,\delta q_i(t) \tag{73}$$

We now calculate the variation of the action in going from the path Γ to the path Γ':

$$\begin{aligned}\delta S &= \int_{t_1}^{t_2} \mathrm{d}t\; \delta\mathscr{L} \\ &= \int_{t_1}^{t_2} \mathrm{d}t \left[\sum_i \frac{\partial\mathscr{L}}{\partial q_i}\,\delta q_i + \sum_i \frac{\partial\mathscr{L}}{\partial \dot{q}_i}\,\delta\dot{q}_i\right] \\ &= \int_{t_1}^{t_2} \mathrm{d}t \left[\sum_i \frac{\partial\mathscr{L}}{\partial q_i}\,\delta q_i + \sum_i \frac{\partial\mathscr{L}}{\partial \dot{q}_i}\frac{\mathrm{d}}{\mathrm{d}t}\,\delta q_i\right]\end{aligned} \tag{74}$$

according to (73). If we integrate the second term by parts, we obtain:

$$\delta S = \left[\sum_i \frac{\partial \mathscr{L}}{\partial \dot{q}_i} \delta q_i \right]_{t_1}^{t_2} + \int_{t_1}^{t_2} \mathrm{d}t \sum_i \delta q_i \left[\frac{\partial \mathscr{L}}{\partial q_i} - \frac{\mathrm{d}}{\mathrm{d}t} \frac{\partial \mathscr{L}}{\partial \dot{q}_i} \right]$$

$$= \int_{t_1}^{t_2} \mathrm{d}t \sum_i \delta q_i \left[\frac{\partial \mathscr{L}}{\partial q_i} - \frac{\mathrm{d}}{\mathrm{d}t} \frac{\partial \mathscr{L}}{\partial \dot{q}_i} \right] \tag{75}$$

since the integrated term is zero, because of conditions (71).

If Γ is the path in space-time which is actually followed during the real motion of the system, the increment δS of the action is zero, according to the principle of least action. For this to be so, it is necessary and sufficient that:

$$\frac{\mathrm{d}}{\mathrm{d}t} \frac{\partial \mathscr{L}}{\partial \dot{q}_i} - \frac{\partial \mathscr{L}}{\partial q_i} = 0 \quad ; \quad i = 1, 2, ..., N \tag{76}$$

It is obvious that this condition is sufficient. It is also necessary, since, if there existed a time interval during which expression (76) were non-zero for a given value k of the index i, the $\delta q_i(t)$ could be chosen so as to make the corresponding increment δS different from zero. (It would suffice, for example, to choose them so as to make the product $\delta q_k \left[\frac{\partial \mathscr{L}}{\partial q_k} - \frac{\mathrm{d}}{\mathrm{d}t} \frac{\partial \mathscr{L}}{\partial \dot{q}_k} \right]$ always positive or zero). Consequently, the principle of least action is equivalent to Lagrange's equations.

References and suggestions for further reading:

See section 6 of the bibliography, in particular Marion (6.4), Goldstein (6.6), Landau and Lifshitz (6.7).

For a simple presentation of the use of variational principles in physics, see Feynman II (7.2), chap. 19.

For Lagrangian formalism applied to a classical field, see Bogoliubov and Chirkov (2.15), chap. I.

Bibliography

1. QUANTUM MECHANICS: GENERAL REFERENCES

INTRODUCTORY TEXTS

Quantum physics

(1.1) E. H. WICHMANN, *Berkeley Physics Course, Vol. 4: Quantum Physics*, McGraw-Hill, New York (1971).

(1.2) R. P. FEYNMAN, R. B. LEIGHTON and M. SANDS, *The Feynman Lectures on Physics, Vol. III: Quantum Mechanics*, Addison-Wesley, Reading, Mass. (1965).

(1.3) R. EISBERG and R. RESNICK, *Quantum Physics of Atoms, Molecules, Solids, Nuclei and Particules*, Wiley, New York (1974).

(1.4) M. ALONSO and E. J. FINN, *Fundamental University Physics, Vol. III: Quantum and Statistical Physics*, Addison-Wesley, Reading, Mass. (1968).

(1.5) U. FANO and L. FANO, *Basic Physics of Atoms and Molecules*, Wiley, New York (1959).

(1.6) J. C. SLATER, *Quantum Theory of Matter*, McGraw-Hill, New York (1968).

Quantum mechanics

(1.7) S. BOROWITZ, *Fundamentals of Quantum Mechanics*, Benjamin, New York (1967).

(1.8) S. I. TOMONAGA, *Quantum Mechanics, Vol. 1: Old Quantum Theory*, North Holland, Amsterdam (1962).

(1.9) L. PAULING and E. B. WILSON JR., *Introduction to Quantum Mechanics*, McGraw-Hill, New York (1935).

(1.10) Y. AYANT et E. BELORIZKY, *Cours de Mécanique Quantique*, Dunod, Paris (1969).

(1.11) P. T. MATTHEWS, *Introduction to Quantum Mechanics*, McGraw-Hill, New York (1963).

(1.12) J. AVERY, *The Quantum Theory of Atoms, Molecules and Photons*, McGraw-Hill, London (1972).

MORE ADVANCED TEXTS :

(1.13) P. A. M. DIRAC, *The Principles of Quantum Mechanics*, Oxford University Press (1958).

(1.14) R. H. DICKE and J. P. WITTKE, *Introduction to Quantum Mechanics*, Addison-Wesley, Reading, Mass. (1966).

(1.15) D. I. BLOKHINTSEV, *Quantum Mechanics*, D. Reidel, Dordrecht (1964).

(1.16) E. MERZBACHER, *Quantum Mechanics*, Wiley, New York (1970).

(1.17) A. MESSIAH, *Mécanique Quantique*, Vols 1 and 2, Dunod, Paris (1964). English translation : *Quantum Mechanics*, North Holland, Amsterdam (1961).

(1.18) L. I. SCHIFF, *Quantum Mechanics*, McGraw-Hill, New York (1968).

(1.19) L. D. LANDAU and E. M. LIFSHITZ, *Quantum Mechanics, Nonrelativistic Theory*, Pergamon Press, Oxford (1965).

(1.20) A. S. DAVYDOV, *Quantum Mechanics*, Translated, edited and with additions by D. Ter HAAR, Pergamon Press, Oxford (1965).

(1.21) H. A. BETHE and R. W. JACKIW, *Intermediate Quantum Mechanics*, Benjamin, New York (1968).

(1.22) H. A. KRAMERS, *Quantum Mechanics*, North Holland, Amsterdam (1958).

PROBLEMS IN QUANTUM MECHANICS

(1.23) *Selected Problems in Quantum Mechanics*, Collected and edited by D. Ter HAAR, Infosearch, London (1964).

(1.24) S. FLÜGGE, *Practical Quantum Mechanics*, I and II, Springer-Verlag, Berlin (1971).

ARTICLES

(1.25) E. SCHRÖDINGER, "What is Matter?", *Scientific American*, **189**, 52 (Sept. 1953).

(1.26) G. GAMOW, "The Principle of Uncertainty", *Scientific American*, **198**, 51 (Jan. 1958).

(1.27) G. GAMOW, "The Exclusion Principle", *Scientific American*, **201**, 74 (July 1959).

(1.28) M. BORN and W. BIEM, "Dualism in Quantum Theory", *Physics Today*, **21**, p. 51 (Aug. 1968).

(1.29) W. E. LAMB JR., "An Operational Interpretation of Nonrelativistic Quantum Mechanics", *Physics Today*, **22**, 23 (April 1969).

(1.30) M. O. SCULLY and M. SARGENT III, "The Concept of the Photon", *Physics Today*, **25**, 38 (March 1972).

(1.31) A. EINSTEIN, "Zur Quantentheorie der Strahlung", *Physik. Z.*, **18**, 121 (1917).

(1.32) A. GOLDBERG, H. M. SCHEY and J. L. SCHWARTZ, "Computer-Generated Motion Pictures of One-Dimensional Quantum-Mechanical Transmission and Reflection Phenomena", *Am. J. Phys.*, **35**, 177 (1967).

(1.33) R. P. FEYNMAN, F. L. VERNON JR. and R. W. HELLWARTH, "Geometrical Representation of the Schrödinger Equation for Solving Maser Problems", *J. Appl. Phys.*, **28**, 49 (1957).

(1.34) A. A. VUYLSTEKE, "Maser States in Ammonia-Inversion", *Am. J. Phys.*, **27**, 554 (1959).

2. QUANTUM MECHANICS: MORE SPECIALIZED REFERENCES

COLLISIONS

(2.1) T. Y. WU and T. OHMURA, *Quantum Theory of Scattering*, Prentice Hall, Englewood Cliffs (1962).

(2.2) R. G. NEWTON, *Scattering Theory of Waves and Particles*, McGraw-Hill, New York (1966).

(2.3) P. ROMAN, *Advanced Quantum Theory*, Addison-Wesley, Reading, Mass. (1965).

(2.4) M. L. GOLDBERGER and K. M. WATSON, *Collision Theory*, Wiley, New York (1964).

(2.5) N. F. MOTT and H. S. W. MASSEY, *The Theory of Atomic Collisions*, Oxford University Press (1965).

RELATIVISTIC QUANTUM MECHANICS

(2.6) J. D. BJORKEN and S. D. DRELL, *Relativistic Quantum Mechanics*, McGraw-Hill, New York (1964).

(2.7) J. J. SAKURAI, *Advanced Quantum Mechanics*, Addison-Wesley, Reading, Mass. (1967).

(2.8) V. B. BERESTETSKII, E. M. LIFSHITZ and L. P. PITAEVSKII, *Relativistic Quantum Theory*, Pergamon Press, Oxford (1971).

FIELD THEORY. QUANTUM ELECTRODYNAMICS

(2.9) F. MANDL, *Introduction to Quantum Field Theory*, Wiley Interscience, New York (1959).

(2.10) J. D. BJORKEN and S. D. DRELL, *Relativistic Quantum Fields*, McGraw-Hill, New York (1965).

(2.11) E. A. POWER, *Introductory Quantum Electrodynamics*, Longmans, London (1964).

(2.12) R. P. FEYNMAN, *Quantum Electrodynamics*, Benjamin, New York (1961).

(2.13) W. HEITLER, *The Quantum Theory of Radiation*, Clarendon Press, Oxford (1954).

(2.14) A. I. AKHIEZER and V. B. BERESTETSKII, *Quantum Electrodynamics*, Wiley Interscience, New York (1965).

(2.15) N. N. BOGOLIUBOV and D. V. SHIRKOV, *Introduction to the Theory of Quantized Fields*, Interscience Publishers, New York (1959); *Introduction à la Théorie des Champs*, Dunod, Paris (1960).

(2.16) S. S. SCHWEBER, *An Introduction to Relativistic Quantum Field Theory*, Harper and Row, New York (1961).

(2.17) M. M. STERNHEIM, "Resource Letter TQE-1 : Tests of Quantum Electrodynamics", *Am. J. Phys.*, **40**, 1363 (1972).

ROTATIONS AND GROUP THEORY

(2.18) P. H. E. MEIJER and E. BAUER, *Group Theory*, North Holland, Amsterdam (1962).

(2.19) M. E. ROSE, *Elementary Theory of Angular Momentum*, Wiley, New York (1957).

(2.20) M. E. ROSE, *Multipole Fields*, Wiley, New York (1955).

(2.21) A. R. EDMONDS, *Angular Momentum in Quantum Mechanics*, Princeton University Press (1957).

(2.22) M. TINKHAM, *Group Theory and Quantum Mechanics*, McGraw-Hill, New York (1964).

(2.23) E. P. WIGNER, *Group Theory and its Application to the Quantum Mechanics of Atomic Spectra*, Academic Press, New York (1959).

(2.24) D. PARK, "Resource Letter SP-I on Symmetry in Physics", *Am. J. Phys.*, **36**, 577 (1968).

MISCELLANEOUS

(2.25) R. P. FEYNMAN and A. R. HIBBS, *Quantum Mechanics and Path Integrals*, McGraw-Hill, New York (1965).

(2.26) J. M. ZIMAN, *Elements of Advanced Quantum Theory*, Cambridge University Press (1969).

(2.27) F. A. KAEMPFFER, *Concepts in Quantum Mechanics*, Academic Press, New York (1965).

ARTICLES

(2.28) P. MORRISON, "The Overthrow of Parity", *Scientific American*, **196**, 45 (April 1957).

(2.29) G. FEINBERG and M. GOLDHABER, "The Conservation Laws of Physics", *Scientific American*, **209**, 36 (Oct. 1963).

(2.30) E. P. WIGNER, "Violations of Symmetry in Physics", *Scientific American*, **213**, 28 (Dec. 1965).

(2.31) U. FANO, "Description of States in Quantum Mechanics by Density Matrix and Operator Techniques", *Rev. Mod. Phys.*, **29**, 74 (1957).

(2.32) D. Ter HAAR, "Theory and Applications of the Density Matrix", *Rept. Progr. Phys.*, **24**, 304 (1961).

(2.33) V. F. WEISSKOPF and E. WIGNER, "Berechnung der Natürlichen Linienbreite auf Grund der Diracschen Lichttheorie", *Z. Physik*, **63**, 54 (1930).

(2.34) A. DALGARNO and J. T. LEWIS, "The Exact Calculation of Long-Range Forces between Atoms by Perturbation Theory", *Proc. Roy. Soc.*, **A 233**, 70 (1955).

(2.35) A. DALGARNO and A. L. STEWART, "On the Perturbation Theory of Small Disturbances", *Proc. Roy. Soc.*, **A 238**, 269 (1957).

(2.36) C. SCHWARTZ, "Calculations in Schrödinger Perturbation Theory", *Annals of Physics* (New York), **6**, 156 (1959).

(2.37) J. O. HIRSCHFELDER, W. BYERS BROWN and S. T. EPSTEIN, "Recent Developments in Perturbation Theory", in *Advances in Quantum Chemistry*, P. O. LOWDIN ed., Vol. I, Academic Press, New York (1964).

(2.38) R. P. FEYNMAN, "Space Time Approach to Nonrelativistic Quantum Mechanics", *Rev. Mod. Phys.*, **20**, 367 (1948).

(2.39) L. VAN HOVE, "Correlations in Space and Time and Born Approximation Scattering in Systems of Interacting Particles", *Phys. Rev.*, **95**, 249 (1954).

3. QUANTUM MECHANICS: FUNDAMENTAL EXPERIMENTS

Interference effects with weak light:

(3.1) G. I. TAYLOR, "Interference Fringes with Feeble Light", *Proc. Camb. Phil. Soc.*, **15**, 114 (1909).

(3.2) G. T. REYNOLDS, K. SPARTALIAN and D. B. SCARL, "Interference Effects Produced by Single Photons", *Nuovo Cimento*, **61 B**, 355 (1969).

Experimental verification of Einstein's law for the photoelectric effect; measurement of h :

(3.3) A. L. HUGHES, "On the Emission Velocities of Photoelectrons", *Phil. Trans. Roy. Soc.*, **212**, 205 (1912).

(3.4) R. A. MILLIKAN, "A Direct Photoelectric Determination of Planck's h", *Phys. Rev.* 7, **355** (1916).

The Franck-Hertz experiment :

(3.5) J. FRANCK und G. HERTZ, "Über Zusammenstöße Zwischen Elecktronen und den Molekülen des Quecksilberdampfes und die Ionisierungsspannung desselben", *Verhandlungen der Deutschen Physikalischen Gesellschaft*, **16**, 457 (1914).
"Über Kinetik von Elektronen und Ionen in Gasen", *Physikalische Zeitschrift*, **17**, 409 (1916).

The proportionality between the magnetic moment and the angular momentum :

(3.6) A. EINSTEIN und J. W. DE HAAS, "Experimenteller Nachweis der Ampereschen Molekularströme", *Verhandlungen der Deutschen Physikalischen Gesellschaft*, **17**, 152 (1915).

(3.7) E. BECK, "Zum Experimentellen Nachweis der Ampereschen Molekularströme", *Annalen der Physik* (Leipzig), **60**, 109 (1919).

The Stern-Gerlach experiment:

(3.8) W. GERLACH und O. STERN, "Der Experimentelle Nachweis der Richtungsquantelung im Magnetfeld", *Zeitschrift für Physik*, **9**, 349 (1922).

The Compton effect:

(3.9) A. H. COMPTON, "A Quantum Theory of the Scattering of X-Rays by Light Elements", *Phys. Rev.*, **21**, 483 (1923).
"Wavelength Measurements of Scattered X-Rays", *Phys. Rev.*, **21**, 715 (1923).

Electron diffraction:

(3.10) C. DAVISSON and L. H. GERMER, "Diffraction of Electrons by a Crystal of Nickel", *Phys. Rev.*, **30**, 705 (1927).

The Lamb shift:

(3.11) W. E. LAMB JR. and R. C. RETHERFORD, "Fine Structure of the Hydrogen Atom",
I – *Phys. Rev.*, **79**, 549 (1950),
II – *Phys. Rev.*, **81**, 222 (1951).

Hyperfine structure of the hydrogen ground state:

(3.12) S. B. CRAMPTON, D. KLEPPNER and N. F. RAMSEY, "Hyperfine Separation of Ground State Atomic Hydrogen", *Phys. Rev. Letters*, **11**, 338 (1963).

Several fundamental experiments are described in:

(3.13) O. R. FRISCH, "Molecular Beams", *Scientific American*, **212**, 58 (May 1965).

4. QUANTUM MECHANICS: HISTORY

(4.1) L. DE BROGLIE, "Recherches sur la Théorie des Quanta", *Annales de Physique*, **3**, 22, Paris (1925).

(4.2) N. BOHR, "The Solvay Meetings and the Development of Quantum Mechanics", *Essays 1958-1962 on Atomic Physics and Human Knowledge*, Vintage, New York (1966).

(4.3) W. HEISENBERG, *Physics and Beyond: Encounters and Conversations*, Harper and Row, New York (1971).
La Partie et le Tout, Albin Michel, Paris (1972).

(4.4) *Niels Bohr, His life and work as seen by his friends and colleagues*, S. ROZENTAL, ed., North Holland, Amsterdam (1967).

(4.5) A. EINSTEIN, M. and H. BORN, *Correspondance 1916-1955*, Editions du Seuil, Paris (1972). See also *La Recherche*, **3**, 137 (Feb. 1972).

(4.6) *Theoretical Physics in the Twentieth Century*, M. FIERZ and V. F. WEISSKOPF eds., Wiley Interscience, New York (1960).

(4.7) *Sources of Quantum Mechanics*, B. L. VAN DER WAERDEN ed., North Holland, Amsterdam (1967); Dover, New York (1968).

(4.8) M. JAMMER, *The Conceptual Development of Quantum Mechanics*, McGraw-Hill, New York (1966). This book traces the historical development of quantum mechanics. Its very numerous footnotes provide a multitude of references. See also (5.12).

ARTICLES

(4.9) K. K. DARROW, "The Quantum Theory", *Scientific American*, **186**, 47 (March 1952).

(4.10) M. J. KLEIN, "Thermodynamics and Quanta in Planck's work", *Physics Today*, **19**, 23 (Nov. 1966).

(4.11) H. A. MEDICUS, "Fifty years of Matter Waves", *Physics Today*, **27**, 38 (Feb. 1974).

Reference (5.11) contains a large number of references to the original texts.

5. QUANTUM MECHANICS: DISCUSSION OF ITS FOUNDATIONS

GENERAL PROBLEMS:

(5.1) D. BOHM, *Quantum Theory*, Constable, London (1954).

(5.2) J. M. JAUCH, *Foundations of Quantum Mechanics*, Addison-Wesley, Reading, Mass. (1968).

(5.3) B. D'ESPAGNAT, *Conceptual Foundations of Quantum Mechanics*, Benjamin, New York (1971).

(5.4) Proceedings of the International School of Physics "Enrico Fermi" (Varenna), Course IL; *Foundations of Quantum Mechanics*, B. D'ESPAGNAT ed., Academic Press, New York (1971).

(5.5) B. S. DEWITT, "Quantum Mechanics and Reality", *Physics Today*, **23**, 30, (Sept. 1970).

(5.6) "Quantum Mechanics debate", *Physics Today*, **24**, 36 (April 1971).

See also (1.28).

MISCELLANEOUS INTERPRETATIONS

(5.7) N. BOHR, "Discussion with Einstein on Epistemological Problems in Atomic Physics", in *A. Einstein: Philosopher-Scientist*, P. A. SCHILPP ed., Harper and Row, New York (1959).

(5.8) M. BORN, *Natural Philosophy of Cause and Chance*, Oxford University Press, London (1951); Clarendon Press, Oxford (1949).

(5.9) L. DE BROGLIE, *Une Tentative d'Interprétation Causale et Non Linéaire de la Mécanique Ondulatoire: la Théorie de la Double Solution*, Gauthier-Villars, Paris (1956); *Etude Critique des Bases de l'Interprétation Actuelle de la Mécanique Ondulatoire*, Gauthier-Villars, Paris (1963).

(5.10) *The Many-Worlds Interpretation of Quantum Mechanics*, B. S. DEWITT and N. GRAHAM eds., Princeton University Press (1973).

A very complete set of references, classified and annotated, can be found in:

(5.11) B. S. DEWITT and R. N. GRAHAM, "Resource Letter IQM-1 on the Interpretation of Quantum Mechanics", *Am. J. Phys.* **39**, 724 (1971).

(5.12) M. JAMMER, *The Philosophy of Quantum Mechanics*, Wiley-Interscience, New York (1974). A general presentation of the different interpretations of the Quantum Mechanics formalism. Gives numerous references.

MEASUREMENT THEORY

(5.13) K. GOTTFRIED, *Quantum Mechanics*, Vol. I, Benjamin, New York (1966).

(5.14) D. I. BLOKHINTSEV, *Principes Essentiels de la Mécanique Quantique*, Dunod, Paris (1968).

(5.15) A. SHIMONY, "Role of the Observer in Quantum Theory", *Am. J. Phys.*, **31**, 755 (1963).

HIDDEN VARIABLES AND "PARADOXES":

(5.16) A. EINSTEIN, B. PODOLSKY and N. ROSEN, "Can Quantum-Mechanical Description of Physical Reality Be Considered Complete?", *Phys. Rev.* **47**, 777 (1935).
N. BOHR, "Can Quantum Mechanical Description of Physical Reality Be Considered Complete?", *Phys. Rev.* **48**, 696 (1935).

(5.17) *Paradigms and Paradoxes, the Philosophical Challenge of the Quantum Domain*, R. G. COLODNY ed., University of Pittsburg Press (1972).

(5.18) J. S. BELL, "On the Problem of Hidden Variables in Quantum Mechanics", *Rev. Mod. Phys.* **38**, 447 (1966).

See also reference (4.8), as well as (5.11) and chap. 7 of (5.12).

6. CLASSICAL MECHANICS

INTRODUCTORY LEVEL

(6.1) M. ALONSO and E. J. FINN, *Fundamental University Physics, Vol. I: Mechanics*, Addison-Wesley, Reading, Mass. (1967).

(6.2) C. KITTEL, W. D. KNIGHT and M. A. RUDERMAN, *Berkeley Physics Course, Vol. 1: Mechanics*, McGraw-Hill, New York (1962).

(6.3) R. P. FEYNMAN, R. B. LEIGHTON and M. SANDS, *The Feynman Lectures on Physics, Vol. I: Mechanics, Radiation, and Heat*, Addison-Wesley, Reading, Mass. (1966).

(6.4) J. B. MARION, *Classical Dynamics of Particles and Systems*, Academic Press, New York (1965).

MORE ADVANCED LEVEL:

(6.5) A. SOMMERFELD, *Lectures on Theoretical Physics, Vol. I: Mechanics*, Academic Press, New York (1964).

(6.6) H. GOLDSTEIN, *Classical Mechanics*, Addison-Wesley, Reading, Mass. (1959).

(6.7) L. D. LANDAU and E. M. LIFSHITZ, *Mechanics*, Pergamon Press, Oxford (1960).

7. ELECTROMAGNETISM AND OPTICS

INTRODUCTORY LEVEL

(7.1) E. M. PURCELL, *Berkeley Physics Course, Vol. 2: Electricity and Magnetism*, McGraw-Hill, New York (1965).
F. S. CRAWFORD JR., *Berkeley Physics Course, Vol. 3: Waves*, McGraw-Hill, New York (1968).

(7.2) R. P. FEYNMAN, R. B. LEIGHTON and M. SANDS, *The Feynman Lectures on Physics, Vol. II: Electromagnetism and Matter*, Addison-Wesley, Reading, Mass. (1966).

(7.3) M. ALONSO and E. J. FINN, *Fundamental University Physics, Vol. II: Fields and Waves*, Addison-Wesley, Reading, Mass. (1967).

(7.4) E. HECHT and A. ZAJAC, *Optics*, Addison-Wesley, Reading, Mass. (1974).

MORE ADVANCED LEVEL

(7.5) J. D. JACKSON, *Classical Electrodynamics*, 2d ed. Wiley, New York (1975).

(7.6) W. K. H. PANOFSKY and M. PHILLIPS, *Classical Electricity and Magnetism*, Addison-Wesley, Reading, Mass. (1964).

(7.7) J. A. STRATTON, *Electromagnetic Theory*, McGraw-Hill, New York (1941).

(7.8) M. BORN and E. WOLF, *Principles of Optics*, Pergamon Press, London (1964).

(7.9) A. SOMMERFELD, *Lectures on Theoretical Physics, Vol. IV: Optics*, Academic Press, New York (1964).

(7.10) G. BRUHAT, *Optique*, 5e Edition revised and completed by A. KASTLER, Masson, Paris (1954).

(7.11) L. LANDAU and E. LIFSHITZ, *The Classical Theory of Fields*, Addison-Wesley, Reading, Mass. (1951); Pergamon Press, London (1951).

(7.12) L. D. LANDAU and E. M. LIFSHITZ, *Electrodynamics of Condinuous Media*, Pergamon Press, Oxford (1960).

(7.13) L. BRILLOUIN, *Wave Propagation and Group Velocity*, Academic Press, New York (1960).

8. THERMODYNAMICS. STATISTICAL MECHANICS

INTRODUCTORY LEVEL

(8.1) F. REIF, *Berkeley Physics Course, Vol. 5: Statistical Physics*, McGraw-Hill, New York (1967).

(8.2) C. KITTEL, *Thermal Physics*, Wiley, New York (1969).

(8.3) G. BRUHAT, *Thermodynamique*, 5e Edition revised by A. KASTLER, Masson, Paris (1962).

See also references (1.4), part. 2, and (6.3).

MORE ADVANCED LEVEL

(8.4) F. REIF, *Fundamentals of Statistical and Thermal Physics*, McGraw-Hill, New York (1965).

(8.5) R. CASTAING, *Thermodynamique Statistique*, Masson, Paris (1970).

(8.6) P. M. MORSE, *Thermal Physics*, Benjamin, New York (1964).

(8.7) R. KUBO, *Statistical Mechanics*, North Holland, Amsterdam and Wiley, New York (1965).

(8.8) L. D. LANDAU and E. M. LIFSHITZ, *Course of Theoretical Physics, Vol. 5: Statistical Physics*, Pergamon Press, London (1963).

(8.9) H. B. CALLEN, *Thermodynamics*, Wiley, New York (1961).

(8.10) A. B. PIPPARD, *The Elements of Classical Thermodynamics*, Cambridge University Press (1957).

(8.11) R. C. TOLMAN, *The Principles of Statistical Mechanics*, Oxford University Press (1950).

9. RELATIVITY

INTRODUCTORY LEVEL

(9.1) J. H. SMITH, *Introduction to Special Relativity*, Benjamin, New York (1965).

See also references (6.2) and (6.3).

MORE ADVANCED LEVEL

(9.2) J. L. SYNGE, *Relativity: The Special Theory*, North Holland, Amsterdam (1965).

(9.3) R. D. SARD, *Relativistic Mechanics*, Benjamin, New York (1970).

(9.4) J. AHARONI, *The Special Theory of Relativity*, Oxford University Press, London (1959).

(9.5) C. MØLLER, *The Theory of Relativity*, Oxford University Press, London (1972).

(9.6) P. G. BERGMANN, *Introduction to the Theory of Relativity*, Prentice Hall, Englewood Cliffs (1960).

(9.7) C. W. MISNER, K. S. THORNE and J. A. WHEELER, *Gravitation*, Freeman, San Francisco (1973).

See also the electromagnetism references, in particular (7.5) and (7.11).

Also valuable:

(9.8) A. EINSTEIN, *Quatre Conférences sur la Théorie de la Relativité*, Gauthier-Villars, Paris (1971).

(9.9) A. EINSTEIN, *La Théorie de la Relativité Restreinte et Générale. La Relativité et le Problème de l'Espace*, Gauthier-Villars, Paris (1971).

(9.10) A. EINSTEIN, *The Meaning of Relativity*, Methuen, London (1950).

(9.11) A. EINSTEIN, *Relativity, the Special and General Theory, a Popular Exposition*, Methuen, London (1920); H. Holt, New York (1967).

A much more complete list of references can be found in :

(9.12) G. HOLTON, Resource Letter SRT-1 on Special Relativity Theory, *Am. J. Phys.* **30**, 462 (1962).

10. MATHEMATICAL METHODS

ELEMENTARY GENERAL TEXTS

(10.1) J. BASS, *Cours de Mathématiques*, Vols. I, II and III, Masson, Paris (1961).

(10.2) A. ANGOT, *Compléments de Mathématiques*, Revue d'Optique, Paris (1961).

(10.3) T. A. BAK and J. LICHTENBERG, *Mathematics for Scientists*, Benjamin, New York (1966).

(10.4) G. ARFKEN, *Mathematical Methods for Physicists*, Academic Press, New York (1966).

(10.5) J. D. JACKSON, *Mathematics for Quantum Mechanics*, Benjamin, New York (1962).

MORE ADVANCED GENERAL TEXTS

(10.6) J. MATHEWS and R. L. WALKER, *Mathematical Methods of Physics*, Benjamin, New York (1970).

(10.7) L. SCHWARTZ, *Mathematics for the Physical Sciences*, Hermann, Paris (1968). *Méthodes mathématiques pour les sciences physiques*, Hermann, Paris (1965).

(10.8) E. BUTKOV, *Mathematical Physics*, Addison-Wesley, Reading, Mass. (1968).

(10.9) H. CARTAN, *Elementary Theory of Analytic Functions of One or Several Complex Variables*, Addison-Wesley, Reading, Mass. (1966). *Théorie élémentaire des fonctions analytiques d'une ou plusieurs variables complexes*, Hermann, Paris (1961).

(10.10) J. VON NEUMANN, *Mathematical Foundations of Quantum Mechanics*, Princeton University Press (1955).

(10.11) R. COURANT and D. HILBERT, *Methods of Mathematical Physics*, Vols. I and II, Wiley, Interscience, New York (1966).

(10.12) E. T. WHITTAKER and G. N. WATSON, *A Course of Modern Analysis*, Cambridge University Press (1965).

(10.13) P. M. MORSE and H. FESHBACH, *Methods of Theoretical Physics*, McGraw-Hill, New York (1953).

LINEAR ALGEBRA. HILBERT SPACES

(10.14) A. C. AITKEN, *Determinants and Matrices*, Oliver and Boyd, Edinburgh (1956).

(10.15) R. K. EISENSCHITZ, *Matrix Algebra for Physicists*, Plenum Press, New York (1966).

(10.16) M. C. PEASE III, *Methods of Matrix Algebra*, Academic Press, New York (1965).

(10.17) J. L. SOULE, *Linear Operators in Hilbert Space*, Gordon and Breach, New York (1967).

(10.18) W. SCHMEIDLER, *Linear Operators in Hilbert Space*, Academic Press, New York (1965).

(10.19) N. I. AKHIEZER and I. M. GLAZMAN, *Theory of Linear Operators in Hilbert Space*, Ungar, New York (1961).

FOURIER TRANSFORMS; DISTRIBUTIONS

(10.20) R. STUART, *Introduction to Fourier Analysis*, Chapman and Hall, London (1969).

(10.21) M. J. LIGHTHILL, *Introduction to Fourier Analysis and Generalized Functions*, Cambridge University Press (1964).

(10.22) L. SCHWARTZ, *Théorie des Distributions*, Hermann, Paris (1967).

(10.23) I. M. GEL'FAND and G. E. SHILOV, *Generalized Functions*, Academic Press, New York (1964).

(10.24) F. OBERHETTINGER, *Tabellen zur Fourier Transformation*, Springer-Verlag, Berlin (1957).

PROBABILITY AND STATISTICS

(10.25) J. BASS, *Elements of Probability Theory*, Academic Press, New York (1966). *Éléments* de *Calcul des Probabilités*, Masson, Paris (1974).

(10.26) P. G. HOEL, S. C. PORT and C. J. STONE, *Introduction to Probability Theory*, Houghton-Mifflin, Boston (1971).

(10.27) H. G. TUCKER, *An Introduction to Probability and Mathematical Statistics*, Academic Press, New York (1965).

(10.28) J. LAMPERTI, *Probability*, Benjamin, New York (1966).

(10.29) W. FELLER, *An Introduction to Probability Theory and its Applications*, Wiley, New York (1968).

(10.30) L. BREIMAN, *Probability*, Addison-Wesley, Reading, Mass. (1968).

GROUP THEORY

Applied to physics:

(10.31) H. BACRY, *Lectures on Group Theory*, Gordon and Breach, New York (1967).

(10.32) M. HAMERMESH, *Group Theory and its Application to Physical Problems*, Addison-Wesley, Reading, Mass. (1962).

See also (2.18), (2.22) and (2.23) or reference (16.13) which gives a simple introduction to continuous groups in physics.

More mathematical:

(10.33) G. PAPY, *Groups*, Macmillan, New York (1964).

(10.34) A. G. KUROSH, *The Theory of Groups*, Chelsea, New York (1960).

(10.35) L. S. PONTRYAGIN, *Topological Groups*, Gordon and Breach, New York (1966).

SPECIAL FUNCTIONS AND TABLES

(10.36) A. GRAY and G. B. MATHEWS, *A Treatise on Bessel Functions and their Applications to Physics*, Dover, New York (1966).

(10.37) E. D. RAINVILLE, *Special Functions*, Macmillan, New York (1965).

(10.38) W. MAGNUS, F. OBERHETTINGER and R. P. SONI, *Formulas and Theorems for the Special Functions of Mathematical Physics*, Springer-Verlag, Berlin (1966).

(10.39) BATEMAN MANUSCRIPT PROJECT, *Higher Transcendental Functions*, Vols. I, II and III, A. ERDELYI ed., McGraw-Hill, New York (1953).

(10.40) M. ABRAMOWITZ and I. A. STEGUN, *Handbook of Mathematical Functions*, Dover, New York (1965).

(10.41) L. J. COMRIE, *Chambers's Shorter Six-Figure Mathematical Tables*, Chambers, London (1966).

(10.42) E. JAHNKE and F. EMDE, *Tables of Functions*, Dover, New York (1945).

(10.43) V. S. AIZENSHTADT, V. I. KRYLOV and A. S. METEL'SKII, *Tables of Laguerre Polynomials and Functions*, Pergamon Press, Oxford (1966).

(10.44) H. B. DWIGHT, *Tables of Integrals and Other Mathematical Data*, Macmillan, New York (1965).

(10.45) D. BIERENS DE HAAN, *Nouvelles Tables d'Intégrales Définies*, Hafner, New York (1957).

(10.46) F. OBERHETTINGER and L. BADII, *Tables of Laplace Transforms*, Springer-Verlag, Berlin (1973).

(10.47) BATEMAN MANUSCRIPT PROJECT, *Tables of Integral Transforms*, Vols. I and II, A. ERDELYI ed., McGraw-Hill, New York (1954).

(10.48) M. ROTENBERG, R. BIVINS, N. METROPOLIS and J. K. WOOTEN JR., *The 3-j and 6-j symbols*, M.I.T. Technology Press (1959); Crosby Lockwood and Sons, London.

11. ATOMIC PHYSICS

INTRODUCTORY LEVEL

(11.1) H. G. KUHN, *Atomic Spectra*, Longman, London (1969).

(11.2) B. CAGNAC and J. C. PEBAY-PEYROULA, *Physique Atomique*, Vols. 1 and 2, Dunod, Paris (1971).
English translation : *Modern Atomic Physics*, Vol. 1 : *Fundamental Principles*, and 2: *Quantum Theory and its Application*, Macmillan, London (1975).

(11.3) A. G. MITCHELL and M. W. ZEMANSKY, *Resonance Radiation and Excited Atoms*, Cambridge University Press, London (1961).

(11.4) M. BORN, *Atomic Physics*, Blackie and Son, London (1951).

(11.5) H. E. WHITE, *Introduction to Atomic Spectra*, McGraw-Hill, New York (1934).

(11.6) V. N. KONDRATIEV, *La Structure des Atomes et des Molécules*, Masson, Paris (1964).

See also (1.3) and (12.1).

MORE ADVANCED LEVEL

(11.7) G. W. SERIES, *The Spectrum of Atomic Hydrogen*, Oxford University Press, London (1957).

(11.8) J. C. SLATER, *Quantum Theory of Atomic Structure*, Vols. I and II, McGraw-Hill, New York (1960).

(11.9) A. E. RUARK and H. C. UREY, *Atoms, Molecules and Quanta*, Vols. I and II, Dover, New York (1964).

(11.10) *Handbuch der Physik, Vols. XXXV and XXXVI, Atoms*, S. FLÜGGE ed., Springer-Verlag Berlin (1956 and 1957).

(11.11) N. F. RAMSEY, *Molecular Beams*, Oxford University Press, London (1956).

(11.12) I. I. SOBEL'MAN, *Introduction to the Theory of Atomic Spectra*, Pergamon Press, Oxford (1972).

(11.13) E. U. CONDON and G. H. SHORTLEY, *The Theory of Atomic Spectra*, Cambridge University Press (1953).

ARTICLES

Numerous references to articles and books, classified and discussed, can be found in:

(11.14) J. C. ZORN, "Resource Letter MB-1 on Experiments with Molecular Beams, *Am. J. Phys.* **32**, 721 (1964).

See also: (3.13).

(11.15) V. F. WEISSKOPF, "How Light Interacts with Matter", *Scientific American*, **219**, 60 (Sept. 1968).

(11.16) H. R. CRANE, "The g Factor of the Electron", *Scientific American*, **218**, 72 (Jan. 1968).

(11.17) M. S. ROBERTS, "Hydrogen in Galaxies", *Scientific American*, **208**, 94 (June 1963).

(11.18) S. A. WERNER, R. COLELLA, A. W. OVERHAUSER and C. F. EAGEN, "Observation of the Phase Shift of a Neutron due to Precession in a Magnetic Field", *Phys. Rev. Letters*, **35**, 1053 (1975).

EXOTIC ATOMS

(11.19) H. C. CORBEN and S. DE BENEDETTI, "The Ultimate Atom", *Scientific American*, **191**, 88 (Dec. 1954).

(11.20) V. W. HUGHES, "The Muonium Atom", *Scientific American*, **214**, 93, (April 1966). "Muonium", *Physics Today*, **20**, 29 (Dec. 1967).

(11.21) S. DE BENEDETTI, "Mesonic Atoms", *Scientific American*, **195**, 93, (Oct. 1956).

(11.22) C. E. WIEGAND, "Exotic Atoms", *Scientific American*, **227**, 102, (Nov. 1972).

(11.23) V. W. HUGHES, "Quantum Electrodynamics: experiment", in *Atomic Physics*, B. Bederson, V. W. Cohen and F. M. Pichanick eds., Plenum Press, New York (1969).

(11.24) R. DE VOE, P. M. MC INTYRE, A. MAGNON, D, Y. STOWELL, R. A. SWANSON and V. L. TELEGDI, "Measurement of the muonium Hfs Splitting and of the muon moment by double resonance, and new value of α", *Phys. Rev. Letters*, **25**, 1779 (1970).

(11.25) K. F. CANTER, A. P. MILLS JR. and S. BERKO, "Observations of Positronium Lyman-Radiation", *Phys. Rev. Letters*, **34**, 177 (1975). "Fine-Structure Measurement in the First Excited State of Positronium" *Phys. Rev. Letters*, **34**, 1541 (1975).

12. MOLECULAR PHYSICS

INTRODUCTORY LEVEL

(12.1) M. KARPLUS and R. N. PORTER, *Atoms and Molecules*, Benjamin, New York (1970).

(12.2) L. PAULING, *The Nature of the Chemical Bond*, Cornell University Press (1948).
See also (1.3), chap. 12; (1.5) and (11.6).

MORE ADVANCED LEVEL

(12.3) I. N. LEVINE, *Quantum Chemistry*, Allyn and Bacon, Boston (1970).

(12.4) G. HERZBERG, *Molecular Spectra and Molecular Structure*, Vol. I: *Spectra of Diatomic Molecules*, and Vol. II: *Infrared and Raman Spectra of Polyatomic Molecules*, D. Van Nostrand Company, Princeton (1963 and 1964).

(12.5) H. EYRING, J. WALTER and G. E. KIMBALL, *Quantum Chemistry*, Wiley, New York (1963).

(12.6) C. A. COULSON, *Valence*, Oxford at the Clarendon Press (1952).

(12.7) J. C. SLATER, *Quantum Theory of Molecules and Solids*, Vol. 1: *Electronic Structure of Molecules*, McGraw-Hill, New York (1963).

(12.8) *Handbuch der Physik, Vol. XXXVII, 1 and 2, Molecules*, S. FLÜGGE, ed., Springer Verlag, Berlin (1961).

(12.9) D. LANGBEIN, *Theory of Van der Waals Attraction*, Springer Tracts in Modern Physics, Vol. 72, Springer Verlag, Berlin (1974).

(12.10) C. H. TOWNES and A. L. SCHAWLOW, *Microwave Spectroscopy*, McGraw-Hill, New York (1955).

(12.11) P. ENCRENAZ, *Les Molécules interstellaires*, Delachaux et Niestlé, Neuchâtel (1974).

See also (11.9), (11.11) and (11.14).

ARTICLES

(12.12) B. V. DERJAGUIN, "The Force Between Molecules", *Scientific American*, **203**, 47 (July 1960).

(12.13) A. C. WAHL, "Chemistry by Computer", *Scientific American*, **222**, 54 (April 1970).

(12.14) B. E. TURNER, "Interstellar Molecules", *Scientific American*, **228**, 51 (March 1973).

(12.15) P. M. SOLOMON, "Interstellar Molecules", *Physics Today*, **26**, 32 (March 1973).

See also (16.25).

13. SOLID STATE PHYSICS

INTRODUCTORY LEVEL

(13.1) C. KITTEL, *Elementary Solid State Physics*, Wiley, New York (1962).

(13.2) C. KITTEL, *Introduction to Solid State Physics*, 3e ed., Wiley, New York (1966).

(13.3) J. M. ZIMAN, *Principles of the Theory of Solids*, Cambridge University Press, London (1972).

(13.4) F. SEITZ, *Modern Theory of Solids*, McGraw-Hill, New York (1940).

MORE ADVANCED LEVEL

General texts:

(13.5) C. KITTEL, *Quantum Theory of Solids*, Wiley, New York (1963).

(13.6) R. E. PEIERLS, *Quantum Theory of Solids*, Oxford University Press, London (1964).

(13.7) N. F. MOTT and H. JONES, *The Theory of the Properties of Metals and Alloys*, Clarendon Press, Oxford (1936); Dover, New York (1958).

More specialized texts:

(13.8) M. BORN and K. HUANG, *Dynamical Theory of Crystal Lattices*, Oxford University Press, London (1954).

(13.9) J. M. ZIMAN, *Electrons and Phonons*, Oxford University Press, London (1960).
(13.10) H. JONES, *The Theory of Brillouin Zones and Electronic States in Crystals*, North Holland, Amsterdam (1962).
(13.11) J. CALLAWAY, *Energy Band Theory*, Academic Press, New York (1964).
(13.12) R. A. SMITH, *Wave Mechanics of Crystalline Solids*, Chapman and Hall, London (1967).
(13.13) D. PINES and P. NOZIERES, *The Theory of Quantum Liquids*, Benjamin, New York (1966).
(13.14) D. A. WRIGHT, *Semiconductors*, Associated Book Publishers, London (1966).
(13.15) R. A. SMITH, *Semiconductors*, Cambridge University Press, London (1964).

ARTICLES

(13.16) R. L. SPROULL, "The Conduction of Heat in Solids", *Scientific American*, **207**, 92 (Dec. 1962).
(13.17) A. R. MACKINTOSH, "The Fermi Surface of Metals", *Scientific American*, **209**, 110 (July 1963).
(13.18) D. N. LANGENBERG, D. J. SCALAPINO and B. N. TAYLOR, "The Josephson Effects", *Scientific American* **214**, 30 (May 1966).
(13.19) G. L. POLLACK, "Solid Noble Gases", *Scientific American*, **215**, 64 (Oct. 1966).
(13.20) B. BERTMAN and R. A. GUYER, "Solid Helium", *Scientific American*, **217**, 85 (Aug. 1967).
(13.21) N. MOTT, "The Solid State", *Scientific American*, **217**, 80 (Sept. 1967).
(13.22) M. Ya. AZBEL', M. I. KAGANOV and I. M. LIFSHITZ, "Conduction Electrons in Metals", *Scientific American*, **228**, 88 (Jan. 1973).
(13.23) W. A. HARRISON, "Electrons in Metals", *Physics Today*, **22**, 23 (Oct. 1969).

14. MAGNETIC RESONANCE

(14.1) A. ABRAGAM, *The Principles of Nuclear Magnetism*, Clarendon Press, Oxford (1961).
(14.2) C. P. SLICHTER, *Principles of Magnetic Resonance*, Harper and Row, New York (1963).
(14.3) G. E. PAKE, *Paramagnetic Resonance*, Benjamin, New York (1962).
See also Ramsey (11.11), Chaps. V, VI and VII.

ARTICLES

(14.4) G. E. PAKE, "Fundamentals of Nuclear Magnetic Resonance Absorption, I and II, *Am. J. Phys.*, **18**, 438 and 473 (1950).
(14.5) E. M. PURCELL, "Nuclear Magnetism", *Am. J. Phys.*, **22**, 1 (1954).
(14.6) G. E. PAKE, "Magnetic Resonance", *Scientific American*, **199**, 58 (Aug. 1958).
(14.7) K. WÜTHRICH and R. C. SHULMAN, "Magnetic Resonance in Biology", *Physics Today*, **23**, 43 (April 1970).
(14.8) F. BLOCH, "Nuclear Induction", *Phys. Rev.* **70**, 460 (1946).
Numerous other references, in particular to original articles, can be found in:
(14.9) R. E. NORBERG, "Resource Letter NMR-EPR-1 on Nuclear Magnetic Resonance and Electron Paramagnetic Resonance", *Am. J. Phys.*, **33**, 71 (1965).

15. QUANTUM OPTICS; MASERS AND LASERS

OPTICAL PUMPING; MASERS AND LASERS

(15.1) R. A. BERNHEIM, *Optical Pumping: An Introduction*, Benjamin, New York (1965).
This book contains many references. In addition, several important original papers are reprinted.

(15.2) *Quantum Optics and Electronics, Les Houches Lectures 1964*, C. DE WITT, A. BLANDIN and C. COHEN-TANNOUDJI eds., Gordon and Breach, New York (1965).

(15.3) *Quantum Optics, Proceedings of the Scottish Universities Summer School 1969*, S. M. KAY and A. MAITLAND eds., Academic Press, London (1970).
These two summer-school books contain several useful texts related to optical pumping and quantum electronics.

(15.4) W. E. LAMB JR., *Quantum Mechanical Amplifiers*, in *Lectures in Theoretical Physics*, Vol. II, W. BRITTIN and D. DOWNS eds., Interscience Publishers, New York (1960).

(15.5) M. SARGENT III, M. O. SCULLY and W. E. LAMB JR., *Laser Physics*, Addison-Wesley, New York (1974).

(15.6) A. E. SIEGMAN, *An Introduction to Lasers and Masers*, McGraw-Hill, New York (1971).

(15.7) L. ALLEN, *Essentials of Lasers*, Pergamon Press, Oxford (1969). This small book contains several reprints of original papers on lasers.

(15.8) L. ALLEN and J. H. EBERLY, *Optical Resonance and Two-Level Atoms*, Wiley Interscience, New York (1975).

(15.9) A. YARIV, *Quantum Electronics*, Wiley, New York (1967).

(15.10) H. M. NUSSENZVEIG, *Introduction to Quantum Optics*, Gordon and Breach, London (1973).

ARTICLES

Two "Resource Letters" give, discuss and classify many useful references:

(15.11) H. W. MOOS, "Resource Letter MOP-1 on Masers (Microwave through Optical) and on Optical Pumping", *Am. J. Phys.*, **32**, 589 (1964).

(15.12) P. CARRUTHERS, "Resource Letter QSL-1 on Quantum and Statistical Aspects of Light", *Am. J. Phys.*, **31**, 321 (1963).

Reprints of many important papers on Lasers have been collected in:

(15.13) *Laser Theory*, F. S. BARNES ed., I.E.E.E. Press, New York (1972).

(15.14) H. LYONS, "Atomic Clocks", *Scientific American*, **196**, 71 (Feb. 1957).

(15.15) J. P. GORDON, "The Maser", *Scientific American*, **199**, 42 (Dec. 1958).

(15.16) A. L. BLOOM, "Optical Pumping", *Scientific American*, **203**, 72 (Oct. 1960).

(15.17) A. L. SCHAWLOW, "Optical Masers", *Scientific American*, **204**, 52 (June 1961). "Advances in Optical Masers", *Scientific American*, **209**, 34 (July 1963). "Laser Light", *Scientific American*, **219**, 120 (Sept. 1968).

(15.18) M. S. FELD and V. S. LETOKHOV, "Laser Spectroscopy", *Scientific American*, **229**, 69 (Dec. 1973).

NON-LINEAR OPTICS

(15.19) G. C. BALDWIN, *An Introduction to Non-Linear Optics*, Plenum Press, New York (1969).

(15.20) F. ZERNIKE and J. E. MIDWINTER, *Applied Non-Linear Optics*, Wiley Interscience, New York (1973).

(15.21) N. BLOEMBERGEN, *Non-Linear Optics*, Benjamin, New York (1965).
See also this author's lectures in references (15.2) and (15.3).

ARTICLES

(15.22) J. A. GIORDMAINE, "The Interaction of Light with Light", *Scientific American*, **210**, 38 (Apr. 1964).
"Non-Linear Optics", *Physics Today*, **22**, 39 (Jan. 1969).

16. NUCLEAR PHYSICS AND PARTICLE PHYSICS

INTRODUCTION TO NUCLEAR PHYSICS

(16.1) L. VALENTIN, *Physique Subatomique: Noyaux et Particules*, Hermann, Paris (1975).
(16.2) D. HALLIDAY, *Introductory Nuclear Physics*, Wiley, New York (1960).
(16.3) R. D. EVANS, *The Atomic Nucleus*, McGraw-Hill, New York (1955).
(16.4) M. A. PRESTON, *Physics of the Nucleus*, Addison-Wesley, Reading, Mass. (1962).
(16.5) E. SEGRE, *Nuclei and Particles*, Benjamin, New York (1965).

MORE ADVANCED NUCLEAR PHYSICS TEXTS

(16.6) A. DESHALIT and H. FESHBACH, *Theoretical Nuclear Physics, Vol. 1: Nuclear Structure*, Wiley, New York (1974).
(16.7) J. M. BLATT and V. F. WEISSKOPF, *Theoretical Nuclear Physics*, Wiley, New York (1963).
(16.8) E. FEENBERG, *Shell Theory of the Nucleus*, Princeton University Press (1955).
(16.9) A. BOHR and B. R. MOTTELSON, *Nuclear Structure*, Benjamin, New York (1969).

INTRODUCTION TO PARTICLE PHYSICS

(16.10) D. H. FRISCH and A. M. THORNDIKE, *Elementary Particles*, Van Nostrand, Princeton (1964).
(16.11) C. E. SWARTZ, *The Fundamental Particles*, Addison-Wesley, Reading, Mass. (1965).
(16.12) R. P. FEYNMAN, *Theory of Fundamental Processes*, Benjamin, New York (1962).
(16.13) R. OMNES, *Introduction à l'Etude des Particules Elémentaires*, Ediscience, Paris (1970).
(16.14) K. NISHIJIMA, *Fundamental Particles*, Benjamin, New York (1964).

MORE ADVANCED PARTICLE PHYSICS TEXTS

(16.15) B. DIU, *Qu'est-ce qu'une Particule Elémentaire?* Masson, Paris (1965).
(16.16) J. J. SAKURAI, *Invariance Principles and Elementary Particles*, Princeton University Press (1964).
(16.17) G. KÄLLEN, *Elementary Particle Physics*, Addison-Wesley, Reading, Mass. (1964).
(16.18) A. D. MARTIN and T. D. SPEARMAN, *Elementary Particle Theory*, North Holland, Amsterdam (1970).
(16.19) A. O. WEISSENBERG, *Muons*, North Holland, Amsterdam (1967).

ARTICLES

(16.20) M. G. MAYER, "The Structure of the Nucleus", *Scientific American*, **184**, 22 (March 1951).
(16.21) R. E. PEIERLS, "The Atomic Nucleus", *Scientific American*, **200**, 75 (Jan. 1959).
(16.22) E. U. BARANGER, "The present status of the nuclear shell model", *Physics Today*, **26**, 34 (June 1973).

(16.23) S. DE BENEDETTI, "Mesonic Atoms", *Scientific American*, **195**, 93 (Oct. 1956).
(16.24) S. DE BENEDETTI, "The Mössbauer Effect", *Scientific American*, **202**, 72 (April 1960).
(16.25) R. H. HERBER, "Mössbauer Spectroscopy", *Scientific American*, **225**, 86 (Oct. 1971).
(16.26) S. PENMAN, "The Muon", *Scientific American*, **205**, 46 (July 1961).
(16.27) R. E. MARSHAK, "The Nuclear Force", *Scientific American*, **202**, 98 (March 1960).
(16.28) M. GELL-MANN and E. P. ROSENBAUM, "Elementary Particles", *Scientific American*, **197**, 72 (July 1957).
(16.29) G. F. CHEW, M. GELL-MANN and A. H. ROSENFELD, "Strongly Interacting Particles", *Scientific American*, **210**, 74 (Feb. 1964).
(16.30) V. F. WEISSKOPF, "The Three Spectroscopies", *Scientific American*, **218**, 15 (May 1968).
(16.31) U. AMALDI, "Proton Interactions at High Energies", *Scientific American*, **229**, 36 (Nov. 1973).
(16.32) S. WEINBERG, "Unified Theories of Elementary-Particle Interaction", *Scientific American*, **231**, 50 (July 1974).
(16.33) S. D. DRELL, "Electron-Positron Annihilation and the New Particles", *Scientific American*, **232**, 50 (June 1975).
(16.34) R. WILSON, "Form Factors of Elementary Particles", *Physics Today*, **22**, 47 (Jan. 1969).
(16.35) E. S. ABERS and B. W. LEE, "Gauge Theories", *Physics Reports*, **9C**, 1, Amsterdam (1973).

Index

N.B. *References followed by* (e) *refer to an exercise*

Coordinate systems

	Cartesian	Cylindrical	Spherical
DEFINITIONS	$U = U(x, y, z)$ $\mathbf{A} = A_x\mathbf{e}_x + A_y\mathbf{e}_y + A_z\mathbf{e}_z$ $A_x = A_x(x, y, z)$ $A_y = A_y(x, y, z)$ $A_z = A_z(x, y, z)$	$U = U(\rho, \varphi, z)$ $\mathbf{A} = A_\rho\mathbf{e}_\rho + A_\varphi\mathbf{e}_\varphi + A_z\mathbf{e}_z$ $A_\rho = A_x \cos\varphi + A_y \sin\varphi$ $A_\varphi = -A_x \sin\varphi + A_y \cos\varphi$	$U = U(r, \theta, \varphi)$ $\mathbf{A} = A_r\mathbf{e}_r + A_\theta\mathbf{e}_\theta + A_\varphi\mathbf{e}_\varphi$ $A_r = A_\rho \sin\theta + A_z \cos\theta$ $A_\theta = A_\rho \cos\theta - A_z \sin\theta$ $A_\varphi = -A_x \sin\varphi + A_y \cos\varphi$
GRADIENT	$\nabla U = (\partial U/\partial x)\mathbf{e}_x$ $+ (\partial U/\partial y)\mathbf{e}_y$ $+ (\partial U/\partial z)\mathbf{e}_z$	$(\nabla U)_\rho = \partial U/\partial \rho$ $(\nabla U)_\varphi = [\partial U/\partial \varphi]/\rho$ $(\nabla U)_z = \partial U/\partial z$	$(\nabla U)_r = \partial U/\partial r$ $(\nabla U)_\theta = [\partial U/\partial \theta]/r$ $(\nabla U)_\varphi = [\partial U/\partial \varphi]/(r \sin\theta)$
LAPLA-CIAN	$\Delta U = \frac{\partial^2 U}{\partial x^2} + \frac{\partial^2 U}{\partial y^2} + \frac{\partial^2 U}{\partial z^2}$	$\Delta U = \frac{1}{\rho}\frac{\partial}{\partial \rho}\left(\rho \frac{\partial U}{\partial \rho}\right) + \frac{1}{\rho^2}\frac{\partial^2 U}{\partial \varphi^2} + \frac{\partial^2 U}{\partial z^2}$	$\Delta U = \frac{1}{r}\frac{\partial^2}{\partial r^2}(rU) + \frac{1}{r^2 \sin\theta}\frac{\partial}{\partial \theta}\left(\sin\theta \frac{\partial U}{\partial \theta}\right) + \frac{1}{r^2 \sin^2\theta}\frac{\partial^2 U}{\partial \varphi^2}$
DIVER-GENCE	$\nabla . \mathbf{A} = \frac{\partial A_x}{\partial x} + \frac{\partial A_y}{\partial y} + \frac{\partial A_z}{\partial z}$	$\nabla . \mathbf{A} = \frac{1}{\rho}\frac{\partial}{\partial \rho}(\rho A_\rho) + \frac{1}{\rho}\frac{\partial A_\varphi}{\partial \varphi} + \frac{\partial A_z}{\partial z}$	$\nabla . \mathbf{A} = \frac{1}{r^2}\frac{\partial}{\partial r}(r^2 A_r) + \frac{1}{r \sin\theta}\frac{\partial}{\partial \theta}(\sin\theta\, A_\theta) + \frac{1}{r \sin\theta}\frac{\partial A_\varphi}{\partial \varphi}$
CURL	$\nabla \times \mathbf{A} = (\partial A_z/\partial y - \partial A_y/\partial z)\mathbf{e}_x$ $+ (\partial A_x/\partial z - \partial A_z/\partial x)\mathbf{e}_y$ $+ (\partial A_y/\partial x - \partial A_x/\partial y)\mathbf{e}_z$	$(\nabla \times \mathbf{A})_\rho = (\partial A_z/\partial \varphi)/\rho - \partial A_\varphi/\partial z$ $(\nabla \times \mathbf{A})_\varphi = \partial A_\rho/\partial z - \partial A_z/\partial \rho$ $(\nabla \times \mathbf{A})_z = [\partial(\rho A_\varphi)/\partial \rho - \partial A_\rho/\partial \varphi]/\rho$	$(\nabla \times \mathbf{A})_r = [\partial(\sin\theta\, A_\varphi)/\partial \theta - \partial A_\theta/\partial \varphi]/(r \sin\theta)$ $(\nabla \times \mathbf{A})_\theta = [\partial A_r/\partial \varphi - \sin\theta\, \partial(r A_\varphi)/\partial r]/(r \sin\theta)$ $(\nabla \times \mathbf{A})_\varphi = [\partial(r A_\theta)/\partial r - \partial A_r/\partial \theta]/r$